W9-BOB-476

www.wadsworth.com

wadsworth.com is the World Wide Web site for Wadsworth Publishing Company and is your direct source to dozens of online resources.

At *wadsworth.com* you can find out about supplements, demonstration software, and student resources. You can also send e-mail to many of our authors and preview new publications and exciting new technologies.

wadsworth.com
Changing the way the world learns®

About the Authors

David Knox, Ph.D., is Professor of Sociology at East Carolina University, where he teaches courses in Courtship and Marriage, Marraige and the Family, and Sociology of Human Sexuality. He is a marriage and family therapist and the author or coauthor of ten books and fifty professional articles. He and Caroline Schacht are married.

Caroline Schacht, M.A. in Sociology and M.A. in Family Relations, is Instructor of Sociology at East Carolina University and teaches Courtship and Marriage, Introduction to Sociology, Deviance, and The Individual and Society. Her clinical work includes marriage and family relationships. She is also a divorce mediator and the coauthor of several books, including *Understanding Social Problems* (Wadsworth, 2005).

Choices in Relationships

An Introduction to Marriage and the Family • EIGHTH EDITION

David Knox
East Carolina University

Caroline Schacht
East Carolina University

THOMSON
WADSWORTH

Australia • Canada • Mexico • Singapore • Spain • United Kingdom • United States

Sociology Editor: Robert Jucha
Development Editor: Julie Sakaue
Assistant Editor: Stephanie Monzon
Editorial Assistant: Melissa Walter
Technology Project Manager: Dee Dee Zobian
Marketing Manager: Matthew Wright
Marketing Assistant: Tara Pierson
Advertising Project Manager: Linda Yip
Project Manager, Editorial Production: Katy German
Print Buyer: Barbara Britton
Permissions Editor: Kiely Sexton
Production Service: Greg Hubit Bookworks
Copy Editor: Jamie Fuller
Cover Designer: Larry Didona
Cover Images: Darren Modricker, Paul Barton, Corbis; Kevin Fitzgerald, Simon Watson, Kaz Chiba, Getty Images
Cover Printer: Coral Graphic Services, Inc.
Compositor: G&S Typesetters
Printer: Courier Corporation/Kendallville

COPYRIGHT © 2005 Wadsworth, a division of Thomson Learning, Inc. Thomson Learning™ is a trademark used herein under license.

ALL RIGHTS RESERVED. No part of this work covered by the copyright hereon may be reproduced or used in any form or by any means—graphic, electronic, or mechanical, including but not limited to photocopying, recording, taping, Web distribution, information networks, or information storage and retrieval systems—without the written permission of the publisher.

Printed in the United States of America
1 2 3 4 5 6 7 08 07 06 05 04

For more information about our products, contact us at:
Thomson Learning
Academic Resource Center
1-800-423-0563
For permission to use material from this text or product, submit a request online at:
http://www.thomsonrights.com
Any additional questions about permissions can be submitted by email to:
thomsonrights@thomson.com

ExamView® and ExamView Pro® are registered trademarks of FSCreations, Inc. Windows is a registered trademark of the Microsoft Corporation used herein under license. Macintosh and Power Macintosh are registered trademarks of Apple Computer, Inc. Used herein under license.

Library of Congress Control Number: 2003116253
ISBN 0-534-62523-1
Instructor's Edition: ISBN 0-534-62524-X

COPYRIGHT 2005 Thomson Learning, Inc. All Rights Reserved. Thomson Learning WebTutor™ is a trademark of

Thomson Learning, Inc.
Thomson Wadsworth
10 Davis Drive
Belmont, CA 94002-3098
USA

Asia
Thomson Learning
5 Shenton Way #01-01
UIC Building
Singapore 068808

Australia/New Zealand
Thomson Learning
102 Dodds Street
Southbank, Victoria 3006
Australia

Canada
Nelson
1120 Birchmount Road
Toronto, Ontario M1K 5G4
Canada

Europe/Middle East/Africa
Thomson Learning
High Holborn House
50/51 Bedford Row
London WC1R 4LR
United Kingdom

Latin America
Thomson Learning
Seneca, 53
Colonia Polanco
11560 Mexico D.F.
Mexico

Spain/Portugal
Paraninfo
Calle Magallanes, 25
28015 Madrid, Spain

To Isabelle,

who now confronts her most important choices.

Contents in Brief

Chapters

Contents

CHAPTER 1 Choices in Relationships: An Introduction 1

CHAPTER 2 Gender in Relationships 31

CHAPTER 5

Sexuality in Relationships *99*

CHAPTER 6

Singlehood and Same-Sex Relationships *132*

CHAPTER 7

Mate Selection 159

CHAPTER 8

Marriage Relationships 182

CHAPTER 9

Communication in Relationships 205

CHAPTER 14

CHAPTER 15

CHAPTER 16

Remarriage and Stepfamilies 382

CHAPTER 17

Aging in Marriage and Family Relationships 407

Preface

We chose the theme of *choices* because of an important, incontestable fact: the choices one makes in regard to relationships have a long-term impact on the individuals, their partners, their marriages, and their children. By being attentive to the importance of making conscious, deliberate, research-based choices, everyone wins (including society). Not to take one's relationship choices seriously is to limit one's ability to have and to share a fulfilling emotional life—the only game in town that matters.

NEW TO THE EIGHTH EDITION: ORGANIZATION AND CHAPTER-BY-CHAPTER CHANGES

The chapters have been rearranged to present chapter topics according to the more common progression of actual occurrence (e.g., sexuality is now addressed before lifestyle diversity, mate selection, and marriage relationships, and stress and crisis in relationships are now addressed before violence and abuse in relationships). A few of the chapters have also been renamed to better reflect chapter content and use more current terminology.

In addition to new census data throughout the text and a new Special Topic, Careers in Marriage and the Family, at the end of the text, examples of new content in each of the respective chapters of this edition include the following:

Chapter 1 Choices in Relationships: An Introduction

New Self-Assessment: Relationship Involvement Scale

New section on correcting bad choices

New theoretical frameworks: Family life course development framework; human ecological framework

Chapter 2 Gender in Relationships

Gender roles in Afghanistan under the Taliban

Gender postmodernism

Emotional stereotypes of women and men

Life goals of college students

Nontraditional career choices of men and women

Chapter 3 **Love in Relationships**

Meaning of phrase "I Love You" when said by college men and women

Compersion

Stalking

Sociological theory of love

Biological wiring, love, and sexual orientation

Chapter 4 **Hanging Out, Pairing Off, and Cohabitation**

Finding a partner online

Speed dating

Video chatting

Eight types of cohabitation

Effects of living together on children

Chapter 5 **Sexuality in Relationships**

Hooking up

Friends with benefits

Meanings of sexual intimacy for women and men

Primary sexual dysfunctions of women

Primary sexual dysfunctions of men

Chapter 6 **Singlehood and Same-Sex Relationships**

New scale assessing attitudes toward homosexuals

Multidimensional model of sexual orientation

New information on the legalization of same-sex marriage

Fluidity of sexual orientation

Twin Oaks Intentional Community

Chapter 7 **Mate Selection**

Manners in dating relationships

Religious homogamy and mate selection

Sexual satisfaction and relationship satisfaction

Ending or continuing an unsatisfactory relationship

Perfectionism and relationship problems

Chapter 8 **Marriage Relationships**

Covenant marriages

Weddings: some college student data

New Self-Assessment: Attitudes toward Interracial Dating Scale

Mixed marriage in music obsession

Attitudes toward interreligious marriages

Chapter 9 **Communication in Relationships**

Verbal and nonverbal communication

Effect of obsessive Internet use on communication with partner

Positives and negatives of Internet communication

Top relationship problem of persons in their first marriage

Top relationship problem among remarrieds

Chapter 10 **Planning Children and Contraception**

New Self-Assessment: Attitudes toward Parenthood Scale

New Self-Assessment: Attitudes toward Transracial Adoption Scale

Physical and psychological effects of abortion

Foster parenting

Chapter 11 **Parenting**

New section on factors involved in parenting

New Self-Assessment on attitudes toward motherhood

New Self-Assessment on attitudes toward fatherhood

New section on attachment theory

New section on bed-sharing with parents

Chapter 12 **Balancing Work and Family Life**

Poverty: definitions, levels, and effects on families

Day-care costs

Effect of wife's earning more than husband on stability of marriage

Women with young children—is home or job their preference?

Effect of having days off on marital interaction

Chapter 13 **Stress and Crisis in Relationships**

Resiliency in families

Biofeedback as a mechanism to help cope with crisis events

Spirituality and coping with crisis

Physical and mental disabilities of children as crisis events for parents

Chapter 14 **Violence and Abuse in Relationships**

Battered woman's syndrome

Treatment of partner abusers

Sibling abuse

Munchausen's syndrome by proxy

Shaken baby syndrome

Chapter 15 **Divorce**

Divorce rates, ratios, estimates

"Starter marriages"

Shared parenting dysfunction

Family relations doctrine as a criterion for awarding custody

Divorce prevention

Chapter 16 **Remarriage and Stepfamilies**

Who remarries first and why

Monitoring of children in stepfather and stepmother families

Stages of remarriage

Moving between households by children after divorce

Ambiguity of extended kin for stepchildren

Chapter 17 **Aging in Marriage and Family Relationships**

Interest of the elderly in family life education topics

Care of the elderly in Spain

Sexual orientation as a factor in entering a nursing home

Effect of divorce on grandparenting

The elderly and driving

Epilogue

Special Topics

Careers in Marriage and the Family

Evaluating Research in Marriage and the Family

Appendix

Family Autobiography Outline

FEATURES OF THE TEXT

Pedagogical features of previous editions identified by adopters and reviewers as valuable have been retained in this eighth edition. Beyond the choices theme, theoretical grounding, and up-to-date research, the features of this edition include the following.

Social Policy Boxes

These identify social policy issues related to marriage and the family with which U.S. society is confronted. Examples include legalization of same-sex marriage, marriage education in the public schools, and divorce law reform. A number of these social policy sections include OVRC (Opposing Viewpoints Resource Center) exercises to encourage students to hone their research and critical thinking skills. These exercises include relevant search terms and questions to guide students' reading. OVRC is more fully described in the supplements section.

Self-Assessment Scales

These are new scales (developed exclusively for this text) that students can complete to measure a particular aspect of themselves and/or their relationship with their partner. Examples include the Relationship Involvement Scale, Traditional Motherhood Scale, Traditional Fatherhood Scale, and Attitudes Toward Transracial Adoption Scale.

Personal Choices Boxes

Embedded in each chapter are detailed discussions of one or two personal choices relevant to the content of the chapter. Examples include "Who Is the Best Person for You to Marry?," "Should I Get Involved in a Long-Distance Dating Relationship?," and "Should I Put My Parents in a Long-Term-Care Facility?" In the previous edition these were the Choices feature at the end of the chapter.

Up Close Boxes

These examine pertinent issues on a personal level (e.g., "Divorce—Uniquely Yours: You *Can* Do It Your Way"—how a couple divorced as friends) or on a conceptual level (e.g., "Theoretical Views of Marital Happiness and Success").

National Data and International Data Features

To replace speculation with facts, we present data from national samples as well as data from around the world.

Diversity in the United States Features

These emphasize race, age, religious, economic, and educational differences in regard to marriage and family in the United States.

Diversity in Other Countries Features

These provide a glimpse of gender roles, love, courtship, and marriage and family patterns in other societies and cultures.

Summary

We changed the format for the summary from a paragraph format to a question-and-answer format with each question highlighting a major point in the chapter. This new format provides students with a more effective, concise review of major concepts and themes covered in the chapter.

Key Terms

Boldfaced in the text, listed at the end of each chapter, and defined in the glossary at the end of the text.

Researching Marriage and the Family with InfoTrac® College Edition

We added new InfoTrac College Edition exercises to the end of each chapter. This new feature includes two search terms relevant to the chapter with instructions for each on the type of article(s) to select and read, as well as questions to help guide students' reading.

Weblinks

Internet addresses are provided at the end of each chapter. We have checked each of these links to ensure that each site was operative at the time the text went to press.

SUPPLEMENTS AND RESOURCES

Choices in Relationships, Eighth Edition, is accompanied by a wide array of supplements prepared for both the instructor and student. Some new resources have been created specifically to accompany the eighth edition, and all of the continuing supplements have been thoroughly revised and updated.

Supplements for the Instructor

Instructor's Edition of *Choices in Relationships: An Introduction to Marriage and the Family,* Eighth Edition The Instructor's Edition contains a Visual Preface, a walk-through of the text that provides an overview of the text's key features, themes, and supplements.

Instructor's Resource Manual with Test Bank (with the Multimedia Manager CD-ROM) Written by Knox and Schacht, this manual provides instructors with learning objectives, a list of major concepts and terms (now with page references), detailed lecture outlines, extensive student projects and classroom activities, *InfoTrac* and Internet exercises, and self-assessment handouts for each chapter. Also included are concise user guides for both *InfoTrac* and *WebTutor* and a table of contents for the *CNN Today* Marriage and Family Video Series. The Test Bank contains forty to fifty multiple-choice, ten to fifteen true/false, ten short answer/discussion questions (including suggested topics for WebTutor online discussion threads), and three to five essay questions per chapter and the Special Topics. The Test Bank items are also available electronically on the *ExamView* Computerized Testing CD-ROM and on the Multimedia Manager CD-ROM. The Multimedia Manager CD-ROM, which now accompanies the Instructor's Resource Manual with Test

Bank, includes book-specific PowerPoint Lecture Slides, graphics from the book itself, the IRM/TB Word documents, CNN Video Clips, and links to many of Wadsworth's important Sociology resources in addition to the Test Bank items.

ExamView Computerized Testing Create, deliver, and customize tests and study guides (both print and online) in minutes with this easy-to-use assessment and tutorial system. *ExamView* offers both a Quick Test Wizard and an Online Test Wizard that guide you step by step through the process of creating tests. The test appears on screen exactly as it will print or display online. Using *ExamView*'s complete word processing capabilities, you can enter an unlimited number of new questions or edit existing questions included with ExamView.

Wadsworth's Marriage and Family 2005 Transparency Acetates A selection of four-color acetates consisting of tables and figures from Wadsworth's marriage and family texts is available to help prepare lecture presentations. Free to qualified adopters.

CNN Today: Marriage and Family Video Series, Volumes I–VI Illustrate how the principles that students learn in the classroom apply to the stories they see on television with the *CNN Today* Marriage and Family Video Series, an exclusive series jointly created by Wadsworth and CNN. Each video consists of approximately forty-five minutes of footage originally broadcast on CNN and selected specifically to illustrate concepts relevant to the marriage and the family course. The videos are broken into short two- to five-minute segments, perfect for classroom use as lecture launchers or to illustrate key concepts. An annotated table of contents accompanies each video with descriptions of the segments. Special adoption conditions apply.

Wadsworth Sociology Video Library This large selection of thought-provoking films, including some from the Films for the Humanities collection, is available to adopters based on adoption size.

Supplements for the Student

Study Guide Written by Knox and Schacht, each chapter of the Study Guide includes learning objectives, key terms (now with page references), a detailed chapter outline, InfoTrac College Edition exercises, Internet exercises, and a personal application section. Students can test and apply their knowledge of concepts with chapter practice tests including fifteen to twenty multiple-choice, ten true/false, ten to fifteen completion, and five to ten answer questions (new), now all with page references, as well as five to ten essay questions. The Study Guide includes practice tests for not only each of the chapters but the Special Topics sections of the text as well.

Marriage and Family: Using Microcase® ExplorIt, Third Edition Written by Kevin Demmitt of Clayton College, this software-based workbook is an exciting way to get students to view marriage and the family from the sociological perspective. With this workbook and accompanying ExplorIt software and data sets, your students will use national and cross-national surveys to examine and actively learn marriage and family topics. This inexpensive workbook will add an exciting dimension to your marriage and family course.

Families and Society, First Edition Written by Scott L. Coltrane, University of California, Riverside, this reader is designed to promote a sociological understanding of families, while at the same time demonstrating the diversity and complexity of contemporary family life. The different parts or sections of the reader are designed to "map" onto most sociology of the family textbooks and course syllabi. The articles emphasize a social constructionist and a sociological view of families. The reader is thus designed to dispel the myth that families are separate from society. Virtually every selection illustrates the myriad links that exist between families and their various social, cultural, economic, and political contexts. The first reading in each section provides a classic theoretical overview of the topic. These classic articles provide students with a historical understanding of the development of the field and offer insight into some of the enduring sociological facts about families.

Readings in the Marriage and Family Experience, Third Edition Written by Bryan Strong, Christine DeVault, and Barbara W. Sayad, this *free* reader contains articles, essays, and excerpts from books, journals, magazines, and newspapers on the topic of marriage and the family. They have all been carefully selected for their diverse and stimulating content. Critical thinking questions are included for each reading in order to encourage classroom discussion and student participation.

The Marriages and Families Activities Workbook What are your risks of divorce? Do you have healthy dating practices? What is your cultural and ancestral heritage and how does it affect your family relationships? The answers to these and many more questions are found in this workbook of nearly a hundred interactive self-assessment quizzes designed for students studying marriage and the family. These self-awareness instruments, all based on known social science research studies, can be used as in-class activities or homework assignments to help students learn more about themselves and their family experience.

Marriage and Family Case Studies CD-ROM This unique student CD-ROM includes a series of ten interactive case study videos that provide a dramatic enactment illustrating key topics and concepts from the text. Students watch each video and answer critical thinking questions, applying marriage and family theories to the case study videos. Students then compare their analysis with that of a marriage and family expert. Also on the CD-ROM is a direct link to InfoTrac College Edition, where students can search for related articles from sociology periodicals and a link to Wadsworth's Virtual Society where the companion Web site for *Choices in Relationships* offers a wide range of book-specific study tools.

Online Resources

Wadsworth's Virtual Society: The Wadsworth Sociology Resource Center
http://www.wadsworth.com/sociology

Here you will find a wealth of resources for students and instructors, including the Marriage and Family Resource Center. A fun, interactive site filled with additional data, exercises, and enriching resources. For example:

GSS Activities—Students can compare their attitudes with GSS data in this interactive exercise

Marriage and Family Activities—Fun self-quizzes for students to learn more about themselves

Marriage and Family Links—Online resources

Resources and Organizations—A library of useful resources.

Also from Wadsworth's Virtual Society, students will find the companion Web site for *Choices in Relationships: An Introduction to Marriage and the Family,* Eighth Edition at http://sociology.wadsworth.com/knox_schacht/choices8e

Access useful learning resources for each chapter of the book. Some of these resources include:

- *Tutorial practice quizzes that can be scored and e-mailed to the instructor*
- *Internet exercises and Web links*
- *Video exercises*
- *InfoTrac College Edition exercises*
- *Flashcards of the text's glossary*
- *Crossword puzzles*
- *Essay questions*
- *Learning objectives*
- *And much more!*

WebTutor™ Toolbox for Webct or Blackboard Preloaded with content and available free via pincode when packaged with this text, WebTutor toolbox pairs all the content of this text's rich book companion Web site with all the sophisticated course management functionality of a Webct or Blackboard product. You can assign materials (including online quizzes) and have the results flow automatically to your grade book. Toolbox is ready to use as soon as you log on—or you can customize its preloaded content by uploading images and other resources, adding Web links, or creating your own practice materials. Students have access only to student resources on the Web site. Instructors can enter a pincode for access to password-protected instructor resources.

InfoTrac® College Edition Give your students anytime, anywhere access to reliable resources with InfoTrac College Edition, the online library. This fully searchable database offers twenty years' worth of full-text articles from almost five thousand diverse sources, such as academic journals, newsletters, and up-to-the-minute periodicals including *Time, Newsweek, Science, Forbes,* and *USA Today*. The incredible depth and breadth of material—available twenty-four hours a day from any computer with Internet access—makes conducting research so easy, your students will want to use it to enhance their work in every course! Through InfoTrac's InfoWrite, students now also have instant access to critical thinking and paper writing tools. Both adopters and their students receive unlimited access for four months.

Opposing Viewpoints Resource Center (OVRC) Newly available from Wadsworth, this online center allows you to expose your students to all sides of today's most compelling issues! The Opposing Viewpoints Resource Center draws on Greenhaven Press's acclaimed social issues series, as well as core reference content from other Gale and Macmillan reference USA sources. The result is a dynamic online library of current event topics—the facts as well as the arguments of each topic's proponents and detractors. Special sections focus on critical thinking (walks student through how to critically evaluate point-counterpoint arguments) and researching and writing papers.

ACKNOWLEDGMENTS

Texts are always collaborative efforts. This text reflects the commitment of a new editor for the eighth edition, Bob Jucha, who came to East Carolina University to meet with the authors and guide/launch the new edition. We thank him for his valuable insight and guidance as we worked on this edition. We would also like to thank Julie Sakaue, our developmental editor (we felt this was not just another text for her, but her "baby"; she read each word and made valuable suggestions throughout); Stephanie Monzon, our assistant editor; Matthew Wright, our marketing manager; Dee Dee Zobian, our technology project manager; Katy German, our production project manager; Jamie Fuller, our copy editor (her keen eye never ceased to amaze us); Greg Hubit of Bookworks, our production service (always a man "on time"); and Kiely Sexton, our permissions editor. All were superb, and we appreciate their professionalism and attention to detail.

We would also like to thank Novita Leasure, who coded the manuscript for production; Kristen McGinty, Britt Sholar, Angela DeCuzzi, Sandi Alessio, Brian Greenberg, and Bill Robbins for researching various topics; Beth Credle for updating the information on contraception/sexually transmitted diseases; and Horanose Raji for formatting the various chapters and typing new sections. Dr. Sharon Ballard, Assistant Professor in the Department of Family Relations at East Carolina University, wrote Special Topics 1, Careers in Marriage and the Family, and provided valuable input in Chapters 1 (Introduction), 11 (Parenting), and 17 (Aging).

REVIEWERS FOR THE EIGHTH EDITION

Von Bakanic, College of Charleston

Donna Crossman, Ohio State University

Karen Dawes, Wake Technical Community College

Doug Dowell, Heartland Community College

Norman Goodman, State University of New York at Stony Brook

Ted Greenstein, North Carolina State University

Cherylon Robinson, University of Texas at San Antonio

Scott Smith, Stanly Community College

Beverly Stiles, Midwestern State University, Wichita Falls, Texas

REVIEWERS FOR THE PREVIOUS EDITIONS

Grace Auyang, University of Cincinnati

Rosemary Bahr, Eastern New Mexico University

Mary Beaubien, Youngstown State University

Sampson Lee Blair, Arizona State University

Mary Blair-Loy, Washington State University

David Daniel Bogumil, Wright State University

Elisabeth O. Burgess, Georgia State University

Craig Campbell, Weber State University

Michael Capece, University of South Florida

Lynn Christie, Baldwin-Wallace College

Laura Cobb, Purdue University and Illinois State University

Jean Cobbs, Virginia State University

Susan Brown Donahue, Pearl River Community College

John Engel, University of Hawaii

Jerry Ann Harrel-Smith, California State University, Northridge

Rudy Harris, Des Moines Area Community College

Sheldon Helfing, College of the Canyons

Mary Ann Gallagher, El Camino College

Shawn Gardner, Genesee Community College

Heidi Goar, St. Cloud State University

Gerald Harris, University of Houston

Terry Hatkoff, California State University, Northridge

Tonya Hilligoss, Sacramento City College

Rick Jenks, Indiana University

Richard Jolliff, El Camino College

Diane Keithly, Louisiana State University

Steve Long, Northern Iowa Area Community College

Carol May, Illinois Central College

Patricia B. Maxwell, University of Hawaii

Scott Potter, Marion Technical College

Eileen Shiff, Paradise Valley Community College

Cynthia Schmiege, University of Idaho

Tommy Smith, Auburn University

Dawood H. Sultan, Louisiana State University

Elsie Takeguchi, Sacramento City College

Myrna Thompson, Southside Virginia Community College

Janice Weber-Breaux, University of Southwestern Louisiana

Loreen Wolfer, University of Scranton

We love the study, writing, and teaching of marriage and the family and welcome your comments for improvement in the ninth edition of this text. We check our e-mail frequently and invite you to write us.

David Knox, E-mail: *Knoxd@mail.ecu.edu*

Caroline Schacht, E-mail: *Schachtc@mail.ecu.edu*

Authors

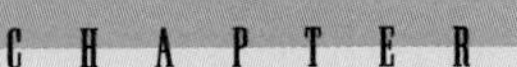

CHAPTER 1

Marriage is one of the most highly prized of all human relationships and a central life goal of most Americans.

Barbara Dafoe Whitehead and David Popenoe

Choices in Relationships: An Introduction

Contents

True or False?

1. In a national sample, first-year female and male college students identified "raising a family" among their top two values.
2. In spite of all the talk about singlehood, only about 5 percent of U.S. adults never marry.
3. Single people are happier and healthier and live longer than married people.
4. Whether marriage education courses in the public school system reduce divorce rates of the students who take them is unknown.
5. Not to make a decision is to make a decision.

Answers: **1.** T **2.** T **3.** F **4.** T **5.** T

Americans (particularly college students) consistently regard marriage and family as among their top values. In a national survey of over a quarter of a million (282,549) first-year undergraduates at 437 two- and four-year universities in the United States, "raising a family" was listed among the top two values by both women and men (women rated "family" as first; men rated "financial security" first) (American Council on Education and University of California, 2002). Other random samples (Abowitz & Knox, 2003) confirm the general importance (and gender differences) college students place on relationships and financial security.

Although the age at which people marry is increasing (people are delaying marriage in favor of career commitments), marriage continues to be a life goal for most people. Indeed, marriage is here to stay, and your choices in reference to it will be the most important in your life. Our goal in this text is to emphasize the various choices we make in regard to marriage and relationships. We begin by focusing on the meanings of marriage and the family.

MARRIAGE

In light of the Supreme Court decision striking down laws that ban homosexual acts, Vermont's offering "civil unions," and the Massachusetts Supreme Court's decision paving the way for gay marriage, the legal status of gay marriage throughout the United States remains uncertain (Biskupic, 2003). For now, **marriage** is viewed as a system of binding a man and a woman together for the reproduction, care (physical/emotional), and socialization of offspring.

Each society works out its own details of what marriage is. Marriage in the United States is a legal contract between a heterosexual couple and the state in which they reside that regulates their economic and sexual relationship. Most marriages not only are a legal contract but also involve emotion, sexual monogamy, protection for future children, and a formal ceremony. These various elements of marriage are discussed in the following paragraphs.

NATIONAL DATA

Over 2.3 million marriages occur annually (Sutton, 2003). Over 95 percent of U.S. adult women and men aged 65 and older have married at least once (*Statistical Abstract of the United States: 2003*, Table 37).

Elements of Marriage

Several elements comprise the meaning of marriage in the United States.

Legal Contract Marriage in our society is a legal contract that may be entered into only by two people of different sexes of legal age (usually 18 or older) who are not already married to someone else. The marriage license certifies that the individuals were married by a legally empowered representative of the state, often with two witnesses present. Though common-law marriages exist in ten states and the District of Columbia, they are the exception.

Under the laws of the state, the license means that all future property acquired by the spouses will be jointly owned and that each will share in the estate of the other. In most states, whatever the deceased spouse owns is legally transferred to the surviving spouse at the time of death. In the event of divorce and unless there was a prenuptial agreement, the property is usually divided equally regardless of the contribution of each partner. The license also implies the expectation of sexual fidelity in the marriage. Through less frequent because of no-fault divorce, infidelity is a legal ground for both divorce and alimony in some states.

The marriage license is also an economic license that entitles a spouse to receive payment by a health insurance company for medical bills if the partner is insured, to collect Social Security benefits at the death of the other spouse, and to inherit from the estate of the deceased. Gay rights advocates seek the legalization of marriage between homosexuals so that they will have the same rights and benefits as heterosexuals. In 2003 two provinces in Canada recognized gay marriages. Movement toward the legalization of gay marriage in the United States is continuing.

Though the definition of what constitutes a "family" is being reconsidered by the courts, the law is currently designed to protect spouses, not lovers or cohabitants. An exception is **common-law marriage,** which means that if a heterosexual couple cohabit and present themselves as married, they will be regarded as legally married in those states that recognize such marriages.

Emotional Relationship "Being in love" was a top life goal (second only to being happy in life) identified by a random sample of students at Bucknell University (Abowitz & Knox, 2003). Indeed, most people in the United States regard being in love with the person they marry as an important prerequisite. The presence of love in a marriage is a primary reason people stay together (Previti & Amato, 2003). However, individuals in other cultures (e.g., India, Iran) do not require love feelings to marry—love is expected to follow, not precede, marriage.

Sexual Monogamy Marital partners are expected to be sexually faithful to each other, and most partners are faithful most of the time. Indeed, over two-thirds (69%) of 620 undergraduates reported that they would end a relationship with a partner who cheated on them and almost half (45%) said that they had done so (Knox, Zusman, Kaluzny, & Sturdivant 2000).

Legal Responsibility for Children Although individuals marry for love and companionship, one of the most important reasons for the existence of marriage from the viewpoint of society is to legally bond a male and a female for the nurture and support of any children they may have. In our society, childrearing is the primary responsibility of the family, not the state.

Marriage is a relatively stable unit that helps to ensure that children will have adequate care and protection, will be socialized for productive roles in society, and will not become the burden of those who did not conceive them. Thus, there is tremendous social pressure for individuals to be married at the time they have children. Even at divorce, the legal obligation of the father and mother to the child is theoretically maintained through child-support payments.

A couple's wedding day reflects great love, anticipation, and hope for a successful future.

Authors

Announcement/Ceremony The legal bonding of a couple is often preceded by an announcement in the local newspaper and a formal ceremony in a church or synagogue. Such a ceremony reflects the cultural importance of the event. Telling parents, siblings, and friends about wedding plans helps to verify the commitment of the partners and also helps to marshal the social and economic support to launch the couple into marital orbit.

Most people in our society decide to marry, and the benefits of doing so are enormous (Wilmoth & Koso, 2002;

Table 1.1 Benefits of Marriage and the Liabilities of Singlehood

	Benefits of Marriage	Liabilities of Singlehood
Health	Spouses have fewer hospital admissions, see a physician more regularly, are "sick" less often	Singles are hospitalized more often, have fewer medical checkups, and are "sick" more often.
Longevity	Spouses live longer than singles.	Singles die sooner than marrieds.
Happiness	Spouses report being happier than singles.	Singles report less happiness than marrieds.
Sexual satisfaction	Spouses report being more satisfied with their sex lives, both physically and emotionally.	Singles report being less satisfied with their sex lives, both physically and emotionally.
Money	Spouses have more economic resources than singles.	Singles have fewer economic resources than marrieds.
Lower expenses	Two can live more cheaply together than separately.	Cost is greater for two singles than one couple.
Drug use	Spouses have lower rates of drug use and abuse.	Singles have higher rates of drug use and abuse.
Connectedness	Spouses are connected to more individuals who provide a support system—partner, in-laws, etc.	Singles have fewer individuals upon whom they can rely for help.
Children	Rates of high school dropouts, teen pregnancies, and poverty are lower among children reared in two-parent homes.	Rates of high school dropouts, teen pregnancies, and poverty are higher among children reared by single parents.
History	Spouses develop a shared history across time with significant others.	Singles may lack continuity and commitment across time with significant others.
Crime	Spouses are less likely to be involved in crime.	Singles are more likely to be involved in crime.
Loneliness	Spouses are less likely to report loneliness.	Singles are more likely to report being lonely.

Kim & McKenry, 2002; Walen & Lachman, 2000; Arrindell & Luteijn, 2000; Waite & Gallagher, 2000). When married persons are compared with singles, the differences are striking (see Table 1.1.).

Types of Marriage

While we think of marriage in the United States as involving one man and one woman, other societies view marriage differently. **Polygamy** is a form of marriage in which there are more than two spouses. Polygamy occurs in societies or subcultures whose norms sanction multiple partners. Polygamous marriages/families have been associated with lower levels of education, socioeconomic status, and family functioning (Al-Krenawi, Graham, & Slonim-Nevo, 2002). There are three forms of polygamy: polygyny, polyandry, and pantogamy.

Polygyny **Polygyny** involves one husband and two or more wives and is practiced illegally in the United States by some religious fundamentalist groups in Arizona, New Mexico, and Utah that have splintered off from the Church of Jesus Christ of Latter-day Saints (commonly known as the Mormon Church) into what is called the Fundamentalist Church of the Latter-day Saints (FLDS) (it is estimated that there are over 40,000 polygamists in Utah alone). Polygyny serves a religious function in that large earthly families are believed to result in large heavenly families. Notice that polygynous sex is only a means to accomplish another goal—large families. Polygynous spouses are rarely prosecuted because polygyny is typically viewed as a "victimless crime." Since neither spouse complains, and there is no apparent victim. However, in some cases the wives have not reached the age of majority, so husbands may be prosecuted for child sex abuse.

Diversity in the United States

The "one size fits all" model of relationships and marriage is nonexistent. Individuals may be described as existing on a continuum from heterosexuality to homosexuality, from rural to urban dwellers, and from being single and living alone to being married and living in communes. Emotional relationships range from being close and loving to being distant and violent. Family diversity includes two parents (other or same sex), single-parent families, blended families, families with adopted children, multigenerational families, extended families, and families representing different racial, religious, and ethnic backgrounds. *Diverse* is the term that accurately describes marriage and family relationships today.

It is often assumed that polygyny exists to satisfy the sexual desires of the man, that the women are treated like slaves, and that jealousy among the wives is common. In most polygynous societies, however, polygyny has a political and economic rather than a sexual function. Polygyny is a means of providing many male heirs to continue the family line. In addition, when a man has many wives, a greater number of children for domestic/farm labor can be produced. Wives are not treated like slaves (although women have less status than men in general), as all household work is evenly distributed among the wives and each wife is given her own house or own sleeping quarters. Jealousy is minimal because the husband often has a rotational system for conjugal visits, which ensures that each wife has equal access to sexual encounters.

Polyandry The Buddhist Tibetans foster yet another brand of polygamy, referred to as **polyandry,** in which one wife has two or more (up to five) husbands. These husbands, who may be brothers, pool their resources to support one wife. Polyandry is a much rarer form of polygamy than polygyny.

The major reason for polyandry is economic. A family that cannot afford wives or marriages for each of its sons may find a wife for the eldest son only. Polyandry allows the younger brothers to also have sexual access to the one wife or marriage that the family is able to afford.

Polyamory **Polyamory** (also known as **open relationships**) is a lifestyle in which two lovers do not limit each other in having other lovers. By agreement, each partner may have numerous other emotional and sexual relationships. Some (about 5%) of the 100 members of Twin Oaks Intentional Community in Louisa, Virginia, are polyamorous in that they have a number of emotional/physical relationships with others. While not legally married, the adults view themselves as emotionally bonded to each other and may even rear children together. Polyamory is not swinging, as polyamorous lovers are concerned about enduring, intimate relationships that include sex (McCullough & Hall, 2003).

Pantogamy This describes a group marriage in which each member of the group is married to the others. Pantogamy is a more formal arrangement than polyamory and is reflected in communes (e.g.,Oneida) of the nineteenth and twentieth centuries.

Our culture emphasizes monogamous marriage and values stable marriages. One expression of this value is the concern for marriage education (see the Social Policy feature on the next page).

FAMILY

Most people who marry choose to have children and become a family. But the definition of what constitutes a family is sometimes unclear. This section examines how families are defined, their numerous types, and how marriages and families have changed in the past fifty years.

Definitions of Family

As with marriage, there are variations in defining families. The United States Bureau of the Census defines **family** as a group of two or more persons related by blood, marriage, or adoption. According to this definition, married couples with or without biological or adopted children, one parent or grandparent and a child or children, and two or more siblings or cousins who live together are the only

Call it a clan, call it a network, call it a tribe, call it a family. Whatever you call it, whoever you are, you need one.

Jane Howard, *Author*

Marriage Education in Public Schools

SOCIAL POLICY

The marriage education movement dates back to the 1930s (Bailey, 2001). Indeed, marriage and family courses have been taught in schools for over a hundred years . However, a new marriage movement in the twenty-first century composed of a group of researchers, academics, grassroots activists, clergy, and disillusioned marriage/family therapists is reemphasizing the need for marriage preparation. Sociologist Marline Pearson emphasized that teenagers "need to hear about relationships before they hit college" (and she has developed a curriculum for high school students) (Madison, 2003). President Bush joined the movement with his proposal to spend as much as $300 million on marriage education and promotion programs (Serafini & Zeller, 2002). The motivation for encouraging marriage is partly economic since divorce often leads to poverty. According to Dr. Steven Noch "for every three divorces, one woman and her children end up below the poverty line. Now that states have more responsibility for welfare, they've hopped on the marriage bandwagon" (Sarifini & Zeller, 2002). Licata (2002) noted that divorce also has negative effects for spouses, marriages, and families.

In addition, with almost half of new marriages ending in divorce, politicians ask whether social policies designed to educate youth about the realities of marriage might be beneficial in promoting marital quality and stability. Might students profit from education about marriage, before they get married, if such relationship skill training is made mandatory in the school curriculum? Over two thousand public schools nationwide are betting on a positive effect.

The philosophy behind premarriage education is that it is better to build a fence at the top of a cliff than to put an ambulance at the bottom. While other states are making it more difficult to divorce (divorce law reform), Florida is leading the way in premarriage public school marriage education. All public school high school seniors are required to take a marriage and relationship-skills course. Utah has budgeted over $600,000 for marriage-promotion education and trains teachers on marriage issues (Serafini & Zeller, 2002).

Beyond the rationale for better marriages is the suggestion that to be good citizens, social and emotional skills are necessary and that these should be taught in the public schools. "If schools are to be charged with producing a responsible, productive citizenry, then they have to teach social and emotional skills," says Nancy McLaren, a former high school instructor (*Curriculum Review,* 2002).

Opponents question using school time for relationship courses. Teachers are already seen as overworked, and an additional course on marriage seems to press the system to the breaking point. In addition, some teachers lack the training to provide relationship courses. Although training teachers would stretch already thin budgets, many schools already have programs in family and consumer sciences and teachers in these programs are trained in teaching about marriage and the family.

A related concern with teaching about marriage and the family in high school is the fear on the part of some parents that the course content may be too liberal. Some parents who oppose teaching sex education in the public schools fear that such courses lead to increased sexual activity.

Whether relationship courses will reduce divorce is unknown. Denial is operative among most lovers who believe that their relationship is unique, that only "others" get divorced, and that they know all they need to know about relationships since their love will see them through any crisis. Nevertheless, high school students might profit from knowing that certain characteristics are associated with spouses who stay together in loving, fulfilling relationships. These include having known each other at least two years before marriage, having similar values/interests/goals, and having parents who approved of their partner and relationship. Individuals who meet and marry within a short time, who have little in common, and whose parents disapprove of their relationship are more likely to divorce. In addition, the relationship skills they learn can provide a lifetime of benefits.

Beyond teaching marriage skills in high school, other efforts are being made to get information to lovestruck altar-bound couples. Florida has a "look before you leap law," which requires either four hours of premarital counseling or a waiting period of three days before a license will be issued. And in Grand Rapids, Michigan, judges can refuse to wed couples who have not had premarital counseling.

Sources

Bailey, B. L. 2001. Scientific truth . . . and love: The marriage education movement in the United States. *Journal of Social History* 87: 711–32

Curriculum Review. 2002. Adding marriage and relationships to the curriculum. Vol 41, Issue 8, 9-11. (*www.curriculumreview.com*) April.

Licata, N. 2002 Should premarital counseling be mandatory as a requisite to obtaining a marriage license? *Family Court Review* 40: 518–32.

Madison, R. C. 2003. Relationships 101. *Time.* 24 November, 63.

Serafini, M. W., and S. Zeller. 2002. Get hitched, stay hitched. *National Journal* 34:694–99.

groups that constitute a family. Couples (heterosexual or homosexual) who live together and foster families are excluded.

Henley-Walters, Warzywoda-Kruszynska, and Gurko (2002) noted that in many respects,

> *the experience of living in a family is the same in all cultures. For example, relationships between spouses and parents and children are negotiated; most relationships within families are hierarchical; the work of the home is primarily the responsibility of the wife; . . . destructive conflict between spouses or parents and children is damaging to children. . . .* (p.449)

Following the tragedy of September 11, 2001, the partners of gay and lesbian individuals were deemed eligible for the same Crime Victims Board benefits as surviving spouses. The executive order signed by New York Governor George Pataki included gay partners as family members, and their relationships were defined as family. However, New York is only one of a dozen states that recognize gay couples and their children as family. The American Red Cross also took the position that families "come in different forms," which included persons who had been living together as domestic partners. In effect, the definition of family is increasingly becoming two adult partners whose interdependent relationship is long-term and characterized by an emotional and financial commitment.

Rather than view family members as partners exclusively related by blood, marriage, or adoption, courts now more often look at the nature of the relationship between the partners. What is the level of emotional and financial commitment and interdependence? How long have they lived together? Do the partners view themselves as a family? Increasingly, families are being defined by function rather than by structure.

In 2000, Vermont was the first state to recognize committed gay relationships as **civil unions** (Massachusetts may be the first to recognize gay legal marriage). While other states may not recognize the civil unions of Vermont (and thus persons moving from Vermont to another state lose the privileges available to them in Vermont), over twenty-four cities and counties (including Canada) recognize some form of **domestic partnership.** In addition, some corporations, such as Disney, are recognizing the legitimacy of such relationships by providing medical coverage for partners of employees. Domestic partnerships are considered an alternative to marriage by some individuals and tend to reflect more egalitarian relationships than those between traditional husbands and wives.

One's family of origin may involve a close and loving relationship with a sibling.

Authors

Types of Families

There are various types of families.

Family of Origin Also referred to as the **family of orientation,** this is the family into which you were born or the family in which you were reared. It involves you, your parents, and your siblings. When you go to your parents' home for the holidays, you return to your **family of origin.** Your experiences in your family of origin have an impact on subsequent outcome behavior (Shaw et al., 2003). For example, if you are born into a family on welfare, you are less likely to attend college than if your parents are both educated and affluent.

Diversity in the United States

African-American families are characterized by their extended nature, multiple parenting, informal adoption practices, familism, supportive social networks, and flexible relationships and roles within the family units (McAdoo, 2000).

Diversity in Other Countries

Asians are also more likely than Anglo-Americans to live with their extended families. Among Asians, the status of the elderly in the extended family derives from religion. Confucian philosophy prescribes that all relationships are of the subordinate-superordinate type—husband-wife, parent-child, and teacher-pupil. For traditional Asians to abandon their elderly rather than include them in larger family units would be unthinkable. However, commitment to the elderly may be changing as a result of the Westernization of Asian countries such as China, Japan, and Korea.

In addition to being concerned for the elderly, Asians are socialized to subordinate themselves to the group. **Familism** (what is best for the group) and group identity are valued over **individualism** (what is best for the individual) and independence. Divorce is not prevalent because Asians are discouraged from bringing negative social attention to the family. In addition, the relationship that is emphasized in Asian families (particularly among the Japanese) is the mother-child relationship, not the husband-wife relationship (Tamura & Lau, 1992).

Family of Procreation The **family of procreation** represents the family that you will begin when you marry and have children. Over 95 percent of U.S. citizens living in the United States marry and establish their own family of procreation (*Statistical Abstract of the United States: 2003*). Across the life cycle, individuals move from the family of orientation to the family of procreation.

Nuclear Family The **nuclear family** refers to either a family of origin or a family of procreation. In practice, this means that your nuclear family consists of you, your parents, and your siblings, or you, your spouse, and your children. Generally, one-parent households are not referred to as nuclear families. They are binuclear families if both parents are involved in the child's life or single-parent families if only one parent is involved in the child's life and the other parent is totally out of the picture.

Sociologist George Peter Murdock (1949) emphasized that the nuclear family is a "universal social grouping" found in all of the 250 societies he studied. Not only does it channel sexual energy between two adult partners who reproduce, but these partners cooperate in the care for and socialization of offspring to be productive members of society. "This universal social structure, produced through cultural evolution in every human society, as presumably the only feasible adjustment to a series of basic needs, forms a crucial part of the environment in which every individual grows to maturity" (p. 11). Neyer and Lang (2003) emphasized that closeness to one's kinship kin continues throughout one's life and found some evidence for "blood being thicker than water" (p. 310).

Binuclear Family A **binuclear family** is a family in which the members live in two separate households. It is created when the parents of the children divorce and live separately, setting up two separate units, with the children remaining a part of each unit. Each of these units may also change again when the parents remarry and bring additional children into the respective units (**blended family**). Hence, the children go from a nuclear family with both parents to a binuclear unit with parents living in separate homes to a blended family when parents remarry and bring additional children into the respective units.

Extended Family The **extended family** includes not only your nuclear family but other relatives as well. These relatives include your grandparents, aunts, uncles, and cousins. An example of an extended family living together would be a husband and wife, their children, and the husband's parents (the children's grandparents). Family reunions are a good place to see several extended families.

DIFFERENCES BETWEEN MARRIAGE AND FAMILY

The concepts of marriage and the family are often used in tandem. Marriage can be thought of as a set of social processes that lead to the establishment of family. Indeed, every society/culture has mechanisms (from "free" dating to arranged

Table 1.2 Differences Between Marriage and the Family in the United States

Marriage	Family
Usually initiated by a formal ceremony.	Formal ceremony not essential.
Involves two people.	Usually involves more than two people.
Ages of the individuals tend to be similar.	Individuals represent more than one generation.
Individuals usually choose each other.	Members are born or adopted into the family.
Ends when spouse dies or is divorced.	Continues beyond the life of the individual.
Sex between spouses is expected and approved.	Sex between near kin is neither expected nor approved.
Requires a license.	No license needed to become a parent.
Procreation expected.	Consequence of procreation.
Spouses are focused on each other.	Focus changes with the addition of children.
Spouses can voluntarily withdraw from marriage with approval of state.	Spouses/parents cannot easily withdraw voluntarily from obligations to children.
Money in unit is spent on the couple.	Money is used for the needs of children.
Recreation revolves around adults.	Recreation revolves around children.

marriages) of guiding their youth into permanent emotionally/legally/socially bonded heterosexual relationships that are designed to lead to the reproduction and care of offspring (children). Although the concepts of marriage and the family are closely related, they are distinct. Some of these differences are identified by sociologist Dr. Lee Axelson in Table 1.2.

CHANGES IN MARRIAGE AND THE FAMILY

Whatever family we experience today was different previously and will change yet again. A look back at some changes in marriage and the family follow.

The Industrial Revolution and Family Change

The Industrial Revolution refers to the social and economic changes that occurred when machines and factories, rather than human labor, became the dominant mode for the production of goods. Industrialization occurred in the United States during the early and mid-1800s and represents one of the most profound influences on the family.

Before industrialization, families functioned as an economic unit that produced goods and services for its own consumption. Parents and children worked together in or near the home to meet the survival needs of the family. As the United States became industrialized, more men and women left the home to sell their labor for wages. The family was no longer a self-sufficient unit that determined its work hours. Rather, employers determined where and when family members would work. Whereas children in preindustrialized America worked on farms and contributed to the economic survival of the family, children in industrialized America became economic liabilities rather than assets. Child labor laws and mandatory education removed children from the labor force and lengthened their dependence on parental support. Eventually, both parents had to work away from the home to support their children. The dual-income family had begun.

During the Industrial Revolution, urbanization occurred as cities were built around factories and families moved to the city to work in the factories. Living space in cities was crowded and expensive, which contributed to a decline in the birthrate and to smaller families.

The development of transportation systems during the Industrial Revolution made it possible for family members to travel to work sites away from the home and to move away from extended kin. With increased mobility, many extended families became separated into smaller nuclear family units consisting of parents and their children. As a result of parents' leaving the home to earn wages and the absence of extended kin in or near the family household, children had less adult supervision and moral guidance. Unsupervised children roamed the streets, increasing the potential for crime and delinquency.

Industrialization also affected the role of the father in the family. Employment outside the home removed men from playing a primary role in child care and in other domestic activities. The contribution men made to the household became primarily economic.

Finally, the advent of industrialization, urbanization, and mobility is associated with the demise of familism and the rise of individualism. When family members functioned together as an economic unit, they were dependent on one another for survival and were concerned about what was good for the family. This familistic focus on the needs of the family has since shifted to a focus on self-fulfillment—individualism. Families from familistic cultures such as China who immigrate to the United States soon discover that their norms, roles, and values begin to alter in reference to the industrialized, urbanized, individualistic patterns and thinking. Individualism and the quest for personal fulfillment are thought to have contributed to high divorce rates, absent fathers, and parents' spending less time with their children.

Hence, although the family is sometimes blamed for juvenile delinquency, violence, and divorce, it is more accurate to emphasize changing social norms and conditions of which the family is a part. When industrialization takes parents out of the home so that they can no longer be constant nurturers, socializers, and supervisors, the likelihood of aberrant acts by children and adolescents increases. One explanation for school violence is that absent, career-focused parents have failed to provide close supervision for their children.

Changes in the Last Half Century

There have been enormous changes in marriage and family in the last fifty years. Changes include divorce replacing death as the endpoint for the majority of marriages, marriage and intimate relations as legitimate objects of scientific study, and the rise of feminism/changes in gender roles in marriage (Pinsof, 2002). Other changes include a delay in age at marriage, increased acceptance of singlehood, cohabitation, and childfree marriages. Table 1.3 reflects these and other changes. Skolnick (2000) reviewed the transition of families across time. She observed, "Some researchers have attacked the 'myth of family decline' . . . that family life is changing, not declining, that family values remain strong, and the new family forms reflect a healthy environment" (p. 11). Skolnick also emphasized the need for a return to intimacy and connectedness of family members.

Table 1.3 Changes in Marriages and Families from 1950 to 2005

	1950	2005
Family Relationship Values	Strong values for marriage and the family. Individuals who wanted to remain single or childless were considered deviant, even pathological. Husband and wife should not be separated by jobs or careers.	Individuals who remain single or childfree experience social understanding and sometimes encouragement. The single and child-free are no longer considered deviant or pathological but are seen as self-actuating individuals with strong job or career commitments. Husband and wife can be separated for job or career reasons and live in a commuter marriage. Married women in large numbers have left the role of full-time mother and housewife to join the labor market.
Gender Roles	Rigid gender roles, with men earning income and wives staying at home, taking care of children.	Fluid gender roles, with most wives in workforce, even after birth of children.
Sexual Values	Marriage was regarded as the only appropriate context for intercourse in middle-class America. Living together was unacceptable, and a child born out of wedlock was stigmatized. Virginity was sometimes exchanged for marital commitment.	For many, concerns about safer sex have taken precedence over the marital context for sex. Virginity is no longer exchanged for anything. Living together is regarded as not only acceptable but sometimes preferable to marriage. For some, unmarried single parenthood is regarded as a lifestyle option. It is certainly less stigmatized.
Homogamous Mating	Strong social pressure existed to date and marry within one's own racial, ethnic, religious, and social class group. Emotional and legal attachments were heavily influenced by obligation to parents and kin.	Dating and mating have become more heterogamous, with more freedom to select a partner outside one's own racial, ethnic, religious, and social class group. Attachments are more often by choice.
Cultural Silence on Intimate Relationships	Intimate relationships were not an appropriate subject for the media.	Talk shows, interviews and magazine surveys are open about sexuality and relationships behind closed doors.
Divorce	Society strongly disapproved of divorce. Familistic values encouraged spouses to stay married for the children. Strong legal constraints kept couples together. Marriage was forever.	Divorce has replaced death as the endpoint of a majority of marriages. Less stigma is associated with divorce. Individualistic values lead spouses to seek personal happiness. No-fault divorce allows for easy divorce. Marriage is tenuous. Increasing numbers of children are being reared in single-parent households apart from other relatives.
Familism versus Individualism	Families were focused on the needs of children. Mothers stayed home to ensure that the needs of their children were met. Adult concerns were less important.	Adult agenda of work and recreation has taken on increased importance, with less attention being given to children. Children are viewed as more sophisticated and capable of thinking as adults, which frees adults to pursue their own interests. Day care used regularly.
Homosexuality	Same-sex emotional and sexual relationships were a culturally hidden phenomenon. Gay relationships were not socially recognized.	Gay relationships are increasingly a culturally open phenomenon. Some definitions of the family include same-sex partners. Domestic partnerships are increasingly given legal status in some states. Same-sex marriage is a hot social/political issue.
Scientific Scrutiny	Aside from Kinsey, few studies on intimate relationships.	Acceptance of scientific study of marriage and intimate relationships.

Reprinted by permission of Dr. Lee Axelson.

CHOICES IN RELATIONSHIPS—THE VIEW OF THIS TEXT

We are not free to choose what we cannot think of.

Tom Peters, *In Search of Excellence*

A central theme of this text is choices in relationships. Although there are over a hundred such choices (see Table 1.4), five major choices include whether to marry, whether to have children, whether to have one or two earners in one marriage, whether to remain emotionally and sexually faithful to one's partner, and

Table 1.4 One Hundred and One Choices in Relationships

Lifestyle

- Marriage?
- Singlehood?
- Live together?
- Communal living?

Marriage

- Age to marry?
- Partner to marry?
- Bicultural marriage?
- Marry while in college?
- Reasons to marry?
- Know partner how long before marriage?
- Live together first?
- Consider parents' approval?
- Prenuptial agreement?
- Premarital counseling?
- Traditional or egalitarian relationship?
- One or two checking accounts?
- Live with parents?
- Partner's night out?
- Separate or joint vacations?
- Who has what roles in relationship?

Communication

- How much about the past to disclose?
- Talk about issues or avoid them?
- Consult a marriage therapist?
- Attend marriage encounter weekend?
- How honest to be?
- Secrets?
- Emotionally open/closed?
- Discuss previous sexual abuses?

Sex

- Sexual values—absolutism, relativism, hedonism?
- How much sex how soon in a relationship?
- Number of sexual partners?
- Require condom for intercourse or oral sex?
- Require HIV test before sexual behavior with new partner?
- Monogamy in relationships?
- Polyamorous relationships? Open relationships?
- Homosexual? Heterosexual? Bisexual? (may not be "choices")
- Consult sex therapist to resolve problems?
- Masturbate with one's partner?
- Disclose sexual fantasies?

Love

- Prerequisite for marriage?
- Prerequisite for sex?
- Romantic or realistic?
- Several loves at one time?
- Tolerance of partner's loving another?
- Divorce if love dies?
- Style (pragmatic, ludic, etc.) of love?

Employment

- Job or career?
- Part-time or full-time career?
- One or two earners in marriage?
- Move if one career requires it?
- Chores after both workers return home?
- Commuter marriage?
- Hire outside help?
- Age of child when return to employment?
- Age to retire?

Birth Control

- Use of contraceptive?
- Which contraceptive?
- How often to use contraceptives?
- Partner responsible for contraceptives?
- Sterilization?
- If pregnant, keep child, abort, place for adoption?

Children

- Whether to have children?
- How many children?
- Age when having first and subsequent children?
- Adopt children?
- Adopt child of another race?
- Single parent by choice?
- Artificial insemination by husband?
- Artificial insemination by donor?
- Test-tube fertilization?
- Amniocentesis?
- Home or hospital birth?
- Day care for child?
- Public or private school?
- Method of childbirth?
- Method of discipline?

Divorce

- Stay married or divorce if unhappy?
- Tolerate violence and abuse?
- Tolerate sexual abuse of one's child by partner?
- When to divorce?
- Divorce mediation?
- Shared parenting with ex-partner after divorce?
- Shared parenting with ex-partner's new partner after divorce?
- When and how to tell parents?
- Take responsibility or blame partner?
- When and what to tell children?
- Relationship with ex-spouse?
- Amount of child support?
- Seek an annulment?

Remarriage

- Whether to remarry?
- How soon to remarry?
- Remarry a person with or without children?
- Remarry against children's wishes?
- Have baby with new partner?
- Live in one of partner's houses or get another house?
- Type of relationship to establish with partner's child?

Widowhood

- Power of attorney?
- Living will?
- Develop a will?
- Amount of life insurance?
- Remarry after death of spouse? How soon?
- Live with one's children?
- Live with a companion?

whether to use a condom. Though structural and cultural influences are operative, a choices framework emphasizes that individuals have some control over their relationship destiny by making deliberate choices to initiate, respond to, nurture, or terminate intimate relationships.

Facts about Choices in Relationships

The facts to keep in mind when making relationship choices include the following.

The three biggest mistakes in life are failing to make wise choices, failing to recognize bad choices, and failing to correct bad choices (if possible) once they have been made.

Marty E. Zusman, *Sociologist*

Not to Decide Is to Decide Not making a decision is a decision by default. If you are sexually active and decide not to use a condom, you have made a decision to increase your risk for contracting a sexually transmissible infection, including HIV. If you don't make a deliberate choice to end a relationship that is going nowhere, you will continue in that relationship and have little chance of getting into a relationship that is going somewhere. If you don't make a decision to be faithful to your partner, you are vulnerable to cheating.

PERSONAL CHOICES

Will You Make Choices Deliberately or by Default?

Some of us believe we can avoid making decisions in reference to relationships, marriage, and the family. We cannot, because not to decide is to decide by default. Some examples follow:

- If we don't make a decision to pursue a relationship with a particular person, we have made a decision (by default) to let that person drift out of our life.
- If we don't decide to do the things that are necessary to keep or improve the relationships we have, we have made a decision to let the relationships slowly disintegrate.
- If we don't make a decision to be faithful to our dating partner or spouse, we have made a decision to be open to situations and relationships in which we are likely to be unfaithful.
- If we don't make a decision to avoid having intercourse early in a new relationship, we have made a decision to let intercourse occur at any time.
- If we are sexually active and don't make a decision to use birth control or a condom, we have made a decision for pregnancy or a sexually transmitted infection.
- If we don't make a decision to break up with our dating partner or spouse, we have made a decision to continue the relationship with him or her.

Throughout the text, as we discuss various relationship choices, consider that you automatically make a choice by being inactive—that not to make a choice is to make one.

Some Choices Require Corrections Some of our choices, while correct at the time, turn out to be disasters. Once it is evident that a choice is having consistently negative consequences, it is important to reverse positions, make new choices, and move forward. Otherwise, one remains consistently locked in the continued negative outcomes of a "bad" choice. For example, the choice of a partner who was once loving and kind but who has turned out to be abusive and dangerous requires correcting that choice. To stay in the abusive relationship will have predictable disastrous consequences—to make the decision to disengage and to move on opens the opportunity for a better life with another partner. Other examples of making a correction may involve ending a dead relationship (we are not suggesting that this be done without investing time and effort to improve one's relationship), returning to school for an advanced degree, or changing one's job or career. Having "regrets" over not correcting an earlier choice is not unusual (Crosnoe & Elder, 2002).

Choices Involve Trade-offs By making one choice you relinquish others. Every relationship choice you make will have a downside and an upside. If you decide to get married, you will give up your freedom of pursuing other emotional/sexual relationships and deciding how you spend your money—but you may get a wonderful companion for life if you marry. If you decide to have children, you will have less money to spend on yourself and your partner; if you don't have children, you might miss what could be a fulfilling parent-child relationship. If you maintain a long-distance relationship with someone, you continue involvement in a relationship that is obviously important to you. But you may spend many weekends alone when you could be discovering new relationships. This chapter's self-assessment provides a way to identify where you are on the continuum of relationship involvement (see p. 16).

Choices Include Selecting a Positive or Negative View Regardless of your circumstances, you can choose to view a situation in positive terms. A terminated relationship can be viewed as an opportunity to become involved in a new, more fulfilling relationship. A financial setback can be viewed as an opportunity to learn how to live on less and as an opportunity to reevaluate the importance of relationships versus money as the ultimate source of personal contentment. The discovery of infidelity on the part of one's partner can be viewed as an opportunity to open up communication channels with one's partner and develop a stronger relationship. Finally, a debilitating illness (e.g., need for kidney transplant or Lou Gehrig's Disease) can be viewed as a challenge to face adversity together. Jones et al. (2002) found that impoverished mothers who have an optimistic outlook believe that they can make a difference in the lives of their children even under difficult living conditions. One's point of view does make a difference—it is the one thing we have control over.

It takes no more time to see the good side of life than to see the bad.

Jimmy Buffet, *Tales from Margaritaville*

Choices Produce Ambivalence Choosing among options and trade-offs often creates ambivalence—conflicting feelings that produce uncertainty or indecisiveness as to what course of action to take. There are two forms of ambivalence: sequential and simultaneous. In sequential ambivalence, the individual experiences one wish and then another. For example, a person may vacillate between wanting to stay with his or her partner and wanting to break up. In simultaneous ambivalence, the person experiences two conflicting wishes at the same time. For example, the individual may at the same time feel both the desire to stay with the partner and the desire to break up. The latter dilemma is reflected in the saying, "You can't live with them and you can't live without them."

Most Choices Are Revocable; Some Are Not Most choices are revocable; that is, they can be changed. For example, a person who has chosen to be sexually active with multiple partners can later decide to be monogamous or to abstain from sexual relations. Or individuals who in the past have chosen to emphasize career (money and advancement) over marriage and family can choose to emphasize relationships over economic and career-climbing considerations.

Other choices are less revocable. For example, backing out of the role of spouse is much easier than backing out of the role of parent. Whereas the law permits disengagement from the role of spouse (formal divorce decree), the law ties parents to dependent offspring (e.g., through child support). Hence the decision to have a child is usually permanent.

Choices Are Influenced by the Stage in the Family Life Cycle The choices a person makes tend to be individualistic or familistic, depending on which stage in the family life cycle the person is in. Before marriage, individualism characterizes most thinking and decisions. People are delaying marriage in favor of completing school, becoming established in a career, and enjoying the freedom of singlehood. Once married, and particularly when they have children, their values and choices become more familistic. Evidence for familism in this regard is the fact that the divorce rate has decreased in recent years (*Statistical Abstract of the United States, 2003*). Also, couples with children have much lower divorce rates than those without children.

Making Choices Is Facilitated with Decision-Making Skills Choices occur at the individual, couple, and family level. Deciding to transfer to another school or take a job out of state may involve all three levels, whereas the decision to lose weight is more likely to be an individual decision. Regardless of the level, the steps in decision making include setting aside enough time to evaluate the issues involved in making a choice, identifying alternative courses of action and carefully weighing the consequences for each choice, and being attentive to your own inner state and motivations. The goal of most people is to make relationship choices that result in the most positive and least negative consequences. Choices are also difficult. We asked our students to identify their "most difficult" choices. Their answers included "continuing a relationship with someone I was intensely in love with but who was abusive," "leaving my partner to come to college . . . I thought we had a solid relationship . . . I was wrong," and "forgiving my partner for cheating on me."

The Relationships Involvement Scale on the next page will help you to identify the level of involvement in your current relationship.

Structural and Cultural Influences on Choices

Choices are influenced by structural and cultural factors. This section reviews the ways in which social structure and culture impact choices in relationships. Although a major theme of this book is the importance of taking active control of your life in making relationship choices, it is important to be aware that such choices are restricted and channeled by the social world in which you live. For example, enormous social disapproval for marrying someone of another race is part of the reason 95 percent of all individuals marry someone of the same race.

Social Structure The social structure of a society consists of institutions, social groups, statuses, and roles.

1. Institutions. The largest elements of society are social institutions, which may be defined as established and enduring patterns of social relationships. Major institutions include the family, economy, education, and religion. Institutions affect individual decision making. For example, most religions encourage their members to marry someone of their own faith.

2. Social groups. Institutions are made up of social groups, defined as two or more people who have a common identity, interact, and form a social relationship. Most individuals spend their day going between social groups. Your family is a social group that is part of the institution of the family. The members of your marriage and family class also form a social group that is part of the educational institution. Your closest friends influence your relationship choices.

The groups an individual belongs to influence that person's choices. In a study of 403 randomly selected college students, those who perceived their peers as drinking alcohol were more likely to drink themselves and to have problems with alcohol (Clapp & McDonnell, 2000). Falstaff, one of Shakespeare's characters, said, "Company, villainous company, hath been the spoil of me."

SELF-ASSESSMENT

The Relationship Involvement Scale

This scale is designed to assess the level of your involvement in a current relationship. Please read each statement carefully, and write the number next to the statement that reflects your level of disagreement to agreement, using the following scale.

1	2	3	4	5	6	7
Strongly Disagree						Strongly Agree

___ **1.** I have told my friends that I love my partner.

___ **2.** My partner and I have discussed our future together.

___ **3.** I have told my partner that I want to marry him/her.

___ **4.** I feel happier when I am with my partner.

___ **5.** Being together is very important to me.

___ **6.** I cannot imagine a future with anyone other than my partner.

___ **7.** I feel that no one else can meet my needs as well as my partner.

___ **8.** When talking about my partner and me, I tend to use the words "us," "we," and "our."

___ **9.** I depend on my partner to help me with many things in life.

___ **10.** I want to stay in this relationship no matter how hard times become in the future.

Scoring

Add the numbers you assigned to each item. A 1 reflects the least involvement and a 7 reflects the most involvement. The lower your total score (10 is the lowest possible score), the lower your level of involvement; the higher your total score (70), the greater your level of involvement. A score of 40 places you at the midpoint between a very uninvolved and very involved relationship.

Other Students Who Completed the Scale

Valdosta State University. The participants were 31 male and 86 female undergraduate psychology students haphazardly selected from Valdosta State University. They received course credit for their participation. These participants ranged in age from 18 to 59 with a mean age of 20.25 (SD = 4.52). The ethnic background of the sample included 70.9% white, 23.9% Black, 1.7% Hispanic, and 3.4% from other ethnic backgrounds. The college classification level of the sample included 46.2% freshmen, 36.8% sophomores, 14.5% juniors, and 2.6% seniors.

East Carolina University. Also included in the sample were 60 male and 129 female undergraduate students haphazardly selected from East Carolina University. These participants ranged in age from 18 to 43 with a mean age of 20.40 (SD = 3.58). The ethnic background of the sample included 76.2% white, 14.3% Black, 0.5% Hispanic, 1.6% Asian, 2.6% American Indian, and 4.8% from other ethnic backgrounds. The college classification level of the sample included 40.7% freshmen, 19.0% sophomores, 19.6% juniors, and 20.6% seniors. All participants were treated in accordance with the ethical guidelines of the American Psychological Association (1992).

Scores of Participants

When students from both universities were combined, the average score of the men was 50.06 (SD = 14.07) and the average score of the women was 52.93 (SD = 15.53), reflecting moderate involvement for both sexes. There was no significant difference between men and women in level of involvement. However, there was a significant difference ($p < .05$) between whites and non-whites, with whites reporting greater relationship involvement (M = 53.37; SD = 14.97) than non-whites (M = 48.33; SD = 15.14).

In addition, there was a significant difference between the level of relationship involvement of seniors compared with juniors ($p < .05$) and freshmen ($p < .01$). Seniors reported more relationship involvement (M = 57.74; SD = 12.70) than did juniors (M = 51.57; SD = 15.37) or freshmen (M = 50.31; SD = 15.33).

Source

"The Relationship Involvement Scale" 2004 by Mark Whatley, Ph.D. Department of Psychology, Valdosta State University, Valdosta, Georgia 31698. Information on validity and reliability may be obtained from Dr. Whatley. The scale is used by permission of Dr. Whatley. Other uses of this scale by written permission only; e-mail *mwhatley@valdosta.edu*

Social groups may be categorized as primary or secondary. **Primary groups,** which tend to involve small numbers of individuals, are characterized by interaction that is intimate and informal. **Secondary groups,** on the other hand, may involve small or large numbers of individuals and are characterized by interaction that is impersonal and formal. Most people live in primary group contexts with family and friends but work in secondary group contexts and interact with employers and coworkers. Membership in a group often involves commitment and sacrifice. Individuals who elect to marry and have children give up some of their private time, interests, and money and spend them in reference to their spouses and children.

3. Statuses. Just as institutions consist of social groups, social groups consist of statuses. A **status** is a position a person occupies within a social group. The statuses we occupy largely define our social identity. The statuses in a family may consist of mother, father, child, sibling, stepparent, etc. In discussing family issues we refer to statuses such as teenager, cohabitant, and spouse. Statuses are relevant to choices in that many choices are those that significantly change one's status. Making decisions that change one's status from single person to spouse to divorcee can influence how people feel about themselves and how others treat them.

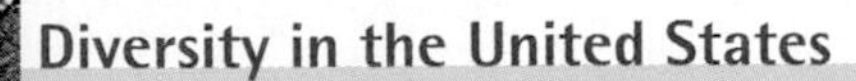

Diversity in the United States

Gay and lesbian relationships are developed and maintained in the context of disapproving institutions (most religions disapprove of homosexuality) and primary groups (family members often react with shock and grief when their child "comes out"). In addition, while the status terms in marriage are husband and wife, gay and lesbian individuals struggle with which term to use (lover, partner, significant other, companion, spouse, lifemate, etc.), and their roles as partners in the relationship are not socially scripted.

*4. **Roles.*** Every status is associated with many *roles,* or sets of rights, obligations, and expectations associated with a status. Our social statuses identify who we are; our **roles** identify what we are expected to do. Roles guide our behavior and allow us to predict the behavior of others. Spouses have adopted a set of obligations and expectations associated with their status. By doing so, they are better able to influence and predict each other's behavior than if they were strangers.

Because we occupy a number of statuses and roles simultaneously, we often experience role conflict. For example, the role of the parent may conflict with the role of the spouse, employee, or student. If your child needs to be driven to the math tutor and your spouse needs to be picked up at the airport and your employer wants you to work late and you have a final exam all at the same time, you are experiencing massive role conflict. Researchers have found that having multiple roles and balancing these roles actually have a positive effect on one's health and ability to handle stress (Stuart & Garrison, 2002). Hence, while having multiple roles may cause stress since there is a lot to do, multiple roles also provide diversion and distraction if one role becomes problematic.

Culture Just as social structure refers to the parts of society, *culture* refers to the meanings and ways of living that characterize persons in a society. Two central elements of culture are beliefs and values.

*1. **Beliefs.*** **Beliefs** refer to definitions and explanations about what is true. Hunt (2003) noted that certain beliefs are operative about why there is a gap in black/white socioeconomic status ("innate inferiority versus motivation") and that these beliefs are changing (decline in inferiority thinking).

The beliefs of an individual or couple influence the choices they make. Dual-career couples who believe that young children flourish best with a full-time parent in the home make different child-care decisions than do couples who believe that day care offers opportunities for enrichment. The belief that children are best served by being reared with two parents will influence a decision regarding what to do about a premarital pregnancy or an unhappy marriage that is different from a decision where there is a belief that single-parent families can provide an enriching context for rearing children.

*2. **Values.*** **Values** are standards regarding what is good and bad, right and wrong, desirable and undesirable. Values influence choices. Individualism, in contrast to familism, reflects the values of the society for the individual or the family. Individualism involves making decisions that are more often based on what serves the individual's rather than the family's interests (familism). Those who remain single, who live together, who seek a childfree lifestyle, and who divorce are more likely to be operating from an individualistic philosophical perspective than those who marry, do not live together before marriage, rear children, and stay married.

We baby boomers invented the concept. We have a right to be happy. Americans now believe in the happiness of entitlement. We are entitled to self realization with generally no strings attached.

Charles Sykes, *A Nation of Victims*

The elements of social structure and culture that we have just discussed play a central role in the decisions individuals make about relationships. One of the goals of this text is to encourage awareness of how powerful social structure and culture are in influencing decision making. Sociologists refer to this awareness as the **sociological imagination** (or sociological mindfulness). For example, though most people in the United States assume that they are free to select their own sex partner, this choice (or lack of it) is in fact heavily influenced by structural and cultural factors. Most people date, have sex with, and marry a person of the same racial background. Structural forces influencing race relations include segregation in housing, religion, and education. The fact that Blacks and whites live in different neighborhoods, worship in different churches, and often attend different schools makes meeting a person of another race less likely. And when encounters occur, they are so colored by the prejudices and bias brought into the interaction that they are hardly "free." Hence, cultural values (transmitted by and through parents and peers) generally do not promote mixed racial interaction, dating, and marriage. Consider the last three relationships in which you were involved, the level of racial similarity, and the structural and cultural influences on those relationships.

Other Influences on Relationship Choices

Aside from structural and cultural influences on relationship choices, other influences include your **family of origin** (the family in which you were reared), unconscious motivations, habit patterns, individual personality, and previous experiences. A dramatic example is the practice of arranged marriage, in which the marriage of two individuals is seen as the merger of two families.

Your family of origin is a major influence on your subsequent choices and relationships. For instance, individuals reared in homes fostering close infant-parent attachment tend to be attracted to and form intimate relationships as adults (Crowell et al., 2002), and "they do not tend to worry about being abandoned or of becoming too intimate" (Sasaki, 2002, 177). Adolescent girls (ages 14 to 15) who have close relationships with their mothers report a later sexual debut (McNeely et al., 2002).

Being reared in an economically impoverished family/neighborhood is also associated with deviant peers and behavior (Barrera et al., 2002). Living apart from both parents whether as a result of divorce or being born out of wedlock is associated with an increased risk of divorce (Teachman, 2002). Humor in one's family of origin (among women) is associated with one's emotional and physical health (Nelson, 2002).

The marriage of these spouses who were brought up in India was arranged by their parents. The couple now live in the United States with their daughters. While they do not require that their daughters marry a person they select, their daughters are very emotionally close to their parents and value their approval. The eldest daughter is already married—to a man from India. The influence of one's family of origin is enormous.

Authors

Another example of the impact of the family of origin on relationship choices is the Chinese family, in which one's parents "continue to have a powerful effect on their children's marriages" (Pimentel, 2000, 44). Their direct involvement in helping to select a suitable partner and their approval of the selection are related to the subsequent reported marital happiness of their children.

Unconscious motivations may also be operative in making choices. A person reared in a lower-class home without adequate food and shelter may become overly concerned about the accumulation of money and make all decisions in reference to gaining higher income. The biographer of Groucho Marx suggested that his life was indelibly affected by his "being pushed out of the nest before he was ready," with the result that he internalized his lost childhood and remained forever "immature in matters of women, money and power" (Kanfer, 2000).

Habit patterns also influence choices. A person who is accustomed to and enjoys spending a great deal of time alone may be reluctant to make a commitment to live with a person who makes demands on his or her time. A person who is a workaholic is unlikely to allocate time to a relationship to make it flourish. Frequent alcohol consumption is associated with a higher number of sexual partners and not using condoms to avoid pregnancy and contraction of sexually transmitted infections (LaBrie, Schiffman, and Earleywine, 2002).

Personalities (e.g., introvert, extrovert; passive, assertive) also influence choices. For example, a person who is assertive is more likely than someone who is passive to initiate talking with someone he or she is attracted to at a party. A person who is very quiet and withdrawn may never choose to initiate a conversation even though he or she is attracted to someone. A divorced woman who is very individualistic in her thinking is less likely to remarry (DeGraaf & Kalmijn, 2003). Finally, a person with a bipolar disorder who is manic one part of the semester (or relationship) and depressed the other part is likely to make different choices when each process is operative (Young 2003).

Current and past relationship experiences also influence one's perceptions and choices. Individuals in a current relationship are more likely to hold relativistic sexual values (choose intercourse over abstinence). And individuals who have had prior sexual experience are more likely to do so in the future (Gillmore et al., 2002).

The life of Ann Landers illustrates how life experience changes one's view and choices. After her death, her only child, Margo Howard (2003), noted that her mother was once against intercourse before marriage, divorce, and involvement with a married man. But when Margo told her that unmarried youth *were* having intercourse, Landers shifted her focus to the use of contraception. When Margo divorced, Landers began to say that ending an unfulfilling marriage is an option. And when Landers was divorced and fell in love with a married man, she said that you can't control who you fall in love with.

THEORETICAL FRAMEWORKS FOR VIEWING MARRIAGE AND THE FAMILY

Although we emphasize choices in relationships as the framework for viewing marriage and the family, other conceptual theoretical frameworks are helpful in understanding the context of relationship decisions. All **theoretical frameworks** are the same in that they provide a set of interrelated principles designed to explain a particular phenomenon and provide a point of view. In essence, theories are explanations (White & Klein, 2002).

Structural-Functional Framework

Just as the human body is made up of different parts that work together for the good of the individual, society is made up of different institutions (family, education, economics, etc.) that work together for the good of society. **Functionalists** view the family as an institution with values, norms, and activities meant to provide stability for the larger society. Such stability is dependent on families' serving various functions for society.

First, families serve to replenish society with socialized members. Since our society cannot continue to exist without new members, we must have some way of ensuring a continuing supply. But just having new members is not enough. We need socialized members—those who can speak our language and know the norms and roles of our society. The legal bond of marriage and the obligation to nurture and socialize offspring help to assure that this socialization will occur.

Second, marriage and the family promote the emotional stability of the respective spouses. Society cannot provide enough counselors to help us whenever we have problems. Marriage ideally provides an in-residence counselor who is a loving and caring partner with whom a person shares his or her most difficult experiences.

Children also need people to love them and to give them a sense of belonging. This need can be fulfilled in a variety of family contexts (two-parent family, single-parent family, extended family). The affective function of the family is one of its major offerings. No other institution focuses so completely on fulfilling the emotional needs of its members as do marriage and the family.

Third, families provide for the economic support of their members. Although modern families are no longer self-sufficient economic units, they provide food, shelter, and clothing for their members. One need only consider the homeless in our society to be reminded of this important function of the family.

In addition to the primary functions of replacement, emotional stability, and economic support, other functions of the family include the following:

- *Physical care—Families provide the primary care for their infants, children, and aging parents. Other agencies (neonatal units, day care, assisted living residences) may help, but the family remains the primary and recurring caretaker.*
- *Regulation of sexual behavior—Spouses are expected to confine their sexual behavior to each other, which reduces the risk of having children who do not have socially and legally bonded parents and of contracting/spreading HIV or other sexually transmitted infections.*
- *Status placement—Being born in a family provides social placement of the individual in society. One's social class, religious affiliation, and future occupation are largely determined by one's family of origin. Prince William, the son of Prince Charles and the late Diana, by virtue of being born into a political family, was automatically in the upper class and destined to be in politics.*
- *Social control—Spouses in high-quality durable marriages provide social control for each other that results in less criminal behavior. Parole boards often note that the best guarantee that a person released from prison will not return to prison is a spouse who expects the partner to get a job and avoid criminal behavior and who reinforces these goals.*

Conflict Framework

Conflict theorists recognize that family members have different goals and values that result in conflict. Conflict is inevitable between social groups (e.g., parents and children) (White & Klein, 2002). Conflict theory provides a lens through which to view these differences. Whereas functionalists look at family practices as good for the whole, conflict theorists recognize that not all family decisions are good for every member of the family. Indeed, some activities that are good for one member are not good for others. For example, a woman who has devoted her life to staying home and taking care of the children may decide to return to school or to seek full-time employment. This may be a good decision for her personally, but her husband and children may not like it. Similarly, divorce may have a positive outcome for spouses in turmoil but a negative outcome for children, whose standard of living and access to the noncustodial parent are likely to decrease.

Conflict theorists also view conflict not as good or bad but as a natural and normal part of relationships. They regard conflict as necessary for change and growth of individuals, marriages, and families. Cohabitation relationships, marriages, and families all have the potential for conflict. Cohabitants are in conflict about commitment to marry, spouses conflict about the division of labor, and parents are in conflict with their children over rules such as curfew, chores, and homework. These three units may also be in conflict with other systems. For example, cohabitants are in conflict with the economic institution for health benefits for their partners. Similarly, employed parents are in conflict with their employers for flexible work hours, maternity/paternity benefits, and day-care/elder-care facilities.

Karl Marx emphasized that conflict emanates from struggles over scarce resources and for power. Though Marxist theorists viewed these sources in terms of the conflict between the owners of production (bourgeoisie) and the workers (proletariat), they are also relevant to conflicts within relationships. The first of these concepts, conflict over scarce resources, reflects the fact that spouses, parents, and children compete for scarce resources such as time, affection, and space. Spouses may fight with each other over how much time should be allocated to one's job, friends, or hobbies. Parents are sometimes in conflict with each other over who does what housework/child care. Children are in conflict with their parents and with each other over time, affection, what programs to watch on television, and money.

Conflict theory is also helpful in understanding choices in relationships with regard to mate selection and jealousy. Unmarried individuals in search of a partner are in competition with other unmarried individuals for the scarce resources of a desirable mate. Such conflict is particularly evident in the case of older women in competition for men. At age 85 and older, there are twice as many women as there are men (*Statistical Abstract of the United States: 2003* Table 12). Jealousy is also sometimes about scarce resources. People fear that their "one and only" will be stolen by someone else who has no partner.

Conflict theorists also emphasize conflict over power in relationships. Premarital partners, spouses, parents, and teenagers also use power to control each other. The reluctance of some courtship partners to make a marital commitment is an expression of wanting to maintain their autonomy, since marriage implies a relinquishment of power from each partner to the other. Spouse abuse is sometimes the expression of one partner trying to control the other through fear, intimidation, or force. Divorce may also illustrate control. The person who executes the divorce is often the person with the least interest in the relationship. Having the least interest gives that person power in the relationship. Parents and adolescents are also in a continuous struggle over power. Parents attempt to use privileges and resources as power tactics to bring compliance in their adolescent. But adolescents may use the threat of suicide as their ultimate power ploy to bring their parents under control.

Feminist Framework

Feminist thought "originated with a social movement for change" (White & Klein 2002, 171) and is embodied by no one feminist framework. Lorber (1998) identified eleven perspectives, including lesbian feminism (oppressive heterosexuality and men's domination of social spaces), psychoanalytic feminism (cultural domination of men's phallic-oriented ideas and repressed emotions), and standpoint feminism (neglect of women's perspective and experiences in the production of knowledge). Regardless of which feminist framework is being discussed, all feminist frameworks have the themes of inequality and oppression. According to feminist theory, gender structures our experiences (i.e., women and men will experience life differently because there are different expectations for the respective genders) (White & Klein, 2002). Up Close 1.1 provides a wider perspective on feminist theory.

Up Close 1.1 Feminist Theories*

There are two widely believed misconceptions about feminism and feminist theory. The first misconception is that feminism seeks to replace men's domination of women with women's domination of men. To the contrary, all feminist theories advocate gender equality. Feminist theories may differ about the sources of gender inequality and its possible solutions, but none seeks simply to replace one form of gender domination with another.

The second misconception is that feminist theory is a unified single theory. Nothing could be further from the truth. Feminist theory encompasses many different ideas and explanations. Feminism is as diverse as any other area of social theory. Many feminist theories differ on the scope of their explanation. Some—such as socialist feminism—are macro approaches, while other theorists such as Arlie Hochschild (1997, 1989) take a more micro perspective. In addition to differences in scope, feminist theories differ in the aspects of social life to which they apply. The substantive application of feminist theory encompasses topics ranging from domestic labor and child care to the effects of patriarchal structures found in religion and government. Like all social theory, feminist theories encompass the full range of human activities. And like other areas of social theory, there are competing perspectives and theoretical differences between scholars.

Sometimes students ask, "How many feminist theories are there?" The answer to this question changes rapidly as the field of feminist scholarship grows. Feminist theory developed rapidly during the second half of the twentieth century, although feminism is hardly a new area of scholarship. In the United States, feminist writings and activism began gaining momentum in the early nineteenth century (see the works of Frances Wright, 1795–1852, and Angelina Grimké 1805–79). Today, despite misrepresentation of feminism and feminists in the popular press, feminist theory is one of the most dynamic and promising areas of development in social science.

Among the areas of feminist research with applications to the study of family are liberal feminism, radical feminism, cultural feminism, and socialist feminism. Liberal feminism locates the source of contemporary gender oppression in the history of exclusion. Until recently women have been denied the opportunity to participate equally in all spheres of social life. Throughout the nineteenth and twentieth centuries feminist activism secured the rights to vote, to own property, to sue for divorce, to have custody of their own children, to have equal access to public education, and to be employed, paid, and promoted without discrimination. While complete gender equality has yet to be realized, liberal feminists argue that opening up the opportunity structure to include women will slowly decrease gender inequality. Thus, the structure of society need not change; it need only include women as unfettered participants.

Radical feminists locate the roots of gender oppression in patriarchy. Patriarchy is a form of organization that originated with family. In this form of organization men are seen as the legitimate and natural rulers, while women and children must be subservient to the father. While patriarchy originated as a family structure, it has been adapted to every other social institution. Many religions worship a "father god." Schools use a hierarchical structure with students subject to teachers, and teachers subject to a principal or head master. Government is similarly patterned on a patriarchal model. In fact, we call loyalty to our government "patriotism." Radical feminists believe that until we remove this form of organization from all our social institutions, there cannot be gender equality. This theory calls for the radical reorganization of social structure— hence the term *radical feminism.*

Socialist feminism takes a position between liberal and radical feminism. Its adherents locate the roots of gender oppression in the economy. Like conflict theorists, they believe that the private ownership of the means of production has created a grossly unequal class system. The capitalist class system creates both fabulous wealth and persistent poverty. Women and their dependent children make up the bulk of the impoverished class. Socialist feminists believe that everyone deserves enough not only to subsist but also to thrive (Kourany, Sterba, & Tong, 1999). They contend that it is our collective social responsibility to guarantee that everyone has access to adequate food, shelter, education, and health care. Like the liberal feminists, they believe that in a social structure freed from gender barriers, women would succeed equally with men. Like the radical feminists, they don't think it can be accomplished with the present structure. But the socialist feminists see capitalism rather than patriarchy as the root of gender oppression.

Cultural feminism contends that by either nature, nurture, or some combination thereof, women and men have developed different cultures within society. Theories within this perspective have also been called "difference feminism." Women, they argue, have developed an emphasis on creating and maintaining cooperative intimate relationships. Men have developed ways of controlling their environment and asserting their own authority over others via intense competition. Cultural feminists study these different cultures in an attempt to transform our competitive and violent culture into a more cooperative social environment.

Sources

Grimké, Angelina E. 1836. *Appeal to the Christian women of the South.* New York.

Hochschild, Arlie Russell. 1989. *The second shift: Working parents and the revolution at home.* New York: Viking.

——. 1997. *The Time bind: When work becomes home and home becomes work.* New York: Metropolitan Books.

Kourany, Janet A., James P. Sterba, and Rosemarie Tong. 1999. *Feminist philosophies.* 2d ed. Upper Saddle River, N.J.: Prentice Hall.

Wright, Frances. 1821. *Views of society and manners in America in a series of letters from the United States to a friend in England 1818, 1819 and 1820.* New York: E. Bliss and E. White.

——. 1829. *Course of popular lectures delivered by Frances Wright in New York, Philadelphia and other cities of the United States.* New York: G.W. & A.J. Matsell.

*Von Bakanic, Department of Sociology, College of Charleston, Charleston, South Carolina.

Symbolic Interaction Framework

Marriages and families represent symbolic worlds in which the various members give meaning to each other's behavior. Human behavior can be understood only by the meaning attributed to behavior (White & Klein, 2002). Herbert Blumer (1969) used the term *symbolic interaction* to refer to the process of interpersonal interaction. Concepts inherent in this framework include the definition of the situation, the looking-glass self, and the self-fulfilling prophecy.

Definition of the Situation Two people who have just spotted each other at a party are constantly defining the situation and responding to those definitions. Is the glance from the other person (1) an invitation to approach, (2) an approach, or (3) a misinterpretation—the other person was looking at someone behind the person? The definition a person arrives at will affect subsequent interaction.

Looking-Glass Self The image people have of themselves is a reflection of what other people tell them about themselves (Cooley, 1964). The people at the party referred to above develop an idea of who they are by the way others act toward them. If no one looks at or speaks to them, according to Cooley, they will begin to feel unsettled. Similarly, family members constantly hold up social mirrors for one another into which the respective members look for definitions of self.

The importance of the family (and other caregivers) as an influence on the development and maintenance of a positive self-concept cannot be overemphasized. Orson Welles, known especially for his film *Citizen Kane,* once said that he was taught that he was wonderful and that everything he did was perfect. He never suffered from a negative self-concept. Cole Porter, known for creating such memorable songs as "I Get a Kick Out of You," had a mother who held up social mirrors offering nothing but praise and adoration. Since children spend their formative years surrounded by their family, the self-concept they develop in that setting is important to their feelings about themselves and their positive interaction with others.

G. H. Mead (1934) believed that people are not passive sponges but evaluate the perceived appraisals of others, accepting some opinions and not others. Although some children are taught by their parents that they are worthless, they may eventually overcome the definition.

Self-Fulfilling Prophecy Once people define situations and the behaviors they are expected to engage in, they are able to behave toward one another in predictable ways. Such predictability of behavior also tends to exert influence on subsequent behavior. If you feel that your partner expects you to be faithful to him or her, your behavior is likely to conform to these expectations. The expectations thus create a self-fulfilling prophecy.

Symbolic interactionism as a theoretical framework helps to explain various choices in relationships. Individuals who decide to marry have defined their situation as a committed reciprocal love relationship. This choice is supported by the belief that the partners will view each other positively (looking-glass self) and be faithful spouses and cooperative parents (self-fulfilling prophecies).

Family Life Course Development Framework

The **family life course development framework** emphasizes the process of how families change over time (White & Klein, 2002). Family development is a process that follows distinct, norm-related stages (e.g., parenthood). As families move systematically through these stages, they change. The family's life course, or **family career,** is comprised of all the stages and events that have occurred within the family (White & Klein, 2002). At different stages there may be developmental tasks or expectations of what typically occurs during this stage. The timing of the movement across stages and the completion of developmental tasks is important. If developmental tasks at one stage are not accomplished, functioning in subsequent stages will be impaired. For example, one of the developmental tasks of early marriage is to emotionally and financially separate from one's family of origin. If such separation does not take place, independence as individuals and as a couple is impaired. LaSala (2002) noted that having distance from one's parents may be particularly functional for gay couples if the parents of one of the gay partners disapprove of the couple's relationship.

Tasks are sometimes completed out of order or simply at the wrong time. Becoming a grandparent at age 45 may seem "off-time" and may make it difficult to accept this new role. Social timing also states that at a certain age, particular events should follow. Off-time, or out-of-sequence, events also can lead to disruptions later in life. For example, having a baby typically follows marriage. This sequence typically occurs after graduating from high school and perhaps from college. If a young teenager has a baby prior to marriage, her education may well be disrupted.

The family life course development framework may also help to identify the choices with which many individuals are confronted throughout life. Each family stage presents choices. For example, the never-married are choosing partners, the newly married are making choices about careers and when to begin their family, the soon-to-be-divorced are making decisions about custody/child support/division of property, and the remarried are making choices with regard to stepchildren and ex-spouses. Grandparents are making choices about how much child care they want to commit to and widows/widowers are concerned with where to live (children, retirement home, with a friend, alone).

Social Exchange Framework

The social exchange framework operates from a premise of **utilitarianism**—that individuals rationally weigh the rewards and costs associated with behavioral choices. Each interaction between spouses, parents, and children can be understood in terms of each individual's seeking the most "benefits" at the least "cost" so as to have the highest "profit" and avoid a "loss" (White & Klein, 2002). Both men and women marry because they perceive more benefits than costs for doing so. Similarly, those who remain single or divorce perceive fewer benefits and more costs for marriage. We examine how the social exchange framework is operative in mate selection later in the text.

A social exchange view of marital roles emphasizes that spouses negotiate the division of labor on the basis of exchange. For example, he participates in child care in exchange for her earning an income, which relieves him of the total financial responsibility. Social exchange theorists also emphasize that power in relationships is the ability to influence, and avoid being influenced by, the partner. The various bases of power, such as money, the need for a partner, and brute force, may be expressed in various ways, including withholding resources, decreasing investment in the relationship, and violence.

Systems Framework

Systems theory is the most recent of all the theories for understanding family interaction (White & Klein, 2002). Its basic premise is that each member of the family is part of a system and the family as a unit develops norms of interacting, which may be explicit (e.g., parents specify chores for the children) or implied (e.g., spouses expect fidelity from each other). These rules serve various functions such as allocating the resources (e.g., money for vacation), specifying the division of power (e.g., who decides how money is spent), and defining closeness and distance between systems (e.g., seeing or avoiding parents/grandparents). Rules are most efficient if they are flexible. For example, they should be adjusted over time in response to children's growing competence. A rule about not leaving the yard when playing may be appropriate for a 4-year-old but inappropriate for a 15-year-old. The rules and individuals can be understood only by recognizing that "all parts of the system are interconnected" (White & Klein, 2002, 122).

Family systems also have subsystems, such as the marital relationship and the parent-child relationship, with one having an influence on the other. A team of researchers (Frosch, Mangelsdorf, & McHale, 2000) observed that children who experience their parents' marriage as positive feel more secure in their relationship with their parents than do children who experience that marriage as negative and conflictual. The findings of Kitzmann (2000) provided further support that couple negativity is correlated with family negativity in that spouses who fight with each other are more likely to be combative with their children.

Family members also develop boundaries that define the individual and the group and separate one system or subsystem from another. A boundary is a "border between the system and its environment that affects the flow of information and energy between the environment and the system" (White & Klein, 2002, p. 124). A boundary may be physical, such as a closed bedroom door, or social, such as expectations that family problems will not be aired in public. Boundaries may also be emotional, such as communication, which maintains closeness or distance in a relationship. Some family systems are cold and abusive; others are warm and nurturing.

In addition to rules and boundaries, family systems have roles (leader, follower, scapegoat) for the respective family members. These roles may be shared by more than one person or may shift from person to person during an interaction or across time. In healthy families, individuals are allowed to alternate roles rather than being locked into one role. In problem families, one family member is often allocated the role of scapegoat, or the cause of all the family's problems (e.g., alcoholic spouse).

Family systems may be open, in that they are open to information and interaction with the outside world, or closed, in that they feel threatened by such contact. The Amish have closed family systems and minimize contact with the outside world. Some communes also encourage minimal outside exposure. Twin Oaks Intentional Community of Louisa, Virginia, does not permit any of its almost one hundred members to own or keep a television in their rooms. Exposure to the negative drumbeat of the evening news is seen as harmful.

Human Ecological Framework

Human ecology is the study of **ecosystems,** or the interaction of families with their environment (Bubolz & Sontag, 1993). Individuals, couples, and families are dependent on the environment for air, food, and water and on other human beings for social interaction (White & Klein, 2002). The well-being of individuals and families cannot be considered apart from the well-being of the ecosystem. This framework emphasizes the importance of examining families within multiple contexts. For example, nutrition and housing are important to the functioning of families. If a family does not have enough to eat or adequate housing, it will not be able to function at an optimal level.

Human beings and their environment also are interdependent in that they depend on each other and they influence each other. Individuals and families do not operate in isolation but interact with multiple environments—schools, workplace, and neighborhoods (Bronfenbrenner,1979; Jenkins, Rasbash,

Table 1.5 Comparison of Theoretical Frameworks

Theory	Description	Concepts	Level of Analysis	Strengths	Weaknesses
Structural-Functional	The family has several important functions within society; within the family, individual members have certain functions.	Structure Function	Institution	Emphasizes the relation of family to society, noting how families affect and are affected by the larger society.	Families with nontraditional structures (single-parent, same-sex couples) are seen as dysfunctional.
Conflict	Conflict in relationships is inevitable, due to competition over resources and power.	Conflict Resources Power	Institution	Views conflict as a normal part of relationships and as necessary for change and growth.	Sees all relationships as conflictual, and does not acknowledge cooperation.
Symbolic Interaction	People communicate through symbols and interpret the words and actions of others.	Definition of the situation Looking-glass self Self-fulfilling prophecy	Couple	Emphasizes the perceptions of individuals, not just objective reality or the viewpoint of outsiders.	Ignores the larger social context and minimizes the influence of external forces.
Family Life Course Development	All families have a life course or family career that is composed of all the stages and events that have occurred within the family.	Family career Stages Transitions Timing	Institution Individual Couple Family	Families are seen as dynamic rather than static. Useful in working with families who are facing transitions in their life courses.	Difficult to adequately test the theory through research.
Social Exchange	In their relationships, individuals seek to maximize their benefits and minimize their costs.	Benefits Costs Profit Loss	Individual Couple Family	Provides explanations for human behavior based on outcome.	Assumes that people always act rationally and all behavior is calculated.
Family Systems	The family is a system of interrelated parts that function together to maintain the unit.	Subsystem Roles Rules Boundaries Open system Closed system	Couple Family	Very useful in working with families who are having serious problems (violence, alcoholism). Describes the effect family members have on each other.	Based on work with troubled families and may not apply to nonproblem families.
Feminism	Women's experience is central and different from man's experience of social reality.	Inequality Power Oppression	Institution Individual Couple Family	Exposes inequality and oppression as explanations for frustrations experienced by women.	Multiple branches of feminism may inhibit central accomplishment of increased equality.
Human Ecological	Families interact with and are interdependent with their environment.	Ecosystem Interdependence Environment	Institution Individual Couple Family	Can be applied to families of different structures and ethnic or racial backgrounds.	Scope of the theory may be too broad.

Appreciation is expressed to Esther Devall of New Mexico State University for her assistance in the development of this table.

Authors

Blacks in the United States are often assumed to be African-American. This couple lives in the United States but is from Panama. They report that their Panamanian heritage and ethnic identification are rarely considered when others meet them for the first time.

& O'Connor, 2003). These interactions are reciprocal. For example, stress at work will influence one's family life and vice versa.

Stratification/Race Framework

Though not formal theoretical frameworks, stratification and race provide ways of viewing and understanding choices in relationships.

After all, there is but one race—humanity.
George Moore

Stratification **Stratification** refers to the ranking of people into layers or strata according to their socioeconomic status or social class, usually indexed according to income, occupation, and educational attainment. Passengers on the *Titanic* were stratified and assigned to different decks, which influenced who they would interact with. Individuals who occupy different socioeconomic statuses are less likely to meet and marry.

Marriages and families are also stratified into different social classes, such as the upper, middle, working, or lower social class. Families in these various social classes reflect dramatic differences in their attitudes, values, and behavior. For example, individuals from the lower class are more likely to divorce than individuals from the higher social classes. Parents in lower socioeconomic classes are also more likely to discuss personal and financial problems with their children than parents in higher socioeconomic groups. The former feel that the sooner their children become aware of the harsh realities of life, the better. Middle-class parents, on the other hand, believe that they should protect their children from the realities that lie ahead. One's social class and, by implication, occupation also influence the time available for leisure activities. Persons in low-skilled occupations have long hours and low pay with little leisure time (Salmon et al., 2000).

Race Racial heritage also influences choices in relationships. For example, Blacks are more likely than whites to report having had sexual intercourse earlier. In a national longitudinal study, almost half (45%) of blacks reported having had sexual intercourse by age 15–16, as compared with 31 percent for other racial groups (Cooksey, Mott, & Neubauer, 2002).

The term **race** is a social construct the meaning of which has less to do with biological differences than with social, cultural, political, economic, behavioral, fertility, and health differences. With regard to the latter, the life expectancy of white men born in 2005 is projected to be 75.4 compared with 69.9 for Black men; the figures for white women and Black women are projected to be 81.1 and 76.8, respectively (*Statistical Abstract of the United States: 2003,* Table 91).

A person's membership in a particular racial or ethnic group is sometimes not as clear-cut as might be expected. For example, not all Black people are African-Americans. Black immigrants from the Caribbean have a strong ethnic identity that is not African-American. In addition, differences that appear to be racial may actually be social class differences. For example, individuals in the lower class (whether white or Black) have higher rates of unemployment, premarital pregnancies, divorce, and crime. In looking at the comparisons between Blacks and whites throughout this text, it is important to keep in mind that many presumed racial differences are really those of social class. However, racism still exists in the form of discrimination against minorities in education, employment, and housing, which affects spouses, parents, and children.

The major theories are summarized in Table 1.5. In addition to the use of theory in studying marriage and the family, a review of how marriage/family research is conducted is provided in Special Topic 2, Evaluating Research in Marriage and the Family. Knowledge of sampling, control groups, etc. will assist you in evaluating the various research referred to in this text.

SUMMARY

What is marriage?

Marriage is a system of binding a man and a woman together for the reproduction, care (physical/emotional), and socialization of offspring. Marriage in the United States is a legal contract between a heterosexual couple and the state in which they reside that regulates their economic and sexual relationship. Other elements of marriage involve emotion, sexual monogamy, and a formal ceremony.

What is family?

The Census Bureau defines family as a group of two or more persons related by blood, marriage, or adoption. In recognition of the diversity of families, the definition of family is increasingly becoming two adult partners whose interdependent relationship is long-term and characterized by an emotional and financial commitment.

How do the concepts of marriage and the family differ?

Marriage involves a license; having a child does not. Marriage ends when one person dies or the couple divorce; relationships between the family members of parent and child continue. Marriages involve spouses who are expected to have sex with each other; families involve parents and children, and sex between them is prohibited.

How have marriage and the family changed?

The advent of industrialization, urbanization, and mobility involved the demise of familism and the rise of individualism. When family members functioned together as an economic unit, they were dependent on one another for survival and were concerned about what was good for the family. This familistic focus on the needs of the family has since shifted to a focus on self-fulfillment—individualism. Other changes in the last fifty years include divorce replacing death as the endpoint for the majority of marriages, marriage and intimate relations as legitimate objects of scientific study, the rise of feminism/changes in gender roles in marriage, a delay in age at marriage, and increased acceptance of singlehood, cohabitation, and childfree marriages.

What is the theme of this text?

A central theme of this text is choices in relationships. Five major choices include whether to marry, whether to have children, whether to have one or two earners in one marriage, whether to remain emotionally and sexually faithful to one's partner, and whether to use a condom. Though structural and cultural influences are operative, a choices framework emphasizes that individuals have some control over their relationship destiny by making deliberate choices to initiate, respond to, nurture, or terminate intimate relationships. Important issues to keep in mind about a choices framework for viewing marriage and the family are that not to decide is to decide, some choices require correcting, all choices involve trade-offs, choices include selecting a positive or negative view, making choices produces ambivalence, and some choices are not revocable.

What are the structural and cultural influences of choices?

Interpersonal choices are restricted and channeled by the social world in which a person lives. This involves the social structure of a society, which consists of institutions (e.g., education, economy), social groups (e.g., primary groups, secondary groups), statuses (e.g., parent, child), and roles (e.g., provider, nurturer). Culture refers to the meanings and ways of living that characterize persons in a society. Central elements of culture are beliefs (e.g., marriage is better than singlehood) and values (e.g., familism versus individualism).

What theoretical frameworks are used to study the family?

Theoretical frameworks provide a set of interrelated principles designed to explain a particular phenomenon and provide a point of view. Those used to study the family are the structural-functional framework (how the family functions to serve society), conflict framework (family members are in conflict over scarce resources of time and money), feminist framework (inequality and oppression), symbolic interaction framework (symbolic worlds in which the various family members give meaning to each other's behavior), family life course development framework (the process of how families change over time), social exchange framework (spouses exchange resources and decisions are made on the basis of perceived profit and loss), systems framework (each member of the family is part of a system and the family as a unit develops norms of interaction), the human ecological framework (individuals, couples, and families are dependent on the environment for air, food, and water and on other human beings for social interaction), and the stratification/race framework(families operate in different social classes and reflect divergent racial identities/histories).

KEY TERMS

beliefs
binuclear family
blended family
civil union
common-law marriage
domestic partnership
ecosystem
extended family
familism
family
family career
family life course development
family of orientation
family of origin
family of procreation
functionalists
human ecology
individualism
marriage
nuclear family
open relationship
polyamory
polyandry
polygamy
polygyny
primary group
race
role
secondary group
sociological imagination
status
stratification
theoretical framework
utilitarianism
values

RESEARCHING MARRIAGE AND THE FAMILY WITH INFOTRAC COLLEGE EDITION

InfoTrac College Edition, an online library, allows you to perform research online anywhere, anytime. Following are two suggested search terms and related questions to help you extend your understanding of the topics covered in this chapter. Go to www.infotrac-college.com to begin your search.

Keyword: **Marriage license.** Locate the "Marriage 101" article that discusses the current trend to reduce the marriage license fee if the couple agrees to a short marriage education course. Do you feel that the government should encourage marriage education via reduced fees or should avoid "meddling" in interpersonal relationships?

Keyword: **Arranged marriage.** Locate pro and con articles on arranged marriages. What are the principal advantages and disadvantages of arranged marriages?

The Companion Web Site for Choices in Relationships: An Introduction to Marriage and the Family, Eighth Edition

http://sociology.wadsworth.com/knox_schacht/choices8e

Supplement your review of this chapter by going to the companion Web site to take one of the Tutorial Quizzes, use the flash cards to master key terms, and check out the many other study aids you'll find there. You'll also find special features such as the Marriage and Family Resource Center, Census 2000 information, and other data and resources at your fingertips to help you with that special project or to do some research on your own.

WEBLINKS

Gilder Lehrman Institute of American History—History of the Family
http://www.digitalhistory.uh.edu/historyonline/familyhistory.cfm

National Center for Health Statistics
http://www.cdc.gov/nchs/

National Council on Family Relations
http://www.ncfr.org/

National Marriage Project
http://marriage.rutgers.edu/about.htm

National Survey of Families and Households
http://www.ssc.wisc.edu/nsfh/home.htm

U.S. Census Bureau
http://www.census.gov/

Authors

CHAPTER 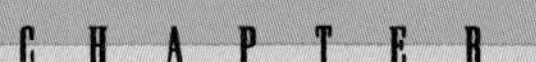 2

Man has his will—but woman has her way.

Oliver Wendell Holmes, Sr., Nineteenth-century physician and author

Gender in Relationships

Contents

True or False?

1. The case of David Reimer, who was reared as a girl, provides convincing evidence that people become the gender they are socialized to be, that biology has a limited impact on our becoming a man or a woman.
2. In contemporary U.S. society, dating interaction norms and scripts continue to be traditional (with the man asking the woman out, deciding on plans, etc. and the woman waiting to be picked up, introducing the male to her family, etc.).
3. Up until age 6 or 7 children think that they can change their gender and become someone of the opposite sex.
4. The majority of the women in rural Afghanistan have little opportunity to escape their house, village, and province.
5. Over half of first-year college men in a national survey agreed that the activities of married women are best confined to the home.

Answers: **1.** F **2.** T **3.** T **4.** T **5.** F

One of the outcomes of 9/11 has been to sensitize the United States to life in other countries. Afghanistan quickly became a national focus and the status of women dominated by the Taliban became visible. The oppression of these women was evident in their being banned from school, required to wear full-length *burqa,* and prohibited from appearing in public without a male chaperon (Skaine, 2002). Their oppression due to gender is not unique to societies formerly controlled by the Taliban. Inequality between women and men continues in industrialized nations. Our focus in this chapter is the importance of gender roles as they influence a person's life and relationship choices. We begin by looking at the terms used to discuss gender issues.

TERMINOLOGY OF GENDER ROLES

In common usage the terms *sex* and *gender* are often interchangeable, but to sociologists, family/consumer science educators, human development specialists, and health educators, these terms are not synonymous. After clarifying the distinction between sex and gender, we discuss other relevant terminology, including *gender identity, gender role,* and *gender role ideology.*

Sex

Sex refers to the biological distinction between females and males. The primary sex characteristics that differentiate women and men include external genitalia (vulva and penis), gonads (ovaries and testes), sex chromosomes (XX and XY), and hormones (estrogen, progesterone, and testosterone). The presence of gonads is particularly tied to one's self-concept as a woman. Elson (2003) found that women who had had their ovaries removed viewed themselves as "less of a woman." Aside from primary sex characteristics are secondary sex characteristics, which include the larger breasts of women and the deeper voice and growth of facial hair in men.

Even though we commonly think of biological sex as consisting of two dichotomous categories (female and male), biological sex exists on a continuum. This view is supported by the existence of individuals with mixed or ambiguous genitals (**hermaphrodites** or **intersexed individuals**). Hence, the genitals are not clearly male or female. In addition, some males produce fewer male hormones (androgens) than some females, just as some females produce fewer female hormones (estrogens) than some males.

Gender

The woman most in need of liberation is the woman in every man and the man in every woman.

William Sloan Coffin, *Clergyman and political activist*

Gender refers to the social and psychological characteristics associated with being female or male. For example, women see themselves (and men agree) as moody and easily embarrassed; men see themselves (and women agree) as competitive, sarcastic, and sexual (Knox, Zusman, & Thompson, in press). In popular usage, gender is dichotomized as an either/or concept (feminine or masculine). But gender may also be viewed as existing along a continuum of femininity and masculinity.

There is an ongoing controversy about whether gender differences are innate as opposed to learned or socially determined. Just as sexual orientation may be best explained as an interaction of biological and social/psychological variables, gender differences may also be seen as a consequence of both biological and social/psychological influences.

While some researchers emphasize the role of social influences, others emphasize a biological imperative as the basis of gender role behavior. As evidence for the latter, John Money, psychologist and former director of the now defunct Gender Identity Clinic at Johns Hopkins University School of Medicine, encouraged the parents of a boy (Bruce) to rear him as a girl (Brenda) because of a botched circumcision that rendered the infant without a penis. Money argued that social mirrors dictate one's gender identity, and thus if the parents treated the child as a girl (name, dress, toys, etc.), the child would adopt the role of a girl and later that of a woman. The child was castrated and sex reassignment began.

But the experiment failed miserably; the child as an adult (David Reimer—his real name) reported that he never felt comfortable in the role of a girl and had always viewed himself as a boy. Today he is married and the adoptive father of two children. In the book *As Nature Made Him: The Boy Who Was Raised as a Girl* (Colapinto, 2000), David worked with a writer to tell his story. His courageous decision to make his poignant personal story public has shed light on scientific debate on the "nature/nurture" question. In the past, David's situation was used as a textbook example of how "nurture" is the more important influence in gender identity, if a reassignment is done early enough. Today, his case makes the point that one's biological wiring dictates gender outcome (Colapinto, 2000).

Nevertheless, **socialization** (the process through which we learn attitudes, values, beliefs, and behaviors appropriate to the social positions we occupy) does impact gender role behaviors, and social scientists tend to emphasize the role of social influences in gender differences. Margaret Mead (1935) provided powerful evidence for the social learning of gender roles in her study of three cultures. She visited three New Guinea tribes in the early 1930s and observed that the Arapesh socialized both men and women to be feminine by Western standards. The Arapesh person was taught to be cooperative and responsive to the needs of others. In contrast, the Tchambuli were known for dominant women and submissive men—just the opposite of our society. And both of these societies were unlike the Mundugumor, which socialized only ruthless, aggressive, "masculine" personalities. The inescapable conclusion of this cross-cultural study is that human beings are products of their social and cultural environment and that gender roles are learned. As Peoples (2001, 18) observed, "cultures construct gender in different ways."

Gender Identity

Gender identity is the psychological state of viewing oneself as a girl or a boy, and later as a woman or a man. Such identity is largely learned and is a reflection of society's conceptions of femininity and masculinity. Some individuals experience **gender dysphoria,** a condition in which one's gender identity does not match one's biological sex. An example of gender dysphoria is transsexualism (discussed in the next section).

Transgenderism

Transgenderism is a political movement seeking to challenge the belief that every person can be categorized simply as a woman or as a man (Roen, 2001). The term **transgendered** refers to individuals who express some characteristics other than those of their assigned gender, which is usually based on their biological sex (male or female). **Cross-dresser** is a broad term for individuals who may dress or present themselves in the gender of the other sex. Some cross-dressers are heterosexual adult males who enjoy dressing and presenting themselves as women. Cross-dressers may also be women who dress as men and present themselves as men. Some cross-dressers are bisexual or homosexual. Another term for cross-dresser is **transvestite,** although the latter term is commonly associated with homosexual men who dress provocatively as women to attract men—sometimes as sexual customers.

Transsexuals are persons with the biological/anatomical sex of one gender (e.g., male) but the self-concept of the other sex (e.g., female). "I am a woman trapped in a man's body" reflects the feelings of the male-to-female transsexual, who may take hormones to develop breasts and reduce facial hair, and may have surgery to artificially construct a vagina. This person lives full-time as a woman. The female-to-male transsexual is one who is a biological/anatomical female but who feels "I am a man trapped in a female's body." This person may take male hormones to grow facial hair and deepen her voice and may have surgery to create an artificial penis. This person lives full-time as a man.

This cross-dresser is on an Alaskan cruise with his wife and teenage daughters. He is an active member of the Society for the Second Self, an organization for heterosexual men who like to dress in women's clothes.

Authors

Individuals need not take hormones or have surgery to be regarded as transsexuals. The distinguishing variable is living full-time in the role of the gender opposite one's biological sex. A man or woman who presents himself or herself full-time as a woman or man is a transsexual by definition. Some transsexuals prefer the term **transgenderist,** which means an individual who lives in a gender role that does not match his or her biological sex but who has no desire to surgically alter his or her genitalia (as does a transsexual). Another variation is the she/male, who looks like a woman and has the breasts of a woman yet has the genitalia and reproductive system of a male.

Gender Roles

Gender roles are the social norms that dictate what is socially regarded as appropriate female and male behavior. All societies have expectations of how boys and girls, men and women "should" behave. Gender roles influence women and men in virtually every sphere of life, including family and occupation. For example, traditional gender role expectations have influenced women to be homemakers, day-care workers, and nurses. Martin (2003) noted that women during childbirth are expected to be "nice," "polite," "kind," and "selfless" (even when they are hurt, bleeding, or dying). Her study emphasized how women are compelled to act in gender-normative ways. McNeely et al. (2004) studied gender differences in beliefs about men and found that women are more likely to view men as thinking about sex more often, cheating in relationships, and preferring to live together rather than to marry.

National Data

Sixteen percent of a random sample of all first-year college undergraduate women and 28 percent of a random sample of all first-year undergraduate men in colleges and universities in the United States agreed that the "activities of married women are best confined to the home" (American Council on Education & University of California, 2004).

The term **sex roles** is often confused with and used interchangeably with the term *gender roles*. However, whereas gender roles are socially defined and can be enacted by either women or men, sex roles are defined by biological constraints and can be enacted by members of one biological sex only—for example, wet nurse, sperm donor, child bearer.

Gender Role Ideology

Gender role ideology refers to beliefs about the proper role relationships between women and men in any given society. In spite of the rhetoric regarding egalitarian interaction between women and men, Laner and Ventrone (2000) documented that dating interaction/scripts continue to be traditional (with the man asking the woman out, deciding on plans, and preparing the car for the date while the woman waits to be picked up, introduces the male to her family, and eats lightly at dinner). However, Donaghue and Fallon (2003) found that individuals in less traditional and more equitable relationships reported higher levels of relationship satisfaction.

Traditional American gender role ideology has perpetuated and reflected patriarchal male dominance and male bias in almost every sphere of life. Even our language reflects this male bias. For example, the words *man* and *mankind* have traditionally been used to refer to all humans. There has been a growing trend away from using male-biased language. Dictionaries have begun to replace *chairman* with *chairperson* and *mankind* with *humankind.*

THEORIES OF GENDER ROLE DEVELOPMENT

Various theories attempt to explain why women and men exhibit different characteristics and behaviors.

Sociobiology

Sociobiology emphasizes that social behavior and therefore gender roles have a biological basis in terms of being functional in human evolution. Theodosius Dobzhansky, the eminent biologist, noted, "Nothing in biology makes sense except in the light of evolution" (cited in Thornhill & Palmer, 2000). In effect, given an appreciation of evolution by natural selection (certain behaviors occurred that contributed to the survival of the animal and its offspring), we can look at social behaviors as having an evolutionary survival function. Joseph (2000) emphasized that the differences between women and men (women nurtured children; men hunted/killed animals for food) were functional for survival. Women stayed in the nest or gathered food nearby, while men could go afar to find food. Such a conceptualization focuses on the division of labor between women and men as functional for the survival of the species.

Although there is little agreement (even among sociologists) on the merits of sociobiology (Alcock, 2001; Miller & Costello, 2001), the theory emphasizes that biological differences (such as hormonal and chromosomal differences) between men and women account for the social and psychological differences in female and male characteristics, behaviors, and roles. For example, testosterone is a male hormone associated with aggression; progesterone is a female hormone associated with nurturance. Such hormonal differences are used to help explain that men have more sexual partners than women, that men are more likely to engage in casual sex, and that men are the perpetrators of most acts of sexual coercion as well as sexual harassment. However, critics of sociobiology point out that a particular dose of a particular hormone does not translate into a specific behavior (Miller & Costello, 2001).

In mate selection, heterosexual men tend to seek and mate with women who are youthful and attractive. These characteristics are associated with fertility, health, and lifetime reproductive potential for women. Alcock (2001) notes that

> *unwrinkled, unblemished skin is far more likely to be possessed by young, healthy women than by older (less fertile) or less healthy (less fertile) women. Are young, healthy women more likely to become pregnant and sustain a pregnancy successfully than older or less healthy women? The answer is yes. Is there any species of animal on earth in which males are more likely to mate with infertile females than with fertile ones, if given the opportunity to choose between the two? The answer is obvious.* (p. 137)

Similarly, women tend to select and mate with men whom they deem will provide the maximum parental investment in their offspring. The term **parental investment** refers to any investment by a parent that increases the offspring's chance of surviving and thus increases reproductive success. Parental investments require time and energy. Women have a great deal of parental investment in their offspring (nine months' gestation, taking care of dependent offspring) and tend to mate with men who have high status, economic resources, and a willingness to share economic resources.

The sociobiological explanation for mate selection is extremely controversial. Critics argue that women may show concern for the earning capacity of a potential mate because women have been systematically denied access to similar economic resources, and selecting a mate with these resources is one of their remaining options. In addition, it is argued that both women and men, when selecting a mate, think more about their partners as companions than as future parents of their offspring. Finally, the sociobiological perspective fails to acknowledge the degree to which social and psychological factors influence our behavior. For example, Miller & Costello (2001) emphasized that biological determinists should "familiarize themselves with the sociological research on the cultural construction of gender" (p. 597).

Diversity in Other Countries

David Buss (1989) found that the pattern of men seeking physically attractive young women and women seeking economically ambitious men was true in 37 groups of women and men in 33 different societies.

Identification

Freud was one of the first researchers to study gender role development. He suggested that children acquire the characteristics and behaviors of their same-sex parent through a process of identification. Boys identify with their fathers; girls identify with their mothers.

In *The Reproduction of Mothering*, Nancy Chodorow uses Freudian identification theory as a basis for her theory that gender role specialization occurs in the family because of the "asymmetrical organization of parenting" (1978, 49).

> *Women, as mothers, produce daughters with mothering capacities and the desire to mother. These capacities and needs are built into and grow out of the mother-daughter relationship itself. By contrast, women as mothers (and men as not-mothers) produce sons whose nurturing capacities and needs have been systematically curtailed and repressed.* (p. 7)

In other words, all activities associated with nurturing and child care are identified as female activities because women are the primary caregivers of young children. This one-sidedness (or asymmetry) of nurturing by women increases the likelihood that females, because they identify with their mothers, will see their own primary identity and role as those of mother. When women have and take care of children, they are "doing gender," just as men are when they avoid child care (West & Zimmerman 1991). In regard to "doing gender," Fox and Murry (2001) noted that "men and women have to be constantly persuaded or reminded to be masculine and feminine. That is, men and women have to 'do' gender rather than 'be' gender" (p. 383).

Social Learning

Derived from the school of behavioral psychology, social learning theory emphasizes the roles of reward and punishment in explaining how a child learns gender role behavior. For example, two young brothers enjoyed playing "lady." Each of them would put on a dress, wear high-heeled shoes, and carry a pocketbook. Their father came home early one day and angrily demanded that they "take those clothes off and never put them on again. Those things are for women," he said. The boys were punished for playing lady but rewarded with their father's approval for playing cowboys, with plastic guns and "Bang! You're dead!" dialogue.

Reward and punishment alone are not sufficient to account for the way in which children learn gender roles. Direct instruction ("girls wear dresses," "a man stands up and shakes hands") by parents or peers is another way children learn. In addition, many of society's gender rules are learned through modeling. In modeling, the child observes another's behavior and imitates that behavior. Gender role models include parents, peers, siblings, and characters portrayed in the media.

The impact of modeling on the development of gender role behavior is controversial. For example, a modeling perspective implies that children will tend to imitate the parent of the same sex, but children in all cultures are usually reared mainly by women. Yet this persistent female model does not seem to interfere with the male's development of the behavior that is considered appropriate for his gender. One explanation suggests that boys learn early that our society generally grants boys and men more status and privileges than girls and women. Therefore, boys devalue the feminine and emphasize the masculine aspects of themselves.

Cognitive-Developmental Theory

The cognitive-developmental theory of gender role development reflects a blend of biological and social learning views. According to this theory, the biological readiness, in terms of cognitive development, of the child influences how the child responds to gender cues in the environment (Kohlberg, 1966). For example, gender discrimination (the ability to identify social and psychological characteristics associated with being female or male) begins at about age 30 months. However, at this age, children do not view gender as a permanent characteristic. Thus, even though young children may define people who wear long hair as girls and those who never wear dresses as boys, they also believe they can change their gender by altering their hair or changing clothes.

Not until age 6 or 7 does the child view gender as permanent (Kohlberg, 1966, 1969). In Kohlberg's view, this cognitive understanding involves the development of a specific mental ability to grasp the idea that certain basic characteristics of people do not change. Once children learn the concept of gender permanence, they seek to become competent and proper members of their gender group. For example, a child standing on the edge of a school playground may observe one group of children jumping rope while another group is playing football. That child's gender identity as either a girl or a boy connects with the observed gender-typed behavior, and the child joins one of the two groups. Once in the group, the child seeks to develop the behaviors that are socially defined as appropriate for her or his gender.

AGENTS OF SOCIALIZATION

Three of the four theories discussed in the preceding section emphasize that gender roles are learned through interaction with the environment. Indeed, though biology may provide a basis for some roles (being seven feet five inches tall is helpful for a basketball player), cultural influences in the form of various socialization agents (parents, peers, religion, and the media) shape the individual toward various gender roles. These powerful influences in large part dictate what a person thinks, feels, and does in his or her role as a woman or a man. Laner and Ventrone (2000) observed that traditional gender roles continue because male and female children continue to be socialized in different environments. After assessing your view of feminism (see p. 40), we look at the different sources influencing your gender socialization.

Family

The family is a gendered institution with female and male roles highly structured by gender. Lorber (2001) noted that parents "create a gendered world for their newborns by naming, birth announcements, and dress" (p. 23). Parents may also relate differently to their children on the basis of gender. Lindsey and Mize (2001) observed videotapes of parent-child play behavior of thirty-three preschool children (eighteen boys, European-American, middle- and upper-middle-class families) and observed that during the physical play session, father-son dyads engaged in more physical play than did father-daughter dyads.

The importance of the father in the family was noted in Pollack's (2001) study of adolescent boys. "America's boys are crying out for a new gender revolution that does for them what the last forty years of feminism has tried to do for girls and women" (p. 18). This new revolution will depend on fathers who teach their sons that feelings and relationships are important. How equipped do you feel today's fathers are to provide these new models for their sons?

Siblings also influence gender role learning. Growing up in a family of all sisters or all brothers intensifies social learning experiences toward femininity or masculinity. A male reared with five sisters and a single parent mother is likely to reflect more feminine characteristics than a male reared in a home with six brothers and a stay-at-home dad.

Race/Ethnicity

The race and ethnicity of one's family also influence gender roles. While African-American families are often stereotyped as being matriarchal, the more common pattern of authority in these families is egalitarian (Taylor, 2002). However, the fact that Black women have increased economic independence provides a powerful role model for young Black women. A similar situation exists among Hispanics, who represent the fastest-growing segment of the U.S. population. While great variability exists among Mexican-American marriages, the employment opportunities of Mexican-American women provide them with both resources and autonomy (Baca Zinn & Pok, 2002). The result is a less than docile model of the emerging Hispanic female (e.g., Jennifer Lopez).

Peers

Though parents are usually the first socializing agents that influence a child's gender role development, peers become increasingly important during the school years. The gender role messages from adolescent peers are primarily traditional. Boys are expected to play sports and be career-oriented. Female adolescents are under tremendous pressure to be physically attractive and thin, popular, and achievement-oriented. Female achievements may be traditional (cheerleading) or nontraditional (sports or academics). Adolescent females are sometimes in conflict because high academic success may be viewed as being less than feminine.

Peers also influence gender roles throughout the family life cycle. In chapter 1 we discussed the family life cycle and noted the various developmental tasks throughout the cycle. With each new stage, role changes are made and one's peers influence those role changes. For example, when a couple move from being childfree to being parents, peers quickly socialize them into the role of parent and the attendant responsibilities. This process of peer socialization continues throughout life. Senator Orin Hatch was seen on *60 Minutes* reporting that he had told Senator Edward Kennedy to "grow up" (translation: stop the wild drinking and partying). The seasoned Senator Kennedy noted that the chiding was warranted and that he had indeed changed his behavior (e.g., allowed his peer to influence his role behavior).

Religion

While the percentage of American adults claiming "no religious preference" continues to increase (Hout & Fischer, 2002), religion remains an influence on gender roles. Indeed, over 85 percent of adults do acknowledge a religious preference (Hout & Fischer, 2002). Since women (particularly white women) are "socialized to be submissive, passive, and nurturing," they may be predisposed to greater levels of religion and religious influence (Miller &Stark, 2002). Such exposure includes a traditional framing of gender roles. "Wives, be subject to your husband, as to the Lord. . . . Husbands, love your wives, even as Christ also loved the church" (Ephesians 5:22–2). Though the Bible has been interpreted in both sexist and nonsexist terms, male dominance is indisputable in the hierarchy of religious organizations, where power and status have been accorded mostly to men. The Roman Catholic Church does not have female clergy, and men dominate the nineteen top positions in the U.S. dioceses. Popular books marketed to the Christian right also emphasize traditional gender roles.

SELF-ASSESSMENT

Attitudes toward Feminism Scale

Following are statements on a variety of issues. At the left of each statement is a place for indicating how much you agree or disagree. Please respond as you personally feel and use the following letter code for your answers:

A Strongly Agree
B Agree
C Disagree
D Strongly Disagree

___ **1.** It is naturally proper for parents to keep a daughter under closer control than a son.

___ **2.** A man has the right to insist that his wife accept his view as to what can or cannot be afforded.

___ **3.** There should be no distinction made between woman's work and man's work.

___ **4.** Women should not be expected to subordinate their careers to home duties to any greater extent than men do.

___ **5.** There are no natural differences between men and women in sensitivity and emotionality.

___ **6.** A wife should make every effort to minimize irritation and inconvenience to her husband.

___ **7.** A woman should gracefully accept chivalrous attentions from men.

___ **8.** A woman generally needs male protection and guidance.

___ **9.** Married women should resist enslavement by domestic obligations.

___ **10.** The unmarried mother is more immoral and irresponsible than the unmarried father.

___ **11.** Married women should not work if their husbands are able to support them.

___ **12.** A husband has the right to expect that his wife will want to bear children.

___ **13.** Women should freely compete with men in every sphere of economic activity.

___ **14.** There should be a single standard in matters relating to sexual behavior for both men and women.

___ **15.** The father and mother should have equal authority and responsibility for discipline and guidance of the children.

___ **16.** Regardless of sex, there should be equal pay for equal work.

___ **17.** Only the very exceptional woman is qualified to enter politics.

___ **18.** Women should be given equal opportunities with men for all vocational and professional training.

___ **19.** The husband should be regarded as the legal representative of the family group in all matters of law.

___ **20.** Husbands and wives should share in all household tasks if both are employed an equal number of hours outside the home.

___ **21.** There is no particular reason why a girl standing in a crowded bus should expect a man to offer her his seat.

___ **22.** Wifely submission is an outmoded virtue.

___ **23.** The leadership of a community should be largely in the hands of men.

___ **24.** Women who seek a career are ignoring a more enriching life of devotion to husband and children.

___ **25.** It is ridiculous for a woman to run a locomotive and for a man to darn socks.

___ **26.** Greater leniency should be adopted toward women convicted of crime than toward male offenders.

___ **27.** Women should take a less active role in courtship than men.

A gender-neutral Bible (both Old and New Testament) is available as of 2005. Examples of changes: "sons of God" will become "children of God" (Matthew 5:9), and "a man is justified by faith" will become "a person is justified by faith" (Romans 3:28).

Education

The educational institution serves as an additional socialization agent for gender role ideology. But such an effect must be considered in the context of the society/culture in which the "school" exists and of the school itself. Schools are basic cultures of transmission in that they make deliberate efforts to reproduce the culture from one generation to the next. Blair (2002) emphasized how schools reflect the broader U.S. culture "and its patriarchal gender roles in their structure, organization, curriculum, and interaction" (p. 22).

___ **28.** Contemporary social problems are crying out for increased participation in their solution by women.

___ **29.** There is no good reason why women should take the name of their husbands upon marriage.

___ **30.** Men are naturally more aggressive and achievement-oriented than women.

___ **31.** The modern wife has no more obligation to keep her figure than her husband to keep down his waistline.

___ **32.** It is humiliating for a woman to have to ask her husband for money.

___ **33.** There are many words and phrases which are unfit for a woman's lips.

___ **34.** Legal restrictions in industry should be the same for both sexes.

___ **35.** Women are more likely than men to be devious in obtaining their needs.

___ **36.** A woman should not expect to go to the same places or to have quite the same freedom of action as a man.

___ **37.** Women are generally too nervous and high-strung to make good surgeons.

___ **38.** It is insulting to women to have the "obey" clause in the marriage vows.

___ **39.** It is foolish to regard scrubbing floors as more proper for women than mowing the lawn.

___ **40.** Women should not submit to sexual slavery in marriage.

___ **41.** A woman earning as much as her male date should share equally in the cost of their common recreation.

___ **42.** Women should recognize their intellectual limitations as compared with men.

Scoring:

Score your answers as follows: A = 1, B = 2, C = 3, D = 4. Because half the items were phrased in a profeminist and half in an antifeminist direction, you will need to reverse the scores (+2 becomes −2, etc.) for the following items: 1, 2, 6, 7, 8, 10, 11, 12, 17, 19, 21, 23, 25, 26, 27, 30, 33, 35, 36, 37, and 42. Now sum your scores for all the items. Scores may range from +84 to −84.

Interpreting your score: The higher your score, the higher your agreement with feminist (Lott used the term *women's liberation*) statements. You may be interested in comparing your score, or that of your classmates, with those obtained by Lott (1973) from undergraduate students at the University of Rhode Island. The sample was composed of 109 men and 133 women in an introductory psychology class, and 47 additional older women who were participating in a special Continuing Education for Women (CEW) program. Based on information presented by Lott (1973), the following mean scores were calculated: men = 13.07, women = 24.30, and continuing education women = 30.67.

Biaggio, Mohan, and Baldwin (1985) administered Lott's questionnaire to 76 students from a University of Idaho introductory psychology class and 63 community members randomly selected from the local phone directory. Although they did not present the scores of their respondents, they reported that they did not find differences between men and women. Unlike Lott's students, in the Biaggio sample women were not more proliberation than men. Biaggio et al. (1985, 61) stated, "It seems that some of the tenets of feminism have taken hold and earned broader acceptance. These data also point to an intersex convergence of attitudes, with men's and women's attitudes toward liberation and child rearing being less disparate now than during the period of Lott's study." It would be interesting to determine whether there are differences in scores between members of each sex in your class.

Sources

Biaggio, M. K., P. J. Mohan, and C. Baldwin. 1985. Relationships among attitudes toward children, women's liberation, and personality characteristics. *Sex Roles* 12:47–62.

Lott, B. E. 1973. Who wants the children? Some relationships among attitudes toward children, parents, and the liberation of women. *American Psychologist* 28:573–82.

Reproduced by permission of Bernice Lott, Department of Psychology, University of Rhode Island.

Economy

The economy of the society influences the roles of the individuals in the society. The economy is a very gendered institution. **Occupational sex segregation** denotes the fact that women and men are employed in gender-segregated occupations, that is, occupations in which workers are either primarily male or primarily female. Examples include a preponderance of men as mechanical and electrical engineers and of females as flight attendants and interior designers (Lippa, 2002). Female-dominated occupations tend to require less education, have lower status, and pay lower salaries than male-dominated occupations. If men typically occupy the role, it tends to pay more. For example, the job of child-care attendant requires more education than the job of dog-pound attendant. However, dog-pound attendants are more likely to be male and earn more than child-care attendants, who are more likely to be female.

PERSONAL CHOICES

Do You Want a Nontraditional Occupational Role?

The concentration of women in certain occupations and men in others is referred to as **occupational sex segregation.** But some jobs traditionally occupied by one gender are now open to the other. Men may become nurses and librarians, and women may become construction workers and lawyers. Increasingly, occupations are becoming less segregated on the basis of gender, and social acceptance of nontraditional career choices has increased. The trend continues; women now fly jet aircraft on military combat missions and school at the previously all-male West Point and Virginia Military Institute. However, only 15 percent and 3 percent, respectively, of the cadets at West Point and Virginia Military Institute are female. Diamond (2003) noted that the military remains a less viable option for many women.

Choosing nontraditional occupational roles may have benefits both for the individual and for society. On the individual level, women and men can make career choices on the basis of their personal talents and interests rather than on the basis of arbitrary social restrictions regarding who can and cannot have a particular job or career. Because traditionally male occupations are generally higher-paying than traditionally female occupations, women who make nontraditional career choices can gain access to higher-paying and higher-status jobs.

On the societal level, an increase in nontraditional career choices reduces gender-based occupational segregation, thereby contributing to social equality among women and men. In addition, women and men who enter nontraditional occupations may contribute greatly to the field that they enter. For example, such traditionally male-dominated occupations as politics, science and technology, and medicine may benefit greatly from increased involvement of women in these fields. Similarly, among preschool/kindergarten teachers (98% of whom are female) there are not enough male role models for children in school. Some parents now ask for male nannies as role models for their male children (Cullen, 2003).

Sources

Cullen, L. T. 2003. I want your job, lady! *Time.* 12 May, 52 et passim.

Diamond, D. A. 2003. Breaking down the barricades: The admission of women at Virginia Military Institute and the United States Military Academy at West Point. Paper presented at the Annual Meeting of the Eastern Sociological Society, Philadelphia, February.

Mass Media

Mass media, such as movies, television, magazines, newspapers, books, music, and computer games both reflect and shape gender roles. Media images of women and men typically conform to traditional gender stereotypes, and media portrayals depicting the exploitation, victimization, and sexual objectification of women are common. *Sex in the City,* produced by HBO, portrayed "Mr. Big" (the on-and-off-again love interest of Carrie) as cool and in control, while Carrie and her three girlfriends (Miranda, Samantha, and Charlotte) were continually frazzled about their relationships with men (until the last episode).

Another form of media is comic strips. Berglund and Inman (2000) reviewed cartoon strips over a twenty-six-week period and observed a reflection of traditional role relationships in regard to leisure, child care, home care, and helping spouse. They found that male characters spent more time in leisure activities than did women, who spent more time in child care, home care, and helping the spouse than men did.

Popular self-help books give advice to both genders to behave consistently with traditional gender socialization. Zimmerman, Holm, and Haddock (2000) conducted a content analysis of the four best-selling self-help books (including *Men Are from Mars, Women Are from Venus,* a 7 million-copy seller) and found that women were encouraged to define themselves in relationship to their male partners. Furthermore, female readers were told that female independence and assertiveness might jeopardize relationships with their partners.

Content on the Internet is becoming a major source of media. Eighty-two percent of almost 300,000 first-year undergraduates reported that they "frequently use a personal computer" (American Council on Education and University of California, 2002). Kayany and Yelsma (2000) noted that information obtained from online viewing is increasing while information obtained from television viewing is decreasing.

The cumulative effect of family, peers, religion, education, the economy, and mass media perpetuate gender stereotypes. Each agent of socialization reinforces gender roles that are learned from other agents of socialization, thereby creating a gender role system that is deeply embedded in our culture.

GENDER ROLES IN OTHER SOCIETIES

Since gender roles are largely influenced by culture, individuals reared in different societies typically display the gender role patterns of those societies. The following subsections discuss how gender roles differ in Afghanistan, China, Pakistan, Sweden, Japan, and Africa.

Gender Roles of Women in Afghanistan under the Taliban

It took bin Laden and al-Qaeda to grab the attention of the American public to the virtual enslavement of Afghan women.

Jennifer Seymour Whitaker, *Director of the Council on Foreign Relations Project on Women's Human Rights*

Afghanistan is a country about the size of Texas with an estimated population of 26 million. The Taliban reached the peak of their dominance in 1996. The Revolutionary Association of the Women of Afghanistan (RAWA) compiled an "abbreviated" list of restrictions against women (including "creating noise when they walk" and "wearing white socks"). The life of women and their children (since what happens to women affects their children) in Afghanistan under the Taliban was cruel, demeaning, and often fatal. Some women drank household bleach rather than continue to endure their plight. They were not allowed to go to school or work and thus were completely dependent economically. Indeed, they were required to stay in the house, to paint the windows black, and to leave the house only if they were fully clothed (wearing a burqa) and accompanied by a male relative. Some could not afford burqas and had no living male relatives. According to Skain, "There are two places for women: one is the husband's bed and the other is the graveyard" (2002, p. 64). One mother reported that the Taliban came to her house, dragged her 19-year-old daughter out of the house, and drove away with her. "They sell them" she lamented (p. 116).

Subsequent to 9/11 the United States attacked the Taliban in Afghanistan with the goal of removing them, and in doing so it improved the lives of Afghan women. But the role of women in Afghanistan, particularly in the rural areas, has typically been one of domination. "Most of the women in rural areas (which comprises over 80% of Afghanistan) have never had the opportunity to get out of their own little house, little village, little province" (Consolatore, 2002, p.13). Hence, outside of Kabul, Afghan women go uneducated, become child brides, produce children, and rarely expect their daughters' lives to be different. The patriarchal social structure and the absence of a centralized and modernized state in Afghanistan predict that changes will be limited for Afghan women (Moghadam, 2002).

Gender Roles among Chinese-Americans

Gender roles in traditional China were very unequal and characterized by female subordination with "close controls over women and wives' responsibility for housework" (Glenn & Yap, 2002). As a result of immigration, industrialization, and urbanization, gender roles of Chinese-Americans have become more egalitarian. Both women and men are now breadwinners, which has resulted in the "downward shift of the husband's occupational status" with more sharing of housework. Pimentel (2000) found that among urban Chinese couples, those spouses who evidenced an egalitarian outlook and who shared household responsibilities and decision making reported higher levels of marital happiness. And this was true of both women and men.

Gender Roles in Punjab, Pakistan

In some cultures of the world, men dominate and control women. Two researchers (Winkvist & Akhtar, 2000) interviewed forty-two women in Punjab, Pakistan, conducted four focus groups of additional women, and interviewed eight mothers-in-law about their perceptions and experiences of bearing sons and daughters. The respondents reported that they had limited control over their lives, as evidenced by the expectation of early marriage, quick conception, and limited access to contraceptives. The women also expressed a strong preference for sons, which reflected women's subordinate position in society. Mothers of daughters and women without children spoke of harassment in the family as well as society.

Gender Roles in Sweden

The Swedish government is strongly concerned with equality between women and men. In 1974 Sweden became the first country in the world to introduce a system that enables mothers and fathers to share parental leave (paid by the government) from their jobs in any way they choose. Furthermore, Swedish law states that employers may not penalize the career of a working parent because he or she has used parental rights. By encouraging fathers to participate more in childrearing, the government aims to provide more opportunities for women to pursue other roles. Women hold about a quarter of the seats in the Swedish Parliament. However, few Swedish women are in high-status positions in business, and governmental efforts to reduce gender inequality are weak compared with the power of tradition. Nevertheless, a comparison of Swedish and U.S. students revealed greater acceptance of gender egalitarianism (Weinberg, Lottes, & Shaver, 2000).

Gender Roles in Africa

Africa is a diverse continent with over fifty nations. The cultures range from Islamic/Arab cultures of Northern Africa to industrial and European influences in South Africa. In some parts of east Africa (e.g., Kenya) gender roles are in flux. Meredith Kennedy (2000) has lived in East Africa and makes the following observations of gender roles:

> *The roles of men and women in most African societies tend to be very separate and proscribed, with most authority and power in the men's domain. For instance, Masai wives of East Africa do not travel much, since when a husband comes home he expects to find his wife (or wives) waiting for him with a gourd of sour milk. If she is not, he has the right to beat her when she shows up. As attempts are made to soften these boundaries and equalize the roles, the impacts are very visible and cause a lot of*

reverberations throughout these communal societies. Many African women who believe in and desire better lives will not call themselves "feminists" for fear of social censure. Change for people whose lives are based on tradition and "fitting in" can be very traumatic.

CONSEQUENCES OF TRADITIONAL GENDER ROLE SOCIALIZATION

This section discusses different consequences, both negative and positive, for women and men, of traditional female or male socialization in the United States.

Consequences of Traditional Female Role Socialization

Table 2.1 summarizes some of the negative and positive consequences of being socialized as a woman in U.S. society. Each consequence may or may not be true for a specific woman. For example, though women in general have less education and income, a particular woman may have more education and a higher income than a particular man.

Negative Consequences of Traditional Female Role Socialization There are several consequences of being socialized as a woman in our society.

1. Less Education/Income. Ross and Van Willigen (1997) emphasized the value of education to one's subjective quality of life—lower levels of depression, anxiety, and anger and lower levels of physical problems. Women earn fewer advanced degrees beyond a master's degree than do men. For the year 2001, women earned 44 percent of the Ph.D.s (National Opinion Research Center, 2002).

The strongest explanation for why women earn fewer advanced degrees and are more likely than men to have jobs rather than careers is their value for and commitment to family. In a study of 102 seniors and 504 alumni from a mid-sized midwestern public university who rated forty-eight job characteristics, women gave significantly higher ratings to family life accommodations, pleasant working conditions, travel, and interpersonal relationships. Men were significantly more concerned about the pay and promotion facts than the women (Heckert et al., 2002). The lack of focus on money and promotions by women is related to the feminization of poverty (see Up Close 2.1).

Less education is associated with lower income. Women still earn about two-thirds of what men earn, even when the level of educational achievement is identical (see Table 2.2).

Women also seem to be invisible in high-paid executive jobs. Of *Fortune* 500 companies, 393 have no women among their top executives. Even in those com-

Table 2.1 Consequences of Traditional Female Role Socialization

Negative Consequences	Positive Consequences
Less education/income (more dependent)	Longer life
Higher HIV infection risk	Closer mother-child bond
Negative self-concept	Greater emotionality
Value defined by youthfulness and beauty	Identity not tied to job
Less marital satisfaction; no "wife" at home	Greater relationship focus

Up Close 2.1 The Feminization of Poverty

The term **feminization of poverty** refers to the disproportionate percentage of poverty experienced by women living alone or with their children. Single mothers are particularly associated with poverty. When head-of-household women are compared with married couple households, the median income is $28,142 versus $60,471 (*Statistical Abstract of the United States:* 2003, Table 685). The process is cyclical—poverty contributes to teenage pregnancy, since teens have limited supervision and few alternatives to parenthood. Such early childbearing interferes with educational advancement and restricts their earning capacity, which keeps them in poverty. Their offspring are born into poverty, and the cycle begins anew.

Even if they get a job, women tend to be employed fewer hours than men, and even when they work full-time, they earn less money. Not only is discrimination in the labor force operating against women, but women usually make their families a priority over their employment, which translates into less income. Such prioritization is based on the patriarchal family, which ensures that women stay economically dependent on men and are relegated to domestic roles. Such dependence limits the choices of many women.

Low pay for women is also related to the fact that they tend to work in occupations that pay relatively low incomes. Indeed, women's lack of economic power stems from the relative indispensability of women's labor (it is easy to replace) and how work is organized (men control positions of power) (Wermuth & Ma'At-Ka-RE Monges, 2002).

When women move into certain occupations, such as teaching, there is a tendency in the marketplace to segregate these occupations from men's, and the result is a concentration of women in lower-paid occupations. The salaries of women in these occupational roles increase at slower rates. For example, salaries in the elementary and secondary teaching profession, which is predominately female, have not kept pace with inflation.

Conflict theorists assert that men are in more powerful roles than women and use this power to dictate incomes and salaries of women and "female professions." Functionalists also note that keeping salaries low for women keeps women dependent and in child-care roles so as to keep equilibrium in the family. Hence, for both conflict and structural reasons, poverty is primarily a feminine issue. One of the consequences of being a woman is to have an increased chance of feeling economic strain throughout life.

Sources

Statistical Abstract of the United States: 2003. 123rd ed. Washington, D.C.: U.S. Bureau of the Census.

Wermuth, L., and M. Ma'At-Ka-Re Monges. 2002. Gender stratification: A structural model for examining case examples of women in less developed countries. *Frontiers* 23:1–22.

Table 2.2 Women's and Men's Median Income with Similar Education

	Bachelor's Degree	Master's Degree	Doctoral Degree
Men	$49,985	$61,960	$72,642
Women	$30,973	$40,744	$52,181

Source: *Statistical Abstract of the United States: 2003.* 123rd ed. Washington, D.C.: U.S. Bureau of the Census, Table 664.

panies that have women CEOs, only three are among the best paid in those firms (Jones, 2003).

With divorce being a close to 50 percent probability for marriages begun in the nineties, the likelihood of being a widow for seven or more years, and the almost certain loss of her parenting role midway through her life, a woman without education and employment skills is often left high and dry. As one widowed mother of four said, "The shock of realizing you have children to support and no skills to do it is a worse shock than learning that your husband is dead." In the words of a divorced, 40-year-old mother of three, "If young women think it can't happen to them, they are foolish." (See Up Close 2.1 for a discussion of the feminization of poverty.)

*2. **Higher STD and HIV Infection Risk.*** Gender roles influence a woman's vulnerability to STDs and HIV infection not only because women receive more bodily fluids from men, who have a greater number of partners (and are therefore more likely to be infected), but because some women feel limited power to influence their partners to wear condoms. Moreover, when a woman possesses a condom, she is viewed as sexually willing, a fact that could undermine any claim of sexual assault (Hynie, Schuller, & Couperthwaite, 2003). A team of researchers (Salgado de Snyder et al., 2000) studied three hundred Mexican women and found that they were at increased risk for HIV/AIDS because of their passive behavior in not insisting that their sexual partners wear condoms.

In developing countries, other negative health outcomes of being socialized as a woman are evident—unsafe abortion, maternal mortality, depression, and psychosomatic symptoms. The progress toward relieving these issues is "uneven and slow" (Murphy, 2003).

3. Negative Self-Concept. Some research suggests that women have more negative self-concepts than men. In a study of 184 British adults, Furham and Gasson (1998) found that the women estimated themselves to have lower intelligence than the men, who estimated that their intelligence was higher than the women's.

Women in the United States live in a society that devalues them. Their lives and experiences are not taken as seriously. **Sexism** is defined as an attitude, action, or institutional structure that subordinates or discriminates against an individual or group because of their sex. Sexism against women reflects the tradition of male dominance and presumed male superiority in our society. It is reflected in the fact that women are rarely found in power positions in our society. In the 108th Congress, of the 435 members of the House of Representatives only 62 are women. There are 14 women and 86 men senators.

4. Negative Body Image. There are over 3,800 beauty pageants annually. The effect for many women who do not match the cultural ideal is to have a negative body image, which is associated with lower rates of intimate partner experiences (Wiederman, 2000).

5. Less Marital Satisfaction. Researchers disagree over whether wives are less happy than husbands (Waite, 2000). Basow (1992) found that wives reported less marital satisfaction than husbands and attributed this to role overload: women are expected to keep their husbands, children, parents, and employers happy and to be good homemakers (cook, clean, wash dishes, do laundry). The results of coping with these unrealistic expectations internalized by or imposed on women are, by contrast with men, more frequent nervous breakdowns, greater psychological anxiety, increased self-blame for not living up to these expectations, and poorer physical health (Bird & Fremont, 1991).

In direct contrast to these findings, Aldous and Woodberry (1994) analyzed General Social Survey data covering nineteen years and found that women reported being happier than men. However, the researchers attributed this finding to the tendency of women to give socially desirable responses.

6. No "Wife" at Home. As we will discuss in Chapter 12, Balancing Work and Family Life, both spouses may need a "wife" at home to take care of the house and children to free them to pursue their jobs or careers. Since wives typically do more housework and child care, they have a double burden, which most husbands do not.

Before leaving this section on negative consequences of being socialized as a woman, we look at the issue of **female genital mutilation** in this chapter's Social Policy feature. This is more of an issue for females born in some African, Middle Eastern, and Asian countries than for women in the United States. But the practice continues even here.

Positive Consequences of Traditional Female Role Socialization We have discussed the negative consequences of being born and socialized as a woman. But there are also decided benefits. Three of these are the potential to live longer, a stronger relationship focus, and a closer emotional bond with children.

Female Genital Mutilation

SOCIAL POLICY

Although **female genital operations** (also referred to as **female circumcision**) occur primarily in African and some Middle Eastern and Asian countries, the practice also occurs in the United States among immigrant families who bring their cultural traditions with them. Each year, about seven thousand immigrants to the United States (from countries that practice various forms of female genital operations) undergo the procedure either in the United States or during a visit to their homeland. The practice has become visible as a result of the increasing number of physicians involved in international assistance and travel (Bosch, 2001). The reason for the practice is cultural—parents believe that female circumcision makes their daughters marketable for marriage. Many daughters view it as a rite of passage and of improving their chance for marriage.

Changing a country's deeply held beliefs and values concerning this practice cannot be achieved by denigration. More effective approaches to discouraging the practice include the following (James & Robertson, 2002; Yount, 2002):

1. Respect the beliefs and values of countries that practice female genital operations. Calling the practice "genital mutilation" and "a barbaric practice" and referring to it as a form of "child abuse" and "torture" convey disregard for the beliefs and values of the cultures where it is practiced. In essence, we might adopt a culturally relativistic point of view (without moral acceptance of the practice). However, Kennedy (2002) traveled extensively in Africa and emphasized that the term *female circumcision* should not be used as the equivalent of female genital operations in general, or infibulation (stitching together of the labia majora so as to leave a tiny hole for menses and urine) specifically. Female genital operations, she noted, are not equivalent to the removal of foreskin.

2. Remember that genital operations are arranged and paid for by loving parents who deeply believe that the surgeries are for their daughters' welfare.

3. It is important to be culturally sensitive to the meaning of being a woman. Indeed, genital cutting is mixed up with how a woman sees herself; thus Westerners are becoming involved in her identity when attempting to alter long-held historical practices.

4. Raising the access of a female to education is related to a decline in female genital cutting. Indeed, educated women more often "oppose circumcision" and give prevention of sexual satisfaction as their reason (Yount, 2002, 352).

Sources

Bosch, X. 2001. Female genital mutilation in developed countries. *The Lancet* 358: 1177–82.

James, S. M., and C. C. Robertson, eds. 2002. *Genital cutting and transnational sisterhood: Disputing U.S. polemics.* Urbana, Ill.: University of Illinois Press

Kennedy, M. 2002. Presentation on infibulation. Department of Sociology, East Carolina University, Greenville, N.C, January.

Yount, K. M. 2002. Like mother, like daughter? Female genital cutting in Minia, Egypt. *Journal of Health and Social Behavior* 43:336–58.

1. Longer Life Expectancy. Women have a longer life expectancy than men. Greater life expectancy may be related more to biological than to social factors.

NATIONAL DATA

Females born in the year 2005 are expected to live to the age of 81, in contrast to men, who are expected to live to the age of 75 (*Statistical Abstract of the United States: 2003*, Table 105).

Diversity in Other Countries

Worldwide, women are more involved in relationships than are men. Two primary groups are their parents and same-sex peers. Williams and Best (1990) collected data from university students in twenty-eight countries (including England, Australia, Nigeria, Japan, and Brazil) and observed that women were more likely than men to have certain relationship characteristics:

- Succoring—soliciting sympathy, affection, or emotional support from others
- Nurturance—engaging in behavior that extends material or emotional benefits to others
- Affiliation—seeking and sustaining numerous personal relationships.

2. Stronger Relationship Focus. Research supports the idea that women value close friends and relatives more than men do (Abowitz & Knox, 2003), are more aware of emotional relationship issues (Croyle & Waltz, 2002), and both express and want higher levels of affection than men do (Kissee et al., 2000). Women also seek to resolve conflict through negotiation rather than attempting to become dominant (Brewer et al., 2002).

3. Bonding with Children. Another advantage of being socialized as a woman is the potential to have a closer bond with children. In general, women tend to be more emotionally bonded with their children than men do. Although the new cultural image of the father is of one who engages emotionally with his children, many fathers continue to be content for their wives to take care of their children, with the result that mothers, not fathers, become more bonded with the children.

Authors

Women typically spend more time with their children and develop strong emotional bonds with them.

Consequences of Traditional Male Role Socialization

Male socialization in our society is associated with its own set of consequences. Both the negative and positive consequences are summarized in Table 2.3. As with women, each consequence may or may not be true for a specific man.

Negative Consequences of Traditional Male Role Socialization There are several negative consequences associated with being socialized as a man in U.S. society.

1. Identity Synonymous with Occupation. Ask men who they are, and many will tell you what they do. Society tends to equate a man's identity with his occupational role. Male socialization toward greater involvement in the labor force is evident in governmental statistics.

National Data

Seventy-five percent of men compared with 60 percent of women were in the civilian work force in 2002 (*Statistical Abstract of the United States: 2003*, Table 587).

Work is the principal means by which men confirm their masculinity and success. Men are also more likely than women to value working more hours to make more money. Indeed, 76 percent of all first-year undergraduate male students, in contrast to 71 percent of female students, said that "being well off financially" was an "essential or very important" objective (American Council on Education & University of California, 2002).

The "work equals identity" equation may be changing. Increasingly, there are more stay-at-home dads, more fathers seeking full custody in divorce litigation, and more men willing to take "time off to be with the family" in lieu of a salary increase. These changes challenge cultural notions of masculinity

2. Limited Expression of Emotions. Some men feel caught between society's expectations that they be competitive, aggressive, and unemotional and their own desire to be more cooperative, passive, and emotional. Indeed, men are pressured to disavow any expression that could be interpreted as feminine (e.g., be emotional). Vogel et al. (2003) confirmed that men, when compared with women, exhibited fewer emotionally expressive behaviors and were more withdrawn. Notice that men are repeatedly told to "prove their manhood" (which implies not being emotional), while there is no dictum in our culture for women "to 'prove their womanhood'—the phrase itself sounds ridiculous" (Kimmel, 2001, 33).

How men hate waiting while their wives shop for clothes and trinkets; how women hate waiting, often for much of their lives, while husbands shop for fame and glory.

Thomas Szaz

Therapists report that the most common complaint of women in distressed marriages is that their husbands are too withdrawn and don't share openly enough.

Howard Markman, *Marriage therapist*

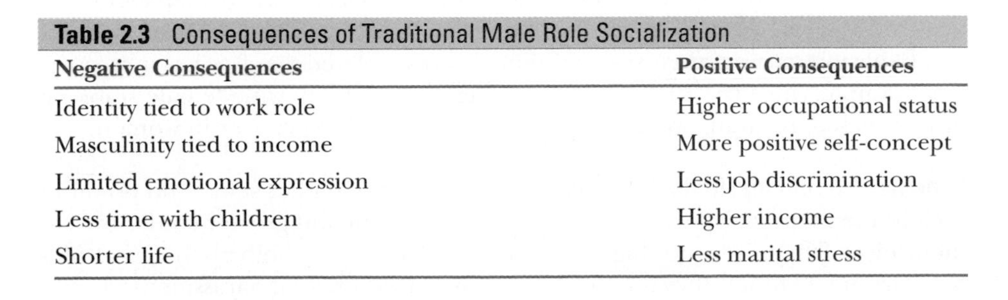

Table 2.3 Consequences of Traditional Male Role Socialization

Negative Consequences	Positive Consequences
Identity tied to work role	Higher occupational status
Masculinity tied to income	More positive self-concept
Limited emotional expression	Less job discrimination
Less time with children	Higher income
Shorter life	Less marital stress

Up Close 2.2 Effects of Gender Role Socialization on Relationship Choices

Women

1. A woman who is not socialized to pursue advanced education (which often translates into less income) may feel pressure to stay in an unhappy relationship with someone on whom she is economically dependent.
2. Women who are socialized to play a passive role and not initiate relationships are limiting interactions that could develop into valued relationships.
3. Women who are socialized to accept that they are less valuable and important than men are less likely to seek or achieve egalitarian relationships with men.
4. Women who internalize society's standards of beauty and view their worth in terms of their age and appearance are likely to feel bad about themselves as they age. Their negative self-concept, more than their age or appearance, may interfere with their relationships.
5. Women who are socialized to accept that they are solely responsible for taking care of their parents, children, and husband are likely to experience role overload. Potentially, this could result in feelings of resentment in their relationships.
6. Women who are socialized to emphasize the importance of relationships in their lives will continue to seek relationships that are emotionally satisfying.

Men

1. Men who are socialized to define themselves more in terms of their occupational success and income and less in terms of positive individual qualities leave their self-esteem and masculinity vulnerable should they become unemployed or work in a low-status job.
2. Men who are socialized to restrict their experience and expression of emotions are denied the opportunity to discover the rewards of emotional interpersonal sharing.
3. Men who are socialized to believe it is not their role to participate in domestic activities (childrearing, food preparation, housecleaning) will not develop competencies in these life skills. Potential partners often view domestic skills as desirable qualities.
4. Heterosexual men who focus on cultural definitions of female beauty overlook potential partners who might not fit the cultural beauty ideal but who would nevertheless be good life companions.
5. Men who are socialized to view women who initiate relationships in negative ways are restricted in their relationship opportunities.
6. Men who are socialized to be in control of relationship encounters may alienate their partners, who may desire equal influence in relationships.

3. Fear of Intimacy. Men are also less likely to feel comfortable being intimate with a partner. Thelen et al. (2000) studied the fear of intimacy among 243 college-aged heterosexual dating couples by asking them to complete a Fear of Intimacy scale. Results showed that men had a greater fear of intimacy than females did.

4. Custody Disadvantages. Courts are sometimes biased against divorced men who want custody of their children. Because divorced fathers are typically regarded as career-focused and uninvolved in child care, some are relegated to seeing their children on a limited basis, such as every other weekend or four evenings a month.

5. Shorter Life Expectancy. Men typically die six years sooner (at age 75) than women (*Statistical Abstract of the United States: 2003,* Table 105). Although biological factors may play a role in greater longevity for women than for men, traditional gender roles play a major part. For example, the traditional male role emphasizes achievement, competition, and suppression of feelings, all of which may produce stress. Not only is stress itself harmful to physical health, but it may also lead to compensatory behaviors, such as smoking, alcohol and other drug abuse, and dangerous risk-taking behavior.

In sum, the traditional male gender role is hazardous to men's physical health. However, as women have begun to experience many of the same stresses and behaviors as men, their susceptibility to stress-related diseases has increased. For example, since the 1950s, male smoking has declined while female smoking has increased, resulting in an increased incidence of lung cancer in women.

Benefits of Traditional Male Socialization As a result of higher status and power in society, men tend to have a more positive self-concept and greater confidence in themselves. They also enjoy higher incomes and an easier climb up the good-old-boy corporate ladder; they are rarely confronted with sexual harassment.

We have been discussing the respective ways in which women and men are affected by traditional gender role socialization. Up Close 2.2 summarizes twelve implications that traditional gender role socialization has for the relationships of women and men.

CHANGING GENDER ROLES

Imagine a society in which women and men each develop characteristics, lifestyles, and values that are independent of gender role stereotypes. Characteristics such as strength, independence, logical thinking, and aggressiveness are no longer associated with maleness, just as passivity, dependence, showing emotions, intuitiveness, and nurturing are no longer associated with femaleness. Both sexes are considered equal, and women and men may pursue the same occupational, political, and domestic roles. Some gender scholars have suggested that persons in such a society would be neither feminine nor masculine but would be described as androgynous. The next subsections discuss androgyny, gender role transcendence, and gender postmodernism.

Androgyny

Androgyny refers to a blend of traits that are stereotypically associated with both masculinity and femininity. It may also imply flexibility of traits; for example, an androgynous individual may be emotional in one situation, logical in another, assertive in another, and so forth. Ward (2001) classified 311 (159 male/152 female) undergraduates at the National University of Singapore as: androgynous (33.8% men and 16.0% women), feminine (11.0% men and 39.6% women), masculine (35.7% men and 13.9% women), and undifferentiated (19.5% men and 30.6% women). For men, masculinity and androgyny were associated with self-acceptance and psychological well-being. For women, masculinity and androgyny were associated with self-acceptance only. Neither masculinity nor androgyny was associated with psychological well-being. Skoe et al. (2002) found that androgynous individuals were much more likely to report "helpful behaviors" than persons scoring high on femininity or masculinity. Stake (2000) analyzed data on interviews of 106 males and 197 females at a community college and found that the capacity to respond androgynously (not be rigid but be both instrumental and expressive) when coping skills were required was associated with positive mental health benefits.

Ward (2001) noted the problems with previous research on androgyny—specifically, the presumption that masculine traits are more desirable than feminine ones. Feminine traits were, on the whole, evaluated as less desirable than masculine qualities. Some researchers suggested this reflected a cultural bias toward male dominance. Woodhill and Samuels (2003) emphasized the need to differentiate between positive and negative androgyny. **Positive androgyny** is devoid of the negative traits associated with masculinity (aggression, hard-heartedness, indifference, selfishness, showing off, and vindictiveness) and femininity (being passive, submissive, temperamental, and fragile). Beyond the concept of androgyny is that of gender role transcendence. The researchers also found that positive androgyny is associated with psychological health and well-being.

Gender Role Transcendence

Beyond the concept of androgyny is that of gender role transcendence. We associate many aspects of our world, including colors, foods, social/occupational roles, and personality traits, with either masculinity or femininity. The concept of

Authors

This 13-year-old felled this blesbok with one shot on a South African hunting trip with her dad. Hunting is traditionally a male-only sport.

gender role transcendence involves abandoning gender schema (i.e., becoming "gender aschematic," Bem, 1983) so that personality traits, social/occupational roles, and other aspects of our lives become divorced from gender categories. But such transcendence is not equal for women and men. Although females are becoming more masculine, in part because our society values whatever is masculine, men are not becoming more feminine. Indeed, adolescent boys may be described as very gender entrenched.

Two researchers identified fifteen couples they regarded as "post-gender," or beyond gender, in that the spouses had dual careers, shared a dual-nurturer relationship, and had a post-gender ideology (Riseman & Johnson-Sumerford, 1998).

"The women and men in this sample generally have rejected gender as an ideological justification for inequality . . . . They do not consider that wifehood involves a script of doing domestic service or that breadwinning is an aspect of successful masculinity" (p. 38). Beyond these changes is gender postmodernism.

Gender Postmodernism

Monro (2000) predicted a new era of gender postmodernism whereby there would be a dissolution of male and female categories as currently conceptualized in Western capitalist society. In essence, people would no longer be categorized as male or female but be recognized as capable of many identities—"a third sex" (p. 37). A new conceptualization of "trans" people would call for new social structures "based on the principles of equality, diversity and the right to self determination" (p. 42). No longer would our society reflect transphobia but would embrace pluralization "as an indication of social evolution, allowing greater choice and means of self-expression concerning gender" (p. 42).

However, Miller (2000) emphasized that a movement away from a dimorphic sex classification is not likely to take hold as "the predominant worldview revolves around the existence of two sexes. The first thing

> *[p]arents want to hear from the obstetrician after birth is "it's a boy" or "it's a girl." Legally, every adult is either male or female. Although it might be a bit dramatic to drop the two-sex model, the existence of individuals who fall outside the two current gender identities deserves some thought* (p. 152).

SUMMARY

What are the important terms related to gender?

Sex refers to the biological distinction between females and males. *Gender* refers to the social and psychological characteristics often associated with being female or male. Other terms related to gender include *gender identity* (one's self-concept as a girl or boy), *gender role* (social norms of what a girl or boy "should" do), *gender role ideology* (how women and men "should" interact), *transgendered* (expressing characteristics different from one's biological sex), and *transgenderist* (person who lives in a role other than one's biological sex).

What theories explain gender role development?

Sociobiology emphasizes social behavior (e.g., gender roles) as having an evolutionary survival function. Women stayed in the nest or gathered food nearby, while men could go afar to find food. Such a conceptualization focuses on the division of labor between women and men as functional for the survival of the species. Identification theory says that children acquire the characteristics and behaviors of their same-sex parent through a process of identification. Boys identify with their fathers; girls identify with their mothers. Social learning theory emphasizes the roles of reward and punishment in explaining how a child learns gender role behavior. Cognitive-developmental theory emphasizes biological readiness, in terms of cognitive development, of the child's responses to gender cues in the environment. Once children learn the concept of gender permanence, they seek to become competent and proper members of their gender group.

What are the various agents of socialization?

Various socialization influences include parents (representing different races and ethnicities), peers, religion, the economy, education, and mass media. These shape the individual toward various gender roles and influence what a person thinks, feels, and does in his or her role as a woman or a man.

How are gender roles expressed in other societies?

The lives of women and their children (since what happens to women affects their children) in Afghanistan under the Taliban were cruel, demeaning, and often fatal. Some women drank household bleach rather than continue to endure their plight. They were not allowed to go to school or work and thus were completely dependent economically. Gender roles in traditional China were very unequal and were characterized by female subordination with "close controls over women and wives' responsibility for housework." Women in Punjab, Pakistan, reported that they had limited control over their lives, as evidenced by the expectation of early marriage, quick conception, and limited access to contraceptives. The women also expressed a strong preference for sons, which reflected women's subordinate position in society. Swedish women hold about a quarter of the seats in the Swedish Parliament. However, few Swedish women are in high-status positions in business, and governmental efforts to reduce gender inequality are weak compared with the power of tradition. Gender roles in Africa reflect a diverse continent with over fifty nations. The cultures range from Islamic/Arab cultures of Northern Africa to industrial and European influences in South Africa. In some parts of east Africa (e.g., Kenya) gender roles are in flux.

What are the consequences of traditional gender role socialization?

Traditional female role socialization may result in less education, less income, greater dependence, lower marital satisfaction, a longer life, a closer emotional bond with children, and a larger number of quality relationships for women. Traditional male role socialization may result in the fusion of self and occupation, a more limited expression of emotion, disadvantages in child custody disputes, a shorter life, higher income, and higher status.

How are gender roles changing?

Androgyny refers to a blend of traits that are stereotypically associated with both masculinity and femininity. It may also imply flexibility of traits; for example, an androgynous individual may be emotional in one situation, logical in another, assertive in another, and so forth. The concept of gender role transcendence involves abandoning gender schema (i.e., becoming "gender aschematic") so that personality traits, social/occupational roles, and other aspects of our lives become divorced from gender categories. But such transcendence is not equal for women and men. Although females are becoming more masculine, in part because our society values whatever is masculine, men are not becoming more feminine. Indeed, adolescent boys may be described as very gender entrenched. A new era of gender postmodernism would involve a dissolution of male and female categories as currently conceptualized in Western capitalist society. In essence, people would no longer be categorized as male or female but be recognized as capable of many identities—"a third sex."

KEY TERMS

androgyny
cross-dresser
female circumcision
female genital mutilation
female genital operations
feminization of poverty
gender
gender dysphoria
gender identity
gender role ideology
gender role transcendence
gender roles
hermaphrodites
intersexed individuals
occupational sex segregation
parental investment
positive androgyny
sex
sex roles
sexism
socialization
sociobiology
transgendered
transgenderism
transgenderist
transsexual
transvestite

RESEARCHING MARRIAGE AND THE FAMILY WITH INFOTRAC COLLEGE EDITION

InfoTrac College Edition, an online library, allows you to perform research online anywhere, anytime. Following are two suggested search terms and related questions to help you extend your understanding of the topics covered in this chapter. Go to www.infotrac-college.com to begin your search.

***Search keyword:* Transgender.** Locate articles on transgenderism and identify the meaning of transgender in the professional literature. Also, identify some of the unique problems these transgenderists face.

***Search keyword:* Gender in marriage.** Locate articles that discuss the advantages and disadvantages of marriage for women and men.

The Companion Web Site for Choices in Relationships: An Introduction to Marriage and the Family, Eighth Edition

http://sociology.wadsworth.com/knox_schacht/choices8e

Supplement your review of this chapter by going to the companion Web site to take one of the Tutorial Quizzes, use the flash cards to master key terms, and check out the many other study aids you'll find there. You'll also find special features such as the Marriage and Family Resource Center, Census 2000 information, and other data and resources at your fingertips to help you with that special project or to do some research on your own.

WEBLINKS

Equal Employment Opportunity Commission
http://www.eeoc.gov/

Men's Issues
http://www.vix.com/pub/men/index.html

National Organization for Women (NOW)
http://www.now.org/

Transgender Forum
http://www.tgforum.com/

Tri-Ess
http://www.triess-outreach.org/

WWWomen
http://www.wwwomen.com

Authors

CHAPTER 3

We love because it's the only true adventure.

Nikki Giovanni

Love in Relationships

Contents

True or False?

1. College women (when compared with college men) report that in a new relationship they are the first one to tell their partner, "I love you."
2. Men and women are biologically wired to fall in love with the opposite sex.
3. A love partner who is very controlling is likely to become a stalker when the relationship ends.
4. People in love report that they are less likely to use a condom with their beloved.
5. Compersion is experiencing intense feelings of jealousy when one's partner shows emotional and sexual interest in another.

Answers: **1.** F **2.** F **3.** T **4.** T **5.** F

When the news media revealed in 2003 that Clara Harris ran over (and killed) her husband with her Mercedes-Benz after she discovered him in a Houston hotel with another woman, we were reminded of how important love feelings are and how devastating the loss of love (particularly to another) can become. Not only is being in love associated with our individual happiness (Gross & Simmons, 2002), but having a strong-intimacy focused relationship is associated with relationship happiness (Sanderson & Karetsky, 2002). Indeed, the absence of love in a relationship paves the way for considering divorce. Over two-thirds (67%) of 620 university students reported that they would divorce if they fell out of love with their spouse and were unhappy (Knox & Zusman, 1998). Because of the importance of love to one's life and relationships, we examine this elusive phenomenon.

This chapter is concerned with the nature of love (both ancient and modern views), various theories of the origin of love, how love develops in a new relationship, and problems associated with love. Since jealousy in love relationships is common, we examine its causes and consequences.

DESCRIPTIONS OF LOVE

Love remains an elusive and variable phenomenon and continues to be thought of as emotional and irrational (Bulcroft et al., 2000). Nevertheless, researchers have conceptualized love as a dichotomy (i.e., "romantic or conjugal") or as a vague construct involving relationship dynamics (i.e., "meeting of needs") (Meyers & Shurts, 2002).

Sternberg (1986) developed the "triangular" view of love as consisting of three basic elements—intimacy, passion, and commitment. The presence or absence of these three elements provides a description of various types of love experienced between individuals regardless of their sexual orientation.

1. *Nonlove—the absence of intimacy, passion, and commitment. Two strangers looking at each other from afar have a nonlove.*
2. *Liking—intimacy without passion or commitment. A new friendship may be described in these terms of the partners liking each other.*
3. *Infatuation—passion without intimacy or commitment. Two persons flirting with each other in a bar may be infatuated with each other.*
4. *Romantic love—intimacy and passion without commitment. Love at first sight reflects this type of love.*
5. *Companionate love—intimacy and commitment without passion. A couple who have been married for fifty years are said to have a companionate love.*
6. *Fatuous love—passion and commitment without intimacy. A couple who are passionately wild about each other and talk of the future but do not have an intimate connection with each other have a fatuous love.*
7. *Empty love—commitment without passion or intimacy. A couple who stay together for social and legal reasons but who have no spark or emotional sharing between them have an empty love.*
8. *Consummate love—combination of intimacy, passion, and commitment. Sternberg's view of the ultimate, all-consuming love.*

Individuals bring different combinations of the elements of intimacy, passion, and commitment (the triangle) to the table of love. One lover may bring a predominance of passion with some intimacy but no commitment (romantic love), while the other person brings commitment but no passion or intimacy (empty love). The triangular theory of love allows lovers to see the degree to which they are matched in terms of passion, intimacy, and commitment in their relationship.

Up Close 3.1 When College Students Say, "I Love You"

What do college students mean when they say, "I love you?" To find out, researchers at a large southeastern university asked 147 undergraduates (72% female; 28% male) to complete an anonymous twenty-nine-item survey designed to reveal the timing and meaning of their telling a new partner, "I love you."

Analysis of the data revealed several findings:

1. *Males say "I love you" first.* Males were significantly more likely to report that, of the mutual love relationships in which they had been involved, they were the first to disclose their love to their partner. The finding that men express love first is supported by previous research. Sharp & Ganong (2000) also found that men fall in love more quickly and have higher levels of romantic beliefs than women.

2. *Males say "I love you" for sex.* Males were significantly more likely than females to report that they say, "I love you" if they think it will increase the chance that their partner will have sex with them. Female students now have empirical verification that hearing "I love you" may mean little more than "I am saying this just so you will have sex with me."

Previous research has confirmed that men are more likely to seek sex early in the relationship (indeed, within hours) than women (Knox, Sturdivant, & Zusman, 2001).

Abridged and adapted from A. Brantley, D. Knox, and M. E. Zusman. 2002. When and why gender differences in saying "I love you" among college students. *College Student Journal* 36: 614–15.

Sources

Knox, D., L. Sturdivant, and M.E. Zusman. 2001. College student attitudes toward sexual intimacy. *College Student Journal* 35:241–43.

Sharp, E. A., and L. H. Ganong. 2000. Raising awareness about marital expectations: An unrealistic beliefs change by integrative teaching? *Family Relations* 49: 71–76.

A common class exercise among professors who teach marriage and the family is to randomly ask class members to identify one word they most closely associate with love. Invariably, students identify different words (commitment, feeling, trust, altruism, etc.), which suggests that there is great variability in the way we think about love. Indeed, just the words "I love you" have different meanings depending on whether they are said by a man or a woman (see Up Close 3.1).

Diversity in Other Countries

Although love is an experience most people throughout the world are capable of, its importance as a prerequisite for marriage varies throughout the world. In the United States individuals think of falling in love and then marriage. In Iran, love follows marriage. "I had a marriage arranged and approved by my parents," said one Iranian woman. "I did not have time for love to develop before the wedding but expected it to develop after I got to know him" (authors' files).

LOVE IN SOCIETAL CONTEXT

Though we think of love as an individual experience, the society in which we live exercises considerable control over our love object/choice and conceptualizes it in various ways.

Social Control of Love

Love may be blind, but it knows what color a person's skin is. Indeed, one might ask, "Is love color blind or is love blinded by color?" The data are clear—love seems to see only people of similar color, as over 95 percent of people marry someone of their own racial background (*Statistical Abstract of the United States: 2003,* Table 62). Hence, parents and peers may approve of their offspring's and friend's love choice when it locks on a same-race partner and disapprove of the selection when it does not. In a larger sense, our society may seek to encourage love bonds since pair-bonded individuals provide mutual aid and are less of a drain on a society's human services resources. Our society cannot provide counselors for the mental health roller coaster of life. But a love partner who is caring and empathetic is just what the doctor ordered.

Love in the Workplace

SOCIAL POLICY

With an increase of women in the workforce, an increase in the age at first marriage, and longer work hours, the workplace has become a common place for romantic relationships to develop. More future spouses may meet at work than at school, social, or neighborhood settings.

Pros and Cons of Office Romances

The energy that both fuels and results from intense love feelings can also fuel productivity on the job. And if the coworkers eventually marry or enter a nonmarital but committed and long-term relationship, they may be more satisfied with and committed to their jobs than spouses whose partners work elsewhere. Working at the same location enables married couples to commute together, go to company-sponsored events together, and talk shop together.

Recognizing the potential benefits of increased job satisfaction, morale, productivity, creativity, and commitment, some companies even look favorably upon love relationships among employees. Prior to the economic downturn following 9/11, Apple Computer in Cupertino, California, encouraged socializing among employees by sponsoring get-togethers every Friday afternoon with beer, wine, food, and, on occasion, live bands. The company also had ski clubs, volleyball clubs, and Frisbee clubs, providing employees with opportunities to meet and interact socially. Some companies hire two employees who are married, reflecting a focus on the value of each employee to the firm rather than on their love relationship outside work.

However, workplace romances can also be problematic for the individuals involved as well as for their employers. When a workplace romance involves a supervisor/subordinate relationship, other employees might make claims of favoritism or differential treatment. In a typical differential-treatment allegation, an employee (usually a woman) claims that the company denied her a job benefit because her supervisor favored a female coworker—who happens to be the supervisor's girlfriend.

If a workplace relationship breaks up, it may be difficult to continue to work in the same environment (and others at work may experience the fallout). A breakup that is less than amicable may result in efforts by partners to sabotage each other's work relationships and performance, incidents of workplace violence, harassment, and/or allegations of sexual harassment. In a survey of 1,221 human-resource managers conducted by the Society for Human Resource Management and CareerJournal.com, 81 percent of the HR professionals and 76 percent of the executives saw office romances as "dangerous" (Franklin, 2002).

Workplace Policies on Intimate Relationships

Some companies such as Disney, Universal, and Columbia have "anti-fraternization" clauses that impose a cap on workers' talking about private issues or sending personal E-mails. Some British firms have "love-contracts" that require workers to tell their managers if they are involved with anyone from the office. While these restrictions are rare, they seem to have the desired effect of curtailing office romances (Cooper, 2003).

Most companies (Wal-Mart is an example) do not prohibit romantic relationships among employees. However, the company may have a policy prohibiting open displays of affection between employees in the workplace and romantic relationships between supervisor and subordinate. Most companies have no policy regarding love relationships at work and generally regard romances between coworkers as "none of their business."

There are some exceptions to the general permissive policies regarding workplace romances. Many companies have written policies prohibiting intimate relationships when one member of the couple is in a direct supervisory position over the other. These policies may be enforced by transferring or dismissing employees who are discovered in romantic relationships.

As the debate continues over whether or not office romances are a good idea, lawsuits and money are at stake. Go to http://sociology.wadsworth.com/knox_schacht/choices8e , select the Opposing Viewpoints Resource Center on the left navigation bar, and enter your passcode. Conduct a search by subject, using the search term "love in the workplace," and read the article on what corporations need to know before they forbid office romances. What policies have been operative in the various places you have worked?

Sources

Cooper, C. 2003. Office affairs are hard work. *The Australian.* 5 March.

Franklin, R. 2002. Office romances: Conduct unbecoming? *Business Week Online.* 14 February, 1.

NATIONAL DATA

In terms of the black-white marriages fewer than 1 percent of the almost 60 million married couples in the United States include a Black and a white spouse (*Statistical Abstract of the United States: 2003*, Table 62).

Another example of the social control of love is that individuals attracted to someone of the same sex quickly feel the social and cultural disapproval of this attraction. Some feign heterosexual attraction, and may even marry and have children, to meet the social requisite of an "appropriate" love choice while harboring secret feelings for same-sex individuals. Diamond (2003) emphasized that women and men are biologically wired and capable of falling in love with and establishing intense emotional bonds with members of their own or opposite sex (hence, one's love-desire and sexual-desire partner can be different). Her notion further illustrates the power of social influence that typically disapproves of such love relationships.

Romantic love is such a powerful emotion and marriage such an important relationship connecting an outsider into an existing family and peer network that mate selection is not left to chance. Parents inadvertently influence the mate choice of their children by moving to certain neighborhoods, joining certain churches, and enrolling their children in certain schools. Doing so increases the chance that their offspring will "hang out," fall in love, and marry persons who are similar in race, education, and social class. Although twentieth-century parents normally do not have large estates and are not concerned about economic transfers, they are concerned about their offspring's meeting someone who will "fit in" and with whom they will feel comfortable.

Peers exert a similar influence toward homogamous mating by approving of certain partners and disapproving of others. Their motive is similar to that of parents—they want to feel comfortable around the people their peers bring with them to social encounters. Both parents and peers are influential, as most offspring and friends end up falling in love with and marrying persons of the same race, education, and social class.

Minority groups are sometimes even more intent on controlling who falls in love with whom. Since individuals typically marry only those with whom they fall in love, minority groups may feel the need to ensure that they fall in love only with those within the group. Not to do so results in marrying outside the group, leaving fewer members of the group to preserve its heritage.

Love is also an issue of control in the workplace (see Social Policy on p. 58).

Ancient Views of Love

Many of our present-day notions of love stem from early Buddhist, Greek, and Hebrew writings.

Buddhist Conception of Love The Buddhists conceived of two types of love—an "unfortunate" kind of love (self-love) and a "good" kind of love (creative spiritual attainment). Love that represents creative spiritual attainment was described as "love of detachment," not in the sense of withdrawal from the emotional concerns of others but in the sense of accepting people as they are and not requiring them to be different from their present selves as the price of friendly affection. To a Buddhist, the best love is one in which you accept others as they are without requiring them to be like you.

Greek and Hebrew Conceptions of Love Three concepts of love introduced by the Greeks and reflected in the New Testament are phileo, agape, and eros. *Phileo* refers to love based on friendship and can exist between family members, friends, and lovers. The city of Philadelphia was named after this phileo type of love. Another variation of phileo love is *philanthropia,* the Greek word meaning "love of humankind."

Agape refers to a love based on a concern for the well-being of others. Agape is spiritual, not sexual, in nature. This type of love is altruistic and requires nothing in return. "Whatever I can do to make your life happy" is the motto of the agape lover, even if this means giving up the beloved to someone else. Such love is not always reciprocal.

Eros refers to sexual love. This type of love seeks self-gratification and sexual expression. In Greek mythology, Eros was the god of love and the son of Aphrodite. Plato described "true" eros as sexual love that existed between two men. According to Plato's conception of eros, homosexual love was the highest form of love because it existed independent of the procreative instinct and free from the bonds of matrimony. Also, women had low status and were uneducated and were therefore not considered ideal partners for the men. By implication, love and marriage were separate.

Love in Medieval Europe—From Economics to Romance

Love in the 1100s was a concept influenced by economic, political, and family structure. In medieval Europe, land and wealth were owned by kings controlling geographical regions—kingdoms. When so much wealth and power were at stake, love was not to be trusted as the mechanism for choosing spouses for royal offspring. Rather, marriages of the sons and daughters of the aristocracy were arranged with the heirs of other states with whom an alliance was sought. Love was not tied to marriage but was conceptualized, even between people not married or of the same sex, as an adoration of physical beauty (often between a knight and his beloved), and as spiritual and romantic.

The presence of kingdoms and estates and the patrimonial households declined with the English revolutions of 1642 and 1688 and the French Revolution of 1789. No longer was power held by aristocratic families; it was transferred to individuals through parliaments or other national bodies. Even today, English monarchs are figureheads, with the real business of international diplomacy being handled by parliament. Since wealth and power were no longer in the hands of individual aristocrats, the need to control mate selection decreased and the role of love changed. Marriage became less of a political and business arrangement and more of a mutually desired emotional union between spouses. Just as partners in medieval society were held together by bureaucratic structure, a new mechanism—love—would now provide the emotional and social bonding. Hence, love in medieval times changed from a feeling irrelevant to marriage, since individuals (representing aristocratic families) were to marry even though they were not in love, to a feeling that bonded a woman and a man together for marriage.

Modern Conceptions of Love—Love Styles

Theorist John Lee (1973, 1988) identified a number of styles of love that describe the way lovers relate to each other. Keep in mind that the same individual may view love in more than one way at a time or may view love in different ways at different times. These love styles are also independent of one's sexual orientation—no one love style is characteristic of heterosexuals or homosexuals.

If you have love in your life it can make up for a great many things you lack. If you don't have it, no matter what else there is, it's not enough.

Ann Landers

1. Ludus. Country-and-western singer George Strait's song "She'll Leave You with a Smile" reflects involvement with a ludic lover."You're gonna give her all your heart/ Then she'll tear your world apart./ You're gonna cry a little while/ Still she'll leave you with a smile." The ludic lover views love as a game, refuses to become dependent on any one person, and does not encourage another's intimacy. Two essential skills of the ludic lover are to juggle several partners at the same time and to manage each relationship so that no one partner is seen too often. These strategies help to ensure that the relationship does not deepen into an all-consuming love. Don Juan represented the classic ludic lover. "Love 'em and leave 'em" is the motto of the ludic lover. Neto, Deschamps, & Barros (2000) also found that men are more likely than women to be ludic lovers.

In a study (Paul, McManus, & Hayes 2000) of "hookups" between college students, certain love styles were characteristic of students who hooked up. Distinguishing features of those who had noncoital hookups were a ludic love style and high concern for personal safety. These individuals may have been participating in collegiate cultural expectations by engaging in "playful" sexual exploration but refraining from intercourse out of their personal safety concern. Indeed, those who engaged in coital hookups were also characterized by ludic love styles, along with symptoms of alcohol intoxication. The researchers worried that the combination of ludic orientation (motivated by the thrill of the game) and alcohol intoxication could be a precursor to sexual experiences that were forced or unwanted by a partner.

The **ludic love style** is sometimes characterized as manipulative and noncaring. But ludic lovers may also be compassionate and very protective of another's feelings. For example, some uninvolved soon-to-graduate seniors avoid involvement with anyone new and become ludic lovers so as not to encourage anyone.

2. Pragma. The **pragma love style** is the love of the pragmatic, who is logical and rational. The pragma lover assesses his or her partner on the basis of assets and liabilities. Economic security may be regarded as very important. The pragma lover does not become involved in interracial, long-distance, or age-discrepant partners, because logic argues against doing so. Bulcroft et al. (2000) noted that, increasingly, individuals are becoming more pragmatic about their love choices.

Authors

The love of this couple is of the eros variety—they are consumed by passion and romance.

3. Eros. Just the opposite of the pragmatic love style, the **eros love style** is one of passion and romance. Intensity of both emotional and sexual feelings dictates one's love involvements. The love between Rose and Jack in the film *Titanic* was of the eros love style. Inman-Amos, Hendrick, and Hendrick (1994) observed that eros was *the* most common love style of women and men in a sample of college students. Hendrick, Hendrick, and Adler (1988) found that couples who were more romantically and passionately in love were more likely to remain together than couples who avoided intimacy by playing games with each other.

4. Mania. The person with **mania love style** feels intense emotion and sexual passion but is out of control. The person is possessive and dependent and "must have" the beloved. Persons who are extremely jealous and controlling reflect manic love. We noted earlier that Clara Harris was convicted in 2003 for running over (and killing) her husband with her Mercedes-Benz. She reported being in a jealous rage when she discovered him in a Houston hotel with another woman. Her love style for him may be characterized as manic and reflects the attitude "If I can't have you, no one else will." Stalking is an expression of love gone wild (see Up Close 3.2 on p. 62)

5. Storge. The **storge love style** is a calm, soothing, nonsexual love devoid of intense passion. Respect, friendship, commitment, and familiarity are characteristics that help to define the relationship. The partners care deeply about each other but not in a romantic or lustful sense. Their love is also more likely to endure than fleeting romance. Spouses who have been married fifty years and who still love and enjoy each other are likely to have a storge type of love.

6. Agape. One of the forms of love identified by the ancient Greeks, the **agape love style** is selfless and giving, expecting nothing in return. The nurturing and caring partners are concerned only about the welfare and growth of each other. The love parents have for their children is also often described as the agape love style. Inman-Amos, Hendrick, and Hendrick (1994) observed that the predominant love style of married couples is agape.

Diversity in Other Countries

Love among Chinese couples is of the storge variety. Pimentel (2000) studied a large representative sample of married couples from urban China and found that "Chinese couples have what Westerners might characterize as a relatively unromantic vision of love, more like companionship. The words most often accompanying remarks about love were "respect," "mutual understanding," and "support." Expressions of passion, of "sparks flying," or similar phrases, were not present.

Up Close 3.2 Stalking—When Love Goes Mad

In the name of love, people have stalked the beloved. **Stalking** is defined as the repeated, willful, and malicious following or harassment by one person of another. It may involve watching a victim, property damage, threats, home invasion, or threats of physical harm, and it usually causes great emotional distress and impairs the recipient's social and work activities. Over a million women and 370,000 men are stalked each year; 8 percent of women and 2 percent of men are stalked at some time during their life (King, 2003). Patriarchal, social learning, and social conflict theories provide explanations for men stalking women—as an expression of male authority, men learn to be dominant, and they use power to control women.

Persons who stalk are obsessional and very controlling. They are typically mentally ill and have one or more personality disorders involving paranoid, antisocial, or obsessive-compulsive behaviors. Exercising a great deal of control in an existing relationship is predictive that the controlling partner will become a stalker when the other partner ends the relationship (King, 2003).

Although various coping strategies have been identified, additional research is needed on how to manage unwanted attention (Regan, 2000). A survey (Spitzberg & Cupach, 1998) of young adults identified five general coping categories:

1. Making a direct statement to the person ("I am not interested in dating you, my feelings about you will not change, and I know that you will respect my decision and direct your attention elsewhere." Regan, 2000, 266).
2. Seeking protection through formal channels (police involvement, restraining order).
3. Avoiding the perpetrator (ignoring, not answering phone).
4. Retaliating.
5. Using informal coping methods (using telephone caller identification; seeking advice of others).

Direct statements and actions that unequivocally communicate lack of interest are probably the most effective types of intervention.

Sources

King, P. A. 2003. Stalking: A control factor. Paper delivered at the 73rd Annual Meeting of the Eastern Sociological Society, Philadelphia, February 28.

Regan, P. 2000. Love relationships. In *Psychological perspectives on human sexuality,* edited by L. T. Szuchman and F. Muscarella. New York: Wiley, 232–82.

Spitzberg, B. H., and W. R. Cupach, eds. 1998. *The dark side of close relationships.* Mahway, N.J.: Erlbaum.

When two people are under the influence of the most violent, most insane, most delusive, and most transient of passions, they are required to swear that they will remain in that excited, abnormal, and exhausting condition continuously, until death do them part.

George Bernard Shaw, *Playwright*

Romantic versus Realistic Love

Love may also be described as being on a continuum from romanticism to realism. For some people, love is romantic; for others, it is realistic. **Romantic love** is characterized by such beliefs as love at first sight, there is only one true love, and love conquers all. The symptoms of romantic love include drastic mood swings, palpitations of the heart, and intrusive thoughts about the partner.

International Data

Based on data from a survey of 641 young adults at three international universities, American young adults were the most romantic, followed by the Turkish students, with Indians having the lowest romanticism scores (Medora et al., 2002).

Whether men or woman are more romantic varies by study. Sharp and Ganong (2000) found that men were more likely than women to fall in love quickly. However, Medora et al. (2002) found that American, Turkish, and Indian women tended to be more romantic than men. In a study of 197 emotional perceptions of college students, women saw themselves (and men agreed) as more romantic than men (Knox, Zusman, & Thompson, in press). Huston et al. (2001) found that after two years of marriage, the couples who had fallen in love more slowly were just as happy as couples who fell in love at first sight.

Diversity in Other Countries

We have already noted the importance of romantic love in the United States. But what about other societies? Jankowiak and Fischer (1992) found evidence of passionate love in 147 out of 166 (88.5%) of the societies they studied. Passionate love was defined by the presence of at least one of the following: accounts depicting personal anguish and longing, love songs, elopement due to mutual affection, or native accounts affirming the existence of passionate love. The researchers' study stands "in direct contradiction to the popular idea that romantic love is essentially limited to or the product of Western culture. Moreover, it suggests that romantic love constitutes a human universal, or at the least a near-universal" (p. 154).

Infatuation is sometimes regarded as synonymous with romantic love. **Infatuation** comes from the same root word as *fatuous,* meaning "silly" or "foolish," and refers to a state of passion or attraction that is not based on reason. Infatuation is characterized by the tendency to idealize the love partner. People who are infatuated magnify their lover's positive qualities ("My partner is always happy") and overlook or minimize their negative qualities ("My partner doesn't have a problem with alcohol; he just likes to have a good time").

In contrast to romantic love is realistic love. Realistic love is also known as conjugal love. **Conjugal (married) love** is less emotional, passionate, and exciting than romantic love and is characterized by companionship, calmness, comfort, and security. *Companionate love* is a term often used for conjugal love. Sprecher and Regan (1998) found that companionate love was more satisfying than passionate love. However, individuals may experience both companionate and passionate love for the same partner. The Love Attitudes Scale offers a way for you to assess the degree to which you tend to be romantic or realistic in your view of love (see p. 64).

When you determine your love attitudes score, be aware that your tendency to be a romantic or a realist is neither good nor bad. Both romantics and realists can be happy individuals and successful relationship partners.

Authors

Men tend to see women as romantic, and women see themselves that way.

PERSONAL CHOICES

Do You Make Relationship Choices with Your Heart or Your Head?

Lovers are frequently confronted with the need to make decisions about their relationships, but they are divided on whether to let their heart or head rule in such decisions. In a marriage and family class, 120 students were asked whether they used their hearts or heads in making such decisions. Some of their answers follow.

Heart Those who relied on their heart (women more than men) for making decisions felt that emotions were more important than logic and that listening to your heart made you happier. One woman said:

> In deciding on a mate, my heart would rule because my heart has reasons to cry and my head doesn't. My heart knows what I want, what would make me most happy. My head tells me what is best for me. But I would rather have something that makes me happy than something that is good for me.

Some men also agreed that your heart should rule. One said:

> I went with my heart in a situation, and I'm glad I did. I had been dating a girl for two years when I decided she was not the one I wanted and that my present girlfriend was. My heart was saying to go for the one I loved, but my head was telling me not to because if I broke up with the first girl, it would hurt her, her parents, and my parents. But I decided I had to make myself happy and went with the feelings in my heart and started dating the girl who is now my fiancée.

Relying on one's emotions does not always have a positive outcome, as the following experience illustrates:

> Last semester, I was dating a guy I felt more for than he did for me. Despite that, I wanted to spend any opportunity I could with him when he asked me to go somewhere with him. One day he had no classes, and he asked me to go to the park by the river for a picnic. I had four classes that day and exams in two of

SELF-ASSESSMENT

The Love Attitudes Scale

This scale is designed to assess the degree to which you are romantic or realistic in your attitudes toward love. There are no right or wrong answers.

Directions

After reading each sentence carefully, circle the number that best represents the degree to which you agree or disagree with the sentence.

1	2	3	4	5
Strongly agree	**Mildly agree**	**Undecided**	**Midly disagree**	**Strongly disagree**

	SA	MA	U	MD	SD
1. Love doesn't make sense. It just is.	1	2	3	4	5
2. When you fall "head over heels" in love, it's sure to be the real thing.	1	2	3	4	5
3. To be in love with someone you would like to marry but can't is a tragedy.	1	2	3	4	5
4. When love hits, you know it.	1	2	3	4	5
5. Common interests are really unimportant; as long as each of you is truly in love, you will adjust.	1	2	3	4	5
6. It doesn't matter if you marry after you have known your partner for only a short time as long as you know you are in love.	1	2	3	4	5
7. If you are going to love a person, you will "know" after a short time.	1	2	3	4	5
8. As long as two people love each other, the educational differences they have really do not matter.	1	2	3	4	5
9. You can love someone even though you do not like any of that person's friends.	1	2	3	4	5
10. When you are in love, you are usually in a daze.	1	2	3	4	5
11. Love "at first sight" is often the deepest and most enduring type of love.	1	2	3	4	5
12. When you are in love, it really does not matter what your partner does because you will love him or her anyway.	1	2	3	4	5
13. As long as you really love a person, you will be able to solve the problems you have with the person.	1	2	3	4	5
14. Usually you can really love and be happy with only one or two people in the world.	1	2	3	4	5
15. Regardless of other factors, if you truly love another person, that is a good enough reason to marry that person.	1	2	3	4	5
16. It is necessary to be in love with the one you marry to be happy.	1	2	3	4	5
17. Love is more of a feeling than a relationship.	1	2	3	4	5
18. People should not get married unless they are in love.	1	2	3	4	5

> them. I let my heart rule and went with him. Nothing ever came of the relationship and I didn't do well in those classes.

Head Most of the respondents (men more than women) felt that it was better to be rational than emotional.

> In deciding on a mate, I feel my head should rule because you have to choose someone that you can get along with after the new wears off. If you follow your heart solely, you may not look deep enough into a person to see what it is that you really like. Is it just a pretty face or a nice body? Or is it deeper than that, such as common interests and values? After the new wears off, it's the person inside the body that you're going to have to live with. The "heart" sometimes can fog up this picture of the true person and distort reality into a fairy tale.

Another student said:

> Love is blind and can play tricks on you. Two years ago, I fell in love with a man who I later found out was married. Although my heart had learned to love this

19. Most people truly love only once during their lives.	**1**	**2**	**3**	**4**	**5**
20. Somewhere there is an ideal mate for most people.	**1**	**2**	**3**	**4**	**5**
21. In most cases, you will "know it" when you meet the right partner.	**1**	**2**	**3**	**4**	**5**
22. Jealousy usually varies directly with love; that is, the more you are in love, the greater your tendency to become jealous will be.	**1**	**2**	**3**	**4**	**5**
23. When you are in love, you are motivated by what you feel rather than by what you think.	**1**	**2**	**3**	**4**	**5**
24. Love is best described as an exciting rather than a calm thing.	**1**	**2**	**3**	**4**	**5**
25. Most divorces probably result from falling out of love rather than failing to adjust.	**1**	**2**	**3**	**4**	**5**
26. When you are in love, your judgment is usually not too clear.	**1**	**2**	**3**	**4**	**5**
27. Love often comes only once in a lifetime.	**1**	**2**	**3**	**4**	**5**
28. Love is often a violent and uncontrollable emotion.	**1**	**2**	**3**	**4**	**5**
29. When selecting a marriage partner, differences in social class and religion are of small importance compared with love.	**1**	**2**	**3**	**4**	**5**
30. No matter what anyone says, love cannot be understood.	**1**	**2**	**3**	**4**	**5**

Scoring

Add the numbers you circled. 1 (strongly agree) is the most romantic response and 5 (strongly disagree) is the most realistic response. The lower your total score (30 is the lowest possible score), the more romantic your attitudes toward love. The higher your total score (150 is the highest possible score), the more realistic your attitudes toward love. A score of 90 places you at the midpoint between being an extreme romantic and an extreme realist. Both men and women undergraduates typically score above 90, with men scoring closer to 90 than women.

A team of researchers (Medora et al., 2002) gave the scale to 641 young adults at three international universities in America, Turkey, and India. Female respondents in all three cultures had higher romanticism scores than male respondents (reflecting their higher value for, desire for, and thoughts about marriage). When the scores were compared by culture, American young adults were the most romantic, followed by Turkish students, with Indians having the lowest romanticism scores.

Reference

Medora, N. P., J. H. Larson, N. Hortacsu, and P. Dave. 2002 Perceived attitudes towards romanticism: A cross-cultural study of American, Asian-Indian, and Turkish young adults. *Journal of Comparative Family Studies* 33:155–78.

Source

Knox, D. 1969. Conceptions of love at three developmental levels. Ph.D. dissertation, Florida State University. Permission to use the scale for research available from David Knox at davidknox2@prodigy.net

> man, my mind knew the consequences and told me to stop seeing him. My heart said, "Maybe he'll leave her for me," but my mind said, "If he cheated on her, he'll cheat on you." I got out and am glad that I listened to my head.

Some individuals feel that both the head and the heart should rule when making relationship decisions.

> When you really love someone, your heart rules in most of the situations. But if you don't keep your head in some matters, then you risk losing the love that you feel in your heart. I think that we should find a way to let our heads and hearts work together.

There is an old saying, "Don't wait until you can find the person you can live with; wait and find the person that you can't live without!" One individual hearing this quote said, "I think both are important. I want my head to let me know it 'feels' right" (authors' files).

THEORIES ON THE ORIGINS OF LOVE

Love—it's everything I understand and all the things I never will.

Mary Chapin Carpenter, *Singer*

Various theories have been suggested with regard to the origins of love.

Evolutionary Theory

Love has an evolutionary purpose by providing a bonding mechanism between the parents during the time their offspring are dependent infants. Love's strongest bonding lasts about four years, the time when children are most dependent and two parents can cooperate in handling their new infant. "If a woman was carrying the equivalent of a 12-lb bowling ball in one arm and a pile of sticks in the other, it was ecologically critical to pair up with a mate to rear the young," observes anthropologist Helen Fisher (Toufexis, 1993). The "four-year itch" is Fisher's term for the time at which parents with one child are most likely to divorce—the time when the woman can more easily survive without parenting help. If the couple have a second child, doing so resets the clock, and "the seven-year itch" is the next most vulnerable time.

Learning Theory

Unlike evolutionary theory, which views the experience of love as innate, learning theory emphasizes that love feelings develop in response to certain behaviors occurring in certain contexts. Individuals on a date who look at each other, smile at each other, compliment each other, touch each other endearingly, do things for each other, and do enjoyable things together are engaging in behaviors that make it easy for love feelings to develop.

During a developing relationship, couples also have a high frequency of reinforcing each other for behaviors they enjoy in each other. The continuation of love feelings depends on each partner's continuing to reinforce the desirable behaviors of the other so that these behaviors (and the love feelings they elicit) continue.

Sociological Theory

Forty-five years ago Ira Reiss (1960) suggested the wheel model as an explanation for how love develops. Basically there are four stages of the wheel—rapport, self-revelation, mutual dependency, and personality need fulfillment. In the rapport stage, each partner has the feeling of having known the partner before, feels comfortable with the partner, and wants to deepen the relationship.

Such desire leads to self-revelation or self-disclosure whereby each reveals intimate thoughts to the other about oneself, the partner, and the relationship. Such revelations deepen the relationship since it is assumed that the confidences are shared only with special people and each partner feels special when listening to the revelations of the other.

As the level of self-disclosure becomes more intimate, a feeling of mutual dependency develops. Each partner is happiest in the presence of the other and begins to depend on the other for creating the context of these euphoric feelings. "I am happiest when I am with you" is the theme of this stage.

The feeling of mutual dependency involves the fulfillment of personality needs. The desires to love and be loved, to trust and be trusted, and to support and be supported are met in the developing love relationship.

Psychosexual Theory

According to psychosexual theory, love results from blocked biological sexual desires. In the sexually repressive mood of his time, Sigmund Freud ([1905] 1938), referred to love as "aim-inhibited sex." Love was viewed as a function of the sexual desire a person was not allowed to express because of social restraints. In Freud's era, people would meet, fall in love, get married, and have sex. Freud felt that the socially required delay from first meeting to having sex resulted in the development of "love feelings." By extrapolation, Freud's theory of love suggests that love dies with marriage (access to one's sexual partner).

Ego-Ideal Theory

Theodore Reik (1949) suggested that love springs from a state of dissatisfaction with oneself and represents a vain urge to reach one's "ego-ideal." He believed that love is a projection of one's ideal image of himself/herself onto another person. For example, suppose you are a shy, passive, dependent person but wish that you were assertive, outgoing, and independent. According to Reik's theory, you will probably fall in love with a person who has the qualities you admire but lack in yourself.

Ontological Theory

Ontology is a branch of philosophy that is concerned with being. Love from an ontological perspective arises from a lack of wholeness in our being. Such lack of wholeness is implied by the division of humans into males and females. From an ontological perspective, love represents women's desire to be united with their other half (i.e., men) and men's desire to be united with their other half (i.e., women). Eric Fromm (1963) viewed love as a means of overcoming the "separateness" of an individual and of quelling the anxiety associated with being lonely. When men and women develop a love relationship, they become whole.

Biochemical Theory

There may be a biochemical basis for love feelings. Oxytocin is a hormone that encourages contractions during childbirth and endears the mother to the suckling infant. It has been referred to as the "cuddle chemical" because of its significance in bonding. Later in life, oxytocin seems operative in the development of love feelings between lovers during sexual arousal. Oxytocin may be responsible for the fact that more women than men prefer to continue cuddling after intercourse.

PEA (phenylethylamine) is a natural amphetamine-like substance that makes the lovers feel euphoric and energized. The high that they report feeling just being with each other is the PEA in their bloodstream released by their brain. The natural chemical high associated with love may explain why the intensity of passionate love decreases over time. As with any amphetamine, the body builds up a tolerance to PEA, and it takes more and more to produce the special kick. Hence, lovers develop a tolerance for each other. "Love junkies" are those who go from one love affair to the next in rapid succession to maintain the high. Alternatively, some lovers break up and get back together frequently as a way of making the relationship new again and keeping the high going.

Attachment Theory

The attachment theory of love emphasizes that a primary motivation in life is to be connected with other people. Persons who were in a secure attachment love relationship with a parent as an infant report less stress and find it is easier to love as an adult. However, an intense adult love relationship (independent of one's love attachment as a child) has similar benefits of reducing stress and making subsequent love relationships easier (Moller, McCarthy, & Fouladi, 2002). Persons who evidence a secure attachment to a love partner also report higher levels

Table 3.1 Love Theories and Criticisms

Theory	Criticism
Evolutionary	Assumption that women and children need men for survival is not necessarily true today. Women can have and rear children without male partners.
Learning	Does not account for (1) why some people will share positive experiences yet will not fall in love and (2) why some people stay in love despite negative behavior.
Psychosexual	Does not account for people who report intense love feelings yet are having sex regularly.
Ego-Ideal	Does not account for the fact that people of similar characteristics fall in love.
Ontological	The focus of an ontological view of love is the separation of women and men from each other as love objects. Homosexual love is unaccounted for.
Biochemical	Does not specify how much of what chemicals result in the feeling of love. Chemicals alone cannot create the state of love; cognitions are also important.
Attachment	Not all people feel the need to be emotionally attached to others. Some prefer to be detached.

of commitment/dedication (Pistole & Vocaturo, 2000). Hence, the benefits of a secure love attachment are enormous.

Each of the theories of love presented in this section has its critics (see Table 3.1).

HOW LOVE DEVELOPS IN A NEW RELATIONSHIP

The development of love relationships is affected by various social, psychological, physiological, and cognitive conditions.

To cheat oneself out of love is the most terrible deception. It is an eternal loss for which there is no reparation, neither in time or in eternity.

Søren Kierkegaard, *Danish philosopher*

Social Conditions for Love

Love is a label given to an internal feeling. Our society promotes love through popular music, movies, television, and novels. These media convey the message that love is an experience to enjoy and to pursue and that you are missing something if you are not in love.

More traditional societies attempt to directly influence one's love choice. In India and other countries (e.g., Iran) where marriages have been arranged, the development of romantic love relationships is tightly controlled. For example, parents select the mate for their child in an effort to prevent any potential love relationship from forming with the "wrong" person and to ensure that the child marries the "right" person. Such a person must belong to the desired social class and have the economic resources desired by the parents. Marriage is regarded as the linking of two families; the love feelings of the respective partners are irrelevant. Love is expected to follow marriage, not precede it.

Psychological Conditions for Love

Two psychological conditions associated with the development of healthy love relationships are high self-esteem and self-disclosure.

Self-Esteem High self-esteem is important for developing healthy love relationships because it enables an individual to feel worthy of being loved. Feeling good about yourself allows you to believe that others are capable of loving you. Individuals with low self-esteem doubt that someone else can love and accept them (DeHart, 2002).

Having self-esteem provides other benefits:

1. *It allows one to be open and honest with others, about both strengths and weaknesses.*
2. *It allows one to feel generally equal to others.*
3. *It allows one to take responsibility for one's own feelings, ideas, mistakes, and failings.*
4. *It allows for the acceptance of both strengths and weaknesses in oneself and others.*
5. *It allows one to validate oneself and not to expect the partner to do this.*
6. *It permits one to feel empathy—a very important skill in relationships.*
7. *It allows separateness and interdependence, as opposed to fusion and dependence.*

Positive physiological outcomes also follow from high self-esteem. People who feel good about themselves are likely to develop fewer ulcers and to cope with anxiety better than those who don't. In contrast, low self-esteem has devastating consequences for individuals and the relationships in which they become involved. Not feeling loved as a child and, worse, feeling rejected and abandoned create the context for the development of a negative self-concept and mistrust of others. People who have never felt loved and wanted may require constant affirmation from their partner as to their worth and may cling desperately to that person out of fear of being abandoned. Such dependence (the modern term is *codependency*) may also encourage staying in unhealthy (abusive and alcoholic) relationships, since the person may feel, "This is all I deserve." Fuller and Warner (2000) studied 257 college students and observed that women had higher codependency scores than men. Codependency was also associated with being reared in families that were stressful and alcoholic.

One characteristic of individuals with low self-esteem is that they may love too much and be addicted to unhealthy love relationships. Petrie and colleagues studied fifty-two women who reported that they were involved in unhealthy love relationships in which they had selected men with problems (such as alcohol/other drug addiction) they attempted to solve at the expense of neglecting themselves. "Their preoccupation with correcting the problems of others may be an attempt to achieve self-esteem" (Petrie, Giordano, & Roberts, 1992, 17).

Although it helps to have positive feelings about oneself going into a love relationship, sometimes these develop after one becomes involved in the relationship. "I've always felt like an ugly duckling," said one woman. "But once I fell in love with him and he with me, I felt very different. I felt very good about myself then because I knew that I was somebody that someone else loved." High self-esteem, then, is not necessarily a prerequisite for falling in love. People who have low self-esteem may fall in love with someone else as a result of feeling deficient (ego-ideal theory of love). The love they perceive the other person having for them may compensate for the deficiency and improve their self-esteem.

Self-Disclosure Having high self-esteem enables one to be able to disclose to others. Such disclosure is necessary if one is to love and be loved by another. Disclosing yourself is a way of investing yourself in another. Once the other person knows some of the intimate details of your life, you will tend to feel more positively about that person because a part of you is now a part of him or her. Open communication in this sense tends to foster the development of an intense love relationship.

Authors

"The more we knew about each other, the more we loved each other," said this couple. Their words reflect the principle that disclosure increases love involvement.

It is not easy for some people to let others know who they are, what they feel, or what they think. They may fear that if others really know them, they will be rejected as a friend or lover. To guard against this possibility, they may protect themselves and their relationships by allowing only limited information about their past behaviors and present thoughts and feelings.

Trust is the condition under which people are most willing to disclose themselves. When people trust someone, they tend to feel that whatever feelings or information they share will not be judged and will be kept safe with that person. If trust is betrayed, a person may become bitterly resentful and vow never to disclose herself or himself again. One woman said: "After I told my partner that I had had an abortion, he told me that I was a murderer and he never wanted to see me again. I was devastated and felt I had made a mistake telling him about my past. You can bet I'll be careful before I disclose myself to someone else" (authors' files).

Gallmeier et al. (1997) studied the communication patterns of 360 undergraduates at two universities and found that women were significantly more likely to disclose information about themselves. Specific areas of disclosure included previous love relationships, what they wanted for the future of the relationship, and what their partners did that they did not like.

Physiological and Cognitive Conditions for Love

Physiological and cognitive variables are also operative in the development of love. The individual must be physiologically aroused and interpret this stirred-up state as love (Walster & Walster, 1978).

> *Suppose, for example, that Dan is afraid of flying, but his fear is not particularly extreme and he doesn't like to admit it to himself. This fear, however, does cause him to be physiologically aroused. Suppose further that Dan takes a flight and finds himself sitting next to Judy on the plane. With heart racing, palms sweating, and breathing labored, Dan chats with Judy as the plane takes off. Suddenly, Dan discovers that he finds Judy terribly attractive, and he begins to try to figure out ways that he can continue seeing her after the flight is over. What accounts for Dan's sudden surge of interest in Judy? Is Judy really that appealing to him, or has he taken the physiological arousal of fear and mislabeled it as attraction?* (Brehm, 1992, 44) [An uncannily similar experience occurred in the life of model Christie Brinkley, who fell in love with and married a man with whom she had survived a helicopter crash in Colorado in the early nineties].

Although most people who develop love feelings are not aroused in this way, they may be aroused or anxious about other issues (being excited at a party, feeling apprehensive about meeting someone) and mislabel these feelings as those of attraction when they meet someone.

In the absence of one's cognitive functioning, love feelings are impossible. Individuals with brain cancer who have had the front part of their brain between the eyebrows removed are incapable of love. Indeed, emotions are not present in them at all (Ackerman, 1994). The social, psychological, physiological, and cognitive conditions are not the only factors important for the development of love feelings. The timing must also be right. There are only certain times in your life when you are seeking a love relationship. When those times occur, you are likely to fall in love with the person who is there and who is also seeking a love relationship. Hence, many love pairings exist because each of the individuals is available to the other at the right time—not because they are particularly suited for each other.

LOVE AS A CONTEXT FOR PROBLEMS

Though love may bring great joy, it also creates a context for problems. Three such problems are simultaneous loves, involvement in an abusive relationship, and making risky/dangerous choices.

Simultaneous Loves

Some people (particularly those separated from the love partner, who may meet and fall in love with another) report the dilemma of being in love with two people at the same time. One solution is to try to reduce their feelings for one of them. This may be accomplished by deciding to see only one of the persons and by thinking negative thoughts about the other. For example, an individual who was in love with two partners decided to see only one of them while focusing on the fact that the other drank heavily, smoked, and had been unfaithful.

Abusive or Unfulfilling Love Relationships

Another problem associated with love is being in love with someone who may be emotionally or physically abusive (see Chapter 14). Someone who beats you, who criticizes you ("you're ugly, stupid, pitiful"), or who is dishonest with you (sexually unfaithful) may create a great deal of pain and disappointment. Nevertheless, you might love that person and feel emotionally drawn to him or her. Most marriage therapists would suggest that you examine why you love and continue to stay with such a person. Do you feel that you deserve this treatment because you are "no good"? Do you feel pity for the person and feel responsible for rescuing him or her? Do you feel you would not be able to find a better alternative? Do you feel you would rather be with a person who treats you badly than be alone?

Another explanation for why some people who are abused by their partners continue to be in love with them is that the abuse is only one part of the relationship. When the partner is not being abusive, he or she may be kind, loving, and passionate. It is these latter qualities that overshadow the occasional abuse and keep the love alive. Love stops when the extent of the abuse is so great that there are insufficient positive behaviors to counteract the abusive behavior.

Love relationships that do not involve emotional or physical abuse may be unfulfilling for other reasons. Partners in love relationships may experience lack of fulfillment if they have radically different values, religious beliefs, role expectations, recreational or occupational interests, sexual needs, or desires concerning family size.

Context for Risky/Dangerous/Questionable Choices

Individuals are aware that love may cause them problems (Bulcroft et al., 2000). Plato said that "love is a grave mental illness," and some research suggests that individuals in love make risky/dangerous/questionable decisions. In a study on "what I did for love," college students reported that "driving drunk," "dropping out of school to be with my partner," and "having sex without protection" were

If you think you have fallen in love, sit down for six months and it will pass.
Paul Pearsall, *Psychologist*

among the more risky/dangerous/questionable choices they had made while they were under the spell of love (Knox , Zusman, & Nieves, 1998). Bankole et al. (2000) confirmed that the longer individuals are in a love relationship, the less likely they are to use a condom.

NATIONAL DATA

Anderson et al. (2000) analyzed national data of U.S. adults and noted that only 19 percent of those in an ongoing (love) relationship used a condom, whereas 62 percent used a condom if their partner was not a person that they were seeing regularly.

JEALOUSY IN RELATIONSHIPS

Jealousy can be defined as an emotional response to a perceived or real threat to an important or valued relationship. People experiencing jealousy feel excluded or they fear being abandoned. Buss (2000) emphasized that "Jealousy is an adaptive emotion, forged over millions of years. . . . It evolved as a primary defense against threats of infidelity and abandonment" (p. 56). Persons become jealous when they fear replacement. Although jealousy does not occur in all cultures (polyandrous societies value cooperation, not sexual exclusivity, Cassidy & Lee, 1989), it does occur in our society and among both heterosexuals and homosexuals. Thirteen percent of 620 university students reported that jealousy was the most frequent problem that they encountered in their current or most recent relationship (Knox & Zusman, 1998).

One hundred and eighty-five students provided information about their experience with jealousy (Knox et al., 1999). On a continuum of 0 ("no jealousy") to 10 ("extreme jealousy"), with 5 representing "average jealousy," these students reported feeling jealous at a mean level of 5.3 in their current or last relationship. Regarding timing, students who had been dating their partner a year or less were significantly more likely to report higher levels of jealousy (mean = 4.7) than those who had dated thirteen months or more (mean = 3.3).

Pines and Friedman (1998) found no gender differences in likelihood, frequency, duration, or intensity of jealousy. However, Buunk et al. (1996) found that men were significantly more jealous when imagining their partner trying new sexual positions with a new partner. This finding was true not only in the United States but in Germany and the Netherlands as well. Researchers have suggested an evolutionary explanation of paternal certainty in that men want to ensure that the child their partner has is their (the father's) biological child. This territorial, possessiveness form of jealousy that men experience may be different from the jealousy as envy that women may experience. Indeed, while a man may want to possess his partner and keep her away from another lover, a woman might feel envious that another lover has the attention of her partner.

Whereas men are more likely to be jealous when their partners engage in sexual behavior, women are more likely to be jealous when their partners become involved in an emotional relationship with someone else. Harris and Christenfeld (1996) explained that the woman's need for her partner's investment in her offspring is threatened by her mate's emotional (not sexual) involvement with another female.

Causes of Jealousy

Jealousy can be triggered by external or internal factors.

External Causes External factors refer to behaviors the partner engages in that are interpreted as (1) an emotional and/or sexual interest in someone (or something) else or (2) a lack of emotional and/or sexual interest in the primary partner. In the study of 185 students (Knox et al., 1999) referred to above, the respondents identified "actually talking to a previous partner" (34%) and "talked about a previous partner" (19%) as the most common sources of their jealousy. Also, men were more likely to report feeling jealous when their partner talked to a previous partner. Women were more likely to report feeling jealous when their partner danced with someone else.

Internal Causes Jealousy may also exist even when there is no external behavior that indicates the partner is involved or interested in an **extradyadic relationship**—an emotional/sexual involvement between a member of a pair and someone other than the partner. Internal causes of jealousy refer to characteristics of individuals that predispose them to jealous feelings, independent of their partner's behavior. Examples include being mistrustful, having low self-esteem, being highly involved in and dependent on the relationship, and having no perceived alternative partners available (Pines, 1992). These internal causes of jealousy are explained below.

1. Mistrust. If an individual has been deceived or cheated on in a previous relationship, that individual may learn to be mistrustful in subsequent relationships. Such mistrust may manifest itself in jealousy.

2. Low self-esteem. Individuals who have low self-esteem tend to be jealous because they lack a sense of self-worth and hence find it difficult to believe anyone can value and love them. Feelings of worthlessness may contribute to suspicions that someone else is valued more.

3. Being involved and dependent. In general, individuals who are more involved in the relationship than their partner or who are more dependent on the relationship than their partner are prone to jealousy (Radecki Bush, Bush, & Jennings, 1988). The person who is more involved in or dependent on the relationship not only is more likely to experience jealousy but also may intentionally induce jealousy in the partner. Such attempts to induce jealousy may involve flirting, exaggerating or discussing an attraction to someone else, and spending time with others. According to White (1980), individuals may try to make their partner jealous as a way of testing the relationship (e.g., to see whether the partner still cares) and/or increasing specific rewards (e.g., to get more attention or affection). White found that women, especially those who thought they were more involved in the relationship than their partners, were more likely than men to induce jealousy in a relationship.

4. Lack of perceived alternatives. Individuals who have no alternative person or who feel inadequate in attracting others may be particularly vulnerable to jealousy. They feel that if they do not keep the person they have, they will be alone.

5. Insecurity. Individuals who feel insecure about the relationship with their partner may experience higher levels of jealousy (Attridge, Berscheid, & Sprecher, 1998).

Consequences of Jealousy

Jealousy can have both desirable and undesirable consequences.

Desirable Outcomes Jealousy may be functional if it occurs at a low level and results in open and honest discussion about the relationship. Not only may jealousy keep the partner aware that he or she is cared for (the implied message is "I love you and don't want to lose you to someone else"), but also the partner may learn that the development of other romantic and sexual relationships is unacceptable. One wife said:

> *When I started spending extra time with this guy at the office my husband got jealous and told me he thought I was getting in over my head and asked me to cut back on the relationship because it was "tearing him up." I felt he really loved me when he told me this and I chose to stop having lunch with the guy at work.* (Authors' files)

According to Buss (2000), the evoking of jealousy also has the positive functions of assessing the partner's commitment and of alerting the partner that one could leave for greener mating pastures. Hence, one partner may deliberately evoke jealousy to solidify commitment and ward off being taken for granted. In addition, sexual passion may be reignited if one partner perceives that another would take his or her love object away. That people want what others want is an adage that may underlie the evocation of jealousy.

Undesirable Outcomes Shakespeare referred to jealousy as the "green-eyed monster," suggesting that it sometimes has undesirable outcomes for relationships. The emotional torment for oneself and one's partner when feelings of jealousy are obsessive is evident. In addition, one's partner can tire of unwarranted jealous accusations and end the relationship.

In its extreme form, jealousy may have devastating consequences. In the name of love, people have stalked the beloved (see Up Close 3.1 earlier in chapter), shot the beloved, and killed themselves in reaction to rejected love. Indeed "I was in a jealous rage" was used as a defense for Clara Harris, who ran over her husband four times and killed him when she discovered that he was involved in an affair.

Compersion

Compersion, describing the situation in which an individual feels positive about a partner's emotional and sexual enjoyment with another person, is the "opposite" of jealousy. Persons in polyamorous relationships strive to rid themselves of jealous feelings and to increase their level of compersion. To feel happy for a partner who delights in the attention and affection of, and sexual involvement with, another is the epitome of compersion.

SUMMARY

What are some ways love has been described?

Love remains an elusive and variable phenomenon and continues to be thought of as emotional and irrational. Researchers have conceptualized love as a dichotomy (i.e., "romantic or conjugal") or as a vague construct involving relationship dynamics (i.e., "meeting of needs"). The "triangular" view of love identifies it as consisting of three basic elements—intimacy, passion, and commitment.

How does societal context influence the development of love?

The society in which we live exercises considerable control over our love object/choice and conceptualizes it in various ways. Love may be blind, but it knows what color a person's skin is. Romantic love is such a powerful emotion and marriage such an important relationship, connecting an outsider into an existing family and peer network, that mate selection is not left to chance. Parents inadvertently influence the mate choice of their children by moving to certain neighborhoods, joining certain churches, and enrolling their children in certain schools. Doing so increases the chance that their offspring will "hang out," fall in love, and marry persons who are similar in race, education, and social class.

What are the various theories of love?

Theories of love include evolutionary (love provides the social glue needed to bond parents with their dependent children and spouses with each other to care for their dependent offspring), social learning (positive experiences create love feelings), sociological (Reiss's "wheel" theory), psychosexual (love results from a blocked biological drive), ego-ideal (love springs from a dissatisfaction with oneself), ontological (love represents people's urge to be reunited with the other half), biochemical (love involves feelings produced by biochemical events), and attachment (a primary motivation in life is to be connected with other people).

How does love develop in a new relationship?

Love occurs under certain conditions. Social conditions include a society that promotes the pursuit of love, peers who enjoy it, and a set of norms that link love and marriage. Psychological conditions involve high self-esteem and a willingness to disclose oneself to others. Physiological and cognitive conditions imply that the individual experiences a stirred-up state and labels it "love."

How is love a context for problems?

Love sometimes provides a context in which lovers experience problems such as being in love with two people at the same time, being in love with someone who is abusive, or making risky/dangerous/questionable choices while in love (e.g., not using a condom).

How do jealousy and love interface?

Jealousy is an emotional response to a perceived or real threat to a valued relationship. Jealous feelings may have both internal and external causes and may have both positive and negative consequences for a couple's relationship.

What is compersion?

Compersion is the "opposite" of jealousy and involves feeling positive about a partner's emotional and physical involvement with another person.

KEY TERMS

agape love style
compersion
conjugal love
eros love style
extradyadic relationship
infatuation
jealousy
ludic love style
mania love style
pragma love style
romantic love
stalking
storge love style

RESEARCHING MARRIAGE AND THE FAMILY WITH INFOTRAC COLLEGE EDITION

InfoTrac College Edition, an online library, allows you to perform research online anywhere, anytime. Following are two suggested search terms and related questions to help you extend your understanding of the topics covered in this chapter. Go to www.infotrac-college.com to begin your search.

Search keyword: **Romantic love.** Locate articles that focus on romantic love and identify whether the women or men in the articles are represented as being more romantic. What gave you this impression?

Search keyword: **Jealousy.** Locate articles that discuss jealousy and argue for and against the position that "the more jealous, the more in love."

The Companion Web Site for Choices in Relationships: An Introduction to Marriage and the Family, Eighth Edition

http://sociology.wadsworth.com/knox_schacht/choices8e

Supplement your review of this chapter by going to the companion Web site to take one of the Tutorial Quizzes, use the flash cards to master key terms, and check out the many other study aids you'll find there. You'll also find special features such as the Marriage and Family Resource Center, Census 2000 information, and other data and resources at your fingertips to help you with that special project or to do some research on your own.

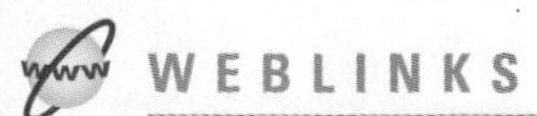

WEBLINKS

Looking for a New Love Partner
http://www.people2people.com/default.asp?connect=rightmate

The Loving Center
http://www.consciouslovingtlc.com/

Marriage Builders
http://marriagebuilders.com/

Authors

CHAPTER 4

Cohabitation appears not to be very helpful and may be harmful as a try-out for marriage.

David Popenoe, Barbara Dafoe Whitehead, Sociologists

Hanging Out, Pairing Off and Cohabitation

Contents

True or False?

1. Couples (even the engaged) no longer go out as a couple but "hang out" in groups.
2. Women are more likely than men to initiate a breakup.
3. Men are more likely to report that "a new partner" helped them to recover from a relationship that ended, while women are more likely to report that "time" helped them to recover from a broken relationship.
4. Students who have been involved in a long-distance dating relationship are more likely to believe "out of sight, out of mind" than those who have not been separated.
5. In a study on Internet dating, the predominant motive for meeting a partner online was romance and sex.

Answers: **1.** F **2.** T **3.** T **4.** T **5.** F

Reality TV shows such as *Elimidate, Fifth Wheel,* and *Blind Date* focus on people meeting, having dinner, and sharing an activity (e.g., horseback riding, playing pool, dancing) to assess the degree to which they feel comfortable, have fun, and want to see each other for a second date. The popularity of these shows reflects the cultural interest in the pairing-off process and the interest of individuals in being involved in this process.

While not all societies have reality TV shows on **dating,** all societies have a way of moving women and men from same-sex groups into pair-bonded legal relationships for the reproduction and socialization of children. In Islamic countries, the family is intimately involved in arranged marriages; in the United States, adolescents and young adults begin to hang out with their friends and eventually pair off into couple relationships. About 60 percent will live together before getting married. In this chapter, we review the pattern of pairing off in this and other societies and discuss the phenomenon of living together.

In their study on "going out or hanging out" Gallmeier, Knox, and Zusman (2002) distinguished between "group daters" and "couple daters." The first group consisted of younger, casual, and first-time individuals who went to a party and "hung out"; by contrast, couple daters were more likely to be older and engaged and to go out to dinner alone. Hence, there was a developmental sequence whereby individuals went to parties to meet each other in group contexts. While they hung out to find each other, later they paired off and saw each other exclusively as a couple. Over two-thirds (69.4%) of Gallmeier's 226 undergraduates reported that they usually went out as a couple (hence, hanging out has *not* replaced couple interaction). Of course, some couples will spend a portion of their time with the group as well as be alone.

FUNCTIONS OF HANGING OUT IN THE UNITED STATES

Because most people regard hanging out and pairing off as a natural part of getting to know someone else, the other functions of these phenomena are sometimes overlooked. There are at least six of these—confirmation of a social self, recreation, companionship/intimacy/sex, anticipatory socialization, status achievement, and mate selection.

Confirmation of a Social Self

In Chapter 1 we noted that symbolic interactionists emphasize the development of the self. Parents are usually the first social mirrors in which we see ourselves and receive feedback about who we are, and hanging-out partners continue the process. When you are hanging out with a person, you are continually trying to assess how that person sees you: Does the person like me? Will the person want to be with me again? When the person gives you positive feedback through speech and gesture, you feel good about yourself and tend to view yourself in positive terms. Hanging out provides a context for the confirmation of a strong self-concept in terms of how you perceive your effect on other people.

Recreation

The focus of hanging out and pairing off is fun. The reality TV programs mentioned earlier always have a recreation/fun focus for the interaction of the individuals who are meeting/pairing off. "She/he is fun" seems to be a criterion for being selected. The couple may make only small talk and learn very little about each other—what seems important is that they "have fun."

Companionship/Intimacy/Sex

Beyond fun, major motivations for hanging out and pairing off are companionship, intimacy, and sex. The impersonal environment of a large university makes a secure relationship very appealing. "My last two years have been the happiest ever," remarked a senior in interior design. "But it's because of the involvement with my partner. During my freshman and sophomore years, I felt alone. Now I feel loved, needed, and secure." In Chapter 5, Sexuality in Relationships, we examine the progress of sexual intimacy in a developing relationship and discover that the longer a couple is together, and the more intimate their relationship, the greater their physical intimacy. Some individuals "hook up" on their first encounter (discussed in Chapter 5).

Anticipatory Socialization

Before puberty, boys and girls interact primarily with same-sex peers. A sixth-grade boy or girl may be laughed at if he or she shows an interest in someone of the other sex. Even when boy-girl interaction becomes the norm at puberty, neither sex may know what is expected of the other. Hanging out provides the first opportunity for individuals to learn how to interact with other-sex partners. Though the manifest function of hanging out is to teach partners how to negotiate differences (e.g., how much sex and how soon), the latent function is to help them learn the skills necessary to maintain long-term relationships (empathy, companionship). In effect, hanging out and pairing off are forms of socialization that anticipate a more permanent union in one's life. Individuals may also try out different role patterns, like dominance or passivity, and try to assess the feel and comfort level of each.

One function of hanging out is to learn how to interact with the other sex and to meet potential partners to pair off with.

Authors

Status Achievement

Being involved with someone is usually associated with more status than being unattached and alone. Some may seek such involvement because of the associated higher status. Others may become involved for peer acceptance and conformity to gender roles, not for emotional reasons. Though the practice is becoming less common, a gay person may pair off with someone of the other sex so as to provide a heterosexual cover for his or her sexual orientation.

Mate Selection

Finally, hanging out and paring off eventually lead to marriage, which remains a major goal in our society (American Council on Education and University of California, 2002). Selecting a mate has become big business. B. Dalton, one of the largest bookstore chains in the United States, carries about two hundred titles on relationships, about fifty of which are specifically geared toward finding a mate.

MEETING A NEW PARTNER

One of the unique qualities of universities is that they provide an environment in which to meet hundreds or even thousands of potential partners of similar age, education, social class, and general goals. This opportunity will probably not be matched later in the workplace or where one lives following graduation.

Individuals today are becoming increasingly rational (using advertisements, agencies, the Internet) about their love relationships with the goal of minimizing the risks (such as potential divorce) associated with such relationships (Bulcroft et al., 2000). While people often meet through friends or on their own through school, work, or recreation contexts, an increasing number are open to a range of alternatives such as personal ads in magazines or on the Internet.

Personal Ads in Magazines and Newspapers

Beyond meeting through friends, some individuals meet a new partner through personal ads in magazines. Some magazines feature ads marketed to a particular group of singles. For example, *Cherry Blossoms* (now in its thirty-first year) is designed for the individual seeking a partner from another country. Each issue features ads from more than five hundred women from forty countries. A twelve-month subscription costs over $500. Some individuals also place ads in local newspapers with the goal of identifying someone locally with mutual interests. But magazine and newspaper ads are being superseded by finding a partner on the Internet.

The Internet—Meeting Online

Cotton (2003) and colleagues (Rohall, Cotton, & Morgan, 2002; Morgan & Cotton, 2003) studied a random sample of residential college freshmen at a medium-sized public university and found that 95 percent reported having been online more than three years. E-mail was their most frequent reason for using the Internet, with four hours the average weekly use. Students who used the Internet for communication purposes reported lower levels of depression and higher levels of self-esteem. Internet usage for noncommunication purposes was associated with higher depression and lower self-esteem.

E-mail romance can be wonderful for the Cyranos of the world, people whose strengths lie less in physical charm than in verbal seduction.

Peter Kramer, M.D.

If you are looking for love on the Internet, you better look here first.

Lisa, *founder of WildXangel.com*

Meeting a partner online is also a reason college students use the Internet. There are over two hundred Web sites designed for this purpose. (RightMate at Heartchoice.com is one of them and offers not only a way to meet others but a free "RightMate Checkup" to evaluate whether the person is right for you.) To find out information about Internet dating, 191 never-married undergraduate university students completed an anonymous twenty-eight-item questionnaire designed to assess their attitudes toward and involvement in use of the Internet to find a mate (Knox et al., 2001). The data revealed that among these college students, friendship, not romance or sex, was the primary interest when going on the Internet to meet new people. The survey also found that over 60 percent of these respondents were successful in establishing an online friendship and almost half felt more comfortable meeting a person online than in person; 40 percent reported that they had lied online.

The advantages of such online meetings include the ability to sift through large numbers of potential partners quickly (by eliminating persons who do not have the basic qualities you require), the potential to exchange information quickly to assess if you have similar interests, and the opportunity to develop a relationship with another on the basis of content independent of visual distraction. Some Web sites reflect certain communities such as BlackPlanet.com (black community), Jdate.com (Jewish community), Gay.com (gay community), and Nerve.com (young, hip, well-educated professionals).

The disadvantages of online meeting include the potential to fall in love quickly as a result of intense mutual disclosure, not being able to observe nonverbal cues/gestures, the tendency to move too quickly from E-mail to phone to meeting to first date to marriage without providing the requisite two years to get to know someone before marrying him or her, and deception. Forty percent of the respondents in the Internet study referred to above reported that they had lied online. Men tend to lie about their economic status and women tend to lie about their weight or age.

Although some relationships begun online result in lifetime marriages, one should be cautious of meeting someone online. See the Web site WildXAngel (the address is in the Weblinks at the end of this chapter) for horror stories of online dating.

Authors

The Internet has become a principal means of meeting a new person. The coed is checking her E-mail from a person she met online.

Video Chatting

Video chatting moves beyond the traditional Internet –typing –of words to each other and allows the partners to see each other while chatting online. ISpQ or Eye Speak is one of the largest downloadable online communities where people visually meet with others from all over the world. Hodge (2003) noted, "Unlike conventional chat rooms, iSpQ does not have a running dialogue or conversation for anyone to view. Video chatting allows users to have personal or private conversation with another user. Hence an individual can not only write information but see the person with whom he or she is interacting online." Half of the respondents in Hodge's study of video chat users reported "meeting people and having fun" as their motivation for video chatting.

PERSONAL CHOICES

Should I Get Involved in a Long-Distance Dating Relationship?

As a result of individuals' meeting online, the delay of marriage, and the respective desires of women and men to finish their educations or launch their careers (which may take them to different states), long-distance dating relationships are increasing. Although these relationships benefit from the partners' insistence that "we care about each other and want to keep our relationship intact even though we are separated," from the fact that the separation may allow the partners to stay "high" on each other, and by permitting personal time and space, there are problems.

Knox et al. (2002) analyzed a sample of 438 undergraduates at a large southeastern university on their attitudes and involvement in a long-distance dating relationship (LDDR)—defined as being separated from a love partner by at least two hundred miles for a period of not less than three months. The median number of miles these LDDR respondents had been separated was 300 to 399 (about a six-hour drive), and the median length of time they were separated was five months. Of the total sample, 20 percent were currently involved in an LDDR, with 37 percent reporting having ever been in an LDDR relationship that had ended.

Being separated was a strain on the couple's relationship. One in five (21.5%) broke up and another one in five (20%) said that the separation made their relationship worse. Only 18 percent reported that the separation improved their relationship (other responses included 33% mixed effect, 9% no effect). Does absence make the heart grow fonder for the beloved? Most of those who have not been separated seem to think so. But 40 percent of those who had experienced a long-distance dating relationship believed that "out of sight, out of mind" was more accurate (Knox et al. 2002). One respondent said, "I got tired of being lonely and the women around me started looking good."

Van Horn et al. (1997) compared eighty college-student, long-distance romantic relationships with eighty-two campus love relationships where distance was not a factor and found that the former were characterized by lower relationship satisfaction and less certainty about the future.

Being frustrated over not being able to be with the partner, loneliness, and feeling as though one is missing out on other activities/relationships are among the disadvantages of being involved in a long-distance dating relationship. Another is that your partner may actually look better from afar than up close. One respondent noted that he and his partner could not wait to live together after they had been separated—but "when we did, I found out I liked her better when she wasn't there."

Nevertheless, here are some issues to consider in making a long-distance dating relationship manageable and keeping the relationship together:

1. Maintain daily contact. In the Knox et al. (2002) study referred to above, actual contact between the lovers during the period of separation was limited. Only 11 percent reported seeing each other weekly, with 16 percent reporting that they never saw each other. However, over three-fourths (77%) reported talking with each other by phone several times each week (22% daily), and over half (53%) E-mailed the partner several times each week (18% daily). Some partners maintain daily contact by Web cams.

> *My fiancé and I had only been dating a month when I went back to college. We weren't sure how the relationship would work out because we lived five hours away from each other. Being in a new relationship is hard enough without the added stress of not being able to see each other every day. For the first month or so, we talked on the telephone every night. That got very expensive very quickly! We had to figure out a way to remedy the problem of talking every day, but not spending so much money. Both of us were typically on the computer every day, so we decided that we should get some information on Web cams. Web cams are small video cameras that plug into a USB port on the computer. You install the software and the camera is up and running in about five minutes. Since both of us were usually on the computer, and we both had Internet connections that didn't use the phone lines, the Web cams have worked perfectly for us.*
>
> *We get to see each other every day, in real time, whenever we want to. Because the connection doesn't interfere with the telephone, we stay connected 24 hours a day. It has been a big help in keeping our relationship going strong. If we need to talk about an important issue, we can do it face to face without worrying about time or money. Also, because we can't physically be together, this device has helped our personal lives as well. We can see each other whenever we want in whatever way that we want. We have been together for over a year, and during that year we have been connected by the Web cams for over eleven months. Technology has certainly helped our relationship last!* (Brooks, 2003)

2. Enjoy/use the time when apart. Since being separated is often for a period of months (Guldner's (2003) sample expected to be separated for twenty-six months), it is important to be involved in worthwhile activities with study, friends, work, sports, and personal projects when apart. Otherwise, resentments may spill over into the interaction with your partner.

3. Avoid conflictual phone conversations. Talking on the phone should involve the typical sharing of events. When the need to discuss a difficult topic arises, the phone is not the best place for such a discussion. Rather, it may be wiser to wait and have the discussion face to face. If you decide to settle a disagreement over the phone, stick at it until you have a solution acceptable to both of you.

4. Stay monogamous. Agreeing not to be open to other relationships is crucial to maintaining a long-distance relationship. This translates into not dating others while apart. Individuals who say, "Let's date others to see if we are really meant to be together" often discover that they are capable of being attracted to and becoming involved with others. Such involvement usually predicts the end of the long-distance dating relationship. Lydon, Pierce, & O'Regan (1997) studied sixty-nine undergraduates who were involved in LDDRs and found that "moral commitment" predicted the survival of the relationships. Individuals committed to maintaining their relationships are often successful in doing so.

Sources

Brooks, S. 2003. Personal communication.

Guldner, G. T. 2003. *Long distance relationships: The complete guide.* Corona, Calif.: JFMilne Publications.

Knox, D., M. Zusman, V. Daniels, and A. Brantley. 2002. Absence makes the heart grow fonder? Long-distance dating relationships among college students. *College Student Journal* 36:365–67.

Lydon, J., T. Pierce, and S. O'Regan. 1997. Coping with moral commitment to long-distance dating relationships. *Journal of Personality and Social Psychology* 73:104–13.

Van Horn, K. R., A. Arone, K. Nesbitt, L. Desilets, T. Sears, M. Griffin, and R. Brundi. 1997. Physical distance and interpersonal characteristics in college students' romantic relationships. *Personal Relationships* 4:25–34.

Video Dating Service

Still another method of finding a partner is through a video dating service. This is an agency that interviews you on videotape and lets others watch your tape in exchange for your watching videotapes already on file. Once you have seen someone you like, that person is contacted by the agency and invited in to review your tape. If the interest is mutual, you and the other person will meet. *Great Expectations* is the largest video dating service in the United States. It has over twenty-one centers and 175,000 clients nationwide. The cost for a lifetime membership (until you get married) in Los Angeles is about $2,000.

Innovations in Dating—Speed Dating

Dating innovations that involve the concept of speed include the eight-minute date. Chetty Red is a bar in Manhattan that features the eight-minute date whereby forty-eight men and forty-eight women pay to sit down with someone new every eight minutes (a bell rings when it is time to go to the next table). The singles then rate their dates and find out two days later via E-mail who wants to see them again (Barker, 2002). "It's cost effective" since you can meet a lot of people in one night. "The idea is to meet a lot of people in a short time and decide whether someone is worth a second look" (Peterson, 2003).

Used by couples involved in long-distance dating relationships, Web cams allow individuals to see and talk with each other over the Internet.

Authors

DATING AFTER DIVORCE

Nothing grows again more easily than love.
Seneca, *Roman philosopher*

Over 2 million Americans get divorced each year. As evidenced by the fact that over three-quarters of the divorced remarry within five years, most of the divorced are open to a new relationship. But there are differences between this single-again population and those becoming involved for the first time.

1. Older population. Divorced individuals are, on the average, ten years older than persons in the marriage market who have never been married before. Hence the divorced are in their mid- to late thirties. Widows and widowers are usually forty and thirty years older, respectively (hence ages 65 and 55), when they begin to date the second time around.

2. Fewer potential partners. Most men and women who are dating the second time around find fewer partners from whom to choose than when they were dating before their first marriage. The large pool of never-marrieds (22% of the population) and currently marrieds (61% of the population) is usually not considered an option (*Statistical Abstract of the United States: 2002,* Table 61). Most divorced persons date and marry others who have been married before.

3. Increased HIV risk. The older an unmarried person, the greater the likelihood that he or she has had multiple sexual partners (Michael et al., 1994), which is associated with increased risk of contracting HIV and other sexually transmitted diseases (STDs). Therefore, individuals entering the dating market for the second time are advised to be more selective in choosing their sexual partners because of the higher likelihood of those partners' having had more sexual partners. In addition, the divorced are much less likely than married adults to be monogamous (Michael et al.1994), so dating an older divorced person may involve even greater risk of contracting an STD. An increasing number of individuals are becoming concerned about HIV and other infections and are discussing HIV testing with new partners before having sex with them. Such a discussion may be particularly important for those reentering the dating market.

4. Children. More than half of those dating again have children from a previous marriage. How these children feel about their parents' dating, how the partners feel about each other's children, and how the partners' children feel about each other are complex issues. Deciding whether to have intercourse when one's children are in the house, when a new partner should be introduced to the children, and what the children should call the new partner are other issues familiar to parents dating for the second time.

5. Ex-spouse issues. Previous ties to the ex-spouse in the form of child support or alimony, phone calls, and the psychological effects of the partner's first marriage will have an influence on the new dating relationship. If the separation/divorce was bitter, the partner may be preoccupied or frustrated in his or her attempts to cope with the harassing ex-spouse (Knox & Zusman, 2001).

6. Brief courtship. Divorced people who are dating again tend to have a shorter courtship period than first-marrieds. In a study of 248 individuals who remarried, the median length of courtship was nine months as opposed to seventeen months the first time around (O'Flaherty & Eells, 1988). A shorter courtship may mean that sexual decisions are confronted more quickly—timing of first intercourse, discussing the use of condoms/contraceptives, and clarifying whether the relationship is to be monogamous.

CULTURAL AND HISTORICAL BACKGROUND OF DATING

Any consideration of current pairing-off patterns must take into account a historical view. Contemporary hanging-out patterns in the United States today are radically different from courtship and dating in other cultures and times.

Traditional Chinese "Dating" Norms

The freedom with which U.S. partners today select each other on the basis of love is a relatively recent phenomenon. At most times and in most cultures, marriages were more often arranged by parents (Hatfield & Rapson, 1996). The "love feelings" between the partners were given either no or limited consideration. In traditional China, **blind marriages,** wherein the bride and groom were prevented from seeing each other for the first time until their wedding day, were the norm. The marriage of two individuals was seen as the linking of two families. Though the influence of parents in the mate selection of their offspring is decreasing (such arranged marriages are no longer the norm), Chinese parents continue to be involved. Indeed, China's dating culture is very different from that of the United States, where individuals meet and date new partners. In a large representative study of couples in urban China, over three-fourths (77%) of the women and two-thirds (66%) of the men reported dating no one or only their spouse prior to marriage (Pimentel, 2000).

Dating during the Puritan Era in the United States

Although less strict than traditional courtship norms among the Chinese, the European marriage patterns brought to America were conservative. The Puritans who settled on the coast of New England in the seventeenth century were radical Protestants who had seceded from the Church of England. They valued marriage and fidelity, as reflected in a very rigid pattern of courtship.

Bundling, also called *tarrying,* was a courtship custom commonly practiced among the Puritans. It involved the would-be groom's sleeping in the girl's bed in her parents' home. But there were rules to restrict sexual contact. Both partners had to be fully clothed, and a board was placed between them. In addition, the young girl might be encased in a type of long laundry bag up to her armpits, her clothes might be sewn together at strategic points, and her parents might be sleeping in the same room.

The justifications for bundling were convenience and economics. Aside from meeting at church, bundling was one of the few opportunities a couple had to get together to talk and learn about each other. Since daylight hours were consumed by heavy work demands, night became the only time for courtship. But how did bed become the courtship arena? New England winters were cold. Firewood, oil for lamps, and candles were in short supply. By talking in bed, the young couple could come to know each other without wasting valuable sources of energy. Although bundling flourished in the middle of the eighteenth century, it provoked a great deal of controversy. By about 1800, the custom had virtually disappeared.

Effects of the Industrial Revolution on Dating

The transition from a courtship system controlled by parents to the relative freedom of mate selection experienced today occurred in response to a number of social changes. The most basic change was the Industrial Revolution, which began in England in the middle of the eighteenth century. No longer were women needed exclusively in the home to spin yarn, make clothes, and process food

SELF-ASSESSMENT

Relationships Dynamics Scale

Please answer each of the following questions in terms of your relationship with your "mate" if married, or your "partner" if dating or engaged. We recommend that you answer these questions by yourself (not with your partner), using the ranges following for your own reflection.

Use the following 3-point scale to rate how often you and your mate or partner experience the following:

1 = almost never
2 = once in a while
3 = frequently

1 2 3 Little arguments escalate into ugly fights with accusations, criticisms, name calling, or bringing up past hurts.
1 2 3 My partner criticizes or belittles my opinions, feelings, or desires.
1 2 3 My partner seems to view my words or actions more negatively than I mean them to be.
1 2 3 When we have a problem to solve, it is like we are on opposite teams.
1 2 3 I hold back from telling my partner what I really think and feel.
1 2 3 I think seriously about what it would be like to date or marry someone else.
1 2 3 I feel lonely in this relationship.
1 2 3 When we argue, one of us withdraws . . . that is, doesn't want to talk about it anymore; or leaves the scene.

Who tends to withdraw more when there is an argument?
Male
Female
Both Equally
Neither Tend to Withdraw

Where Are You in Your Marriage
We devised these questions based on seventeen years of research at the University of Denver on the kinds of communication and conflict management patterns that predict if a relationship is headed for trouble. We have recently completed a nationwide, random phone survey using these questions. The average score was 11 on this scale. While you should not take a higher score to mean that your relationship is somehow destined to fail, higher scores can mean that your relationship may be in greater danger unless changes are made. (These ranges are based only on your individual ratings—not a couple total.)

8 to 12 "Green Light"
If you scored in the 8–12 range, your relationship is probably in good or even great shape at *this time*, but we emphasize "at *this time*" because relationships don't stand still. In the next twelve months, you'll either have a stronger, happier relationship, or you could head in the other direction.

To think about it another way, it's like you are traveling along and have come to a green light. There is no need to stop, but it is probably a great time to work on making your relationship all it can be.

13 to 17 "Yellow Light
If you scored in the 13–17 range, it's like you are coming to a "yellow light." You need to be cautious. While you may be happy now in your relationship, your score reveals warning signs of patterns you don't want to let get worse. You'll want to be taking action to protect and improve what you have. Spending time to strengthen your relationship now could be the best thing you could do for your future together.

Stanley, S. M., and H. J. Markman. 1997. *Marriage and Family: A Brief Introduction.* Reprinted with permission of PREP, Inc.

from garden to table. Commercial industries had developed to provide these services, and women transferred their activities in these areas from the home to the factory. The result was that women had more frequent contact with men.

Women's involvement in factory work decreased parental control, since parents were unable to dictate the extent to which their offspring could interact with those they met at work. Hence, values in mate selection shifted from the parents to the children. In the past, the "good wife" was valued for her domestic aptitude—her ability to spin yarn, make clothes, cook meals, preserve food, and care for children. The "good husband" was evaluated primarily in terms of being an economic provider. Though these issues may still be important, contemporary mates are more likely to be selected on the basis of personal qualities, particularly for love and companionship, than for either utilitarian or economic reasons.

Changes in Dating in the Last Fifty Years

The Industrial Revolution had a profound effect on courtship patterns, but these patterns have continued to change in the past fifty years. The changes include an increase in the age at marriage, which has been accompanied by each person's having a longer period of time during which he or she becomes involved with more people. Marrying at age 29 rather than 24 provides more time and opportunity to date more people.

The dating pool today also includes an increasing number of individuals in their 30s who have been married before. These individuals often have children, which changes the nature of a date from two adults going out alone to see a movie to renting a movie and baby-sitting in the apartment or home of one of the partners.

18 to 24 "Red Light"

Finally, if your scored in the 18–24 range, it's like approaching a red light. Stop, and think about where the two of you are headed. Your score indicates the presence of patterns that could put your relationship at significant risk. You may be heading for trouble—or already be there. But there is *good news.* You can stop and learn ways to improve your relationship now!

For more information on danger signs and constructive tools for strong marriages, see: Markman, H. J., Stanley, S. M., & Blumberg, S. L. (1994). *Fighting for Your Marriage: Positive Steps for a Loving and Lasting Relationship.* San Francisco: Jossey Bass, Inc. (PREP 1-800-366-0611)

To: Those interested in using the Relationships Dynamics Scale From: PREP, Inc.

1. We wrote these items based on an understanding of many key studies in the field. The content or themes behind the questions are based on numerous in-depth studies on how people think and act in their marriages. These kinds of dynamics have been compared with patterns on many other key variables, such as satisfaction, commitment, problem intensity, etc. Because the kinds of methods researchers can use in their laboratories are quite complex, this actual measure is far simpler than many of the methods we and others use to study marriages over time. But the themes are based on many solid studies. Caution is warranted in interpreting scores.

2. The discussion of the Relationships Dynamics Scale gives rough guidelines for interpreting the meaning of the scores. The ranges we suggest for the measure are based on results from a nationwide, random phone survey of 947 people (85% married) in January 1996. These ranges are meant as a rough guideline for helping couples assess the degree to which they are experiencing key danger signs in their marriages. The measure as you have it here powerfully discriminated between those doing well in their marriages/relationships and those who were not doing well on a host of other dimensions (thoughts of divorce, low satisfaction, low sense of friendship in the relationship, lower dedication, etc.). Couples scoring more highly on these items are truly more likely to be experiencing problems (or, based on other research, are more likely to experience problems in the future).

3. This measure in and of itself should not be taken as a predictor of couples who are going to fail in their marriages. No couple should be told they will not "make it" based on a higher score. That would not be in keeping with our intention in developing this scale or with the meaning one could take from it for any one couple. While the items are based on studies that assess such things as the likelihood of a marriage working out, we would hate for any one person to take this and assume the worst about their future based on a high score. Rather, we believe that the measure can be used to motivate high and moderately high scoring people to take a serious look at where their marriages are heading—and take steps to turn such negative patterns around for the better.

For more information on constructive tools for strong marriages, see: Markman, H. J., Stanley, S. M., & Blumberg, S. L. (1994). *Fighting for Your Marriage: Positive Steps for a Loving and Lasting Relationship.* If you have questions about the measure and the meaning of it, please write to us at:

PREP, Inc.
P. O. Box 102530
Denver, Colorado 80250-2530
E-mail: PREPinc@AOL.com

Scott M. Stanley, Ph.D.
Howard J. Markman, Ph.D.

To order books, audio or videotapes: Call 1-800-366-0166

As we will note later in this chapter, cohabitation has become more normative. For some couples, the sequence of date, fall in love, and get married has been replaced by date, fall in love, and live together. Such a sequence results in the marriage of couples who are more relationship-savvy than those who dated and married out of high school.

Not only do individuals now date more partners and more often live together, but also gender role relationships have become more egalitarian. Though the double standard still exists, women today are more likely than women in the 1950s to ask men to go out, to have sex with them without requiring a commitment, and to postpone marriage until their own educational and career goals have been met. Women no longer feel desperate to marry but consider marriage one of many goals they have for themselves.

Unlike during the fifties, both sexes today are aware of and somewhat cautious of becoming HIV-infected. Sex has become potentially deadly, and condoms are being used more frequently. The 1950s fear of asking a druggist for a condom has been replaced by the confidence and mundaneness of buying condoms along with one's groceries.

Finally, couples of today are more aware of the impermanence of marriage. However, most couples continue to feel that divorce will not happen to them, and they remain committed to domestic goals. Almost three-quarters (72%) of all first-year college students in the United States reported that "an essential or very important goal" for them is "raising a family" (American Council on Education & University of California, 2002).

To assess the relationship with your partner at this time, complete the Relationships Dynamics Scale, which begins on page 86.

COHABITATION

Cohabitation, also known as living together, is becoming a "normative life experience" (Raley, 2000, 19), with almost 60 percent of U.S. women who married in the 1990s reporting that they had cohabited before marriage (Bachrach, Hindin, & Thomson, 2000). Persons who live together before marriage are more likely to be high school dropouts than college graduates (60% versus 37%), to have been married, to be less religious/traditional, and to be supportive of egalitarian gender roles. Half of previously married cohabiters have children and 35 percent of never-married cohabiters have children (Smock, 2000; Moors, 2000). Half of new unwed mothers are cohabiting with the fathers at the time their children are born (Sigle-Rushton & McLanahan, 2002). There are few racial and ethnic differences in the demographics of cohabiters (Smock, 2000; Moors, 2000).

NATIONAL DATA

There are over 5 million unmarried couple households in the United States (*Statistical Abstract of the United States, 2003*, Table 69). In a national random sample of 947 individuals, 60 percent of couples who had been married five years or less reported that they had cohabited before they married (Stanley & Markman, 1997).

In research, over twenty definitions of **cohabitation** (also referred to as **living together**) have been used. These various definitions involve variables such as duration of the relationship, frequency of overnight visits, emotional/sexual nature of the relationship, and sex of the partners. Most research on cohabitation has been conducted on heterosexual live-in couples. We define cohabitation as two unrelated adults involved in an emotional and sexual relationship who sleep overnight in the same residence on a regular basis. The terms used to describe live-ins include cohabitants and **POSSLQs** (people of the opposite sex sharing living quarters), the latter term used by the U.S. Bureau of the Census.

In the 1920s, Judge B. B. Lindsey, who witnessed an endless stream of couples coming through his court to divorce, suggested that couples live together rather than marry. His suggestion grew out of concern for the parade of divorcing couples he saw in his courtroom. He reasoned that if couples lived together before marriage, they might be better able to assess their compatibility. Similarly, Margaret Mead suggested a **two-stage marriage.** The first stage would involve living together without having children. If the partners felt that their relationship was stable and durable, they would get married and have children.

Although Judge Lindsey's suggestion was made in the 1920s, cultural support for living together has not flourished until recently. Former President and Mrs. Clinton lived together before they were married.

Reasons for the increase in cohabitation include fear of marriage; career or educational commitments; increased tolerance from society, parents, and peers; improved birth-control technology; and the desire for a stable emotional and sexual relationship without legal ties. Cohabitants also regard living together as a vaccination against divorce. Later, we will review studies emphasizing that this hope is more often an illusion.

Diversity in Other Countries

Iceland is a homogeneous country of 250,000 descendants of the Vikings. Their sexual norms include early (age 14) protected intercourse, nonmarital parenthood, and living together before marriage. Indeed, a wedding photo often includes not only the couple but the children they have already had. One American woman who was involved with an Icelander noted, "My parents were upset with me because Ollie and I were thinking about living together, but his parents were upset that we were not already living together" (authors' files).

Eight Types of Cohabitation Relationships

Various types of cohabitation follow:

1. Here and now.—The new partners have an affectionate relationship and are focused on the here and now, not the future of the relationship. Only a small proportion of persons living together report that the "here and now" type characterizes their relationship (Jamieson et al., 2002).

2. Testers. The couples are involved in a relationship and want to assess whether they have a future together. As in the case of here-and-now cohabitants, only a small proportion of cohabitants characterize their relationship as "testers" (Jamieson et al., 2002).

3. Engaged. These couples are in love and are planning to marry. While not all cohabitants consider marriage their goal, most view themselves as committed to each other (Jamieson et al., 2002). Oppenheimer (2003) studied a national sample of cohabitants and found that cohabiting whites were much more likely to have married the partner than cohabiting blacks—51 percent versus 22 percent.

4. Money savers. The couples live together primarily out of economic convenience. They are open to the possibility of a future together but regard such a possibility as unlikely.

5. Pension partners. A variation of the money savers category is that of pension partners. These individuals are older, have been married before, still derive benefits from their previous relationships, and are living with someone new. Getting married would mean giving up their pension benefits from the previous marriage. An example is a widow from the war with Iraq who was given military benefits due to her husband's death. If she remarries, she forfeits both health and pension benefits; she is now living with a new partner but getting the benefits from the previous marriage.

6. Security blanket cohabiters. Also known as "Linus Blanket" cohabiters, some of the individuals in these relationships are drawn to each other out of a need for security rather than mutual attraction.

7. Rebellious cohabiters. Some couples use cohabitation as a way of making a statement to their parents that they are independent and can make their own choices. Their cohabitation is more about rebelling from parents than being drawn to each other.

8. Marriage never. These couples feel that a real relationship is a commitment of the heart, not a legal document. Living together provides both companionship and sex without the responsibilities of marriage. Skinner et al. (2002) found that individuals in long-term cohabiting relationships scored the lowest in terms of relationship satisfaction when compared with married and remarried couples.

The "marriage never" couples are rare. However, women and men sometimes have different definitions of the living-together experience, with women viewing it more as a sign of a committed relationship moving toward marriage and men viewing it as an alternative to marriage or as a test to see whether future commitment is something to pursue.

There are various reasons and motivations for living together as a permanent alternative to marriage. Some may have been married before and don't want the entanglements of another marriage. Others feel that the real bond between two people is (or should be) emotional. They contend that many couples stay together because of the legal contract, even though they do not love each other any longer. "If you're staying married because of the contract," said one partner, "you're staying for the wrong reason." Some couples feel that they are "married" in their hearts and souls and don't need or want the law to interfere with what they feel is a private act of commitment. For most couples, living together is a short-lived experience. About 55 percent will marry and 40 percent will break up within five years of beginning cohabitation (Smock, 2000). Some couples who view their living together as "permanent" seek to have it defined as a domestic partnership (see Social Policy, p. 90).

Diversity in Other Countries

Over 90 percent of first marriages in Sweden are preceded by cohabitation; however, only 12 percent of first marriages in Italy are preceded by cohabitation (Kiernan, 2000).

Diversity in the United States

Though cohabitation as a permanent alternative to marriage is relatively rare on the U.S. mainland, it is common and expected among the Hopi Native Americans living in central Arizona. After the respective families express approval of the marriage (this is done by an exchange of food/gifts), the man lives with the woman in her parents' home, has children with her, and later marries her (Knox & Schacht, 2000).

Domestic Partnerships

SOCIAL POLICY

Domestic partnerships refer to two adults who have chosen to share each other's lives in an intimate and committed relationship of mutual caring. Such cohabitants, both heterosexual and homosexual, want their employers, whether governmental or corporate, to afford them the same rights as spouses. Specifically, an employed person who pays for health insurance would like his or her domestic partner to be covered in the same way that one's spouse would be. Employers have been reluctant to legitimize domestic partners as qualifying for benefits money because of the additional expense. One reason for the reluctance is the fear that a higher proportion of partners may be HIV-infected, which would involve considerable medical costs. Aside from the economic issue, domestic partner benefits have been criticized by fundamentalist religious groups as eroding family values by giving nonmarital couples the same rights as married couples.

California leads the way in domestic partner benefits (Lisotta, 2003), with the law providing rights and responsibilities in areas as varied as child custody, legal claims, housing protections, bereavement leave, and state government benefits.

To receive benefits, domestic partners must register, which involves signing an affidavit of domestic partnership verifying that they are a nonmarried cohabitating couple 18 years of age or older and unrelated by blood close enough to bar marriage in the state of residence. Other criteria typically used to define a domestic partnership include that the individuals must be jointly responsible for debts to third parties, they must live in the same residence, they must be financially interdependent, and they must intend to remain in the intimate committed relationship indefinitely. Should they terminate their domestic partnership, they are required to file notice of such termination.

The Netherlands allows the legal registration of both heterosexual and homosexual partnerships, making a couple's relationship the functional equivalent of marriage. The one exception is that cohabiting couples do not have the right to adopt children (Kiernan, 2000).

Sources

Kiernan, K. 2000. European perspectives on union formation. In *The ties that bind*, edited by Linda J. Waite. New York: Aldine de Gruyter, 40–58.

Lisotta, C. 2003. Toward perfect unions: California's new domestic partnership law is second only to Vermont's legislation. *The Advocate*. 14 October, 17–18.

Consequences of Cohabitation

McGinnis (2003) noted that one of the effects of cohabitation was to increase the chance that a couple would end up getting married. Although living together before marriage does not ensure a happy, stable marriage, it has some potential advantages.

Advantages of Cohabitation Many unmarried couples who live together report that it is an enjoyable, maturing experience. Other potential benefits of living together include the following:

*1. **Sense of well-being.*** Cohabitants are likely to report a sense of well being. They are in love, the relationship is new, and the disenchantment that frequently occurs in long-term relationships has not had time to surface. One student reported, "We have had to make some adjustments in terms of moving all our stuff into one place, but we very much enjoy our life together."

*2. **Delayed marriage.*** Another advantage of living together is that those who live together are remaining unmarried. And the longer one waits to marry, the better. Being older at the time of marriage is predictive of marital happiness and stability, just as being young (particularly 18 and below) is associated with marital unhappiness and divorce. Hence, if a young couple who have known each other for a short time is faced with the choice of living together or getting married, their delaying marriage while they live together seems to be the better choice. Also, If they break up, the split will not go on their record as will a "divorce" if they get married at a young age.

*3. **Learning about self and partner.*** Living with an intimate partner provides an opportunity for individuals to learn more about themselves and their partner. For example, individuals in living-together relationships may find that their role expectations are more (or less) traditional than they had previously thought. Learning more about one's partner is a major advantage of living together. A person's values (calling parents daily), habits (leaving the lights on), and relationship expectations (how emotionally close or distant) are sometimes more fully revealed in a living-together context than in a traditional dating context.

PERSONAL CHOICES

Will Living Together Ensure a Happy, Durable Marriage?

Couples who live together before getting married assume that doing so will increase their chances of having a happy and durable marriage relationship. But will it? In a word, no. Researchers refer to the **cohabitation effect** as the tendency for couples who cohabit to end up in less happy and shorter-lived marriages (more likely to divorce). Cohabitants are more likely not only to divorce but to report more disagreements, more violence, lower levels of happiness, and lower levels of ability to negotiate conflict (Cohan & Kleinbaum, 2002). However, Skinner et al. (2002) compared those who had cohabited and married and those who married but did not cohabit and found no distinguishing characteristics. They concluded that "cohabiting couples may not be stigmatized if there is an expectation that marriage will occur."

While cohabitation does not "cause" divorce, there are several reasons why cohabitation is associated with subsequent divorce (Smock, 2000). One, the selection hypothesis, suggests that cohabitation may select or draw people with nontraditional values (who are not ready to commit to each other or to the institution of marriage) and who may have poor relationship skills. Cohan and Kleinbaum (2002) suggest that couples who cohabit are less effective in soliciting support from and providing support to their partners and that this deficit makes them vulnerable as spouses to unhappiness and divorce. Indeed, they may look for an out when there is conflict rather than work things out. They may be ludic lovers looking to play at love, not engage in a committed relationship.

Cohabitation has been shown to attract a different type of couple than marriage, and to foster attitudes that contribute to divorce.
Steven Nock, *Sociologist*

A second explanation is that cohabitants tend to be people who are willing to violate social norms and live together before marriage. Once they marry, they may be more willing to break another social norm and divorce if they are unhappy than are unhappily married persons who tend to conform to social norms and have no history of unconventional behavior. Kamp Dush et al. (2003) suggested that it is not the types of people that cohabitation attracts but the experience of cohabitation that is associated with subsequent divorce. "According to this perspective, cohabitation changes people and their relationships in ways that undermine later marital quality and commitment" (p. 545). For example, being in a relationship with an uncertain commitment to the future may make people less invested and committed and therefore most vulnerable to divorce. Whatever the reason, cohabitants should not assume that cohabitation will make them happier spouses or insulate them from divorce.

Sources

Cohan, C. L., and S. Kleinbaum. 2002. Toward a greater understanding of the cohabitation effect: Premarital cohabitation and marital communication. *Journal of Marriage and the Family* 64:180–92.

Kamp Dush, C. M. K., C. L. Cohan, and P. R. Amato. 2003. The relationship between cohabitation and marital quality and stability: Change across cohorts? *Journal of Marriage and the Family* 65:539–49.

Skinner, K. B., S. J. Bahr, D. R. Crane, and V. R. A. Call. 2002. Cohabitation, marriage, and remarriage. *Journal of Family Issues* 23:74–90.

Smock, P. J. 2000. Cohabitation in the United States: An appraisal of research themes, findings, and implications. *Annual Review of Sociology* 26:1–20.

Disadvantages of Cohabitation There is a downside for individuals and couples who live together.

1. Feeling used or tricked. We have mentioned that women are more prone than men to view cohabitation as reflective of a more committed relationship. When expectations differ, the more invested partner may feel used or tricked if the relationship does not progress toward marriage. One partner said, "I always felt we would be getting married, but it turns out that he never saw a future for us."

2. Problems with parents. Some cohabiting couples must contend with parents who disapprove of or do not fully accept their living arrangement. For example, cohabitants sometimes report that when visiting their parents' homes, they are required to sleep in separate beds in separate rooms. Some cohabitants who have parents with traditional values respect these values, and sleeping in separate rooms is not a problem. Other cohabitants feel resentful of parents who require them to sleep separately.

Some parents express their disapproval of their child's cohabiting by cutting off communication, as well as economic support, from their child. Other parents display lack of acceptance of cohabitation in more subtle ways. One woman who had lived with her partner for two years said that her partner's parents would not include her in the family's annual photo portrait. Emotionally, she felt very much a part of her partner's family and was deeply hurt that she was not included in the family portrait (authors' files). Still other parents are completely supportive of their children's cohabiting and support their doing so. "I'd rather my kid live together than get married, and besides it is safer for her and she's happier," said one father.

3. Economic disadvantages. Some economic liabilities exist for those who live together instead of getting married. In the Social Policy section on domestic partnerships, we noted that cohabitants typically do not benefit from their partner's health insurance, Social Security, or retirement benefits. In most cases only spouses qualify for such payoffs. We have already noted that cohabitants tend to be from lower socioeconomic strata than noncohabitants (Smock, 2000).

Given that most living-together relationships are not long-term relationships and that breaking up is not uncommon, cohabitants might develop a written and signed legal agreement should they purchase a house, car, or other costly items together. The written agreement should include a description of the item, to whom it belongs, how it will be paid for, and what will happen to the item if the relationship terminates. Purchasing real estate together may require a separate agreement, which should include how the mortgage, property taxes, and repairs will be shared. The agreement should also specify who gets the house if the partners break up and how the value of the departing partner's share will be determined. If the couple have children, another agreement may be helpful in defining custody, visitation, and support issues in the event the couple terminates their relationship. Such an arrangement may take some of the romance out of the cohabitation relationship, but it can save a great deal of frustration should the partners decide to go their separate ways.

In addition, couples who live together instead of marrying can protect themselves from some of the economic disadvantages of living together by specifying their wishes in wills; otherwise, their belongings will go to next of kin or to the state. They should also own property through joint tenancy with rights of survivorship. This means that ownership of the entire property will revert to one partner if the other partner dies. In addition, the couple should save for retirement, since Social Security benefits may not be accessed by live-in companions, and some company pension plans bar employees from naming anyone other than a spouse as the beneficiary.

4. Effects on children. About 40 percent of children will spend some time in a home where the adults are cohabitating. In addition to being disadvantaged in terms of parental income and education, they are likely to experience more disruptions in family structure. Smock (2000, 11) notes that "there is evidence that the number of changes in family structure is particularly important." The more changes, the more negative the outcome (e.g., children separated from biological parents).

Legal Aspects of Living Together

In recent years, the courts and legal system have become increasingly involved in living-together relationships. Some of the legal issues concerning cohabiting partners include common-law marriage, palimony, child support, and child inheritance. Lesbian and gay couples also confront legal issues when they live together .

Technically, cohabitation is against the law in some states. For example, in North Carolina, cohabitation is a misdemeanor punishable by a fine not to exceed $500, imprisonment for not more than six months, or both. Most law enforcement officials view cohabitation as a victimless crime and feel that the general public can be better served by concentrating upon the crimes that do real damage to citizens and their property.

Common-Law Marriage The concept of **common-law marriage** dates back to a time when couples who wanted to be married did not have easy or convenient access to legal authorities (who could formally sanction their relationship so that they would have the benefits of legal marriage). Thus, if the couple lived together, defined themselves as husband and wife, and wanted other people to view them as a married couple, they would be considered married in the eyes of the law.

Despite the assumption by some that heterosexual couples who live together a long time have a common-law marriage, only eleven jurisdictions recognize such marriages. In ten states (Alabama, Colorado, Idaho, Iowa, Kansas, Rhode Island, South Carolina, Montana, Pennsylvania, and Texas) and the District of Columbia, a heterosexual couple may be considered married if they are legally competent to marry, if there is an agreement between the partners that they are married, and if they present themselves to the public as a married couple. A ceremony or compliance with legal formalities is not required.

In common-law states, individuals who live together and who prove that they were married "by common law" may inherit from each other or receive alimony and property in the case of "divorce." They may also receive health and Social Security benefits as would other spouses who have a marriage license. In states not recognizing common-law marriages, the individuals who live together are not entitled to benefits traditionally afforded married individuals. Over three-quarters of the states have passed laws prohibiting the recognition of common-law marriages within their borders.

Palimony A take-off on the word *alimony,* **palimony** refers to the amount of money one "pal" who lives with another "pal" may have to pay if the partners end their relationship. Technically it means legally arranged support payments between unmarried partners. Claire and Tony Maglica lived together for twenty-three years. Although they were never legally married, they would often present themselves as husband and wife. When they broke up, Claire said that she and Tony had a "private marriage ceremony," and she sued Tony for $150 million (half) of Mag Instruments, the business she contended she had helped to build.

Tony said that they were never married and that he was responsible for the success of the $300 million business. A jury in California, which does not recognize common-law marriage, awarded her $84 million. Tony appealed the decision.

Child Support Individuals who conceive children are responsible for those children whether they are living together or married. In most cases, the custody of young children will be given to the mother, and the father will be required to pay child support. In effect, living together is irrelevant with regard to parental obligations.

Couples who live together or who have children together should be aware that laws traditionally applying only to married couples are now being applied to many unwed relationships. Palimony, distribution of property, and child support payments are all possibilities once two people cohabit or parent a child.

Child Inheritance Children born to cohabitants who view themselves as spouses and who live in common-law states are regarded as legitimate and can inherit from their parents. However, children born to cohabitants who do not hold themselves out to the public as married or who do not live in common-law states are also able to inherit. A biological link between the parent and the offspring is all that needs to be established.

ENDING AN UNSATISFACTORY RELATIONSHIP

Endings are as common as beginnings in relationships. Why they end and considerations in ending one's own relationship are the concerns of this section.

Why College Students End Relationships

Table 4.1 shows the percentage of 185 undergraduates who gave the reasons for ending their last relationship (Knox et al. 1997).

Although 43 percent of the respondents noted that "too many differences/different values" was the reason their last relationship ended, closer examination reveals that "another person" was the culprit in the majority of endings. Respondents mentioned this in various ways: "cheating" (18%), "I met someone new" (15%), "My partner met someone new" (13%), "I went back to a previous lover" (6%), and "My partner went back to a previous lover" (5%). When these percentages are totaled, the major reason relationships ended in almost 60 percent of the cases (57%) was "someone else." If we include the 18 percent who reported "dishonesty" as the reason, and assume that such dishonesty was in reference to another person, the percentage goes to almost 75 percent.

The negative effect of an external person on a couple's relationship has been documented. Miller (1997) studied attentiveness to desirable alternatives to one's partner in a sample of 99 men and 147 women undergraduates and found that there was no better predictor of relationship failure than high attentiveness to alternatives.

Table 4.1 Why 185 College Students Ended Their Last Love Relationship*

Reason	Percentage
Too many differences/different values	43%
Got tired of each other	27%
Cheating	18%
Dishonesty	18%
I met someone new	15%
Separation	15%
My partner met someone new	13%
Parental disapproval	13%
Violence/abuse	9%
Alcohol/drugs	7%
I went back to previous lover	6%
My partner went back to previous lover	5%

*Respondents could give more than one reason.

Source: D. Knox, L. Gibson, M. E. Zusman, and C. Gallmeier. 1997. Why college students end relationships. *College Student Journal* 31:449–52.

Considerations in Ending a Relationship

All relationships have difficulties, and all necessitate careful consideration of various issues before they are ended. These considerations include the following:

1. Is there any desire/hope to revive and improve the relationship? In some cases, people end relationships and later regret having done so. Setting unrealistically high standards may eliminate an array of individuals who might be superb part-

Up Close 4.1 College Student Recovery from a Broken Heart

A sample of 410 freshmen and sophomore undergraduates at a large southeastern university completed a confidential survey revealing their recovery from a previous love relationship. Some of the findings were:

1. Sex differences in relationship termination. Women were significantly more likely (50%) than men (40%) to report that they initiated the breakup. Sociologists explain such a phenomenon as parental investment in a potential good father for their offspring. One student recalled, "I got tired of his lack of ambition—I just thought I could do better. He's a nice guy but living in a trailer is not my idea of a life."

2. Sex differences in relationship recovery. Though recovery was not traumatic for either men or women, men reported more difficulty than women did in adjusting to a breakup. When respondents were asked to rate their level of difficulty from "no problem"(0) to "complete devastation" (10) on a ten-point continuum, women reported 4.35 and men scored 4.96. In explaining why men might have more difficulty adusting to terminated relationships, some of the students said, "Men have such inflated egos they can't believe that a woman would actually dump them." Others said, "Men are oblivious to what is happening in a relationship and may not have a clue that it is heading toward an abrupt end. When it does end, they are in shock."

3. Time/new partner as factors in recovery. The passage of time and involvement with a new partner were identified as the most helpful factors in getting over a love relationship that ended. Though the difference was not statistically significant, men more than women reported "a new partner" was more helpful in relationship recovery (34% versus 29%). Similarly, women more than men reported that "time" was more helpful in relationship recovery (34% versus 29%).

4. Other findings. Other factors associated with recovery for women and men were "moving to a new location" (13% versus 10%) and recalling that "the previous partner lied to me" (7% versus 5%). Men more than women were much more likely to use alcohol to help them get over a previous partner (9% versus 2%). Neither men nor women reported using therapy with any frequency to help them get over a partner (1% versus 2%).

Shakespeare noted that "true love never did run smooth." These data suggest that breaking up is not terribly difficult for most individuals (more difficult for men than women) and involves both time to recover and a new partner.

Based on original data collected for and adapted from Knox, D., M. E. Zusman, M. Kaluzny, and C. Cooper. 2000. College student recovery from a broken heart. *College Student Journal* 34: 322–24.

ners, companions, and mates. If the reason for ending a relationship is conflict over an issue or set of issues, an alternative to ending the relationship is to attempt to resolve the issues through negotiating differences, compromising, and giving the relationship more time. (We do not recommend giving an abusive relationship more time, as abuse, once started, tends to increase in frequency and intensity.)

Independent of abuse, some couples elect to end a scarred relationship. As Rhett Butler says to Scarlett O'Hara in *Gone with the Wind,* "I was never one to patiently pick up broken fragments and glue them together and tell myself that the mended whole was as good as new. What's broken is broken—I'd rather remember it as it was at its best than mend it and see the broken pieces as long as I lived" (Mitchell, [1936] 1977, 945).

2. Acknowledge and accept that terminating a relationship may be painful for both partners. There may be no way you can stop the hurt. One person said, "I can't live with him anymore, but I don't want to hurt him either." The two feelings are incompatible. To end a relationship with someone who loves you is usually hurtful to both partners. Persons who end a relationship usually conclude that the pain and suffering of staying in a relationship is more than they will experience from leaving

3. Blame yourself for the end. One way to end a relationship is to blame yourself by giving a reason that is specific to you ("I need more freedom," "I want to go to graduate school in another state," "I'm not ready to settle down," etc.). If you blame your partner or give your partner a way to make things better, the relationship may continue because *you* may feel obligated to give your partner a second chance if he or she promises change.

4. Cut off the relationship completely. If you are the person ending the relationship, it will probably be easier for you to continue to see the other person without feeling too hurt. But the other person will probably have a more difficult time and will heal faster if you stay away completely. Alternatively, some people are skilled at ending love relationships and turning them into friendships. Though this is difficult and infrequent, it can be rewarding across time for the respective partners.

5. Learn from the terminated relationship. Stets (1993) observed that when past issues involved in terminated relationships remain unresolved, they reappear in later relationships and lead to consecutive relationship breakups. Issues that remain unresolved may include problems of being oversensitive, jealous, too picky, cheating, fearing commitment, and being unable to compromise and negotiate conflict. Recognizing one's own contribution to the breakup and working on any characteristics that might be a source of future relationship problems are some of the benefits of terminating a relationship. Otherwise, one might repeat the process.

6. Allow time to grieve over the end of the relationship. Ending a love relationship is painful. It is okay to feel this pain, to hurt, to cry. Allowing yourself to experience such grief may help you get over a relationship so as to become healed for the next one. The passage of time also aids in recovery. Up Close 4.1 emphasizes the factors associated with relationship recovery.

SUMMARY

What are the functions of hanging out in the United States?

Confirmation of a social self, recreation, companionship/intimacy/sex, anticipatory socialization, status achievement, and mate selection are the functions of dating/hanging out in the United States.

How do individuals go about meeting a new partner?

Besides the traditional way of meeting people at work/school or through friends, other ways new people meet each other include through personal ads in magazines/newspapers, on line, video chatting, video dating, and the "eight-minute date." Regardless of how people meet, the top problem they identify on dates is communication.

What issues do divorced persons face when they start dating again?

The divorced are older and select from an older, more limited population who have a higher chance of having an STD. Most have children from a previous marriage, and are dealing with an ex-spouse.

What is the cultural and historical background of dating?

The freedom with which U.S. partners today select each other on the basis of love is a relatively recent phenomenon. At most times and in most cultures, marriages were more often arranged by parents. Unlike during the fifties, both sexes today are aware of and somewhat cautious of becoming STD/HIV-infected. Sex has become potentially deadly, and condoms are being used more frequently. Couples of today are also more aware of the impermanence of marriage.

What is cohabitation like among today's youth?

Cohabitation, also known as living together, is becoming a "normative life experience," with almost 60 percent of U.S. women who married in the 1990s reporting that they had cohabited before marriage. Reasons for an increase in living together since 1970 include a delay of marriage for educational or career commitments, fear of marriage, increased tolerance from society for living together, and a desire to avoid the legal entanglements of marriage. Types of living-together relationships include the here and now, testers (testing the relationship), engaged couples (planning to marry), and cohabitants forever (never planning to marry). Most people who live together eventually get married but not necessarily to each other.

Domestic partnerships refer to two adults who have chosen to share each other's lives in an intimate and committed relationship of mutual caring. Such cohabitants, both heterosexual and homosexual, want their employers, whether governmental or corporate, to afford them the same rights as spouses. Only about 10 percent of firms recognize domestic partners and offer any benefits

What are the consequences of cohabitation?

Although living together before marriage does not ensure a happy, stable marriage, it has some potential advantages. These include a sense of well-being, delayed marriage, learning about self/partner, and being able to disengage with minimal legal hassle. Disadvantages include feeling exploited, feeling guilty about lying to parents, and not having the same economic benefits as those who are married. Social Security and retirement benefits are paid to spouses, not live-in partners.

Why do college students end relationships?

Although 43 percent of the respondents noted that "too many differences/different values" was the reason their last relationship ended, closer examination reveals that "another person" was the culprit in the majority of endings. Considerations in ending a relationship include improving it, acknowledging that endings are often painful, and taking the blame. Recovery involves time and a new partner.

KEY TERMS

blind marriage
bundling
cohabitation
cohabitation effect
common-law marriage
dating
domestic partnership
living together
palimony
POSSLQ
two-stage marriage

RESEARCHING MARRIAGE AND THE FAMILY WITH INFOTRAC COLLEGE EDITION

InfoTrac College Edition, an online library, allows you to perform research online anywhere, anytime. Following are two suggested search terms and related questions to help you extend your understanding of the topics covered in this chapter. Go to www.infotrac-college.com to begin your search.

Keyword: **Hanging out.** Locate articles that discuss this pattern of meeting a new partner. To what degree do individuals "hang out" versus date in pairs?

Keyword: **Internet dating.** Locate articles that discuss the use of the Internet in meeting a new person. What are the advantages and disadvantages of meeting someone on the Internet?

The Companion Web Site for Choices in Relationships: An Introduction to Marriage and the Family, Eighth Edition

http://sociology.wadsworth.com/knox_schacht/choices8e

Supplement your review of this chapter by going to the companion Web site to take one of the Tutorial Quizzes, use the flash cards to master key terms, and check out the many other study aids you'll find there. You'll also find special features such as the Marriage and Family Resource Center, Census 2000 information, and other data and resources at your fingertips to help you with that special project or to do some research on your own.

WEBLINKS

Right Mate at Heartchoice
http://www.heartchoice.com/rightmate/

WildXAngel
http://www.wildxangel.com/

CHAPTER 5

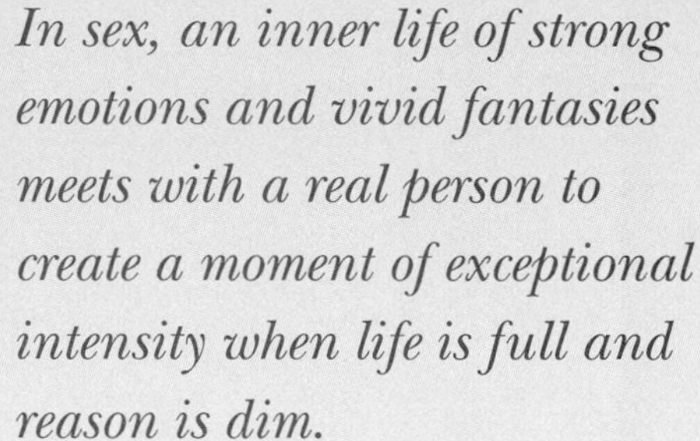

In sex, an inner life of strong emotions and vivid fantasies meets with a real person to create a moment of exceptional intensity when life is full and reason is dim.

Thomas Moore, The Soul of Sex

Authors

Sexuality in Relationships

Contents

TRUE OR FALSE?

1. Students exposed to abstinence sex education in the public schools are more likely to report an intention to delay premarital intercourse.
2. Hooking up, particularly for females, has become a way to attract men and initiate relationships where love and commitment eventually develop.
3. The longer a couple has been dating, the less likely they are to use a condom.
4. Sexual but not relationship satisfaction is adversely affected by the man's premature ejaculation.
5. Women and men view "friends with benefits" relationships differently—for women the focus is friends; for men the focus is benefits.

Answers: **1.** F **2.** F **3.** T **4.** T **5.** T

Sex contributes to the excitement in a new relationship. The partners can't seem to get enough of each other. Over time, sex is no longer the focus of a couple's time together and becomes a part of the fabric of their overall relationship. Although the frequency of intercourse does decrease the longer the partners are in a sexual relationship with each other, the reported emotional and physical satisfaction may remain high. Sexuality in relationships is the focus of this chapter, and we begin by discussing the sexual values the partners bring to each other.

SEXUAL VALUES

The following are some examples of choices (reflecting sexual values) with which individuals in a new relationship are confronted:

How much sex and how soon in a relationship is appropriate?

Require a condom for vaginal or anal intercourse?

Require a condom and/or dental dam for oral sex?

Require STD/HIV test before becoming sexually active with a new partner?

Tell partner actual *number of previous sexual partners?*

Tell partner of sexual fantasies (which may include other people)?

Reveal same-sex previous/current behavior/interests?

But what are sexual values and what is their function for individuals and relationships?

Definition of Sexual Values

Sexual values are moral guidelines for making sexual choices in nonmarital, marital, heterosexual, and homosexual relationships. Attitudes and values predict sexual behavior (Meier, 2003). Among the various sexual values are absolutism, relativism, and hedonism. One's sexual values may be identical to one's sexual choices. For example, a person who values abstinence until marriage may choose to remain a virgin until marriage. Or a person who values being in a stable love relationship before having sex may choose not to have a one-night encounter of casual sex. Finally, a person who values monogamy may have sex only with his or her partner.

But one's behavior does not always correspond with one's values. Persons who express a value of waiting until marriage sometimes have intercourse during their engagement. Those who value love as a condition of sex will sometimes have sex with a person they have just met. And persons who value sexual monogamy may sometimes have sex with someone other than their primary partner. This struggle between values and behavior is not new. St. Paul wrote in about A.D. 53, "For I do not do what I want, but I do the very thing I hate" (Romans 7:15).

One explanation for the discrepancy between values and behavior is that a person may engage in a sexual behavior, then decide the behavior was wrong and adopt a sexual value against it. Fifty-eight percent of 3,432 respondents who said that premarital sex is always wrong also said that they themselves had sex before they were married (Michael et al., 1994, 239). Nevertheless, most people behave most of the time in reference to their sexual values. Most spouses value fidelity, and their behavior reflects it . . . most of the time—less than 5 percent of both husbands and wives in a national sample report that they have been unfaithful in the past twelve months (Wiederman, 1997).

Functions of Sexual Values

We have already mentioned one function of sexual values—to serve as a moral compass. Other functions include solidifying one's self-concept, sorting potential partners, and helping to avoid sexually transmissible diseases (STDs).

Solidifying Self-Identity Sexual values help to establish and maintain one's self-identity. Our self-identity refers to who we are—that is, how we perceive that others view us and how we view ourselves. Our sexual values are a reflection of how we view ourselves. Persons who value sex without love, sex with casual partners, extramarital sex, and sex with children have a different view of themselves than those who value sex with love, sex in long-term, committed, monogamous relationships, and sex with consenting adults.

Scripting of Sexual Behavior **Sexual scripts** are shared interpretations and expected behaviors in sexual situations. Sexual values script the sexual behaviors for oneself and one's partner. These scripts reflect the respective values of absolutism, relativism, and hedonism.

Selecting Dating/Marriage Partners Sexual values are also an important factor in selecting dating partners and mates. The value theory of mate selection suggests that individuals with radically dissimilar sexual values are unlikely to consider each other as potential dating or marriage partners. Those who value monogamous sexual relationships tend not to select mates who approve of extradyadic sexual relationships.

Premarital sexual behavior may be punished in one society, tolerated in a second, and rewarded in a third. In general, the sexual behavior of women has been subject to greater social control than that of men. Traditionally **patriarchy** (rule by the father) helped to ensure that women were faithful to their husbands and remained in the home as economic assets, childbearers, and child-care workers.

Reducing STD/HIV Risk One's sexual values (and the behavior that reflects these values) also affect one's risk of acquiring or transmitting sexually transmissible diseases. Persons who have sexual values that dictate having one sexual partner and using a condom have a lower risk of contracting sexually transmissible

Figure 5.1
Sexual Values and Risk of Contracting an STD

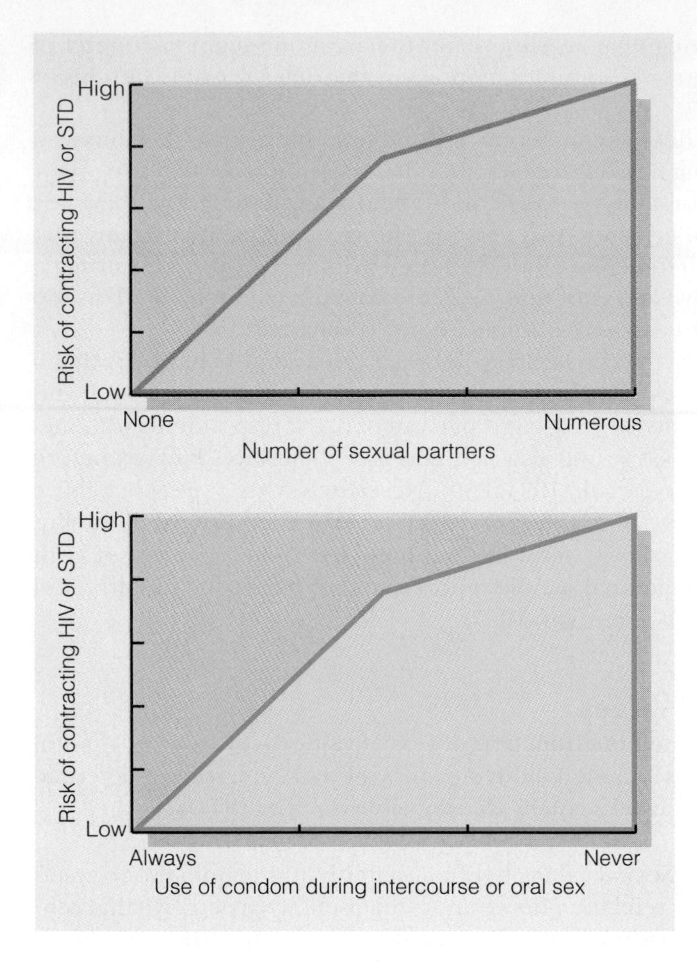

diseases than those whose sexual values permit numerous sexual partners without using a condom (see Figure 5.1). As noted earlier, the context of the relationship influences condom use. Sixty-two percent of adults reported using a condom at last intercourse outside an ongoing relationship, while only 19 percent reported using condoms when the most recent intercourse occurred within a steady relationship (Anderson et al., 2000).

ALTERNATIVE SEXUAL VALUES

There are at least three sexual value perspectives that guide choices in sexual behavior: absolutism, relativism, and hedonism. People sometimes have different sexual values at different stages of the family life cycle. For example, elderly individuals are more likely to be absolutist, whereas those in the middle years are more likely to be relativistic. Young unmarried adults are more likely than the elderly to be hedonistic.

Absolutism

Absolutism refers to a belief system based on unconditional allegiance to the authority of science, law, tradition, or religion. A religious absolutist makes sexual choices on the basis of moral considerations. To make the correct moral choice is to comply with God's will, and not to comply is a sin. A legalistic absolutist makes sexual decisions on the basis of a set of laws. People who are guided by absolutism in their sexual choices have a clear notion of what is right and wrong.

The official creeds of fundamentalist Christian and Islamic religions encourage absolutist sexual values. Intercourse is solely for procreation, and any sexual acts (masturbation, oral sex, homosexuality) that do not lead to procreation are immoral and regarded as sins against God, Allah, self, and community. Waiting until marriage to have intercourse is also an absolutist sexual value.

Diversity in Other Countries

In an effort to ascertain the importance of virginity in mate selection in China, Zambia, Sweden, and the United States, two researchers asked college students in these countries, "How desirable is chastity in potential long-term mates or marriage partners?" Respondents were asked to rate the level of desirability on a scale from 0 (irrelevant or unimportant) to 3 (indispensable). Chinese students were the most concerned (2.5 out of 3), and Swedish students were the least (less than 0.5). Male students in the United States rated their concern slightly less than 1, with women reporting a 0.5 level of concern (Buss & Schmitt, 1993).

"True Love Waits" is an international campaign designed to challenge teenagers and college students to remain sexually abstinent until marriage. Under this program, created and sponsored by the Baptist Sunday School Board, young people are asked to agree to the absolutist position and sign a commitment to the following: "Believing that true love waits, I make a commitment to God, myself, my family, my friends, my future mate, and my future children to be sexually abstinent from this day until the day I enter a biblical marriage relationship."

Being an absolutist and not planning to have intercourse as an adolescent may have implications for the well-being of young people. A team of researchers (Whitaker, Miller, & Clark, 2000) studied a cross-sectional sample of 982 adolescents aged 14–16 and found that delayers (those who had not had intercourse and who rated their chance of having intercourse in the next 12 months as less than 50%) reported greater psychological health (e.g., greater self-esteem, less hopelessness) than anticipators (those who had not had intercourse but rated their chance of intercourse in the next 12 months as greater than 50%).

VIRGIN: Teach your kid it's not a dirty word.

Billboard in Baltimore

Carpenter (2003) discussed in her paper "Like a Virgin . . . Again?" the concept of **secondary virginity**—the conscious decision of a sexually active person to refrain from intimate encounters for a specified period of time. Secondary virginity closely resembles a pattern scholars have called "regretful" nonvirginity, the chief difference being the adoption of the label "virgin" by the nonvirgin in question. Secondary virginity may be a result of physically painful, emotionally distressing, or romantically disappointing sexual encounters. Of the sixty-one young adults Carpenter interviewed, over half (women more than men) believed that a person could, under some circumstances, be a virgin more than once. Fifteen people contended that a person could resume her or his virginity in an emotional, psychological, or spiritual sense. Terence Deluca (27, heterosexual, white, Roman Catholic), explained:

> *There is a different feeling when you love somebody and when you just care about somebody. So I would have to say if you feel that way, then I guess you could be a virgin again. Christians get born all the time again, so. . . . When there's true love involved, yes, I believe that.*

Some individuals still define themselves as virgins even though they have engaged in oral sex. In a study of 353 female and 245 male college students, only 40 percent said that oral sex constituted "having sex," and only 15 percent said that manual genital stimulation met the criteria of "having sex" (Sanders & Reinisch, 1999).

A subcategory of absolutism is **asceticism.** The ascetic believes that giving in to carnal lust is unnecessary and attempts to rise above the pursuit of sensual pleasure into a life of self-discipline and self-denial. Accordingly, spiritual life is viewed as the highest good, and self-denial helps one to achieve it. Catholic priests, monks, nuns, and some other celibates have adopted the sexual value of asceticism.

Abstinence Sex Education in the Public Schools

SOCIAL POLICY

Sexuality education was introduced in the American public school system in the late nineteenth century with the goal of combating STDs and instilling sexual morality. "The strong emphasis on abstinence-only sexuality education has been the most common approach and has become formalized over the past 20 years" (Elia, 2000, p. 123). The federal government supports sex education in the public school system that is focused on abstinence education and has appropriated $50 million a year administered by the Maternal and Child Health Bureau. To be eligible for this funding, programs must focus *exclusively* on "teaching the social, psychological, and health gains to be realized by abstaining from sexual activity" (Sonfield & Gold, 2001, 170). Programs that emphasize abstinence but also discuss contraception and other means of protection are not eligible.

Sather and Zinn (2002) compared the values and attitudes of two groups of seventh- and eighth-grade adolescents toward premarital sexual activity. One group received state-funded, abstinence-only education; the other group did not receive that education. Abstinence-only education did not significantly change adolescents' values and attitudes about premarital sexual activity, or their intentions to engage in premarital sexual activity. The majority of both the treatment and control group subjects expressed disagreement with the statement: "It is okay for people my age to have sexual intercourse," and they did not intend to have sexual intercourse while they were unmarried teenagers.

In contrast to the belief that sex education should be restrictive, focused on biomedical/hygienic aspects, and should promote abstinence is the belief that it should be "broad based covering the biological, ethical, psychological, sociocultural, and spiritual dimensions of sexuality" (Elia, 2000, 123). Wiley (2002) argues that presenting abstinence-only information may harm secondary students by not providing them with potential lifesaving information. The substantial protection provided by condoms against certain sexually transmitted diseases, including human immunodeficiency virus (HIV), represents scientifically valid information that should be provided to students.

Public education must also be realistic. Young and Goldfarb (2000) note that abstinence education imposes a value or standard to which the vast majority of adults in the United States have not adhered—and simply do not support. Given that young adolescents are sexually active, it may be prudent to provide broad-based sex education classes to encourage protected sex. In addition, sex education in whatever format is most effective if it promotes sex-positive information and views rather than "fear and morality-based messages" (Bay-Cheng, 2001, 249). Balanko (2002) reemphasized the need for "sex-positive programs" including the fact that sex is pleasurable.

Sources

Balanko, S. L. 2002. Good sex? A critical review of school sex education. *Guidance & Counseling* 17:117–214.

Bay-Cheng, L. Y. 2001. SexEd.com: Values and norms in web-based sexuality education. *Journal of Sex Research* 38:241–251.

Elia, J. P. 2000. Democratic sexuality education: A departure from sexual ideologies and traditional schooling. *Journal of Sex Education and Therapy* 25:122–29.

Sather, L., and K. Zinn. 2002. Effects of abstinence-only education on adolescent attitudes and values concerning premarital sexual intercourse. *Family and Community Health* 25:15–20.

Sonfield, A., and R. B. Gold. 2001. States' implementation of the section 510 abstinence education program, FY 1999. *Family Planning Perspectives* 33:166–71.

Wiley, D. C. 2002. The ethics of abstinence-only and abstinence-plus sexuality education. *Journal of School Health* 72:164–68.

Young, M., and E. S. Goldfarb. 2000. The problematic (a)–(h) in abstinence education. *Journal of Sex Education and Therapy* 25:156–61.

Relativism

In contrast to absolutism is **relativism**—a value system emphasizing that decisions should be made in the context of a particular situation (hence, values are relative). Most college students feel that the context of mutual agreement, affection or love, comfort, security, and care provide a justifiable context for sexual behavior. "Permissiveness with affection" is the term used by Reiss (1967) over thirty years ago to describe this sexual value.

Sex cannot be contained within a definition of physical pleasure; it cannot be understood as merely itself for it has stood for too long as a symbol of profound connection between human beings.

Elizabeth Janeway, ***Between Myth and Morning***

NATIONAL DATA

Fifty-five percent of undergraduate men versus 32 percent of undergraduate women in over 460 colleges and universities agreed that "if two people really like each other, it's all right for them to have sex if they've known each other for only a very short time" (American Council on Education & University of California, 2002).

A disadvantage of relativism as a sexual value is the difficulty of making sexual decisions on a relativistic case-by-case basis. The statement "I don't know

Up Close 5.1 Friends with Benefits

A new trend is emerging in relational/sexual behavior. "Friends with benefits" is a relationship consisting of nonromantic friends who also have a sexual relationship. Previous research on "friends with benefits" (FWB) relationships has focused on the transition from romantic to platonic relationships (Schneider & Kenny, 2000), the perceived costs/benefits of opposite-sex friendships (Bleske & Buss, 2000), and the impact of sexual activity in cross-sex friendships. This study assessed the prevalence, attitudes, and sex differences of college student involvement in an FWB relationship.

Sample

The sample consisted of 170 undergraduates at a large southeastern university who responded to an anonymous twenty-three-item questionnaire designed to assess prevalence, attitudes, and sex differences of involvement in an FWB relationship. Seventy-five percent of the respondents were female; 25 percent were male. The median age of the respondents was 20, with most (86.5%) reporting that they were white (13.5% nonwhite).

Findings and Discussion

Almost 60 percent of these undergraduates (57.3%) reported that they were or had been involved in a friends with benefits relationship. There were no significant differences between the percentages of women and men reporting involvement in an FWB relationship. This is one of the few studies finding no difference in sexual behavior between women and men (e.g., one would expect men to have more friends with benefits relationships than women). However, the percentages of women and men in our sample were very similar in their reported rates of FWB involvement—57.1% and 57.9%. Is a new sexual equality in FWB operative?

Continued analysis of the data did reveal some other significant differences between women and men college students in regard to various aspects of the FWB relationship.

1. Women are more emotionally involved. Women were significantly ($p < .006$) more likely than men (62.5% vs. 38.1%) to view their current FWB relationship as an emotional relationship. In addition, women were significantly ($p < .021$) more likely than men to be perceived as being more emotionally involved in the FWB relationship. Over 40 percent (43.5%) of the men compared with 13.6 percent of the women reported that "my partner is more emotionally involved than I am." Previous research has confirmed that women evidence more emotional than sexual interest in relationships (Knox, Cooper & Zusman, 2001).

2. Men are more sexually focused. As might be expected from the above finding, men were significantly ($p < .05$) more likely than women to agree with the statement, "I wish we had sex more often than we do" (43.5% versus 13.6%). Previous research has reported that men are more hedonistic than women and more focused on sexual pleasure (Michael et al., 1994; Knox, Cooper, & Zusman, 2001).

3. Men are more polyamorous. With *polyamorous* defined as desiring to be involved in more than one emotional/sexual relationship at the same time, men were significantly ($p < .000$) more likely than women to agree that "I would like to have more than one FWB relationship going on at the same time" (34.8% versus 4.5%). Serial FWB relationships may already be occurring. Over half of the men (52.2%) compared with almost a quarter of the women (24.6%) reported that they had been involved in more than one FWB relationship. This finding is not surprising, since previous research has revealed men's penchant for multiple sexual encounters (Wiederman, 1977).

Implications

University students who consider involvement in a FWB relationship might be aware that women and men bring different perceptions, expectations, and definitions to the relationship table. These data suggest that women tend to view such a relationship as emotional with the emphasis on friends while men tend to view the relationship as more casual with an emphasis on benefits (as in sexual benefits). Indeed, when the women and men who were involved in a friends with benefits relationship were asked if they were more friends than lovers, almost 85 percent (84.4%) of the women compared with almost 15 percent (14.8%) of the men reported that they were more friends than lovers.

From Schacht C., D. Knox, M. E. Zusman, and K. McGinty. 2004. Friends with benefits: Some college student data. Presentation at the National Council on Family Relations, November 19, Orlando, Florida.

Sources

Bleske, A. L., and D. M. Buss. 2000. Can men and women be just friends? *Personal Relationships* 7:131–51.

Knox, D., C. Cooper, and M. E. Zusman. 2001. Sexual values of college students. *College Student Journal* 35:24–27.

Schneider, C. S., and D. A. Kenny. 2000. Cross-sex friends who were once romantic partners: Are they platonic friends now? *Journal of Social and Personal Relationships* 17:451–66.

Wiederman, M. W. 1977. Extramarital sex: Prevalence and correlates in a national survey. *Journal of Sex Research* 34: 167–74.

what's right anymore" reflects the uncertainty of a relativistic view. Once a person decides that mutual love is the context justifying intercourse, how often and how soon is it appropriate for the person to fall in love? Can love develop after two hours of conversation? How does one know that love feelings are genuine? The freedom that relativism brings to sexual decision making requires responsibility, maturity, and judgment. In some cases, individuals may convince themselves that they are in love so that they will not feel guilty about having intercourse. Though one may feel "in love," "secure," and "committed" as a prerequisite for a first intercourse experience, of all first intercourse experiences reported by women, only 17 percent of these are with the person they eventually marry (Raley, 2000).

Whether or not two unmarried people should have intercourse would be viewed differently by absolutists and relativists. Whereas an absolutist would say that it is wrong for unmarried people to have intercourse and right for married people to do so, a relativist would say, "It depends on the situation." Suppose, for example, that a married couple do not love each other and intercourse is an abusive, exploitative act. Suppose also that an unmarried couple love each other and their intercourse experience is an expression of mutual affection and respect. A relativist might conclude that in this particular situation, it is "more right" for the unmarried couple than the married couple to have intercourse.

A specific expression of relativism is reflected in students who become involved in a "friends with benefits" relationship (See Up Close, p. 105).

Hedonism

Hedonism is the belief that the ultimate value and motivation for human actions lie in the pursuit of pleasure and the avoidance of pain. The hedonistic value is reflected in the statement "If it feels good, do it." Hedonism assumes that sexual desire, like hunger and thirst, is an appropriate appetite and its expression is legitimate. "Permissiveness without affection" is the term used by Reiss (1967) over thirty years ago to describe a similar sexual value. Data were collected on the sexual values of 618 undergraduates (Knox, Zusman, & Cooper, 2001) showing that close to 20 percent of college males but less than 4 percent of college females regard their own sexual values as hedonistic. Traditionally, women have been socialized to be more concerned than men about the context (love, commitment, security) as a condition for sexual expression. By contrast, men have traditionally been more sexually aggressive and pleasure-focused independent of the relationship context.

When I'm good, I'm very good, but when I'm bad I'm better.
Mae West, *Actress*

In addition to sex of respondent, other dependent variables found to be significantly related to one's sexual values include age (younger, more absolutist), nature of the relationship (more involved, more relativistic), religion (more religious, more absolutist), and race. East (1998) found that young (sixth- through eighth-graders) Southeast Asian females were the most likely to expect that they would delay sexual relations, and young Black females were the least likely to predict that they would delay sexual relations. Realo and Goodwin (2003) also found that familism, characteristic of eastern European families, is more likely to be associated with conservative sexual values.

Another factor related to sexual values is the willingness to have intercourse during menstruation. Rempel and Baumgartner (2003) found that women who were comfortable having intercourse during menstruation were also more likely to report

> a wider range of sexually stimulating acts, particularly acts that many might regard as shocking or immoral (e.g., viewing or performing in live sex shows, group sex, and spanking or punishment). Thus it seems that women who have engaged in menstrual sex not only experience greater sexual desire but they are also uniquely aroused by sexual acts that push the boundaries of social convention. By implication, sexual activity during menses may carry with it an element of being unconventional or avant-garde (p. 162).

The nature of an individual's relationship involvement is also related to his or her sexual values (see Table 5.1). Respondents who reported that they were

Table 5.1 Sexual Value by Level of Involvement

	Absolutism	Relativism	Hedonism
Involved	8.8% (26)	85.5% (272)	5.7% (18)
Casual	12.4% (37)	72.2% (223)	12.4% (37)

Source: Knox, D., C. Cooper, & M. E. Zusman. 2001. Sexual values of college students. *College Student Journal* 35:26. Used by permission.

not dating or were casually dating different people were categorized as being in "casual" relationships, in contrast to those who were emotionally involved, living together, or engaged, who were categorized as being in "involved" relationships.

The emotionally involved were significantly more likely to identify themselves as believing in relativism than those who were involved in casual relationships. Similarly, casual daters were more than twice as likely as the involved to report being hedonistic in their sexual values (12.4% versus 5.7%). Previous researchers have found one's relationship context tends to define one's sexual values. Michael et al. (1994) found that noncohabiting individuals who reported having sex with two or more partners in the past twelve months were twice as likely to have a hedonistic (recreational) view of sex as individuals who reported having only one partner in the last year.

Hooking Up

The sexual values of relativism ("we are hitting it off . . . this feels right") and hedonism ("this is going to be some good sex . . . I'll never see the person again") provide the context in which "hooking up" may occur. The Institute for American Values (a conservative think tank) collected data from one thousand undergraduates, 40 percent of whom reported that they had "hooked up" at least once (Glenn & Marquardt, 2001). Similarly, Bogle (2003) interviewed fifty-seven college students and alumni at two universities in the eastern United States about their experiences in dating and sex. She found that "**hooking up**" (defined as meeting someone and becoming sexually involved that same evening with no commitment or expectation beyond the encounter) had become the primary means for heterosexuals to get together on campus. Bogle's respondents perceived that "others" were hooking up more often ("everybody's doing it") and were going further sexually ("intercourse") than were the respondents. Researchers (Bogle, 2003; Glenn & Marquardt, 2001) note that while hooking up may be an exciting sexual adventure, it is fraught with feelings of regret. Some of the women in their studies were particularly disheartened to discover that hooking up usually did not result in the development of a relationship that went beyond a one-night encounter. However, the hope that something will develop is what keeps the "hooking up" system in tact.

There is no sense of true intimacy. We are back to "hooking up," back to the casual sex of the 60s.

Howard Markman, *Center for Marital and Family Studies*

Bogle (2002) noted three outcomes of hooking up. In the first, described above, nothing results from the first-night sexual encounter. In the second, the college students will repeatedly hook up with each other on subsequent "hanging out" occasions. However, this type of relationship is characterized by a low level of commitment in that each is still open to hooking up with someone else. A third outcome of hooking up— the least likely is that the relationship becomes one of going out or being together or with the other person in an exclusive relationship. The hope that a hookup is going to lead to something more (i.e., some version of a relationship) is what keeps the hooking-up system intact.

Diversity in Other Countries

The double standard in other countries may be particularly pronounced. In Turkey women are "significantly disrespected and penalized for having sex before marriage" (Sakalh-Ugurlu & Glick, 2003).

SEXUAL DOUBLE STANDARD

The **sexual double standard**—the view that encourages and accepts sexual expression of men more than women—is reflected in Table 5.2. Indeed, men were almost six times more hedonistic than women.

Kimmel (2000) emphasized that the double standard continues to be reflected in today's sexual norms:

Men are not going to move women who have sex with them on the first date from the "good for now woman" status to the "wife potential" status.

Bradley Gerstman, Christopher Pizzo, Rich Seldes, *What Men Want*

The double standard persists today—perhaps less in what we actually do, and more in the way we think about it. Men still stand to gain status and women to lose it from sexual experience: he's a stud who scores; she's a slut who "gives it up." Boys are taught to try to get sex; girls are taught strategies to foil the boys' attempts. "The whole game was to get a girl to give out," one man told sociologist Lillian Rubin. "You expected her to resist; she had to if she wasn't going to ruin her reputation. But you kept pushing" (p. 222).

Table 5.2 Sexual Value by Sex of Respondent

	Hedonism	Absolutism	Relativism
Male Students	18.7% (43)	8.3% (19)	73% (168)
Female Students	3.6% (14)	11.9% (46)	84.5% (328)

Source: Knox, D., C. Cooper, & M. E. Zusman. 2001. Sexual values of college students. *College Student Journal* 35:24. Used by permission.

SOURCES OF SEXUAL VALUES

The sources of one's sexual values are numerous and include one's school, family, religion, and peers, as well as technology, television, social movements, and the Internet.

We have noted that public schools in the United States promote absolutist sexual values through abstinence education. Specifically, under Title V, Section 510, of the Social Security Act, the federal government funds states that offer abstinence promotion programs meeting the following eight-point definition of abstinence education (Sonfield & Gold, 2001). Sex education in the public school system

1. *has as its exclusive purpose, teaching the social, psychological, and health gains to be realized by abstaining from sexual activity;*
2. *teaches abstinence from sexual activity outside marriage as the expected standard for all school-age children;*
3. *teaches that abstinence from sexual activity is the only certain way to avoid out-of-wedlock pregnancy, STDs, and other associated health problems;*
4. *teaches that a mutually faithful monogamous relationship in the context of marriage is the expected standard of human sexual activity;*
5. *teaches that sexual activity outside the context of marriage is likely to have harmful psychological and physical effects;*
6. *teaches that bearing children out of wedlock is likely to have harmful consequences for the child, the child's parents, and society;*
7. *teaches young people how to reject sexual advances and how alcohol and drug use increases vulnerability to sexual advances;*
8. *teaches the importance of attaining self-sufficiency before engaging in sexual activity. (P. 170)*

We noted in the Social Policy section (p. 104) that the effectiveness of this program has been questioned.

In regard to family influences, sexual attitudes of parents may provide a model for sexual values of their children. Talking with one's children about sex is not easy. A team of researchers (Jaccard, Dittus, & Gordon, 2000) reported that mothers are reluctant to discuss sex and birth control with their teenagers out of fear that they will embarrass the teens or that their children might ask them something they do not know. Teenagers avoid involvement in such conversations not only because of being embarrassed but also out of fear that their mother will ask too many personal questions. However, Miller (2002) noted that parent-child closeness or connectedness and parental supervision or regulation of children, in combination with parents' values against teen intercourse (or unprotected intercourse), decrease the risk of adolescent pregnancy.

Authors

This father is fostering specific values in his son (a high school senior) by having him consider such colleges as Oral Roberts University, which encourages and supports absolutist sexual values. His son did attend and graduated from ORU.

NATIONAL DATA

Seven percent of 1,011 adult Americans noted that their parents had been their most important source of sexual information (Stodgill, 1998).

Family influences on sexual values include sibling influences. Kornreich et al. (2003) found that girls who had older brothers held more conservative sexual values. "Those with older brothers in the home may be socialized more strongly to adhere to these traditional standards in line with power dynamics believed to shape and reinforce more submissive gender roles for girls and women" (p. 197).

Religion also is an important influence. Over half (52.3%) of the respondents in the nationwide study agreed with the statement "My religious beliefs have guided my sexual behavior" (Michael et al., 1994, 234). Weinberg, Lottes, & Shaver (2000) compared undergraduates at an American and a Swedish university and found that the religiosity among Americans was higher and contributed to their more restrictive views toward sexuality. Buddhism, Hinduism, and Islam all encourage waiting until marriage for sexual intercourse.

Peers are an important source of sex education and are important influences on one's sexual values (O'Sullivan & Meyer-Bahlburg, 2003). Somers and Gleason (2001) evaluated the comparative contribution that multiple sources of education about sexual topics (family, peers, media, school, and professionals) made on teen sexual knowledge, attitudes, and behavior and found that, in general, teens tended to get less of their sex education from schools and more of their sex education from nonsibling family.

Reproductive technologies such as birth-control pills, the morning-after pill, and condoms influence sexual values by affecting the consequences of behavior. Being able to reduce the risk of pregnancy and HIV infection with the pill and condoms allows one to consider a different value system than if these methods of protection did not exist.

Sexuality is a major theme of television in the United States. However, as a source of sexual values and responsible treatments on contraception, condom us age, abstinence, or consequences of sexual behavior, it is woefully inadequate. Indeed, the television viewer learns that sex is romantic and exciting but learns nothing about discussing the need for contraception or HIV and STD protection. With few exceptions (*Sex in the City,* a Home Box Office sitcom, has featured condom usage in a sexual encounter), viewers are inundated with role models who engage in casual sex without protection.

NATIONAL DATA

In 248 television episodes involving 3,228 scenes showing sexual activity in 1989–99, including 135 scenes of simulated on-screen sexual intercourse, only eight pregnancies and no case of sexually transmitted diseases resulted (Lichter, Lichter, & Amundson, 2001).

Social movements such as the women's movement affect sexual values by empowering women with an egalitarian view of sexuality. This translates into encouraging women to be more assertive about their own sexual needs and giving them the option to experience sex in a variety of contexts (e.g., without love or commitment) without self-deprecation. The net effect is an increase in the frequency of recreational, hedonistic sex. The gay liberation movement has also been influential in encouraging values that are accepting of sexual diversity.

Another influence on sexual values is the Internet. Sexually explicit dialogue and photos are now available on one's home computer. In a study of 191 undergraduates, over three-quarters (77.4 percent) (more men than women) reported that they had visited a sexually explicit site (Knox, Daniels, Sturdivant, & Zusman, 2001). A unanimous Supreme Court has taken the position that a broad restriction on posting "indecent" material is inconsistent with free speech.

Authors

University sexuality classes may provide explicit information on masturbation and sex toys.

SEXUAL BEHAVIORS

We have been discussing the various sources of sexual values. We now focus on what people report that they do sexually in terms of masturbation, oral sex, vaginal intercourse, and anal sex. Our discussion of sexual behavior ends with an examination of gender differences in sexual behavior.

Masturbation

Masturbation involves stimulating one's own body with the goal of experiencing pleasurable sexual sensations. Alternative terms for masturbation include *autoeroticism, self-pleasuring, solo sex,* and *sex without a partner.* Several older, more pejorative terms for masturbation are *self-pollution, self-abuse, solitary vice, sin against nature, voluntary pollution,* and *onanism.* The negative connotations associated with these terms are a result of various myths about masturbation (e.g., that it causes insanity, blindness, and hair growth on the palms of the hands). Replacing these myths are new attitudes toward the practice. Although some religious leaders still view it in negative terms, most health care providers and therapists today regard masturbation as normal and healthy sexual behavior. Furthermore, masturbation has become known as a form of safe sex in that it involves no risk of transmitting diseases (such as HIV) or producing unintended pregnancy. Thomsen and Chang (2000) surveyed the first intercourse experiences of 292 university undergraduates and found that prior experience with masturbation was the strongest single predictor of orgasm and emotional satisfaction with first intercourse.

Masturbation—it's sex with someone I love.
Woody Allen

Oral Sex

Fellatio is oral stimulation of the man's genitals by his partner. In many states, legal statutes regard fellatio as a "crime against nature." "Nature" in this case refers to reproduction, and the "crime" is sex that does not produce babies. Nevertheless, most men have experienced fellatio.

With regard to the "last sexual event," 28 percent of men reported that the last time they had sex they experienced fellatio (Michael et al., 1994). A team of researchers noted that the presence of the AIDS virus in the semen of infected men is a well-established medical fact (Eyre, Zheng, & Kiessling, 2000). Use of a condom to prevent semen in the mouth is recommended.

Cunnilingus is oral stimulation of the woman's genitals by her partner.

With regard to the last sexual event, 20 percent of women reported that the last time they had sex their partner performed cunnilingus on them (Michael et al., 1994). Women who were neither married nor living together reported the highest frequency of cunnilingus as their last sexual event (26%, noncohabiting; 22%, cohabiting; 17%, married).

African-Americans typically have lower rates of oral sex than whites. Gagnon (2004) noted:

> *I think that these differences are mostly the result of education and religion—but there may be other factors relating to the symbolic meanings of oral sex in Western cultures. As a result of the economic and racial oppression of African-Americans,*

potential threats to men's power and masculinity are often more present when African-American men have sex. There is always the threat of symbolic subordination when men perform oral sex, particularly if it is not identified by the woman as masculinity-enhancing and it is not reciprocated. This is a parallel to the symbolic subordination of women when they perform unreciprocated oral sex. However, if the men will not do it, then the women will not either, since they view fellatio without reciprocation as simply servicing the man.

Increasingly, youth who have oral sex regard themselves as virgins, believing that only sexual intercourse constitutes "having sex." One teen reported, "The consensus in my high school is that oral sex makes girls popular, whereas intercourse would make them outcasts" (Peterson, 2000, D1).

Vaginal Intercourse

Vaginal intercourse, or coitus, refers to the sexual union of a man and woman by insertion of the penis into the vagina. Kaestle, Morisky, & Wiley (2002) noted that having an older partner as a female is associated with intercourse. Data from 1,975 females revealed that if a 17-year-old female was romantically involved with a partner six years older, the female was twice as likely to report having had intercourse as was a female with a partner the same age.

PERSONAL CHOICES

Should You Have Intercourse with a New Partner?

Sexual values are operative in choosing whether to engage in sexual intercourse with a partner in a new relationship. The following might be considered in making such a decision.

1. Personal consequences. How do you predict you will feel about yourself after you have had intercourse? An increasing percentage of college students are relativists and feel that if they are in love and have considered their decision carefully, the outcome will be positive. (The quotes in this section are from students in the authors' classes.)

> I believe intercourse before marriage is OK under certain circumstances. I believe that when a person falls in love with another, it is then appropriate. This should be thought about very carefully for a long time, so as not to regret engaging in intercourse.

Those who are not in love and have sex in a casual context sometimes feel bad about their decision:

> I viewed sex as a new toy—something to try as frequently as possible. I did my share of sleeping around, and all it did for me was to give me a total loss of self-respect and a bad reputation. Besides, guys talk. I have heard rumors that I sleep with guys I have never slept with.

The effect intercourse will have on you personally will be influenced by your personal values, your religious values, and the emotional involvement with your partner. Some people prefer to wait until they are married to have intercourse and feel that this is the best course for future marital stability and happiness. Abstinence is an appropriate alternative choice. There is often, but not necessarily, a religious basis for this value. Strong personal and religious values against nonmarital intercourse plus a lack of emotional involvement usually result in guilt and regret following an intercourse experience. In contrast, values that regard intercourse as appropriate within the context of a love relationship are likely to result in feelings of personal satisfaction and contentment after intercourse.

2. Partner consequences. Because a basic moral principle is to do no harm to others, it may be important to consider the effect of intercourse on your partner. Whereas intercourse may be a pleasurable experience with positive consequences for you, your partner may react differently. What are your partner's feelings about nonmarital intercourse and her or his ability to handle the experience? If you suspect that your partner will regret having intercourse, you might reconsider whether intercourse would be appropriate with this person.

3. Relationship consequences. What is the effect of intercourse on a couple's relationship? A team of researchers (Rostosky et al., 1999) studied the effect of various levels of sexual intimacy (from kissing to intercourse) on relationship commitment and found that "the milder, but more affectionate, behaviors of holding hands and kissing seemed to be more important indicators of couples' commitment than the more 'intense' behaviors of fondling and sexual intercourse" (p. 331). However, if this was the first partner with whom one had sexual intercourse, the commitment to the partner was much greater than if the person had had intercourse with previous partners (p. 331).

4. Contraception. Another potential consequence of intercourse is pregnancy. Once a couple decide to have intercourse, a separate decision must be made as to whether intercourse should result in pregnancy. Some couples want children and make a mutual commitment to love and care for their offspring. Other couples do not want children. If the couple want to avoid pregnancy, they must choose and effectively use a contraceptive method (discussed in detail in Chapter 10). But many do not. In a study of almost 73,000 pregnancies, 45 percent were unintended (Naimi et al. 2003).

Some students have a hard time understanding that the consequence of one unprotected sexual encounter may not be reversible.

American College Health Association

Thomsen and Chang (2000) noted in their study of first intercourse experiences that not using contraception was associated with labeling the experience as negative. Not only might a pregnancy result, but the worry over such a pregnancy contributed to the evaluation. One woman recalled: "It was the first time I had intercourse, so I didn't really think I would get pregnant my first time. But I did. And when I told him I was pregnant, he told me he didn't have any money and couldn't help me pay for the abortion. He really wanted nothing to do with me after that" (authors' files).

5. HIV and other sexually transmissible infections. Engaging in casual sex has potentially fatal consequences. Avoiding HIV infection and other sexually transmissible infections (STDs) is an important consideration in deciding whether to have intercourse in a new relationship. The increase in the number of people having more partners results in the rapid spread of the bacteria and viruses responsible for numerous varieties of STDs.

Serovich and Mosack (2003) reported that only 37 percent of men with HIV told a casual partner of their positive HIV status. The reason they did so was that of personal responsibility: "I thought he had a right to know." While depending on a person's integrity is laudable, it may not be wise. Indeed, a condom should be used that is 87 percent effective in providing protection from HIV (Davis & Weller, 2000).

About a half of male college students and approximately 40 percent of female college students report consistent condom use (Eisenberg, 2002). Civic (2000) found that the primary reasons for not using a condom were knowledge that the partner was on contraception and the feeling that "my partner does not have an STD" (Civic, 2000).

6. Influence of Alcohol and Other Drugs. A final consideration with regard to the decision to have intercourse in a new relationship is to be aware of the influence of alcohol and other drugs on such a decision. Naimi et al. (2003) noted that unintended pregnancy was much more likely among women who consumed alcohol and women who binge-drank. Coren (2003) also found that marijuana use was associated with more sexual partners and lower condom use.

The mere drinking of alcohol by a woman may be perceived as indicating her availability for sex. This is the conclusion of Mulligan Rauch and Bryant (2000), who studied men, women, and alcohol on a first date. And it is not just any alcohol but a particular type of alcohol. Osman and Davis (2000) noted that though a woman's drinking alcohol per se is not indicative of her perceived availability for sex, her drinking a type of alcohol that is out of the ordinary (e.g., switching from wine to liquor) or reaching a particular volume (e.g., four tequila shots) might make her vulnerable to such a perception.

Sources

Civic, D. 2000. College students' reasons for nonuse of condoms within dating relationships. *Journal of Sex and Marital Therapy* 26:95–105.

Coren, C. 2003. Timing, amount of teenage alcohol or marijuana use may make future risky sex more likely. *Perspectives on Sexual and Reproductive Health* 25:49–51.

Davis, K. R., and S. C. Weller. 2000. The effectiveness of condoms in reducing transmission of HIV. *Family Planning Perspectives* 31:272–79.

Eisenberg, M. E. 2002. The association of campus resources for gay, lesbian, and bisexual students with college students' condom use. *Journal of American College Life* 51:109–16.

Mulligan Rauch, S. A., and B. Bryant. 2000. Gender and context differences in alcohol expectancies. *Journal of Social Psychology* 140:240–53.

Naimi, T. S., L. E. Lipscomb, R. D. Brewer, and B. C. Gilbert. 2003. Binge drinking in the preconception period and the risk of unintended pregnancy: Implications for women and their children. *Pediatrics* 111:1136–41.

Osman, S. L., and C. M. Davis. 2000. Predicting perceptions of date rape based on individual beliefs and female alcohol consumption. *Journal of College Student Development* 40:701–9.

Rostosky, S. S., D. Welsh, M. C. Kawaguchi, and R. V. Galliher. 1999. Commitment and sexual behaviors in adolescent dating relationships. In *Handbook of interpersonal commitment and relationship stability,* edited by J. M. Adams and W. H. Jones. New York: Academic/Plenum Publishers, 323–38.

Serovich, J. M., and K. E. Mosack. 2003. Reasons for HIV disclosure or nondisclosure to casual sexual partners. *AIDS Education and Prevention* 15:70–81.

Thomsen, D., and I. J. Chang. 2000. Predictors of satisfaction with first intercourse: A new perspective for sexuality education. Poster presentation at the 62nd Annual Conference of the National Council on Family Relations, Minneapolis, November.

First Intercourse

Thomsen and Chang (2000) surveyed 292 undergraduates (60 percent women) in regard to their first intercourse experience. Those who reported the highest level of satisfaction expected the event to occur (they were not drunk and just "let it happen"), were in love with the partner, discussed the use of condoms and birth control ahead of time, actually used birth control and/or condoms, and experienced an orgasm. Indeed, the two most frequent comments made by the respondents about their first intercourse experience were that they wished they had been older before their first experience and wished they had been in a committed relationship with their partner. This study suggests that negative first intercourse experiences are those that occur at a young age in an uncommitted relationship without protection and with no previous plan or discussion.

Anal Sex

Whereas vaginal intercourse was reported as "very appealing" by 78 percent of women aged 18–44 and by 83 percent of men aged 18–44, only 1 percent of the women and 5 percent of the men reported that anal sex was very appealing (Michael et al., 1994, 146, 147). Frequency of anal sex may also vary by race and national origin. Latinos in Mediterranean and Latin American cultures have higher rates than persons in either northern Europe or the United States (Gagnon, 2004). Anal (not vaginal) intercourse is the sexual behavior associated with the highest risk of HIV infection. The potential to tear the rectum so that blood contact becomes possible presents the greatest danger. AIDS is lethal. Partners who use a condom during anal intercourse reduce their risk of infection not only from HIV but also from other STDs.

Gender Differences in Sexual Behavior

Do differences exist in the reported sexual behaviors of women and men? Yes. In national data based on interviews with 3,432 adults, women reported thinking about sex less often than men (19% vs. 54% reported thinking about sex several times a day), reported having fewer sexual partners than men (2% vs. 5% reported having had five or more sexual partners in the last year), and reported having orgasm during intercourse less often (29% vs. 75%) (Michael et al., 1994, 102, 128, 156).

Sociobiologists explain males' higher rates of thinking about and engaging in sex with multiple partners as biologically based (higher testosterone levels). Social learning theorists, on the other hand, emphasize that men are socialized by the media and peers to think about and to seek sexual experiences. Men are also accorded social approval and called "studs" for their sexual exploits. Women, on the other hand, are more often punished and labeled "sluts" if they have many sexual partners. Because sexual behavior is socially scripted, what individuals think, do, and experience is a reflection of what they have learned (Simon & Gagnon, 1998). Hynie et al. (1997), for example, found that women perceived other women less favorably if they had sex in an uncommitted relationship than if they had sex in a committed relationship. Such a perception is a learned social script.

Men and women differ in their motivations for sexual intercourse. Men report the desire for sexual pleasure, conquest, and relief of sexual tension more often than women, who emphasize emotional closeness and affection (Michael et al., 1994). Men are also more likely than women to approve of sex earlier in a relationship.

Men and women also differ in the way they read cues to assess each other's interest in physical intimacy (Reiss & Reiss, 1990). For example, a man might assess his date's interest in physical intimacy by looking to see whether she had another drink, agreed to go to his apartment, laughed at a dirty joke, and permitted him to touch her breast. A woman might assess her date's interest in physical intimacy according to whether he smiled at her in a friendly way, laughed at her jokes, listened to her opinions, and kissed her tenderly. Both may guess the intentions of the other without engaging in direct, clear communication about physical intimacy.

SEXUALITY IN RELATIONSHIPS

Sexuality occurs in a social context that influences its frequency and perceived quality.

Sexual Relationships among the Never-Married

The never-married and not living together report more sexual partners than those who are married or living together. In one study, 9 percent of the never-married and not living together reported having had five or more sexual partners in the previous twelve months; 1 percent of the marrieds and 5 percent of cohabitants reported the same (Michael et al., 1994). However, the unmarried, when compared with marrieds and cohabitants, report having the lowest level of sexual satisfaction. One-third of a national sample of persons who were not married and not living with anyone reported that they were emotionally satisfied with their sexual relationships. In contrast, 85 percent of the married and pair-bonded reported emotional satisfaction. Up Close 5.2 details how unmarried college students view sexual intimacy (Michael et al., 1994).

Up Close 5.2 College Student Attitudes toward Sexual Intimacy

"Look inside a college student's head and you'll find six parts sex, three parts alchohol/drugs, and one part academics," lamented a university professor. This pessimistic and disgruntled estimate reflects the belief that college students are preoccupied with sex. Though genital sex is often presumed to be the referent of the term *sex,* this study focused on sexual intimacy—the conditions under which individuals are willing to have intercourse with someone and their definition of the term *sexual intimacy.*

Sample

Data for the present study consisted of ninety-nine never-married undergraduates at a large southeastern university who voluntarily completed an anonymous twenty-item questionnaire designed to assess the respondents' conditions for having intercourse and their definitions of sexual intimacy. Among the respondents, 79 percent were women and 21 percent were men. The median age was 19.

Findings and Discussion

Analysis of the data revealed some significant differences.

1. Sex differences in attitudes toward sexual intimacy. Men were significantly ($p < .01$) more likely than women to report that they were willing to have intercourse with someone they had known for three hours, to have intercourse with two different people within a six-hour period, to have intercourse with someone they did not love, and to have intercourse with someone with whom they did not have a good relationship. Previous researchers using national samples have documented that men are more hedonistic than women in their willingness to have intercourse devoid of relationship considerations (Michael et al., 1994). Researchers using smaller, nonrepresentative samples have reached similar conclusions. Townsend and Levy (1990) studied 382 respondents and found that men decide on the basis of physical attractiveness alone whether they want to have intercourse with a particular person. In contrast, women consider a number of factors in making a decision to have intercourse, including affection ("Does he love me?"), commitment ("Is he interested in a continued relationship with me?"), and socioeconomic resources ("Does he have money?").

Sociobiologists have emphasized that women decide to have intercourse with a man on the basis of his potential for investing in her subsequent offspring and that they are more likely to have sex with men who demonstrate such investment (Ellis & Symons, 1990). Pepper and Weiss (1987) and Motley and Reeder (1995) demonstrated that women limit and slow the movement toward increasing levels of sexual intimacy in a relationship so as to give them time to assess the man's parental investment potential. Traditionally, women have been socialized to be more concerned than men about the relationship context (commitment and security) as a condition for sexual expression.

2. Relationship differences in attitudes toward sexual intimacy. Respondents who were dating different people were significantly ($p < .04$) more likely than persons involved in a relationship to be willing to have intercourse without being in love and to have intercourse with someone with whom they did not have a good relationship. Previous research supports the notion that involvement in a relationship is associated with movement away from hedonistic sexual values (Knox, Zusman, & Cooper, 2001). Michael et al. (1994) also found that individuals who reported having sex with two or more partners in the previous twelve months were twice as likely to have a hedonistic (recreational) view of sex as individuals who reported having only one partner in the previous year.

In addition to observing that significant gender and relationship differences are associated with the conditions under which our respondents expressed a willingness to have intercourse, we conducted a content analysis of the respective definitions of sexual intimacy provided by the women and men and developed a summary profile of these definitions. According to the male students surveyed, sexual intimacy was "sexual activity of any kind—particularly sexual intercourse and oral sex."

In contrast, according to the female students surveyed, sexual intimacy was "any kind of sexual activity with one person with whom there is mutual emotional involvement." The definition provided by women included more than sexual intercourse; it included touching, foreplay, and being close to that special person in a sexual way, particularly through intercourse and oral sex.

Summary

This study confirmed previous research findings that men and persons not involved in a relationship, in contrast to women and persons involved in a relationship, are more likely to approve of intercourse devoid of relationship considerations. Similarly, men tend to define sexual intimacy as sexual activity of any kind, while women regard it in terms of its emotional commitment and relationship connotations. Sociobiological differences between men and women involving concern for parental investment were suggested as the theoretical explanation for these differing attitudes.

Abridged and adapted from "College Student Attitudes toward Sexual Intimacy" by D. Knox, L. Sturdivant and M. E. Zusman. 2001. *College Student Journal* 35:241–43. Used by permission of *College Student Journal.*

Sources

Ellis, B. J., and D. Symons. 1990. Sex differences in sexual fantasy: An evolutionary psychological approach. *Journal of Sex Research* 27:527–56.

Knox, D., M. E. Zusman, and C. Cooper. 2001. Sexual values of college students. *College Student Journal* 35:24–27.

Michael, R. T., J. H. Gagnon, E. O. Laumann, and G. Kolata. 1994. *Sex in America.* Boston: Little, Brown.

Motley, M. T., and H. M. Reeder. 1995. Unwanted escalation of sexual intimacy: Male and female perceptions of connotations and relational consequences of resistance messages. *Communication Monographs* 62:355–82.

Pepper, T., and D. L. Weiss. 1987. Proceptive and rejective strategies of U.S. and Canadian college women. *Journal of Sex Research* 23:455–80.

Townsend, J. M., and G. D. Levy. 1990. Effects of potential partners' physical attractiveness and socioeconomic status on sexuality and partner selection. *Archives of Sexual Behavior* 19:149–64.

Sexual Relationships among the Married

Marital sex is distinctive for its social legitimacy, declining frequency, and satisfaction (both physical and emotional).

1. Social legitimacy. In our society, marital intercourse is the most legitimate form of sexual behavior. Homosexual, premarital, and extramarital intercourse do not enjoy as high a level of social approval as does marital sex. It is not only okay to have intercourse when married, it is expected. People assume that married couples make love and that something is wrong if they do not.

NATIONAL DATA

The number of times per month a national sample of married adults reported having sex was 4.79 (Waite & Joyner, 2001).

2. Declining frequency. Sexual intercourse often occurs more than once a week, particularly among relatively young married couples. A sample of 261 wives interviewed twelve months after the birth of their babies noted that they had had intercourse about six times (5.82) in the past 30 days (Hyde, DeLamater, & Durik, 2001). However, this frequency of six times a month declines as the age of the spouses increases.

In addition to biological changes due to aging, reasons for declining frequency are related to the demands of employment, the care of children, and satiation. Psychologists use the term **satiation** to mean that repeated exposure to a stimulus results in the loss of its ability to reinforce. For example, the first time you listen to a new CD, you derive considerable enjoyment and satisfaction from it. You may play it over and over during the first few days. But after a week or so, listening to the same music is no longer new and does not give you the same level of enjoyment that it first did. So it is with intercourse. The thousandth time that a person has intercourse with the same partner is not as new and exciting as the first few times.

NATIONAL DATA

When cohabitants, marrieds, widowed, singles, and divorced were asked whether intercourse was occurring less frequently than they desired, the respective yes answers were 38 percent, 49 percent, 60 percent, 65 percent, and 74 percent. Hence, those most dissatisfied with frequency are the divorced and those most satisfied are cohabitants (Dunn, Croft, & Hackett, 2000, 145).

Some spouses do not have intercourse at all. One percent of husbands and 3 percent of wives in a nationwide study of sexuality reported that they had not had intercourse in the past twelve months (Michael et al., 1994, p. 116). Health, age, sexual orientation, stress, depression, and conflict were some of the reasons given for not having intercourse with one's spouse. Such an arrangement may be accompanied by either limited or extensive affection.

3. Satisfaction (emotional and physical). Despite declining frequency, marital sex remains a richly satisfying experience. On a scale from 1 (not at all satisfying) to 4 (extremely satisfying), 3.8 was the average score reported by a sample of 261 wives (Hyde, DeLameter, & Durik al., 2001). The husbands of the wives estimated the satisfaction of their wives at 3.6. Contrary to the popular belief that unattached singles have the best sex, it is the married and pair-bonded adults who enjoy the most satisfying sexual relationships. In a national sample, 88 percent of married people said they received great physical pleasure from their sexual lives, and almost 85 percent said they received great emotional satisfaction (Michael et al., 1994). Married women are particularly likely to report emotional satisfaction with sex (Waite & Joyner, 2001). Individuals least likely to report being physically and emotionally pleased in their sexual relationships are those who are not married, not living with anyone, and not in a stable relationship with one person (Michael et al., 1994).

Sexual Relationships among the Divorced

Of the almost 2 million people getting divorced, most will have intercourse within one year of being separated from their spouses. The meanings of intercourse for the separated or divorced vary. For many, intercourse is a way to reestablish—indeed, repair—their crippled self-esteem. Questions like "What did I do wrong?" "Am I a failure?" and "Is there anybody out there who will love me again?" loom in the minds of the divorced. One way to feel loved, at least temporarily, is through sex. Being held by another and being told that it feels good gives people some evidence that they are desirable. Because divorced people may be particularly vulnerable, they may reach for sexual encounters as if for a lifeboat. "I felt that as long as someone was having sex with me, I wasn't dead and I did matter," said one recently divorced person.

Before getting remarried, most divorced people seem to go through predictable stages of sexual expression. The initial impact of the separation is followed by a variable period of emotional pain. During this time, the divorced may turn to sex for intimacy to soothe some of the pain, although this is rarely achieved.

This stage of looking for intimacy through intercourse overlaps with the divorced person's feeling of freedom and the desire to explore a wider range of sexual partners and behaviors than marriage provided. "I was a virgin at marriage and was married for twelve years. I've never had sex with anyone but my spouse, so I'm curious to know what other people are like sexually," one divorced person said.

Since the divorced are usually in their 30s or older, they may not be as sensitized to the danger of contracting HIV as persons in their 20s. Yet AIDS is not a youth disease. In the United States, 60 percent of the HIV diagnoses are among individuals aged 35 and older (Centers for Disease Control and Prevention, 2003). Divorced individuals should always use a condom.

SEXUAL FULFILLMENT: SOME PREREQUISITES

There are several prerequisites for having a good sexual relationship.

Self-Knowledge, Self-Esteem, Health

Sexual fulfillment involves knowledge about yourself and your body. Such information not only makes it easier for you to experience pleasure but also allows you to give accurate information to a partner about pleasing you. It is not possible to teach a partner what you don't know about yourself.

Sexual fulfillment also implies having a positive self-concept. To the degree that you have positive feelings about yourself and your body, you will regard yourself as a person someone else would enjoy touching, being close to, and making love with. If you do not like yourself or your body, you might wonder why anyone else would.

Effective sexual functioning also requires good physical and mental health. This means regular exercise, good nutrition, lack of disease, and lack of fatigue. Performance in all areas of life does not have to diminish with age—particularly if people take care of themselves physically (see Chapter 17, Aging in Marriage and Family Relationships).

Good health also implies being aware that some drugs may interfere with sexual performance. Alcohol is the drug most frequently used by American adults. Although a moderate amount of alcohol can help a person become aroused through a lowering of inhibitions, too much alcohol can slow the physiological processes and deaden the senses. Shakespeare may have said it best: "It [alcohol] provokes the desire, but it takes away the performance" (*Macbeth,* act 2, scene 3). The result of an excessive intake of alcohol for women is a reduced chance of orgasm; for men, overindulgence results in a reduced chance of attaining an erection.

The reactions to marijuana are less predictable than the reactions to alcohol. Though some individuals report a short-term enhancement effect, others say that marijuana just makes them sleepy. In men, chronic use may decrease sex drive because marijuana may lower testosterone levels.

A Good Relationship

A guideline among therapists who work with couples who have sexual problems is to treat the relationship before focusing on the sexual issue. The sexual relationship is part of the larger relationship between the partners, and what happens outside the bedroom in day-to-day interaction has a tremendous influence on what happens inside the bedroom. The statement "I can't fight with you all day and want to have sex with you at night" illustrates the social context of the sexual experience.

McGuirl and Wiederman (2000) asked 185 men and 244 women to identify their preferences in an ideal sex partner. Men most valued a physically attractive partner, one who was open to discussing sex, who communicated her desires clearly, who was easily aroused, and who was uninhibited. Paying him compliments during sex and experiencing orgasm easily were also important.

Women most valued a partner who was open to discussing sex, who was knowledgeable about sex, who clearly communicated his desires, who was physically attractive, and who paid her compliments during sex. Being easily sexually aroused and being uninhibited were also important.

Sexual interaction communicates how the partners are feeling and acts as a barometer for the relationship. Each partner brings to a sexual encounter, sometimes unconsciously, a motive (pleasure, reconciliation, procreation, duty), a psychological state (love, hostility, boredom, excitement), and a physical state (tense, exhausted, relaxed, turned on). The combination of these factors will change from one encounter to another. Tonight the wife may feel aroused and loving and seek pleasure, but her husband may feel exhausted and hostile and have sex only out of a sense of duty. Tomorrow night, both partners may feel relaxed and have sex as a means of expressing their love for each other.

Open Sexual Communication

Sexually fulfilled partners are comfortable expressing what they enjoy and do not enjoy in the sexual experience. Unless both partners communicate their needs, preferences, and expectations to each other, neither is ever sure what the other wants. In essence, the Golden Rule ("Do unto others as you would have them do unto you") is not helpful, because what you like may not be the same as what your partner wants. A classic example of the uncertain lover is the man who picks up a copy of *The Erotic Lover* in a bookstore and leafs through the pages until the topic on how to please a woman catches his eye. He reads that women enjoy having their breasts stimulated by their partner's tongue and teeth. Later that night in bed, he rolls over and begins to nibble on his partner's breasts. Meanwhile, she wonders what has possessed him and is unsure what to make of this new (possibly unpleasant) behavior.

SELF-ASSESSMENT

Student Sexual Risks Scale

The following self-assessment allows you to evaluate the degree to which you may be at risk for engaging in behavior that exposes you to HIV. Safer sex means sexual activity that reduces the risk of transmitting the AIDS virus. Using condoms is an example of safer sex. Unsafe, risky, or unprotected sex refers to sex without a condom, or to other sexual activity that might increase the risk of AIDS virus transmission. For each of the following items, check the response that best characterizes your option.

A = Agree
U = Undecided
D = Disagree

		A	U	D
1.	If my partner wanted me to have unprotected sex, I would probably give in.	___	___	___
2.	The proper use of a condom could enhance sexual pleasure.	___	___	___
3.	I may have had sex with someone who was at risk for HIV/AIDS.	___	___	___
4.	If I were going to have sex, I would take precautions to reduce my risk of HIV/AIDS.	___	___	___
5.	Condoms ruin the natural sex act.	___	___	___
6.	When I think that one of my friends might have sex on a date, I ask him/her if he/she has a condom.	___	___	___
7.	I am at risk for HIV/AIDS.	___	___	___
8.	I would try to use a condom when I had sex.	___	___	___
9.	Condoms interfere with romance.	___	___	___
10.	My friends talk a lot about safer sex.	___	___	___
11.	If my partner wanted me to participate in risky sex and I said that we needed to be safer, we would still probably end up having unsafe sex.	___	___	___
12.	Generally, I am in favor of using condoms.	___	___	___
13.	I would avoid using condoms if at all possible.	___	___	___
14.	If a friend knew that I might have sex on a date, he/she would ask me whether I was carrying a condom.	___	___	___
15.	There is a possibility that I have HIV/AIDS.	___	___	___
16.	If I had a date, I would probably not drink alcohol or use drugs.	___	___	___
17.	Safer sex reduces the mental pleasure of sex.	___	___	___
18.	If I thought that one of my friends had sex on a date, I would ask him/her if he/she used a condom.	___	___	___
19.	The idea of using a condom doesn't appeal to me.	___	___	___
20.	Safer sex is a habit for me.	___	___	___
21.	If a friend knew that I had sex on a date, he/she wouldn't care whether I had used a condom or not.	___	___	___
22.	If my partner wanted me to participate in risky sex and I suggested a lower-risk alternative, we would have the safer sex instead.	___	___	___
23.	The sensory aspects (smell, touch, etc.) of condoms make them unpleasant.	___	___	___
24.	I intend to follow "safer sex" guidelines within the next year.	___	___	___
25.	With condoms, you can't really give yourself over to your partner.	___	___	___
26.	I am determined to practice safer sex.	___	___	___

Sexually fulfilled partners take the guesswork out of their relationship by communicating preferences and giving feedback. This means using what some therapists call the touch-and-ask rule. Each touch and caress may include the question "How does that feel?" It is then the partner's responsibility to give feedback. If the caress does not feel good, the partner can say what does feel good. Guiding and moving the partner's hand or body are also ways of giving feedback.

27. If my partner wanted me to have unprotected sex and I made some excuse to use a condom, we would still end up having unprotected sex. ___ ___ ___

28. If I had sex and I told my friends that I did not use condoms, they would be angry or disappointed. ___ ___ ___

29. I think safer sex would get boring fast. ___ ___ ___

30. My sexual experiences do not put me at risk for HIV/AIDS. ___ ___ ___

31. Condoms are irritating. ___ ___ ___

32. My friends and I encourage each other before dates to practice safer sex. ___ ___ ___

33. When I socialize, I usually drink alcohol or use drugs. ___ ___ ___

34. If I were going to have sex in the next year, I would use condoms. ___ ___ ___

35. If a sexual partner didn't want to use condoms, we would have sex without using condoms. ___ ___ ___

36. People can get the same pleasure from safer sex as from unprotected sex. ___ ___ ___

37. Using condoms interrupts sex play. ___ ___ ___

38. It is a hassle to use condoms. ___ ___ ___

(To be read after completing the scale)

Scoring

Begin by giving yourself eighty points. Subtract one point for every undecided response. Subtract two points every time that you disagreed with odd-numbered items or with item number 38. Subtract two points every time you agreed with even-numbered items 2 through 36.

Interpreting Your Score

Research shows that students who make higher scores on the SSRS are more likely to engage in risky sexual activities, such as having multiple sex partners and failing to consistently use condoms during sex. In contrast, students who practice safer sex tend to endorse more positive attitudes toward safer sex, and tend to have peer networks that encourage safer sexual practices. These students usually plan on making sexual activity safer, and they feel confident in their ability to negotiate safer sex even when a dating partner may press for riskier sex. Students who practice safer sex often refrain from using alcohol or drugs, which may impede negotiation of safer sex, and often report having engaged in lower-risk activities in the past. How do you measure up?

(Below 15) Lower Risk

(Of 200 students surveyed by DeHart and Birkimer, 16 percent were in this category.) Congratulations! Your score on the SSRS indicates that, relative to other students, your thoughts and behaviors are more supportive of safer sex. Is there any room for improvement in your score? If so, you may want to examine items for which you lost points and try to build safer sexual strengths in those areas. You can help protect others from HIV by educating your peers about making sexual activity safer.

(15 to 37) Average Risk

(Of 200 students surveyed by DeHart and Birkimer, 68 percent were in this category.) Your score on the SSRS is about average in comparison with those of other college students. Though it is good that you don't fall into the higher-risk category, be aware that "average" people can get HIV, too. In fact, a recent study indicated that the rate of HIV among college students is ten times that in the general heterosexual population. Thus, you may want to enhance your sexual safety by figuring out where you lost points and work toward safer sexual strengths in those areas.

(38 and Above) Higher Risk

(Of 200 students surveyed by DeHart and Birkimer, 16 percent were in this category.) Relative to other students, your score on the SSRS indicates that your thoughts and behaviors are less supportive of safer sex. Such high scores tend to be associated with greater HIV-risk behavior. Rather than simply giving in to riskier attitudes and behaviors, you may want to empower yourself and reduce your risk by critically examining areas for improvement. On which items did you lose points? Think about how you can strengthen your sexual safety in these areas. Reading more about safer sex can help, and sometimes colleges and health clinics offer courses or workshops on safer sex. You can get more information about resources in your area by contacting the CDC's HIV/AIDS Information Line at 1-800-342-2437.

Source

DeHart, D. D., and Birkimer, J. C. 1997. The Student Sexual Risks Scale (modification of SRS for popular use; facilitates student self-administration, scoring, and normative interpretation). Developed specifically for this text by Dana D. DeHart, College of Social Work at the University of South Carolina; John C. Birkimer, University of Louisville. Used by permission of Dana DeHart.

Addressing Safer Sex Issues

Sexually transmitted diseases (STDs) have become a concern of sexual involvement, yet consistent condom use is lacking. Civic (2000) found that in her sample of 210 heterosexual college students in dating relationships, half reported consistent condom use in the first month of their dating relationship, while only 34 percent reported consistent condom use in the last month. Subjective assessments of the partner's safety and the belief that sufficient measures were being taken to avoid pregnancy were the identified reasons for condom nonuse.

A team of researchers (Katz et al., 2000) studied 297 persons aged 13–24 who had a sexually transmitted disease and identified the relationship characteristics associated with consistent condom use. These included lower relationship quality, lower emotional reasons for sex, lower coital frequency, sex with a new partner, and not being in a cohabitation relationship. Conversely, those persons least likely to use a condom for safe sex are involved in a high-quality emotional cohabitation relationship in which frequent sex occurs with an established partner.

The need to address safe sex issues is not specific to heterosexuals. A team of researchers examined the conditions under which gay men address safe sex issues and found that those who regarded themselves as emotionally involved were less likely to use a condom since doing so would imply less trust. Men with negative moods and poor self-images were also less likely to use a condom (Adam et al., 2000).

Sexuality in an age of HIV and STDs demands talking about safer sex issues with a new potential sexual partner. Bringing up the issue of condom use should be perceived as caring for oneself, the partner and the relationship, not a sign of distrust. Some individuals routinely have a condom available, and it is a "given" in any sexual encounter.

The self-assessment on page 120 allows you to assess the degree to which you take sexual risks.

Having Realistic Expectations

To achieve sexual fulfillment, expectations must be realistic. A couple's sexual needs, preferences, and expectations may not coincide. It is unrealistic to assume that your partner will want to have sex with the same frequency and in the same way that you do on all occasions. It may also be unrealistic to expect the level of sexual interest and frequency of sexual interaction in long-term relationships to remain consistently high.

Sexual fulfillment means not asking things of the sexual relationship that it cannot deliver. Failure to develop realistic expectations will result in frustration and resentment. One's health, feelings about the partner, age, previous sexual experiences (including child sexual abuse, rape, etc.) will have an affect on one's sexuality and one's sexual relationship.

Avoiding Spectatoring

One of the obstacles to sexual functioning (we discuss sexual dysfunctions in detail at the end of this chapter) is **spectatoring,** which involves mentally observing your sexual performance and that of your partner. When the researchers in one extensive study observed how individuals actually behave during sexual intercourse, they reported a tendency for sexually dysfunctional partners to act as spectators by mentally observing their own and their partners' sexual performance. For example, the man would focus on whether he was having an erection, how complete it was, and whether it would last. He might also watch to see whether his partner was having an orgasm (Masters & Johnson, 1970).

Spectatoring, as Masters and Johnson conceived it, interferes with each partner's sexual enjoyment because it creates anxiety about performance, and anxiety blocks performance. A man who worries about getting an erection reduces his chance of doing so. A woman who is anxious about achieving an orgasm probably will not. The desirable alternative to spectatoring is to relax, focus on and enjoy your own pleasure, and permit yourself to be sexually responsive.

Table 5.3 Common Sexual Myths

Masturbation is sick.
Masturbation will make you go blind and grow hair on your palm.
Sex education makes children promiscuous.
Sexual behavior usually ends after age 60.
People who enjoy pornography end up committing sexual crimes.
Most "normal" women have orgasms from penile thrusting alone.
Extramarital sex always destroys a marriage.
Extramarital sex will strengthen a marriage.
Simultaneous orgasm with one's partner is the ultimate sexual experience.
My partner should enjoy the same things that I do sexually.
A man cannot have an orgasm unless he has an erection.
Most people know a lot of accurate information about sex.
Using a condom ensures that you won't get HIV.
Most women prefer a partner with a large penis.
Few women masturbate.
Women secretly want to be raped.
An erection is necessary for good sex.
An orgasm is necessary for good sex.

Spectatoring is not limited to sexually dysfunctional couples and is not necessarily associated with psychopathology. It is a reaction to the concern that the performance of one's sexual partner is consistent with his or her expectations. We all probably have engaged in spectatoring to some degree. It is when spectatoring is continual that performance is impaired.

Debunking Sexual Myths

Sexual fulfillment also means not being victim to sexual myths. Some of the more common myths include that sex equals intercourse and orgasm, that women who love sex don't have values, and that the double standard is dead. Table 5.3 presents some other sexual myths.

SEXUAL DYSFUNCTIONS OF WOMEN

Either women or men may experience sexual problems in the marriage. Also referred to as sexual dysfunctions, these may be classified by time of onset, situations in which they occur, and cause. A **primary sexual dysfunction** is one that a person has always had. A **secondary sexual dysfunction** is one that a person is currently experiencing, after a period of satisfactory sexual functioning. For example, a woman who has never had an orgasm with any previous sexual partner has a primary dysfunction, whereas a woman who has been orgasmic with previous partners but not with a current partner has a secondary sexual dysfunction.

A **situational dysfunction** occurs in one context or setting and not in another, while a **total dysfunction** occurs in all contexts or settings. For example, a man who is unable to become erect with one partner but who can become erect with another has a situational dysfunction. Finally, a sexual dysfunction can be classified according to whether it is caused primarily by biological (or organic) factors, such as insufficient hormones or physical illness, or by psychosocial or cultural factors, such as negative learning, guilt, anxiety, or an unhappy relationship.

Both women and men report experiencing sexual problems. In a study of 1,768 adults, about a third (34%) of the women and four in ten men (41%) reported having a current sexual problem (Dunn, Croft, & Hackett, 2000). Sexual problems become visible in societies where specialists (sexologists, sex therapists, marriage and family therapists) emphasize the importance of problem-free sexual relationships and offer (for a price) a therapeutic remedy.

The national sex study reported by Dunn, Croft, & Hackett (2000) identified the most frequent sexual problems creating dissatisfaction in women as arousal problems (51%), unpleasurable sex (47%), and inability to achieve orgasm (39%). A fourth of the women in Bancroft, Loftus, & Long's (2003) study reported marked distress over their sexual relationship, with the greatest predictor of such distress being the woman's emotional and relationship distress. We will examine the potential causes and alternative solutions for the top three problems.

Arousal Problems

Also referred to as hypoactive sexual desire, lack of interest in sex is the most frequent sexual problem reported by women in the United States. Lack of interest may be caused by one or more factors, including restrictive upbringing, nonacceptance of one's sexual orientation, learning a passive sexual role, and physical factors such as stress, illness, drug use, and fatigue. More often, lack of interest in sex and difficulty in becoming aroused can be explained by the woman's emotional relationship with her partner. In addition to a negative emotional context, there may be lower testosterone levels in women reporting low sexual desire.

The treatment for lack of interest in sex depends on the underlying cause or causes of the problem. The following are some of the ways in which lack of sexual desire can be treated:

Couples might consider increased intimacy, rather than sexual performance, as the goal of sexual therapy.
Clark Christensen, *Sex therapist*

1.Improve relationship satisfaction. Treating the relationship before treating the sexual problem is standard therapy in treating any sexual dysfunction, including lack of interest in sex. A prerequisite for being interested in sex with a partner, particularly from the viewpoint of a woman, is to be in love and feel comfortable and secure with the partner. Sammons (2003) emphasized the role of subjective mental factors in a woman's sexual arousal. Indeed, the Masters and Johnson sexual response model may not apply to women in that subjective mental factors such as feeling loved, trust, commitment and security may be a prerequisite to arousal. Couple therapy focusing on a loving egalitarian relationship becomes the focus of therapy.

2. Practice sensate focus. *Sensate focus* is a series of exercises developed by Masters and Johnson used to treat various sexual dysfunctions. Sensate focus may also be used by couples who are not experiencing sexual dysfunction but who want to enhance their sexual relationship. In carrying out the sensate focus exercise, the couple take turns pleasuring each other in nongenital ways, with each taking turns giving and receiving pleasure (while getting feedback from the partner about what is and is not pleasurable). On subsequent occasions, genital touching is allowed, but orgasm is not the goal. Indeed, the goal of sensate focus is to help the partners learn to give and receive pleasure by promoting trust and communication and by reducing anxiety related to sexual performance.

3. Be open to reeducation. Reeducation involves being open to examining and reevaluating the thoughts, feelings, and attitudes learned in childhood. The goal is to redefine sexual activity so that it is viewed as a positive, desirable, healthy, and pleasurable experience.

Diversity in Other Countries

A national study of Finnish women shows dramatic changes in their reported increases in sexual satisfaction. Women from unreserved and nonreligious homes who had high education and who were sexually assertive reported the greatest pleasure in sexual intercourse (Haavio-Mannila & Kontula, 1997). The authors suggested that increased emancipation of women will be associated with increases in sexual pleasure experienced by women.

4. Consider other treatments. Other treatments for lack of sexual desire include rest and relaxation. This is particularly indicated where the culprit is chronic fatigue syndrome (CFS), the symptoms of which are overwhelming fatigue, low-grade fever, and sore throat. Still other treatments for lack of sexual desire include hormone treatment and changing medications (if possible) in cases where medication interferes with sexual desire. In addition, sex therapists often recommend that people who are troubled by a low level of sexual desire engage in sexual fantasies and masturbation as a means of developing positive sexual images and feelings.

Unpleasurable Sex

Sex that is not pleasurable may be both painful and aversive. Pain during intercourse, or dyspareunia, occurs in about 10 percent of gynecological patients and may be caused by vaginal infection, lack of lubrication, a rigid hymen, or an improperly positioned uterus or ovary. Because the causes of dyspareunia are often medical, a physician should be consulted. Sometimes surgery is recommended to remove the hymen.

Dyspareunia may also be psychologically caused. Guilt, anxiety, or unresolved feelings about a previous trauma, such as rape or childhood molestation, may be operative. Therapy may be indicated.

Some women report that they find sex aversive. Sexual aversion, also known as sexual phobia and sexual panic disorder, is characterized by the individual's wanting nothing to do with genital contact with another person. The immediate cause of sexual aversion is an irrational fear of sex. Such fear may result from negative sexual attitudes acquired in childhood or sexual trauma such as rape or incest. Some cases of sexual aversion may be caused by fear of intimacy or hostility toward the other sex.

Treatment for sexual aversion involves providing insight into the possible ways in which the negative attitudes toward sexual activity developed, increasing the communication skills of the partners, and practicing sensate focus. Understanding the origins of the sexual aversion may enable the individual to view change as possible. Through communication with the partner and through sensate focus exercises, the individual may learn to associate more positive feelings with sexual behavior.

Inability to Achieve Orgasm

Orgasmic difficulty, also referred to as **inhibited female orgasm,** or orgasmic dysfunction, occurs when a woman is unable to achieve orgasm after a period of continuous stimulation. Difficulty achieving orgasm can be primary, secondary, situational, or total. Situational orgasmic difficulties, in which the woman is able to experience orgasm under some circumstances but not others, are the most common. Many women are able to experience orgasm during manual or oral clitoral stimulation but are unable to experience orgasm during intercourse (i.e., in the absence of manual or oral stimulation).

Biological factors associated with orgasmic dysfunction can be related to fatigue, stress, alcohol, and some medications, such as antidepressants and antihypertensives. Diseases or tumors that affect the neurological system, diabetes, and radical pelvic surgery (e.g., for cancer) may also impair a woman's ability to experience orgasm.

Psychosocial and cultural factors associated with orgasmic dysfunction are similar to those related to lack of sexual desire. Causes of orgasm difficulties in women include restrictive childrearing and learning a passive female sexual role. Guilt, fear of intimacy, fear of losing control, ambivalence about commitment, and spectatoring may also interfere with the ability to experience orgasm. Other women may not achieve orgasm because of their belief in the myth that women are not supposed to enjoy sex.

Relationship factors, such as anger and lack of trust, can also produce orgasmic dysfunction. For some women, the lack of information can result in orgasmic difficulties (e.g., some women do not know that clitoral stimulation is important for orgasm to occur). Some women might not achieve orgasm with their partners because they do not tell their partners what they want in terms of sexual stimulation out of shame and insecurity. Or, even in those cases where the woman is open about her sexual preferences, the partner may be unwilling to provide the necessary stimulation. Hence, a cooperative partner rather than a nonorgasmic woman should be the focus for resolution.

Since the causes for primary and secondary orgasm difficulties vary, the treatment must be tailored to the particular woman. Treatment can include rest and relaxation, testosterone injections, or limiting alcohol consumption prior to sexual activity. Sensate focus exercises might help a woman explore her sexual feelings and increase her comfort with her partner. Treatment can also involve improving relationship satisfaction and teaching the woman how to communicate her sexual needs. Teaching the woman how to masturbate is also a frequent therapeutic option. The rationale behind masturbation as a therapeutic technique for a nonorgasmic woman is that masturbation is the behavior that is most likely to produce orgasm and can enable her to show her partner what she needs. Masturbation gives the individual complete control of the stimulation, provides direct feedback to the woman of the type of stimulation she enjoys, and eliminates the distraction of a partner.

SEXUAL DYSFUNCTIONS OF MEN

Men also report sexual problems. The national sex study reported by Dunn et al. (2000) identified some of the most frequent sexual problems creating dissatisfaction in men as erectile problems (48%) and premature ejaculation (43%).

Erectile Dysfunction

Also referred to as **erectile dysfunction,** loss of erection involves the man's inability to get and maintain an erection. Like other sexual dysfunctions, erectile dysfunction can be primary, secondary, situational, or total. Occasional, isolated episodes of the inability to attain or maintain an erection are not considered dysfunctional; these are regarded as normal occurrences. To be classified as an erectile dysfunction, the erection difficulty should last continuously for a period of at least three months.

Erectile dysfunction may be caused by physiological conditions. Such biological causes include blockage in the arteries, diabetes, neurological disorders, alcohol/other drug abuse, chronic disease (kidney or liver failure), pelvic surgery, and neurological disorders. Smoking is also related to erectile dysfunction. Indeed, cigarette smoking almost doubles the likelihood of moderate or complete erectile dysfunction (Feldman et al., 2000).

Psychosocial factors associated with erectile dysfunction include depression, fear (e.g., of unwanted pregnancy, intimacy, HIV, or other STDs), guilt, and relationship dissatisfaction. For example, the man who is having an extradyadic sexual relationship may feel guilty. This guilt may lead to difficulty in achieving or maintaining an erection in sexual interaction with the primary partner and/or the extradyadic partner.

Anxiety may also inhibit the man's ability to create and maintain an erection. One source of anxiety is performance pressure, which may be self-imposed or imposed by a partner. In self-imposed performance anxiety, the man constantly checks (mentally or visually) to see that he is erect. Such self-monitoring creates anxiety, since the man fears that he may not be erect.

Partner-imposed performance pressure involves the partner's communicating to the man that he must get and stay erect to be regarded as a good lover. Such pressure usually increases the man's anxiety, thus ensuring no erection. Whether self- or partner-imposed, the anxiety associated with performance pressure results in a vicious cycle—anxiety, erectile difficulty, embarrassment, followed by anxiety, erectile difficulty, and so on.

Performance anxiety may also be related to alcohol use. After consuming more than a few drinks, the man may initiate sex but may become anxious after failing to achieve an erection (too much alcohol will interfere with erection). Although alcohol may be responsible for his initial failure, his erection difficulties continue because of his anxiety.

Authors

This retired 70-year-old says, "Hefner is right. Viagra is the greatest legal drug available today."

Treatment of erectile dysfunction (like treatment of other sexual dysfunctions) depends on the cause(s) of the problem. When erection difficulties are caused by psychosocial factors, treatment may include improving the relationship with the partner and/or removing the man's fear, guilt, or anxiety (i.e., performance pressure) about sexual activity. These goals may be accomplished through couple counseling, reeducation, and sensate focus exercises. A sex therapist would instruct the man and his partner to temporarily refrain from engaging in intercourse so as to remove the pressure to attain or maintain an erection. During this period, the man is encouraged to pleasure his partner in ways that do not require him to have an erection (e.g., cunnilingus, manual stimulation of partner). Once the man is relieved of the pressure to perform and learns alternative ways to satisfy his partner, his erection difficulties (if caused by psychosocial factors) often disappear. Most therapists bypass the use of these exercises in favor of Viagra, Cialis, or Levitra, discussed below.

Treatment for erectile dysfunction related to biological factors can include modification of medication, alcohol, or other drugs. Increasingly, physicians are prescribing sildenafil (marketed as Viagra), which increases blood flow to the penis and results in an erection when the penis is stimulated. Though a physician should be consulted before the medication is taken, about 80 percent of men experiencing erectile dysfunction report restored potency as a consequence of Viagra (Sammons, 2003). Two new FDA-approved drugs are Cialis and Levitra. The newer medications can be taken twelve hours prior to sex and last for twenty-four

Diversity in Other Countries

Seventy-one percent of 964 male patients at the All India Institute of Medical Sciences complained of "nocturnal emission" as a sexual problem. These men felt that the loss of semen could lead to a loss of virility and manhood (Verma, Khaitan, & Singh,1998).

Diversity in Other Countries

In China, men with erectile dysfunction are regarded as "suffering from deficiency of Yang elements in the kidney" and are treated with a solution prepared with water and several chemicals designed to benefit kidney function. They may also be given acupuncture therapy.

to thirty-six hours. The man does not have a constant erection but may become erect with stimulation.

Viagra is having some unanticipated effects. Men in their 80s are beginning to take Viagra, which does produce an erection. But in some cases, the couple may not have had intercourse in ten years, and the man's partner is no longer interested. A reshuffling of the sexual relationship of a couple in their later years may be indicated with medications such as Viagra, Levitra, or Cialis.

Men with low libido as well as erectile dysfunction may also benefit from testosterone replacement therapy (TRT). Once a low testosterone level (20% of men over age 60 have low levels) is confirmed by a blood test, testosterone supplements in the form of a gel, patch, or injection may begin. Potential risks include an increase in prostate size and changes in blood levels of cholesterol (AFUD, 2003).

Rapid Ejaculation

Also referred to as premature ejaculation, **rapid ejaculation** is defined as recurrent ejaculation with minimal sexual stimulation before, upon, or shortly after penetration and before the person wishes it. Whether a man ejaculates too soon is a matter of definition, depending on his and his partner's desires. Some partners define a rapid ejaculation in positive terms. One woman said she felt pleased that her partner was so excited by her that he "couldn't control himself." Another said, "The sooner he ejaculates, the sooner it's over with, and the sooner the better." Other women prefer that their partner delay ejaculation. Some women regard a pattern of rapid ejaculation as indicative of selfishness in their partner. This feeling can lead to resentment and anger. Regardless of the definition, rapid ejaculation, or failing to control the timing of his ejaculation, is a man's most common sexual problem (Metz & Pryor, 2000) and can lead to other sexual dysfunctions, such as female inorgasmia, low sexual desire, and sexual aversion.

The cause of rapid ejaculation may be biological, psychogenic, or both (Metz & Pryor, 2000). Some men are thought to have a constitutionally hypersensitive sympathetic nervous system that predisposes them to rapid ejaculation. Psychogenic factors include psychological distress, such as shame, or psychological constitution, such as being obsessive-compulsive. Rapid ejaculation is less of a problem for the woman than for her male partner. And, while lower sexual satisfaction for both the woman and the man is associated with rapid ejaculation, relationship satisfaction is not affected (Byers & Grenier, 2003).

Treatment depends on the cause. If a hypersensitive neurologic constitution seems to be the primary culprit, pharmacologic intervention is used. For example, 50 mg of clomipramine hydrochloride taken four to six hours before sex has been found to delay orgasm in 30 percent of the cases. Where the cause is identified as psychogenic, cognitive-behavioral-sex therapy is indicated. This may include the stop-start technique or frequent ejaculations.

The pause, or stop-start, involves the man's stopping penile stimulation (or signaling his partner to stop stimulation) at the point that he begins to feel the urge to ejaculate. After the period of preejaculatory sensation subsides, stimulation resumes. This process may be repeated as often as desired by the partners.

Still another method of increasing the delay of ejaculation is for the man to ejaculate often. In general, the greater the number of ejaculations a man has in one twenty-four-hour period, the longer he will be able to delay each subsequent ejaculation. The man's relationship with his partner is also important. Success of the respective treatments in reference to the cause awaits further research (Metz & Pryor, 2000).

The buyer should beware, and couples who decide to seek sex therapy should see only a credentialed therapist. The American Association of Sex Educators, Counselors, and Therapists (AASECT) maintains a list of certified sex therapists throughout the country. The association's address is 435 N. Michigan Ave., Suite 1717, Chicago, IL 60611-4067. Phone: 312-644-0828.

SUMMARY

What are sexual values and what are their functions?

Sexual values are moral guidelines for making sexual choices in nonmarital, marital, heterosexual, and homosexual relationships. Their functions include solidifying one's self-concept, scripting sexual behavior, sorting potential partners, and helping to avoid sexually transmissible infections (STDs).

What are three alternative sexual values?

Three sexual values are absolutism (rightness is defined by an official code of morality), relativism (rightness depends on the situation—who does what, with whom, in what context), and hedonism ("if it feels good, do it"). Relativism is the sexual value held by most college students, with women being more relativistic than men and men being more hedonistic than women. About 60 percent of men and women college students in one study reported a "friends with benefits" relationship. The sources of one's sexual values are numerous and include one's school, family, religion, and peers, as well as technology, television, social movements, and the Internet.

What is the sexual double standard?

The sexual double standard is the view that encourages and accepts sexual expression of men more than women. For example, men may have more sexual partners than women.

What are the various sources of sexual values?

The sources of sexual values include one's school, family, religion, and peers, as well as technology, television, social movements, and the Internet.

What are various sexual behaviors?

Masturbation involves stimulating one's own body with the goal of experiencing pleasurable sexual sensations. Fellatio is oral stimulation of the man's genitals by his partner. In many states, legal statutes regard fellatio as a "crime against nature." "Nature" in this case refers to reproduction, and the "crime" is sex that does not produce babies. Cunnilingus is oral stimulation of the woman's genitals by her partner. Increasingly, youth who have oral sex regard themselves as virgins, believing that only sexual intercourse constitutes "having sex." Vaginal intercourse, or coitus, refers to the sexual union of a man and woman by insertion of the penis into the vagina. Anal (not vaginal) intercourse is the sexual behavior associated with the highest risk of HIV infection. The potential to tear the rectum so that blood contact becomes possible presents the greatest danger. AIDS is lethal. Partners who use a condom during anal intercourse reduce their risk of infection not only from HIV but also from other STDs.

Do differences exist in the reported sexual behaviors of women and men? Yes. In national data based on interviews with 3,432 adults, women reported thinking about sex less often than men (19% vs. 54% reported thinking about sex several times a day), reported having fewer sexual partners than men (2% vs. 5% reported having had five or more sexual partners in the last year), and reported having orgasm during intercourse less often (29% vs. 75%).

What is the sexual behavior of singles, marrieds, and divorced?

The never-married and not living together report more sexual partners than those who are married or living together. Marital sex is distinctive for its social legitimacy, declining frequency, and satisfaction (both physical and emotional). The divorced have a lot of sexual partners but are the least sexually fulfilled.

What are the prerequisites for sexual fulfillment?

Fulfilling sexual relationships involve self-knowledge, self-esteem, health, a good nonsexual relationship, open sexual communication, safer sex practices, and making love with, not to, one's partner. Other variables include realistic expectations ("my partner will not always want what I want") and not buying into sexual myths ("masturbation is sick").

What are the sexual dysfunctions of women?

The most frequent sexual problems creating dissatisfaction in women are arousal problems, unpleasurable sex, and inability to achieve orgasm.

What are the sexual dysfunctions of men?

The most frequent sexual problems creating dissatisfaction in men are erectile problems and premature ejaculation.

KEY TERMS

absolutism
asceticism
cunnilingus
double standard
erectile dysfunction
fellatio
hedonism
hooking up
inhibited female orgasm
patriarchy
primary sexual dysfunction
rapid ejaculation
relativism
satiation
secondary sexual dysfunction
secondary virginity
sexual script
sexual values
spectatoring

RESEARCHING MARRIAGE AND THE FAMILY WITH INFOTRAC COLLEGE EDITION

InfoTrac College Edition, a n online library, allows you to perform research online anywhere, anytime. Following are two suggested search terms and related questions to help you extend your understanding of the topics covered in this chapter. Go to www.infotrac-college.com to begin your search.

Keyword: **Sex education.** Locate articles that discuss sex education in the public school system and determine the degree to which sex education is "abstinence education." In addition, find research that reflects the degree to which abstinence education is effective as defined by higher abstinence rates among students who take the abstinence education course.

Keyword: **Hooking up.** Locate articles that point out how men and women define hooking up differently and the degree to which their expectations are fulfilled by hooking up.

The Companion Web Site for Choices in Relationships: An Introduction to Marriage and the Family, Eighth Edition

http://sociology.wadsworth.com/knox_schacht/choices8e

Supplement your review of this chapter by going to the companion Web site to take one of the Tutorial Quizzes, use the flash cards to master key terms, and check out the many other study aids you'll find there. You'll also find special features such as the Marriage and Family Resource Center, Census 2000 information, and other data and resources at your fingertips to help you with that special project or to do some research on your own.

WEBLINKS

Body Health: A Multimedia AIDS and HIV Information Resource
http://www.thebody.com

Centers for Disease Control and Prevention (CDC)
http://www.cdc.gov

Go Ask Alice: Sexuality
http://www.goaskalice.columbia.edu/

Sexual Intimacy
http://www.heartchoice.com/sex_intimacy/

Sexual Health
http://www.sexualhealth.com/

Sexuality Information and Education Council (SIECUS)
http://www.siecus.org

Authors

CHAPTER 6

Singlehood and Same-Sex Relationships

We recognize today that children can be effectively raised in many different family systems and that it is the emotional climate of the family, rather than its kinship structure, that primarily determines a child's emotional well being and healthy development.

David Elkind,
Child development specialist

Contents

True or False?

1. Fewer people are marrying; only about three-fourths of U.S. adults eventually marry.
2. A fifteen-year longitudinal study of 24,000 individuals found that singles reported being happier than marrieds.
3. Children reared by lesbians when compared with children reared by heterosexual couples have lower emotional well-being and are more likely to be homosexual as adults.
4. The average length of stay of a person in the Twin Oaks Intentional Community (a commune in existence for thirty years) is eight months.
5. Sexual identity may be more fluid (changeable) than commonly thought.

Answers: **1.** F **2.** F **3.** F **4.** F **5.** T

One of the hallmarks of American culture is the increasing social acceptance of a range of choices. While the overwhelming majority of people elect both marriage (95%) and parenthood (90%), no longer is heterosexual marriage with children the only status sought by everyone. "Diverse" is the most accurate way to describe relationship, marriage, and family structures today. Although heterosexual marriage with children is the relationship context of most people, other lifestyles include singlehood, single parenthood, intentional community living, and homosexuality. In this chapter we explore these options, beginning with singlehood as an alternative to marriage.

SINGLE LIFESTYLE

In this section we discuss the effect of social movements on the acceptance of singlehood, the various categories of singles, the choice to be permanently unmarried, and the HIV infection risk associated with this choice.

Social Movements and Single Lifestyle

Though over 95 percent of U.S. adults eventually marry, more people are delaying marriage and living the single lifestyle. The percentage of all adults of all ages in our society who are never-married increased from 22 percent in 1990 to 25 percent in 2002 (*Statistical Abstract of the United States: 2003,* Table 61). Part of the increase can be attributed to social movements—the sexual revolution, the women's movement, and the gay liberation movement. The sexual revolution involved openness about sexuality and permitted intercourse outside the context of marriage. No longer did people feel compelled to wait until marriage for involvement in a sexual relationship. Hence, the sequence changed from dating, love, maybe intercourse with a future spouse, then marriage and parenthood to hanging out, hooking up with numerous partners, maybe living together (in one or more relationships), marriage and children.

The women's movement emphasized equality in education, employment, and income for women. As a result, rather than get married and depend on a husband for income, women earned higher degrees in school, sought career opportunities, and earned their own income. This economic independence brought with it independence of choice. Women could afford to remain single or to leave an unfulfilling or abusive relationship.

The gay liberation movement increased the visibility of gay people and increased the acceptance of homosexuality as an alternative lifestyle. Though some gays still marry heterosexuals to provide a traditional social front, the gay liberation movement has provided support for a lifestyle consistent with one's sexual orientation. This includes rejecting traditional heterosexual marriage. Today, some gay pair-bonded couples regard themselves as "married" even though they are not legally wed. Some gay couples have formal "wedding" ceremonies in which they exchange rings and vows of love and commitment.

In effect, there is a new wave of youth who feel that their commitment is to themselves in early adulthood and to marriage in their late 20s and 30s, if at all. This translates into getting a job or staying in school, establishing oneself in a career, and becoming economically and emotionally independent from one's parents. The old pattern was to leap from high school into marriage. The new pattern is to wait until after college or after being established in a career. Some individuals opt for remaining single forever.

Categories of Singles

Three categories of singles of the U.S. adult male and female population are the never- married, the divorced, and the widowed.

Never-Married Singles The social support for singlehood has increased and the stigma of singlehood has decreased. Kevin Eubanks (*Tonight Show* music director), Oprah Winfrey, Diane Keaton, and Drew Carey are examples of persons who have never married. Nevertheless, it is rare for a person to remain single his or her entire life. Even Gloria Steinem, the ardent feminist who earlier in her career had suggested that marriage was a prison for women, married at age 66.

Most women find themselves never-married by accident, not design. Almost all are surprised by the fact that they're not married. It's not something they dreamed of.
Carol Anderson, *Psychologist*

NATIONAL DATA

By age 75, only 3.6 percent of U.S. women and 3.8 percent of U.S. men have never married (*Statistical Abstract of the United States: 2003*, Table 63)

Though the never-married singles consist mostly of those who want to marry someday, increasingly these individuals are comfortable delaying marriage to pursue educational and career opportunities. Others, such as African-American women who have never married, note a lack of potential marriage partners. Educated Black women report a particularly difficult time finding eligibles from which to choose.

NATIONAL DATA

Considering all adult females at all ages, about 38 percent of adult Black women have never married, in contrast to about 17 percent of adult white women who have never married (*Statistical Abstract of the United States: 2003*, Table 61).

Never-married Chinese-American and Japanese-American women also perceive a lack of suitable marriage partners (Ferguson, 2000). Indeed, some view the men available to them as too traditional and restrictive in their expectations of a traditional wife. They would rather remain single than be in a traditional marriage like the one their parents have.

In spite of the viability of singlehood as a lifestyle, representative data from national samples have been almost nonexistent. One exception is research by Lucas et al. (2003), who analyzed data from a fifteen-year longitudinal study of over 24,000 individuals. They found that married people were happier than single people and hypothesized that marriage may draw persons who are already more satisfied than average (e.g., unhappy people may be less sought after as marriage partners).

As we leave this section on never-married singles, we look at the single life of one woman (see Up Close 6.1 on p. 136).

Divorced Singles People who are divorced from their spouses are also among the single. For many of the divorced, the return to singlehood is not an easy transition. The separated and divorced are the least likely to say that they are "very happy" with their life: only 18 percent of divorced men compared with 36 percent of married men said they were very happy. Only 19 percent of divorced women compared with 40 percent of married women said they were very happy (Glenn & Weaver, 1988).

NATIONAL DATA

There are almost 21 million divorced adults in the United States, representing about 10 percent of our adult population. This percentage has increased from 8.3 percent in 1990 (*Statistical Abstract of the United States: 2003*, Table 61).

Divorced persons tend not to live as long as marrieds or are more likely to die at a younger age. On the basis of a study of 44,000 deaths, Hemstrom (1996) observed that, "on the whole, marriage protects both men and women from the higher mortality rates experienced by unmarried groups" (p. 376). One explanation is the protective aspect of marriage. "The protection against diseases and mortality that marriage provides may take the form of easier access to social support, social control, and integration, which leads to risk avoidance, healthier lifestyles, and reduced vulnerability" (p. 375).

Widowed Singles Whereas some divorced people choose to be single rather than remain in an unhappy marriage, the widowed are forced into singlehood.

NATIONAL DATA

There are about 14 million widowed adults in the United States, representing about 7 percent of the adult population (*Statistical Abstract of the United States: 2003*, Table 61).

As a group, the widowed are happier than the divorced but not as happy as the married. Twenty-one percent of widowed men in contrast to 18 percent of divorced men and 36 percent of married men reported that they were very happy. For women, 29 percent of the widowed compared with 19 percent of the divorced and 40 percent of the married reported that they were very happy (Glenn & Weaver, 1988). Lucas et al (2003) also found that widows are struggling to regain their previous level of satisfaction.

Singlehood as a Lifestyle Choice

Singlehood may have different meanings to the never-married, the divorced, and the widowed. Whereas some view it as a lifestyle, others view it as a stage leading to marriage or remarriage. Meredith Kennedy is a never-married veterinarian who has found that singlehood works very well for her. She writes:

> *As a little girl, I always assumed I'd grow up to be swept off my feet, get married, and live happily ever after. Then I hit my 20s, became acquainted with reality, and discovered that I had a lot of growing up to do. I'm still working on it now, in my late 30s. Growing up for me has become an ongoing adventure in individuality and creating my own path. This is a very personal journey, and remaining single has helped me envision who I am and who I'm evolving into. Marriage for a lot of people (it seems to me) brings an end, or a plateau, to individual development, leading to boredom and disenchantment. Life is not a dress rehearsal, so I want to continue to learn and grow as much as I can, and not settle for any situation that hinders this. A marriage in which both partners could really continue to evolve as individuals would be ideal, but this takes a lot of maturity and self-awareness, and I'm still working on growing up.*
>
> *I've gotten a lot out of my relationships with men over the years, some serious and some not so serious, and they've all left their impression on me. But gradually I've moved away from considering myself "between boyfriends" to getting very comfortable with being alone, and finding myself good company. The thought of remaining single for the rest of my life doesn't bother me, and the freedom that comes with it is very precious. I've worked and traveled all over the world, and my schedule is my own. I've lived in West Africa (learning to play drums), East Africa (teaching marine biology), and southern Africa (wildlife biology), and now I'm taking six months off to write a book. Currently I have two professions and I'm working on a transition to a third, all of which fill my life with passion, intensity, and creativity. I don't think this could have come about with the responsibilities of marriage and a family, and the time and space I have as a single woman have allowed me to really explore who I am in this life.*

Up Close 6.1 Why Singlehood? A Woman's View

A never-married woman, 40 years of age, spoke to the authors' marriage and family class about her experience as a single woman. The following is from the outline she developed and the point she made in regard to each topic.

Stereotypes about Single Women
Various assumptions are made about the never-married woman and why she is single. These include the following:

Unattractive—she's either overweight or homely or she would have a man.

Lesbian—she has no real interest in men and marriage because she is homosexual.

Workaholic—she's career-driven and doesn't make time for relationships.

Poor interpersonal skills—she has no social skills and men are embarrassed by her.

History of abuse—she has been turned off to men by sexual abuse (father, brother, or date).

Negative previous relationships—she's been rejected again and again and can't hold a man.

Man hater—deep down she hates men.

Frigid—she hates sex and avoids men and intimacy.

Promiscuous—she is indiscriminate in her sexuality so that no man respects or wants her.

Too picky—she always finds something wrong with each partner and is never satisfied.

Too weird—she would win the Miss Weird contest and no man wants her.

Dependent on parents—she is still bonded to her family and still lives with them. She has not become independent and can't function in an adult-adult relationship.

Societal Expectations of the Adult Woman

1. Get married.
2. If married, have children.
3. If not married or without children, always be in a relationship.
4. Compromise career goals since marriage and children are more important.

Translation of Societal Expectations into Comments Made to a Single Woman

Family: "Why haven't you found a nice young man yet?" "All of your friends have boyfriends and are getting married—can't one of them introduce you to someone else?"

Friends: "I've got a friend you may be interested in; can I fix you up with him?"

Work Associates: "Are you married?" "What does your spouse do?" "Do you have children?"

Social Functions (mostly for couples): Invitations read, "You and your spouse are invited. . ."

Marketing (sometimes for couples): "Companion fares"; "Coupons for dinners for two."

Media: Magazine articles on finding a man and getting him to make a commitment.

Authors

Having been to China, Africa, and Australia, Meredith Kennedy notes that freedom to travel is among the benefits of choosing to remain single. In addition to being a published novelist, she is a veterinarian (notice her love for dogs) and enjoys a network of friendships.

I love children, and I'm fortunate to have a number of friends with families, so I'm "auntie" to several wonderful little people. Parenting is a very difficult job, and I think there's a need for parents and their children to have relationships with single adults who can add their love, time, and perspective. Traditionally the demands of parenthood would have been spread out among an extended family, but that doesn't happen much anymore, so I see myself (and other single people) filling an important role in helping kids grow up. This then becomes my contribution to society, and I don't need to have my own children to feel complete.

It's not always easy to explain why I'm single in a culture that expects women to get married and have children, but the freedom and independence I have allow me to lead a unique and interesting life. (Written exclusively for this text by Meredith Kennedy.

Positive Aspects of Being Single

1. Freedom to define self in reference to own accomplishments, not in terms of attachments (e.g., spouse).
2. Freedom to pursue own personal and career goals and advance with limited restrictions on time by spouse and children.
3. Freedom to come and go as you please and to do what you want, when you want.
4. Freedom to establish relationships with members of both sexes at desired level of intensity.
5. Freedom to travel and explore new cultures, ideas, values.

Negative Aspects of Being Single

1. Increased extended family responsibilities. The unmarried sibling is assumed to have the time to care for elderly parents.
2. Increased job expectations. The single employee does not have marital or family obligations so can be expected to work at night and on weekends.
3. Isolation. Too much time alone does not allow others to give feedback: "Are you drinking too much?" "Have you had a checkup lately?" "Are you working too much?"
4. Decreased privacy. Others assume the single person is always at home and always available. They may call late at night or drop in whenever they feel like it. They tend to ask personal questions freely.
5. Less safety. A single woman living alone is more vulnerable than a married woman with a man in the house.
6. Feeling different. Many work-related events are for couples, husbands, and wives. A single woman sticks out.
7. Lower income. Single women have much lower incomes than married couples.
8. Less psychological intimacy. The single woman may have no context of psychological intimacy waiting for her at the end of the workday.
9. Negotiation skills lie dormant. Since there is no one with whom the single woman must negotiate issues on a regular basis, she may become deficient in compromise and negotiation skills.
10. Rigidity patterns become entrenched. Since there is no other person to express preferences in terms of what to do, the single person may establish a very repetitive lifestyle.

Maximizing One's Life as a Single

1. Frank discussion. Talk with parents about your commitment to and enjoyment of the single lifestyle and request that they drop marriage references. Talk with siblings about joint responsibility for aging parents and your willingness to do your part (but not all elderly care). Talk with employers about spreading workload among all workers, not just those who are unmarried and childfree.
2. Relationships. Develop and nurture close relationships with parents, siblings, extended family, and friends to have a strong and continuing support system.
3. Participate in social activities. Go to social events with or without a friend. Avoid becoming a social isolate.
4. Be cautious. Be selective in giving out name, address, phone number, etc.
5. Money. Pursue education to maximize income; set up retirement plan.
6. Health. Exercise, have regular checkups, and eat healthy food. Take care of yourself.

People elect to remain single for different reasons. Table 6.1 reflects some of the perceived benefits of remaining single and the costs associated with getting married. The primary advantage of remaining single is freedom and control over one's life. Once a decision has been made to involve another in one's life, it follows that one's choices become vulnerable to the influence of that other person. The person who chooses to remain single may view the needs/influence of another as something to avoid.

Diversity in the United States

Ferguson (2000) studied sixty-two never-married Chinese-American and Japanese-American women and found that 40 percent expressed some regret at not being married. However, for most of these women, their regret was primarily about not having children and not about being never-married per se. Instead, most of these never-married women were happy with the decisions they made and were living rich and fulfilling lives. Most were economically successful, immersed in a community of friends and family, and actively involved in their work or community projects (p. 155).

Table 6.1 Reasons to Remain Single

Benefits of Singlehood	Limitations of Marriage
Freedom to do as one wishes	Restricted by spouse or children
Responsible for oneself only	Responsible for spouse and children
Close friends of both sexes	Pressure to avoid close other-sex friendships
Spontaneous lifestyle	Routine, predictable lifestyle
Feeling of self-sufficiency	Potential to feel dependent
Spend money as one wishes	Expenditures influenced by needs of spouse/children
Freedom to move as career dictates	Restrictions on career mobility
Avoid being controlled by spouse	Potential to be controlled by spouse
Avoid emotional/financial stress of divorce	Possibility of divorce

PERSONAL CHOICES

Is Singlehood for You?

Singlehood is not a unidimensional concept. While some are committed to singlehood, others enjoy it for now but intend to eventually marry, and still others are conflicted about it. There are many styles of singlehood from which to choose. As a single person, you may devote your time and energy to career, travel, privacy, heterosexual or homosexual relationships, living together, communal living, or a combination of these experiences over time. An essential difference between traditional marriage and singlehood is the personal and legal freedom to do as you wish.

Although singlehood offers freedom, single people are sometimes challenged by such issues as loneliness, less money, and establishing an identity.

1. Loneliness. For some singles, being alone is a desirable and enjoyable experience. "The major advantage of being single," said one 29-year-old man, "is that I don't have to deal with another person all the time. I like my privacy." Henry David Thoreau, who never married, spent two years alone on fourteen acres bordering Walden Pond in Massachusetts. He said of his experience, "I love to be alone. I never found the companion that was so companionable as solitude."

While unmarried single men report more loneliness than single women (Pinquart, 2003), one never-married-by-choice 77-year-old woman spoke to the authors' classes and noted, "It hit me at age 60 that I was alone and that the status and money of my very successful career were of less importance than the love and companionship of a man that I now wish I had married. I am here tonight to tell you that I made a mistake."

Some single women want a family even though they are not married, and they decide to have a baby without a husband. "Single women express more regret about not having children than about not being married" (Toufexis, 1996, 79). We discuss single parenthood beginning on page 139.

2. Less money. Married couples who combine their incomes usually have more income than single people living alone. The median income of a married couple is $60,471, compared with $40,715 for a male householder with no wife and $28,482 for a female householder with no husband (*Statistical Abstract of the United States: 2003,* Table 685). Lichter, Graefe, & Brown (2003) also confirmed higher rates of poverty among the never-married and divorced.

Individuals who want to remain single might consider the advantages of increased education, since it is associated with increased income. The median income of a householder with a master's degree is about $10,000 more per year than that for a householder with a bachelor's degree ($40,744 versus $30,973). Persons with a Ph.D. typically earn $12,000 more than those with a master's ($52,181 versus $40,744). (*Statistical Abstract of the United States: 2003*, Table 693). "The more you learn, the more you earn" is a phrase used to emphasize the value of education.

3. Identity. Single people must establish an identity—a role—that helps to define who they are and what they do. Couples eat together, sleep together, party together, and cooperate economically. They mesh their lives into a cooperative relationship that gives them the respective identity of being a married couple. On the basis of their spousal roles, we can predict what they will be doing most of the time. For example, at noon on Sunday, they are most likely to be having lunch together. Not only can we predict what they will be doing, but their roles as spouses tell them what they will be doing—interacting with each other.

The single person finds other roles and avenues to identity. A meaningful career is the avenue most singles pursue. A career provides structure, relationships with others, and a strong sense of identity ("I am a veterinarian and love my work" said one woman). To the degree that singles find meaning in their work, they are successful in establishing autonomous identities independent of the marital role.

In evaluating the single lifestyle, to what degree, if any, do you feel that loneliness is or would be a problem for you? What are your economic needs, and to what degree might education help to meet those needs? The old idea that you can't be happy unless you are married is no longer credible. Whereas marriage will be the first option for some, it will be the last option for others. One 76-year-old single-by-choice said, "A spouse would have to be very special to be better than no spouse at all."

Sources

Lichter, D. T., D. R. Graefe, and J. B. Brown. 2003. Is marriage a panacea? Union formation among economically disadvantaged unwed mothers. *Social Problems* 50:60–86.

Statistical Abstract of the United States: 2003, 123rd ed. Washington, D.C.: U.S. Bureau of the Census.

Toufexis, A. 1966. *When the ring doesn't fit.* Psychology Today 29(6): 52 passim.

Singlehood and HIV Infection Risk

Unmarried individuals who are not married or not living with someone are at greater risk for contracting HIV and other STDs. In a study comparing single and married white women, the former report feeling more vulnerable to HIV and STD infection than do married white women, and they are more likely to engage in risky sexual behavior (Wayment et al., 2003). Though women typically report having had fewer sexual partners than men, the men they had sex with usually had had multiple sexual partners. Hence, women are more likely to get infected from men than men are from women. In addition to the social reason for increased risk of infection, there is a biological reason—sperm, which may be HIV-infected, is deposited into the woman's body.

SINGLE PARENTHOOD

Most (55%) single parents enter the role through divorce or separation (e.g., Nicole Kidman) and 4 percent enter the role via death of a spouse. Ten percent of single parents are fathers. Unmarried late-night TV host David Letterman and his longtime girlfriend, Regina Lasko, have a child. Still another category of single parents is represented by Susan Sarandon (actress), who lives with but is not married to Tim Robbins, the father of their two children. Nevertheless, the stereotype of the single parent is the unmarried Black single mother. In reality, 40 percent of single mothers are white and only 33 percent are Black (Sugarman, 2003).

It is important to distinguish between a single-parent "family" and a single-parent "household." A **single-parent family** is one in which there is only one parent—the other parent is completely out of the child's life through death, sperm donation, or complete abandonment, and no contact is ever made with the other parent. In contrast, a **single-parent household** is one in which one parent typically has primary custody of the child or children but the parent living out of the house is still a part of the child's family. This is also referred to as a binuclear family. In most divorce cases where the mother has primary physical custody of the child, the child lives in a single-parent household, since he or she is still connected to the father, who remains part of the child's family. In cases in which one parent has died, the child or children live with the surviving parent in a single-parent family, since there is only one parent.

Single Mothers by Choice

Although women have traditionally had their children after marriage, some have grown tired of waiting for the "right" partner to come along. Jodie Foster, Academy Award-winning actress, has elected to have children without a husband. She now has two children and smiles when asked, "Who's the father?" The implication is that she has a right to her private life and that choosing to have a single-parent family is a viable option.

NATIONAL DATA

There are over a million births to unmarried women annually in the United States, representing about a third (33 %) of all births (Curtin & Martin, 2000). Half of all nonmarital births are planned: 39 percent among single women and 54 percent among women who are cohabiting (Musick, 2002).

An organization for women who want children and who may or may not marry is Single Mothers by Choice.

> *Typically, we are career women in our thirties and forties. The ticking of our biological clocks has made us face the fact that we could no longer wait for marriage before starting our families. Some of us went to a doctor for donor insemination or adopted in the United States or abroad. Others accidentally became pregnant and discovered we were thrilled . . . Most of us would have preferred to bring a child into the world with two loving parents, but although we have a lifetime to marry or find a partner, nature is not as generous in allotting child-bearing years.* (Mattes, 1994)

Bock (2000) noted that single mothers by choice (SCMs) are, for the most part, in the middle to upper class, mature, well-employed, politically aware, and dedicated to motherhood. Interviews with twenty-six SCMs revealed their struggle to avoid stigmatization and to seek legitimization for their choice. Most felt that their age (older), sense of responsibility, maturity, and fiscal capability justified their choice. Their self-concepts were those of competent, ethical, and mainstream mothers.

Challenges Faced by Single Parents

The single-parent lifestyle involves numerous challenges, including some of the following issues.

1. Responding to the demands of parenting with limited help. Perhaps the greatest challenge for single parents is taking care of the physical, emotional, and disciplinary needs of their children—alone. Many single parents resolve this problem by getting help from their parents or extended family.

2. Adult emotional needs. Single parents have emotional needs of their own that children are often incapable of satisfying. The unmet need to share an emotional relationship with an adult can weigh heavily on the single parent. One single mother said, "I'm working two jobs, taking care of my kids, and trying to go to school. Plus my mother has cancer. Who am I going to talk to about my life?" (authors' files).

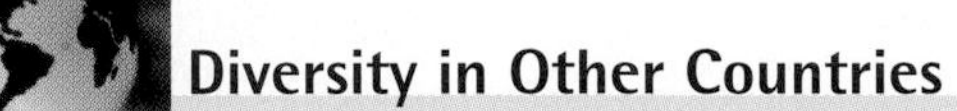

Diversity in Other Countries

There are wide variations in the percentage of children born to unmarried mothers. One half of Hawaiian mothers are not married when they give birth to their children. In contrast, only 8 percent of Chinese women and 10 percent of Japanese women are not married when they give birth to their children (*Statistical Abstract of the United States: 2003,* Table 91).

3. Adult sexual needs. Some single parents regard their parental role as interfering with their sexual relationships. They may be concerned that their children will find out if they have a sexual encounter at home or be frustrated if they have to go away from home to enjoy a sexual relationship. Some choices with which they are confronted include "Do I wait until my children are asleep and then ask my lover to leave before morning?" "Do I openly acknowledge my lover's presence in my life to my children and ask them not to tell anybody?" "Suppose my kids get attached to my lover, who may not be a permanent part of our lives?"

4. Lack of money. Single-parent families, particularly those headed by women, report that money is always lacking.

NATIONAL DATA

The median income for a female householder with no husband present is $28,142 (male householder with no wife present is $40,715), compared with $60,471 for two-parent households (*Statistical Abstract of the United States: 2003,* Table 685).

5. Guardianship. If the other parent is completely out of the child's life, the single parent needs to appoint a guardian to take care of her or his child in the event of her or his own death or disability.

6. Prenatal care. Single women who decide to have a child have poorer pregnancy outcomes than married women. The reason for such an association may be the lack of economic funds (no male partner with economic resources available) as well as the lack of social support for the pregnancy or the working conditions of the mothers, all of which result in less prenatal care for their babies.

7. Absence of a father. Another consequence for children of single-parent mothers is that they often do not have the opportunity to develop an emotionally supportive relationship with their father. Knox (2000) reported that children who grow up without fathers are more likely to drop out of school, be unemployed, abuse drugs, experience mental illness, and be a target of child sexual abuse. Conversely, those with fathers in their lives report higher life satisfaction, more stable marriages, and closer friendships. Sociologist David Popenoe observed that "fatherlessness is a major cause of the degenerating conditions of our young" (Peterson, 1995, 6d). Ansel Adams, the late great photographer, attributes his personal and life success to his father, who steadfastly guided his development. One positive aspect of father absence is that children in homes where the single parent is the mother experience less physical abuse than those in two-parent homes (Nobes & Smith, 2002).

> *If you think dads are not important, ask any child who doesn't have one.*
>
> Robert Sammons, M.D., *Psychiatrist*

8. Negative life outcomes for the child in a single-parent family. Researcher Sarah McLanahan, herself a single mother, set out to prove that children reared by single parents were just as well off as those reared by two parents. McLanahan's data on 35,000 children of single parents led her to a different conclusion—children of only one parent were twice as likely to drop out of high school, get pregnant before marriage, have drinking problems, and experience a host of other difficulties (including getting divorced themselves) as were children reared by two married parents (McLanahan & Booth, 1989; McLanahan, 1991). Lack of supervision, fewer economic resources, and less extended family support were among the culprits. Other research suggests that negative outcomes are reduced or eliminated when income levels remain stable (Pong & Dong, 2000).

Though the risk of negative outcomes is higher for children in single-parent homes, most are happy and well adjusted. Benefits to single parents themselves include a sense of pride and self-esteem that results from being independent.

INTENTIONAL COMMUNITY LIVING: TWIN OAKS

In addition to remaining single or opting for parenthood without a spouse, another lifestyle is to live in the context of an intentional community. Twin Oaks is one such community of ninety adults and fifteen children living together on 450 acres of land in Louisa, Virginia (about forty-five minutes east of Charlottesville and one hour west of Richmond).* Known as an **intentional community** (a group of people who choose to live together on the basis of a set of shared values), the commune (an older term now replaced by the newer term) was founded in 1967 and is one of the oldest nonreligious intentional communities in the United States.

The membership is 55 percent male and 45 percent female. Most of the members are white, but there is a wide range of people from different racial, ethnic, and social class backgrounds. Membership includes gay, straight, bisexual, and transgender people; most are single never-married adults, but there are married couples and families. Sexual values range from celibate to monogamous to polyamorous (involvement in more than one emotional/sexual relationship at the same time). The age range of the members is newborn to 78 with an average age of 40. The average length of stay for current members is about eight years.

Making hammocks is a major industry at Twin Oaks. Here one of the 100 members is working on a hammock.

Kate Adamson

There are no officially sanctioned religious beliefs at Twin Oaks—it is not a "spiritual" community, though individual members represent various religious values (Jewish, Christian, pagan, atheist, etc.). The core values of the community are nonviolence (no guns, low tolerance for violence of any kind in the community, no parental use of violence against children), egalitarianism (no leader, with everyone having equal political power and the same access to resources), and environmental sustainability (the community endeavors to live off the land, growing its own food and heating its buildings with wood from the forest).

Another core value is income sharing. All the members work together to support everyone in the community instead of working individually to support themselves. Money earned from the three community businesses (weaving hammocks, making tofu, and writing indexes for books) is used to provide food and other basic needs for all members. No one needs to work outside the community. Each member works in the community in a combination of income-producing and domestic jobs. An hour of cooking or gardening receives the same credit as an hour of fixing computers or business management. No matter what work one chooses to do, each member's commitment to the community is forty-two hours of work a week. This includes preparing meals, taking care of children, clean-

*This section is based on information provided by Kate Adamson and Ezra Freeman, members of Twin Oaks, and used with their permission.

ing bathrooms, and maintaining buildings—activities not considered in the typical mainstream forty-hour workweek.

The community places high value on actively creating a "homegrown" culture. Members provide a large amount of their own entertainment, products, and services. They create homemade furniture, present theatrical and musical performances, and enjoy innovative community holidays such as Validation Day (a distant relative of Valentine's Day minus the commercialism). The community also encourages participation in activism outside the community. Many members are activists for peace-and-justice, feminist, and ecological organizations.

SAME-SEX RELATIONSHIPS

Same-sex relationships continue to be a controversial issue in our society. The debate over same-sex marriage rages, and the outcome is uncertain (see Epilogue, The Future of Marriage and the Family). In this section we review the nature, origins, and types of same-sex relationships (male, lesbian, bisexual). We also note the heterosexist bias in our society and the prejudice and discrimination gay individuals are subjected to.

The Nature of Sexual Orientation

Sexuality is much more than the biological act of procreation. It involves values, emotions, thoughts, lifestyles, identities, behaviors, and relationships. The term **sexual orientation** refers to the direction of one's thoughts, feelings, and sexual interactions toward members of the same sex, the other sex, or both sexes.

Although the terms *sexual preference* and *sexual orientation* are often used interchangeably, the term *sexual orientation* avoids the implication that homosexuality, heterosexuality, and bisexuality are determined voluntarily. Hence, *orientation* is used more often by those who believe that sexual orientation is inborn, and *preference* is used more often by those who think that individuals choose their sexual orientation. Money (1987) suggests that "[p]olitically, sexual preference is a dangerous term, for it implies that if homosexuals choose their preference, then they can be legally forced, under threat of punishment, to choose to be heterosexual" (p. 385). Persons who feel that homosexuality is genetically determined are more likely to be in favor of homosexual rights (Tyagart, 2002).

Homosexuality refers to the predominance of cognitive, emotional, and sexual attraction to those of the same sex. The term *gay* is synonymous with the term *homosexual* and may refer to either males or females who have a same-sex orientation. More often *gay* is used to refer to male homosexuals and *lesbian* is used to refer to homosexual women. **Heterosexuality** refers to the predominance of cognitive, emotional, and sexual attraction to those of the other sex.

Bisexuality involves the cognitive, emotional, and sexual attraction to both men and women. While some bisexuals express an equal attraction to members of both sexes, others note a preference for one sex but a capacity to love and to enjoy sex with either. Lesbians, gays, and bisexuals are sometimes referred to collectively as **LesBiGays.**

In addition to being defined in terms of cognitive, emotional, and sexual attraction, sexual orientation is defined by one's sexual self-identity. This complicates the definition of homosexuality, heterosexuality, and bisexuality because attractions and sexual behavior are not always consistent with sexual self-identity. For example, individuals may feel emotionally drawn to same-sex members but not be sexually attracted (Diamond, 2003b). In a study of 6,982 men, of those who reported having had prior sexual experiences with both men and women,

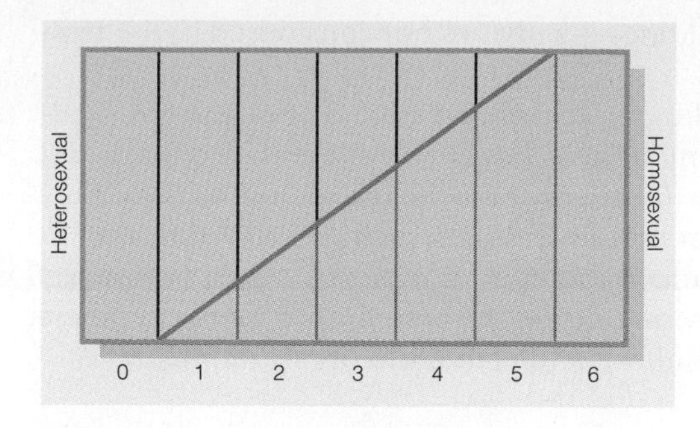

Based on both psychologic reactions and overt experience, individuals rate as follows:

0. Exclusively heterosexual with no homosexual
1. Predominantly heterosexual, only incidentallly homosexual
2. Predominantly heterosexual, but more than incidentally homosexual
3. Equally heterosexual and homosexual
4. Predominantly homosexual, but more than incidentally heterosexual
5. Predominantly homosexual, but incidentally heterosexual
6. Exclusively homosexual

Figure 6.1
The Heterosexual-Homosexual Rating Scale

Source: Kinsey, A. C., W. B. Pomeroy, C. H. Martin, and P. H. Gebhard. 1953. *Sexual behavior in the human female.* Philadelphia: W. B. Saunders. Reproduced by permission of the Kinsey Institute for Research in Sex, Gender, and Reproduction, Inc.

69 percent described themselves as heterosexual, 29 percent as bisexual, and 2 percent as homosexual (Lever et al., 1992).

Although most people view heterosexuality and homosexuality as discrete categories (the dichotomous model), sex researchers have suggested that sexual orientation exists on a continuum (the unidimensional continuum model) (see Figure 6.1).

On the Heterosexual-Homosexual Rating Scale developed by Kinsey and his colleagues (1953), individuals with ratings of 0 or 1 are entirely or largely heterosexual; 2, 3, or 4 are more bisexual; and 5 or 6 are largely or entirely homosexual. Kinsey and his colleagues developed the continuum of sexual orientation after finding that many women and men reported having had sexual experiences involving both sexes. According to this continuum, few people are entirely either heterosexual or homosexual but are somewhere in between. In addition, individuals with either sexual orientation have more similarities than differences—they both have the capacity to love and to feel jealous.

The Prevalence of Homosexuality, Heterosexuality, and Bisexuality

The prevalence of homosexuality, heterosexuality, and bisexuality is difficult to determine. Because of embarrassment, a desire for privacy, or fear of social disapproval, many individuals are not willing to answer questions about their sexuality honestly (Smith & Gates, 2001). Indeed, a team of researchers (Black et al., 2000) estimated that only about one-third of gay men and lesbian couples reported themselves as such in the latest census.

Nevertheless, research data have yielded rough estimates of prevalence rates of homosexuality and bisexuality.

National Data

It is estimated that there are more than 10 million (10,456,405) gay and lesbian individuals in the United States, which represents between 4 and 5 percent of the total U.S. population aged 18 and over (200 million) (Smith & Gates, 2001). According to the 2000 U.S. Census, 601,209 gay and lesbian families were reported, which comprise 304,148 gay male families and 297,061 lesbian families. The human rights campaign estimates that these 2000 census figures on gay and lesbian families represent a 62 percent undercount due to antigay sentiment (Smith & Gates, 2001).

International Data

Of one thousand adults born in 1972–73 in New Zealand, 1.6 percent of men and 2.1 percent of women report being currently attracted to a person of the same sex. However, by age 26, 10.7 percent of the men and 24.5 percent of the women report having been attracted to someone of the same sex in the past (Dickson, Paul, & Herbison, 2003).

The Origins of Sexual Orientation: Nature or Nurture?

One of the prevailing questions raised regarding homosexuality and bisexuality centers on their origin or cause. Gay and bisexual people are often irritated by the fact that the same question is rarely asked about heterosexuality ("How did you get to be heterosexual?"), since it is assumed that this sexual orientation is normal and needs no explanation.

It may be an academically interesting puzzle as to why we are gay . . . but it is much more interesting and important to find out why people are homophobic.

Peter Nardi

Despite the growing research on this topic, a concrete cause of sexual orientation diversity has yet to be discovered. After presenting research findings on the biological bases of sexual orientation, we look briefly at environmental explanations. Most researchers acknowledge an interaction of biological and social factors.

Biological Origins of Sexual Orientation Is sexual orientation an inborn trait that is transmitted genetically, like eye color? There does seem to be a genetic influence, although unlike with eye color, a single gene for sexual orientation has not been confirmed. Dawood et al. (2000) studied thirty-seven gay male sibling pairs and suggested that genetic biological etiology was operative. Cantor et al. (2002) also noted that men with older homosexual brothers are more likely to be homosexual themselves. "[R]oughly one gay man in 7 owes his sexual orientation to the fraternal birth order effect" (p. 63).

LeVay (1991) suggested a biological component of homosexuality as a result of scanning the brains of forty-one cadavers (nineteen homosexual men, sixteen heterosexual men, and six heterosexual women). He found that the portion of the brain thought to be involved in the regulation of sexual activity (the third interstitial nucleus of the anterior hypothalamus known as INAH 3) was half as large in homosexual men as in heterosexual men. LeVay's research is controversial and raises more questions than it answers; for example, does that brain structure produce sexual orientation, or were the differences a result of adult behavior? LeVay (1994, p. 108) observed the complexities of biological factors (genes, level of sex steroids before birth, brain structure) and environmental events, and concluded, "I do not know—nor does anyone else—what makes a person gay, bisexual, or straight. I do believe, however, that the answer to this question will eventually be found by doing biological research in laboratories and not by simply talking about the topic, which is the way most people have studied it up to now."

Environmental Explanations of Sexual Orientation According to environmental social/cultural explanations for sexual orientation, while adrenal androgens provide the fuel for the sex drive (around age 10), they do not provide the direction for sexual orientation that sociocultural forces such as the media, one's peer group, and parents (Hyde, 2000) may provide. While the dominant marital and family models of U.S. society are heterosexual, same-sex relationships are becoming more visible (e.g., TV sitcom *Will and Grace)* and increase the acceptability of these relationships as part of our diverse culture.

The degree to which early sexual experiences have been negative or positive has also been hypothesized as influencing sexual orientation. Having pleasurable same-sex experiences would be likely to increase the probability of a homosexual orientation. By the same token, early sexual experiences that are either unsuccessful or traumatic have been suspected as causing fear of heterosexual activity. One lesbian in the authors' classes explained her attraction to women as a result of being turned off to men—that her uncle molested her regularly and often for four years. However, one study that compared sexual histories of lesbian and heterosexual women found no difference in the incidence of traumatic experiences with men (Brannock & Chapman, 1990).

Complicating the understanding of the origins of sexual orientation are studies suggesting that sexual identity is less rigid, more flexible, and more fluid than previously thought. Diamond (2003a) conducted a longitudinal study of eighty lesbians, bisexuals, and "unlabeled women," first when they were between the ages of 16 and 23 and then over a five-year follow-up. During the period of study, one fourth relinquished their lesbian or bisexual identities; half returned to heterosexual identifications; half stopped labeling their sexuality altogether. Those who relinquished their sexuality-minority-status identity reported significant declines in their same-sex behavior, but their attractions did not change significantly. Hence, women who returned to heterosexual behavior maintained emotional attractions to same-sex individuals. This finding reflects the unidimensional theory of sexual orientation.

GAY RELATIONSHIPS

Whatever the origins of sexual orientation, there is a mystique about gay relationships and how they differ from heterosexual relationships. On the basis of a review of gay and lesbian relationships, Peplau, Veniegas, & Campbell (1996) concluded that there are many similarities in the relationship experiences of same-sex and different-sex couples. They noted: "That which most clearly distinguishes same-sex from heterosexual couples is the social context of their lives. Whereas heterosexuals enjoy many social and institutional supports for their relationships, gay and lesbian couples are the object of prejudice and discrimination" (p. 268).

Gay Male Relationships

A common stereotype of gay men is that they do not develop close, intimate relationships with their partners. A team of researchers (Green, Bettinger, & Sacks, 1996) compared 50 gay male couples with 218 married couples and found the former almost twice as likely to report the highest levels of cohesiveness (closeness) in their relationships.

Another stereotype is that gay men do not seek monogamous long-term relationships. While part of this stereotype may be true (Dreher, 2002), most gay men prefer long-term relationships. When sex outside the homosexual relationship does occur, it is usually infrequent and not emotionally involving (Green, Bettiner, & Sacks, 1996). Edmund White (1994) described the typical long-term gay male couple:

> *If all goes well, two gay men will meet through sex, become lovers, weather the storms of jealousy and diminution of lust, develop shared interests (a hobby, a business, a house, common friends), and end up with a long-term . . . camaraderie that is not as disinterested as friendship or as seismic as passion. . . . Younger couples feel that this sort of relationship, when it happens to them, is incomplete, a compromise, and they break up in order to find total fulfillment (i.e., tireless passion) elsewhere. But older gay couples stay together, cultivate their mild, reasonable love, and defend it against the ever-present danger of the sexual allure exercised by a newcomer* (p. 164).

NATIONAL DATA

In a national survey of homosexual men, 71 percent reported that they preferred long-term monogamous relationships to other arrangements (Lever, 1994).

A third stereotype of gay men is in reference to their roles in a relationship. Contrary to stereotypical beliefs, same-sex couples (male or female) do not typically adopt "husband" and "wife" roles (Peplau, Veniegas, & Campbell, 1996). Gay male couples also report having more flexibility in their roles than heterosexual couples, and their level of relationship satisfaction is roughly similar to that of heterosexual couples.

One unique aspect of gay male relationships in the United States involves coping with the high rate of HIV and AIDS. Many gay men have lost a love partner to this disease; some have experienced multiple losses. Those still in relationships with partners who are HIV-positive experience profound changes such as developing a sense of urgency to "speed up" their relationship because they may not have much time left together (Palmer & Bor, 2001).

Lesbian Relationships

Like many heterosexual women, most gay women value stable, monogamous relationships that are emotionally as well as sexually satisfying. Women in U.S. society, gay and "straight," are taught that sexual expression should occur in the context of emotional or romantic involvement. Transitory sexual encounters among gay women do occur, but they are the exception, not the rule. In a study of long-term gay relationships, five years was the average length of the relationship of 706 lesbian couples; 18 percent reported that they had been together eleven or more years. Ninety-one percent reported that they were sexually monogamous (National survey results, 1990). The majority (57 percent) of the women in these lesbian relationships noted that they wore a ring to symbolize their commitment to each other. Some (19 percent) acknowledged their relationship with a ceremony. Most had met through friends or at work. Only 4 percent had met at a bar. For most gay (and straight) women the formula is love first.

Although gay female relationships normally last longer than gay male relationships, long-term relationships (twenty years or more) are rare. Of 706 lesbian couples, only 1 percent had been in relationships lasting over twenty years (National survey results, 1990). Serial monogamy—one relationship at a time—was the predominant pattern, and 6.6 years was the average relationship duration.

Kurdek (1994) reviewed the literature on lesbian and gay couples and concluded, "The most striking finding regarding the factors linked to relationship satisfaction is that they seem to be the same for lesbian couples, gay couples, and heterosexual couples" (p. 251). These factors include having equal power and control, being emotionally expressive, perceiving many attractions and few alternatives to the relationship, placing a high value on attachment, and sharing decision making. Also, as is true with heterosexual couples, relationship satisfaction among gay couples depends on the emotional intimacy of the partners (Deenan, Gijiis, & Van Naerssem, 1994). In another study, Kurdek (2003) compared lesbian and gay male partners and found that the former reported stronger liking for their partners, more trust, and more equality.

Green, Bettinger, & Sacks (1996) compared 52 lesbian couples with 50 gay male couples and 218 married couples. They found that the lesbian couples were the most cohesive (closest), the most flexible in terms of roles, and the most satisfied in their relationships. Means-Christensen, Snyder, & Negy (2003) compared lesbians with heterosexual women and gay men and found that lesbians reported greater satisfaction with relationship quality and shared leisure time.

Authors

These lesbians, like most couples, prefer a stable monogamous relationship.

Bisexual Relationships

Contrary to the common myth that bisexuals are, by definition, nonmonogamous, some bisexuals prefer monogamous relationships (especially considering the widespread concern about HIV). In one study, 16.4 percent of bisexuals reported being in monogamous relationships with no desire to stray (Rust, 1996).

Monogamous bisexual women and men find that their erotic attractions can be satisfied through fantasy and their affectional needs through nonsexual friendships (Paul, 1996). Even in a monogamous relationship, "the partner of a bisexual person may feel that a bisexual person's decision to continue to identify as bisexual . . . is somehow a withholding of full commitment to the relationship. The bisexual person may be perceived as holding onto the possibility of other relationships by maintaining a bisexual identity and, therefore, not fully committed to the relationship" (Ochs, 1996, 234). However, this perception overlooks the fact that one's identity is separate from one's choices about relationship involvement or monogamy. Ochs (1996) notes that "a heterosexual's ability to establish and maintain a committed relationship with one person is not assumed to falter, even though the person retains a sexual identity as 'heterosexual' and may even admit to feeling attractions to other people despite her or his committed status" (p. 234).

Sexual Orientation and HIV Infection

Most worldwide HIV infection occurs through heterosexual transmission. However, in the United States, HIV infection remains the most threatening sexually transmitted disease (STD) for male homosexuals and bisexuals. Men who have sex with men account for more cases of AIDS in the United States than persons in any other transmission category. Women who have sex exclusively with other women have a much lower rate of HIV infection than men (both gay and straight) and women who have sex with men (Centers for Disease Control and Prevention, 2004).

HETEROSEXISM, HOMONEGATIVITY, HOMOPHOBIA, AND BIPHOBIA

Heterosexism

Our society is heterosexist, homophobic, and biphobic. The United States, along with many other countries throughout the world, is predominantly heterosexist. **Heterosexism** refers to the denigration and stigmatization of any behavior, person, or relationship that is not heterosexual. Heterosexism says that to be heterosexual is good; to be homosexual is bad. Heterosexism is based on the belief that heterosexuality is superior to homosexuality and results in prejudice and discrimination against homosexuals and bisexuals. *Prejudice* refers to negative attitudes, and *discrimination* refers to behavior that denies individuals or groups equality of treatment. Homosexuals are often victims of social disapproval, isolation, and ridicule.

Homonegativity and Homophobia

Negative attitudes toward homosexuality are reflected in the high percentage of the U.S. population that disapproves of homosexuality.

NATIONAL DATA

Seventy-one percent of white men and 63 percent of white women report that "homosexual activity" is wrong (Michael, 2001).

Other countries also do not approve of homosexuality.

INTERNATIONAL DATA

Homosexuality is illegal in about fifty countries; eight countries have the death penalty for homosexuality—Afghanistan, Iran, Mauritania, Pakistan, Saudi Arabia, Sudan, United Arab Emirates, and Yemen (Mackay, 2001).

Homonegativity, the construct that refers to antigay responses, is multidimensional and includes negative feelings (fear, disgust, anger), thoughts (homosexuals are HIV carriers), and behavior (homosexuals deserve a beating). The affective component, **homophobia,** refers to emotional responses toward and aversion to homosexuals. Homophobia may be cultural (e.g., cultural stereotypes that lesbian/gay individuals are inherently immoral, abnormal, dangerous, and/or predatory), social (fear that one will be perceived by others as gay), or psychological (irrational fear of lesbian/gay individuals and phobic reaction to them—a psychological defense mechanism against the fear that they themselves are gay). To assess your attitudes toward homosexuals, you might complete this chapter's self-assessment on page 150.

Homophobia is learned early, beginning in primary school (Plummer, 2001). Homophobia is not necessarily a clinical phobia (that is, one involving a compelling desire to avoid the feared object in spite of recognizing that the fear is unreasonable) but may involve distress that spreads from the original source to related objects or situations. For example, homophobics often feel discomfort not only with homosexuals, but also with traits or behaviors they associate with gay men or lesbians (such as effeminate behavior in men). Homophobics are also often reluctant to express affection toward someone of the same sex for fear of being labeled gay.

Males (particularly Black men, Lewis, 2003) tend to be more homophobic than females (Herek, 2002). Indeed, traditional male role socialization burdens men with the expectation that they display traditional masculine heterosexual traits or else be regarded as feminine and socially unacceptable. Theodore and Basow (2000) studied homophobia in a group of college-age men and found that the more the men believed that having traditionally prescribed masculine traits was important but did not see themselves as measuring up, the more homophobic they were.

A unique finding in research is that college-educated men exhibit homonegative characteristics toward gay men but not toward lesbians. In a study of heterosexual college students, Louderback and Whitley (1997) replicated previous research showing that heterosexual women and men hold similar attitudes toward lesbians but that men are more negative toward gay men. The researchers explained that heterosexual men attribute a high erotic value to lesbianism and that this erotic value lessens their negative view of lesbians. The erotic value placed on lesbianism most likely stems from female-female sexual themes in erotic materials marketed to heterosexual men.

Biphobia

Just as the term *homophobia* is used to refer to negative attitudes and emotional responses and discriminatory behavior toward gay men and lesbians, **biphobia** refers to similar reactions and discrimination toward bisexuals. Eliason (2001) noted that bisexual men are viewed more negatively than bisexual women, gay men, or lesbians. Bisexuals are thought to be really homosexuals afraid to acknowledge their real identity or homosexuals maintaining heterosexual relationships to avoid rejection by the heterosexual mainstream. In addition, bisexual individuals are sometimes viewed as heterosexuals who are looking for exotic sexual experiences. Bisexuals may experience double discrimination in that they are accepted by neither the heterosexual nor the homosexual community.

SELF-ASSESSMENT

The Self-Report of Behavior Scale (Revised)

This questionnaire is designed to examine which of the following statements most closely describes your behavior during past encounters with people you thought were homosexuals. Rate each of the following self-statements as honestly as possible by choosing the frequency that best describes your behavior.

1. I have spread negative talk about someone because I suspected that he or she was gay.

   A. Never B. Rarely C. Occasionally D. Frequently E. Always

2. I have participated in playing jokes on someone because I suspected that he or she was gay.

   A. Never B. Rarely C. Occasionally D. Frequently E. Always

3. I have changed roommates and/or rooms because I suspected my roommate was gay.

   A. Never B. Rarely C. Occasionally D. Frequently E. Always

4. I have warned people who I thought were gay and who were a little too friendly with me to keep away from me.

   A. Never B. Rarely C. Occasionally D. Frequently E. Always

5. I have attended anti-gay protests.

   A. Never B. Rarely C. Occasionally D. Frequently E. Always

6. I have been rude to someone because I thought that he or she was gay.

   A. Never B. Rarely C. Occasionally D. Frequently E. Always

7. I have changed seat locations because I suspected the person sitting next to me was gay.

   A. Never B. Rarely C. Occasionally D. Frequently E. Always

8. I have had to force myself to keep from hitting someone because he or she was gay and very near me.

   A. Never B. Rarely C. Occasionally D. Frequently E. Always

9. When someone I thought to be gay has walked toward me as if to start a conversation, I have deliberately changed directions and walked away to avoid him or her.

   A. Never B. Rarely C. Occasionally D. Frequently E. Always

10. I have stared at a gay person in such a manner as to convey to him or her my disapproval of his or her being too close to me.

    A. Never B. Rarely C. Occasionally D. Frequently E. Always

11. I have been with a group in which one (or more) person(s) yelled insulting comments to a gay person or group of gay people.

    A. Never B. Rarely C. Occasionally D. Frequently E. Always

12. I have changed my normal behavior in a restroom because a person I believed to be gay was in there at the same time.

    A. Never B. Rarely C. Occasionally D. Frequently E. Always

13. When a gay person has checked me out, I have verbally threatened him or her.

    A. Never B. Rarely C. Occasionally D. Frequently E. Always

14. I have participated in damaging someone's property because he or she was gay.

    A. Never B. Rarely C. Occasionally D. Frequently E. Always

15. I have physically hit or pushed someone I thought was gay because he or she brushed his or her body against me when passing by.

    A. Never B. Rarely C. Occasionally D. Frequently E. Always

16. Within the past few months, I have told a joke that made fun of gay people.

    A. Never B. Rarely C. Occasionally D. Frequently E. Always

Lesbians seem to exhibit greater levels of biphobia than do gay men. This is because many lesbian women associate their identity with a political stance against sexism and patriarchy. Some lesbians view heterosexual and bisexual women who "sleep with the enemy" as traitors to the feminist movement.

DISCRIMINATION AGAINST HOMOSEXUAL AND BISEXUAL RELATIONSHIPS

There is enormous discrimination against homosexual and bisexual relationships. Almost 70 percent (68%) of 655 first-year undergraduates at a southeastern university reported that they had "heard peers make insensitive or disparaging remarks" about a person's sexual orientation (Mather, 2000). In addition,

17. I have gotten into a physical fight with a gay person because I thought he or she had been making moves on me.

A. Never B. Rarely C. Occasionally D. Frequently E. Always

18. I have refused to work on school and/or work projects with a partner I thought was gay.

A. Never B. Rarely C. Occasionally D. Frequently E. Always

19. I have written graffiti about gay people or homosexuality.

A. Never B. Rarely C. Occasionally D. Frequently E. Always

20. When a gay person has been near me, I have moved away to put more distance between us.

A. Never B. Rarely C. Occasionally D. Frequently E. Always

The Self-Report of Behavior Scale (Revised) (SBS-R) is scored by totaling the number of points endorsed on all items (Never = 1; Rarely = 2; Occasionally = 3; Frequently = 4; Always = 5), yielding a range from 20 to100 total points. The higher the score, the more negative the attitudes toward homosexuals.

Comparison Data

The SBS was originally developed by Sunita Patel (1989) in her thesis research in her clinical psychology master's program at East Carolina University. College men (from a university campus and from a military base) were the original participants (Patel et al., 1995). The scale was revised by Shartra Sylivant (1992), who used it with a coed high school student population, and by Tristan Roderick (1994), who involved college students to assess its psychometric properties. The scale was found to have high internal consistency. Two factors were identified: a passive avoidance of homosexuals and active or aggressive reactions.

In a study by Roderick et al. (1998) the mean score for 182 college women was 24.76. The mean score for 84 men was significantly higher, at 31.60. A similar sex difference, although with higher (more negative) scores, was found in Sylivant's high school sample (with a mean of 33.74 for the young women, and 44.40 for the young men).

The following table provides detail for the scores of the college students in Roderick's sample (from a mid-sized state university in the southeast):

	N	Mean	Standard Deviation
Women	182	24.76	7.68
Men	84	31.60	10.36
Total	266	26.91	9.16

Sources

Patel, S. 1989. Homophobia: Personality, emotional, and behavioral correlates. Master's thesis, East Carolina University.

Patel, S., T. E. Long, S. L. McCammon, and K. L. Wuensch. 1995. Personality and emotional correlates of self reported antigay behaviors. *Journal of Interpersonal Violence* 10: 354–66.

Roderick, T. 1994. Homonegativity: An analysis of the SBS-R. Master's thesis, East Carolina University.

Roderick, T., S. L. McCammon, T. E. Long, and L. J. Allred. 1998. Behavioral aspects of homonegativity. *Journal of Homosexuality* 36:79–88.

Sylivant, S. 1992. The cognitive, affective, and behavioral components of adolescent homonegativity. Master's thesis, East Carolina University.

The SBS-R is reprinted by the permission of the students and faculty who participated in its development: S. Patel, S. L. McCammon, T. E. Long, L. J. Allred, K. Wuensch, T. Roderick, & S. Sylivant.

29 percent of the men and 17 percent of the women reported that they would feel "discomfort" in interacting with students who had a same-sex sexual orientation. Such discrimination may be a factor in the higher incidence of psychological distress experienced by homosexuals and the greater use of mental health services (Cochran, Sullivan, & Mays, 2003).

Discrimination against homosexuals sometimes results in murder. Matthew Shepard was a 21-year-old gay college student at the University of Wyoming. He was beaten and lashed to a split-rail fence by homophobic and heterosexist Russell Henderson and Aaron McKinney, who face the death penalty for their hate crime against Shepard. Their crime reflects the extent of heterosexism in our society.

Heterosexism is also reflected in the cultural discrimination against same-sex relationships. The Social Policy section on the next page reveals that gays continue to struggle for recognition of their relationships.

I know what it feels like to try to blend in so that everybody will think that you are OK and they won't hurt you.

Ellen DeGeneres, *Actress*

Should Legal Marriage Be Granted to Same-Sex Couples?

SOCIAL POLICY

In 2003, Massachusetts's highest court lifted a ban on same-sex marriages, paving the way for the nation's first gay marriage licenses. Previously, in 2000, Vermont had become the first state to legalize "civil unions" for same-sex couples. At the time this text goes to press, Vermont is the only state to legalize same-sex civil unions. While not equal to traditional heterosexual "marriage," civil unions are devoid of the gendered and sexist history of traditional marriage. However, the benefits of having one's relationship legally identified as a civil union include joint property rights, inheritance rights, shared health-care benefits, hospital visitation rights, and immunity from being compelled to testify against a partner. In Vermont, gay couples may obtain a civil union license (not a marriage license) for $20 from the town clerk and have the union certified by a judge or member of the clergy. In case of a breakup of the civil union, the couple would have to go through family court to obtain a legal dissolution. Civil unions do not include federal government recognition—translation, no Social Security benefits for partners. All the same-sex "marriages" performed in cities throughout the United States have *no* legal value.

Another difference between civil unions and traditional marriage unions is that the former may be recognized only in the state of Vermont. Hence, the civil union of a Vermont gay couple (unlike the marriage of a heterosexual couple) may not be recognized when they cross state lines. At least thirty-six states have banned gay marriages, and Congress has passed the **Defense of Marriage Act** denying federal recognition of homosexual marriage and allowing states to ignore same-sex marriages licensed elsewhere. Hawaii also recognizes same-sex relationships and provides some benefits to same-sex partners. To be clear, there are no gay *marriages,* only civil unions. No state (as of April, 2004) recognizes the concept of gay marriage. Indeed, Senate Majority leader Bill Frist has gone on record supporting a constitutional amendment to ban homosexual marriage in the United States (Ritter, 2003). One half of a national sample of adults for a *USA Today* poll said that they favored "a constitutional amendment that would define marriage as between a man and a woman, thus barring marriages between gay or lesbian couples" (Page, 2003). President Bush emphasized in his State of the Union message in January 2004 that "our nation must preserve the sanctity of Marriage."

High school students are much more accepting of gay marriage. Two-thirds of a national sample of high school students agreed that gay marriages should be legal (Michand, 2001). In regard to university students, female and male college students differ in their support for gay marriages. Nearly two-thirds (66.9%) of female university students and over half (50.2%) of male university students agree or strongly agree that same-sex couples should have the right to legal marital status (American Council on Education and University of California, 2004).

Antigay advocates view homosexuality as unnatural and against our country's moral standards and do not want their children to learn that homosexuality is an accepted, "normal" lifestyle. Indeed, the most common argument against same-sex marriage is that it subverts the stability and integrity of the heterosexual family. The continued ban on sanctioning same-sex relationships "clearly reinforces the already omnipresent stereotypes and misconceptions burdening their community" (Moss, 2002, 107).

Advocates of same-sex marriage argue that permitting such marriages would encourage many lesbians and gays to live within long-term, committed relationships. Such relationships would benefit individuals, couples, and society. Not only would there be greater emotional stability resulting from stable units, but monogamy would encourage fewer sexual partners, thus reducing the spread of STDs and AIDS. Everyone wins. The acceptance of same-sex marriages would also help to "eliminate heterosexism and homophobia by elevating homosexuality to the level of acceptability" (Bolte, 2001, p. 40).

Despite the fact that that there are as many as 3 million children born into families headed by lesbians (Cornelius-Cozzi, 2002), children reared by homosexuals, it is argued, will not be "normal" and will become homosexual. However, Allen and Burrell (2002) compared the impact of homosexual and heterosexual parents on children in regard to the emotional

Discrimination in Marriage and Family Relationships

The rights afforded married heterosexuals but denied pair-bonded homosexuals who love and care for each other are numerous:

assumption of spouse's pension
automatic housing lease transfer
automatic inheritance
bereavement leave
burial determination
child custody
confidentiality of conversations
crime victim's recovery benefits
domestic violence protection
family leave to care for sick partner

well-being and sexual orientation of the child and found no differences on any measures of emotional well-being or sexual orientation. In other words, the data failed to support the continuation of the bias against homosexual parents.

The fear that children would be damaged by gay marriage is also debatable. Patterson (2001) noted that not only do most children who are reared by gay parents have a heterosexual orientation as an adult but the "home environments provided by lesbian and gay parents are just as likely as those provided by heterosexual parents to enable psychosocial growth among family members" (p. 283).

In sum, advocates suggest that gay marriage would strengthen, not weaken, the family. A comparison between groups of homosexuals, bisexuals, and heterosexuals found little evidence that the dimension of sexual orientation is related to quality of life (sense of well-being/happiness), lifestyle (smoking/drinking), or health indicators (exercise) (Horowitz, Weis, & Laflin, 2001).

The arguments above reflect that considerable debate has raged over whether homosexual couples should be allowed to marry. Go to http://sociology.wadsworth.com/knox_schacht/choices8e, select the Opposing Viewpoints Resource Center on the left navigation bar, and enter your pass code. Conduct a search by subject, using the search term "same-sex marriage," and read the articles on the Viewpoints tab that support each position. Do you find yourself agreeing with one perspective more than the other? Why?

In 2001, the Netherlands became the first country in the world to offer legal marriage to same-sex couples, and in 2003, Belgium became the second (Demian, 2003). The Canadian provinces of Ontario and British Columbia also ruled in 2003 to extend the option of marriage to same-sex couples. And, in a number of countries (Denmark, Finland, Germany, Iceland, Greenland, Netherlands, Norway, and Sweden), same-sex couples can sign as "Registered Partners" to claim a status and benefits similar to those of marriage. In 2003, Quebec approved civil unions for same-sex couples, which confer some but not all of the rights, benefits, and responsibilities of marriage.

Attitudes toward homosexual relationships are becoming more accepting. According to a Gallup poll, 60 percent of adult Americans accept the idea that same-sex relations between consenting adults should be legal and that homosexuality is an acceptable way of life. This is up from 52 percent in 2002 and up from 43 percent in 1977 (Lisotta, 2003). And, with the Supreme Court's ruling that homosexual acts are no longer illegal and the legalization of gay marriage in two Canadian provinces (Ritter, 2003), there is increased movement toward the recognition of gay marriage in the United States.

Sources

Allen, M., and N. Burrell. 2002. Sexual orientation of the parent: The impact on the child. In *Interpersonal communication research: Advances through meta-analysis,* edited by M. Allen, R. Preiss, B. Gayle, and N. Burrell. Mahwah, N.J.: Lawrence Erlbaum, 111–24.

American Council on Education and University of California. 2004. The American freshman: National norms for fall, 2003. Los Angeles: Higher Education Research Institute. U.C.L.A. Graduate School of Education and Information Studies.

Bolte, A. 2001. Do wedding dresses come in lavender? The prospects and implications of same-sex marriage. In *The gay and lesbian marriage and family reader,* edited by J. M. Lehmann. Lincoln, Neb.: Gordian Knot Books, 25–46.

Cornelius-Cozzi, T. 2002. Effects of parenthood on the relationships of lesbian couples. *PROGRESS: Family Systems Research and Therapy* 11:8594.

Demian. 2003. Legal marriage report: Global status of legal marriage. Partners Task Force for Gay & Lesbian Couples. www.buddybuddy.com

Horowitz, S. M., D. L. Weis, and M. T. Laflin, 2001. Differences between sexual orientation behavior groups and social background, quality of life, and health behaviors. *Journal of Sex Research* 38: 205–18.

Lisotta, C. 2003. Poll: 6 in 10 Americans OK gay unions. Gay.com 15 May.

Michand, C. 2001. Survey: Students hold mostly pro-gay views. http://www.hrc.org/familynet

Moss, K. 2002. Legitimizing same-sex marriages. *Peace Review* 14:101–8.

Page, S. 2003. Gay rights tough to sharpen into political "wedge issue." *USA Today.* 28 July, 10A.

Patterson, C. J. 2001. Family relationships of lesbians and gay men. In *Understanding families into the new millennium: A decade in review,* edited by R. M. Milardo. Minneapolis: National Council on Family Relations, 271–88.

Ritter, J. 2003. Canada gives gays hope for change. *USA Today.* 30 June, 3A

insurance breaks | *joint bankruptcy*
medical decisions on behalf of partner | *visitation of partner in hospital*
reduced-rate memberships | *Social Security benefits*
tax advantages | *visitation of partner's children*
wrongful death benefits | *no property tax on partner's death*

Oswald (2000) and Oswald and Culton (2003) analyzed surveys of 527 rural gay, lesbian, bisexual, transgendered people and found that the respondents reported low levels of connections with their parents and siblings. In addition, the

Authors

Gay men often labor under the stereotype of being promiscuous and having no interest in stable relationships. These gay men with their children from previous heterosexual marriages have been in a stable monogamous relationship with each other for sixteen years.

respondents reported that they made family out of close-knit friendship networks. Living in rural areas was an advantage in that it fostered close relationships and a high quality of life. However, there was the feeling of living in a homophobic environment.

Actor Richard Chamberlain revealed his homosexuality after keeping it a secret until he was 69. He noted, "I would have been in trouble with my career" had he been open about his lifestyle. He reveals his ordeal of fear of being discovered and his 27-year closeted relationship with Martin in *Shattered Love* (Chamberlain, 2003).

Discrimination in Child Custody, Visitation, and Adoption Rights

Lesbians and gay men are discriminated against in family matters (Richman et al., 2002). Although between 6 and 14 million children are being reared by gay men and lesbians (Duran-Aydintug & Causey, 2001), there is a general belief that homosexuals should not rear children. Two grounds are typically used. The first, per se, "presumes the parent to be unfit merely based on the existence of same-sex orientation. Lesbian mothers are thought to be 'unfit parents, emotionally unstable, or unable to assume a maternal role'" (Erera & Fredriksen, 2001, 88). This cultural bias exists despite the fact that research on children of lesbian mothers has found that they are just as likely to be well adjusted as the children of heterosexual mothers (Golombok et al., 2003).

The "gayby boom" is underway as gay adults create families. Lesbians are conceiving with donor sperm, gay men are employing surrogate mothers, and both sexes are adopting.

Marily Elias

The second ground is the nexus approach, which is used to deny custody to homosexual parents on the basis of their being a negative influence on the sexual development of the child, the social stigmatization of the child who is being parented by homosexual parents, and the potential sexual molestation of children by homosexual parents (Duran-Aydintug & Causey, 2001). Although the effect that homosexual parents have on a child's sexual development is still being debated (Stacey & Biblarz, 2001), public concern remains. Social stigmatization is a fact of life, but this is also true of biracial children. Cases of documented sexual abuse by homosexual parents of their children are virtually "nonexistent" (Duran-Aydintug & Causey, 2001).

COMING OUT TO FAMILY MEMBERS

Family members sometimes come out to each other.

Coming Out to One's Spouse and Children

Coming out can be an emotionally charged experience. One instructor asked his students to participate in an experimental written assignment whereby they wrote a "coming out" letter to a person of their choice. "Many students expressed great difficulty—fear, humiliation, shame, embarrassment, intimidation, nervousness, frustration, shock and loneliness—in both their letters and the analysis of their letters" (Welde & Hubbard, 2003, 79–89)

Up to 2 million gay, lesbian, or bisexual persons in the United States have been in heterosexual marriages at some point (Buxton, 2004). Some of these individuals married as a "cover" for their homosexuality; others discovered their attraction/interest in same-sex relationships after they married. Issues that surface include sexual behavior ("Do I have HIV?" "Do we still have sex?"), marriage ("Do we separate or divorce, as two-thirds do, or cope, as a third do?"), and children ("What do we tell them?"). Issues to consider in regard to the latter issue include the prospect that the children may find out from another source, the age of the child (the teen years may be a bad time to make such a disclosure since children are confronting their own sexuality), the likelihood that the child will tell peers, and whether therapy would help the child adjust to this information.

Coming Out to One's Parents

Deciding whether to tell one's parents of one's homosexuality is very difficult. Evans and Broido (2000) found that coming out is related to the perceived risks and reactions of doing so. Not to do so is to hide one's true self from one's parents and to feel alienated, afraid, and alone. To tell them is to risk rejection and disapproval. Some parents of homosexuals suspect that their children are gay, even before they are told. Of 402 parents of gay and lesbian children, 26 percent stated that they suspected their offspring's homosexuality (Robinson, Walters, & Skeen, 1989).

In general, young adults are more likely to come out to mothers than to fathers (Cohen & Savin-Williams, 1996). Parents may be told face to face, through an intermediary (a sibling, another relative, or a counselor), or in a letter. There is probably no best way to disclose the information to parents. Mary Cheney, daughter of Vice President (in 2004) Dick Cheney, and Chrissy Gephardt, daughter of Senator Richard Gephardt, have also made public their homosexuality and been embraced by their parents.

Parental reaction to a child's disclosure of homosexuality is diverse and may not follow the assumed shock, denial, and isolation pattern (Savin-Williams & Dube, 1997). Though the majority of families are neither totally rejecting nor totally accepting (Beeler & DiProva, 1999), parents most often grieve (which may take five years) as they obtain accurate information about gay lifestyles. However, "a complete acceptance of a gay son or lesbian daughter's sexual orientation may be impossible for most parents" (LaSala, 2000, p.79). The result is that some gays (men more than women) maintain their distance from parents (LaSala, 2002).

Some parents find ways to look beyond cultural prejudices. One parent who became aware that her son was gay said:

> *I had my suspicions, but when he told me he was gay I cried anyway. It was over the phone, and I think this was more difficult because I couldn't hug him. At the time I think the tears shed were more for me and the dreams I was losing than for him and the prejudice and battles he would have to endure. I found an organization called PFLAG—Parents and Friends of Lesbians and Gays—to be a lifesaver to both my husband and myself. It's a support group mostly of parents but is also attended by siblings, friends, and gays themselves. They meet once a month and make you feel safe to share your feelings and concerns about being a parent of a gay. They say when the child comes out of the closet, the parent goes in, and there is some truth to that. They have a speaker and group discussions and also work toward changing legislation that is antigay. PFLAG also helped us to realize that it wasn't anything we said or did that made our son gay—it is how he was born. Needless to say, we love our son just as much as before we knew.* (Authors' files)

Chastity Bono, a lesbian, is the daughter of the famous singing duo Sonny and Cher. She has written *Family Outing* (Bono & Fitzpatrick, 1998), designed to serve as a guide to coming out for gay and lesbian people and their families.

SUMMARY

What social movements created the single lifestyle, and what are its primary advantages?

As a result of the sexual revolution, the women's movement, and the gay liberation movement, there is increased social approval for being unmarried. The primary benefits of singlehood include freedom to do as one wishes and to have responsibility for oneself only. Those who choose single parenthood report a sense of pride and well-being that results from being independent. Singles include not only the never-married but also the divorced and widowed.

What are the factors to be considered in becoming a single parent?

Most (55%) single parents enter the role through divorce or separation, and 4 percent enter the role via death of a spouse. Some, such as Jodie Foster, elect to have a child as an unmarried person. Single Mothers by Choice is an organization of women who embrace this approach.

The challenges of single parenthood include taking care of the emotional and physical needs of a child alone, meeting one's own adult emotional and sexual needs, money, and rearing a child without a father (the influence of whom can be positive and beneficial).

What is living in Twin Oaks, an intentional community, like?

Twin Oaks is a community of ninety adults and fifteen children living together on 450 acres of land in Louisa, Virginia (about forty-five minutes east of Charlottesville and one hour west of Richmond). Known as an intentional community (a group of people who choose to live together on the basis of a set of shared values), the commune (an older term now replaced by the newer term) was founded in 1967 and is one of the oldest nonreligious intentional communities in the United States. Membership includes gay, straight, bisexual, and transgender people; most are single never-married adults, but there are married couples and families. Sexual values range from celibate to monogamous to polyamorous (involvement in more than one emotional/sexual relationship at the same time). The age range of the members is newborn to 78 with an average age of 40. The average length of stay for current members is about eight years.

The core values of the community are nonviolence (no guns, low tolerance for violence of any kind in the community, no parental use of violence against children), egalitarianism (no leader, with everyone having equal political power and the same access to resources), and environmental sustainability (the community endeavors to live off the land, growing its own food and heating its buildings with wood from the forest). The community also encourages participation in activism outside the community. Many members are activists for peace-and-justice, feminist, and ecological organizations.

What is the nature of the homosexual lifestyle, and how prevalent is it?

Fewer than 3 percent of adult men and fewer than 2 percent of adult women report that their self-identity is homosexual. Theories of sexual orientation focus on biological and environmental influences. Most researchers agree that sexual orientation results from the interaction of biological, social, and cultural influences.

What are gay relationships like?

There are many similarities in the relationship experiences of same-sex and different-sex couples. That which most clearly distinguishes same-sex from heterosexual couples is the social context of their lives. Whereas heterosexuals enjoy many social and institutional supports for their relationships, gay and lesbian couples are sometimes the object of prejudice and discrimination.

A common stereotype of gay men is that they do not seek monogamous long-term relationships. However, most gay men prefer long-term relationships. Like many heterosexual women, most gay women value stable, monogamous relationships that are emotionally as well as sexually satisfying.

What are heterosexism, homonegativity, homophobia, and biphobia?

Heterosexism refers to the denigration and stigmatization of any behavior, person, or relationship that is not heterosexual. Heterosexism is based on the belief that heterosexuality is superior to homosexuality and results in prejudice and discrimination against homosexuals and bisexuals. Homonegativity includes negative feelings (fear, disgust, anger), thoughts (homosexuals are HIV carriers), and behavior (homosexuals deserve a beating). The affective component, homophobia, refers to emotional responses toward and aversion to homosexuals. Demographic characteristics associated with homonegativity include being less educated, being "born again" or being an evangelical Christian, being a right-wing authoritarian, being African-American or Hispanic, and being male. Biphobia expresses negative reactions to bisexuals.

What are examples of discrimination against homosexual and bisexual relationships?

Gay individuals are discriminated against in terms of custody and adoption decisions and are not allowed to legally marry. They are also victims of insensitive and disparaging remarks. In some cases they are murdered because they are gay.

KEY TERMS

biphobia
bisexuality
civil union
Defense of Marriage Act
heterosexism
heterosexuality
homonegativity
homophobia
homosexuality
intentional community
sexual orientation
single-parent family
single-parent household

RESEARCHING MARRIAGE AND THE FAMILY WITH INFOTRAC COLLEGE EDITION

InfoTrac College Edition, an online library, allows you to perform research online anywhere, anytime. Following are two suggested search terms and related questions to help you extend your understanding of the topics covered in this chapter. Go to www.infotrac-college.com to begin your search.

Keyword: **Single parenthood.** Locate articles that review children reared in single-parent homes. Identify some of the advantages and disadvantages of single parent families.

Keyword: **Homosexuality.** Locate articles that discuss attitudes toward homosexuality. What are the sources of such negative attitudes and how do they impact the lives of those involved?

The Companion Web Site for Choices in Relationships: An Introduction to Marriage and the Family, Eighth Edition

http://sociology.wadsworth.com/knox_schacht/choices8e

Supplement your review of this chapter by going to the companion Web site to take one of the Tutorial Quizzes, use the flash cards to master key terms, and check out the many other study aids you'll find there. You'll also find special features such as the Marriage and Family Resource Center, Census 2000 information, and other data and resources at your fingertips to help you with that special project or to do some research on your own.

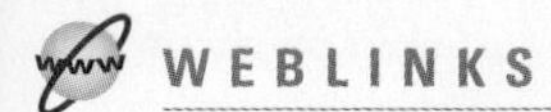

WEBLINKS

Advocate (Online Newspaper)
http://www.advocate.com/

Alternatives to Marriage Project
http://www.unmarried.org

Bisexual Resource Center
http://www.biresource.org

Gay and Lesbian Support Groups for Parents
http://www.gayparentmag.com/29181.html

National Gay and Lesbian Task Force
http://www.ngltf.org

The Single Parent Network
http://www.makinglemonade.com

Single Mothers by Choice
http://mattes.home.pipeline.com/

Twin Oaks Community
http://www.twinoaks.org/

Authors

CHAPTER 7

Marry yourself—people who marry someone just like them find that they enjoy the person and it is easy to be best friends. Opposites may attract but few relationships are sustained where the partners are completely different.

Jack Wright Jr., Sociologist and attorney

Mate Selection

Contents

True or False?

1. In a study of 700 undergraduates, extrinsic characteristics such as physical appearance and wealth were more important than intrinsic characteristics such as being warm, kind, and open.
2. Persons who are casually dating different people are better mannered with their partners than persons who are involved with and dating only one person.
3. Being a perfectionist and always requiring oneself to be perfect is also associated with relationship problems.
4. The long-term effectiveness of premarital education remains unknown.
5. Beginning in 2004 over 10 percent of couples married someone outside their own race.

Answers: **1.** F **2.** T **3.** T **4.** T **5.** F

The plethora of television programs featuring boy meets girl scenarios (*Bachelor, Bachelorette, Blind Date, Dating Story, Elimidate*) reflects America's fascination with the mate selection process. The most desirable partner seems to be one who is good-looking and fun to be with. These shows provide no information or emphasis on the qualities associated with a durable marriage partner. Indeed, no other choice in life is as important as your choice of a marriage partner. Your day-to-day happiness, health, and economic well-being will be significantly influenced by the partner with whom you choose to share your life. Most have high hopes and even believe in the perfect mate. An anonymous comic said, "I married Miss Right. . . . I just didn't know her first name was Always."

In this chapter we examine the cultural influences and pressures that influence the choice of one's mate, the tendency for individuals to select a partner with similar characteristics, and the effect of psychological factors on mate selection. Since the heart of this text is about mate choice, we provide an inventory to help them examine their relationship. We also submit for a couple's consideration the wisdom of calling off the wedding if several factors predictive of an unhappy and unstable relationship are present in their relationship.

CULTURAL ASPECTS OF MATE SELECTION

Individuals are not free to marry whomever they please. Indeed, university students routinely assert, "I can marry whomever I want!" Hardly. Rather, their culture and society radically restrict and influence their choice. The best example of mate choice being culturally and socially controlled is the fact that *fewer* than *1 percent* of persons marry someone outside their race (details later in the chapter). Two forms of cultural pressure operative in mate selection are endogamy and exogamy.

Endogamy

Endogamy is the cultural expectation to select a marriage partner within one's own social group, such as race, religion, and social class. Endogamous pressures involve social approval and disapproval to encourage you to select a partner within your own group. The pressure toward an endogamous mate choice is especially strong when race is concerned. Killian (1997) studied twenty Black-white married couples who reported that public reaction to their relationship was often negative. Almost all of the respondents had experienced stares, disapproving expressions, and harassment at the work site. The effect of such relentless disapproval is reflected in the percentage of individuals who marry someone of the same race. Love may be blind, but it knows the color of one's partner.

Diversity in Other Countries

Endogamous pressures remain strong in some families. The mate "choice" of Muslim females in traditional Muslim cultures is strongly controlled because of the influence of the family, particularly the father. Indeed, Muslim female dating behavior is prohibited. However, Muslim women socialized in the United States are questioning the normative patterns of marriage of the old country. Nevertheless, many are willing to have an arranged marriage despite their lack of faith in the arranged marriage system (Zaidi & Shuraydi, 2002).

Exogamy

In addition to the cultural pressure to marry within one's social group, there is also the cultural expectation that one will marry outside his or her own family group. This expectation is known as **exogamy.** Incest taboos are universal. In no society are children permitted to marry the parent of the other sex. In the United States, siblings and (in some states) first cousins are also prohibited from marrying each other. The reason for such restrictions is fear of genetic defects in children whose parents are too closely related.

Once cultural factors have identified the general pool of eligibles, individual mate choice becomes more operative. However, even when individuals feel that they are making their own choices, social influences are still operative.

Diversity in the United States

Exogamy is operative among the Hopi Indians, about 10,000 of whom live on reservations in Arizona. The Hopi belong to different clans (Bear clan, Badger clan, Rain clan, etc.), and the youth are expected to marry someone of a different clan. By doing so they bring resources from another clan into their own.

SOCIOLOGICAL FACTORS OPERATIVE IN MATE SELECTION

Numerous sociological factors are at work in bringing two people together who eventually marry.

Homogamy

Whereas endogamy is a concept that refers to cultural pressure, **homogamy** refers to individual initiative toward sameness. The homogamy theory of mate selection states that we tend to be attracted to and become involved with those who are similar to ourselves in such characteristics as age, race, religion, and social class. Jepsen and Jepsen (2002) found homogamy operative in both same-sex and opposite-sex couples, whether married or unmarried.

Love does not consist in gazing at each other, but in looking together in the same direction.

Antoine de Saint-Exupáry, *French philosopher*

Some data suggest (and new data confirm, Jepsen & Jepsen, 2002) that homogamy is more important in selecting a mate than choosing a date (Knox, Zusman, & Nieves, 1997). Two hundred seventy-eight undergraduates were asked to reveal the degree to which it was important to them to (1) date and (2) marry partners who had similar characteristics (race, religion, education, age, etc.) on a ten-point continuum from 0 (not important) to 10 (very important). The respondents averaged 7.4 per item when asked about the importance of a future mate (in contrast to a mean of 6.8 reflecting importance of similarity in a dating partner). Hence, the more involved the relationship, the greater the importance of finding a partner with whom they had a lot in common. Some of the homogamous factors operative in mate selection include the following.

Race As noted above, racial homogamy operates strongly in selecting a living-together or marital partner (with greater homogamy for marital partners) (Jepsen & Jepsen, 2002; Blackwell & Lichter, 2000).

NATIONAL DATA

Of the almost 58 million married couples in the United States, fewer than 1 percent (.006) consist of a Black and white spouse. Three percent consist of a Hispanic and a non-Hispanic (*Statistical Abstract of the United States: 2003*, Table 62).

Some evidence suggests that openness among college students to dating/involvement with those across racial lines may be increasing. In a study of 620 university students (Knox et al., 2000), almost half (49.6%) of the respondents reported that they were open to involvement in an interracial relationship. Almost a quarter (24.2%) said that they had dated someone of another race.

Blacks were twice as likely as whites (83% versus 43%) to report that they were open to involvement in an interracial relationship. Explanations include more benefits to Blacks if they join the majority than vice versa, the greater number of whites available to Blacks than

The principle of homogamy is operative in the emergence of this couple's relationship—they are the same age (17) and share the same religion (Catholic) and racial/cultural background (Hispanic).

Authors

Authors

The principle of homogamy is reflected in the mutual value these individuals have for biking and traveling. Most research suggests that people of similar interests find each other and their similarity is a major factor in their staying together.

vice versa, and the greater exposure of Blacks to the white culture than vice versa. Finally, Black mothers and white fathers have different roles in the respective Black and white communities in terms of setting the norms of interracial relationships.

> *In many Black families, mothers play the key role in accepting or not accepting an interracial relationship, much more than in white families. In white families, fathers were more often major players, as were siblings, grandparents, and other kin. . . . If women respond more often with openness and efforts to relate and less often with prejudice to the relationship choices of sons and daughters, then the fact that the crucial person in Black families is most often a woman means there may be more acceptance of a family member's entry into an interracial couple* (Rosenblatt, Karis, & Powell, 1995, 118).

Age Most individuals select someone who is relatively close in age. And men tend to select women three to five years younger than themselves. The result is the **marriage squeeze,** which is the imbalance of the ratio of marriageable-age men to marriageable-age women. In effect, women have fewer partners to select from since men have not only their same-age partners but those younger than themselves. One 40-year-old recently divorced woman said, "What chance do I have with all these guys looking at these younger women?"

Education Educational homogamy also operates strongly in selecting a living-together and marital partner (with greater homogamy for marital partners) (Kalmijn & Flap, 2001). Not only does college provide an opportunity to meet, date, live together, and marry another college student, but it also increases one's chance that only a college-educated partner becomes acceptable as a potential cohabitant or spouse. The very pursuit of education becomes a value to be shared. However, Lewis and Oppenheimer (2000) observed that when persons of similar education are not available, women are particularly likely to marry someone with less education. And the older the woman, the more likely she is to marry a partner with less education.

Social Class You have been reared in a particular social class that reflects your parents' occupations, incomes, and educations as well as your residence, language, and values. If you were brought up in a home in which both parents were physicians, you probably lived in a large house in a nice residential area—summer vacations and a college education were givens. Alternatively, if your parents dropped out of high school and worked at Wal Mart, your home would be smaller, in a less expensive part of town, and your opportunities (e.g., education) would be more limited. Social class affects one's comfort in interacting with others—we tend to feel more comfortable with others from our same social class.

Diversity in Other Countries

Educational homogamy is particularly prominent in Asian countries, where there is a strong emphasis on formal education as a channel for social mobility (Smits, Ultee, & Lammers, 1998).

The **mating gradient** refers to the tendency for husbands to be more advanced than their wives with regard to age, education, and occupational success. Two researchers assessed the expectations of 131 single female and 103 male college students and found that most women expected their husbands to be "superior in intelligence, ability, success, income, and education. Less than 10 percent

Up Close 7.1 Spirituality and Religion and a Couple's Relationship

Religion may be broadly defined as a specific fundamental set of beliefs (in reference to a supreme being, etc.) and practices generally agreed upon by a number of persons or sects. Similarly, some individuals view themselves as "not religious" but "spiritual," with spirituality defined as belief in the spirit as the seat of the moral or religious nature that guides one's decisions and behavior. Because religious or spiritual views reflect, in large part, who the person is, they have an enormous impact on one's attraction to a partner, the level of emotional engagement with that partner, and the durability of the relationship.

Religious homogamy is operative in that persons of similar religion sometimes seek out each other. Almost a third (27%) of undergraduate women and 15 percent of undergraduate men in one study reported that they would only marry someone of their same religion (Knox, Zusman, & Daniels, 2002). Baptists were significantly more likely than Methodists or Catholics to believe that persons marrying outside their religion would eventually divorce.

Couples with a homogeneous religious marriage may have greater marital stability and a lower chance of divorce due to the value of religion for resolving conflicts. Butler, Stout, & Gardner (2002) found that couples who prayed in reference to their marital conflicts reported a softening of their positions and a feeling of "healing." Indeed the authors of the study suggested that marriage counselors should be aware of this spiritual source of healing.

It appears that those couples with religious or spiritual ties—especially shared beliefs—are more inclined to try every means possible to make their marriage work. They may see themselves as letting down the community and their Higher Power (God, divine power, source of energy), not just separating themselves from each other, if the marriage fails. At the same time, such couples tap into a power beyond themselves to help them handle both daily life and unexpected crises.

Religious literature often provides practical, down-to-earth suggestions for relationship enhancement. For example, forgiveness involves recognizing the offense but choosing to learn from it, releasing the partner from any penalty, and moving on.

Spiritual involvement brings with it guidelines designed to lessen stress and conflict in relationships. The general outcome includes enhanced interest in the partner and greater desire to be with that beloved. Examples abound in religious writings on how to demonstrate romantic love and desire in healthy, growth-producing ways. (One prime example is the Song of Solomon—Song of Songs—found in the Christian and Jewish Bibles.)

The phrase "the couple that prays together, stays together" is more than just a cute cliché. However, it should also be pointed out that religion can serve as a divisive force. For example, when one partner becomes "born again" or "saved," unless the other partner shares the experience, the relationship can be dramatically altered and eventually terminated. A former rock-and-roll hard-drinking, drug-taking wife noted that when her husband "got saved," it was the end of their marriage. "He gave our money to the church as the 'tithe' when we couldn't even pay the light bill," she said. "And when he told me I could no longer wear pants or lipstick, I left."

Sources

Butler, M. H., J. A. Stout, & B. C. Gardner. 2002. Prayer as a conflict resolution ritual: Clinical implications of religious couple's report of relationship softening, healing perspective, and change responsibility. *American Journal of Family Therapy* 30:19–37.

Knox, D., M. E. Zusman, and V. W. Daniels. 2002. College student attitudes toward interreligious marriage. *College Student Journal* 31:445–48.

Appreciation is expressed to Steven O. Langehough for his participation in the development of this section.

of the women in this sample expected to exceed their marriage partner on any of the variables measured" (Ganong & Coleman, 1992, 61).

Physical Appearance Marcus and Miller (2003) noted, "Beauty is quite clearly not entirely in the eye of the beholder. Instead, some of us are judged by almost everyone we meet as handsome or lovely, whereas others of us nearly always seem plain" (p. 333). In their study of physical attractiveness, they found that "people know they are pretty or handsome" (p. 325).

In addition, people tend to become involved with those who are similar in physical attractiveness. However, a partner's attractiveness may be a more important consideration for men than for women. In a study of homogamous preferences in mate selection, men and women rated physical appearance an average of 7.7 and 6.8 (out of 10) in importance, respectively (Knox, Zusman, & Nieves, 1997).

Marital Status The never-married tend to select as marriage partners the never-married, the divorced tend to select the divorced, and the widowed tend to select the widowed. Similar marital status may be more important to women than to men. In the study of homogamous preferences in mate selection, women and men rated similarity of marital status an average of 7.2 and 6.3 (out of 10) in importance, respectively (Knox, Zusman, & Nieves, 1997).

Religion Religion is not just a creed, thought, experience, or affiliation, and it may impact individuals in various ways (Brodsky, 2000). Although "religious" is not a characteristic typically associated with college students, over 83 percent of all first-year college students reported that they had attended a religious service in the last year (American Council on Education & University of California, 2002). Up Close 7.1 on page 163 details the influence of spiritual/religious homogamy on a couple's relationship.

Personality Conservatives, liberals, risk takers, etc. tend to select each other as marital partners. And their doing so has positive consequences for them and their relationship. Partners who select others with similar personalities report high subjective well-being (Arrindell & Luteijn, 2000).

Propinquity

Almost seventy years ago, sociologist James Bossard found that one-sixth of five thousand individuals applying for marriage licenses in Philadelphia lived within one city block of their partners. One-third of the individuals lived within five blocks of each other, and half lived within twenty blocks (Bossard, 1932). His finding illustrates the principle of **residential propinquity,** or the tendency to select marriage partners from among those who live nearby.

The more generic term **propinquity** refers to the tendency to marry someone who lives, works, or goes to school in the same area. Propinquity is clearly evident among college students who meet in class or on campus, date, and marry. Such propinquity also involves sharing similar interests, frustrations, values, and life experiences.

PSYCHOLOGICAL FACTORS OPERATIVE IN MATE SELECTION

Psychologists have focused on complementary needs, exchanges, parental characteristics, and personality types with regard to mate selection.

Complementary-Needs Theory

"In spite of the women's movement and a lot of assertive friends, I am a shy and dependent person," remarked a transfer student. "My need for dependency is met by Warren, who is the dominant, protective type." The tendency for a submissive person to become involved with a dominant person (one who likes to control the behavior of others) is an example of attraction based on complementary needs. **Complementary-needs theory** states that we tend to select mates whose needs are opposite and complementary to our own. Partners can also be drawn to each other on the basis of nurturance versus receptivity. These complementary needs suggest that one person likes to give and take care of another, while the other likes to be the benefactor of such care. Other examples of complementary needs may involve responsibility versus irresponsibility, peacemaker versus troublemaker, and disorder versus order. The idea that mate selection is based on complementary needs was suggested by Winch (1955), who noted that needs can be complementary if they are different (for example, dominant and submissive) or if the partners have the same need at different levels of intensity.

As an example of the latter, two individuals may have a complementary relationship when they both want to do advanced graduate study but do not both have a need to get a Ph.D. The partners will complement each other if one is comfortable with his or her level of aspiration as represented by a master's degree but still approves of the other's commitment to earn a Ph.D.

Winch's theory of complementary needs, commonly referred to as "opposites attract," is based on the observation of twenty-five undergraduate married couples at Northwestern University. The findings have been criticized by other researchers who have not been able to replicate Winch's study (Saint, 1994). Two researchers said, "It would now appear that Winch's findings may have been an artifact of either his methodology or his sample of married people" (Meyer & Pepper, 1977).

Three questions can be raised about the theory of complementary needs:

1. Couldn't personality needs be met just as easily outside the couple's relationship as through mate selection? For example, couldn't a person who has the need to be dominant find such fulfillment in a job that involved an authoritative role, such as being a supervisor?

2. What is a complementary need as opposed to a similar value? For example, is the desire to achieve at different levels a complementary need or a shared value?

3. Don't people change as they age? Could a dependent person grow and develop self-confidence so that he or she might no longer need to be involved with a dominant person? Indeed, the person might no longer enjoy interacting with a dominant person.

Exchange Theory

Exchange theory emphasizes that mate selection is based on assessing who offers the greatest rewards at the lowest cost. Five concepts help to explain the exchange process in mate selection.

If you would be married fitly, wed your equal.

Ovid, *Roman poet*

1. Rewards. Rewards are the behaviors (your partner looking at you with the eyes of love), words (saying "I love you"), resources (being beautiful or handsome, having a car, condo, and money), and services (cooking for you, typing for you) your partner provides for you that you value and that influence you to continue the relationship. Increasingly, men are interested in women who offer "financial independence." In a study of Internet ads placed by women, the woman who described herself as "financially independent . . . successful and ambitious" produced 50 percent more responses than the next most popular ad, in which the woman described herself as "lovely . . . very attractive and slim" (Strassberg & Holty, 2003).

2. Costs. Costs are the unpleasant aspects of a relationship. A woman identified the costs associated with being involved with her partner: "He abuses drugs, doesn't have a job, and lives nine hours away." The costs her partner associated with being involved with this woman included "she nags me," "she doesn't like sex," and "she wants her mother to live with us if we marry." Ingoldsby, Schvaneveldt, & Uribe (2003) assessed the degree to which various characteristics were associated with reducing one's attractiveness on the marriage market. The most damaging traits were not being heterosexual, having alcohol/drug problems, having a sexually transmitted disease, and being lazy.

3. Profit. Profit occurs when the rewards exceed the costs. Unless the couple referred to above derive a profit from staying together, they are likely to end their relationship and seek someone else with whom there is a higher profit margin.

4. Loss. Loss occurs when the costs exceed the rewards.

5. Alternative. No other person is currently available who offers a higher profit margin.

Most people have definite ideas about what they are looking for in a mate. For example, Xie et al. (2003) found that men with good incomes were much more likely to marry than men with no or low incomes. The currency used in the marriage market consists of the socially valued characteristics of the persons involved, such as age, physical characteristics, and economic status. In our free-choice system of mate selection, we typically get as much in return for our social attributes as we have to offer or trade. An unattractive, drug-abusing high school dropout with no job has little to offer an attractive, drug-free, 3.5 GPA college student who has just been accepted to graduate school.

Once you identify a person who offers you a good exchange for what you have to offer, other bargains are made about the conditions of your continued relationship. Waller and Hill (1951) observed that the person who has the least interest in continuing the relationship could control the relationship. This **principle of least interest** is illustrated by the woman who said, "He wants to date me more than I want to date him, so we end up going where I want to go and doing what I want to do." In this case, the woman trades her company for the man's acquiescence to her recreational choices.

Parental Characteristics

Whereas the complementary needs and exchange theories of mate selection are relatively recent, Freud suggested that the choice of a love object in adulthood represents a shift in libidinal energy from the first love objects—the parents. Role theory and modeling theory emphasize that a son or daughter models after the parent of the same sex by selecting a partner similar to the one the parent selected. This means that a man looks for a wife who has similar characteristics to those of his mother and that a woman looks for a husband who is very similar to her father.

Desired Personality Characteristics for a Potential Mate

In a study of seven hundred undergraduates, both men and women reported that the personality characteristics of being warm, kind, open, and having a sense of humor were very important to them in selecting a romantic/sexual partner. Indeed, these intrinsic personality characteristics were rated as more important than physical attractiveness or wealth (extrinsic characteristics) (Sprecher & Regan, 2002). Similarly, adolescents wanted intrinsic qualities such as intelligence and humor in a romantic partner but looked for physical appearance and high sex drive in a casual partner (no sex differences between women and men were found) (Regan & Joshi, 2003).

A man falls in love through his eyes, a woman through her ears.
Woodrow Wyatt

In another study, women were significantly more likely than men to identify "having a good job" and "being well educated" as important attributes in a future mate while significantly more men than women wanted their spouse to be physically attractive (Medora et al., 2002). Toro-Morn and Sprecher (2003) noted that both American and Chinese undergraduate women identified characteristics associated with status (earning potential, wealth) more than physical attractiveness. The behavior that 60 percent of a national sample of adult single women reported as the most serious fault of a man was his being "too controlling" (Edwards, 2000).

Women are also attracted to men who have good manners. In a study of 398 undergraduates, women were significantly more likely than men to report that they "only wanted to date or be involved with someone who had good manners," that "manners are very important," and that "the more well mannered the person, the more I like the person" (Zusman, Knox, Gescheidler, & McGinty, in press).

Personality Characteristics Predictive of Divorce

Researchers have defined several personality factors predictive of divorce.

1. Poor impulse control. Persons who have poor impulse control have little self-restraint and may be prone to aggression and violence (Snyder & Regts, 1990). Lack of impulse control is also problematic in marriage because the person is less likely to consider the consequences of his or her actions. For example, to some people, having an affair might sound like a good idea at the time, but it will have devastating consequences for the marriage.

2. Hypersensitivity. Hypersensitivity to perceived criticism involves getting hurt easily. Any negative statement or criticism is received with a greater impact than intended by the partner. The disadvantage of such hypersensitivity is that the partner may learn not to give feedback for fear of hurting the hypersensitive partner. Such lack of feedback to the hypersensitive partner blocks information about what the person does that upsets the other and what he or she could do to make things better. Hence, the hypersensitive one has no way of learning that something is wrong, and the partner has no way of alerting the hypersensitive partner. The result is a relationship in which the partners can't talk about what is wrong, so the potential for change is limited (Snyder & Regts, 1990).

3. Inflated ego. An exaggerated sense of oneself is another way of saying the person has a big ego and always wants things to be his or her way. A person with an inflated sense of self may be less likely to consider the other person's opinion in negotiating a conflict and prefer to dictate an outcome. Such disrespect for the partner can be damaging to the relationship (Snyder & Regts, 1990).

4. Being neurotic. Such individuals are perfectionists and require of themselves and others that they be perfect. This attitude is associated with relationship problems (Haring, Hewitt, & Flett, 2003).

5. Anxiety. Husbands who report high levels of anxiety tend to report lower marital adjustment. No such association holds for wives. While the explanation is not clear, the authors of one study postulated that "anxious husbands may have more negative interactions with their wives, which may also lower perceived marital quality" (Dehle & Weiss, 2002, 336).

6. Insecurity. Feelings of insecurity also compromise marital happiness. Researchers studied the personality trait of attachment and its effect on marriage in 157 couples at two time intervals and found that "insecure participants reported more difficulties in their relationships. . . . [I]n contrast secure participants reported greater feelings of intimacy in the relationship at both assessments (Crowell, Treboux, & Waters, 2002)

7. Controlled. Individuals who are controlled by their parents, grandparents, former partner, child, or whomever compromise the marriage relationship since their allegiance is external to the couple's relationship. Unless the person is able to break free of such control, the ability to make independent decisions will be thwarted, which will both frustrate the spouse and challenge the marriage.

To summarize, a number of personality factors may be predictive of negative outcomes in marriage—poor impulse control, hypersensitivity, overly inflated ego, being neurotic, perfectionism, high anxiety, insecurity, and being controlled by others. Figure 7.1 summarizes the cultural, sociological, and psychological filters involved in mate selection.

Figure 7.1
Cultural, Sociological, and Psychological Filters in Mate Selection

Cultural Filters

For two people to consider marriage to each other,

Endogamous factors (same race, age) and Exogamous factors (not blood-related)
↓
must be met.
↓

After the cultural prerequisites have been satisfied, sociological and psychological filters become operative.

Sociological Filters

Propinquity = the tendency to select a mate from among those who live, work, or go to school nearby.

Homogamy = the tendency to select a mate similar to oneself with regard to the following:

Race	Physical appearance
Education	Body clock compatibility
Social class	Religion
Age	Marital status
Intelligence	Interpersonal values

Psychological Filters

Complementary needs
Reward-cost ratio for profit
Parental characteristics
Desired personality characteristics

PERSONAL CHOICES

Who Is the Best Person for You to Marry?

In a study commissioned by *Time* magazine of 465 never-married adults, 78 percent of the women and 79 percent of the men thought they would find and marry their perfect mate (Edwards, 2000). Although there is no perfect mate, some individuals are more suited to you as a marriage partner than others. As we have seen in this chapter, persons who have a big ego, poor impulse control, and an oversensitivity to criticism, and who are anxious and neurotic should be considered with great caution.

Equally as important as avoiding someone with problematic personality characteristics is selecting someone with whom you have a great deal in common. "Marry someone just like you" may be a worthy guideline in selecting a marriage partner. Homogamous matings with regard to race, education, age, values, religion, social class, and marital status (e.g., never-marrieds marry never-marrieds; divorced with children marry those with similar experience) tend to result in more durable, satisfying relationships. "Marry your best friend" is another worthy guideline for selecting the person you marry.

Finally, marrying someone with whom you have a relationship of equality and respect is associated with marital happiness. Relationships in which one partner is exploited or intimidated engender negative feelings of resentment and distance. One man said, "I want a co-chair, not a committee member, for a mate." He was saying that he wanted a partner to whom he related as an equal.

Reference

Edwards, T. M. Flying solo. *Time,* August, 47–53.

SOCIOBIOLOGICAL FACTORS OPERATIVE IN MATE SELECTION

In contrast to cultural, sociological, and psychological aspects of mate selection, which reflect a social learning assumption, the sociobiological perspective suggests that biological/genetic factors may be operative in mate selection.

Definition of Sociobiology

Sociobiology suggests a biological basis for all social behavior—including mate selection. Based on Charles Darwin's theory of natural selection, which states that the strongest of the species survive, sociobiology holds that men and women select each other as mates on the basis of their innate concern for producing offspring who are most capable of surviving.

According to sociobiologists, men look for a young, healthy, attractive, sexually conservative woman who will produce healthy children and who will invest in taking care of the children. Women, in contrast, look for an ambitious man with good economic capacity who will invest his resources in her children. Earlier in this chapter, we provided data supporting the idea that men seek attractive women and women seek ambitious, financially successful men.

Criticisms of the Sociobiological Perspective

The sociobiological explanation for mate selection is controversial. Critics argue that women may show concern for the earning capacity of men because women have been systematically denied access to similar economic resources, and selecting a mate with these resources is one of their remaining options. In addition, it is argued that both women and men, when selecting a mate, think about their partners more as companions than as future parents of their offspring.

ENGAGEMENT

Engagement moves the relationship of a couple from a private love-focused experience to a public experience. Family and friends are invited to enjoy the happiness and commitment of the individuals to a future marriage. Unlike casual dating, engagement is a time in which the partners are emotionally committed, are sexually monogamous, and are focused on wedding preparations. The engagement period is your last opportunity before marriage to systematically examine your relationship, ask each other specific questions, find out about the partner's parents and family background, and participate in marriage education or counseling.

Asking Specific Questions

Since partners may neither ask nor reveal information that they feel will be met with disapproval during casual dating, the engagement is a time to get specific about the other partner's thoughts, feelings, values, goals, and expectations. The Involved Couple's Inventory is designed to help individuals in committed relationships learn more about each other by asking specific questions.

Self-Assessment

Involved Couple's Inventory

The following questions are designed to increase your knowledge of how you and your partner think and feel about a variety of issues. Assume that you and your partner have considered getting married. Each partner should ask the other the following questions:

Careers and Money

1. What kind of job or career will you have? What are your feelings about working in the evening versus being home with the family? Where will your work require that we live? How often do you feel we will be moving? How much travel will your job require?
2. What are your feelings about a joint versus a separate checking account? Which of us do you want to pay the bills? How much money do you think we will have left over each month? How much of this do you think we should save?
3. When we disagree over whether to buy something, how do you suggest we resolve our conflict?
4. What jobs or work experience have you had? If we end up having careers in different cities, how do you feel about being involved in a commuter marriage?
5. What is your preference for where we live? Do you want to live in an apartment or a house? What are your needs for a car, large, high-definition TV, cablevision, DVD, and so on?
6. How do you feel about my having a career? Do you expect me to earn an income? If so, how much annually? To what degree do you feel it is your responsibility to cook, clean, and take care of the children? How do you feel about putting young children or infants in day-care centers? When the children are sick and one of us has to stay home, who will that be?
7. To what degree do you want me to account to you for the money I spend? How much money, if any, do you feel each of us should have to spend each week as we wish without first checking with the other partner? What percentage of income, if any, do you think we should give to charity each year?

Religion and Children

1. To what degree do you regard yourself as a religious/spiritual person? What do you think about religion, a Supreme Being, prayer, and life after death?
2. Do you go to religious services? Where? How often? Do you pray? How often? What do you pray about? When we are married, how often would you want to go to religious services? In what religion would you want our children to be reared? What responsibility would you take to ensure that our children had the religious training you wanted them to have?
3. How do you feel about abortion? Under what conditions, if any, do you feel abortion is justified?
4. How do you feel about children? How many do you want? When do you want the first child? At what intervals would you want to have additional children? What do you see as your responsibility for child care—changing diapers, feeding, bathing, playing with children, and taking them to piano lessons? To what degree do you regard these responsibilities as mine?
5. Suppose I did not want to have children or couldn't have them—how would you feel? How do you feel about artificial insemination, surrogate motherhood, in vitro fertilization, and adoption?
6. To your knowledge, can you have children? Are there any genetic problems in your family history that would prevent us from having normal children?
7. Do you want our children to go to public or private schools?
8. How should children be disciplined? How were you disciplined as a child?
9. How often do you think we should go out alone without our children?
10. What are your expectations of me regarding religious participation with you and our children?

Sex

1. How much sexual intimacy do you feel is appropriate in casual dating, involved dating, and engagement?
2. What do you think about masturbation, oral sex, homosexuality, S & M, and anal sex?
3. What type of contraception do you suggest? Why? If that method does not prove satisfactory, what method would you suggest next?
4. What are your values regarding extramarital sex? If I were to have an affair and later told you, what would you do? Why? If I had an affair, would you want me to tell you? Why?
5. What sexual behaviors do you most and least enjoy? How often do you want to have intercourse? How do you want me to turn you down when I don't want to have sex? How do you want me to approach you for sex? How do you feel about just being physical together—hugging, rubbing, holding, but not having intercourse?
6. By what method of stimulation do you experience an orgasm most easily?
7. What are pornography and/or erotica to you? How do you feel about watching X-rated videos?
8. How important is our using a condom to you?
9. Do you want me to be tested for HIV? Are you willing to be tested?
10. What sexually transmitted infections have you had?

11. How much do you want to know about my sexual behavior with previous partners?

12. If we broke up, would you be willing to have a "friends with benefits" relationship?

Relationships with Friends/Coworkers

1. How do you feel about my spending one evening a week with my friends or coworkers?

2. How do you feel about my spending time with friends of the opposite sex during this time?

3. What do you regard as appropriate and inappropriate affectional behaviors with opposite sex friends?

Recreation and Leisure

1. How do you feel about golfing, surfing, swimming, boating, horseback riding, jogging, lifting weights, racquetball, basketball, football, baseball, tennis, soccer, wrestling, fishing, hunting? How often do you engage in each of these activities? What recreational activities would you like me to share with you?

2. What hobbies do you have?

3. What do you like to watch on TV? How often do you watch TV and for what periods of time?

4. What are the amount and frequency of your current alcohol and other drug (e.g., marijuana, cocaine, crack, speed) consumption? What, if any, have been your previous alcohol and other drug behaviors and frequencies? What are your expectations of me regarding the use of alcohol and other drugs?

5. Where did you vacation with your parents? Where will you want us to go? How will we travel? How much money do you feel we should spend on vacations each year?

Partner Feelings

1. If you could change one thing about me, what would it be?

2. What would you like me to do to increase your happiness?

3. What would you like me to say or not say that would make you happier?

4. What do you think of yourself? Describe yourself with three adjectives.

5. What do you think of me? Describe me with three adjectives.

6. What do you like best about me?

7. Do you think I get jealous easily? How will you cope with my jealousy?

8. How do you feel about me emotionally?

9. To what degree do you feel we each need to develop and maintain outside relationships so as not to focus all of our interpersonal expectations on each other?

Feelings about Parents/Family

1. How often do you have contact with your father/mother? How do you feel about your parents? How often do you want to visit your parents and/or siblings? How often would you want them to visit us? Do you want to spend holidays alone?

2. What do you like and dislike about my parents?

3. What is your feeling about living near our parents? How would you feel about my parents living with us? What will we do with our parents if they can't take care of themselves?

4. How do your parents get along? Rate their marriage on a 0–10 scale (0 = unhappy, 10 = happy). What are your parents' role responsibilities in their marriage?

5. To what degree did members of your family enjoy spending their free time together? What are your expectations of our spending free time together?

6. To what degree did members of your family consult one another on their decisions? To what degree do you expect me to consult you on the decisions that I make?

7. How close were your family members to one another? To what degree do you value closeness versus separateness in our relationship?

8. Who was the dominant person in your family? Who had more power? Who do you regard as the dominant partner in our relationship? How do you feel about this power distribution?

9. What "problems" has your family experienced? Is there any history of mental illness, alcoholism, drug abuse, suicide, etc.?

Remarriage Questions

1. How and why did your first marriage end? What are your feelings about your former spouse now? What are the feelings of your former spouse toward you? How much "trouble" do you feel she or he will want to cause us? What relationship do you want with your former spouse?

2. Do you want your children from a previous marriage to live with us? What are your emotional and financial expectations of me in regard to your children? What are your feelings about my children living with us? Do you want us to have additional children? How many? When?

3. When your children are with us, who will be responsible for their food preparation, care, discipline, and driving them to activities?

4. Suppose your children do not like me and vice versa. How will you handle this? Suppose they are against our getting married?

5. Suppose our respective children do not like one another. How will you handle this?

(*continued on next page*)

Involved Couple's Inventory (*continued*)

6. What assets or debts will you bring into the marriage?
7. How much child support/alimony do you get or pay each month? Tell me about your divorce
8. May I read your divorce settlement agreement? When?

Other Questions

1. Do you have any history of abuse or violence, either as an abused child or adult or as the abuser in an adult relationship?
2. If we could not get along, would you be willing to see a marriage counselor? Would you see a sex therapist if we were having sexual problems?
3. What is your feeling about prenuptial agreements?
4. What value do you place on the opinions or values of your parents and friends?
5. What are your feelings about our living together?
6. To what degree do you enjoy getting and giving a massage? How important is it to you that we massage each other regularly?

It would be unusual if you agreed with each other on all of your answers to the previous questions. You might view the differences as challenges and then find out the degree to which the differences are important for your relationship. You might need to explore ways of minimizing the negative impact of those differences on your relationship. It is not possible to have a relationship with someone where there is total agreement. Disagreement is inevitable; the issue becomes how you and your partner manage the differences.

Note: This self-assessment is intended to be thought-provoking and fun. It is not intended to be used as a clinical or diagnostic instrument.

Visiting Partner's Parents

Seize the opportunity to discover the family environment in which your partner was reared and consider the implications for your subsequent marriage. When visiting your partner's parents, observe their standard of living, the way they relate to each other, and the degree to which your partner is similar to the same-sex parent. How does their standard of living compare with that of your own family? How does the emotional closeness (or distance) of your partner's family compare with that of your family? Such comparisons are significant because both you and your partner will reflect your respective home environments to some degree. If you want to know what your partner may be like in the future, look at his or her parent of the same sex. There is a tendency for a man to become like his father and a woman to become like her mother.

And, if you want to know how your partner is likely to treat you in the future, observe the way your partner's parent of the same sex treats and interacts with his or her spouse. Their relationship is the model of a spousal relationship your partner is likely to duplicate in relating to you.

Premarital Programs and Counseling

Some premarital couples attend the Prevention and Relationship Enhancement Program (PREP). PREP is designed for all couples who want to learn the essential skills for a lasting relationship and is offered throughout the United States, Canada, and Europe.

Outcome research over ten years on the effectiveness of PREP is impressive. Not only have couples who learned how to communicate and negotiate conflict been less likely than a control group to divorce or separate (8 percent versus 16 percent), but they reported greater marital satisfaction, fewer conflicts, and less physical violence (Renick, Blumberg, & Markman, 1992). Couples who want to help

Diversity in the United States

Among the Hopi Indians, when a couple think of getting married, the potential bride's family prepares food and takes it to the potential groom's family. If his family members accept the food and approve of the union, they reciprocate by giving food (usually meat) to her family—and the engagement is on.

insulate themselves against divorce might well consider such preventive programs, which focus on communication and negotiation skills

Individuals who want to marry in the Roman Catholic Church are required to take premarital education. Other faiths may also offer premarital sessions (usually three) before the wedding. These sessions consist of information about marriage, an assessment of the couple's relationship, and/or resolving conflicts that have surfaced in the couple's relationship. Although the couple might deny the existence of a problem for fear that looking at it will break up the relationship, the counselor can help them examine the problem and work toward solving it.

Prenuptial Agreement

Some couples (particularly those with considerable assets or those in subsequent marriages) discuss and sign a prenuptial agreement. To reduce the chance that the agreement will later be challenged, each partner should hire an attorney (months before the wedding) to develop and/or review the agreement. The primary purpose of a **prenuptial agreement** (also referred to as a premarital agreement, marriage contract, or antenuptial contract) is to specify ahead of time how property will be divided if the marriage ends in divorce or when it ends by the death of one partner. In effect, the value of what you take into the marriage is the amount you are allowed to take out of the marriage. For example, if you bring $150,000 into the marriage and buy the marital home with this amount, at divorce, your ex-spouse is not automatically entitled to half the house. Some agreements may also contain clauses of no spousal support (no alimony) if the marriage ends in divorce. See Appendix D for an example of a Prenuptial Agreement developed by a husband and wife who had both been married before and had assets and children.

Reasons for a prenuptial agreement include the following:

1. Protecting assets for children from a prior relationship. Persons who are in the middle or later years, who have considerable assets, who have been married before, and who have children are often concerned that money and property be kept separate in a second marriage so that the assets at divorce or death go to the children. Some children encourage their remarrying parent to draw up a prenuptial with the new partner so that their (the offspring's) inheritance, house, or whatever will not automatically go to the new spouse upon the death of their parent.

2. Protecting business associates. A spouse's business associate may want a member of a firm or partnership to draw up a prenuptial agreement with a soon-to-be-spouse to protect the firm from intrusion by the spouse if the marriage does not work out.

Prenuptial contracts do have a value beyond the legal implications. Their greatest value may be that they facilitate the partners' discussing with each other their expectations of the relationship. In the absence of such an agreement, many couples may never discuss the issues they may later face.

There are disadvantages of signing a prenuptial agreement. They are often legally challenged ("My partner forced me to sign it or call off the wedding"), and not all issues can be foreseen ("Who gets the time share vacation property?"). Prenuptial agreements also are not very romantic ("I love you, but sign here and see what you get if you don't please me.") and may serve as a self-fulfilling prophecy ("We were already thinking about divorce.").

Prenuptials are almost nonexistent in first marriages and, while they occur in second marriages, they are rare. Whether or not it is a good idea to sign a

Increasing Requirements for a Marriage License

SOCIAL POLICY

Should marriage licenses be obtained so easily? Should couples be required, or at least encouraged, to participate in premarital education before saying "I do"? Given the high rate of divorce today, policymakers and family scholars are considering this issue. While evidence of long-term effectiveness of premarital education remains elusive (Carroll & Doherty, 2003), some believe that "mandatory counseling will promote marital stability" (Licata, 2002, 518).

Several states have proposed legislation requiring premarital education (also referred to as premarital counseling, premarital therapy, and marriage preparation). For example, an Oklahoma statute provides that parties who complete a premarital education program pay a reduced fee for their marriage license. Also, in Lenawee County, Michigan, local civil servants and clergy have made a pact—they will not marry a couple unless that couple has attended marriage education classes. Other states that are considering policies to require or encourage premarital education include Arizona, Illinois, Iowa, Maryland, Minnesota, Mississippi, Missouri, Oregon, and Washington.

Proposed policies include not only mandating premarital education and lowering marriage license fees for those who attend courses but imposing delays on issuing marriage licenses for those who refuse premarital education. However, "no state mandates premarital counseling as a prerequisite to obtaining a license" (Licata, 2002, 525).

Traditionally, most Protestant pastors and Catholic priests require premarital counseling before they will perform marriage ceremonies. Couples who do not want to participate in premarital education can simply get married in secular ceremonies.

Advocates of mandatory premarital education emphasize that it will reduce marital discord. However, questions remain about who will offer what courses and whether couples will take seriously the content of such courses. Indeed, persons contemplating marriage are often narcotized with love and would doubtless not take any such instruction seriously. Love myths such as "divorce is something that happens to other people" and "our love will overcome any obstacles" work against the serious consideration of such courses.

Sources

Carroll, J. S., and W. J. Doherty. 2003. Evaluating the effectiveness of premarital prevention programs: A meta-analytic review of outcome research. *Family Relations* 52:105–18.

Licata, N. 2002 Should premarital counseling be mandatory as a requisite to obtaining a marriage license? *Family Court Review* 40:518–32

prenuptial agreement depends on the circumstances. Some individuals who do sign an agreement later regret it. Sherry, a never-married 22-year-old at the time, signed such an agreement:

> *Paul was adamant about my signing the contract. He said he loved me but would never consider marrying me unless I signed a prenuptial agreement stating that he would never be responsible for alimony in case of a divorce. I was so much in love, it didn't seem to matter. I didn't realize that basically he was and is a selfish person. Now, five years later after our divorce, I live in a mobile home and he lives in a big house overlooking the lake with his new wife.*

The husband viewed it differently. He was glad that she had signed the agreement and that his economic liability to her was limited. He could afford the new house by the lake with his new wife because he was not sending money to Sherry. Billionaire Donald Trump attributed his economic survival of two divorces to prenuptial agreements with his ex-wives.

Couples who decide to develop a prenuptial agreement need separate attorneys to look out for their respective interests. The laws regulating marriage and divorce vary by state, and only attorneys in those states can help ensure that the document drawn up will be honored. Full disclosure of assets is also important. If one partner hides assets, the prenuptial can be thrown out of court. One husband recommended that the issue of the premarital agreement should be brought up and signed a minimum of six months before the wedding. "This gives the issue time to settle rather than being an explosive emotional issue if it is brought up a few weeks before the wedding." Indeed, as noted above, if a prenuptial agreement is signed within two weeks of the wedding, that is grounds enough for the agreement to be thrown out of court, since it is assumed that the document was executed under pressure.

While individuals are deciding whether to have a prenuptial agreement, states are deciding whether to increase marriage license requirements, this chapter's social policy issue.

CONSIDER CALLING OFF THE WEDDING IF . . .

If your engagement is characterized by the factors identified below, consider prolonging your engagement and delaying the marriage at least until the most distressing issues have been resolved.

A lady took out an ad in the classifieds: "Husband wanted." Next day she received a hundred letters. They all said the same thing: "You can have mine."

Internet humor

Age 18 or Younger

The strongest predictor of getting divorced is getting married as an adolescent. Individuals who marry in their teens have a greater risk of divorce than those who delay marriage into their mid-20s. Teenagers may be more at risk for marrying to escape an unhappy home and may be more likely to engage in impulsive decision making and behavior.

Waiting until one is older and finished with college might not be a bad idea. Research by Meehan and Negy (2003) on being married while in college revealed higher marital distress among spouses who were also students. In addition, when married college students were compared with single college students, the married students reported more difficulty adjusting to the demands of higher education. The researchers conclude that "these findings suggest that individuals opting to attend college while being married are at risk for compromising their marital happiness and may be jeopardizing their education" (p. 688).

Hence, waiting until one is older and through college not only may result in a less stressful marriage, but also may be associated with less economic stress.

NATIONAL DATA

The median annual incomes of women who complete high school, college, and graduate school (master's level) are $15,665, $30,973, and $40,744, respectively. The corresponding salaries for men are $28,343, $49,985, and $61,960 (*Statistical Abstract of the United States: 2003*, Table 693).

Known Partner Less Than Two Years

Impulsive marriages where the partners had known each other for less than a month are associated with a higher-than-average divorce rate. Indeed, partners who date each other for at least two years (25 months is best) before getting married report the highest level of marital satisfaction and are less likely to divorce (Huston et al., 2001). A short courtship does not allow the partners to observe and scrutinize each other's behavior in a variety of settings. Indeed, some individuals may be more prone to fall in love at first sight and to want to hurry the partner into a committed love relationship. Two researchers observed that anxious people "fall in love quickly, making them susceptible to choosing partners who are inappropriate or even dangerous" (Morgan & Shaver, 1999).

Marry in haste and repent at leisure.

John Ray, *English Proverbs*

Suggestions include making a joint decision that getting to know each other over several years is important, taking a five-day "primitive" camping trip, taking a fifteen-mile hike together, wallpapering a small room together, or spending

Authors

One of the best ways to get to know each other is to travel together. This couple took a six-week trip together to the Far East.

several days together when one partner has the flu. If the couple plan to have children, they may want to take care of a six-month-old together for a weekend. Time should also be spent with each other's friends.

Abusive Relationship

As we will discuss in Chapter 14, Violence and Abuse in Relationships, partners who emotionally and/or physically abuse their partners while dating and living together continue these behaviors in marriage. Abusive lovers become abusive spouses with predictable negative outcomes. Though extricating oneself from an abusive relationship is difficult before the wedding, it becomes even more difficult after marriage. The price to oneself of being in an abusive relationship spreads to one's children.

One aspect of abuse to be aware of is a partner who attempts to isolate you from family or friends. Some partners attempt to systematically detach their intended spouse from all other relationships ("I don't want you spending time with your family and friends—you should be here with me"). This is a serious flag of impending relationship doom and should not be overlooked.

Critical Remarks

Stanley and Markman (1997) analyzed the communication patterns of 947 adults and found that those who belittled each other with caustic critical remarks were less likely to stay together. Of course, giving negative feedback or having a disagreement can be functional for a couple's relationship. The best of relationships are those in which the partners are free to express their differences. Partners who hold back what they really feel may harbor resentment that, over time, becomes destructive to the marriage. Being able to discuss differences and resolve issues moves the couple closer together, not farther apart.

Numerous Significant Differences

However, relentless conflict often arises from numerous significant differences. Though all spouses are different from each other in some ways, those who have numerous differences in key areas such as race, religion, social class, education, values, and goals are less likely to report being happy and to have durable relationships. Persons who report the greatest degree of satisfaction in durable relationships have a great deal in common (Houts, Robin, & Huston, 1996; Michael et al., 1994). Conversely, Skowron (2000) found that the less couples had in common, the more their marital distress.

On-and-Off Relationship

A roller-coaster premarital relationship is predictive of a marital relationship that will follow the same pattern. Partners who break up and get back together several times have developed a pattern where the dissatisfactions in the relationship become so frustrating that separation becomes the antidote for relief. In courtship, separations are of less social significance than marital separations. "Breaking up" in courtship is called "divorce" in marriage. Couples who routinely break up and get back together should examine the issues that continue to recur in their relationship and attempt to resolve them.

Dramatic Parental Disapproval

A parent recalled, "I knew when I met the guy it wouldn't work out. I told my daughter and pleaded that she not marry him. She did, and they divorced." Such parental predictions (whether positive or negative) often come true. If the predictions are negative, they sometimes contribute to stress and conflict once the couple marries.

Even though parents who reject the commitment choice of their offspring are often regarded as uninformed and unfair, their opinions should not be taken lightly. The parents' own experience in marriage and their intimate knowledge of their offspring combine to put them in a unique position to assess how their child might get along with a particular mate. If the parents of either partner disapprove of the marital choice, the partners should try to evaluate these concerns objectively. The insights might prove valuable. The value of parental approval is illustrated in a study of Chinese marriages. Pimentel (2000) found that higher marital quality was associated with parents' approving of the mate choice of their offspring.

Low Sexual Satisfaction

Sexual satisfaction is linked to relationship satisfaction, love, and commitment. Sprecher (2002) followed 101 dating couples across time and found that low sexual satisfaction (for both women and men) was related to reporting low relationship quality, less love, lower commitment, and breaking up. Hence, couples who are dissatisfied with their sexual relationship might explore ways of improving it (alone or through counseling) or consider the impact of such dissatisfaction on the future of their relationship.

Marrying for the Wrong Reason

Some reasons for getting married are more questionable than others. These reasons include the following:

1. Rebound. A rebound marriage results when you marry someone immediately after another person has ended a relationship with you. It is a frantic attempt on your part to reestablish your desirability in your own eyes and in the eyes of the partner who dropped you. One man said, "After she told me she wouldn't marry me, I became desperate. I called up an old girlfriend to see if I could get the relationship going again. We were married within a month. I know it was foolish, but I was very hurt and couldn't stop myself." To marry on the rebound is questionable because the marriage is made in reference to the previous partner and not to the partner being married. In reality, you are using the person you intend to marry to establish yourself as the "winner" in the previous relationship. To avoid the negative consequences of marrying on the rebound, wait until the negative memories of your past relationship have been replaced by positive aspects of your current relationship. In other words, marry when the satisfactions of being with your current partner outweigh any feelings of revenge. This normally takes between twelve and eighteen months.

*2. **Escape.*** A person might marry to escape an unhappy home situation in which the parents are often seen as oppressive, overbearing, conflictual, or abusive. The parents' continued bickering might be highly aversive, causing the person to marry to flee the home. A family with an alcoholic parent might also create a context from which a person might want to escape. One woman said, "I couldn't wait to get away from home. Ever since my parents divorced, my mother has been drinking and watching me like a hawk. 'Be home early, don't drink, and watch out for those horrible men,' she would always say. I admit it. I married the first guy that would have me. Marriage was my ticket out of there."

Marriage for escape is a bad idea. It is far better to continue the relationship with the partner until mutual love and respect, rather than the desire to escape an unhappy situation, become the dominant forces propelling you toward marriage. In this way you can evaluate the marital relationship in terms of its own potential and not solely as an alternative to an unhappy situation.

*3. **Unanticipated pregnancy.*** Getting married just because a partner becomes pregnant is usually a bad idea. Indeed, the decision of whether to marry should be kept separate from the fact that there is now a pregnancy. An abortion, single parenthood, and unmarried parenthood (the couple can remain together as an unmarried couple and have the baby) are all alternatives to simply deciding to marry if a partner becomes pregnant. Avoiding feelings of being trapped or "You married me just because of the pregnancy" are among the reasons for not rushing into marriage based on pregnancy.

*4. **Psychological blackmail.*** Some individuals get married because their partner takes the position that "I can't live without you" or "I will commit suicide if you leave me." Because the person fears that the partner may commit suicide, he or she agrees to the wedding. The problem with such a marriage is that one partner has learned to manipulate the relationship to get what he or she wants. Use of such power often creates resentment in the other partner, who feels trapped in the marriage. Escaping from the marriage becomes even more difficult. One way of coping with a psychological blackmail situation is to encourage the person to go with you to a counselor to "discuss the relationship." Once inside the therapy room, you can tell the counselor that you feel pressured to get married because of the suicide threat. Counselors are trained to respond to such a situation.

*5. **Pity.*** Some partners marry because they feel guilty about terminating a relationship with someone whom they pity. The fiancé of one woman got drunk one Halloween evening and began to light fireworks on the roof of his fraternity house. As he was running away from a Roman candle he had just ignited, he tripped and fell off the roof. He landed on his head and was in a coma for three weeks. A year after the accident his speech and muscle coordination were still adversely affected. The woman said she did not love him anymore but felt guilty about terminating the relationship now that he had become physically afflicted. She was ambivalent. She felt it was her duty to marry her fiancé, but her feelings were no longer love feelings. Pity may also have a social basis. For example, a partner may fail to achieve a lifetime career goal (for example, he or she may flunk out of medical school). Regardless of the reason, if one partner loses a limb, becomes brain-damaged, or fails in the pursuit of a major goal, it is important to keep the issue of pity separate from the advisability of contracting the marriage. The decision to marry should be based on factors other than pity for the partner.

6. Filling a void. Julie Bateman is a university student whose father died of cancer. She noted that his death created a vacuum that she felt driven to fill immediately by getting married so that she would have a man in her life. Since she was focused on filling the void, she had paid little attention to the personality characteristics of or the relationship with the man who had asked to marry her. She reported that she discovered on her wedding night that her new husband had several other girlfriends whom he had no intention of giving up. The marriage was annulled.

In deciding whether to continue or terminate a relationship, listen to what your senses tell you ("Does it feel right?"), listen to your heart ("Do you love this person or do you question whether you love this person?"), and evaluate your similarities ("Are we similar in terms of core values, goals, view of life?"). Also, be realistic. It would be unusual if none of the factors listed above applied to you. Indeed, most people have some negative and some positive indicators before they marry. For example, actor Russell Crowe had known his girlfriend, Danielle Spencer, for thirteen years when they married (a positive predictor). However, their relationship had been described as an "on-again, off-again" one (a negative predictor) (Crawley, 2003).

Breaking up is never easy. Almost a third of 279 undergraduates reported that they "sometimes" remained in relationships they thought should end (31%) or those relationships which became "unhappy" (32.3%). This finding reflects the ambivalence students sometimes feel in ending an unsatisfactory relationship and their reluctance to do so (Knox, Zusman, McGinty, & Davis, 2002).

SUMMARY

What are the cultural factors that influence mate selection?

Two types of cultural influences in mate selection are endogamy (marry someone inside one's own social group—race, religion, social class) and exogamy (marry someone outside one's own family).

What are the sociological factors that influence mate selection?

Sociological aspects of mate selection include homogamy (people prefer someone like themselves in race, age, and education) and propinquity (people are more likely to find a mate who attends the same school or works for the same company).

What are the psychological factors operative in mate selection?

Psychological aspects of mate selection include complementary needs, exchange theory, and parental image. Complementary-needs theory suggests that people select others who have characteristics opposite to their own. For example, a highly disciplined, well-organized individual might select a free-and-easy, worry-about-nothing mate. Most researchers find little evidence for complementary-needs theory.

Exchange theory suggests that one individual selects another on the basis of rewards and costs. As long as an individual derives more profit from a relationship with one partner than with another, the relationship will continue. Exchange concepts influence who dates whom, the conditions of the dating relationship, and the decision to marry. Parental image theory suggests that individuals select a partner similar to the opposite-sex parent.

What are the sociobiological factors operative in mate selection?

The sociobiological view of mate selection suggests that men and women select each other on the basis of their biological capacity to produce and support healthy offspring. Men seek young women with healthy bodies, and women seek ambitious men who will provide economic support for their offspring. There is considerable controversy about the validity of this theory.

What factors should be considered when becoming engaged?

The engagement period is the time to ask specific questions about the partner's values, goals, and marital agenda, visit each other's parents to assess parental models, and consider involvement in premarital educational programs and/or counseling. Negative reasons for getting married include being on the rebound, escaping from an unhappy home life, psychological blackmail, and pity.

Some couples (particularly those with children from previous marriages) decide to write a prenuptial agreement to specify who gets what and the extent of spousal support in the event of a divorce. To be valid, the document should be developed by an attorney in accordance with the laws of the state in which the partners reside. Last-minute prenuptial agreements put enormous emotional strain on the couple and are often considered invalid by the courts. Discussing a prenuptial agreement six months in advance is recommended.

What factors suggest you might consider calling off the wedding?

Factors suggesting that a couple may not be ready for marriage include being in their teens, having known each other less than two years, and having a relationship characterized by significant differences and /dramatic parental disapproval. Some research suggests that partners with the greatest number of similarities in terms of values, goals, and common interests are most likely to have happy and durable marriages.

KEY TERMS

complementary-needs theory
endogamy
exchange theory
exogamy
homogamy
mating gradient
prenuptial agreement
principle of least interest
propinquity
residential propinquity
sociobiology

RESEARCHING MARRIAGE AND THE FAMILY WITH INFOTRAC COLLEGE EDITION

InfoTrac College Edition, an online library, allows you to perform research online anywhere, anytime. Following are two suggested search terms and related questions to help you extend your understanding of the topics covered in this chapter. Go to www.infotrac-college.com to begin your search.

Keyword: **Homogamy.** Locate articles that provide evidence for and against homogamy as a factor in lasting marriage relationships. Do you think that the best marriages are those in which the partners are similar or different?

Keyword: **Marriage requirements.** Locate articles that discuss marriage requirements such as premarital education, age, etc. To what degree do you think it is "too easy" to get married? What legal requirements (if any) do you recommend that persons should be required to meet before being allowed to marry?

The Companion Web Site for Choices in Relationships: An Introduction to Marriage and the Family, Eighth Edition

http://sociology.wadsworth.com/knox_schacht/choices8e

Supplement your review of this chapter by going to the companion Web site to take one of the Tutorial Quizzes, use the flash cards to master key terms, and check out the many other study aids you'll find there. You'll also find special features such as the Marriage and Family Resource Center, Census 2000 information, and other data and resources at your fingertips to help you with that special project or to do some research on your own.

WEBLINKS

PAIR Project
www.utexas.edu/research/pair

RightMate
http://www.heartchoice.com/rightmate/

Singles Christian Network
http://www.singlec.com/

Authors

CHAPTER 8

Courtship brings out the best.
Marriage brings out the rest.

Cullen Hightower

Marriage Relationships

Contents

True or False?

1. Black students are more likely than white students to feel that newlyweds should take a honeymoon soon after the wedding.
2. Economic security is the greatest expected benefit of marriage in the United States.
3. About a third of states now offer covenant marriages and a third of people getting married in these states elect the covenant alternative.
4. Most college students now express a preference for a civil ceremony to save money on the wedding and provide a down payment on a house.
5. Marital satisfaction increases across time—the longer spouses are married, the happier they report themselves with each other.

Answers: **1.** F **2.** F **3.** F **4.** F **5.** F

Marriage in our culture is characterized by diversity. There is no one marriage relationship but only marriage relationships. In this chapter we look at why people marry (both macro and micro reasons), the three levels of commitment, the numerous changes in a couple's relationship after marriage, and the characteristics of successful marriages. We also look at how college students view weddings and how marriage might be strengthened by divorce law reform. We begin with why people elect to marry and the social functions of marriage.

MOTIVATIONS FOR AND FUNCTIONS OF MARRIAGE

In this section we discuss both why people marry and the functions that getting married serves for society.

Individual Motivations for Marriage

We have defined marriage as a legal contract between two heterosexual adults that regulates their economic and sexual interaction. However, individuals in the United States tend to think of marriage in more personal than legal terms. The following are some of the reasons people give for getting married.

Love Many couples view marriage as the ultimate expression of their love for each other—the desire to spend their lives together in a secure, legal, committed relationship. In U.S. society love is expected to precede marriage—thus only couples in love consider marriage. Those not in love would be ashamed to admit it.

Personal Fulfillment We marry because we feel a sense of personal fulfillment in doing so. We were born into a family (family of origin) and want to create a family of our own (family of procreation). We remain optimistic that our marriage will be a good one. Even if our parents divorced or we have friends who have done so, we may feel that our relationship will be different.

Companionship Talk show host Oprah Winfrey once said that lots of people want to ride in her limo, but that what she wants is someone who will take the bus when the limo breaks down. One of the motivations for marriage is to enter a structured relationship with a genuine companion, a person who will take the bus with you when the limo breaks down.

Although marriage does not ensure it, companionship is the greatest expected benefit of marriage in the United States. Coontz (2000) noted that it has become "the legitimate goal of marriage" (p. 11). Eating meals together is one of the most frequent normative behaviors of spouses. Indeed, while spouses may eat lunch apart, dinner together becomes an expected behavior. **Commensality** is eating with others, and one of the issues spouses negotiate is "who eats with us" (Sobal et al., 2002).

Parenthood Most people want to have children. Ninety-two percent of 620 undergraduates reported that they wanted to have children (Knox & Zusman, 1998). Although some people are willing to have children outside marriage (in a cohabitating relationship or in no relationship at all), most Americans prefer to have children in a marital context. A relatively strong norm exists in our society (particularly for whites) that individuals should be married before they have children.

Economic Security Married persons report higher household incomes than do unmarried persons. Indeed, national data from the Health and Retirement Survey revealed that individuals who were not continuously married had significantly lower wealth than those who remained married throughout the life course. Remarriage offsets the negative effect of marital dissolution (Wilmoth & Koso, 2002).

Although individuals may be drawn to marriage for the preceding reasons on a conscious level, unconscious motivations may also be operative. Individuals reared in a happy family of origin may seek to duplicate this perceived state of warmth, affection, and sharing. Alternatively, individuals reared in unhappy, abusive, drug-dependent families may inadvertently seek to re-create a similar family because that is what they are familiar with. In addition, individuals are motivated to marry because of the fear of being alone, to better themselves economically, to avoid an out-of-wedlock birth, and to prove that someone wants them.

Just as most individuals want to marry (regardless of the motivation), most parents want their children to marry. If their children do not marry too young and if they marry someone they approve of, parents feel some relief from the economic responsibility of parenting, anticipate that marriage will have a positive, settling effect on their offspring, and look forward to the possibility of grandchildren.

Societal Functions of Marriage

As we noted in Chapter 1, Choices in Relationships: An Introduction, the primary function of marriage is to bind a male and female together who will reproduce, provide physical care for their dependent young, and socialize them to be productive members of society who will replace those who die (Murdock, 1949). Marriage helps protect children by giving the state legal leverage to force parents to be responsible to their offspring whether or not they stay married. If couples did not have children, the state would have no interest in regulating marriage.

Additional functions include regulating sexual behavior (spouses have less STD exposure than singles) and stabilizing adult personalities by providing a companion and "in house" counselor. In the past, marriage and family have served protective, educational, recreational, economic, and religious functions. But as these functions have gradually been taken over by the police/legal system, schools, entertainment industry, workplace, and church/synagogue, only the companionship-intimacy function has remained virtually unchanged.

The emotional support each spouse derives from the other in the marital relationship remains one of the strongest and most basic functions of marriage (Coontz, 2000). In today's social world, which consists mainly of impersonal, secondary relationships, living in a context of mutual emotional support may be particularly important. Indeed, the companionship and intimacy needs of contemporary U.S. marriage have become so strong that many couples consider divorce when they no longer feel "in love" with their partner. Sixty-seven percent of 620 undergraduates reported that they would divorce if they fell out of love and were very unhappy (Knox & Zusman, 1998).

The very nature of the marriage relationship has also changed from being very traditional or male-dominated to being very modern or egalitarian. A summary of these differences is presented in Table 8.1. Keep in mind that these are stereotypical marriages and that only a small percentage of today's modern marriages have all the traditional/egalitarian characteristics that are listed.

Table 8.1 Traditional versus Egalitarian Marriages

Traditional Marriage	Egalitarian Marriage
There is limited expectation of husband to meet emotional needs of wife and children.	Husband is expected to meet emotional needs of wife and to be involved with children.
Wife is not expected to earn income.	Wife is expected to earn income.
Emphasis is on ritual and roles.	Emphasis is on companionship.
Couples do not live together before marriage.	Couples may live together before marriage.
Wife takes husband's last name.	Wife may keep her maiden name.
Husband is dominant; wife is submissive.	Neither spouse is dominant.
Roles for husband and wife are rigid.	Roles for spouses are flexible.
Husband initiates sex; wife complies.	Sex is initiated by either spouse.
Wife takes care of children.	Parents share childrearing.
Education is important for husband, not for wife.	Education is important for both spouses.
Husband's career decides family residence.	Career of either spouse determines family residence.

MARRIAGE AS A COMMITMENT

Whereas love is a private feeling, marriage is a public commitment. It is the second of three times that one's name can be expected to appear in the local newspaper. Marriage represents a multilevel commitment—person-to-person, family-to-family, and couple-to-state.

Person-to-Person Commitment

Individuals commit themselves to someone whom they love, with whom they feel a sense of equality, and who they feel is the best of the alternative persons available to them (Crawford et al., 2003). Beyond the idea of **commitment** as an intent to maintain a relationship, behavioral indexes of commitment (928 of them) were identified by 248 people who were committed to someone. These behaviors were then coded into ten major categories and included providing affection, providing support, maintaining integrity, sharing companionship, making an effort to communicate, showing respect, creating a relational future, creating a positive relational atmosphere, working on relationship problems together, and expressing commitment (Weigel & Ballard-Reisch, 2002).

Family-to-Family Commitment

The movie *My Big Fat Greek Wedding* emphasized the fact that while two people get married, it is also their families who must meet and mesh and negotiate a relationship of understanding with each other. Marriage also involves commitments by each of the marriage partners to the family members of the spouse. Married couples are often expected to divide their holiday visits between both sets of parents.

Strengthening Marriage through Divorce Law Reform

SOCIAL POLICY

Some family scholars and policymakers advocate strengthening marriage by reforming divorce laws to make divorce harder to obtain. Since California became the first state to implement "no-fault" divorce laws in 1969, every state has passed similar laws allowing couples to divorce without proving in court that one spouse was at fault for the marital breakup. The intent of no-fault divorce legislation was to minimize the acrimony and legal costs involved in divorce, making it easier for unhappy spouses to get out of their marriage. Under the system of no-fault divorce, a partner who wanted a divorce could get one, usually by citing irreconcilable differences, even if his or her spouse did not want a divorce.

Other states believe the no-fault system has gone too far and have taken measures designed to make breaking up harder to do by requiring proof of fault (such as infidelity, physical or mental abuse, drug or alcohol abuse, and desertion), or extending the waiting period required before a divorce is granted. In most divorce law reform proposals, no-fault divorces would still be available to couples who mutually agree to end their marriages.

Opponents argue that divorce law reform measures would increase acrimony between divorcing spouses (which harms the children as well as the adults involved), increase the legal costs of getting a divorce (which leaves less money to support any children), and delay court decisions on child support and custody and distribution of assets. In addition, critics point out that ending no-fault divorce would add countless court cases to the dockets of an already overloaded court system. Efforts in many state legislatures to repeal no-fault divorce laws have largely failed.

However, in June 1997, the Louisiana legislature became the first in the nation to pass a law creating a new kind of marriage contract that would permit divorce only in narrow circumstances. Under the Louisiana law, couples can voluntarily choose between two types of marriage contracts: the standard contract that allows a no-fault divorce or a "**covenant marriage**" that permits divorce only under conditions of fault (such as abuse, adultery, or imprisonment on a felony) or after a marital separation of more than two years. Couples who choose a covenant marriage are also required to get premarital counseling from a clergy member or another counselor. While covenant marriage is a unique way to begin a marriage, there has been no rush on the part of couples to seek these marriages. Fewer than 3 percent of couples that marry in Louisiana (and now Arizona) have chosen to take on the extra restrictions of marriage by covenant (Licata, 2002).

There is considerable disagreement on whether it is worse for children to go through the divorce of their parents or continue to live with parents who no longer love each other and are unhappy. Go to http://sociology.wadsworth.com/knox_schacht/choices8e, select the Opposing Viewpoints Resource Center on the left navigation bar and enter your pass code. Conduct a search by subject, using the search term "Divorce," and read the articles on the Viewpoints tab that support each position. Do you find yourself agreeing with one perspective more than the other? Why?

Reference

Licata, N. 2002. Should premarital counseling be mandatory as a requisite to obtaining a marriage license? *Family Court Review* 40:518–32.

PERSONAL CHOICES

Should a Married Couple Have Their Parents Live with them?

Of course, this question is more often asked by individualized westernized couples who live in isolated nuclear units. Asian couples reared in extended family contexts expect to take care of their parents and consider it an honor to do so. As the parents of American spouses get older, a decision must often be made by the spouses about whether to have the parents live with them. Usually the person involved is the mother of either spouse, since the father is more likely to die first. One wife said: "We didn't have a choice. His mother is 82 and has Alzheimer's disease. We couldn't afford to put her in a nursing home at $5,200 a month, and she couldn't stay by herself. So we took her in. It's been a real strain on our marriage, since I end up taking care of her all day. I can't even leave her alone to go to the grocery store." Some elderly persons have resources for nursing home care or their married children can afford such care. But even in these circumstances, some spouses decide to have their parents live with them. "I couldn't live with myself if I knew my mother was propped up in a wheelchair eating Cheerios when I could be taking care of her," said one spouse.

When spouses disagree about parents in the home, the result can be devastating. According to one wife, "I told my husband that Mother was going to live with us. He told me she wasn't and that he would leave if she did. She moved in, and he moved out (we were divorced). Five months later, my mother died."

Couple-to-State Commitment

In addition to making person-to-person and family-to-family commitments, spouses become legally committed to each other according to the laws of the state in which they reside. This means they cannot arbitrarily decide to terminate their own marital agreement.

Just as the state says who (not close relatives, the insane, or the mentally deficient) can marry and when (usually at age 18 or older), legal procedures must be instituted if the spouses want to divorce. The state's interest is that a couple stay married, have children, and take care of them. Should they divorce, the state will dictate how the parenting is to continue, both physically and economically.

Social policies designed to strengthen marriage through divorce law reform reflect the value the state places on stable, committed relationships (see Social Policy on p. 186).

MARRIAGE AS A RITE OF PASSAGE

A **rite of passage** is an event that marks the transition from one social status to another. Starting school, getting a driver's license, and graduating from high school or college are events that mark major transitions in status (to student, to driver, and to graduate). The wedding itself is another rite of passage that marks the transition from fiancé to spouse.

Weddings

The wedding is a rite of passage that is both religious and civil. To the Catholic Church, marriage is a sacrament that implies that the union is both sacred and indissoluble. According to Jewish and most Protestant faiths, marriage is a special bond between the husband and wife sanctified by God, but divorce and remarriage are permitted. Wedding ceremonies still reflect traditional cultural definitions of women as property. For example, the bride is usually walked down the aisle by her father and "handed over to the new husband." Gay weddings are devoid of such traditional cultural scripts.

Whereas love is a private experience, marriage is a public experience in the United States. This is emphasized by the wedding in which family and friends of both parties are invited to participate. The wedding is a time for the respective families to learn how to cooperate with each other for the benefit of the respective daughter and son. Conflicts over number of bridesmaids and ushers, number of guests to invite, and place of the wedding are not uncommon. Though some families harmoniously negotiate all differences, others become so adamant about their preferences that the prospective bride and groom elope to escape or avoid the conflict. However, most families recognize the importance of the event in the life of their daughter or son and try to be helpful and nonconflictual.

Some states require that in order to obtain a marriage license, the partners must have a blood test to certify that neither has a sexually transmitted disease. The document is then taken to the county courthouse, where the couple applies for a marriage license. Two-thirds of the states require a waiting period between the issuance of the license and the wedding. Eighty percent of couples are married by a member of the clergy; 20 percent (primarily in remarriages) go to a justice of the peace, judge, or magistrate.

Some brides wear something old, new, borrowed, and blue. The "old" (e.g., gold locket) is something that represents the durability of the impending marriage. The "new," perhaps in the form of new unlaundered undergarments, emphasizes the new life to begin. The "borrowed" (e.g., a wedding veil) is something that has already been worn by a currently happy bride. The "blue" (e.g., ribbons)

Up Close 8.1 Weddings: Some Data on College Student Perceptions

Once a couple make a commitment to marry and become engaged, they begin to discuss the wedding, which may occasion their first set of realized differences. While previous research on weddings has focused on the role of "white weddings" in popular culture (Ingraham, 1999), on traditional white wedding ideals and aspirations (Abowitz, 2000, 2002), and on celebrity weddings (Boden, 2001), this study documents the ways in which undergraduate women and men differ in their attitudes toward weddings

Sample

The sample consisted of 196 undergraduates from a large southeastern university who responded to an anonymous forty-seven-item questionnaire designed to assess wedding attitudes and perceptions. Seventy percent of the respondents were female; 30 percent were male. The median age was 20 and almost half (47.9%) were involved with someone at the time of the survey.

Findings and Discussion

There are considerable differences between women and men in their attitudes toward weddings.

1. Women prepare more. Women were significantly ($p < .000$) more likely than men (55.9% vs. 17.6 %) to report that they were intent on preparing for their wedding by reading *Modern Bride Magazine,* which they felt would be helpful to them in planning their wedding. College women also prepare by watching programs on television such as "A Wedding Story," which is replacing soap operas in the daytime viewing habits of many such women (Abowitz, 2000).

2. The wedding is for the bride's family. In terms of who the wedding is for, 16.4 percent of the men, compared with only 11.6 percent of the women, see the wedding as being for the bride's family and not for the bride and groom. Hence, while both the potential groom and bride view the wedding as being for them as a couple, male college students are likely to observe the influence of the bride's mother in suggesting certain arrangements and conclude that the wedding is for the bride's family.

3. The bride wants the wedding documented. Female respondents were significantly ($p < .000$) more likely than male respondents (84.8 % vs. 70.6 %) to report their desire to "hire a professional photographer to take photos and videotape the wedding."

4. The bride prefers a formal wedding. Female respondents were significantly ($p < .015$) more likely than male respondents (77.4 % vs. 61.4%) to report that they did not want a civil wedding ceremony with a justice of the peace officiating. A formal wedding was much preferred. Ingraham (1999) emphasized in *White Weddings* the modern-day preference for a formal, traditional wedding complete with formal white gown, several attendants, numerous guests, and some kind of party or reception. There is also enormous cultural pressure on the bride to approximate the "superbride" ideal (Boden, 2001).

In response to the open-ended question, "What is your dream wedding?" "tradition" was the most frequent response (28%), with "beach" and "I don't care as long as I am with the woman I love" written in by 13 and 8 percent of the respondents, respectively. There were some unusual "dream" weddings identified by the respondents: "on a cruise," "on a cliff over the Pacific at sunset," and "in Charlottesville, VA with the Dave Matthews Band." One respondent wrote, "one where everyone attends."

5. Both parents should be invited if they are still married. When asked if they would invite both parents to the wedding, those students with both parents still married to each other answered yes 97.3 percent of the time versus only 77 percent of students whose parents were divorced. One wonders how much the divorce of one's parents influences the wedding preparation and pleasure of the soon-to-be spouses.

6. Racial background affects perception of who should pay for the wedding. One's racial background was also associated with significant differences. White students were significantly ($p < .000$) and dramatically (58.4% versus 21.0%) more likely than nonwhite students to report that parents should be responsible for paying for the wedding expenses. Similarly, white students were significantly ($p < .001$) more likely than nonwhite students (95.6 % versus 76.3%) to report that the couple should take a honeymoon soon after the wedding.

Implications

University students, most of whom will marry, might be reminded to consider viewing conflict in their relationships as an opportunity to learn to tolerate and negotiate differences. One of the first major conflicts an engaged couple may have is the way they view and approach the wedding. These data suggest that women and men college students differ in the degree to which they would "prepare" for their wedding and in their perception of who the wedding is for (the bride's mother?), as well as in their desire for a professional photographer and a formal wedding. Indeed, the bride and groom may also differ on which parents are to be invited to the wedding. None of these issues are important or relevant to a couple's marital happiness and satisfaction unless they are unable to accept or negotiate them.

Sources

Abowitz, D. A. 2000. A "Wedding story," or, A modern American fairy tale: Gender and romance in the construction of white weddings. Paper presented at the Eastern Sociological Society, Baltimore, March 5.

———. 2002. On the road to "happily ever after": A survey of attitudes toward romance and marriage among college students. Paper presented at the Southern Sociological Society, Baltimore, April 6.

Boden, S. 2001. "Superbrides": Wedding consumer culture and the construction of bridal identity. *Sociological Research on Line* www.socresonline.org.uk Vol 6, 1, May.

Ingraham, Chrys. 1999. *White weddings: Romancing heterosexuality in popular culture.* New York: Routledge.

Adapted from an article of the same name by D. Knox, M. E. Zusman, K. McGinty, and D. A. Abowitz 2003. *College Student Journal* 37:197–200. Used by permission of *College Student Journal.*

represents fidelity (those dressed in blue have lovers true). The bride's throwing her floral bouquet signifies the end of girlhood; the rice thrown by the guests at the newly married couple signifies fertility.

It is no longer unusual for couples to have weddings that are neither religious nor traditional. In the exchange of vows, neither partner promises to obey the other, and the couple's relationship is spelled out by the partners rather than by tradition. Vows often include the couple's feelings about equality, individualism, humanism, and openness to change.

Part of the preference for less lavish, less traditional weddings is economic. The average wedding costs between $15,000 and $20,000 and includes 188 guests. A second wedding costs almost as much as the first (Kirn & Cole, 2000). Up Close 8.1 reveals wedding attitudes of a sample of college students.

Honeymoons

Traditionally, another rite of passage follows immediately after the wedding—the honeymoon. The functions of the honeymoon are both personal and social. The personal function is to provide a period of recuperation from the usually exhausting demands of preparing for and being involved in a wedding ceremony and reception. The social function is to provide a time for the couple to be alone to solidify the change in their identity from that of an unmarried to a married couple. And, now that they are married, their sexual expression with each other achieves full social approval and legitimacy. Now the couple can have children with complete societal approval.

Honeymoon costs can be extravagant or inexpensive. Individuals seeking to minimize honeymoon expenses (airfare) can go to a local bed and breakfast. The authors of your text drove forty-five minutes from the town in which they married for their honeymoon at a B and B (bed and breakfast). Total cost was less than $200 (for a million-dollar honeymoon and marriage).

CHANGES AFTER MARRIAGE

After the wedding and honeymoon, the new spouses begin to experience the stark realities of marriage, including changes in their personal, social, legal, and sexual relationship.

All women are married to two men— the one they married and the one they think they married.

Jay Leno, *Comedian and* Tonight Show *host*

Personal Changes

New spouses experience an array of changes in their lives. One initial consequence of getting married may be an enhanced self-concept. Parents and close friends usually arrange their schedules to participate in your wedding and give gifts to express their approval. In addition, the strong evidence that your spouse approves of you and is willing to spend a lifetime with you also tells you that you are a desirable person.

The married person also begins adopting new values and behaviors consistent with the married role. Although new spouses often vow that "marriage won't change me," it does. For example, rather than stay out all night at a party, which is not uncommon for singles who may be looking for a partner, spouses (who are already paired off) tend to go home early. Their roles of spouse, employee, and parent force them to adopt more regular hours. The role of married person implies a different set of behaviors than the role of single person. Although there is an initial resistance to "becoming like old married folks," the resistance soon gives way to the realities of the role.

Another effect of getting married is **disenchantment**—the transition from a state of newness and high expectation to a state of mundaneness tempered by reality. It may not happen in the first few weeks or months of marriage, but it is almost inevitable. Whereas courtship is the anticipation of a life together, marriage is the day-to-day reality of that life together—and reality does not always fit the dream. "Moonlight and roses become daylight and dishes" is an old adage reflecting the realities of marriage. Disenchantment after marriage is also related to the partners' shifting their focus of interest away from each other to work or children; each partner usually gives and gets less attention in marriage than in courtship. If the partners do not discuss these changes, they may define each other's interests that are external to the relationship as betrayal.

Other changes that a couple experience when they marry follow:

1. Change in how money is spent. Entertainment expenses in courtship become allocated to living expenses and setting up a household together.

2. Discovering that one's mate is different from one's date. Courtship is a context of deception. Marriage is one of reality. Spouses sometimes say, "He (she) is not the person I married."

3. Experiencing loss of freedom. Although premarital norms permit relative freedom to move in and out of relationships, marriage involves a binding legal contract. The new sense of confinement to one person and a routine life may bring out the worst in partners who thrived on freedom.

Parents, In-laws, and Friendship Changes

Marriage affects relationships with parents, in-laws, and the friends of both partners. Parents are likely to be more accepting of the partner following the wedding. "I encouraged her not to marry him," said the father of a recent bride, "but once they were married, he was her husband and my son-in-law, so I did my best to get along with him."

Time spent with parents and extended kin radically increases when the couple have children. Indeed, a major difference between couples with and without children is the amount of time they spend with relatives. The resources provided by parents and kin as well as their participation in celebrations such as birthdays help to account for the increased interaction (Miller, 1999).

Two researchers (Serovich & Price, 1992) examined in-law relationships of 309 spouses. They found that most reported high relationship quality with in-laws, that wives reported equally satisfying relationships with both mothers-in-law and fathers-in-law, and that how close one lived to one's in-laws was not a significant factor in satisfaction with in-law relationships. Malia and Blackwell (1997) also found evidence for good in-law relationships. Only four of twenty-two daughters-in-law identified their mothers-in-law as negative forces in their lives. Timmer and Veroff (2000) found that husbands' and wives' closeness to their in-laws predicted higher levels of marital happiness in their first year of marriage.

Emotional separation from one's parents is an important developmental task in building a successful marriage. When choices must be made between one's parents and one's spouse, more long-term positive consequences for the married couple are associated with choosing the spouse over the parents. But such choices become more complicated and difficult when one's parents are old, ill, or widowed.

Marriage also affects relationships with friends of the same and other sex. Less time will be spent with friends because of the new role demands as a spouse. More time will be spent with other married couples, who will become powerful influences on the new couple's relationship.

PERSONAL CHOICES

Is "Partner's Night Out" a Good Idea?

Although spouses may want to spend time together, they may also want to spend time with their friends—shopping, having a drink, fishing, golfing, seeing a movie, or whatever.

Some spouses have a flexible policy based on trust with each other. Other spouses are very suspicious of each other. One husband said, "I didn't want her going out to bars with her girlfriends after we were married. You never know what someone will do when they get three drinks in them."

For partner's night out to have a positive impact on the couple's relationship, it is important that the partners maintain emotional and sexual fidelity to each other, that each partner have a partner's night out, and that the partners spend some nights alone with each other. Friendships can enhance a marriage relationship by making the individual partners happier, but friendships cannot replace the marriage relationship. Spouses must spend time alone to nurture their relationship.

What spouses give up in friendships, they gain in developing an intimate relationship with each other. However, abandoning one's friends after marriage is problematic, since one's spouse cannot be expected to satisfy all of one's social needs. And since many marriages end in divorce, friendships that have been maintained throughout the marriage can become a vital source of support for a person adjusting to a divorce.

Legal Changes

Unless the partners have signed a prenuptial agreement specifying that their earnings and property will remain separate, the wedding ceremony is associated with an exchange of property. Once two individuals become spouses, each automatically becomes part owner of what the other earns in income and accumulates in property. Although the laws on domestic relations differ from state to state, courts typically award to each spouse half of the assets accumulated during the marriage (even though one of the partners may have contributed a smaller proportion). For example, if a couple buy a house together, even though one spouse invested more money in the initial purchase, the other will likely be awarded half of the value of the house if they divorce. (Having children complicates the distribution of assets, since the house is often awarded to the custodial parent.) In the case of death of the spouse, the remaining spouse is legally entitled to inherit between one-third and one-half of the partner's estate, unless a will specifies otherwise.

Should a couple divorce after having children, both are legally responsible to provide for the economic support of their children. In a typical case, the mother is awarded primary physical custody of the children and the father is required by the court to pay about one-half of his gross income if there are two children (until the children graduate from high school). Should the wife have the higher income and the husband be the custodial parent, the courts will require her to pay child support. In cases of joint custody where the children stay with each parent equally, no child support is awarded.

Sexual Changes

The sexual relationship of the couple also undergoes changes with marriage. First, since spouses are more sexually faithful to each other than are dating partners or cohabitants (Treas & Giesen, 2000), their number of sexual partners will decline dramatically. Second, the frequency with which they have sex with each other will decrease. According to one wife:

> *The urgency to have sex disappears after you're married. After a while you discover that your husband isn't going to vanish back to his apartment at midnight. He's going to be with you all night, every night. You don't have to have sex every minute because you know you've got plenty of time. Also, you've got work and children and other responsibilities, so sex takes a lower priority than before you were married.*

Although married couples may have intercourse less frequently than they did before marriage, marital sex is still the most satisfying of all sexual contexts. Eighty-eight percent of married people in a national sample reported that they experienced extreme physical pleasure, and 85 percent reported experiencing extreme emotional satisfaction, with their spouses. In contrast, 54 percent of individuals who were not married or not living with anyone said that they experienced extreme physical pleasure with their partners, and 30 percent said that they were extremely emotionally satisfied (Michael et al., 1994).

Interactional Changes

The way wives and husbands perceive and interact with each other continues to change throughout the course of the marriage. Two researchers studied 238 spouses who had been married over thirty years and observed that (across time) men changed from being patriarchal to collaborating with their wives and that women changed from deferring to their husbands' authority to challenging that authority (Huyck & Gutmann, 1992). We have also noted other changes, including less focus on each other and less sex.

Before marriage, a man declares that he would lay down his life to serve you; after marriage, he won't even lay down his newspaper to talk to you.
Helen Rowland, 1876–1950

RACIAL AND CULTURAL MARITAL DIVERSITY

Racial background affects marital relationships because of the cultural heritage of the spouses. Whereas Anglo-American (Euro-American) marriages are characterized by the values of independence, equality, materialism, and competition, less is known about other racial variations. We begin our discussion with African-American marriages.

African-American Marriages

African-American families have often been described in negative terms, such as being low-income families, having high birthrates among unmarried mothers, being one-parent families, having absentee fathers, and having spouses with limited educations. Such a pathological view of African-American family life is a result of researchers' looking at the African-American family as a deviation from the white norm. More recently, African-American families are being discussed in terms of their uniqueness and resilience as a cultural variant (Taylor, 2002).

While previous historical reviews of Black families emphasized their lack of family cohesiveness due to slave conditions, new versions have emphasized the impressive family structures maintained by Blacks. And, while it was previously suggested that white slave owners routinely sold off family members, new research suggests that only the most dehumanized slaveholder failed to recognize marital and family ties. Similarly, the slavery myths of the invisible male, the Black family headed by a single female, and the controlling Black female have been replaced by information emphasizing the positive visible role of the Black husband/father, the two-parent family, and the egalitarian relationship between the husband and wife (Taylor, 2002).

Browning (1999) argued that today's Black marriages can best be understood as responding to structural conditions such as poor educational systems, high unemployment, and high underemployment and that negative labels placed on Black families are inappropriate. Rather, Black families may more accurately be described in terms of their unique strengths, including strong kinship bonds, favorable attitudes toward their elderly, adaptable roles, strong achievement orientations, strong religious values, and a love of children.

Mexican-American Marriages

The term Mexican-American refers to people of Mexican origin or descent living in America. The breakdown of the Hispanic population in the United States is 59 percent Mexican, 10 percent Puerto Rican, 4 percent Cuban, and 28 percent labeled as "other Hispanic or Latino" (*Statistical Abstract of the United States: 2003,* Table 22).

NATIONAL DATA

Hispanics are the fastest-growing segment of the U.S. population. It is estimated that by the year 2005 there will be 42 million Hispanics (about 14% of the population) living in the United States (*Statistical Abstract of the United States: 2003,* Table16).

The term *Mexican-American* is sometimes used synonymously with *Chicano, Spanish American, Hispanic, Mexican, Californian,* and *Latin American* (*Latino*). It is sometimes regarded as derogatory. The treatment of Mexican-Americans as second-class citizens has historical roots. When America annexed Texas in 1845, Mexico became outraged, and the Mexican War followed (1846–48). In the Treaty of Guadalupe Hidalgo, Mexico recognized the loss of Texas and accepted the Rio Grande as the boundary between Mexico and the United States. Although the war was over, hostilities continued, and the negative stereotyping of Mexican-Americans as a conquered and subsequently inferior people became entrenched. Such stereotyping and discrimination have contributed to the stress to which Mexican-American spouses have been exposed. Compared with Anglos, Mexican-Americans have less education, earn lower incomes, and work in lower-status occupations (Becerra, 1998).

The Husband-Wife Relationship Great variability exists among Mexican-American marriages. What is true in one relationship may not be true in another, and the same relationship may change over time. Nevertheless, some "typical" characteristics of Mexican-American relationships are detailed here.

1. Male dominance. Although role relationships between women and men are changing in all segments of society, traditional role relationships between the Mexican-American sexes are characterized by male domination. "Male dominance is the designation of the father as the head of the household, the major decision maker, and the absolute power holder in the Mexican-American family. In his absence, this power position reverts to the oldest son. All members of the household are expected to carry out the orders of the male head" (Becerra, 1998, 159).

2. Female submissiveness. The complement to the male authority figure in the Mexican-American marriage is the submissive female partner. Traditionally, the Latina is subservient to her husband and devotes her time totally to the roles of homemaker and mother. However, as more wives begin to work outside the home, the nature of the Mexican-American husband-wife relationship will become more egalitarian in terms of joint decision making and joint childrearing.

Strong Familistic Values Mexican-Americans, regardless of their national origin, value having children and having close relationships with both nuclear and extended family members (familism). Mexican-Americans report receiving a high level of family support and desire geographical closeness. "For decades, familism has been considered to be a defining feature of the Mexican-origin population" (Baca Zinn & Pok, 2002, 93).

The relationship between Mexican-American children and their parents has traditionally been one of respect. This respect results not from fear but from the fact that the parents (including fathers) are warm, nurturing, and companionable. In addition, it is common for the younger generation to pay great deference to the older generation. When children speak to their elders, they do so in a formal way.

Native American Marriages

About 2.8 million individuals define themselves as Native Americans (*Statistical Abstract of the United States: 2003,* Table 21). The term refers not only to American Indians but also to Inuit (Eskimos) and Aleuts (native people of the Aleutian Islands). American Indians comprise over 95 percent of all Native Americans and are the group to which we will refer.

Native Americans, the original Indian inhabitants, are an incredibly diverse group. They comprise 510 federally recognized tribes and 278 reservations and speak 187 different languages (Coburn et al., 1995, 226). Family patterns of Native Americans are also diverse. When viewed across time, Native American families have been patrilineal (heritage traced through males), matrilineal (heritage traced through females), monogamous, polygynous, and polyandrous. Tribal identity has consistently taken precedence over family identity, and the values of a family reflect those of the particular tribe or clan.

Given these caveats, Yellowbird & Snipp (2002) and McLain (2000) made the following observations about Native Americans:

1. They tend to marry young. When compared with both Blacks and whites, Native Americans marry at younger ages (median age for males = 22, for females = 20).

2. Little stigma is attached to having a child without being married. Children are highly valued in the Native American community. Indeed, among the Hopi Indians, a couple will live in the home of the woman (with her parents), have children, and marry years later.

3. Intermarriage rates are the highest of any racial group. The most frequent intermarriage involving a Native American is between a white husband and a Native American wife. The reason for such high rates of intermarriage is that "American Indians are perceived, especially by whites, as socially acceptable marriage partners" (Yellowbird & Snipp, 2002, 236). Hence, the stigma and racism associated with Black-white marriages are nonexistent.

4. Divorce among Native Americans is regarded as a less traumatic event and is usually not associated with guilt and recriminations. Indeed, the percentage of divorced persons is higher among Native Americans than among either Blacks or whites.

5. Elders are viewed as important and are looked up to. They are given meaningful economic, political, religious, and familial roles within the tribe.

6. Extended families are the norm.

7. Role relationships between husbands and wives are becoming less traditional as Native American women become increasingly involved in working outside the home.

8. Native American family values include concern for the group, generosity, and disdain for material possessions. These values are counter to the individualism and materialism characteristic of mainstream American culture. But materialism and individualism are making inroads into Native American life. The authors visited in the home of a Hopi elder in north-central Arizona (near Tuba City) and listened to him talk about the changes among the Hopi people. He noted, "Too much TV and whiskey—they are ruining my people."

Asian- and Pacific Islander-American Marriages

It is estimated that by the year 2005 there will be about 13 million Asian and Pacific Islander Americans representing about 5 percent of the U.S. population (*Statistical Abstract of the United States: 2003,* Table 16). These two groups comprise Chinese, Filipino, Japanese, Korean, Vietnamese, Cambodian, Thai, Lao Hmong, Burmese, Samoan, and Guamanian. Each group is different, depending on its cultural heritage, immigration history, American response to its arrival or presence, and its resulting socioeconomic and social adaptation. Immigration history is relevant to family patterns in that, historically, immigration laws restricted whole families from immigrating. Rather, only men were permitted entry, because they were a source of inexpensive labor; this situation resulted in splintered families (wives and children later joined the men).

The cultural heritage brought to the states included Confucian philosophy and the importance of familism. Confucian principles emphasized superiority of husbands and elders over wives and children. These values were in conflict with egalitarian values Asian children socialized in America were taught. Familism as a value also emphasized the importance of the family group over the individual, with the result of a much lower divorce rate among Asians and Pacific Islanders.

Japanese-American marriages differ depending on how long the spouses have lived in the United States. Issei, or first-generation, Japanese-Americans were born in Japan and immigrated to the United States in the early 1900s. They are now in their eighties and live with their children, in nursing homes, or in senior citizen housing projects such as the Little Tokyo Towers in Los Angeles. Their values, beliefs, and patterns reflect those of traditional Japanese families: (1) offspring not allowed to select their own spouse, (2) a stronger parent-child bond than the bond between husband and wife, (3) male dominance, (4) rigid division of labor by sex, and (5) precedence of family values over individual values.

The Nisei, or second-generation, children have been influenced by Japanese-American peers and the American mainstream. The younger Nisei believe in romantic love, select their own mates, regard the husband-wife relationship as more important than the parent-child relationship, and have egalitarian sex roles. However, family gatherings of extended kin are common.

The Sansei are the third-generation children and reflect even greater Americanization than the Nisei. The Sansei marry for love and rarely hesitate to marry someone who is not Japanese. Native-born Japanese women are also willing to delay marriage and, increasingly, to remain single.

Interracial Marriages

Interracial marriages may involve many combinations, including American white, American Black, Indian, Chinese, Japanese, Korean, Mexican, Malaysian, and Hindu mates. Interracial marriages are rare in the United States—less than 5 percent of all marriages in the United States are interracial. Of these, fewer than 1 percent consist of Black/white spouses (*Statistical Abstract of the United States: 2003,* Table 62). Examples of African-American men who are married to Caucasian women are Quincy Jones and Charles Barkley. Segregation in religion (the races worship in separate churches), housing (white and Black neighborhoods), and education (white and Black colleges), not to speak of parental and peer endogamous pressure to marry within one's own race, are factors that help to explain the low percentage of interracial Black/white marriages.

SELF-ASSESSMENT

Attitudes toward Interracial Dating Scale

Interracial dating or marrying is the dating or marrying of two people from different races. The purpose of this survey is to gain a better understanding of what people think and feel about interracial relationships. Please read each item carefully and consider how you feel about each statement. There are no right or wrong answers to any of these statements. Please read each statement carefully, and respond by using the following scale:

1	2	3	4	5	6	7
Strongly Disagree						Strongly Agree

___ **1.** I believe that interracial couples date outside their race to get attention.

___ **2.** I feel that interracial couples have little in common.

___ **3.** When I see an interracial couple I find myself evaluating them negatively.

___ **4.** People date outside their own race because they feel inferior.

___ **5.** Dating interracially shows a lack of respect for one's own race.

___ **6.** I would be upset with a family member who dated outside his/her race.

___ **7.** I would be upset with a close friend who dated outside his/her race.

___ **8.** I feel uneasy around an interracial couple.

___ **9.** People of different races should associate only in non-dating settings.

___ **10.** I am offended when I see an interracial couple.

___ **11.** Interracial couples are more likely to have low self-esteem.

___ **12.** Interracial dating interferes with my fundamental beliefs.

___ **13.** People should date only within their race.

___ **14.** I dislike seeing interracial couples together.

___ **15.** I would not pursue a relationship with someone of a different race regardless of my feelings for him/her.

___ **16.** Interracial dating interferes with my concept of cultural identity.

___ **17.** I support dating between people with the same skin color, but not with a different skin color.

___ **18.** I can imagine myself in a long-term relationship with someone of another race.

___ **19.** As long as the people involved love each other, I do not have a problem with interracial dating.

___ **20.** I think interracial dating is a good thing.

Scoring

Having placed a number representing the continuum from 1 to 7 in each of the twenty spaces above, reverse-score items 18, 19, and 20. For example, if you selected 7 for item 18, replace it with a 1; if you selected 1, replace it with a 7, etc. Next, add your scores and divide by 20. Possible scores range from 1 to 7 with 1 representing the most positive attitudes toward interracial dating and 7 representing the most negative attitudes toward interracial dating.

Norms

The norming sample was based upon 113 male and 200 female students attending Valdosta State University. The participants received no compensation for their participation. All participants were United States citizens. The average age of participants completing the Attitudes toward Interracial Dating Scale was 23.02 years (*SD* = 5.09), and participants ranged in age from 18 to 50. The ethnic composition of the sample was 62.9 percent white, 32.6 percent Black, 1 percent Asian, 0.6 percent Hispanic, and 2.2 percent classified themselves as "Other." The classification of the sample was 9.3 percent freshman, 16.3 percent sophomore, 29.1 percent junior, 37.1 percent senior, and 2.9 percent were graduate students. The average score on the IRDS was 2.88 (*SD* = 1.48) and ranged from 1.00 to 6.60, suggesting very positive views of interracial dating. Men scored an average of 2.97 (SD=1.58); women 2.84 (SD=1.42). There were no significant differences between women and men.

Copyright: "The Attitudes Toward Interracial Dating Scale" 2004 by Mark Whatley, Ph.D. Department of Psychology, Valdosta State University, Valdosta, Georgia 31698. Information on validity and reliability may be obtained from Dr. Whatley. The scale is used by permission of Dr. Whatley. Other uses of this scale by written permission only; e-mail mwhatley@valdosta.edu

Black-white spouses are more likely to have been married before, to be age-discrepant, to live far away from their families of orientation, to have been reared in racially tolerant homes, and to have educations beyond high school. Some may also belong to religions that encourage interracial unions. The Baha'i religion, which has over 6 million members worldwide and 84,000 in the United States, teaches that God is particularly pleased with interracial unions. Finally, interracial spouses may tend to seek contexts of diversity. "I have been reared in a military family, been everywhere and met people of different races and nationalities throughout my life. I seek diversity," noted one student.

Kennedy (2003) identified three reactions to a Black-white couple who cross racial lines to marry—approval (increases racial open-mindedness, decreases social segregation), indifference (interracial marriage is seen as a private choice)

and disapproval (reflects racial disloyalty, impedes perpetuation of Black culture). "The argument that intermarriage is destructive of racial solidarity has been the principal basis of black opposition" (p. 115).

Interracial partners sometimes experience negative reactions to their relationship. Blacks partnered with whites have their Blackness and racial identity challenged by Blacks. Whites partnered with Blacks may lose their white status and have their awareness of whiteness heightened more than ever before. At the same time, one partner is not given full status as a member of the other partner's race (Hill & Thomas, 2000). Gaines and Leaver (2002) also note that the pairing of a Black male and a white female is regarded as "less appropriate" than that of a white male and a Black female. In the former, the Black male "often is perceived as attaining higher social status (i.e. the White woman is viewed as the Black man's 'prize,' stolen from the more deserving white man)" (p. 68). In the latter, when a white male pairs with a Black female, "no fundamental change in power within the American social structure is perceived as taking place" (p. 68).

Black-white interracial marriages are likely to increase . . . slowly. Not only has white prejudice against African-Americans in general declined, but segregation in school, at work, and in housing has decreased, permitting greater contact between the races. The Self-Assessment on page 196 allows you to assess your openness to involvement in an interracial relationship.

Authors

The fact that fewer than 1 percent of the almost 60 million married couples in America consist of a Black and a white spouse reflects that "freedom" to marry whomever we choose is an illusion—that social factors influence and restrict our choices. Those who do cross racial lines to marry have usually been married before and are older, as is true of the couple in this photo.

Interreligious Marriage

My Big Fat Greek Wedding, sleeper film of 2002, detailed the difficulties of a Greek never-married women who fell in love with and married a man of a different religion and culture. Although the couple resolved the dilemma by having him convert to her religion, the movie emphasized that marriage is the merging of two families and these families may represent very different learning histories.

Although religion may be a central focus of some individuals and their marriage, Americans in general have become more secular, and as a result religion has become less influential as a criterion for selecting a partner. In a study on attitudes of college students toward interreligious marriage, over 70 percent of the women and 85 percent of the men reported a willingness to marry someone outside their religion (Knox, Zusman, & Daniels,2002).

Are people in interreligious marriages less satisfied with their marriages than those who marry someone of the same faith? The answer depends on a number of factors. First, people in marriages in which one or both spouses profess "no religion" tend to report lower levels of marital satisfaction than those in which at least one spouse has a religious tie. People with no religion are often more liberal and less bound by traditional societal norms and values—they feel less constrained to stay married for reasons of social propriety.

The impact of a mixed religious marriage may also depend more on the devoutness of the partners than on the fact that the partners are of different religions. If both spouses are devout in their religious beliefs, they may expect some problems in the relationship (although not necessarily). Less problematic is the relationship in which one spouse is devout but the partner is not. If neither spouse

This Christian husband and Jewish wife were married by both a Protestant minister and a rabbi. They wrote their own vows, which included the statement that one of their strengths as a couple was their respect for their connections to their respective faiths.

Authors

in an interfaith marriage is devout, problems regarding religious differences may be minimal or nonexistent.

One interfaith couple who married (he Christian, she Jewish) said in their marriage vows that they viewed their different religions as an opportunity to strengthen their connections to their respective faiths and to each other. "Our marriage ceremony seeks to celebrate both the Jewish and Christian traditions, just as we plan to in our life together" (see photo at bottom of p. 197).

Cross-National Marriages

The number of international students studying at American colleges and universities is large.

National Data

Approximately 548,000 foreign students are enrolled at more than 2,500 colleges and universities in the United States. Most (62%) are from Asia (*Statistical Abstract of the United States: 2003*, Table 281).

Since American students take classes with foreign students, there is the opportunity for dating and romance between the two groups, which may lead to marriage. Since some persons from foreign countries marry an American citizen to gain citizenship in the United States, immigration laws now require the marriage to last two years before citizenship is granted. If the marriage ends before the two years, the foreigner must prove good faith (he or she did not marry just to gain entry into the country) or be asked to leave the country.

When the international student is male, more likely than not his cultural mores will prevail and will clash strongly with his American bride's expectations, especially if the couple should return to his country. One female American student described her experience of marriage to a Pakistani, who violated his parents' wishes by not marrying the bride they had chosen for him in childhood. The marriage produced two children before the four of them returned to Pakistan. The woman felt that her in-laws did not accept her and were hostile toward her. The in-laws also imposed their religious beliefs on her children and took control of their upbringing. When this situation became intolerable, the woman wanted to return to the United States. Because the children were viewed as being "owned" by their father, she was not allowed to take them with her and was banned from even seeing them. Like many international students, the husband was from a wealthy, high-status family, and the woman was powerless to fight the family. The woman has not seen her children in six years.

International students on campus provide an opportunity for the development of relationships with persons from different cultures.

Authors

Cultural differences do not necessarily cause stress in cross-national marriage, and degree of cultural difference is not necessarily related to degree of stress. Much of the stress is related to society's intolerance of cross-national marriages, as manifested in attitudes of friends and family. Japan and Korea place an extraordinarily high value on racial purity. At the other extreme is the racial tolerance evident in Hawaii, where a high level of out-group marriage is normative.

Age-Discrepant Relationships and Marriages

There have been a number of age-discrepant celebrity marriages, including Celine Dion, who is twenty-six years younger than Rene Angelil (in 2005 she is 37 and he is 63). Among all sexually active women aged 15 to 44, 20 percent had a partner who was three to five years older, and 18 percent had a partner who was six or more years older (Darroch, Landry, & Oslak, 2000). Silverthorne and Quinsey (2000) asked 192 adults to express their age preferences for a preferred partner. Both heterosexual and homosexual men preferred younger partners (homosexual women preferred older partners).

When the partners are ten or more years apart in age, the union is regarded as an age-discrepant relationship. A study (Knox, Britton, & Crisp, 1997) of seventy-seven female university faculty and their female students who were involved with men ten to twenty-five years older revealed five themes in these age-discrepant relationships.

1. Age-discrepant relationships are happy. Eighty percent reported that they were happy in their relationships. Forty percent agreed with the statement, "I am happy in my current relationship" and 40 percent reported "strong agreement." Only 4 percent disagreed with the statement. Over 60 percent said that they would become involved in another age-discrepant relationship if their current relationship ended.

2. Age-discrepant relationships lack social approval and support. Only a quarter of the respondents reported that their friends, mothers, and fathers provided clear support for their relationship. Fathers were least approving, with over 40 percent not approving.

3. Age-discrepant relationships are not without problems. In addition to lack of support, the respondents in this study reported a range of problems they attributed to the age difference with their partners, including money, in-laws, and recreation.

4. Women perceive benefits from involvement with older partners. Respondents noted financial security (58%), maturity (58%), and dependability (51%) as the primary advantages of involvement with an older man. Higher status was regarded as less important: only 28 percent of the respondents identified this.

5. Friends of the couple are joint friends. Over 70 percent (71%) of respondents reported that when they did something recreational, the friends were likely to be both of theirs. However, if the friends were friends of only one of them, it was more likely to be the man (22%) than the woman (5%).

Some age-discrepant dating relationships become age-discrepant marriage relationships, also referred to as age-dissimilar marriages (ADM). When the man is considerably older than the woman, such marriages are referred to as **May-December marriages.** Typically, she is in the spring of her youth (May), and he is in the later years of his life (December). Well-known personalities and the number of years they are older than their spouses include Tony Randall, fifty; Tony Bennett, forty; and Michael Douglas, twenty-five. Though the situation is less common, some women are older than their partners. Demi Moore has been linked romantically in the media with Ashton Kutcher—she is sixteen years older.

At the age of 18, she became the fourth wife of 54-year-old Charlie Chaplin. The May-December alliance was expected to last the requisite six months, but they confounded skeptics by staying together, raising eight children, and remaining, in their words, blissfully happy.

Jane Scovell *of Oona Chaplin, wife of Charles Chaplin, from the book* Oona: Living in the Shadows

Mixed Marriage in Music Obsession

While mixed marriages are typically discussed in terms of race, religion, culture and age, some are unique and specific. Adams and Rosen-Grandon (2002) reported on marriages in which one spouse is an obsessed "Deadhead" (follower of the Grateful Dead) and one spouse is not. "The 'Deadhead' spouse is sometimes upset by the non-Deadhead partner's inability to understand why sub cultural membership is important and meaningful. The non-Deadhead spouse is sometimes irritated by the expenditure of family resources for sub cultural activities and illegal behavior. If their values and interests diverge significantly, noncouple socializing can become problematic" (p. 85). Being a Deadhead is also a stigma and a "voluntary identity rather than an inherited disability," resulting in pressure to give it up. And the time and expense can be considerable. The average Deadhead reports having attended sixty-one shows and spends over a thousand dollars a year related to "the Dead."

SUCCESSFUL MARRIAGES

Judith Wallerstein, a wife for fifty years, a clinical psychologist, and coauthor with Sandra Blakeslee of *The Good Marriage* (1995), identified herself as having a successful marriage. Most people who marry have the same goal—to be married successfully. But what is a successful marriage?

Definition of a Successful Marriage

Marital success is measured in terms of marital stability and marital happiness. Stability refers to how long the spouses have been married and how permanent they view their relationship, whereas marital happiness refers to more subjective aspects of the relationship. For a discussion of theoretical views of marital success, see Up Close 8.2 on page 201.

In describing marital success, researchers have used the terms *satisfaction, quality, adjustment, lack of distress,* and *integration.* Marital success is often measured by asking spouses how happy they are, how often they spend their free time together, how often they agree about various issues, how easily they resolve conflict, how sexually satisfied they are, and how often they have considered separation or divorce. The degree to which the spouses enjoy each other's companionship is

Authors

One criterion for defining a marriage as successful is its duration.

Up Close 8.2 Theoretical Views of Marital Happiness and Success

Interactionists, developmentalists, exchange theorists, and functionalists view marital happiness and success differently. Symbolic interactionists emphasize the subjective nature of marital happiness and point out that the definition of the situation is critical. Only when spouses define the verbal and nonverbal behavior of their partner as positive, and only when they label themselves as being in love, may a happy marriage exist. Hence, marital happiness is not defined by the existence of eight or more specific criteria but is subjectively defined by the respective partners.

Family developmental theorists emphasize the developmental tasks that must be accomplished to enable a couple to have a happy marriage. Wallerstein and Blakeslee (1995) identified several of these tasks, including separating emotionally from one's parents, building a sense of "we-ness," establishing an imaginative and pleasurable sex life, and making the relationship safe for expressing differences. A developmental view of marriage and family also suggests that certain problems are more pronounced at certain times of the family life cycle than at others. For example, in a national survey, only one-third of the couples under age 50 reported that they had enough time for each other. These couples were still rearing their children and building careers. But three-fourths of those over 50 said that they had enough time to be with each other (Saad, 1995).

Exchange theorists focus on the exchange of behavior of a kind and at a rate that is mutually satisfactory to both spouses. When spouses exchange positive behaviors at a high rate, they are more likely to feel marital happiness than when the exchange is characterized by high-frequency negative behavior (Turner, 2004).

Structural functionalists see marital happiness as contributing to marital stability, which is functional for society. When two parents are in love and happy, the likelihood that they will stay together to provide physical care and emotional nurturing for their offspring is increased. Furthermore, when spouses take care of their own children, society is not burdened with having to pay for the children's care through welfare payments, paying foster parents, or paying for institutional management (group homes) when all else fails. Happy marriages also involve limiting sex to each other. In their national sex survey, Michael and colleagues reported, "[H]appiness is clearly linked to having just one partner—which may not be too surprising since that is the situation that society smiles upon" (1994, 130). Fewer HIV cases also mean lower medical bills for society. Similarly, marriage is associated with improved health (Stack & Eshleman, 1998), since spouses monitor each other's health and encourage/facilitate medical treatment as indicated.

Sources

Michael, R. T., J. H. Gagnon, E. O. Laumann, and G. Kolata. 1994. *Sex in America: A definitive survey.* Boston: Little, Brown.

Saad, L. 1995. Children, hard work taking their toll on baby boomers. *Gallup Poll Monthly.* April, 21–24.

Stack, S., and J. R. Eshleman. 1998. Marital happiness: A 17-nation study. *Journal of Marriage and the Family* 60:527–36.

Turner, A. J. 2004. Personal communication, Huntsville, Alabama, September. Used by permission.

Wallerstein, J., and S. Blakeslee. 1995. *The good marriage.* Boston: Houghton-Mifflin.

another variable of marital success. Not all couples, even those recently married, achieve high-quality marriages.

Wallerstein and Blakeslee (1995) studied fifty financially secure couples in stable (from ten to forty years) and happy marriages with at least one child. These couples defined marital happiness as feeling respected and cherished. They also regarded their marriage as a work in progress that needed continued attention lest it become stale. No couple said that they were happy all the time. Rather, a good marriage was a process. Frye and Karney (2002) noted that spouses generally perceive themselves to be better off than other couples in terms of having a good marriage and feel that their problems will remain stable (not get worse) over time.

I am certain that I have one of America's better marriages; and yet the challenge of keeping it successful never dims, for Camille and I may be blinded by love, but we have Braille for each other's flaws.

Bill Cosby, *Comedian*

Characteristics of Successful Marriages

Billingsley, Lim, and Jennings (1995) interviewed thirty happily married couples who had been wed an average of thirty-two years and had an average of 2.5 children. Various themes of couples who stay together and who enjoy each other and their relationship include the following:

1. Commitment. Divorce was not considered an option. The couples were committed to each other for personal reasons rather than societal pressure. They also did not

Diversity in Other Countries

Though marital happiness is an important concept in American marriages, this is "not necessarily a salient concept in Chinese society" (Pimentel, 2000, 32). What is important to the Chinese is the allegiance of the husband to his family of origin (his parents). Also, the five basic relationships in order of importance in Confucian philosophy are ruler-minister, father-son, elder brother-younger brother, husband-wife, and friend-friend.

leave themselves open to affairs. One respondent said "on any number of occasions attractive people had given clear signals of availability; however, whenever I sense something like this, I'm thrilled to talk about my wife. Talking about the beauty of our relationship helps to build a hedge in what could be a troublesome situation" (Billingsley et al., 1995, 288).

*2. **Common interests.*** The spouses talked of sharing values, children, work, travel, goals, dependability, and the desire to be together.

*3. **Communication.*** Gottman and Carrere (2000) studied the communication patterns of couples over an eleven-year period and emphasized that those spouses who stay together are five times more likely to lace their arguments with positives ("I'm sorry I hurt your feelings") and to consciously choose to say things to each other that nurture the relationship rather than destroy it.

*4. **Religiosity.*** A strong religious orientation provided the couples with social, spiritual, and emotional support from church members and with moral guidance in working out problems. Knox et al. (1998) found in a sample of 235 undergraduates that high religiosity scores (measured by religious attendance, prayer patterns, and belief in God) were related to higher self-esteem and fewer antisocial behaviors.

*5. **Trust.*** Trust in the partner provided a stable floor of security for the respective partners and their relationship. Neither partner feared that the other partner would leave or become involved in another relationship.

*6. **Finances and work.*** Being nonmaterialistic, being disciplined, and being flexible with each other's work schedules and commitments were characteristic of these happily married couples.

*7. **Role models.*** The couples spoke of positive role models in their parents. Good marriages beget good marriages— good marriages run in families. It is said that the best gift you can give your children is a good marriage.

*8. **Low stress levels.*** A team of researchers (Harper, Schaalje, & Sandberg, 2000) found that low stress levels in one's life were associated with marital quality in mature marriages in which the partners were aged 55 to 75.

*9. **Sexual desire.*** Regan (2000) studied twenty-five men and twenty-five women involved in dating relationships and found that sexual desire was related to a greater desire for the partner, relationship stability, being faithful, and not being attracted to others. It is quite likely that sexual desire would have similar effects on spouses. Earlier, we noted the superiority of marital sex over sex in other relationship contexts in terms of both emotion and physical pleasure.

What can we learn from our knowledge about the characteristics of successful marriages? Commitment, common interests, communication skills, and a nonmaterialistic view of life all seem related to being involved in and maintaining a successful marriage. Couples might strive to include these as part of their relationship.

He who laughs, lasts.
Mary Poolegive

As we have noted, durability is only one criterion for a successful marriage. Satisfaction is another. Researchers in the early nineties analyzed cross-sectional data and concluded that marital satisfaction drops across time, reaches a low point during the years the couple has teens in the house, and returns to preteen satisfaction levels (Vaillant & Vallant, 1993). More recent researchers have studied longitudinal data and found that marital satisfaction consistently drops across time (with the steepest declines in the early and later years) (Vanlaningham, Johnson, & Amato, 2001).

Diversity in Other Countries

Walters et al. (1997) provided cross-cultural data on perceived marital quality by both husbands and wives in 1,718 families in the United States, the republic of Georgia (south of Moscow), Poland, and Samara (south-central Russia). Both spouses had been married seven years or less and had at least one child. Approximately 20 to 30 percent of all the couples in all the countries perceived the quality of their marriages to be low.

SUMMARY

What are individual motivations and social functions of marriage?

Individuals' motives for marriage include personal fulfillment, companionship, legitimacy of parenthood, and the emotional and financial security from marriage. Social functions include continuing to provide society with socialized members, regulating sexual behavior, and stabilizing adult personalities.

What are three levels of commitment in marriage?

Marriage involves a commitment—person-to-person, family-to-family, and couple-to-state.

What are two rites of passage associated with marriage?

The wedding is a rite of passage signifying the change from the role of fiancé to the role of spouse. The honeymoon is a time of personal recuperation and making the transition to a new role of spouse.

What changes might a person anticipate after marriage?

Personal changes include an enhanced self-concept, a satisfying sex life (when compared with that of singles), improved acceptance of the mate by one's parents, less time with one's parents, and less time with one's same-sex friends.

How do marriage patterns differ among racial and cultural groups?

African-American marriages are characterized by strong kinship ties and strong mother-child relationships. Mexican-American marriages have traditionally been characterized by male dominance and female submissiveness. Native American marriages are not so easily categorized. Since tribal identity supersedes family identity and the values and beliefs vary widely among the 510 federally recognized tribes, there are few fixed characteristics of Native American marriages. There are over 10 million Asian and Pacific Islander Americans. Japanese-American marriages differ, depending on the degree of socialization of the spouses in the United States. Issei (first-generation) marriages are very traditional, in contrast to Sansei (third-generation) marriages, in which the spouses are very Americanized.

Some marriages are mixed in that the partners represent different races, religions, cultures, ages, and musical obsessions. Although interracial marriages remain rare, spouses in such marriages have usually been married before, are older, were reared in racially tolerant homes, have parents who live far away, and seek contexts of diversity. They are also self-confident and autonomous and do not see their marriage as different from any other marriage.

An increasing number of marriages are interreligious. Although mixed religious marriages do not necessarily imply a greater risk to marital happiness, marriages in which one or both spouses profess no religion are in the greatest jeopardy. Also, husbands in interreligious marriages seem less satisfied because children are usually reared in the faith (or nonfaith) of the wife.

The mating gradient results in numerous pairings of age-discrepant partners, with the man usually older. Though partners in age-discrepant relationships report satisfaction, there is social disapproval, particularly from the woman's father. May-December marriages may also generate disapproval from one's children if they feel their inheritance is being jeopardized.

What are the characteristics associated with successful marriages?

Marital success is defined in terms of both quality and durability. Characteristics associated with marital success include commitment, common interests, communication, religiosity, trust, being nonmaterialistic, having positive role models, low stress levels, and sexual desire.

KEY TERMS

commensality
commitment
covenant marriage
disenchantment
marital success
May-December marriage
Mexican-American
racism
rite of passage

RESEARCHING MARRIAGE AND THE FAMILY WITH INFOTRAC COLLEGE EDITION

InfoTrac College Edition, an online library, allows you to perform research online anywhere, anytime. Following are two suggested search terms and related questions to help you extend your understanding of the topics covered in this chapter. Go to www.infotrac-college.com to begin your search.

Keyword: **Marriage myths.** Locate articles that discuss myths about marriage. What are the sources of such myths and how prevalent are they?

Keyword: **Interracial marriage.** Locate articles that discuss interracial marriage in the United States. Why do you think the rate of interracial marriage remains low?

The Companion Web Site for Choices in Relationships: An Introduction to Marriage and the Family, Eighth Edition

http://sociology.wadsworth.com/knox_schacht/choices8e

Supplement your review of this chapter by going to the companion Web site to take one of the Tutorial Quizzes, use the flash cards to master key terms, and check out the many other study aids you'll find there. You'll also find special features such as the Marriage and Family Resource Center, Census 2000 information, and other data and resources at your fingertips to help you with that special project or to do some research on your own.

WEBLINKS

African American News
http://www.africanamerican.com/
(this link will take you to The News Channel Network home page. Scroll down and click on African-Americans.com under the list of Other News Channels)

Americans for Divorce Reform
http://www.divorcereform.org/index.html

Asian-American Links
http://www.africanamerican.com/
(this link will take you to The News Channel Network home page. Scroll down and click on Asian-Americans.com under the list of Other News Channels)

Bridal Registry
http://www.theknot.com

Brides and Grooms
http://www.bridesandgrooms.com/

Facts about Marriage
http://www.cdc.gov/nchs/fastats/marriage.htm

Hispanic-American Links
http://www.africanamerican.com/
(this link will take you to The News Channel Network home page. Scroll down and click on Hispanic-Americans.com under the list of Other News Channels)

Smart Marriages
http://www.smartmarriages.com/

CORBIS

CHAPTER 9

It's not the differences between partners that cause problems but how the differences are handled when they arise.

Clifford Notarius and Howard Markman, Marriage therapists

Communication in Relationships

Contents

True or False?

1. In a study of undergraduates, females valued and engaged in more nonverbal communication than did males.
2. Spending an obsessive amount of time on the Internet is sometimes at the expense of communicating with a partner.
3. In some states, not informing a partner of a communicable disease such as HIV is a felony, which may involve a five-year prison term.
4. Lying about the number of previous partners was the most frequently reported lie told by a sample of university students.
5. Being assertive and cooperative (collaborating style) is the style of conflict most associated with marital and spousal satisfaction.

Answers: **1.** T **2.** T **3.** T **4.** T **5.** T

Communication is what turns strangers into lovers ("We talked all night") and its absence ("We have nothing to say to each other") is what sometimes results in spouses drifting apart and seeking a divorce. In this chapter we examine various issues related to communication and identify some factors that might be helpful in improving communication skills. We begin by looking at the nature of interpersonal communication.

I really didn't say everything I said.
Yogi Berra

THE NATURE OF INTERPERSONAL COMMUNICATION

Communication can be defined as the process of exchanging information and feelings between two people. Although most communication is focused on verbal content, much (estimated to be as high as 80%) interpersonal communication is nonverbal. Regardless of what a person says, crossed arms and lack of eye contact will convey a very different meaning than the same words accompanied with a gentle touch and eye-to-eye contact. We often attend to the nonverbal cues in interaction and assign them more importance—we like to hear sweet words but we feel more confident when we see behavior that supports the words. "Show me the money, honey" is a phrase that reflects a partner's focus on behavior rather than words. Up Close 9.1 on the next page features a study on college students' nonverbal communication.

And we'll meet up without even talking and you'll know what I'm saying.
Lila McCann, *"I Wanna Fall in Love"*

Relationship Problems Reported by Casual and Involved Daters

A great deal of communication is devoted to resolving relationship problems. Five hundred twenty-seven never-married undergraduates at a large southeastern university identified the most frequent problem they had experienced in their current or most recent dating relationship (Zusman & Knox, 1998). There were statistically significant differences between those reporting that they were "casual" daters (dating more than one person or just starting to date someone) and those reporting that they were "involved" (mutual emotional involvement). Table 9.1 shows the top ten relationship problems identified by casual and involved daters.

The data in Table 9.1 make it clear that communication is the top problem in dating relationships and suggest that it does not abate as a problem for partners who have become more involved. Indeed, these data suggest that those who regarded themselves as emotionally involved were more likely than those who were dating casually to identify communication as their most frequent problem.

Regardless of what this couple is saying, their nonverbal behavior is communicating delight in being together.

Authors

Lack of commitment was the second most frequently reported problem by casual daters. Presumably, commitment issues have been resolved among involved couples, since this problem dropped to number 10 for the involved partners. Only 2.4 percent regarded this as a relationship issue, which suggests that one of the benefits of being in an involved relationship is feeling more secure about one's own and the partner's commitment.

While casual daters were struggling with commitment issues, the second most frequent problem reported by involved daters was "other problems." A flaw of this study was that respondents had no way to identify what they meant by "other" problems. However, we often ask our students to write down anonymous questions regarding problems they are having in their respective relationships as

Up Close 9.1 Nonverbal and Verbal Communication in "Involved" and "Casual" Relationships

This study examined the nonverbal and verbal communication between partners in two categories of college students—the "involved" (emotionally involved in a reciprocal love relationship with one person) and "casual"daters (dating different people). Two hundred and thirty-three undergraduates at a large southeastern university completed a forty-five-item questionnaire.

Findings and Discussion

Analysis of the data revealed several significant findings:

1. **Involved daters valued nonverbal communication more than casual daters.** When students were asked "How important do you think nonverbal communication is in a relationship?" (with 1 = "very important" and 10 = "not important at all," so that the lower the score the greater the importance), the mean value of the involved and casual daters was 3.08 vs. 3.75, respectively ($p < .03$). Previous research has demonstrated that the more involved the individuals become, the more serious they regard their relationship issues (e.g., requiring a person to have similar values) (Knox, Zusman, & Nieves, 1997). Being sensitive to and concerned about the nuances of nonverbal communication is an extension of being more serious about relationship issues.

2. **Involved daters worked on nonverbal behavior more than casual daters.** Not only did the involved respondents value nonverbal behavior in the abstract, but they were more likely ($p < .005$) than casual daters to "work hard" to ensure that their nonverbal behavior reinforced their verbal behavior. "I try to make sure that what I do backs up what I say," noted one college student.

3. **Females valued nonverbal behavior more than males.** When females in our sample were compared with males, the former were significantly more likely ($p < 0.001$) than males to report that nonverbal behavior "should" be regarded as important. Previous research has demonstrated that women (more than men) are more serious about communication in relationships (Tannen, 1990).

4. **Females engaged in more nonverbal behavior.** When females were compared with males, the former were more likely to look their partner straight in the eye and to nod their heads when their partner spoke. Guerrero (1997) also noted that females were more likely than men to engage in nonverbal behavior when such behavior was defined as displaying "direct body orientation and gaze." Moore (1998) found that when women want to dissuade an aggressor, they display high rates of nonverbal rejection behavior. She documented seventeen such behaviors that signaled noninterest or rejection such as yawning, frowning, and pocketing the hands.

4. **Whites valued nonverbal behavior more.** Whites were significantly ($p < .017$) more likely than Blacks to believe that nonverbal behavior is ($p < .001$) and should be ($p < .019$) important to a relationship. Our hypothesis regarding this finding is that Blacks face enormous pressure to adapt to the mainstream white culture even with little to no attention given to nonverbal communication. Hence, Blacks may feel more predisposed to believe in the verbal, the literal. Meanwhile, whites may feel no such pressure and may feel more "free" to focus on nonverbal aspects of communication. Whites (compared with Blacks) also significantly ($p < .027$) feel that their relationship would be better if their partner would use more nonverbal behavior.

Abridged and adapted from K. McGinty, D. Knox, and M. E. Zusman. 2003. Nonverbal and verbal communication in "involved" and "casual" relationships among college students. *College Student Journal* 37:68–71. Reprints of the original article may be obtained from the senior author at Vedettee@aol.com

Sources

Guerrero, L. K. 1997. Nonverbal involvement across interactions with same-sex friends, opposite-sex friends and romantic partners: Consistency or change? *Journal of Social and Personal Relationships* 14: 31–54.

Knox, D., M. E. Zusman, and W. Nieves. 1997. College students' homogamous preferences for a date and mate. *College Student Journal* 31: 445–48.

Moore, M. M. 1998. Nonverbal courtship patterns in women: Rejection signaling—An empirical investigation. *Semiotica* 118: 201–14.

Tannen, D. 1990. *You just don't understand: Women and men in conversation.* London: Virago.

Table 9.1 Top Ten Problems Experienced By Casual and Involved Daters (N = 527)

Casual Daters (N = 240)	Percentage	Involved Daters (N = 287)	Percentage
Communication	19.6%	Communication	22.3%
Lack of commitment	12.5%	Other problems	15.3%
Jealousy	12.1%	Jealousy	13.9%
Other problems	9.6%	No problems	13.2%
No problems	8.3%	Time for relationship	9.1%
Different values	7.9%	Lack of money	5.2%
Honesty	7.5%	Places to go	4.9%
Shyness	5.4%	Honesty	4.2%
Unwanted sex pressure	2.1%	Different values	3.1%
Acceptance	1.7%	Lack of commitment	2.4%

Source: M. E. Zusman and D. Knox. 1998. Relationship problems of casual and involved university students. *College Student Journal* 32:606–9,

issues to be discussed in class. Among the issues they identify are different interests in sexual involvement, sexual dysfunctions, alcohol/substance abuse, depression, and self-concept issues. It is possible that these are some of the issues with which involved daters were coping.

Twice as many casual daters as involved daters reported different values. The theory of homogamy suggests that new dating partners are in the process of eliminating partners whose values are dissimilar to their own. Those who remain are partners with whom they share similar values. It is not surprising that "involved" daters report fewer differences in values.

Principles and Techniques of Effective Communication

It's more important to be right ***with each other*** *than it is to be right.*

Fraley and Marilyn Bost, *married twenty-seven years*

Persons who are concerned about effective communication in their relationship follow various principles and techniques, including the following.

1. Make communication a priority. Communicating effectively implies making communication an important priority in a couple's relationship. When communication is a priority, partners make time for it to occur in a setting without interruptions—they are alone; they do not answer the phone; they turn the television off. Making communication a priority results in the exchange of more information between the partners, which increases the knowledge each partner has about the other.

Negative relationship outcomes occur when partners do not prioritize communication with each other but are passionately and obsessively interacting with others via the Internet (Seguin-Levesque et al., 2003). These researchers found that it is not use of the Internet per se but the obsessive passion of involvement with the Internet that is destructive.

2. Establish and maintain eye contact. Shakespeare noted that a person's eyes are the "mirrors to the soul." Partners who look at each other when they are talking not only communicate an interest in each other but also are able to gain information about the partner's feelings and responses to what is being said. Not looking at your partner may be interpreted as lack of interest and prevents you from observing nonverbal cues.

Eye-to-eye communication is helpful in connecting with one's partner. What "message" are these spouses communicating to each other?

Authors

3. Ask open-ended questions. When your goal is to find out your partner's thoughts and feelings about an issue, it is best to use open-ended questions. An open-ended question (e.g., "How do you feel about me?") encourages your partner to give an answer that contains a lot of information. Closed-ended questions (e.g., "Do you love me?"), which elicit a one-word answer such as yes or no, do not provide the opportunity for the partner to express a range of thoughts and feelings.

4. Use reflective listening. Effective communication requires being a good listener. One of the skills of a good listener is the ability to use the technique of **reflective listening,** which involves paraphrasing or restating what the person has said to you while being sensitive to what the partner is feeling. For example, suppose you ask your partner, "How was your day?" and your partner responds, "I felt exploited today at work because I went in early and stayed late and a memo from my new boss said that future bonuses would be eliminated because of a company takeover." Listening to what your partner is both saying and feeling, you might respond, "You feel frustrated because you really worked hard and felt unappreciated."

Reflective listening serves the following functions: (1) creates the feeling for the speaker that she or he is being listened to and is being understood and (2) increases the accuracy of the listener's understanding of what the speaker is saying. If a reflective statement does

Table 9.2 Judgmental and Nonjudgmental Responses to your Partner's Saying "I'd Like to Spend One Evening a Week with my Friends"

Nonjudgmental Reflective Statements	Judgmental Statements
It sounds like you really miss your friends.	You only think about what *you* want.
You think it is healthy for us to be with our friends some of the time.	Your friends are more important to you than I am.
You really enjoy your friends and want to spend some time with them.	You just want a night out so that you can meet someone new.
You think it is important that we not abandon our friends just because we are involved.	You just want to get away so you can drink.
You think that our being apart one night each week will make us even closer.	You are selfish.

not accurately reflect what the speaker thinks and feels, the speaker can correct the inaccuracy by restating her or his thoughts and feelings.

An important quality of reflective statements is that they are nonjudgmental. For example, suppose two lovers are arguing about spending time with their respective friends and one says, "I'd like to spend one night each week with my friends and not feel guilty about it." The partner may respond by making a statement that is judgmental (critical or evaluative), such as those exemplified in Table 9.2. Judgmental responses serve to punish or criticize someone for what he or she thinks, feels, or wants and often result in frustration and resentment. Table 9.2 also provides several examples of nonjudgmental reflective statements.

5. Use "I" statements. **"I" statements** focus on the feelings and thoughts of the communicator without making a judgment on others. Because "I" statements are a clear and nonthreatening way of expressing what you want and how you feel, they are likely to result in a positive change in the listener's behavior.

In contrast, **"you" statements** blame or criticize the listener and often result in increasing negative feelings and behavior in the relationship. For example, suppose you are angry with your partner for being late. Rather than say, "You are always late and irresponsible" (which is a "you" statement), you might respond with, "I get upset when you are late and will feel better if you call me when you will be delayed." The latter focuses on your feelings and a desirable future behavior rather than blaming the partner for being late.

6. Avoid brutal criticism. Research on marital interaction has consistently shown that one brutal "zinger" can erase twenty acts of kindness (Notarius & Markman, 1994). Because intimate partners are capable of hurting each other so intensely, be careful in how you communicate disapproval to your partner.

7. Say positive things about your partner. People like to hear others say positive things about them. These positive statements may be in the form of compliments (e.g., "You look terrific!") or appreciation (e.g., "Thanks for putting gas in the car"). Gable et al. (2003) asked fifty-eight heterosexual dating couples to monitor their interaction with each other. The respondents observed that they were overwhelmingly positive at a five-to-one ratio.

8. Tell your partner what you want. Focus on what you want rather than on what you don't want. Rather than say, "You always leave the bathroom a wreck," an alternative might be "Please hang up your towel after you take a shower."

9. Stay focused on the issue. **Branching** refers to going out on different limbs of an issue rather than staying focused on the issue. If you are discussing the overdrawn checkbook, stay focused on the checkbook. To remind your partner that

SELF-ASSESSMENT

Supportive Communication Scale

This scale is designed to assess the degree to which partners experience supportive communication in their relationships. After reading each item, circle the number that best approximates your answer.

0 = strongly disagree (SD)
1 = disagree (D)
2 = undecided (UN)
3 = agree (A)
4 = strongly agree (SA)

	SD	D	UN	A	SA
1. My partner listens to me when I need someone to talk to.	0	1	2	3	4
2. My partner helps me clarify my thoughts.	0	1	2	3	4
3. I can state my feelings without his/her getting defensive.	0	1	2	3	4
4. When it comes to having a serious discussion, it seems we have little in common (reverse scored).	0	1	2	3	4
5. I feel put down in a serious conversation with my partner (reverse scored).	0	1	2	3	4
6. I feel it is useless to discuss some things with my partner (reverse scored).	0	1	2	3	4
7. My partner and I understand each other completely.	0	1	2	3	4
8. We have an endless number of things to talk about.	0	1	2	3	4

Scoring

Look at the numbers you circled. Reverse score the numbers for questions 4, 5, and 6. For example, if you circled a 0, give yourself a 4; if you circled a 3, give yourself a 1, etc. Add the numbers and divide by 8, the total number of items. The lowest possible score would be 0, reflecting the complete absence of supportive communication; the highest score would be 4, reflecting complete supportive communication. The average score of 94 male partners who took the scale was 3.01; the average score of 94 female partners was 3.07. Thirty-nine percent of the couples were married, 38 percent were single, and 23 percent were living together. The average age was just over 24.

Source

Sprecher, Susan, Sandra Metts, Brant Burelson, Elaine Hatfield, and Alicia Thompson. 1995. Domains of expressive interaction in intimate relationships: Associations with satisfaction and commitment. *Family Relations* 44:203–10. Copyright (1995) by the National Council on Family Relations, 3989 Central Ave. NE, Suite 550, Minneapolis, MN 55421. Reprinted with permission.

he or she is equally irresponsible when it comes to getting things repaired or doing housework is to get off the issue of the checkbook. Stay focused.

10. Make specific resolutions to disagreements. To prevent the same issues or problems from recurring, it is important to agree on what each partner will do in similar circumstances in the future. For example, if going to a party together results in one partner's drinking too much and drifting off with someone else, what needs to be done in the future to ensure an enjoyable evening together? (E.g., How many drinks within what time period?)

11. Give congruent messages. **Congruent messages** are those in which the verbal and nonverbal behaviors match. A person who says, "OK. You're right" and smiles as he or she embraces the partner with a hug is communicating a congruent message. In contrast, the same words accompanied by leaving the room and slamming the door communicate a very different message.

12. Share power. One of the greatest sources of dissatisfaction in a relationship is conflict over power (Kurdek, 1994). **Power** is the ability to impose one's will on the partner and to avoid being influenced by the partner. Expressions of power are numerous and include:

Withdrawal (not speaking to the partner)

Guilt induction ("How could you ask me to do this?")

Being pleasant ("Kiss me and help me move the sofa.")

Negotiation ("I'll go with you to your parents if you will let me golf for a week with my buddies.")

Deception (running up bills on charge card)

Blackmail ("I'll tell your parents you do drugs if you . . . ")

Physical abuse or verbal threats ("I'll kill you if you leave.")

Criticism ("I can't think of anything good about you.")

In general, the spouse with the more prestigious occupation, higher income, and more education exerts the greater influence on family decisions. But power may also take the form of love and sex. The person in the relationship who loves less and who needs sex less has enormous power over the partner who is very much in love and who is dependent on the partner for sex. This pattern reflects the principle of least interest we discussed earlier in the text.

13. Keep the process of communication going. Communication includes both content (verbal and nonverbal information) and process (interaction). It is important not to allow difficult content to shut down the communication process (Turner, 2004). To ensure that the process continues, the partners should focus on the fact that the sharing of information is essential and reinforce each other for keeping the process alive. For example, if your partner tells you something that you do that bothers him or her, it is important to thank him or her for telling you that rather than becoming defensive. In this way, your partner's feelings about you stay out in the open rather than hidden behind a wall of resentment. Otherwise, if you punish such disclosure because you don't like the content, subsequent disclosure will stop.

14. Fight fair. When an argument ensues, it is important to establish rules for fighting that will leave the partners and their relationship undamaged after the disagreement. Such fair-fighting guidelines include not calling each other names, not bringing up past misdeeds, not attacking each other, waiting twenty-four hours before saying what's on one's mind, and not beginning a heated discussion late at night. In some cases, a good night's sleep has a way of altering how a situation is viewed and may even result in the problem's no longer being an issue.

An underlying principle in all of these techniques is to engage in supportive communication. Supportive communication exists when the partners feel comfortable discussing a range of issues with each other. The Self-Assessment on page 210 allows you to assess the degree to which your relationship is characterized by supportive communication.

Authors

Whatever this mother is saying, she is communicating absolute delight and pleasure in holding her son. The result for the son is to know that he is loved and valued.

DISCLOSURE, HONESTY, PRIVACY, AND LYING

Hillary Clinton said in a *20/20* interview with Barbara Walters that Bill Clinton awakened her hours before the Monica Lewinsky story broke. She reported that he told her he had lied to her and that what the press had been saying was true. "I was dumbfounded," she said (Clinton, 2003). Bill Clinton is not unique. All of us make choices, consciously or unconsciously, on the degree to which we disclose, are honest, and/or lie.

Self-Disclosure in Intimate Relationships

One aspect of intimacy in relationships is self-disclosure, which involves revealing personal information and feelings about oneself to another person. McKenna, Green, and Gleason (2002) found that a positive function of meeting on-line is that persons were better able to express themselves and were more disclosing on the Internet than in person. In a related study, "when compared to face-to-face

Should One Partner Disclose HIV/STD to the Other?

SOCIAL POLICY

Individuals often struggle over whether, or how, to tell a partner if they have an STD (sexually transmissible disease), including HIV. If a person in a committed relationship becomes infected with an STD, that individual, or his or her partner, may have been unfaithful and have had sex with someone outside the relationship. Thus, disclosure about an STD may also mean confessing one's own infidelity or confronting one's partner about his or her possible infidelity. (However, the infection may have occurred prior to the current relationship but gone undetected.) Individuals who are infected with an STD and who are beginning a new relationship face a different set of concerns. Will their new partners view them negatively? Will they want to continue the relationship?

Although telling a partner about having an STD may be difficult and embarrassing, avoiding disclosure or lying about having an STD represents a serious ethical violation. The responsibility to inform a partner that one has an STD—before having sex with that partner—is a moral one. But there are also legal reasons for disclosing one's sexual health condition to a partner. If you have an STD and you do not tell your partner, you may be liable for damages if you transmit the disease to your partner. In over half the states, transmission of a communicable disease, including many STDs, is considered a crime. Penalties depend on whether the crime is regarded as a felony (which may involve a five-year prison term) or a misdemeanor (which may involve a fine of $100) (Davis & Scott, 1988).

Some states and cities have partner notification laws that require health care providers to advise all persons with serious sexually transmitted diseases about the importance of informing their sex or needle-sharing partner(s). Partner notification laws may also require health care providers to either notify any partners the infected person names or forward the information about partners to the Department of Health, where public health officers notify the partners that they have been exposed to an STD and schedule an appointment for STD testing. The privacy of the infected individual is protected by not revealing his or her name to the partner being notified of potential infection. In cases where the infected person refuses to identify partners, standard partner notification laws require doctors to undertake notification without cooperation if they know who the sexual partner or spouse is (Norwood, 1995).

Sources

Davis, M., and Scott, R. S. 1988. *Lovers, doctors and the law.* New York: Harper & Row.

Norwood, Chris. 1995. Mandated life versus mandatory death: New York's disgraceful partner notification record. *Journal of Community Health* 20 (2): 161–70.

Rothenberg, Karen H., and Stephen J. Paskey. 1995. The risk of domestic violence and women with HIV infection: Implications for partner notification, public policy, and the law. *American Journal of Public Health* 85 (11):1569–76.

Diversity in Other Countries

Individuals in Japan are taught that quick self-disclosure in social relationships is inappropriate. They are much less likely to disclose information about themselves than are individuals socialized in the United States (Nakanishi, 1986).

interactions, people were better able to present, and have accepted by others, their 'true' selves over the Internet" (Bargh, Mckenna, & Fitzsimons, 2002).

Relationships become more stable when individuals disclose themselves—their formative years, relationships, experiences, hopes, and dreams. One way to encourage disclosure in one's partner is to make disclosures about one's own life and then ask the partner about his or her life. Patford (2000) found that the higher the level of disclosure, the more committed the spouses were to each other.

Honesty in Intimate Relationships

Sammy Sosa's cork bat, Martha Stewart's selling stock on inside information, Pete Rose's early denials about betting on baseball, and the Enron scandal reflect a cultural meltdown of honesty. One anonymous saying captures the pervasiveness of dishonesty: "The secret of success is sincerity and once you can fake that you have made it." Sharon Stone, the actress, says, "Yeah, women may fake orgasm but men fake whole relationships." A student in the authors' class wrote:

> *At this moment in my life I do not have any love relationship. I find college dating to be very hard. The guys here lie to you about anything and you wouldn't know the truth. I find it's mostly about sex here and having a good time before you really have to get serious. That is fine, but that is just not what I am all about.*

Forms of Dishonesty and Deception

Dishonesty and deception take various forms. In addition to telling an outright lie, people may exaggerate the truth, pretend, conceal the truth, or withhold information. Regarding the latter, in virtually every relationship, there are things that partners have not shared with each other about themselves or their past. We often withhold information or keep secrets in our intimate relationships for what we believe are good reasons—we believe that we are protecting our partners from anxiety or hurt feelings, protecting ourselves from criticism and rejection, and protecting our relationships from conflict and disintegration. Finkenauer and Hazam (2000) found that happy relationships depend on withholding information. The researchers contend, "Nobody wants to be criticized (e.g., 'You're really too fat') or talk about topics that are known to be conflictive (e.g., 'You should not have spent that much money')."

Extent of Lying among College Students

Over 95 percent of university students in one study reported having lied to their parents when they were living at home (Knox et al., 2001). Lying also occurs in romantic relationships (Miller & Abraham, 1998). In two studies of university dating relationships, both women and men reported altering their self-presentation in an attempt to get a date (Rowatt, Cunningham, & Druen, 1998). Number of previous partners was the most frequently told lie reported by a sample of university students (Knox et al., 1993).

One of the ways in which college students deceive their partners is by failure to disclose that they have a sexually transmitted disease (STD). It is estimated that 25 percent of college students will contract an STD while they are in college. Since the potential to harm an unsuspecting partner is considerable, should we have a national social policy regarding such disclosure?

NATIONAL DATA

Fifteen percent of a national sample of men infected with an STD had had sex while they were infected. Fifteen percent did not inform their partners of their infection before having intercourse (Payn et al., 1997).

PERSONAL CHOICES

How Much Do I Tell My Partner about My Past?

Because of the fear of HIV infection and other STDs, some partners want to know the details of each other's previous sex life, including how many partners they have had sex with and in what contexts. Those who are asked will need to make a decision about whether to disclose the requested information, which may include one's sexual orientation, present or past sexually transmitted diseases, and any sexual proclivities or preferences the partner might find bizarre (e.g., bondage and discipline). Ample evidence suggests that individuals are sometimes dishonest with regard to the sexual information they provide to their partners.

In deciding whether or not to talk honestly about your past to your partner, you may want to consider the following questions: How important is it to your partner to know about your past? Do you want your partner to tell you (honestly) about her or his past?

GENDER DIFFERENCES IN COMMUNICATION

When a woman says, "Sure . . . go ahead," what she means is "I don't want you to." When a woman says, "I'm sorry," what she means is "You'll be sorry." When a woman says, "I'll be ready in a minute," what she means is "Kick off your shoes and find a game on TV."

Internet humor

Beyond issues of disclosure, honesty, and deception, women and men differ in their approach to and patterns of communication. Sollie (2000) noted that although husbands are increasingly expected to become more emotional, they have received less socialization than women to show emotion.

Tannen (1990) observed that men and women, in general, focus on different content in their conversations. Men tend to focus on activities; women, on relationships. To men, talk is information; to women, it is interaction. To men, communication should emphasize what is rational; to women, communication is about emotion. To men, conversations are negotiations in which they try to "achieve and maintain the upper hand if they can, and to protect themselves from others' attempts to put them down and push them around" (p. 25). However, to women, conversations are negotiations for closeness in which they try "to seek and give confirmations and support, and to reach consensus" (p. 25). A woman's goal is to preserve intimacy and avoid isolation. Women are more concerned than men about how their partner is reacting to dialogue and want to avoid hurting the partner's feelings (Kim & Aune, 1998).

A team of researchers reviewed the literature on intimacy in communication and observed that men approach a problem in the relationship cognitively, whereas women approach it emotionally (Derlega et al., 1993). A husband might react to a seriously ill child by putting pressure on the wife to be mature (stop crying) about the situation and by encouraging stoicism (asking her not to feel sorry for herself). Wives, on the other hand, want their husbands to be more emotional (by asking them to cry to show that they really care that their child is ill).

Women are also more likely to initiate discussion of relationship problems than men. In a study of 203 undergraduates, two-thirds of the women, in contrast to 60 percent of the men, reported that they were likely to start a discussion about a problem in their relationship (Knox, Hatfield, & Zusman, 1998). Mackey and O'Brien (1999) found a similar pattern among the twenty-year spouses they interviewed—wives were much more likely to confront their husbands about a relationship problem.

The literature on how men and women respond to crisis events reveals that men more often report controlling their emotions, accepting the problem, not thinking about the situation, and engaging in problem-solving efforts. Women more often report seeking social support, distracting themselves, letting out their feelings, and turning to prayer.

Finally, as noted earlier, women disclose more in their relationships than men do (Gallmeier et al., 1997). In this study of 360 undergraduates, women were more likely to disclose information about previous love relationships, previous sexual relationships, their love feelings for the partner, and what they wanted for the future of the relationship. They also wanted their partners to reciprocate their (the women's) disclosure, but such disclosure was not forthcoming.

PERSONAL CHOICES

How Close Do You Want to Be?

Individuals differ in their capacity for and interest in an emotionally close/disclosing relationship. These preferences may vary over time; the partners may want closeness at some times and distance at other times. Individuals frequently choose partners according to an "emotional fit"—agreement about the amount of closeness they desire in their relationship.

Because courtship is so deceptive and the partners may reveal very little about themselves, and because they may be distracted with having fun in the relationship, little attention may be given to the desired level of emotional closeness. Only later may the partners discover that they have different emotional needs.

Another form of closeness is physical presence. Some partners prefer a pattern of complete togetherness (the current buzz word is co-dependency) in which all of their leisure and discretionary time is spent together. Others enjoy time alone and time with other friends and don't want to feel burdened by the demands of a partner with high companionship needs. Partners might consider their own choices and those of their partners in regard to emotional and spatial closeness.

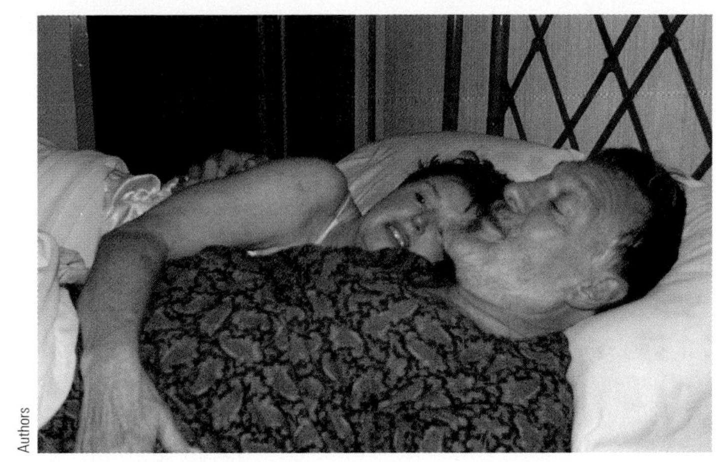

Authors

Couples who enjoy their relationship take time to talk and share their lives with each other.

THEORIES APPLIED TO RELATIONSHIP COMMUNICATION

Symbolic interactionism and social exchange are theories that help to explain the communication process.

Symbolic Interactionism

Interactionists examine the process of communication between two actors in terms of the meanings each attaches to the actions of the other. Definition of the situation, the looking-glass self, and taking the role of the other (discussed in Chapter 1) are all relevant to understanding how partners communicate. With regard to resolving a conflict over how to spend the semester break (e.g., vacation alone or go to see parents), the respective partners must negotiate their definition of the situation (is it about their time together as a couple or their loyalty to their parents?). The looking-glass self involves looking at each other and seeing the reflected image of someone who is loved and cared for and someone with whom a productive resolution is sought. Taking the role of the other involves each partner's understanding the other's logic and feelings about how to spend the break.

Social Exchange

Exchange theorists suggest that the partners' communication can be described as a ratio of rewards to costs. Rewards are positive exchanges, such as compliments, compromises, and agreements. Costs refer to negative exchanges, such as critical remarks, complaints, and attacks. When the rewards are high and the costs are low, the outcome is likely to be positive for both partners (profit). When the costs are high and the rewards low, neither may be satisfied with the outcome (loss).

When discussing how to spend the semester break, the partners are continually in the process of exchange—not only in the words they use but also in the way they use them. If the communication is to continue, each partner

Diversity in Other Countries

The culture in which one is reared will influence the meaning of various words. An American woman was dating a man from Iceland. When she asked him, "Would you like to go out to dinner?" he responded, "Yes, maybe." She felt confused by this response and was uncertain whether he wanted to eat out. It was not until she was visiting in his home in Iceland and asked his mother, "Would you like me to set the table?"—to which his mother replied, "Yes, maybe"—that she discovered that "Yes, maybe" means "Yes, definitely."

needs to feel acknowledged for his or her point of view and to feel a sense of legitimacy and respect. Communication in abusive relationships is characterized by the parties' criticizing and denigrating each other, which usually result in a shutdown of the communication process.

CONFLICTS IN RELATIONSHIPS

My wife said I don't listen—at least I think that's what she said.
Laurence Peter, *Humorist*

Conflict can be defined as the process of interaction that results when the behavior of one person interferes with the behavior of another. A professor in a marriage and family class said, "If you haven't had a conflict with your partner, you haven't been going together long enough." This section explores the inevitability, desirability, sources, and styles of conflict in relationships.

Inevitability of Conflict

If you are alone this Saturday evening from six o'clock until midnight, you are assured of six conflict-free hours. But if you plan to be with your partner, roommate, or spouse during that time, the potential for conflict exists. Whether you eat out, where you eat, where you go after dinner, and how long you stay must be negotiated. Although it may be relatively easy for you and your companion to agree on one evening's agenda, marriage involves the meshing of desires on an array of issues for potentially sixty years or more. Indeed, conflict is inevitable in any intimate relationship because "there are two unique individuals, often with different gender perspectives, changing as they mature but not always in the same direction or at the same rate, and functioning within a social context that is also changing and to which they may respond differently" (Goodman, 2003).

Desirability of Conflict

Conflict can be healthy and productive for a couple's relationship. Indeed, ignoring and resigning oneself to a problem may actually increase stress levels. Although discussing conflictual issues and negotiating differences may not reduce immediate stress, the long-term outcome for the relationship is improvement (Gottman, 1994a). Two researchers found that the presence of conflict in a relationship was predictive of the couple also doing things to help maintain their relationship (self-disclosing and discussing problems) (Sprecher & Felmlee, 1993).

Sources of Conflict

Trying to be right all the time is a very subtle way of being wrong.
Sanford M. Manley

Conflict has numerous sources, some of which are easily recognized, while others are hidden inside the web of marital interaction.

1. Behavior. Stanley, Markman, and Whitton (2002) noted that money was the issue over which a national sample of couples reported that they argued the most. The behavioral expression of a money issue might include how the partner spends money (excessively), the lack of communication about spending (e.g., does not consult the partner), and the target (i.e., items considered unnecessary by the partner). But marital conflict is not limited to behavioral money issues. Stanley, Markman, and Whitton (2002) found that remarried couples argued most about the children (e.g., rules for, discipline of, etc).

2. Cognitions and perceptions. Aside from your partner's actual behavior, your cognitions and perceptions of a behavior can be a source of satisfaction or dissatisfaction. One husband complained that his wife "had boxes of coupons everywhere and always kept the house in a wreck." The wife made the husband aware that she saved $100 on their grocery bill every week and asked him to view the boxes and a mess as "saving money." He changed his view and the clutter ceased to be a problem.

3. Value differences. Because you and your partner have had different socialization experiences, you may also have different values—about religion (one feels religion is a central part of life; the other does not), money (one feels uncomfortable being in debt; the other has the buy-now-pay-later philosophy), in-laws (one feels responsible for parents when they are old; the other does not), and children (number, timing, discipline).

The effect of value differences depends less on the degree of the difference than on the degree of rigidity with which each partner holds his or her values. Dogmatic and rigid thinkers, feeling threatened by value disagreement, may try to eliminate alternative views and thus produce more conflict. Partners who recognize the inevitability of difference may consider the positives of an alternative view and move toward acceptance. When both partners do this, the relationship takes priority and the value differences suddenly become less important.

4. Inconsistent rules. Partners in all relationships develop a set of rules to help them function smoothly. These unwritten but mutually understood rules include what time you are supposed to be home after work, whether you should call if you are going to be late, how often you can see friends alone, and when and how to make love. Conflict results when the partners disagree on the rules or when inconsistent rules develop in the relationship. For example, one wife expected her husband to take a second job so they could afford a new car, but she also expected him to spend more time at home with the family.

5. Leadership ambiguity. Unless a couple has an understanding about which partner will make decisions in which area (for example, the husband will decide when to ground teenage children; the wife will decide how much money to spend on vacations), each partner may continually struggle to control the other. All conflict is seen as an "I win–you lose" encounter because each partner is struggling for dominance in the relationship. "In low-conflict marriages, leadership roles vary and are flexible, but they are definite. Each partner knows most of the time who will make certain decisions" (Scoresby, 1977, 141).

Styles of Conflict

Spouses develop various styles of conflict. If you were watching a videotape of various spouses disagreeing over the same issue, you would notice at least six styles of conflict. These styles have been described by Greeff and De Bruyne (2000).

Competing Style The partners are both assertive and uncooperative. Each tries to force his or her way on the other so that there is a winner and a loser. A couple arguing over whether to discipline a child with a spanking or time out would resolve the argument with the dominant partner's forcing a decision.

Collaborating Style The respective partners are both assertive and cooperative. Each partner expresses his or her view and cooperates to find a solution. The above issue might be resolved by a spanking, time out, or just talking to the child, but both partners would be satisfied with the resolution.

Compromising Style Here there would be an intermediate solution: both partners would find a middle ground they could live with—perhaps spanking the child for serious infractions such as playing with matches in the house and imposing time out for talking back.

Diversity in the United States

African-American husbands are much more likely to report a confrontational style of dealing with marital conflict than either Mexican-American or white husbands. In one study of spouses in twenty-year marriages, 72, 25, and 18 percent, respectively, reported a confrontational style (Mackey & O'Brien, 1998).

Avoiding Style The partners are neither assertive nor cooperative. They would avoid a confrontation and let either parent do what he or she wanted in disciplining the child. Thus the child might be both spanked and put in time out. Marchand and Hock (2000) noted that depressed spouses were particularly likely to use avoidance as a conflict-resolution strategy.

NATIONAL DATA

Forty-three percent of the men, in contrast to 26 percent of the women, in a national random sample reported that they withdrew when a conflict arose (Stanley & Markman, 1997).

Accommodating Style The respective partners are not assertive in their positions but each accommodates to the other's point of view. Each attempts to soothe the other and to avoid conflict. While the goal of this style is to rise above the conflict and keep harmony in the relationship, fundamental feelings about the "rightness" of one's own approach may be maintained.

Parallel Style Both partners deny, ignore, and retreat from addressing a problem issue. "Don't talk about it, and it will go away" is the theme of this conflict style. Gaps begin to develop in the relationship, neither partner feels free to talk, and both partners believe that they are misunderstood. Both eventually become involved in separate activities rather than spending time together.

Greeff and De Bruyne (2000) studied fifty-seven couples who had been married at least ten years and found that the collaborating style was associated with the highest level of marital and spousal satisfaction. The competitive style used by either partner was associated with the lowest level of marital satisfaction. Regardless of the style of conflict, partners who say positive things to each other on a five-to-one (positive-to-negative) ratio seem to stay together (Gottman, 1994b).

FIVE STEPS IN CONFLICT RESOLUTION

It is better to debate a question without settling it than to settle a question without debating it.

Joseph Joubert

Every relationship experiences conflict. If left unresolved, conflict may create tension and distance in the relationship, with the result that the partners stop talking, stop spending time together, and stop being intimate. A conflictual, unsatisfacting marriage is similar to divorce in terms of it's impact on diminished psychological, social, and physical well-being (Hetherington, 2003). Developing and using conflict resolution skills are critical for the maintenance of a good relationship.

Howard Markman is head of the Center for Marital and Family Studies at the University of Denver. He and his colleagues have been studying 150 couples at yearly intervals (beginning before marriage) to determine those factors most responsible for marital success. They have found that communication skills that reflect the ability to handle conflict, which they call "constructive arguing," are the single biggest predictor of marital success over time (Marano, 1992). According to Markman: "Many people believe that the causes of marital problems are the differences between people and problem areas such as money, sex, children. However, our findings indicate it is not the differences that are important, but how these differences and problems are handled, particularly early in marriage" (Marano, 1992, 53).

There is also merit in developing and using conflict negotiation skills before problems develop. Not only are individuals more willing to work on issues when

Up Close 9.2 College Student Discussion of Relationship Problems

Two hundred three undergraduates (72% female, 28% male) from the authors' classes completed a questionnaire concerning discussion of problems in their respective relationships. The following is based on analysis of the original data collected for Knox, Hatfield, and Zusman (1998).

1. **Ease of communication.** The respondents noted a mean level of 7 on a continuum from 0 (very difficult) to 10 (very easy). There were no differences in the mean scores with respect to sex of respondent, racial background, age, or year in school.

2. **Perceived skill level.** These respondents also reported being relatively confident in their skill level, with 7.1 being the average mean (0 = very difficult and 10 = very easy).

3. **Most difficult topic.** Of eighteen relationship topics, "future of the relationship" was the most difficult. It was also the "last" problem the individuals reported having discussed. Married couples identify "money" and "discipline of children" as their most difficult topics (Stanley & Markman, 1997).

4. **Effect of relationship type.** Respondents involved in reciprocal love relationships reported that they were able to resolve their disagreements by compromise, in contrast to those who were in nonreciprocal relationships.

Sources

Knox, D., S. Hatfield, and M. E. Zusman. 1998. College student discussion of relationship problems. *College Student Journal* 32:19–21.

Stanley, S. M., and H. J. Markman. 1997, *Marriage in the 90s: A nationwide random phone survey.* Denver: PREP, Inc.

things are going well, but they have not developed negative patterns of response that are difficult to change. The following sections identify five steps helpful in resolving interpersonal conflict.

Address Recurring, Disturbing Issues

It is important to address issues in the relationship. A helpful ground rule is "Either partner can bring up any issue at any time, but the listener has the right to say 'this is not a good time.' If the listener says, 'this is not a good a good time,' he or she takes the responsibility to find and initiate a good time within 24 hours" (Stanley & Trathen, 1994, 158).

Some couples are uncomfortable talking about issues that plague them. They fear that a confrontation will further weaken their relationship. Pam is jealous that Mark spends more time with other people at parties than with her. "When we go someplace together," she blurts out, "he drops me to disappear with someone else for two hours." Her jealousy is spreading to other areas of their relationship. "When we are walking down the street and he turns his head to look at another woman, I get furious." If Pam and Mark don't discuss her feelings about Mark's behavior, their relationship may deteriorate as a result of a negative response cycle: He looks at another woman, she gets angry, he gets angry at her getting angry and finds that he is even more attracted to other women, she gets angrier because he escalates his looking at other women, and so on.

To bring the matter up, Pam might say, "I feel jealous when you spend more time with other women at parties than with me. I need some help in dealing with these feelings." By expressing her concern in this way, she has identified the problem from her perspective and asked her partner's cooperation in handling it.

When discussing difficult relationship issues, it is important to avoid attacking, blaming, or being negative. Such reactions reduce the motivation of the partner to talk about an issue and thus reduce the probability of a positive outcome (Forgatch, 1989).

It is also important to use good timing in discussing difficult issues with your partner. In general, it is best to discuss issues or conflicts when (1) you are alone with your partner in private rather than in public, (2) you and your partner have ample time to talk, and (3) you and your partner are rested and feeling generally good (avoid discussing conflict issues when one of you is tired, upset, or under unusual stress).

Identify New Desired Behaviors

Dealing with conflict is more likely to result in resolution if the partners focus on what they want rather than what they don't want. For example, rather than tell Mark she doesn't want him to spend so much time with other women at parties, Pam might tell him that she wants him to spend more time with her at parties.

Identify Perceptions to Change

Rather than change behavior, it may be easier and quicker to change one's perception of a behavior. Rather than expect one's partner to always be "on time," it may be easier to drop the expectation that one's partner be on time and to stop being mad about something that doesn't matter. Also recall the earlier example of the husband who changed his view of the clutter in the house rather than demanding that his wife change her pack-rat behavior.

Summarize Your Partner's Perspective

We often assume that we know what our partner thinks and why he or she does things. Sometimes we are wrong. Rather than assume how our partner thinks and feels about a particular issue, we might ask our partner open-ended questions in an effort to get him or her to tell us thoughts and feelings about a particular situation. Pam's words to Mark might be, "What is it like for you when we go to parties?" "How do you feel about my jealousy?"

Once your partner has shared his or her thoughts about an issue with you, it is important for you to summarize your partner's perspective in a nonjudgmental way. After Mark has told Pam how he feels about their being at parties together, she can summarize his perspective by saying, "You feel that I cling to you more than I should, and you would like me to let you wander around without feeling like you're making me angry." (She may not agree with his view, but she knows exactly what it is—and Mark knows that she knows.)

Not summarizing the partner's perspective but instead expressing contempt, belligerence, and defensiveness effectively derails couples from constructive conflict resolution (Gottman et al., 1998). Couples who avoid these destructive behaviors are more likely to stay together.

Generate Alternative Win-Win Solutions

It is imperative to look for win-win solutions to conflicts. Solutions in which one person wins and the other person loses mean that one person is not getting his or her needs met. As a result, the person who loses may develop feelings of resentment, anger, hurt, and hostility toward the winner and may even look for ways to get even. In this way, the winner is also a loser. In intimate relationships, one winner really means two losers.

Generating win-win solutions to interpersonal conflict often requires **brainstorming.** The technique of brainstorming involves suggesting as many alternatives as possible without evaluating them. Brainstorming is crucial to conflict resolution because it shifts the partners' focus from criticizing each other's perspective to working together to develop alternative solutions.

The authors and their colleagues (Knox et al., 1995) studied the degree to which two hundred college students who were involved in ongoing relationships were involved in win-win, win-lose, and lose-lose relationships. Descriptions of the various relationships follow:

Win-win relationships are those in which conflict is resolved so that each partner derives benefits from the resolution. For example, suppose a couple have a limited amount of money and disagree on whether to spend it on eating out or on seeing a current movie. One possible win-win solution might be for the couple to eat a relatively inexpensive dinner and rent a movie.

An example of a **win-lose solution** would be for one of the partners to get what he or she wanted (eat out or go to a movie), with the other partner getting nothing of what he or she wanted. A **lose-lose solution** is one in which both partners get nothing that they want—in the scenario presented, the partners would neither go out to eat nor see a movie and would be mad at each other.

Over three-quarters (77.1%) of the students reported being involved in a win-win relationship, with men and women reporting similar percentages. Twenty percent of the respondents were involved in win-lose relationships. Only 2 percent reported that they were involved in lose-lose relationships. Eighty-five percent of the students in win-win relationships reported that they expected to continue their relationships, in contrast to only 15 percent of students in win-lose relationships. No student in a lose-lose relationship expected the relationship to last.

After a number of solutions are generated, each solution should be evaluated and the best one selected. In evaluating solutions to conflicts, it may be helpful to ask the following questions:

1. *Does the solution satisfy both individuals? (Is it a win-win solution?)*
2. *Is the solution specific? Does it specify exactly who is to do what, how, and when?*
3. *Is the solution realistic? Can both parties realistically follow through with what they have agreed to do?*
4. *Does the solution prevent the problem from recurring?*
5. *Does the solution specify what is to happen if the problem recurs?*

Kurdek (1995) emphasized that conflict resolution styles that stress agreement, compromise, and humor are associated with marital satisfaction, whereas conflict engagement, withdrawal, and defensiveness styles are associated with lower marital satisfaction. In his own study of 155 married couples, the style in which the wife engaged the husband in conflict and the husband withdrew was particularly associated with low marital satisfaction for both spouses.

Be Alert to Defense Mechanisms

Effective conflict resolution is sometimes blocked by **defense mechanisms**—unconscious techniques that function to protect individuals from anxiety and minimize emotional hurt. The following paragraphs discuss some common defense mechanisms:

Escapism is the simultaneous denial of and withdrawal from a problem. The usual form of escape is avoidance. The spouse becomes "busy" and "doesn't have time" to think about or deal with the problem, or the partner may escape into recreation, sleep, alcohol, marijuana, or work. Denying and withdrawing from problems in relationships offer no possibility for confronting and resolving the problems.

Rationalization is the cognitive justification for one's own behavior that unconsciously conceals one's true motives. For example, one wife complained that her husband spent too much time at the health club in the evenings. The underlying reason for the husband's going to the health club was to escape an unsatisfying home life. But the idea that he was in a dead marriage was too painful and difficult for the husband to face, so he rationalized to himself and his wife that he spent so much time at the health club because he made a lot of important business contacts there. Thus, the husband concealed his own true motives from himself (and his wife).

Projection occurs when one spouse unconsciously attributes his or her own feelings, attitudes, or desires to the partner. For example, the wife who desires to have an affair may accuse her husband of being unfaithful to her. Projection may be seen in such statements as "You spend too much money" (projection for "I spend too much money") and "You want to break up" (projection for "I want to break up"). Projection interferes with conflict resolution by creating a mood of hostility and defensiveness in both partners. The issues to be resolved in the relationship remain unchanged and become more difficult to discuss.

Displacement involves shifting your feelings, thoughts, or behaviors from the person who evokes them onto someone else. The wife who is turned down for a promotion and the husband who is driven to exhaustion by his boss may direct their hostilities (displace them) onto each other rather than toward their respective employers. Similarly, spouses who are angry at each other may displace this anger onto someone else, such as the children.

By knowing about defense mechanisms and their negative impact on resolving conflict, you can be alert to them in your own relationships. When a conflict continues without resolution, one or more defense mechanisms may be operating.

PERSONAL CHOICES

Should Parents Argue in Front of the Children?

Parents may disagree about whether to argue in front of their children. One parent may feel that it is best to argue behind closed doors so as not to upset their children, but the other may feel it is best to expose children to the reality of relationships. This includes seeing parents argue and, hopefully, negotiating win-win solutions. Most therapists agree that being open is best. Children need to know that relationships involve conflict and how to resolve it. In the absence of such exposure, children may have an unrealistic view of relationships.

SUMMARY

What is the nature of interpersonal communication?

Communication is the process of exchanging information and feelings between two individuals. It involves both verbal and nonverbal messages. The nonverbal part of a message often carries more weight than the verbal part.

Some basic principles and techniques of effective communication include making communication a priority, maintaining eye contact, asking open-ended questions, using reflective listening, using "I" statements, complimenting each other, and sharing power. Partners must also be alert to keep the dialogue (process) going even when they don't like what is being said (content).

How are relationships affected by dishonesty and lying?

Intimacy in relationships is influenced by the level of self-disclosure and honesty. High levels of self-disclosure are associated with increased intimacy. Most individuals value honesty in their relationships. Honest communication is associated with trust and intimacy.

Despite the importance of honesty in relationships, deception occurs frequently in interpersonal relationships. Partners sometimes lie to each other about previous sexual relationships, how they feel about each other, and how they experience each other sexually. Telling lies is not the only form of dishonesty. People exaggerate, minimize, tell partial truths, pretend, and engage in self-deception. Partners may withhold information or keep secrets in order to protect themselves and/or preserve the relationship. However, the more intimate the relationship, the greater our desire to share our most personal and private selves with our partner and the greater the emotional consequences of not sharing. In intimate relationships, keeping secrets can block opportunities for healing, resolution, self-acceptance, and a deeper intimacy with your partner.

What are gender differences in communication?

Men and women tend to focus on different content in their conversations. Men tend to focus on activities, information, logic, and negotiation and "to achieve and maintain the upper hand." To women, communication is emotion, relationships, interaction, and maintaining closeness. A woman's goal is to preserve intimacy and avoid isolation. Women are also more likely than men to initiate discussion of relationship problems. Women also disclose more than men.

How are interactionist and exchange theories applied to relationship communication?

Symbolic interactionists examine the process of communication between two actors in terms of the meanings each attaches to the actions of the other. Definition of the situation, the looking-glass self, and taking the role of the other are all relevant to understanding how partners communicate.

Exchange theorists suggest that the partners' communication can be described as a ratio of rewards to costs. Rewards are positive exchanges, such as compliments, compromises, and agreements. Costs refer to negative exchanges, such as critical remarks, complaints, and attacks. When the rewards are high and the costs are low, the outcome is likely to be positive for both partners (profit). When the costs are high and the rewards low, neither may be satisfied with the outcome (loss).

What are various issues related to conflict in relationships?

Conflict is both inevitable and desirable. Unless individuals confront and resolve issues over which they disagree, one or both may become resentful and withdraw from the relationship. Conflict may result from one partner's doing something the other does not like, having different perceptions, or having different values. Sometimes it is easier for one partner to view a situation differently or alter a value than for the other partner to change the behavior causing the distress.

What are five steps in conflict resolution?

The sequence of resolving conflict includes deciding to address recurring issues rather than suppressing them, asking the partner for help in resolving the issue, finding out the partner's point of view, summarizing in a nonjudgmental way the partner's perspective, and finding alternative win-win solutions. Defense mechanisms that interfere with conflict resolution include escapism, rationalization, projection, and displacement.

KEY TERMS

accommodating style of conflict
avoiding style of conflict
brainstorming
branching
collaborating style of conflict
communication
competing style of conflict
compromising style of conflict
conflict
congruent message
defense mechanisms
displacement
escapism
"I" statements
lose-lose solution
parallel style of conflict
power
projection
rationalization
reflective listening
win-lose solution
win-win relationships
"you" statements

RESEARCHING MARRIAGE AND THE FAMILY WITH INFOTRAC COLLEGE EDITION

InfoTrac College Edition, an online library, allows you to perform research online anywhere, anytime. Following are two suggested search terms and related questions to help you extend your understanding of the topics covered in this chapter. Go to www.infotrac-college.com to begin your search.

Keyword: **Honesty.** Locate articles that discuss honesty as a value in relationships and in communication. How honest do you think partners are with each other? Are men or women more likely to be dishonest?

Keyword: **Marital conflict.** Locate articles that discuss conflict in marriage. What are the typical sources of marital conflict and how do these differ from the premarital days together?

The Companion Web Site for Choices in Relationships: An Introduction to Marriage and the Family, Eighth Edition

http://sociology.wadsworth.com/knox_schacht/choices8e

Supplement your review of this chapter by going to the companion Web site to take one of the Tutorial Quizzes, use the flash cards to master key terms, and check out the many other study aids you'll find there. You'll also find special features such as the Marriage and Family Resource Center, Census 2000 information, and other data and resources at your fingertips to help you with that special project or to do some research on your own.

WEBLINKS

Department of Communication Resources
http://communication.ucsd.edu/resources/commlinks.html

Guidelines on Effective Communication, Healthy Relationships & Successful Living
http://www.drnadig.com/

Association for Couples in Marriage Enrichment
http://www.bettermarriages.org/

Episcopal Marriage Encounter
http://www.episcopalme.com/

Authors

CHAPTER 10

The reason we have so many children is that I'm deaf. When we would go to bed, my wife would ask, "Do you want to go to sleep or what?" And I would always ask, "What?"

Joe Hancock, Married twenty-four years

Planning Children and Contraception

Contents

True or False?

1. The primary disadvantage of having a child in late life is that limited resources are usually available to an elderly couple.
2. "Double Dutch" in the Netherlands means using both the pill and the condom.
3. In an infertile couple, the man is as likely to feel stressed and depressed as the woman.
4. Fertility clinics can now offer an infertile couple about an 80 percent probability of getting pregnant.
5. Women are more willing than men to adopt transracially.

Answers: **1.** F **2.** T **3.** F **4** F **5.** T

Love is a fourteen letter word—Family Planning.

Planned Parenthood poster

Having children continues to be a major goal of most college students. Over 70 percent of a random sample of all first-year students in U.S. colleges and universities noted that "raising a family" was an "essential" or "very important" life goal (American Council on Education and University of California, 2002). And most want at least two (Townsend, 2003).

As the U.S. population nears 300 million, the world population is moving toward 7 billion (Townsend, 2003). In a study of fifteen states involving 72,907 pregnant women, 45 percent of the pregnancies were unintended (Naimi et al., 2003). Planning when to become pregnant has benefits for both the mother and the child. Having several children at short intervals increases the chances of premature birth, infectious disease, and death of the mother or the baby. Would-be parents can minimize such risks by planning fewer children with longer intervals in between. Women who plan their pregnancies can also modify their behaviors and seek preconception care from a health-care practitioner to maximize their chances of having healthy pregnancies and babies. For example, women planning pregnancies can make sure they eat properly and avoid alcohol and other substances (such as cigarettes) that could harm developing fetuses (Haller, Miles, & Dawson, 2003).

You and your partner may also benefit from family planning by pacing the financial demands of your offspring. Having children four years apart helps to avoid having more than one child in college at the same time.

Conscientious family planning will help to reduce the number of unwanted pregnancies. Two researchers (David, Dytrych, & Matejcek, 2003) studied 220 children who were born to women who were twice denied abortion for the same pregnancy. The children were medically, psychologically, and socially evaluated at ages 9, 14–16, 21–23, 30, and 35 and found to have poorer mental health as adults than children born to mothers who wanted the pregnancy.

International Data

Worldwide 56 million unwanted pregnancies occur annually (Trevor & Althaus, 2002).

National Data

In the United States 3 million unplanned pregnancies occur annually (Burkman, 2002).

Your choices in regard to children and contraception have an important effect on your happiness, lifestyle, and resources. These choices, in large part, are influenced by social and cultural factors that may operate without your awareness. We now discuss these influences.

DO YOU WANT TO HAVE CHILDREN?

Children are an expected part of one's adult life. This section examines the social influences that motivate individuals to have children, the lifestyle changes that result from such a choice, and the costs of rearing children.

Ivor L. Livingston

Most individuals want to have children for psychological reasons, among them the joy of watching happy children develop into adulthood.

Social Influences Motivating Individuals to Have Children

Our society tends to encourage childbearing, an attitude known as **pronatalism.** Our family, friends, religion, and government help to develop positive attitudes toward parenthood. Cultural observances also function to reinforce these attitudes.

Family Our experience of being reared in families encourages us to have families of our own. Our parents are our models. They married; we marry. They had children; we have children. Some parents exert a much more active influence. "I'm 73 and don't have much time. Will I ever see a grandchild?" asked the mother of an only child.

Friends Our friends who have children influence us to do likewise. After sharing an enjoyable weekend with friends who had a little girl, one husband wrote to the host and hostess, "Lucy and I are always affected by Karen—she is such a good child to have around. We haven't made up our minds yet, but our desire to have a child of our own always increases after we leave your home." This couple became parents sixteen months later.

Religion Religion is a powerful influence on the decision to have children. Catholics are taught that having children is the basic purpose of marriage and gives meaning to the union. Mormonism and Judaism also have a strong family orientation. Women who elect not to have children are more likely to report no religious affiliation (Kaufman, 2000).

Race Hispanics have the highest fertility rate, with ninety-six births per thousand women aged 15–44 years of age. Non-Hispanic Blacks have the next highest rate at sixty-nine per thousand (Ventura, Hamilton, & Sutton, 2003).

Government The tax structures imposed by our federal and state governments support parenthood. Married couples without children pay higher taxes than couples with children, although the reduction in taxes is not sufficient to offset the cost of rearing a child and is not large enough to be a primary inducement to have children.

Economy Times of affluence are associated with a high birth rate. Postwar expansion of the fifties resulted in the oft-noted "baby boom" generation. Similarly, couples are less likely to decide to have a child during economically depressed times. In addition, the necessity of two wage earners in our postindustrial economy is associated with a reduction in the number of children.

Diversity in Other Countries

While the average number of children in North America is 1.9, in Europe it is 1.4; East Asia, 1.8; and in sub-Saharan Africa, 6 (Townsend, 2003).

Diversity in Other Countries

In premodern societies (before industrialization and urbanization) and in some parts of China (despite China's one-child policy) and Korea today, spouses try to have as many children as possible because children are an economic asset. At an early age, children work as free labor for parents or for wages, which they give to their parents. Children are also expected to take care of their parents when the parents are elderly. In China and Korea, the eldest son is expected to take care of his aging parents by earning money for them and by marrying and bringing his wife into their home to physically care for his parents. The wife's parents need their own son and daughter-in-law to provide old-age insurance. Beyond its value in providing economic security and old-age insurance, having numerous children is regarded as a symbol of virility for the man, a source of prestige for the woman, and a sign of good fortune for the couple. In modern societies, having large numbers of children is less valued.

How Old Is Too Old to Have a Child?

SOCIAL POLICY

Couples are delaying the age at which they have a child (Wu & Macneill, 2002). The birth rate in 2001 for women aged 45-49 doubled from the 1990 rate (0.5 per 1,000 versus 0.2 per 1,000) (Ventura, Hamilton, & Sutton, 2003). But how old is too old to begin parenthood? Actor Tony Randall had a child at the age of 77, and the late Senator Strom Thurmond was in his 80s when he became a father. Situations like these are becoming more common, and questions are now being asked about the appropriateness of elderly individuals becoming parents. Should social policies on this issue be developed?

There are advantages and disadvantages of having a child as an elderly parent. The primary developmental advantage for the child of retirement-aged parents is the attention the parents can devote to their offspring. Not distracted by their careers, these parents have more time to nurture, play with, and teach their children. And, although they may have less energy, their experience and knowledge are doubtless better. Indeed the "biological disadvantage is to a degree balanced by social advantage" (Stein & Susser, 2000, p. 1682).

The primary disadvantage of having a child in the later years is that the parents are likely to die before, or early in, the child's adult life. When Tony Randall's daughter, Julie, begins college, her dad will be 95. The daughter of James Dickey, the Southern writer, lamented the fact that her late father (to whom she was born when he was in his 50s) would not be present at her graduation or wedding.

There are also medical concerns for both the mother and the baby during pregnancy in later life. They include an increased risk of morbidity (chronic illness and disease) and mortality (death) for the mother. These risks are typically a function of chronic disorders that go along with aging, such as diabetes, hypertension, and cardiac disease. Stillbirths, miscarriage, ectopic pregnancies, multiple births, and congenital malformations are also higher in women with advancing age (Stein & Susser, 2000). However, prenatal testing can identify some potential problems such as the risk of Down syndrome, and any chromosome abnormality and negative neonatal outcomes are not inevitable. However, women who delay the age at which they become pregnant risk not being able to get pregnant. Pregnancy outcomes of seventy-seven postmenopausal women aged 50 to 63 using in vitro fertilization were less than 50 percent (45.5%) (Paulson et al., 2002).

Because an older woman can usually have a healthy baby, government regulations on the age at which a woman can become pregnant are not likely. Indeed, 255 births in 2000 were to women aged 50 to 54, a 50 percent increase over the year before (Regalado, 2002).

Authors

This 50-year-old new father makes time to take care of and be with his son. When his son marries at about age 25, the dad will be in his mid-70s.

Sources

Paulson, R. J. , R. Boostanfar, P. Saadat, E. Mor, D. Tourgeman, C. C. Stater, et al. 2002. Pregnancy in the sixth decade of life: Obstetric outcomes in women in advanced reproductive age. *Journal of the American Medical Association* 288: 2320–24.

Regalado, A. 2002. Age is no barrier to motherhood—Study gives latest proof older women can get pregnant with donor eggs. *Wall Street Journal.* 13 November, D3.

Stein, Z., and M. Susser 2000. The risks of having children in later life. *British Medical Journal* 320: 1681–83.

Ventura, S. J., B. E. Hamilton, and P. D. Sutton. 2003. Revised birth and fertility rates for the United States, 2000 and 2001. *National Vital Statistics Reports* 51 (4). Hyattsville, Md.: National Center for Health Statistics.

Wu, Z, and L. Macneill. 2002. Education, work and childbearing after age 30. *Journal of Comparative Family Studies* 33:191–213.

Cultural Observances Our society reaffirms its approval of parents every year by identifying special days for Mom and Dad. Each year on Mother's Day and Father's Day (and now Grandparents' Day) parenthood is celebrated across the nation with cards, gifts, and embraces. There is no cultural counterpart (e.g., Childfree Day) for persons choosing not to have children.

In addition to influencing individuals to have children, society and culture also influence feelings about the age parents should be when they have children. Recently, couples have been having children at later ages. Is this a good idea? The Social Policy: How Old Is Too Old to Have a Child? (p. 228) addressed this issue.

Individual Motivations for Having Children

Individual motivations, as well as social influences, play an important role in the decision to have children. Some of these are conscious, as in the desire for love and companionship with one's own offspring and in the desire to be personally fulfilled as an adult by having a child. Some also want to recapture their own childhood and youth by having a child.

Cross-cultural research shows that children everywhere are valued mainly for psychological reasons.

James Walters and Linda Walters, *Past presidents, National Council on Family Relations*

The Self-Assessment allows you to assess the degree to which having children is a value for you.

Unconscious motivations for parenthood may also be operative. Examples include wanting a child to avoid career tracking and to gain the acceptance and approval of one's parents and peers. Teenagers sometimes want to have a child to have someone to love them. See Up Close 10.1 on page 232, which details teenage motherhood as a major social issue.

Lifestyle Changes and Economic Costs of Parenthood

Although there are numerous potential positive outcomes for becoming a parent, there are also drawbacks. Every parent knows that parenthood involves difficulties as well as joys. Some of the difficulties associated with parenthood are discussed next.

Lifestyle Changes Becoming a parent often involves changes in lifestyle. Daily living routines become focused around the needs of the children. Living arrangements change to provide space for another person in the household. Some parents change their work schedule to allow them to be home more. Food shopping and menus change to accommodate the appetites of children. A major lifestyle change is the loss of freedom of activity and flexibility in one's personal schedule.

Lifestyle changes are particularly dramatic for women. The time and effort required to be pregnant and rear children often compete with the time and energy needed to finish one's education. Hofferth, Reid, and Mott (2001) observed that early childbearing has a significant negative impact on years of completed schooling. Teenage mothers have about two fewer years of education than do women who delay their first birth until age 30.

Building a career is also negatively impacted by the birth of children. Parents learn quickly that it is difficult to both be an involved on-the-spot parent and climb the career ladder. The careers of women may suffer most.

You can be sort of married, you can sort of be divorced, you can sort of be living together, but you can't sort of have a baby.

David Shire

Financial Costs Meeting the financial obligations of parenthood is difficult for many parents. New parents are often shocked at the relentless expenses. The annual cost of a child less than 2 years old for middle income parents ($39,100 to $65,800)—which includes housing, food, transportation, clothing, health care, and child care—is $9,030. For a 15- to 17-year-old the cost is $10,140 (*Statistical Abstract of the United States: 2003,* Table 675). These costs do not include the wages lost when a parent drops out of the workforce to provide child care.

SELF-ASSESSMENT

Attitudes toward Parenthood Scale

The purpose of this survey is to assess your attitudes toward the role of parenthood. Please read each item carefully and consider what you believe and feel about parenthood. There are no right or wrong answers to any of these statements. After reading each statement, select the number which best reflects your answer using the following scale:

1	2	3	4	5	6	7
Strongly Disagree						Strongly Agree

___ **1.** Parents should be involved in their children's school.

___ **2.** I think children add joy to a parent's life.

___ **3.** Parents attending functions such as sporting events, recitals, etc. of their children help build their child socially.

___ **4.** Parents are responsible for providing a healthy environment for their children.

___ **5.** When you become a parent, your children become your top priority.

___ **6.** Parenting is a job.

___ **7.** I feel that spending quality time with children is an important aspect of child rearing

___ **8.** Mothers and fathers should share equal responsibilities in raising children.

___ **9.** The formal education of children should begin at as early an age as possible.

Scoring

After assigning a number to each item, add the items and divide by 9. The higher the number (7 is the highest number), the more positive your view and the stronger your commitment to the role of parenthood. The lower the number (1 is the lowest number) the more negative your view and the weaker your commitment to parenthood.

Norms

Norms for the scale are based upon twenty-two male and seventy-two female students attending Valdosta State University. The average score on the Attitudes toward Parenthood Scale was 6.36 (SD = 0.65) (hence, the respondents had very positive views) and ranged from 3.89 to 7.00. There was no significant difference between male participants' attitudes toward parenthood (M = 6.15, SD = 0.70) and female participants' attitudes (M = 6.42, SD = 0.63), $F(1, 92) = 2.96$, $p > .05$. There were no significant differences between ethnicities as well.

The average age of participants completing the Attitudes toward Parenthood Scale was 22.09 years (SD = 3.85) and ages ranged from 18 to 39. The ethnic composition of the sample was 73.4% White, 22.3% Black, 2.1% Asian, 1.1% American Indian, and 1.1% Other. The classification of the sample was 20.2% Freshman, 6.4% Sophomore, 22.3% Junior, 47.9% Senior, and 3.2% were graduate students.

Copyright: "The Attitudes Toward Parenthood Scale" 2004 by Mark Whatley, Ph.D. Department of Psychology, Valdosta State University, Valdosta, Georgia 31698. Information on validity and reliability may be obtained from Dr. Whatley. The scale is reprinted here with permission of Dr. Whatley. Other uses of this scale by written permission only from Dr. Whatley—e-mail **mwhatley@valdosta.edu**

HOW MANY CHILDREN DO YOU WANT?

Most people do not regard the lifestyle changes and economic costs of children as important when it comes to deciding to have children. While some opt for a childfree marriage, most (90%) want to have not only one child but several. Procreative liberty is the freedom to decide whether or not to have children.

Childfree Marriage?

Late-night TV talk show host Jay Leno and his wife, Mavis, are an example of a couple who have chosen not to have children. About 20 percent of individuals aged 65 or over report that they have never had children (Wu & Hart, 2002). Typical reasons individual couples give for not having children include the freedom to spend their time and money as they choose, to enjoy their partner without interference, to pursue their career, to avoid health problems associated with pregnancy, and to avoid passing on genetic disorders to a new generation (DeOllos & Kapinus, 2002). Gillespie (2003) noted that childfree women no longer view having a child as evidence of their femininity.

Some people simply do not like children. Aspects of our society reflect **antinatalism** (against children). Indeed, there is a continuous fight to get corporations to implement or enforce any family policies (from family leaves to flex time to on-site day care). Profit and money—not children—are priorities. In addition, although people are generally tolerant of their own children, they often exhibit antinatalistic behavior in reference to the children of others. Notice the unwillingness of some individuals to sit next to a child on an airplane. What other examples illustrate an antinatalistic view?

The social pressure, particularly for women, to have children is enormous. "Motherhood is still considered to be a primary role for women and women who do not mother either biologically or socially are often stereotyped as either desperate or selfish" (Letherby, 2002, 7). Women today who want to remain childfree tend to have egalitarian gender role attitudes (Kaufman, 2000), to evidence greater interest in a career, and to value freedom from the constraints of having children. They also tend to be better educated, to live in urban areas, and to be less affiliated with religion (DeOllos & Kapinus, 2002). Childfree women have life satisfaction and mental health similar to women who have children (Wu & Hart, 2002). Mirowsky and Ross (2003) reviewed the literature on psychological well-being and children in the home and concluded that children "often make it worse" (p. 90). Some of the reasons include economic hardships, decreased support from one's spouse, inequity in child-care responsibilities, tension between work and family, and arranging for child care.

How happy are the marriages of couples who elect to remain childfree compared with the marriages of couples who opt for having children? Though marital satisfaction declines across time for all couples whether or not they have children, children tend to lessen marital satisfaction by decreasing spousal time together, spousal interaction, and agreement over finances. However, DeOllos and Kapinus (2002) reviewed the studies comparing marital satisfaction of childfree and parent couples and found contradictory results. Hence, the jury is still out on whether childfree or parent couples have happier marriages.

Another consequence of not having children is an increased risk of being in a residential care facility in one's old age (Wenger, Scott, & Patterson, 2000). Married childfree women are the most vulnerable because their spouses are likely to die before them. With no children to care for them and siblings likely having their own families, these women are likely to end up in a residential care facility.

One Child?

One in five women aged 40 to 44 has a single child (Roberts and Blanton, 2000). Twenty young adult only children were asked about the recommendations they would give to parents rearing single children today. These suggestions included facilitating connections with age mates, avoiding overindulgence, and encouraging independence (Roberts and Blanton, 2000).

We had a quicksand box in our back yard. I was an only child, eventually.

Steven Wright, *comedian*

Two Children?

The most preferred family size in the United States is the two-child family. Reasons for this preference include feeling that a family is "not complete" without two children, having a companion for the first child, having a child of each sex, and repeating the positive experience of parenthood they had with their first child. Some couples may not want to "put all their eggs in one basket." They may fear that if they have only one child and that child dies or turns out to be disappointing, they will not have another opportunity to enjoy parenting. Kramer and Ramsburg (2002) emphasized the importance of preparing existing children for the advent of new siblings. Such preparation requires the mother to deliberately set aside time to give to existing children as well as to encourage and reinforce them for prosocial behavior with the new infant.

Teenage Motherhood

Extent of Teenage Childbirth
About half a million (445,492) children are born annually to teenage (15–19) mothers in the United States (Ventura, Hamilton, & Sutton, 2003). In one study, not one of thirty-seven teen pregnancies was planned (Bales & Stephens, 2000). In addition to lack of systematic socialization for contraceptive protection and a lack of supervision more prevalent in single-parent homes (Hogan, Sun, & Cornwell, 2000), a perceived lack of future occupational opportunities also contributes to teenage parenthood (Aassve, 2003). Teenage females who do poorly in school may have little hope of success and achievement in pursuing educational and occupational goals. One of the only remaining options for a meaningful role in life is to become a parent. Geronimus (2003) emphasized that teen pregnancy was adaptive, particularly among African-Americans who "must contend with structural constraints that shorten healthy life expectancy" (p. 881). In addition, some teenagers feel lonely and unloved and have a baby to create a sense of being needed and wanted. In contrast, in Sweden, eligibility requirements for welfare payments make it almost necessary to complete an education and get a job before becoming a parent.

Problems Associated with Teenage Motherhood
Teenage parenthood is associated with various negative consequences, including the following:

1. Poverty among single teen mothers and their children. Many teenagers have no means of economic support or have limited earning capacity. Teen mothers and their children often live in substandard housing and have inadequate nutrition and medical care. The public bears some of the economic burden of supporting these mothers and their children, but even with public assistance, many unmarried teenage parents often struggle to survive economically.

2. Poor health habits. Teenage unmarried mothers are less likely to seek prenatal care and more likely than older and married women to smoke, drink alcohol, and take other drugs. These factors have an adverse effect on the health of the baby. Indeed, babies born to unmarried teenage mothers are more likely to have low birth weights (less than five pounds, five ounces) and to be born prematurely. Children of teenage unmarried mothers are also more likely to be developmentally delayed. These outcomes are largely a result of the association between teenage unmarried childbearing and persistent poverty.

3. Poor academic achievement. Poor academic achievement is both a contributing factor and a potential outcome of teenage parenthood. Three-fifths of teenage mothers drop out of school and, as a consequence, have a much higher probability of remaining poor throughout their lives. Since poverty is linked to unmarried parenthood, a cycle of successive generations of teenage pregnancy may develop. Indeed, teens whose parents have less than a high school education are more vulnerable to becoming pregnant (Hogan, Sun, & Cornwell, 2000).

Of course, despite the problems associated with teenage parenthood, many teenage parents provide a safe and nurturing environment for their children. Conversely, many older parents do not provide such an environment. Nevertheless, the difficulties teenage mothers and their children face are enormous.

Teenage Parenthood: Social Strategies and Interventions
Some interventions regarding teenage childbearing are aimed at prevention. Others attempt to minimize the negative effects associated with teenage childbearing. One preventive intervention for unmarried teenage childbearing is to provide sex education and family-planning programs before unwanted or unintended pregnancy occurs. Sex education programs are offered in schools, churches, family-planning clinics, and public health departments. Another strategy aimed at both preventing unmarried teenage childbirth and minimizing the negative effects of such childbearing is to provide unmarried teenage mothers with programs that increase their life options. Such programs include educational programs, job training, and skills-building programs. Other programs designed to help unmarried teenage mothers and their children include public welfare (such as Temporary Assistance to Needy Fami-

International Data

Forty international population experts concluded that in seventy-four "intermediate-fertility" countries where women have between two and five children—fertility levels will fall below 2.1 per family (Alan Guttmacher Institute, 2003).

Three or More Children?

Couples are more likely to have a third child, and to do so quickly, if they already have two girls rather than two boys. They are least likely to bear a third child if they already have a boy and a girl. Some individuals may want three children because they enjoy children and feel that "three is better than two." In some instances, a couple who have two children may simply want another child because they enjoy parenting and have the resources to do so.

lies), prenatal programs to help ensure the health of the mother and baby, and parenting classes for both teenage unmarried fathers and unmarried teenage mothers.

Some programs combine various services. For example, a program at the New Futures School in Albuquerque, New Mexico, offers health care, parenting education, child-care services, and vocational training to teenage parents and parents-to-be. The goal of this program is to help teenage parents have healthy babies, complete their high school education, and become self-sufficient. A team of researchers (Manlove, Mariner, & Papillo 2000) found that teen mothers who stayed in school or earned a high school diploma or even a GED were less likely to have a second teen birth. Another team (Lewis et al., 1999) evaluated 139 community changes such as new policies, programs, and practices (e.g., access to contraceptives) and found that these were associated with a reported reduction in adolescent sexual activity. The researchers noted that this "may have influenced a modest reduction in average estimated pregnancy rate" (p. 16).

Indeed, as a result of increased public awareness of the need for pregnancy prevention via abstinence and responsible behavior, the birthrate for teenagers has been declining since 1991. In 1991 there were 61.8 births per thousand women aged 15–19; in 2002 the rate was 43.0 (Martin et al., 2003). However, the teenage birthrate in the United States, as measured by births per thousand teenagers, is nine times that of teens in European countries (Meschke, Bartholamae, & Zentall, 2000). Teens in countries such as France, Germany, and Denmark are as sexually active as U.S. teenagers, but the former grow up in a society that promotes responsible contraceptive use. In the Netherlands, for example, individuals are taught to use both the pill and the condom for prevention of pregnancy and STDs (an approach called "double Dutch"). A movement is also afoot to focus on teenage males as a way of decreasing teenage motherhood. Prevention programs in high school target at-risk males (e.g., those who are failing a subject or repeating a grade; who have a history of STDs, drug use, cigarette smoking, or alcohol use; and who are from single-parent families) have begun (Smith et al., 2003).

Sources

Assve, A. 2003. The impact of economic resources on premarital childbearing and subsequent marriage among young American women. *Demography* 40: 105–26.

Bales, D., and D. Stephens. 2000. What do teen parents need? Assessing community policies and effective services for pregnant and parenting teens. Poster presentation at the 62nd Annual Conference of the National Council on Family Relations, Minneapolis, November

Geronimus. A. T. 2003. Damned if you do: Culture, identity, privilege, and teenage childbearing in the United States. *Social Science and Medicine* 57:881–93.

Hogan, D. P., R. Sun, and G. T. Cornwell. 2000. Sexual and fertility behaviors of American females aged 15–19 years: 1985, 1990, and 1995. *American Journal of Public Health* 90:1421–25.

Lewis, R. K., A. Paine-Andrews, J. Fisher, C. Custard, M. Fleming-Randle, and S. B. Fawcett. 1999. Reducing the risk for adolescent pregnancy: Evaluation of a school/community partnership in a Midwestern military community. *Family Community Health* 22:16–30.

Manlove, J., C. Mariner, and A. R. Papillo. 2000. Subsequent fertility among teen mothers: Longitudinal analysis of recent national data. *Journal of Marriage and the Family* 62:430–48.

Martin, J. A., B. E. Hamilton, P. D. Sutton, S. J. Ventura, F. Menacker, and M. L. Munson. 2003. Birth: Final data for 2002. *National Vital Statistics Reports* 52 (10). Hyattsville, Md: National Center for Health Statistics.

Meschke, L. L., S. Bartholamae, and S. R. Zentall. 2000. Adolescent sexuality and parent-adolescent processes: Promoting healthy teen choices. *Family Relations* 49:143–54.

Smith, P. B., R. S. Buzi, and M. L. Weinman. 2003. Targeting males for teenage pregnancy prevention in a school setting. *School of Social Work Journal* 27:23–36.

Ventura, S. J., B. E. Hamilton, and P. D. Sutton. 2003. Revised birth and fertility rates for the United States, 2000 and 2001. *National Vital Statistics Reports* 51 (4). Hyattsville, Md.: National Center for Health Statistics.

Having a third child creates a "middle child." This child is sometimes neglected because parents of three children may focus more on the "baby" and the firstborn than on the child in between. However, an advantage to being a middle child is the chance to experience both a younger and an older sibling. Each additional child also has a negative effect on the existing children by reducing the amount of parental time available to existing children. The economic resources for each child are also affected for each subsequent child.

Hispanics are more likely to want larger families than are whites or African-Americans. Larger families have complex interactional patterns and different values. The addition of each subsequent child dramatically increases the possible relationships in the family. For example, in the one-child family, four interpersonal relationships are possible: mother-father, mother-child, father-child, and father-mother-child. In a family of four, eleven relationships are possible; in a family of five, twenty-six; and in a family of six, fifty-seven.

PERSONAL CHOICES

Is Genetic Testing for You?

Since each of us may have flawed genes that may carry increased risk for diseases such as cancer and Alzheimer's, the question of whether to get a genetic test before becoming pregnant becomes relevant. The test involves giving a blood sample. The advantage is the knowledge of what defective genes you may have and what diseases you may pass to your children. The disadvantage is that since no treatment may be available, the knowledge that you may pass diseases to your children can be devastating. In addition, the results of genetic testing may become a part of your medical history, which may be used to deny you health insurance and employment. Finally, genetic testing is expensive, ranging from $200 to $2,400. The National Society of Genetic Counselors offers information about genetic testing. Its address is NSGC, Department P, 233 Canterbury Drive, Wallingford, PA 19086.

INFERTILITY

Infertility is defined as the inability to achieve a pregnancy after at least one year of regular sexual relations without birth control, or the inability to carry a pregnancy to a live birth. Different types of infertility include the following:

1. Primary infertility. The woman has never conceived even though she wants to and has had regular sexual relations for the past twelve months.

2. Secondary infertility. The woman has previously conceived but is currently unable to do so even though she wants to and has had regular sexual relations for the past twelve months.

3. Pregnancy wastage. The woman has been able to conceive but has been unable to produce a live birth.

NATIONAL DATA

About 6 million American women and their partners cope with infertility (Hart, 2002). About 17 percent of couples of childbearing age have not conceived after a year of unprotected intercourse (Peterson, Newton, & Rosen, 2003).

Causes of Infertility

Although popular usage does not differentiate between the terms *fertilization* and the *beginning of pregnancy*, **fertilization** or **conception** refers to the fusion of the egg and sperm, while pregnancy is not considered to begin until five to seven days later, when the fertilized egg is implanted (typically in the uterine wall; Pinon, 2002). Hence, not all fertilizations result in a pregnancy. An estimated 30–40 percent of conceptions are lost prior to or during implantation.

Forty percent of infertility problems are attributed to the woman, 40 percent to the man, and 20 percent to both of them. Some of the more common causes of infertility in men include low sperm production, poor semen motility, effects of sexually transmitted diseases (such as chlamydia, gonorrhea, and syphilis), and interference with passage of sperm through the genital ducts due to an enlarged prostate. The causes of infertility in women include blocked fallopian tubes, endocrine imbalance that prevents ovulation, dysfunctional ovaries, chemically hostile cervical mucus that may kill sperm, and effects of sexually transmitted diseases.

The psychological reaction to infertility is often depression, particularly if individuals link parenthood to happiness (Brothers & Maddux, 2003). When a couple view themselves as infertile, the woman is more likely than the man (21% versus 9%) to view it as a stressful experience and to become depressed. When both partners experience the infertility with the same level of stress, there are higher levels of marital adjustment than when only one partner becomes stressed (Peterson, Newton, & Rosen, 2003).

Assisted Reproductive Technology

A number of technological innovations are available to assist women and couples in becoming pregnant. These include hormonal therapy, artificial insemination, ovum transfer, in vitro fertilization, gamete intrafallopian transfer, and zygote intrafallopian transfer.

Hormone Therapy Drug therapies are often used to treat hormonal imbalances, induce ovulation, and correct problems in the luteal phase of the menstrual cycle. Frequently used drugs include Clomid, Pergonal, and human chorionic gonadotropin (HCG), a hormone extracted from human placenta. These drugs stimulate the ovary to ripen and release an egg. Although they are fairly effective in stimulating ovulation, hyperstimulation can occur, which may result in permanent damage to the ovary.

Hormone therapy also increases the likelihood that multiple eggs will be released, resulting in multiple births. The increase of triplets and higher-order multiple births over the last decade in the United States is largely attributed to the increased use of ovulation-inducing drugs for treating infertility. Infants of higher-order multiple births are at greater risk of having low birth weight and experience higher infant mortality rates. Mortality rates have improved for these babies (the mortality rates in the mid-1980s were only half those in the 1960s), but these low birth weight survivors may need extensive neonatal medical and social services.

Here is a mother with her daughter, who was conceived with donor sperm. The media often focus on fetuses and babies—not full-grown individuals who have positive relationships—when covering artificial insemination. Yet, if all goes as most parents hope, including parents who seek children through AIH or AID, they enjoy positive, loving relationships with their children, as this mother does with her daughter.

Authors

Artificial Insemination When the sperm of the male partner are low in count or motility, sperm from several ejaculations may be pooled and placed directly into the cervix. This procedure is known as *artificial insemination by husband* (AIH). When sperm from someone other than the woman's partner are used to fertilize a woman, the technique is referred to as *artificial insemination by donor* (AID). Lesbians who want to become pregnant may use sperm from a friend or from a sperm bank (some sperm banks cater exclusively to lesbians). Regardless of the source of the sperm, it should be screened for genetic abnormalities/sexually transmitted diseases, quarantined for 180 days, and retested for HIV; also, the donor should be under 50 to diminish hazards related to aging. These precautions are not routinely taken—let the buyer beware!

Artificial Insemination of a Surrogate Mother In some instances, artificial insemination does not help a woman get pregnant. (Her fallopian tubes may be blocked, or her cervical mucus may be hostile to sperm.) The couple who still wants a child and has decided against adoption may consider parenthood through a surrogate mother. There are two types of surrogate mothers. One is the contracted surrogate mother who supplies the egg, is impregnated with the male partner's sperm, carries the child to term, and gives the baby to the man and his partner. A second type is the surrogate mother who carries to term a baby to whom she is not genetically related (a fertilized egg is implanted in her uterus).

As with AID, the motivation of the prospective parents is to have a child that is genetically related to at least one of them. For the surrogate mother, the primary motivation is to help childless couples achieve their aspirations of parenthood and to make money. (The surrogate mother is usually paid about $10,000.)

In Vitro Fertilization About 2 million couples cannot have a baby because the woman's fallopian tubes are blocked or damaged, preventing the passage of eggs to the uterus. In some cases, blocked tubes can be opened via laser-beam surgery or by inflating a tiny balloon within the clogged passage. When these procedures are not successful (or when the woman decides to avoid invasive tests and exploratory surgery), *in vitro* (meaning "in glass") *fertilization (IVF),* also known as test-tube fertilization, is an alternative.

Using a laparoscope (a narrow, telescope-like instrument inserted through an incision just below the woman's naval to view tubes and ovaries), the physician is able to see a mature egg as it is released from the woman's ovary. The time of release can be predicted accurately within two hours. When the egg emerges, the physician uses an aspirator to remove the egg, placing it in a small tube containing stabilizing fluid. The egg is taken to the laboratory, put in a culture Petri dish, kept at a certain temperature-acidity level, and surrounded by sperm from the woman's partner (or donor). After one of these sperm fertilizes the egg, the egg divides and is implanted by the physician in the wall of the woman's uterus. Usually, several eggs are implanted in the hope one will survive.

Occasionally, some fertilized eggs are frozen and implanted at a later time, if necessary. This procedure is known as *cryopreservation.* Separated or divorced couples may disagree over who owns the frozen embryos, and the legal system is still wrestling with the fate of their unused embryos, sperm, or ova in the event of divorce or death.

Ovum Transfer In conjunction with in vitro fertilization is ovum transfer, also referred to as embryo transfer. In this procedure an egg is donated, fertilized in vitro with the husband's sperm, and then transferred to his wife. Alternatively, the sperm of the male partner is placed by a physician in a surrogate woman. After about five days, her uterus is flushed out (endometrial lavage), and the contents are analyzed under a microscope to identify the presence of a fertilized ovum. The fertilized ovum is then inserted into the uterus of the otherwise infertile partner. Although the embryo can also be frozen and implanted at another time, fresh embryos are more likely to result in successful implantation.

Infertile couples who opt for ovum transfer do so because the baby will be biologically related to at least one of them (the father) and the partner will have the experience of pregnancy and childbirth. As noted earlier, the surrogate woman participates out of her desire to help an infertile couple or to make money.

Other Reproductive Technologies A major problem with in vitro fertilization is that only about 15 to 20 percent of the fertilized eggs will implant on the uterine wall. To improve this implant percentage (to between 40 and 50 percent), physicians place the egg and the sperm directly into the fallopian tube, where they meet and fertilize. Then the fertilized egg travels down into the uterus and implants. Since the term for sperm and egg together is *gamete,* this procedure is called *gamete intrafallopian transfer,* or GIFT. This procedure, as well as in vitro fertilization, is not without psychological costs to the couple.

Gestational surrogacy, another technique, involves fertilization in vitro of the wife's ovum and transfer to a surrogate. Trigametic IVF also involves the use of sperm in which the genetic material of another person has been inserted. This technique allows lesbian couples to have a child genetically related to both women

Infertile couples hoping to get pregnant through one of the over three hundred in vitro fertilization clinics should make informed choices by asking questions such as "What is the center's pregnancy rate for women with a similar diagnosis? What percentage of these women have a live birth? How many cycles are attempted per patient?" The most successful reproductive technology programs report live birthrates of 20 percent (Rosenthal, 1997).

ADOPTION

Some would-be parents are interested in adopting a child. The various routes to adoption are public (children from the child welfare system), private agency (children are placed with nonrelatives through agencies), independent adoption (children are placed directly by birth parents or through an intermediary such as a physician or attorney), kinship (children are placed in a family member's home), and stepparent (children are adopted by a spouse). Motives for adopting a child include wanting a child because of their inability to have a biological child (infertility), their desire to give an otherwise unwanted child a permanent loving home, or their desire to avoid contributing to overpopulation by having more biological children. Some couples may seek adoption for all of these motives. While parents can return a child they provisionally adopt, public opinion is against their doing so. Almost sixty (58%) percent of a national sample of U.S. respondents feel that parents who adopt a child should be required to keep that child (Hollingsworth, 2003).

He changed everything, but in the most wonderful way. Everything that should matter, matters. He's absolutely the center of my life.

Angelina Jolie on her 2-year-old adopted son, Maddox

NATIONAL DATA

There are approximately 100,000 adoptions annually in the United States (Simon & Roorda, 2000).

Demographic Characteristics of Persons Seeking to Adopt a Child

Whereas demographic characteristics of those who typically adopt are white, educated, and high-income, increasingly adoptees are being placed in nontraditional families including older, gay, and single individuals; it is recognized that these individuals may also be white, educated, and high-income (Finley, 2000). Rosie O'Donnell represents the unmarried gay adoptive mother. She lives in Florida, where state law (Mississippi and Utah have similar laws) prohibits gay and lesbian individuals and couples from adopting children. Regarding single parents in general, Haugaard, Palmer, and Wojslawowicz (1999) reviewed the research literature on adoption and suggested that there is no indication that adoptions by single parents are more problematic than adoptions by two parents.

Fisher (2003) noted that adoption is typically portrayed negatively, particularly by college textbooks and readers on the family. He noted that the potential problems of adoption are emphasized twice as much as the successes and rewards. Furthermore, he noted that the Evan B. Donaldson Adoption Institute has

SELF-ASSESSMENT

Attitudes toward Transracial Adoption Scale

Transracial adoption is defined as the practice of parents adopting children of another race—for example, a white couple adopting a Korean or African-American child. Please read each item carefully and consider what you believe about each statement. There are no right or wrong answers to any of these statements, so please give your honest reaction and opinion. After reading each statement, select the number which best reflects your answer using the following scale:

1	2	3	4	5	6	7
Strongly Disagree						Strongly Agree

___ **1.** Transracial adoption can interfere with a child's well-being.

___ **2.** Transracial adoption should not be allowed.

___ **3.** I would never adopt a child of another race.

___ **4.** I think that transracial adoption is unfair to the children.

___ **5.** I believe that adopting parents should adopt a child within their own race.

___ **6.** Only same-race couples should be allowed to adopt.

___ **7.** Biracial couples are not well prepared to raise children.

___ **8.** Transracially adopted children need to choose one culture over another.

___ **9.** Transracially adopted children feel as though they are not part of the family they live in.

___ **10.** Transracial adoption should only occur between certain races.

___ **11.** I am against transracial adoption.

___ **12.** A person has to be desperate to adopt a child of another race.

___ **13.** Children adopted by parents of a different race have more difficulty developing socially than children adopted by foster parents of the same race.

___ **14.** Members of multi-racial families do not get along well.

___ **15.** Transracial adoption results in "cultural genocide."

Scoring

After assigning a number to each item, add the items and divide by 15. The lower the number (1 is the lowest number), the more positive one's view of transracial adoptions. The higher the number (7 is the highest number) the more negative one's view of transracial adoptions. The norming sample was based upon thirty-four male and sixty-nine female students attending Valdosta State University. The average score on the Attitudes toward Transracial Adoption Scale was 2.27 ($SD = 1.15$) (suggesting a generally positive view of transracial adoption by the respondents) and scores ranged from 1.00 to 6.60.

The average age of participants completing the Attitudes toward Transracial Adoption Scale was 22.22 years ($SD = 4.23$) and ages ranged from 18 to 48. The ethnic composition of the sample was 74.8% White, 20.4% Black, 1.9% Asian, 1.0% Hispanic, 1.0% American Indian, and one person did not indicate ethnicity. The classification of the sample was 15.5% Freshman, 6.8% Sophomore, 32.0% Junior, 42.7% Senior, and 2.9% were graduate students.

Copyright: "The Attitudes Toward Transracial Adoption Scale" 2004 by Mark Whatley, Ph.D. Department of Psychology, Valdosta State University, Valdosta, Georgia 31698. Information on validity and reliability may be obtained from Dr. Whatley. The scale is used by permission of Dr. Whatley. Other uses of this scale by written permission only—e-mail **mwhatley@valdosta.edu**

conducted an impressive annotated bibliographic resource of over 1,200 citations on adoption; thus would-be adoptive parents have an enormous resource available to them.

Characteristics of Children Available for Adoption

Adoptees in the highest demand are infant white, healthy children. Those who are older, of a racial or ethnic group different from that of the adoptive parents, of a sibling group, or with physical or developmental disabilities have been difficult to place. Flower Kim (2003) noted that since the waiting period for a healthy white infant is from five to ten years, couples are increasingly open to cross-racial adoptions. Of the 1.6 million adopted children under the age of 18 living in U.S. households, the percentages adopted from other countries are Korea , 24 percent; China, 11 percent; Russia , 10 percent; and Mexico, 9 percent. The ratio of females to males is about equal (53% to 47%) (Peterson, 2003). Judge (2003) studied 109 mother-father pairs who adopted children from Eastern Europe and found stress associated with parenthood was similar to what all parents experience, including higher stress scores reported by fathers than by mothers.

Transracial Adoption

Transracial adoption is defined as the practice of adopting children of a race different from that of the parents— for example, a white couple adopting a Korean or African-American child. In a study on transracial adoption attitudes of college students (using the Attitudes toward Transracial Adoption Scale on p. 238), the scores of the 188 respondents reflected overwhelmingly positive attitudes toward transracial adoption. Significant differences included that women, persons willing to adopt a child at all, interracially experienced daters, and those open to interracial dating were more willing to adopt transracially than men, persons rejecting adoption as an optional route to parenthood, persons with no previous interracial dating experience, and persons closed to interracial dating (Ross et al., 2003). About one in five parents (21%) who adopt children from foster care adopt transracially (Urban Institute, 2003).

National Data

Eight percent of all adoptions are transracial; of these 1.2% involve a Black child adopted by a white family (Simon & Roorda, 2000).

Transracial adoptions are controversial. Kennedy (2003) noted, "Whites who seek to adopt black children are widely regarded with suspicion. Are they ideologues, more interested in making a political point than in actually being parents?" (p. 447). Another controversy is whether it is beneficial for children to be adopted by parents of the same racial background (Hollingsworth, 2000). In regard to the adoption of African-American children by same-race parents, the National Association of Black Social Workers (NABSW) passed a resolution against transracial adoptions, citing that such adoptions prevented Black children from developing a positive sense of themselves "that would be necessary to cope with racism and prejudice that would eventually occur" (Hollingsworth, 1997, 44). The counterargument is that healthy self-concepts, an appreciation for one's racial heritage, and coping with racism/prejudice can be learned in a variety of contexts.

Legal restrictions on transracial adoptions have disappeared. The Adoption and Safe Families Act of 1996 prohibited the use of race "to delay or deny the placement of a child for adoption" (Simon & Roorda, 2000, 3). Social approval for transracial adoptions is increasing. Hollingsworth (2000) reported on a telephone survey of 916 individuals and found that 71 percent believed that race should not be a factor in who should be allowed to adopt. Data comparing children reared in transracial and same-race homes show few differences (Simon & Roorda, 2000). However, a substantial number of studies conclude that "same-race placements are preferable and that special measures should be taken to facilitate such placements, even if it means delaying some adoptions" (Kennedy, 2003, 469).

One 26-year-old Black female was asked how she felt being reared by white parents and replied, "Again, they are my family and I love them, but I am black. I have to deal with my reality as a black woman" (Simon & Roorda, 2000, p. 41). A Black man reared in a white home advised white parents considering a transracial adoption, "Make sure they have the influence of blacks in their lives; even if they have to go out and make friends with black families—it's a must" (p. 25). Indeed, Huh and Reid (2000) found that positive adjustment by adoptees was associated with participation in the cultural activities of the race of the parents who adopted them.

Open Versus Closed Adoptions

Another controversy is whether adopted children should be allowed to obtain information about their biological parents. In general, there are considerable benefits from an open adoption—the biological parent has the opportunity to stay involved in the child's life. Adoptees learn early that they are adopted and who their biological parents are. Birth parents are more likely to avoid regret and to be able to stay in contact with their child. Adoptive parents have information about the genetic background of their adopted child. Avery (1998) studied a sample of 1,274 adoptive parents in 743 adoptive homes and found that the majority of adoptive parents (particularly older adoptive mothers) were in favor of information disclosure.

Regardless of the way individuals enter the role of the adoptive parent, those who enroll in a pre-adoption course report being more emotionally ready to adopt, having more parenting knowledge, and knowing more about the adoption process. Such a course is offered through some private nonprofit agencies (Farber et al., 2003).

NATIONAL DATA

Private adoption usually costs from $15,000 to $35,000. A child from foster care costs nothing (Pertman, 2000).

FOSTER PARENTING

Some individuals seek the role of parent via foster parenting. A **foster parent,** also known as a family caregiver, is neither a biological nor an adoptive parent but is a person who takes care of and fosters a child taken into custody. He or she has made a contract with the state for the service, has judicial status, and is reimbursed by the state. In general, the person is not required to have any degree or formal training (Isomaki, 2002).

About 600,000 children are in foster care (Urban Institute, 2003). Children placed in foster care have typically been removed from parents who are abusive, who are substance abusers, and/or who are mentally incompetent. While foster parents are paid an income for taking care of children in their home, they are also motivated by "empathy, love, and generosity, a willingness to help" (p. 629). The goal of placing children in foster care is to remove them from a negative family context, improve that context and return them, or find a more permanent home than foster care.

CONTRACEPTION

Once individuals have decided on whether and when they want children, choices in regard to contraception must be made.* Pregnancy prevention, STD prevention, ease of use, and need to plan ahead are the primary criteria for selection among women. Men report that pregnancy prevention, STD prevention, and sexual pleasure are the primary criteria for their selection of a method (Grady, Kepinger, and Nelson-Wally, 2000). All contraceptive practices have one of two

*Appreciation is expressed to Beth Credle, MAEd, CHES, a health education specialist, Coordinator of Human Sexuality Programs at UNC-Chapel Hill in the Center for Healthy Student Behaviors, who updated this section to provide state of the art information.

common purposes: to prevent the male sperm from fertilizing the female egg or to keep the fertilized egg from implanting itself in the uterus. In this section, we look at the various methods of contraception. Whatever contraception a woman selects, change is likely as 40 percent of married women and 61 percent of unmarried women in one study reported switching their method of contraception over a two-year period. The primary reason for switching was related to level of contraceptive effectiveness, health risks, and sexually transmitted disease prevention (Grady, Billy, & Klepinger, 2002).

Hormonal Contraceptives

Hormonal methods of contraception currently available to women include the "pill," Norplant®, Jadelle®, Depo-Provera®, NuvaRing®, and Ortho Evra®.

Oral Contraceptive Agents (Birth Control Pill) The birth control pill is the most commonly used method of all the nonsurgical forms of contraception. Although 8 percent of women who take the pill still become pregnant in the first year of use (Ranjit et al., 2001), it remains a desirable birth control option.

Oral contraceptives are available in basically two types: the combination pill, which contains varying levels of estrogen and progestin, and the minipill, which is progestin only. Combination pills work by raising the natural level of hormones in a woman's body, inhibiting ovulation, creating an environment where sperm cannot easily reach the egg, and hampering implantation of a fertilized egg.

The second type of birth control pill, the minipill, contains the same progesterone-like hormone found in the combination pill but does not contain estrogen. Progestin-only pills are taken every day with no hormone-free interval. As with the combination pill, the progestin in the minipill provides a hostile environment for sperm and does not allow implantation of a fertilized egg in the uterus, but unlike the combination pill, the minipill does not always inhibit ovulation. For this reason, the minipill is somewhat less effective than other types of birth control pills. The minipill has also been associated with a higher incidence of irregular bleeding.

Neither the combination pill nor the minipill should be taken unless prescribed by a health-care provider who has detailed information about the woman's previous medical history. Contraindications—reasons for not prescribing birth control pills—include hypertension, impaired liver function, known or suspected tumors that are estrogen-dependent, undiagnosed abnormal genital bleeding, pregnancy at the time of the examination, and a history of poor blood circulation or blood clotting. The major complications associated with taking oral contraceptives are blood clots and high blood pressure. Also, the risk of heart attack is increased for those who smoke or have other risk factors for heart disease. If they smoke, women over 35 should generally use other forms of contraception.

Although the long-term negative consequences of taking birth control pills are still the subject of research, short-term negative effects are experienced by 25 percent of all women who use them. These side effects include increased susceptibility to vaginal infections, nausea, slight weight gain, vaginal bleeding between periods, breast tenderness, headaches, and mood changes (some women become depressed and experience a loss of sexual desire). Women should also be aware of situations in which the pill is not effective, such as the first month of use, with certain prescription medications, and when pills are missed. On the positive side, pill use reduces the incidence of ectopic pregnancy and offers noncontraceptive benefits: reduced incidence of ovarian and endometrial cancers, pelvic inflammatory disease, anemia, and benign breast disease.

Finally, women should be aware that pill use is associated with an increased incidence of chlamydia and gonorrhea. One reason for the association of pill use and a higher incidence of STDs is that sexually active women who use the pill sometimes erroneously feel that because they are protected from becoming pregnant, they are also protected from contracting STDs. The pill provides no protection against STDs; the only methods that provide some protection against STDs are the male and female condoms.

Despite the widespread use of birth control pills, many women prefer a method that is longer acting and does not require daily action. Research continues toward identifying safe, effective hormonal contraceptive delivery methods that are more convenient (Schwartz & Gabelnick, 2002).

One such new hormonal contraceptive is the FDA approved Seasonale®, which reduces the number of periods a woman experiences from thirteen to four per year. This hormonal contraceptive manages the menstrual cycles by skipping the hormone-free week and limiting women's menstrual periods to once every three months.

Norplant® A long-acting reversible hormonal contraceptive consisting of six thin flexible silicone capsules (34 mm in length) implanted under the skin of the upper arm, Norplant, which had been used in the United States for ten years, was taken off the market permanently in the summer of 2002 (Wyeth Pharmaceuticals, Inc., 2002).

Jadelle® Like Norplant, **Jadelle** is a system of rod-shaped silicone implants that are inserted under the skin in the upper inner arm and provide time-release progestin into a woman's system for contraception. This method has replaced Norplant, the first hormonal implant system introduced on the market. The difference between Jadelle and Norplant is that Jadelle consists of only two thin flexible silicone rods, whereas Norplant consisted of six capsules, and Jadelle is inserted with a needle rather than through a minor surgical procedure. Jadelle was originally approved for three years of constant contraceptive protection, but in July 2002, the FDA extended approval for use for five years of pregnancy protection. Side effects are similar to those produced by Norplant, with the most common side effect that occurs being irregular menstrual bleeding (Population Council, 2003b). Population Council scientists are also developing a single rod implant containing Nestrone® (a synthetic progestin similar to the natural hormone progesterone). It may be used by lactating women and may protect against pregnancy for two years.

Depo-Provera® Also known as "Depo" and "the shot," **Depo-Provera** is a synthetic compound similar to progesterone injected into the woman's arm or buttock that protects a woman against pregnancy for three months by preventing ovulation. It has been used by 30 million women worldwide since it was introduced in the late 1960s (although it was not approved for use in the United States until 1992).

Side effects of Depo-Provera include menstrual spotting, irregular bleeding, and some heavy bleeding the first few months of use, although eight out of ten women using Depo-Provera will eventually experience amenorrhea, or the absence of a menstrual period. Mood changes, headaches, dizziness, and fatigue have also been observed. Some women report a weight gain of five to ten pounds. Also, after the injections are stopped, it takes an average of eighteen months before the woman will become pregnant at the same rate as women who have not used Depo-Provera. Slightly fewer than 3 percent of U.S. women report using the injectable contraceptive. The reasons they cite for not using "Depo" include lack of knowledge, fear of side effects or health hazards, and satisfaction with their current contraceptive method (Tanfer, Wierzbicki, & Payne, 2000).

Vaginal Rings **NuvaRing®**, which is a soft, flexible, and transparent ring approximately two inches in diameter that is worn inside the vagina, provides month-long pregnancy protection. Like oral contraceptives, NuvaRing is a highly effective contraceptive when used according to the labeling. Out of a hundred women using NuvaRing for a year, one or two will become pregnant. This method is self-administered. NuvaRing is inserted into the vagina and is designed to release hormones that are absorbed by the woman's body for three weeks. The ring is then removed for a week, at which time the menstrual cycle will occur; afterward the ring is replaced with a new ring. Side effects of NuvaRing are similar to those of the birth control pill. In a one-year study by the manufacturer, more than two thousand women tested NuvaRing; 85 percent of the women were satisfied, and 95 percent would recommend it to others (Roumen et al., 2001). In 2001 *Time* magazine recognized NuvaRing as one of the best health inventions of the year. NuvaRing became available in the United States in mid-2002.

Transdermal Applications **Ortho Evra®** is a contraceptive transdermal patch that delivers hormones to a woman's body through skin absorption. The contraceptive patch is worn on the buttocks, abdomen, upper torso (excluding the breasts), or on the outside of the upper arm for three weeks and is changed on a weekly basis. The fourth week is patch-free and the time when the menstrual cycle will occur (Mishell, 2002). Under no circumstances should a woman allow more than seven days to lapse without wearing a patch (Zieman, 2002).

Ortho Evra provides pregnancy protection and has side effects similar to those of the pill. Ortho Evra simply offers a different delivery method of the hormones needed to prevent pregnancy from occurring. However, a major advantage of the patch over the pill is higher compliance rates with the patch (Mishell, 2002). In clinical trials, the contraceptive patch was found to keep its adhesiveness even through showers, workouts, and water activities, such as swimming. Another transdermal delivery system under development is a contraceptive gel that can be applied to a woman's abdomen. Current efforts in the development of this Nestorone-containing gel yielded a product that women apply to the abdomen for contraceptive or hormone therapy. Because marketing studies in a number of locales have demonstrated that gels are more acceptable in some cultures and patches in others, both Nestorone gel and patch formulations are currently being investigated and developed (Population Council, 2003a).

Male Hormonal Methods Because of dissatisfaction with the few contraceptive options available to men, research and development are occurring in this area. Several promising male products under development rely on MENT ™, a synthetic steroid that resembles testosterone. In contrast to testosterone, however, MENT does not have the effect of enlarging the prostate. A MENT implant and MENT transdermal gel and patch formulation are being developed for contraception and hormone therapy (Population Council Annual Report, 1999). When male hormonal methods do become available, men will need access to clinical screening, prescriptions, and monitoring similar to the follow-up of women who are taking the pill (Alan Guttmacher Institute, 2002).

Male Condom

The condom is a thin sheath made of latex, polyurethane, or natural membranes. Latex condoms, which can be used only with water-based lubricants, historically have been more popular. They are also less likely to slip off and have a much lower chance of breakage (Walsh et al., 2003). However, the polyurethane condom, which is thinner but just as durable as latex, is growing in popularity.

Polyurethane condoms can be used with oil-based lubricants, are an option for some people who have latex-sensitive allergies, provide some protection against the HIV virus and other sexually transmitted diseases, and allow for greater sensitivity during intercourse. Condoms made of natural membranes (sheep intestinal lining) are not recommended because they are not effective in preventing transmission of HIV or other STDs. Individuals are more likely to use condoms with casual than with stable partners (Morrison, et al., 2003).

The condom works by being rolled over and down the shaft of the erect penis before intercourse. When the man ejaculates, sperm are caught inside the condom. When used in combination with a spermicidal lubricant that is placed on the inside of the reservoir tip of the condom as well as a spermicidal or sperm-killing agent that the woman inserts into her vagina, the condom is a highly effective contraceptive.

Like any contraceptive, the condom is effective only when used properly. It should be placed on the penis early enough to avoid any seminal leakage into the vagina. In addition, polyurethane or latex condoms with a reservoir tip are preferable, as they are less likely to break. Even when a condom has a reservoir tip, air should be squeezed out of the tip as it is being placed on the penis to reduce the chance of breaking during ejaculation! But such breakage does occur. Finally, the penis should be withdrawn from the vagina immediately after ejaculation, before it returns to its flaccid state. If the penis is not withdrawn and the erection subsides, semen may leak from the base of the condom into the vaginal lips. Alternatively, when the erection subsides, the condom will come off when the man withdraws his penis if he does not hold onto the condom. Either way, the sperm will begin to travel up the vagina to the uterus and fertilization may occur.

In addition to furnishing extra protection, spermicides also provide lubrication, which permits easy entrance of the condom-covered penis into the vagina. If no spermicide is used and the condom is not of the prelubricated variety, a sterile lubricant (such as K-Y Jelly) may be needed. Vaseline or other kinds of petroleum jelly should not be used with condoms because vaginal infections and/or condom breakage may result. Though condoms should also be checked for visible damage and for the date of expiration, this is rarely done. Three-fourths of the respondents in Lane's (2003) study did not check for damage and 61 percent did not check the date of expiration.

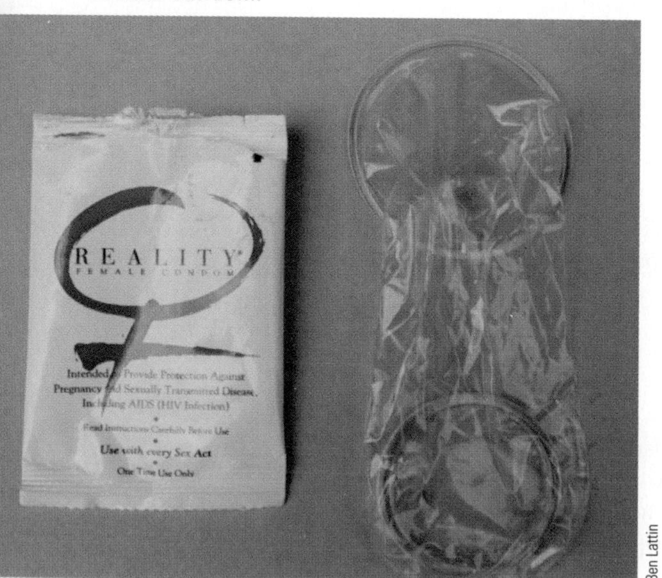

Increasingly women are using the female condom.

Ben Lattin

Female Condom

The **female condom** resembles the male condom except that it fits in the woman's vagina to protect her from pregnancy, HIV infection, and other STDs. The female condom is a lubricated, polyurethane adaptation of the male version. It is about six inches long and has flexible rings at both ends. It is inserted like a diaphragm, with the inner ring fitting behind the pubic bone against the cervix; the outer ring remains outside the body and encircles the labial area. Like the male version, the female condom is not reusable. Female condoms have been approved by the FDA and are being marketed under the brand names Femidom and Reality. The one-size-fits-all device is available without a prescription.

The female condom is durable and may not tear as easily as latex male condoms. Some women may encounter some difficulty with first attempts to insert the female condom. A major advantage of the female condom is that, like the male counterpart, it helps protect against transmission of the HIV virus and other STDs, giving women an option for protection if their partner refuses to wear a condom.

Placement may occur up to eight hours before use, allowing greater spontaneity (Beksinska et al., 2001).

A number of studies that have been conducted internationally have found that women assess the female condom as acceptable and have high rates of trial use with positive reactions (Artz et al., 2000). Women who have had a sexually transmitted disease are more likely to use the female condom. Eighty-five percent of 895 women who attended an STD clinic reported that they had used the female condom during the past six months (Macaluso et al., 2000). If women have instruction and training in the use of the female condom, including a chance to practice using the method, this increases its use. Women who had a chance to practice skills on a pelvic model were more likely to rate the method favorably, use it, and use it correctly. Although not a method completely under the woman's control, the female condom gives women more control than does the male condom (Van Devanter et al., 2002).

There are problems with the female condom however. London (2003) found in a study of 2,232 female condom uses, that slippage occurred in 10 percent of the cases and that women may have been exposed to semen in one in five uses. Problems occur when there is a large disparity between the size of the woman's vagina and the size of her partner's penis and when intercourse is very active.

Vaginal Spermicides

A **spermicide** is a chemical that kills sperm. Vaginal spermicides come in several forms, including foam, cream, jelly, film, and suppository. In the United States, the active agent in most spermicides is nonoxynol-9, which has previously been recommended for added STD protection. However, recent preliminary research in South African countries has shown that women who used nonoxynol-9 became infected with HIV at approximately a 50 percent higher rate than women who used a placebo gel (Van Damme, 2000). Spermicidal creams or gels should be used with a diaphragm. Spermicidal foams, creams, gels, suppositories, and films may be used alone or with a condom.

Spermicides must be applied before the penis enters the vagina (appropriate applicators are included when the product is purchased) no more than twenty minutes before intercourse. Foam is effective immediately, but suppositories, creams, and jellies require a few minutes to allow the product to melt and spread inside the vagina (package instructions describe the exact time required). Each time intercourse is repeated, more spermicide must be applied. Spermicide must be left in place for at least six to eight hours after intercourse; douching or rinsing the vagina should not be done during this period.

One advantage of using spermicides is that they are available without a prescription or medical examination. They also do not manipulate the woman's hormonal system and have few side effects. It was believed that a major noncontraceptive benefit of some spermicides is that they offer some protection against sexually transmitted diseases. However, the 2002 guidelines for prevention and treatment of STDs from the Centers for Disease Control (CDC) do not recommend spermicides for STD/HIV protection. Furthermore, the CDC emphasizes that condoms lubricated with spermicides offer no more protection from STDs than other lubricated condoms, and spermicidally lubricated condoms may have a shorter shelf life, cost more, and be associated with increased urinary tract infections in women (Centers for Disease Control and Prevention, 2002).

Intrauterine Device (IUD)

Although not technically a barrier method, the **intrauterine** device (IUD) is a structural device that prevents implantation. It is inserted into the uterus by a physician to prevent the fertilized egg from implanting on the uterine wall or to dislodge the fertilized egg if it has already implanted. Two common IUDs sold in the United States are ParaGard Copper T 380A and the Progestasert Progesterone T. The Copper T is partly wrapped in copper and can remain in the uterus for ten years. The Progesterone T contains a supply of progestin, which it continuously releases into the uterus in small amounts; after a year the supply runs out and a new IUD must be inserted. The Copper T has a lower failure rate than the Progesterone T. Recently, the FDA also approved an IUD called Mirena, widely used for years in Europe, for use in the United States. Mirena releases tiny amounts of the hormone levonorgestrel into the uterus and protects against pregnancy for five years.

Worldwide, the IUD is a popular method; 60 percent of women using IUDs continue their use after two years (Motamed, 2002). As a result of infertility and miscarriage associated with the Dalkon Shield IUD and subsequent lawsuits against its manufacturer by persons who were damaged by the device, use of all IUDs in the United States declined in the 1980s. However, other IUDs do not share the rates of pelvic inflammatory disease (PID) or resultant infertility associated with the Dalkon Shield. Nevertheless, other manufacturers voluntarily withdrew their IUDs from the U.S. market. In contrast to the 1 percent rate of use by women in the United States, the IUD accounts for 9–24 percent of all contraceptive use in five European countries (Italy, Spain, Poland, Germany, and Denmark). The January 2001 reintroduction of the IUD in the United States and training of health-care providers in its proper screening, insertion, and follow-up care may prompt its revival in this country (Hubacher, 2002). The IUD is often an excellent choice for women who do not anticipate future pregnancies but do not wish to be sterilized or for women who are unable to use hormonal contraceptives. The IUD is not for women who have multiple sex partners.

Diaphragm

The **diaphragm** is a shallow rubber dome attached to a flexible, circular steel spring. Varying in diameter from two to four inches, the diaphragm covers the cervix and prevents sperm from moving beyond the vagina into the uterus. This device should always be used with a spermicidal jelly or cream.

To obtain a diaphragm, a woman must have an internal pelvic examination by a physician or nurse practitioner, who will select the appropriate size and instruct the woman on how to insert the diaphragm. The woman will be told to apply spermicidal cream or jelly on the inside of the diaphragm and around the rim before inserting it into the vagina (no more than two hours before intercourse). The diaphragm must also be left in place for six to eight hours after intercourse to permit any lingering sperm to be killed by the spermicidal agent.

After the birth of a child, a miscarriage, abdominal surgery, or the gain or loss of ten pounds, a woman who uses a diaphragm should consult her physician or health practitioner to ensure a continued good fit. In any case, the diaphragm should be checked every two years for fit.

A major advantage of the diaphragm is that it does not interfere with the woman's hormonal system and has few, if any, side effects. Also, for those couples who feel that menstruation diminishes their capacity to enjoy intercourse, the diaphragm may be used to catch the menstrual flow for a brief time.

On the negative side, some women feel that use of the diaphragm with the spermicidal gel is messy and a nuisance, and it is possible that use of a spermicide may produce an allergic reaction. Furthermore, some partners feel that spermicides make oral genital contact less enjoyable. Finally, if the diaphragm does not fit properly or is left in place too long (more than twenty-four hours), pregnancy or toxic shock syndrome can result.

Cervical Cap

The cervical cap is a thimble-shaped contraceptive device made of rubber or polyethylene that fits tightly over the cervix and is held in place by suction. Like the diaphragm, the cervical cap, which is used in conjunction with spermicidal cream or jelly, prevents sperm from entering the uterus. Cervical caps have been widely available in Europe for some time and were approved for marketing in the United States in 1988. The cervical cap cannot be used during menstruation, since the suction cannot be maintained. The effectiveness, problems, risks, and advantages are similar to those of the diaphragm.

Natural Family Planning

Also referred to as **periodic abstinence,** rhythm method, and fertility awareness, **natural family planning** involves refraining from sexual intercourse during the seven to ten days each month when the woman is thought to be fertile.

Women who use natural family planning must know their time of ovulation and avoid intercourse just before, during, and immediately after that time. Calculating the fertile period involves knowing when ovulation has occurred. This is usually the fourteenth day (plus or minus two days) before the onset of the next menstrual period. Numerous "home ovulation kits" are available without a prescription in a pharmacy; these allow the woman (by testing her urine) to identify twelve to twenty-six hours in advance when she will ovulate. Another method of identifying when ovulation has occurred is by observing an increase in the basal body temperature and a sticky cervical mucus. Calculating the time of ovulation may also be used as a method of becoming pregnant since it helps the couple to know when the woman is fertile.

Nonmethods: Withdrawal and Douching

Because withdrawal and douching are not effective in preventing pregnancy, we call them "nonmethods" of birth control. Also known as **coitus interruptus, withdrawal** is the practice whereby the man withdraws his penis from the vagina before he ejaculates. The advantages of coitus interruptus are that it requires no devices or chemicals, and it is always available. The disadvantages of withdrawal are that it does not provide protection from STDs, it may interrupt the sexual response cycle and diminish the pleasure for the couple, and it is very ineffective in preventing pregnancy.

Withdrawal is not a reliable form of contraception for two reasons. First, a man can unknowingly emit a small amount of pre-ejaculatory fluid, which may contain sperm. One drop can contain millions of sperm. In addition, the man may lack the self-control to withdraw his penis before ejaculation, or he may delay his withdrawal too long and inadvertently ejaculate some semen near the vaginal opening of his partner. Sperm deposited there can live in the moist vaginal lips and make their way up the vagina.

Though some women believe that douching is an effective form of contraception, it is not. Douching refers to rinsing or cleansing the vaginal canal. After intercourse, the woman fills a syringe with water, any of a variety of solutions that can be purchased over the counter, or a spermicidal agent and flushes (so she assumes) the sperm from her vagina. But in some cases, the fluid will actually force sperm up through the cervix. In other cases, a large number of sperm may already have passed through the cervix to the uterus, so the douche may do little good.

Sperm may be found in the cervical mucus within ninety seconds after ejaculation. In effect, douching does little to deter conception and may even encourage it. In addition, douching is associated with an increased risk for pelvic inflammatory disease and ectopic pregnancy.

Emergency Contraception

Also called postcoital contraception, **emergency contraception** refers to various types of morning-after pills that are used primarily in three circumstances: when a woman has unprotected intercourse, when a contraceptive method fails (such as condom breakage or slippage), and when a woman is raped. Emergency contraception methods should be used in emergencies for those times when unprotected intercourse has occurred, and medication can be taken within seventy-two hours of exposure. Planned Parenthood now makes emergency contraception help available by phone. A person can call in, get a prescription, and begin taking the medication immediately.

Fear that emergency contraception is being used as a routine method of contraception is not supported by data. Of 235 women who had received emergency contraception, 70 percent were using the pill and 73 percent were using the condom prior to their need for emergency contraception (Harvey et al., 2000).

Combined Estrogen-Progesterone The most common morning-after pills are the combined estrogen-progesterone oral contraceptives routinely taken to prevent pregnancy. In higher doses, they serve to prevent ovulation, fertilization of the egg, or transportation of the egg to the uterus. They may also make the uterine lining inhospitable to implantation. Known as the "Yuzpe method" after the physician who proposed it, this method involves ingesting a certain number of tablets of combined estrogen-progesterone. *These pills must be taken within seventy-two hours of unprotected intercourse to be effective, and a pregnancy test is required.* Common side effects of combined estrogen-progesterone emergency contraception pills (sold under the trade names Preven and Plan B) include nausea, vomiting, headaches, and breast tenderness, although some women also experience abdominal pain, headache, and dizziness. Side effects subside within a day or two after treatment is completed. The pregnancy rate is 1.2 percent if combined estrogen-progesterone is taken within twelve hours of unprotected intercourse, 2.3 percent if taken within forty-eight hours, and 4.9 percent if taken within forty-eight to seventy-two hours (Rosenfeld, 1997). Although the FDA has not yet approved the sale of this morning-after pill over the counter, the American Medical Association supports this action.

Postcoital IUD Insertion of a copper IUD within five to seven days after ovulation in a cycle when unprotected intercourse has occurred is very effective for preventing pregnancy. This option, however, is used much less frequently than hormonal treatment because women who need emergency contraception often are not appropriate IUD candidates.

Mifepristone (RU-486) **Mifepristone,** also known as **RU-486,** is a synthetic steroid that effectively inhibits implantation of a fertilized egg by making the endometrium unsuitable for implantation. The so-called abortion pill, approved by the FDA in the United States in 2000, is marketed under the name Mifeprex, and can be given to induce abortion within seven weeks of pregnancy. Side effects of RU-486 are usually very severe and may include cramping, nausea, vomiting, and breast tenderness. The pregnancy rate associated with RU-486 is 1.6 percent, which suggests that RU-486 is an effective means of emergency contraception (Rosenfeld, 1997). More than 90 percent of U.S. women who tried RU-486 would recommend it to others and choose it over surgery again. Its use remains controversial.

Diversity in Other Countries

A greater proportion of U.S. women report no contraceptive use at either the first or most recent intercourse (25% and 20%, respectively) than do women in France (11% and 12%, respectively), Great Britain (21% and 4%, respectively) and Sweden (22% and 7%, respectively) (Darroch et al., 2001).

Effectiveness of Various Contraceptives

In Table 10.1 (pp. 250–251), we present data on the effectiveness of various contraceptive methods in preventing pregnancy and protecting against sexually transmitted diseases. Table 10.1 also describes the benefits, disadvantages, and costs of various methods of contraception. Also included in the chart is the obvious and most effective form of birth control: abstinence. Its cost is the lowest, it is 100 percent effective for pregnancy prevention, and it also eliminates the risk of HIV and STD infection from intercourse. Abstinence can be practiced for a week, a month, several years, until marriage, or until someone finds the "right" sexual partner.

Sterilization

Unlike the temporary and reversible methods of contraception already discussed, **sterilization** is a permanent surgical procedure that prevents reproduction. It is the most prevalent method of contraception in the United States (Godecker, Thomson, & Bumpass, 2001). Sterilization may be a contraceptive method of choice when the woman should not have more children for health reasons or when individuals are certain about their desire to have no more children or to remain childfree. Most couples complete their intended childbearing in their late 20s or early 30s, leaving more than fifteen years of continued risk of unwanted pregnancy. Because of the risk of pill use at older ages and the lower reliability of alternative birth control methods, sterilization has become the most popular method of contraception among married women who have completed their families.

Slightly more than half of all sterilizations are performed on women. Although male sterilization is easier and safer than female sterilization, women feel more certain they will not get pregnant if they are sterilized. "I'm the one that ends up being pregnant and having the baby," said one woman. "So I want to make sure that I never get pregnant again."

Female Sterilization Although a woman may be sterilized by removal of her ovaries (**oophorectomy**) or uterus (**hysterectomy**), these operations are not normally undertaken for the sole purpose of sterilization because the ovaries produce important hormones (as well as eggs) and because both procedures carry the risks of major surgery. But sometimes there is another medical problem requiring hysterectomy.

The usual procedures of female sterilization are the salpingectomy and a variant of it, the laparoscopy. **Salpingectomy,** also known as tubal ligation, or tying the tubes (see Figure 10.1, p. 251), is often performed under a general anesthetic while the woman is in the hospital just after she has delivered a baby. An incision is made in the lower abdomen, just above the pubic line, and the fallopian tubes are brought into view one at a time. A part of each tube is cut out, and the ends are tied, clamped, or cauterized (burned). The operation takes about thirty minutes. About 700,000 such procedures are performed annually. The cost is around $2500.

A less expensive and quicker (about fifteen minutes) form of salpingectomy, which is performed on an outpatient basis, is the **laparoscopy.** Often using local anesthesia, the surgeon inserts a small, lighted viewing instrument (laparoscope) through the woman's abdominal wall just below the navel through which the uterus and the fallopian tubes can be seen. The surgeon then makes another

Table 10.1 Methods of Contraception and Sexually Transmitted Disease Protection

Method	Pregnancy Protection[1]	STD Protection	Benefits	Disadvantages	Cost[2]
Oral contraceptive ("The Pill")	95–99.5%	No	High effectiveness rate 24-hour protection. Menstrual regulation.	Daily administration. Side effects possible. Medication interactions.	$10–42 per month
Jadelle® (2 rod 3–5 year implant)	99.95%	No	High effectiveness rate. Long-term protection.	Side effects possible. Menstrual changes.	$300–600 insertion
Depo-Provera® (3-month injection)	99.7%	No	High effectiveness rate. Long-term protection.	Decreases body calcium. Side effects likely.	$45–75 per injection
Ortho Evra® (transdermal patch)	98.5–99.9%	No	Same as oral contraceptives except use is weekly, not daily.	Patch changed weekly. Side effects possible.	$15–32 per month
NuvaRing® (vaginal ring)	99–99.9%	No	Same as oral contraceptives except use is monthly, not daily.	Must be comfortable with body for insertion.	$15–48 per month
Male condom	86–97%	Yes	Few or no side effects. Easy to purchase and use.	Can interrupt spontaneity.	$2–10 a box
Female condom	79–95%	Yes	Few or no side effects. Easy to purchase.	Decreased sensation. Insertion takes practice.	$4–10 a box

small incision in the lower abdomen and inserts a special pair of forceps that carry electricity to cauterize the tubes. The laparoscope and the forceps are then withdrawn, the small wounds are closed with a single stitch, and small bandages are placed over the closed incisions. (Laparoscopy is also known as "the Band-Aid operation.")

As an alternative to reaching the fallopian tubes through an opening below the navel, the surgeon may make a small incision in the back of the vaginal barrel (vaginal tubal ligation).

In late 2002, the FDA approved Essure, a permanent sterilization procedure that requires no cutting and only a local anesthetic in a half-hour procedure that blocks the fallopian tubes. Women typically may return home within 45 minutes (and to work the next day). Essure is already available in Australia, Europe, Singapore, and Canada.

These procedures for female sterilization are over 95 percent effective, but sometimes they have complications. In rare cases, a blood vessel in the abdomen is torn open during the sterilization and bleeds into the abdominal cavity. When

Table 10.1 *(continued)*

Method	Pregnancy Protection[1]	STD Protection	Benefits	Disadvantages	Cost[2]
Spermicide	74–94%	No	Many forms to choose. Easy to purchase and use.	Can cause irritation. Can be messy.	$8–18 per box/tube/can
Diaphragm and cervical cap	80–94%	No	Few side effects. Can be inserted within 2 hours before intercourse.	Can be messy. Increased risk of vaginal/UTI infections.	$50 to $200 plus spermicide
Intrauterine evice (IUD)	97.4–99.2%	No	Little maintenance. Longer term protection.	Risk of PID increased. Chance of expulsion.	$150–300
Withdrawal	81–96%	No	Requires little planning. Always available.	Pre-ejaculatory fluid can contain sperm.	$0
Periodic abstinence	75–91%	No	No side effects. Accepted in all religions/cultures.	Requires a lot of planning. Need ability to interpret fertility signs.	$0
Emergency contraception	75%	No	Provides an option after intercourse has occurred.	Must be taken within 72 hours. Side effects likely.	$10–32
Abstinence	100%	Yes	No risk of pregnancy or STDs.	Partners both have to agree to abstain.	$0

[1] Percentages range from actual/typical use to perfect use rates.

[2] Costs may vary.

Source: Beth Credle, MAEd, CHES, a health education specialist, Coordinator of Human Sexuality Programs at UNC-Chapel Hill in the Center for Healthy Student Behaviors, who updated this section to provide state-of-the-art information as of February 2004, developed this table.

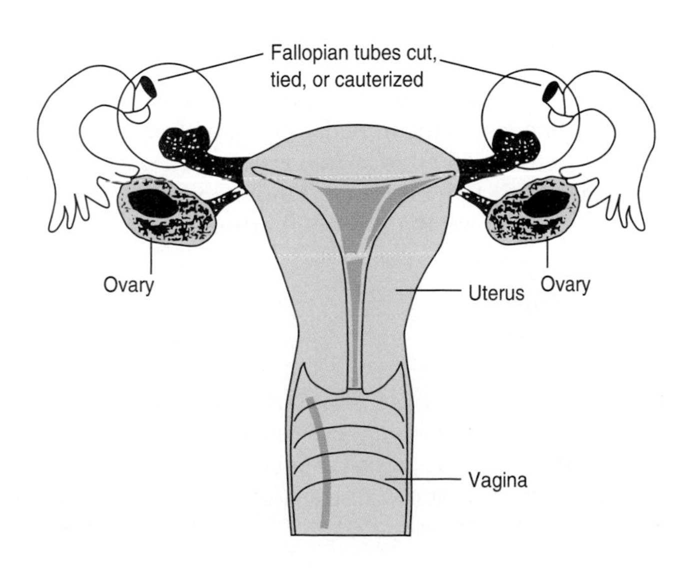

Figure 10.1
Female Sterilization: Tubal Sterilization

this happens, another operation is necessary to find the bleeding vessel and tie it closed. Occasionally, injury occurs to the small or large intestine, which may cause nausea, vomiting, and loss of appetite. The fact that death may result, if only rarely, is a reminder that female sterilization is surgery and, like all surgery, involves some risks.

In addition, although some female sterilizations may be reversed, a woman should become sterilized only if she does not want to have a biological child. Hillis et al. (2000) reported that women who were 30 or younger at the time they were sterilized were twice as likely to report feelings of regret as women who were older than 30.

Male Sterilization **Vasectomies** are the most frequent form of male sterilization. They are usually performed in the physician's office under a local anesthetic. In this operation the physician makes two small incisions, one on either side of the scrotum, so that a small portion of each vas deferens (the sperm-carrying ducts) can be cut out and tied closed. Sperm are still produced in the testicles, but since there is no tube to the penis, they remain in the epididymis and eventually dissolve. The procedure takes about fifteen minutes and costs about $800. The man can leave the physician's office within a short time.

Since sperm do not disappear from the ejaculate immediately after a vasectomy (some remain in the vas deferens above the severed portion), a couple should use another method of contraception until the man has had about twenty ejaculations. In about 1 percent of the cases, the vas deferens grows back and the man becomes fertile again.

A vasectomy does not affect the man's desire for sex, ability to have an erection or an orgasm, amount of ejaculate (sperm comprise only a minute portion of the seminal fluid), health, or chance of prostate cancer. Although in some instances a vasectomy may be reversed, a man should get a vasectomy only if he does not want to have a biological child.

ABORTION

An abortion may be either an **induced abortion,** which is the deliberate termination of a pregnancy through chemical or surgical means, or a **spontaneous abortion (miscarriage),** which is the unintended termination of a pregnancy. In this text, however, we will use the term *abortion* to refer to induced abortion. In general, U.S. law protects the right of the woman to decide to have an abortion. This protection may be threatened by new conservative pro-life appointments to the Supreme Court under the Bush administration. In 2003 Congress passed and President Bush signed a bill banning late-term (also referred to as "partial birth") abortions. Legal challenges were immediately launched. In 2004 the Senate passed the Unborn Victims of Violence Act, which makes it a separate crime to kill a fetus during a violent federal crime. Opponents view the bill as setting a precedent that could lead to restrictions on abortion rights.

Incidence of Abortion

NATIONAL DATA

There are about 1.3 million abortions annually (*Statistical Abstract of the United States: 2003,* Table 104). Among unmarried women, 39 percent who get pregnant have an abortion; 7 percent of married women have an abortion (Ventura et al., 2003). However, the available abortion numbers do not reflect reality because only about 35–50 percent of the abortions that actually occur are reported (Michael, 2001).

Table 10.2 Abortion Rates and Ratios for Selected Years		
Year	**Rate** (Number of abortions per 1,000 women aged 15–44)	**Ratio** (Number of abortions per 1,000 live births)
1990	27.4	389
1995	22.9	351
1999	21.4	327

Source: *Statistical Abstract of the United States: 2003.* 123rd ed. Washington, D.C.: U.S. Bureau of the Census, Table 102.

Table 10.2 reflects the abortion rates and ratios in the United States for selected years. **Abortion rate** refers to the number of abortions per thousand women aged 15–44; **abortion ratio** refers to the number of abortions per thousand live births.

Compared with 1990, the number of abortions per 1,000 women aged 15–44 and the number of abortions per 1,000 live births has dropped. Reasons for the decreasing number of abortions include a reduced rate of unintended pregnancies and more supportive attitudes toward women's becoming single parents. In addition, part of the decline may be due to an increase in restrictive abortion policies, such as those requiring parental consent and mandatory waiting periods. A drop in Medicaid-financed abortions also lowers the abortion rate by as much as 25 percent. "Thus, there is a growing body of evidence of price sensitivity on the part of lower-income women" (Michael, 2001, 380).

Who Gets Abortions and Why

A study in New York City identified women obtaining an abortion as unmarried, young (particularly those under age 16), having no religious affiliation, having had a previous abortion, and white (Michael, 2001; Hollander, 2001). Because there are proportionately more white women than minorities in the United States, about half of all abortions are by white women.

Thirty-two states in the United States have parental notification laws that require explicit permission from one or both parents before an adolescent has an abortion (Adler, Ozer, & Tschann, 2003). The available evidence suggests low psychological risk from having an abortion and that most adolescents make wise choices (Adler, Ozer, & Tschann, 2003).

Nationally, however, the number of abortions to teens has decreased in recent years compared with other age groups, largely because of their use of long-acting (injectable) hormonal contraceptive methods, increased use of contraceptives at first intercourse, and decline in sexual activity (Finer & Henshaw, 2003). About 19 percent of abortions in 2000 and 2001 were obtained by teenagers (Jones, Darroch,& Henshaw, 2002). The majority of abortions (56%) during that time period were obtained by women in their 20s. The abortion rates for women older than their 20s decreases with age because older women are less fertile, may have selected sterilization, and are more likely to be married (Jones, Darroch, & Henshaw, 2002).

Ninety-two pregnant women who sought an abortion reported the following reasons for doing so (Barnett, Freudenberg, & Willie, 1992): fear that the children would cause difficulties with their training or work (46%), pressure from their partners to have an abortion (29%), or concern that their relationships with their partners were unstable (20%). Women also have abortions because of inadequate finances, the feeling they are not ready for the responsibility of having a child, and the perceived effect on existing children.

Diversity in Other Countries

There are wide variations in the range of cultural responses to the abortion issue. On one end of the continuum is the Kafir tribe in Central Asia, where an abortion is strictly the choice of the woman. In this preliterate society, there is no taboo or restriction with regard to abortion, and the woman is free to exercise her decision to terminate her pregnancy. One reason for the Kafirs' approval of abortion is that childbirth in the tribe is associated with high rates of maternal mortality. Since birthing children may threaten the life of significant numbers of adult women in the community, women may be encouraged to abort. Such encouragement is particularly strong in the case of women who are viewed as too young, too sick, too old, or too small to bear children.

Abortion may be encouraged by a tribe or society for a number of other reasons, including practicality, economics, lineage, and honor. Abortion is practical for women in migratory societies. Such women must control their pregnancies, since they are limited in the number of children they can nurse and transport. Economic motivations become apparent when resources (e.g., food) are scarce—the number of children born to a group must be controlled. Abortion for reasons of lineage or honor involves encouragement of an abortion in those cases in which a woman becomes impregnated in an adulterous relationship. To protect the lineage and honor of her family, the woman may have an abortion.

From a worldwide perspective, access to safe, legal abortion is often restricted. Also, many countries do not have the medical technology and facilities to provide women safe abortions.

Abortions are also performed for health reasons, although this is not a frequent reason. Some women choose to abort after learning, through prenatal testing, that their fetus has a serious abnormality. Other women may choose to abort if their physician informs them that continuing the pregnancy would jeopardize their health or life. Only 3 percent of women in one study reported their reason for abortion was a possible fetal problem, and 3 percent aborted because of their own health (Torres & Forrest, 1988). Abortions performed to protect the life or health of the woman are called **therapeutic abortions.** However, there is disagreement over the definition. Garrett, Baillie, and Garrett (2001) noted, "Some physicians argue that an abortion is therapeutic if it prevents or alleviates a serious physical or mental illness, or even if it alleviates temporary emotional upsets. In short, the health of the pregnant woman is given such a broad definition that a very large number of abortions can be classified as therapeutic" (p. 218).

Some women with multifetal pregnancies (a common outcome of the use of fertility drugs) may have a procedure called *transabdominal first trimester selective termination.* In this procedure, the lives of some fetuses are terminated to increase the chance of survival for the others or to minimize the health risks associated with multifetal pregnancy for the woman. For example, a woman carrying five fetuses may elect to abort three of them to minimize the health risks of the other two.

Pro-Life and Pro-Choice Abortion Positions

A dichotomy of attitudes toward abortion is reflected in two opposing groups of abortion activists. Individuals and groups who oppose abortion are commonly referred to as "pro-life" or "antiabortion."

Pro-Life Pro-life groups advocate restrictive abortion policies or a complete ban on abortion. They essentially believe the following:

The unborn fetus has a right to live and that right should be protected.

Abortion is a violent and immoral solution to unintended pregnancy.

The life of an unborn fetus is sacred and should be protected, even at the cost of individual difficulties for the pregnant woman.

Individuals who are over the age of 44, female, mothers of three or more children, married to white-collar workers, affiliated with a religion, and Catholic are most likely to be pro-life (Begue, 2001). About a quarter of women in the world live in countries in which abortion is either completely prohibited or permitted only to save the woman's life (Adler, Ozer, & Tschann, 2003).

Pro-Choice Pro-choice advocates support the legal availability of abortion for all women. They essentially believe the following:

Freedom of choice is a central value.

Those who must personally bear the burden of their moral choices ought to have the right to make these choices.

Procreation choices must be free of governmental control.

In *Breaking the Abortion Deadlock: From Choice to Consent,* Dr. Eileen McDonagh (1996) argues that if a woman has the right to defend herself against a rapist, she also should be able to defend herself against the invasion of a fetus. Although abortion opponents argue that a woman doesn't have the right to terminate the life of a fetus, Dr. McDonagh argues that the fetus doesn't have the right to invade a woman's body. No laws, she says, besides those restricting abortion, allow a person to invade another's body.

No woman can call herself free who does not own and control her own body.

Margaret Sanger

People most likely to be pro-choice have the following characteristics: they are female, are mothers of one or two children, have some college education, are employed, and have annual income of more than $50,000. Although many self-proclaimed feminists and women's organizations, such as the National Organization for Women (NOW), have been active in promoting abortion rights, not all feminists are pro-choice.

Physical Effects of Abortion

Part of the debate over abortion is related to the presumed effects of abortion. Legal abortions, performed under safe conditions, in such countries as the United States are "now so effective and safe that almost always it is safer for the woman to have an abortion than to continue with the pregnancy" (Baird, 2000, 251). The earlier in the pregnancy the abortion is performed, the safer it is.

Postabortion complications include the possibility of incomplete abortion, which occurs when the initial procedure misses the fetus and a repeat procedure must be done. Other possible complications include uterine infection; excessive bleeding; perforation or laceration of the uterus, bowel, or adjacent organs; and an adverse reaction to a medication or anesthetic. After having an abortion women are advised to expect bleeding for up to two weeks (usually not heavy) and to return to their health-care provider thirty days after the abortion to check that all is well (Baird, 2000).

Vacuum aspiration abortions, comprising most U.S. abortions, do not increase the risks to future childbearing. However, late-term abortions do increase the risks of subsequent miscarriages, premature deliveries, and low-birth-weight babies (Baird, 2000).

Psychological Effects of Abortion

What are the data on the psychological effects? Hollander (2001) studied 442 women two years after they had had an abortion and found that 16 percent regretted their decision. This number was up from 11 percent one month after the abortion. However, 69 percent reported that they would make the same decision to abort if faced with the same situation. The predominant emotion in reference to the abortion was relief, rather than either positive or negative emotions. Although 20 percent met the criteria for clinical depression, only 1 percent experienced post-traumatic stress disorder. The researcher concluded "for most women, elective abortion . . . does not pose a risk to mental health" (p. 3).

The finding in the Hollander study is similar to the findings of a panel convened by the American Psychological Association, which reviewed the scientific literature on the mental health impact of abortion. For most women, a legal first-trimester abortion does not create psychological hazards, and symptoms of distress are within normal bounds. Such a conclusion is not to deny that an abortion is a difficult experience for some women and may be associated with increased substance use (to the point of abuse) as a method of coping (Reardon & Ney, 2000).

Younger, unmarried women who have not had a child tend to have more negative responses than older women who have already given birth. Women whose culture or religion prohibits abortion and women with less perceived support for their determination to obtain an abortion tend to experience more distress. Second-trimester abortions, which may reflect greater conflict about the pregnancy and involve more distressing medical procedures, are associated with a higher likelihood of negative response. Nevertheless, "well-designed studies of psychological responses following abortion have consistently shown that risk of psychological harm is low" (Adler, Ozer, & Tschann 2003, 211).

PERSONAL CHOICES

Should You Have an Abortion?

The decision to have an abortion continues to be a complex one. Women who are faced with the issue may benefit by considering the following guidelines:

1. Consider all the alternatives available to you, realizing that no alternative may be all good or all bad. As you consider each alternative, think about both the short-term and the long-term consequences of each course of action.

2. Obtain information about each alternative course of action. Inform yourself about the medical, financial, and legal aspects of abortion, childbearing, parenting, and placing the baby up for adoption.

3. Talk with trusted family members, friends, or unbiased counselors. Consider talking with the man who participated in the pregnancy. If possible, also talk with women who have had abortions as well as with women who have kept and reared their baby or placed their baby for adoption. If you feel that someone is pressuring you in your decision making, look for help elsewhere.

4. Consider your own personal and moral commitments in life. Understand your own feelings, values, and beliefs concerning the fetus and weigh those against the circumstances surrounding your pregnancy.

SUMMARY

Why do people have children?

Having children continues to be a major goal of most college students. Conscientious family planning will help to reduce the number of unwanted pregnancies. Worldwide 56 million unwanted pregnancies occur annually. In the United States 3 million unplanned pregnancies occur annually.

The decision to become a parent is encouraged (sometimes unconsciously) by family, friends, religion, government, favorable economic conditions, and cultural observances. The reasons people give for having children include personal fulfillment and identity and the desire for a close relationship.

How many children do people have?

About 20 percent of individuals never have children. Reasons that spouses elect a childfree marriage include the freedom to spend their time and money as they choose, to enjoy their partner without interference, to pursue their career, to avoid health problems associated with pregnancy, and to avoid passing on genetic disorders to a new generation.

The most preferred type of family in the United States is the two-child family. Some of the factors in a couple's decision to have more than one child are the desire to repeat a good experience, the feeling that two children provide companionship for each other, and the desire to have a child of each sex.

What are the causes of infertility?

Infertility is defined as the inability to achieve a pregnancy after at least one year of regular sexual relations without birth control, or the inability to carry a pregnancy to a live birth. Forty percent of infertility problems are attributed to the woman, 40 percent to the man, and 20 percent to both of them. Some of the more common causes of infertility in men include low sperm production, poor semen motility, effects of sexually transmitted diseases (such as chlamydia, gonorrhea, and syphilis), and interference with passage of sperm through the genital ducts due to an enlarged prostate. The causes of infertility in women include blocked fallopian tubes, endocrine imbalance that prevents ovulation, dysfunctional ovaries, chemically hostile cervical mucus that may kill sperm, and effects of sexually transmitted diseases. The psychological reaction to infertility is often depression over having to give up a lifetime goal. When a couple view themselves as infertile, the woman is more likely than the man to view it as a stressful experience and to become depressed (21% versus 9%).

About one in four women with infertility problems reports receiving some form of infertility service. A number of technological innovations are available to assist women and couples in becoming pregnant. These include hormonal therapy, artificial insemination, ovum transfer, in vitro fertilization, gamete intrafallopian transfer, and zygote intrafallopian transfer.

Why do people adopt?

Motives for adoption include a couple's inability to have a biological child (infertility), their desire to give an otherwise unwanted child a permanent loving home, or their desire to avoid contributing to overpopulation by having more biological children. There are approximately 100,000 adoptions annually in the United States.

Whereas demographic characteristics of those who typically adopt are white, educated, and high-income, increasingly, adoptees are being placed in nontraditional families including older, gay, and single individuals; it is recognized that these individuals may also be white, educated, and high-income. Most college students are open to transracial adoption.

Some individuals seek the role of parent via foster parenting. A foster parent, also known as a family caregiver, is a person who at home either alone or with a spouse takes care and fosters a child taken into custody. He or she has made a contract with the state for the service, has judicial status, and is reimbursed by the state.

What are various methods of contraception?

The primary methods of contraception include hormonal methods (the newest of which are Jadelle®, Depo-Provera®, NuvaRing®, and Ortho Evra®), which prevent ovulation; the IUD, which prevents implantation of the fertilized egg; condoms and diaphragms, which are barrier methods; and vaginal spermicides and the rhythm method. These and numerous new methods vary in effectiveness and safety.

Sterilization is a surgical procedure that prevents fertilization, usually by blocking the passage of eggs or sperm through the fallopian tubes or vas deferens, respectively. The procedure for female sterilization is called salpingectomy, or tubal ligation. Laparoscopy is another method of tubal ligation. The most frequent form of male sterilization is vasectomy.

What are the types of and motives for an abortion?

An abortion may be either an induced abortion, which is the deliberate termination of a pregnancy through chemical or surgical means, or a spontaneous abortion (miscarriage), which is the unintended termination of a pregnancy. In this text we use the term *abortion* to refer to induced abortion. In general, U.S. law protects the right of the woman to decide to have an abortion. Ninety-two pregnant women who sought an abortion reported the following reasons for doing so: fear that the children would cause difficulties with their training or work (46%), pressure from their partners to have an abortion (29%), or concern that their relationships with their partners were unstable (20%). Women also have abortions because of inadequate finances, the feeling they are not ready for the responsibility of having a child, and the perceived effect on existing children.

KEY TERMS

abortion rate
abortion ratio
antinatalism
cervical cap
coitus interruptus
conception
Depo-Provera®
diaphragm
emergency contraception
female condom
fertilization
foster parent
hysterectomy
induced abortion
infertility
intrauterine device
Jadelle®
laparoscopy
Mifepristone
miscarriage
natural family planning
NuvaRing®
oophorectomy
Ortho Evra®
periodic abstinence
pregnancy
pronatalism
RU-486
salpingectomy
spermicide
spontaneous abortion
sterilization
therapeutic abortion
transracial adoption
vasectomy
withdrawal

RESEARCHING MARRIAGE AND THE FAMILY WITH INFOTRAC COLLEGE EDITION

InfoTrac College Edition, an online library, allows you to perform research online anywhere, anytime. Following are two suggested search terms and related questions to help you extend your understanding of the topics covered in this chapter. Go to www.infotrac-college.com to begin your search.

Keyword: **Transracial adoption.** Locate articles that provide different points of view on the benefits/drawbacks for children of being adopted by parents of a different race. How do you evaluate the research and what is your opinion?

Keyword: **Abortion.** Locate articles that reflect a different perspective from your own pro-life or pro-choice position and make a case for the "other" position.

The Companion Web Site for Choices in Relationships: An Introduction to Marriage and the Family, Eighth Edition

http://sociology.wadsworth.com/knox_schacht/choices8e

Supplement your review of this chapter by going to the companion Web site to take one of the Tutorial Quizzes, use the flash cards to master key terms, and check out the many other study aids you'll find there. You'll also find special features such as the Marriage and Family Resource Center, Census 2000 information, and other data and resources at your fingertips to help you with that special project or to do some research on your own.

WEBLINKS

Alan Guttmacher Institute
http://www.agi-usa.org

Engenderhealth
http://www.engenderhealth.org/

The Evan B. Donaldson Adoption Institute
http://www.adoptioninstitute.org/

Fetal Fotos (bonding with your fetus)
http://www.fetalfotosusa.com/

Georgia Reproduction Specialists (male infertility)
http://www.ivf.com/male.html

National Right to Life
http://www.nrlc.org/

Planned Parenthood Federation of America, Inc.
http://www.plannedparenthood.org

NARAL Pro-Choice America (reproductive freedom and choice)
http://www.naral.org/

SafeDreams
http://www.safedreams.com

Authors

CHAPTER 11

Teaching children to count is not as important as teaching them what counts.

E.C. McKenzie

Parenting

CONTENTS

TRUE OR FALSE?

1. There is little difference in the qualities of the mother-infant interaction between mothers whose infants spend a lot of time in day care and mothers who stay at home with their infants.
2. Attachment theorist Robert Sears recommends that children not be allowed to cry themselves to sleep but be picked up and comforted by the parents.
3. Support, monitoring, and avoidance of harsh punishment are the parenting factors that seem crucial to positive child outcomes.
4. Infants who sleep with their own parents in the parents' bed are at significant risk of Sudden Infant Death Syndrome when compared with children who do not share a bed with their parents.
5. In the United States, parents rather than the government will be responsible for monitoring what children are exposed to on the Internet.

Answers: **1.** T **2.** T **3.** T **4.** F **5.** T

American parents feel immediate empathy when they witness other parents on national television reacting to the bizarre behavior of their children who have shot someone or themselves. As they watch the television screen, parents are careful not to assign blame to the distraught and grieving parents—there is the haunting fear "Could this be my child?" or "Is my child capable of this?" All parents wonder whether their parenting choices are encouraging their children toward triumph or tragedy. This chapter focuses on the issue of parenting choices with the goal of facilitating happy, successful, socially contributing members to our society. We begin by looking at the various roles of parenting.

ROLES INVOLVED IN PARENTING

While it is difficult to find one definition of parenting, there is general agreement about the roles parents play in the lives of their children. These include the following.

Caregiver

Sprey (2001) emphasized that "families are responsible for the care of their children" (p. 12). From the moment of birth, when infants draw their first breath and are placed on the mother's stomach, parents stand ready to provide nourishment (milk), cleanliness (diapers), and temperature control (warm blanket). The need for such sustained care continues and becomes an accepted and anticipated role of parents. The parents who excuse themselves from a party early because they "need to check on the baby" are alerting the hostess of their commitment to the role of caregiver.

Emotional Resource

Beyond providing physical care, parents are sensitive to the emotional needs of children in terms of their need to belong, to be loved, and to develop positive self-concepts. Hugging, holding, kissing their infant not only express their love for their infant but also reflect an awareness that such display of emotion is good for the child's sense of self-worth. The security that children feel when they are emotionally attached to their parents cuts across racial and ethnic identities. Arbona and Power (2003) found that securely attached adolescents in African-American, European-American, and Mexican-American families all reported positive self-esteem. In contrast, individuals who reported being emotionally neglected as children reported higher levels of psychological distress as adults (Wark et al., 2003).

Table 11.1 Annual Costs per Child in Three Income Groups

Income Group	Total Annual Cost
Less than $39,100	
Less than 2 years old	$6,490
15–17 years old	$7,480
$39,100–$65,800	
Less than 2 years old	$9,030
15–17 years old	$10,140
More than $65,800	
Less than 2 years old	$13,430
15–17 years old	$14,670

Source: *Statistical Abstract of the United States:* 2003 123rd ed. Washington, D.C.: U.S. Bureau of the Census, Table 675.

Parents are regularly encouraged to spend "quality time" with their children, and it is implied that putting children in day care robs the child of this time. Booth et al. (2002) compared children in day care with those in home care in terms of time mother and child spent together per week. While the mothers of day-care children spent less time with their children than the mothers who cared for their children at home, the authors concluded that the "groups did not differ in the quality of mother-infant interaction" and that the difference in the "quality of the mother-infant interaction may be smaller than anticipated" (p. 16). Other research has suggested that social emotional outcomes may not be compromised by day care (NICHD, 2003).

There is a certain level of confidence in knowing for a fact that someone loves you. And my parents always made us feel like we were good. From that, succeeding is just, all right, how can I let the world know.

Michael Jordan, *Basketball superstar*

Economic Resource

New parents are also acutely aware of the costs for medical care, food, and clothes for infants and seek ways to ensure that such resources are available to their children. Working longer hours, taking second jobs, and cutting back on leisure expenditures are attempts to ensure that money is available to meet the needs of the child. Parents may also take out additional life insurance or begin a college fund for their offspring. Table 11.1 reflects the costs at different ages for three different economic groups. These total annual costs include those for housing (the largest expenditure), food, transportation, clothing, health and child care.

This father (taught by his parents and grandparents to fish) is now modeling this interest and teaching his son the same skill.

Authors

Teacher

All parents think they have a philosophy of life or set of principles they feel their children will benefit from. The plethora of *Life's Little Instruction Books* are bought primarily by parents to give to their children to remind them of basic life lessons the parents want to teach. Parents later discover that their children may not be interested in their religion or philosophy—indeed, may rebel against it. But this possibility does not deter them from their role as teacher.

Protector

Parents also feel the need to protect their children from harm. Insisting that their children wear seat belts reflects this protective role, which extends to protecting them from drugs, from violence

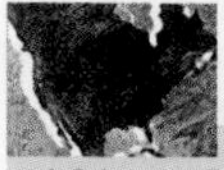

Diversity in the United States

African-American fathers may have a unique role, such as educating their children how to survive in a white world. Greif, Hrabowski, and Maton (1998) noted that high-status Black fathers alert their sons to the importance of learning how to play golf, since "business deals are often completed on the golf course" (p. 50).

Lesbian parents also reflect a unique parenting context where there is a need to buttress children against potential prejudice and discrimination. Findings comparing children reared in heterosexual and lesbian homes show positive mother-child relationships and well-adjusted children in both contexts (Golombok et al. 2003).

Diversity in Other Countries

Six hundred and fifty-four parents in Guyana identified the values they most wanted to teach their children as those of obedience, honesty, and mannerly conduct (Wilson, Wilson, & Berkeley-Caines, 2003).

and nudity in the media, and from strangers. Children are admonished early to be careful whom they interact with—that there are people to seek (family) and people to avoid (strangers). Parents may also try to protect the future of their children by instilling the value of education, helping with homework, etc., to ensure success in the educational system or workplace.

Some parents feel that protecting their children from harm implies appropriate discipline for inappropriate behavior. Galambos, Barker, and Almeida (2003) noted "parents' firm behavioral control seemed to halt the upward trajectory in externalizing problems among adolescents with deviant peers." In other words, parents who intervened when they saw a negative context developing were able to help their children avoid negative peer influences.

Finally, some parents feel it is important to protect their children from certain television content. This ranges from families that do not allow a television in their home to setting the V chip on their TV to allow only "G" rated movies. Research confirms that parents do find media ratings helpful (Bushman and Cantor, 2003).

Authors

These parents are in the role of protecting their children from becoming sunburned.

Ritual Bearer

In order to build a sense of family cohesiveness, parents often foster rituals to bind members together in emotion and in memory. Prayer at meals and before bedtime, celebrating birthdays, and vacationing at the same place (beach, mountains, etc.) provide predictable times of togetherness and sharing.

CHOICES PERSPECTIVE OF PARENTING

Although both genetic and environmental factors are at work, it is the choices parents make that have a dramatic impact on their children. In this section we review the nature of parental choices and some of the basic choices parents make.

Nature of Parenting Choices

Parents might keep the following points in mind when they make choices about how to rear their children.

1. The absence of a parental decision is a decision. Parents are constantly making choices even when they think they are not doing so. When a child is impolite and the parent does not provide feedback and encourage polite behavior, the parent has chosen to teach his or her child that being impolite is acceptable. When a child makes a promise ("I'll call you when I get to my friend's house") and does not do as promised, the parent has chosen to allow the child to not take commitments seriously. Hence, parents cannot choose not to make choices in their parenting since their inactivity is a choice that has as much impact as a deliberate decision to reinforce politeness and responsibility.

2. Parental choices involve trade-offs. Parents are also continually making trade-offs in the parenting choices they make. The decision to take on a second job or to work overtime to afford the larger house will come at the price of having less time to spend with one's children and being more exhausted when such time is available. The choice to enroll one's child in the highest-quality day care (which may also be the most expensive) will mean less money for family vacations. The choice to have an additional child will provide siblings for the existing children but will mean less time and fewer resources for those children. Parents should increase their awareness that no choice is without a tradeoff and should evaluate the costs and benefits in making such decisions.

Diversity in the United States

Parents are diverse. Amish parents rear their children in homes without electricity (no TV, phones, CD players). Charismatics (members of the conservative religious denomination) educate their children in Christian schools emphasizing a biblical and spiritual view of life and the world. More secular parents bring up their children amid cultural cutting-edge technology and encourage individualistic liberal views.

3. View bad choices positively. All parents are plagued by the belief that they made a bad decision—they should have held their child back a year in school (or not done so); they should have intervened in a bad peer relationship; they should have handled their child's drug use differently, etc. Whatever the issue, parents chide themselves for their mistakes. Rather than berate themselves as parents, they might emphasize the positive outcome of their choices: not holding the child back made the child the "first" to experience some things among his or her peers; they made the best decision they could at the time, etc. Children might also be encouraged to view their own decisions positively.

Five Basic Parenting Choices

The five basic choices parents make include deciding whether to have a child, deciding the number of children, deciding the interval between children, deciding one's method of discipline and guidance, and deciding the degree to which one will be invested in the role of parent. Though all of these decisions are important, the relative importance one places on parenting as opposed to one's career will have implications for the parents, their children, and their children's children. While some parents focus their life around their children, others regard the children as only one aspect of their lives. One father of seven noted that while the first five years of parenting involved physical caregiving and intense emotional bonding, the best parenting is to "create the context" and let the children flourish. In effect, he felt the seven children could best learn from one another with parents as the guardrails for safety.

TRANSITION TO PARENTHOOD

Transition to parenthood refers to that period of time from the beginning of pregnancy through the first few months after the birth of a baby. The mother, father, and couple all undergo changes and adaptations during this period.

The best gift you can give your children is your own happy marriage. Your children need to see you closing the bedroom door or going out together without them. That gives a sense of security greater than what they get by just being loved.

Shirley Glass, *Psychologist*

Transition to Motherhood

The Self-Assessment on page 264 examines one's view of traditional motherhood.

Although childbirth is sometimes thought of as a painful ordeal, some women describe the experience as fantastic, joyful, and unsurpassed. A strong emotional bond between the mother and her baby usually develops early, and both the mother and infant resist separation. Sociobiologists suggest that there is a biological basis (survival) for the attachment between a mother and her offspring. The mother alone carries the fetus in her body for nine months, lactates

SELF-ASSESSMENT

The Traditional Motherhood Scale

The purpose of this survey is to assess the degree to which students possess a traditional view of motherhood. Read each item carefully and consider what you believe. There are no right or wrong answers, so please give your honest reaction and opinion. After reading each statement, select the number which best reflects your level of agreement using the following scale:

1	2	3	4	5	6	7
Strongly Disagree						Strongly Agree

___ **1.** The mother has a better relationship with her children.

___ **2.** A mother knows more about her child, therefore being the better parent.

___ **3.** Motherhood is what brings women to their fullest potential.

___ **4.** A good mother should stay at home with her children for the first year.

___ **5.** Mothers should stay at home with the children.

___ **6.** Motherhood brings much joy and contentment to a woman.

___ **7.** A mother is needed in a child's life for nurturance and growth.

___ **8.** Motherhood is an essential part of a female's life.

___ **9.** I feel that all women should experience motherhood in some way.

___ **10.** Mothers are more nurturing.

___ **11.** Mothers have a stronger emotional bond with their children.

___ **12.** Mothers are more sympathetic to children who have hurt themselves.

___ **13.** Mothers spend more time with their children.

___ **14.** Mothers are more lenient toward their children.

___ **15.** Mothers are more affectionate toward their children.

___ **16.** The presence of the mother is vital to the child during the formative years.

___ **17.** Mothers play a larger role in raising children.

___ **18.** Women instinctively know what a baby needs.

Scoring

After assigning a number from 1 (strongly disagree) to 7 (strongly agree), add the numbers and divide by 18. The higher your score (7 is the highest possible score), the stronger the traditional view of motherhood. The lower your score (1 is the lowest possible score), the less traditional the view of motherhood.

Norms

The norming sample of this self-assessment was based upon twenty male and eighty-six female students attending Valdosta State University. The average age of participants completing the scale was 21.72 years (*SD* = 2.98) and ages ranged from 18 to34. The ethnic composition of the sample was 80.2% White, 15.1% Black, 1.9% Asian, 0.9% American Indian, and 1.9% Other. The classification of the sample was 16.0% Freshman, 15.1% Sophomore, 27.4% Junior, 39.6% Senior, and 1.9% were graduate students.

Participants responded to each of eighteen items on a seven-point scale from (1) *Strongly Disagree* to (7) *Strongly Agree.* Higher scores on the scale represented more traditional views toward motherhood. The most traditional score was 6.33; the score reflecting the least support for traditional motherhood was 1.78. The midpoint between the top and bottom score was 4.28 so that persons scoring above this number tended to have a more traditional view of motherhood and persons scoring below this number a less traditional view of motherhood. The average score on the Traditional Motherhood Scale was 4.28 (*SD* = 1.04) and scores ranged from 1.78 to 6.33.

There was a significant difference ($p. < .05$) between female participants' scores ($M = 4.19$, $SD = 1.08$) and male participants' scores ($M = 4.68$, $SD = 0.73$), suggesting that males had more traditional views of motherhood than females.

Copyright: The Traditional Motherhood Scale 2004 by Mark Whatley, Ph.D. Department of Psychology, Valdosta State University, Valdosta, Georgia 31698. Use of this scale is permitted only by prior written permission of Dr. Whatley. His e-mail is **mwhatley@valdosta.edu**

to provide milk, and produces **oxytocin**—a hormone from the pituitary gland during the expulsive stage of labor that has been associated with the onset of maternal behavior in lower animals.

Not all mothers feel joyous after childbirth. Emotional bonding may be temporarily impeded by a mild depression, characterized by irritability, crying, loss of appetite, and difficulty in sleeping. The mother may also feel overwhelmed with the work of caring for an infant and feel a loss of a sense of mastery (Cassidy & Davies, 2003). Many new mothers experience **baby blues**—transitory symptoms of depression twenty-four to forty-eight hours after the baby is born. A few, about 10 percent, experience postpartum depression—a more severe reaction than baby blues. **Postpartum depression** is believed to be a result of the numerous physiological and psychological changes occurring during pregnancy, labor, and delivery. Although the woman may become depressed in the hospital, she more often experiences these feelings within the first month after returning home with her baby. Behaviors associated with postpartum depression include eating a high-fat diet, not exercising, smoking (Gennaro & Fehder, 2000), partner-associated stress, physical abuse, and not breast-feeding (Gross et al., 2002). In addition, women who report that pregnancy was a difficult time for them are more likely to experience postpartum depression (Gross et al., 2002). Most women recover within a short time; some (about 5 percent) seek therapy to speed their recovery. To minimize baby blues and postpartum depression, one must recognize that having misgivings about the new infant is normal and appropriate. In addition, the woman who has negative feelings about her new role as mother should elicit help with the baby from her family or other support network so that she can continue to keep up her social contacts with friends and to spend time by herself and with her partner. Mulsow et al. (2002) found that intimacy with the partner reduced maternal stress.

Is transition to motherhood similar for lesbian and heterosexual mothers? Not according to Cornelius-Cozzi (2002), who interviewed lesbian mothers and found that the egalitarian norm of the lesbian relationship had been altered—e.g., the biological mother became the primary caregiver and that the coparent, who often heard the biological mother refer to the child as "her child," suffered a lack of validation.

Transition to Fatherhood

The Self-Assessment on page 266 examines one's view of traditional fatherhood.

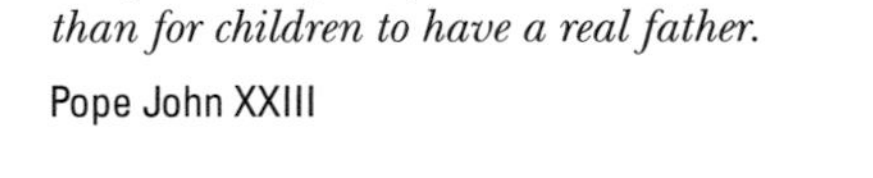

It is far easier for a father to have children than for children to have a real father.

Pope John XXIII

While the likelihood of a male's living with his own children has been in decline (Eggebeen, 2002), the importance of the father in the lives of his children is enormous and goes beyond his economic contribution (Flouri & Buchanan, 2003; Aldous & Mulligan, 2002; Knox, 2000). Children from intact homes or those in which fathers maintained an active involvement in their lives after divorce tend to

make good grades	*be less involved in crime*
have good health/self-concept	*have a strong work ethic*
have durable marriages	*have a strong moral conscience*
have higher life satisfaction	*have higher incomes as adults*
have higher education levels	*form close friendships*
have stable jobs	*have fewer premarital births*
have lower child sex abuse	*exhibit fewer anorectic symptoms*

Strom et al. (2002) emphasized that the amount of time fathers spent with their children was important for being evaluated as a "successful father." How much time fathers actually spend with their children depends on whom you ask: fathers typically report spending more time with their children than mothers report that they do (Coley & Morris, 2002). Gavin et al. (2002) noted that parental involvement was predicted most strongly by the quality of the parent's romantic relationship. If the father was emotionally and physically involved with the mother, he was more likely to take an active role in the child's life. Fathers whose wives worked more hours than the fathers worked also reported more involvement with their children (McBride, Schoppe, & Rane, 2002).

Diversity in the United States

In the District of Columbia, over 10 percent of African-American men between the ages of 18 and 35 are in prison; over half are under some form of correctional supervision; 75 percent can expect to be incarcerated at some point in their lives. The result is less opportunity to function in the role of active father for their children. African-American children are more vulnerable to growing up in homes without a father (Mauer & Chesney-Lind, 2003).

SELF-ASSESSMENT

The Traditional Fatherhood Scale

The purpose of this survey is to assess the degree to which students have a traditional view of fatherhood. Read each item carefully and consider what you believe. There are no right or wrong answers, so please give your honest reaction and opinion. After reading each statement, select the number which best reflects your level of agreement using the following scale:

1	2	3	4	5	6	7
Strongly Disagree						Strongly Agree

___ 1. Fathers do not spend much time with their children.

___ 2. Fathers should be the disciplinarian in the family.

___ 3. Fathers should never stay at home with the children while the mother works.

___ 4. The father's main contribution to his family is giving financially.

___ 5. Fathers are less nurturing.

___ 6. Fathers expect more from children.

___ 7. Most men make horrible fathers.

___ 8. Fathers punish children more than mothers do.

___ 9. Fathers do not take a highly active role in their children's lives.

___ 10. Fathers are very controlling.

Scoring

After assigning a number from 1 (strongly disagree) to 7 (strongly agree), add the numbers and divide by 10. The higher your score (7 is the highest possible score), the stronger the traditional view of fatherhood. The lower your score (1 is the lowest possible score), the less traditional the view of fatherhood.

Norms

The norming sample was based upon twenty-four male and sixty-nine female students attending Valdosta State University. The average age of participants completing the Traditional Fatherhood Scale was 22.15 years ($SD = 4.23$) and ages ranged from 18 to 47. The ethnic composition of the sample was 77.4% White, and 19.4% Black, 1.1% Hispanic, and 2.2% Other. The classification of the sample was 16.1% Freshman, 11.8% Sophomore, 23.7% Junior, 46.2% Senior, and 2.2% were graduate students.

Participants responded to each of ten items on a seven-point scale from (1) *Strongly Disagree* to (7) *Strongly Agree*. The higher the score, the stronger the view of traditional fatherhood. Lower scores on the scale represented weaker support for a traditional view of fatherhood. The most traditional score was 5.50; the score representing the least support for traditional fatherhood was 1.00. The average score on the Traditional Fatherhood Scale was 3.33 ($SD = 1.03$) (suggesting a less than traditional view) and the scores ranged from 1.00 to 5.50.

There was a significant difference ($p < .05$) between female participants' attitudes ($M = 3.20$, $SD = 1.01$) and male participants' attitudes toward fatherhood ($M = 3.69$, $SD = 1.01$) suggesting that males had more traditional views of fatherhood than females. There were no significant differences between ethnicities.

Copyright: "The Traditional Fatherhood Scale" 2004 by Mark Whatley, Ph.D. Department of Psychology, Valdosta State University, Valdosta, Georgia 31698. Use of this scale is permitted only by prior written permission of Dr. Whatley. His e-mail is **mwhatley@valdosta.edu**

Transition from a Couple to a Family

Researchers disagree over whether children have a negative or positive impact on a couple's marital relationship. Some research suggests that parenthood decreases marital happiness. Bost et al. (2002) interviewed 137 couples before the birth of their first child and then at three-, twelve-, and twenty-four-month periods. The spouses consistently reported depression and adjustment through twenty-four months postpartum. Interview data from a probability sample of 3,407 white and African-American adults in twenty-one cities revealed that those who were parents were less happy in their relationships than those who did not have children (Tucker, 1997).

Twenge (2003) reviewed 148 samples representing 47,692 individuals in regard to the effect children have on marital satisfaction. They found that (1) parents (both women and men) report lower marital satisfaction than nonparents; (2) mothers of infants report the most significant drop in marital satisfaction; (3) the higher the number of children, the lower the marital satisfaction; and (4) the factors that depress marital satisfaction are conflict and loss of freedom.

For parents who experience a pattern of decreased happiness, it bottoms out during the teen years and gradually improves. When the children leave home, parents typically report increased relationship satisfaction, though it is not as high as pre-baby levels.

Previous research revealed that children neither increase nor decrease marital happiness. In a study comparing married couples who had children with those who did not, the researchers observed that, over time, the spouses in both groups reported declines in love feelings, marital satisfaction, doing things together, and positive interactions. The parents were no less happy in their marriages than the childfree couples. The researchers concluded, "the transition to parenthood is not an inescapable detriment to marital quality" (MacDermid, Huston, & McHale, 1990, 485).

Regardless of how children affect the feelings spouses have about their marriage, spouses report more commitment to their relationship once they have children (Stanley & Markman, 1992). Figure 11.1 illustrates that the more children a couple has, the more likely the couple will stay married. A primary reason for this increased commitment is the desire on the part of both parents to provide a stable family context for their children. In addition, parents of dependent children may keep their marriage together to maintain continued access to and a higher standard of living for their children. Finally, people (especially mothers) with small children feel more pressure to stay married (if the partner provides sufficient economic resources) regardless of how unhappy they may be. Hence, though children may decrease happiness, they increase stability, since pressure exists to avoid divorce.

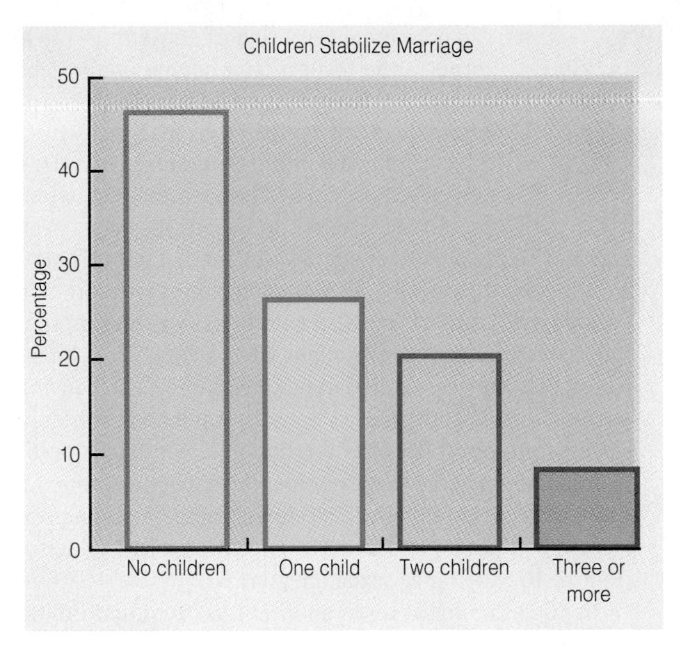

Figure 11.1
Percentage of Couples Getting Divorced by Number of Children

PERSONAL CHOICES

Should Children Bed-Share with Parents?

Frequency of bed-sharing of parents and children ranges from never to every night. Children most likely to sleep in their parents' bed are young (1–12 weeks) and breast-fed. Indeed, most of the infants (60–90%) sharing a bed with their parents do so for ease of nighttime breast-feeding. Other reasons for regular bed sharing include ideological reasons, enjoyment, and lack of space. Children who bed share with their parents on rare or specific occasions do so for reasons of illness, transient lack of space when traveling/visiting, or irritability (Ball, 2002).

One concern about bed-sharing has been the risk of Sudden Infant Death Syndrome (SIDS). Hauck et al. (2003) studied 260 SIDS deaths and found that the usual bed-sharing where one infant shares the bed with a parent is not associated with SIDS. However, where the parent slept on a sofa or where more than one child was in the bed, there was an increased risk of SIDS.

Sources

Ball, H. L. 2002. Reasons to bed-share: Why parents sleep with their infants. *Journal of Reproductive and Infant Psychology* 20:207–22.

Hauck, F. R., S. M. Herman, M. Donovan, C. M. Moore, S. Iyasu, E. Donoghue, R. H. Kirschner, and M. Willinger. 2003. Sleep environment and the risk of sudden death syndrome in an urban population: The Chicago Infant Mortality Study. *Pediatrics* 111:1207–15.

PARENTHOOD: SOME FACTS

Parenting is only one stage in an individual's or couple's life (children typically live with an individual 30% of that person's life and with a couple 40% of their marriage). It involves responding to the varying needs of children as they grow

Should the Government Control Sexual Content on the Internet?

SOCIAL POLICY

Governmental censoring of content on the Internet is the focus of an ongoing public debate that concerns protecting children from sexually explicit content on the Internet. One side of the issue is reflected in the Communications Decency Act, passed by Congress in 1996, which prohibited sending "indecent" messages over the Internet to people under age 18. But the Supreme Court struck down the law in 1997, holding that it was too broadly worded and violated free-speech rights by restricting too much material that adults might want access to. A similar ruling in 2002 has resulted in no government crackdown, even on virtual child pornography. In support of not limiting sexual content on the Internet, the American Civil Liberties Union emphasized that governmental restrictions threaten material protected by the First Amendment, including sexually explicit poetry and material educating disabled persons on how to experience sexual pleasure.

In 1998 Congress passed another law (the Child Online Protection Act), which makes it a crime to knowingly make available to people under age 17 any Web materials that, based on "contemporary community standards," are designed to pander to prurient interests. The law requires commercial operators to verify that a user is an adult through credit card information and adult access codes. Business owners who break the law are subject to a $50,000 fine and six months in jail. An inadvertent effect of the act was to require public libraries to install Internet filters to block access to objectionable sites. A U.S. appeals court and panel of federal judges have also struck down this law on the basis that it violates First Amendment rights.

In essence, many object to government control of sexual content on the Internet on the grounds of First Amendment rights. Many others believe government restrictions are necessary to protect children from inappropriate sexual content. Go to http://sociology.wadsworth.com/knox_schacht/choices8e, select the Opposing Viewpoints Resource Center on the left navigation bar and enter your pass code. Conduct a search by subject, using the search term "Censorship," and read the articles on the Viewpoints tab that support each position. Do you find yourself agreeing with one perspective more than the other? Why?

Other governments have adopted restrictive Internet policies (Casanova et al., 2001). In Singapore, the government requires Internet service providers to block access to certain Web sites that contain pornography or inflame political, religious, or racial sensitivities. China's largest service provider blocks at least one hundred sites, including *Playboy*. Germany is also moving toward government control of the Internet (Newsbytes, 2001).

In the United States, parents, not the government, will be responsible for regulating their children's use of the Internet. Software products, such as Net Nanny, Surfwatch, CYBERsitter, CyberPatrol, and Time's Up, are being marketed to help parents control what their children view on the Internet. These software programs allow parents to block unapproved Web sites and categories (such as pornography), block transmission of personal data (such as address and telephone numbers), scan pages for sexual material before they are viewed, and track Internet usage. A more cumbersome solution is to require Internet users to provide passwords or identification numbers that would verify their ages before allowing access to certain Web sites.

Another alternative is for parents to use the Internet with their children both to monitor what their children are viewing and to teach their children values about what they believe is right and wrong on the Internet. Some parents believe that children must learn how to safely surf the Internet. One parent reported that the Internet is like a busy street, and just as you must teach your children how to safely cross in traffic, you must teach them how to avoid giving information to strangers on the Internet.

Sources

Casanova, M. F., D. Solursh, L. Solursh, E. Roy, and L. Thigpen. 2001. The history of child pornography on the Internet. *Journal of Sex Education and Therapy* 25:245–51.

Newsbytes. 2001. Germans seek to centralize Internet content control. 31 August, NWSBO1, 24300e.

up, and parents require help from family and friends in rearing their children. Three additional facts of parenthood follow.

Each Child Is Unique

Children differ in their genetic makeup, physiological wiring, intelligence, tolerance for stress, capacity to learn, comfort in social situations, and interests. Parents soon become aware of the uniqueness of each child—of her or his difference from every other child they know. Parents of two or more children are often amazed at how children who have the same parents can be so different.

Children also differ in their mental and physical health. There are 5.6 million children with disabilities—over half (50.5%) of these children have learning disabilities, 11 percent mental retardation, and 8 percent serious emotional disturbance (*Statistical Abstract of the United States: 2003,* Table 261). Mental and physical disabilities of children present emotional and financial challenges to their parents. Green (2003) discussed the potential stigma associated with disability and how parents cope with their children who are stigmatized.

Although parents often contend "we treat our children equally," Tucker, McHale, and Crouter (2003) found that parents treat children differently, with firstborns usually receiving more privileges than children born later. However, the assignment of chores seems to be equal.

Parents Are Only One Influence in a Child's Development

Although parents often take the credit—and the blame—for the way their children turn out, they are only one among many influences on child development. Although parents are the first significant influence, peer influence becomes increasingly important during adolescence. Pinquart and Silbereisen (2002) studied seventy-six dyads of mothers and their 11–16-year-old adolescents and observed a decrease in connectedness adolescents had with their mothers and a movement toward their adolescent friends.

Siblings also have an important and sometimes lasting effect on each other's development. Siblings are social mirrors and models (depending on the age) for each other. They may also be sources of competition for each other and be jealous of each other.

Teachers are also a significant influence in the development of a child's values. Some parents send their children to religious schools to ensure teachers with conservative religious values. This may continue into the child's college and university education.

Media in the form of television, replete with MTV and "parental discretion advised" movies, is a major source of language, values, and lifestyles for children that may be different from those of the parents. Parents are also concerned about the violence (e.g., as seen in *The Sopranos*) that television exposes their children to.

Another influence of concern to parents is the Internet. Though parents may encourage their children to conduct research and write term papers using the Internet, they may fear the downloading of pornography and related sex sites (we discussed the degree to which the government should be involved in control of the sexual content in the Social Policy section on page 268). Parental supervision of teens on the Internet and the right of the teen for privacy remain potential conflict issues.

Parenting Styles Differ

In her classic work of the 1960s, Diana Baumrind (1966) developed a typology of parenting styles. She noted that parenting behavior has two dimensions: responsiveness and demandingness. **Responsiveness** refers to the extent to which parents respond to and meet the needs of their children. In other words, how supportive are the parents? Warmth, reciprocity, person-centered communication, and attachment are all aspects of responsiveness. **Demandingness**, on the other hand, is the manner in which parents place demands on children in regard to expectations and discipline. How much control do they exert over their children? Monitoring and confrontation and are also aspects of demandingness.

Categorizing parents in terms of their responsiveness and their demandingness creates four categories of parenting styles: permissive (also known as indulgent), authoritarian, authoritative, and uninvolved.

1. Permissive parents are high on responsiveness and low on demandingness. They are very lenient and allow their children to largely regulate their own behavior.

Diversity in the United States

Supple (2000) studied a sample of 1,206 Euro-American, African-American, and Mexican-American adolescents (ages 10–17) and compared the degree to which their parents talked over important decisions that affected them. Findings revealed more discussion among parents of African-American and Mexican-American adolescents than among Euro-American adolescents.

Diversity in Other Countries

Whereas American parents are the primary disciplinarians of their children, in Asian families the extended family kin may be more involved (Fontes, 1998).

2. ***Authoritarian parents are high on demandingness and low in responsiveness.*** They feel that children should obey their parents no matter what and they provide a great deal of structure in the child's world.

3. ***Authoritative parents are both demanding and responsive.*** They impose appropriate limits on their children's behavior but emphasize reasoning and communication. This style offers a balance of warmth and control.

4. ***Uninvolved parents are low in responsiveness and demandingness.*** These parents are not invested in their children's lives.

Research suggests that authoritative parenting results in more positive child outcomes, including higher levels of social competence and lower levels of problem behavior (Darling, 1999). Amato and Fowler (2002) also noted that a family context of "support, monitoring, and avoidance of harsh punishment" independent of race, ethnicity, family structure, education, income, and gender are the crucial elements of a parenting style that has positive outcomes for children. Hence, as long as these elements are present, the style of parenting is less important and children are more likely to flourish.

PRINCIPLES OF EFFECTIVE PARENTING

Numerous principles are involved in being an effective parent. Some of these follow.

This loving father has made it a point to share fun activities with his children.

Authors

Give Time, Love, Praise, and Encouragement

"Nine years after your death—today . . . I remember you and I remember me. But I don't remember us," wrote Louie Anderson (1989, 8), comedian. His words reflect the longing for his alcoholic father's time and love—the absence of which contributed to Anderson's obesity. "As you drank and got drunk, I ate and got big" (p. 23).

Since children depend first on their parents for the development of their sense of emotional security, it is critical that parents provide a warm emotional context in which the children can develop. Feeling loved as an infant also affects one's capacity to become involved in adult love relationships. The effects "exist well below the threshold of awareness, are neurological and psychological in nature, and are influential in relationships throughout the life span" (Lundeen, 1999, 103).

As the child matures, positive reinforcement for prosocial behavior also helps to encourage desirable behavior and a positive self-concept. Instead of focusing only on correcting or reprimanding bad behavior, parents should frequently comment on and reinforce good behavior. Comments like "I like the way you shared your toys," "You asked so politely," and "You did such a good job cleaning your room" help to reinforce positive social behavior and may enhance a child's self-concept. However, parents need to be careful not to overpraise their children, as too much praise may lead to the child's striving to please others rather than trying to please himself or herself.

Praise focuses on other people's judgments of a child's actions, whereas encouragement focuses more on the child's efforts. For example, telling a child who brings you his painting, "I love your picture; it is the best one that I have ever seen" is not as effective in building the child's confidence as saying, " You worked really hard on your painting. I notice that you used lots of different colors."

PERSONAL CHOICES

Should Parents Reward Positive Behavior?

Most parents agree that some form of punishment is necessary to curb a child's inappropriate behavior, but there is disagreement over whether positive behavior (taking out the trash, cleaning one's room, making good grades) should be rewarded by praise, extra privileges, or money. Some parents feel that children should "do the right things anyway" and that to reward them is to bribe them. One parent said, "My children are going to do what I say because I say so, not because I am going to give them something for doing it."

Other parents feel that both the child and the parent benefit when the parents reward the child for good behavior. Rewarding a child for a behavior will likely result in the child's engaging in that behavior more often and developing a set of positive behaviors. The parents, in turn, feel good about the child.

There is also concern among some professionals that parents praise their children too much. They feel that children develop an inflated ego and do not learn how to be realistic about themselves and to cope with failure.

Monitor Child's Activities

As noted in Chapter 9, Communication in Relationships, 95 percent of the university students in one study reported having lied to their parents as a teenager (Knox et al., 2001). Where they were, whom they were with, alcohol, and sex were issues about which they reported having told lies to their parents the most. Abundant research suggests that parents who monitor their children—know where their children are, who they are with, etc.—are less likely to report that their adolescents are involved in delinquent behavior and drinking alcohol (Laird et al., 2003; Longmore, Manning, & Giordano, 2001), poor academic performance (Jacobson & Crockett, 2000), and sexual activity (Hollander, 2003).

Set Limits and Discipline Children for Inappropriate Behavior

The goal of guidance is self-control. Parents want their children to be able to control their own behavior and to make good decisions without their parents. Guidance may involve reinforcing desired behavior or providing limits to children's behavior. This sometimes involves disciplining children for negative behavior. Unless parents provide negative consequences for lying, stealing, and hitting, children can grow up to be dishonest, to steal, and to be inappropriately aggressive.

Time out (a noncorporal form of punishment that involves removing the child from a context of reinforcement to a place of isolation for one minute for each year of the child's age) has been shown to be an effective consequence for inappropriate behavior. Withdrawal of privileges (watching television, playing with friends), pointing out the logical consequences of the misbehavior ("you were late; we won't go"), and positive language ("I know you meant well but . . .") are also effective methods of guiding children's behavior.

Physical punishment is less effective in reducing negative behavior; it teaches the child to be aggressive and encourages negative emotional feelings toward the parents. When using time out or the withdrawal of privileges, parents should make it clear to the child that they disapprove of the child's behavior, not the child. Some evidence suggests that consistent discipline has positive outcomes for children. Lengua et al. (2000) studied 231 mothers of 9- to 12-year-olds and found that inconsistent discipline was related to adjustment problems, particularly for children high in impulsivity.

PERSONAL CHOICES

Should Parents Use Corporal Punishment?

Parents differ in the type of punishment they feel is appropriate for children. Some parents—particularly religiously conservative parents (Day, Peterson, & McCracken, 1998) or right-wing authoritarian parents (Danso, Hunsberge, & Pratt, 1997)—use corporal punishment as a means of disciplining their children. Half of American children report having experienced corporal punishment in early adolescence (Straus, 1994). Adolescent boys are more likely than girls to report that their parents use corporal punishment (Hay, 2003).

Other parents, particularly older parents (Day, Peterson, & McCracken, 1998), feel that corporal punishment is unnecessary or wrong and elect to put their children in time out. Types of punishment also differ according to race, with more African-Americans than whites reporting support for spanking (Flynn, 1998).

The decision to choose a corporal or noncorporal method of punishment should be based on the consequences of use. In general, the use of time out and withholding of privileges seems to be more effective than corporal punishment in stopping undesirable behavior (Straus, 1994). Though beatings and whippings will temporarily decrease the negative verbal and nonverbal behaviors, they have major side effects. First, punishing children by inflicting violence teaches them that it is acceptable to physically hurt someone you love. Hence, parents may be inadvertently teaching their children to use violence in the family. Second, parents who beat their children should be aware that they are teaching their children to fear and avoid them. Third, children who grow up in homes in which corporal punishment is used are more likely to be aggressive and disobedient (Straus, 1994). Recognizing the negative consequences of corporal punishment, the law in Sweden forbids parents to spank their children.

So what kind of discipline is best? A team of researchers found that parents who reasoned with their children and then backed up their reasoning with punishment reported the most behavior change as compared with parents who just used reasoning or punishment (Larzelere et al., 1998). Their data suggest that words backed up by some type of consequence have the most desirable outcome.

A review of some of the alternatives to corporal punishment include the following:*

*1. **Be a positive role model.*** Children learn behaviors by observing their parents' actions, so parents must model the ways in which they want their children to behave. If a parent yells or hits, the child is likely to do the same.

*2. **Set rules and consequences.*** Make rules that are fair, realistic, and appropriate to a child's level of development. Explain the rules and the consequences of not following them. If children are old enough, they can be included in establishing the rules and consequences of breaking them.

*3. **Encourage and reward good behavior.*** When children are behaving appropriately, give them verbal praise and occasionally reward them with tangible objects, privileges, or increased responsibility.

*Based on National Mental Health Association, 2003. Effective discipline techniques for parents: Alternatives to spanking. *Strengthening Families Fact Sheet.* www.nmha.org

4. Use charts. Charts to monitor and reward behavior can help children learn appropriate behavior. Charts should be simple and focus on one behavior at a time, for a certain length of time.

5. Use time out. "Time out" involves removing children from a situation following a negative behavior. This can help children calm down, end the inappropriate behavior, and re-enter the situation in a positive way. Explain what the inappropriate behavior is, why the time out is needed, when it will begin, and how long it will last. Set an appropriate length of time for the time out based on age and level of development, usually one minute for each year of the child's age.

Sources

Danso, H., B. Hunsberger, and M. Pratt. 1997. The role of parental religious fundamentalism and right-wing authoritarianism child-rearing goals and practices. *Journal for the Scientific Study of Religion* 36:496–502.

Day, R. D., G. W. Peterson, and C. McCracken. 1998. Predicting spanking of younger and older children by mothers and fathers. *Journal of Marriage and the Family* 60:79–94.

Flynn, C. P. 1998. To spank or not to spank: The effect of situation and age of child on support for corporal punishment. *Journal of Family Violence* 13:21–37.

Hay, C. 2003. Family strain, gender, and delinquency. *Sociological Perspectives* 46:107–35.

Larzelere, R. E., P. R. Sather, W. N. Schneider, D. B. Larson, and P. L. Pike. 1998. Punishment enhances reasoning's effectiveness as a disciplinary response to toddlers. *Journal of Marriage and the Family* 60:388–403.

Straus, M. A. 1994. *Beating the devil out of them: Corporal punishment in American families.*, San Francisco: Jossey-Bass.

Provide Security

Predictable responses from parents, a familiar bedroom or playroom, and an established routine help to encourage a feeling of security in children. Security provides children with the needed self-assurance to venture beyond the family. If the outside world becomes too frightening or difficult, a child can return to the safety of the family for support. Knowing it is always possible to return to an accepting environment enables a child to become more involved with the world beyond the family.

One of the ways parents might consider providing security for their children is to nurture their own relationship. Harry and Ainslie (1998) found that lower marital discord was associated with higher-quality parent-child relationships and lower levels of child aggression.

Encourage Responsibility

Giving children increased responsibility encourages the autonomy and independence they need to be assertive and independent. Giving children more responsibility as they grow older can take the form of encouraging them to choose healthy snacks and letting them decide what to wear and when to return from playing with a friend (of course, the parents should praise appropriate choices). Children who are not given any control and responsibility for their own lives remain dependent on others. Successful parents can be defined in terms of their ability to rear children who can function as independent adults. A dependent child is a vulnerable child.

Provide Sex Education

Parents are a powerful influence on the sexual behavior of their children. Although they are reluctant to discuss safe sex, their doing so often has positive consequences. In a sample of 237 Australian university students aged 16–19, researchers Troth and Peterson (2000) found that neither fathers nor mothers engaged in any substantial amount of education or communication with their offspring about the topic of safe sex. However, when parents do have such discussions, which are interpreted by their offspring as emphasizing that safe sex is important, the adolescents are more likely to follow through behaviorally by implementing frequent condom use. Remez (2003) found that mothers have a greater effect on daughters than sons and that mothers' disapproval of their daughters' having sex and satisfaction with the mother-daughter relationship were associated with reduced risk of their daughters' becoming sexually active early in adolescence.

If parents have the goal of delaying the first intercourse experience of their children, they might consider taking the children to church. Karnehm (2000) studied the sexual debut of 815 children of relatively young mothers and found that children who reported at least monthly church attendance with parents at age 10 or 11 were more likely to delay their first sexual intercourse experience until at least age 16.

Express Confidence

"One of the greatest mistakes a parent can make," confided one mother, "is to be anxious all the time about your child, because the child interprets this as your lack of confidence in his or her ability to function independently." Rather, this mother noted that it is best to convey to the child that you know that he or she will be all right and that you are not going to worry about the child because you have confidence in him or her. "The effect on the child," said this mother, "is a heightened sense of self-confidence." Another way to conceptualize this parental principle is to think of the self-fulfilling prophecy as a mechanism that facilitates self-confidence. If the parents show the child that they have confidence in him or her, the child begins to accept these social definitions as real and becomes more self-confident.

Respond to the Teen Years Creatively

Parenting teenage children presents challenges that differ from those in parenting infants and young children. The teenage years have been characterized as a time when adolescents defy authority, act rebellious, and search for their own identity. Teenagers today are no longer viewed as innocent, naive children.

Conflicts between parents and teenagers often revolve around money and independence. The desires for a cell phone, DVD player, and plasma TV can outstrip the budget of many parents. And teens increasingly want more freedom. But neither of these issues need necessarily result in conflicts. And when they do, the effect on the parent-child relationship may be inconsequential. One parent tells his children, "I'm just being the parent, and you're just being who you are; it is OK for us to disagree—but you can't go." The following suggestions can help to keep conflicts with teenagers at a low level.

1. Catch them doing what you like rather than criticizing them for what you don't like. Adolescents are like everyone else—they don't like to be criticized but they do like to be noticed for what they do that is good.

2. Be direct when necessary. Though parents may want to ignore some behaviors of their children, addressing some issues directly may also be effective. Regarding the avoidance of STD/HIV infections, Dr. Louise Sammons (2003) tells her teenagers, "It is utterly imperative to require that any potential sex partner produce a certificate indicating no STDs or HIV infection and to require that a condom or dental dam be used before intercourse or oral sex." Such direct communication is rare. In a study of 163 functional families, the researchers noted that sensitive topics such as sex, alcohol, smoking, and drugs were rarely discussed (Riesch et al., 2000).

*3. **Provide information rather than answers.*** When teens are confronted with a problem, try to avoid making a decision for them. Rather, it is helpful to provide information on which they may base a decision. What courses to take in high school and what college to apply for are decisions that might be made primarily by the adolescent. The role of the parent might best be that of providing information or helping the teenager obtain information.

*4. **Be tolerant of high activity levels.*** Some teenagers are constantly listening to loud music, going to each other's homes, and talking on cell phones for long periods of time. Parents often want to sit in their easy chairs and be quiet. Recognizing that it is not realistic to expect teenagers to be quiet and sedentary may be helpful in tolerating their disruptions.

*5. **Engage in some activity with your teenagers.*** Whether it is renting a video, eating a pizza, or taking a camping trip, it is important to structure some activities with your teenagers. Such activities permit a context in which to communicate with them.

Sometimes teenagers present challenges with which the parents feel unable to cope. Aside from monitoring their behavior closely, family therapy may be helpful. A major focus of such therapy is to increase the emotional bond between the parents and the teenagers and to encourage positive consequences (e.g., concert tickets) for desirable behavior (e.g., good grades) and negative consequences (loss of car privileges) for undesirable behavior (e.g., getting a speeding ticket).

GAY PARENTING ISSUES

Bigner (in press) identified several issues unique to gay parents. These include the following:

*1. **Identity issues.*** The degree to which the gay individual is comfortable with his or her own identity and coming out to children.

*2. **Concerns about parenting effectiveness.*** Gays sometimes question their ability to parent effectively. Research consistently shows that sexual orientation is an irrelevant variable in parental effectiveness.

*3. **New intimate relationships.*** For the individual who has recently come out to self and others, there are general issues of how to be involved in a same-sex intimate relationship and how to make children aware of such relationships.

*4. **Boundary issues.*** Some gays isolate themselves from the gay community out of fear of being found out and losing custody.

Bigner (in press) summarized the needs a family therapist might address with gay parents:

- *Addressing intrapsychic disturbances relating to coming-out issues*
- *Helping clients to resolve internalized homophobic issues and facilitating the coming out process to self, ex-spouses, and children*
- *Helping clients to achieve a healthy new personal identity as a gay father or lesbian mother*
- *Directing and encouraging client's socialization into the new world of homosexuality*
- *Assisting clients in addressing intimate relationship issues*
- *Assisting clients to resolve developmental tasks of queer stepfamily formation and functioning*

Even if she knew her father was gay, she still loved him.

Of Jennifer Grant (daughter), Charles Higham and Roy Moseley in their biography, *Cary Grant: The Lonely Heart*

APPROACHES TO CHILDREARING

Parents are offered advice from professionals, their own parents, siblings, and friends. Early parenting education efforts were directed solely at women. Today, parenting information is targeted to fathers as well as mothers and is readily available to all parents. Sometimes, however, there is conflicting information that can be confusing to parents. Adding to the confusion is the fact that parenting advice may change from time to time. For example, parents in 1914 who wanted to know what to do about their child's thumb sucking were told to try and control such a bad impulse by pinning the sleeves of the child to the bed if necessary. Today, parents are told that thumb sucking meets an important psychological need for security and they should not try to prevent it. If the child's teeth become crooked as a result, an orthodontist should be consulted.

There are several theoretical approaches to rearing children. A review of these approaches can be found in Table 11.2. In examining these approaches, it is important to keep in mind that no single approach is superior to another. What works for one child may not work for another. Any given approach may not even work with the same child at two different times. In addition, parents and the professional caregivers of children often differ in regard to childrearing approaches. It is important for new parents to use a cafeteria approach when examining parenting advice from health care providers, family members, friends, parenting educators, and other well-meaning individuals. Take what makes sense, works, and feels right as a parent and leave the rest behind. Parents know their own child better than anyone else and should be encouraged to combine different approaches to find what works best for them and their unique child.

Developmental-Maturational Approach

For the past sixty years, Arnold Gesell and his colleagues at the Yale Clinic of Child Development have been known for their ages-and-stages approach to childrearing (Gesell, Ilg, & Ames, 1995). The **developmental-maturational approach** has been widely used in the United States. The basic perspective, some considerations for childrearing, and some criticisms of the approach follow.

Basic Perspective Gesell views what children do, think, and feel as being influenced by their genetic inheritance. Although genes dictate the gradual unfolding of a unique person, every individual passes through the same basic pattern of growth. This pattern includes four aspects of development: motor behavior (sitting, crawling, walking), adaptive behavior (picking up objects and walking around objects), language behavior (words and gestures), and personal-social behavior (cooperativeness and helpfulness). Through the observation of hundreds of normal infants and children, Gesell and his coworkers have identified norms of development. Although there may be large variations, these norms suggest the ages at which an average child displays various behaviors. For example, on the average, children begin to walk alone (although awkwardly) at age 13 months and use simple sentences between the ages of 2 and 3. Jenkins and Buccioni (2000) also observed that 9-year-old children are capable of understanding that parental conflict may emanate from divergent goals, whereas 5-year-old children are incapable of this conceptualization.

Considerations for Childrearing Gesell suggested that if parents are aware of their children's developmental clock, they will avoid unreasonable expectations. For example, a child cannot walk or talk until the neurological structures necessary for those behaviors have matured. Also, the hunger of a 4-week-old must be im-

Table 11.1 Theories of Childrearing

Theory	Major Contributor	Basic Perspective	Focal Concerns	Criticisms
Developmental-Maturational	Arnold Gesell	Genetic basis for child passing through predictable stages	Motor behavior Adaptive behavior Language behavior Social behavior	Overemphasis on biological clock Inadequate sample to develop norms Demand schedule questionable Upper-middle-class bias
Behavioral	B. F. Skinner	Behavior is learned through operant and classical conditioning	Positive reinforcement Negative reinforcement Punishment Extinction Stimulus response	Deemphasis on cognitions of child Theory too complex to be accurately/appropriately applied by parent Too manipulative/controlling Difficult to know reinforcers and punishers in advance
Parent Effectiveness Training	Thomas Gordon	The child's worldview is the key to understanding the child	Change the environment before attempting to change the child's behavior Avoid hurting the child's self-esteem Avoid win-lose solutions	Parents must cometimes impose their will on the child's How to achieve win-win solutions is not specified
Socioteleological	Alfred Adler	Behavior is seen as attempt of child to secure a place in the family	Insecurity Compensation Power Revenge Social striving Natural consequences	Limited empirical support Child may be harmed taking "natural consequences"
Attachment	William Sears	Goal is to establish a firm emotional attachment with child	Connecting with baby Responding to cues Breast-feeding Wearing baby Sharing sleep	May result in spoiled, overly dependent child Exhausting for parents

mediately appeased by food, but at 16 to 28 weeks, the child has some capacity to wait because the hunger pains are less intense. In view of this and other developmental patterns, Gesell suggested that the infant's needs be cared for on a demand schedule; instead of having to submit to a schedule imposed by parents, infants are fed, changed, put to bed, and allowed to play when they want. Children are likely to be resistant to a hard-and-fast schedule because they may be developmentally unable to cope with it.

In addition, Gesell emphasized that parents should be aware of the importance of the first years of a child's life. In Gesell's view, these early years assume

The developmental maturational approach encourages parents to feed their babies on demand—when the babies are hungry, not when it is convenient for the parents to feed them.

the greatest significance because the child's first learning experiences occur during this period.

Criticisms of the Developmental-Maturational Approach Gesell's work has been criticized because of (1) its overemphasis on the idea of a biological clock, (2) the deficiencies of the sample he used to develop maturational norms, (3) his insistence on the merits of a demand schedule, and (4) the idea that environmental influences are weak.

Most of the children who were studied to establish the developmental norms were from the upper middle class. Children in other social classes are exposed to different environments, which influence their development. So norms established on upper-middle-class children may not adequately reflect the norms of children from other social classes.

Gesell's suggestion that parents do everything for the infant when the infant wants has also been criticized. Rearing an infant on the demand schedule can drastically interfere with the parents' personal and marital interests. In the United States, with its emphasis on individualism, many parents feed their infants on a demand schedule but put them to bed to accommodate the parents' schedule.

Behavioral Approach

The **behavioral approach** to childrearing, also known as the social learning approach, is based on the work of B. F. Skinner (Skinner et al., 1997). Behavioral approaches to child behavior have received the most empirical study. Public health officials and policymakers are encouraged to promote programs with empirically supported treatments. We now review the basic perspective, considerations, and criticisms of this approach to childrearing.

Basic Perspective Behavior is learned through classical and operant conditioning. Classical conditioning involves presenting a stimulus with a reward. For example, infants learn to associate the faces of their parents with food, warmth, and comfort. Although initially only the food and feeling warm will satisfy the infant, later just the approach of the parent will soothe the infant. This may be observed when a father hands his infant to a stranger. The infant may cry because the stranger is not associated with pleasant events. But when the stranger hands the infant back to the parent, the crying may subside because the parent represents positive events and the stimulus of the parent's face is associated with pleasurable feelings and emotional safety.

Other behaviors are learned through operant conditioning, which focuses on the consequences of behavior. Two principles of learning are basic to the operant explanation of behavior—reward and punishment. According to the reward principle, behaviors that are followed by a positive consequence will increase. If the goal is to teach the child to say "please," doing something the child likes after he or she says "please" will increase the use of "please" by the child. Rewards may be in the form of attention, praise, desired activities, or privileges. Whatever consequence increases the frequency of an occurrence is, by definition, a reward. If a particular reward doesn't change the behavior in the desired way, a different reward needs to be tried.

The punishment principle is the opposite of the reward principle. A negative consequence following a behavior will decrease the frequency of that behavior;

for example, the child could be isolated for five or ten minutes following an undesirable behavior. The most effective way to change behavior is to use the reward and punishment principles together to influence a specific behavior. Praise children for what you want them to do and provide negative consequences for what they do that you do not like.

Considerations for Childrearing Parents often ask, "Why does my child act this way, and what can I do to change it?" The behavioral approach to childrearing suggests the answer to both questions. The child's behavior has been learned through being rewarded for the behavior, and it can be changed by eliminating the reward for or punishing the undesirable behavior and rewarding the desirable behavior.

The child who cries when the parents are about to leave home to go to dinner or see a movie is often reinforced for crying by the parents' staying home longer. To teach the child not to cry when the parents leave, the parents should reward the child for not crying when they are gone for progressively longer periods of time. For example, they might initially tell the child they are going outside to walk around the house and they will give the child a treat when they get back if he or she plays until they return. The parents might then walk around the house and reward the child for not crying. If the child cries, they should be out of sight for only a few seconds and gradually increase the amount of time they are away. The essential point is that children learn to cry or not to cry depending on the consequences of crying. Because children learn what they are taught, parents might systematically structure learning experiences to achieve specific behavioral goals.

Criticisms of the Behavioral Approach Professionals and parents have attacked the behavioral approach to childrearing on the basis that it is deceptively simple and does not take cognitive issues into account. Although the behavioral approach is often presented as an easy-to-use set of procedures for child management, many parents do not have the background or skill to implement the procedures effectively. What constitutes an effective reward or punishment, how it should be presented and in what situation, with what child, to influence what behavior are all decisions that need to be made before attempting to increase or decrease the frequency of a behavior. Parents often do not know the questions to ask or lack the training to make appropriate decisions in the use of behavioral procedures. One parent locked her son in the closet for an hour to punish him for lying to her a week earlier—a gross misuse of learning principles.

Behavioral childrearing has also been said to be manipulative and controlling, thereby devaluing human dignity and individuality. Some professionals feel that humans should not be manipulated to behave in certain ways through the use of rewards and punishments.

Finally, the behavioral approach has been criticized because it deemphasizes the influence of thought processes on behavior. Too much attention, say the critics, has been given to rewarding and punishing behavior and not enough attention has been given to how the child perceives a situation. For example, parents might think they are rewarding a child by giving her or him a bicycle for good behavior. But the child may prefer to upset the parents by rejecting the bicycle and may be more rewarded by their anger than by the gift.

Parent Effectiveness Training Approach

Thomas Gordon (2000) developed a model of childrearing based on **parent effectiveness training** (PET).

Basic Perspective Parent effectiveness training focuses on what children feel and experience in the here and now—how they see the world. The method of trying to understand what the child is experiencing is active listening, in which the parent reflects the child's feelings. For example, the parent who is told by the child, "I want to quit taking piano lessons because I don't like to practice" would reflect, "You're really bored with practicing the piano and would rather have fun doing something else."

PET also focuses on the development of the child's positive self-concept. To foster a positive self-concept in their child, parents should reflect positive images to the child—letting the child know he or she is liked, admired, and approved of.

Considerations for Childrearing To assist in the development of a child's positive self-concept and in the self-actualization of both children and parents, Gordon recommended managing the environment rather than the child, engaging in active listening, using "I" messages, and resolving conflicts through mutual negotiation. An example of environmental management is putting breakables out of reach of young children. Rather than worry about how to teach children not to touch breakable knickknacks, it may be easier to simply move the items out of the children's reach.

The use of active listening becomes increasingly important as the child gets older. When Joanna is upset with her teacher, it is better for the parent to reflect the child's thoughts than to take sides with the child. Saying "You're angry that Mrs. Jones made the whole class miss play period because Becky was chewing gum" rather than saying "Mrs. Jones was unfair and should not have made the whole class miss play period" shows empathy with the child without blaming the teacher.

Gordon also suggested using "I" rather than "you" messages. Parents are encouraged to say "I get upset when you're late and don't call" rather than "You're an insensitive, irresponsible kid for not calling me when you said you would." The former avoids damaging the child's self-concept but still expresses the parent's feelings and encourages the desired behavior.

Gordon's fourth suggestion for parenting is the no-lose method of resolving conflicts. Gordon rejects the use of power by parent or child. In the authoritarian home, the parent dictates what the child is to do and the child is expected to obey. In such a system, the parent wins and the child loses. At the other extreme is the permissive home, in which the child wins and the parent loses. The alternative, recommended by Gordon, is for the parent and the child to seek a solution that is acceptable to both and to keep trying until they find one. In this way, neither parent nor child loses and both win.

Criticisms of the Parent Effectiveness Training Approach Although much is commendable about PET, parents may have problems with two of Gordon's suggestions. First, he recommends that because older children have a right to their own values, parents should not interfere with their dress, career plans, and sexual behavior. Some parents may feel they do have a right (and an obligation) to "interfere."

Second, the no-lose method of resolving conflict is sometimes unrealistic. Suppose a 16-year-old wants to spend the weekend at the beach with her boyfriend and her parents do not want her to. Gordon advises negotiating until a decision is reached that is acceptable to both. But what if neither the daughter nor the parents can suggest a compromise or shift their position? The specifics of how to resolve a particular situation are not always clear.

Socioteleological Approach

Alfred Adler, a physician and former student of Sigmund Freud, saw a parallel between psychological and physiological development. When a person loses her or his sight, the other senses (hearing, touch, taste) become more sensitive—they compensate for the loss. According to Adler, the same phenomenon occurs in the psychological realm (Adler, 1992). When individuals feel inferior in one area, they will strive to compensate and become superior in another. Rudolph Dreikurs, a student of Adler, has developed an approach to childrearing that alerts parents as to how their children might be trying to compensate for feelings of inferiority (Soltz & Dreikurs, 1991). Dreikurs's **socioteleological approach** is based on Adler's theory.

Basic Perspective According to Adler, it is understandable that most children feel they are inferior and weak. From the child's point of view, the world is filled with strong giants who tower above him or her. Because children feel powerless in the face of adult superiority, they try to compensate by gaining attention (making noise, becoming disruptive), exerting power (becoming aggressive, hostile), seeking revenge (becoming violent, hurting others), and acting inadequate (giving up, not trying). Adler suggested that such misbehavior is evidence that the child is discouraged or feels insecure about her or his place in the family. The term *socioteleological* refers to social striving or seeking a social goal. In the child's case, the goal is to find a secure place within the family—the first "society" the child experiences.

Considerations for Childrearing When parents observe misbehavior in their children, they should recognize it as an attempt to find security. According to Dreikurs, parents should not fall into playing the child's game by, say, responding to a child's disruptiveness with anger but should encourage the child, hold regular family meetings, and let natural consequences occur. To encourage the child, the parents should be willing to let the child make mistakes. If John wants to help Dad carry logs to the fireplace, rather than saying, "You're too small to carry the logs," Dad should allow John to try and should encourage him to carry the size limb or stick that he can manage. Furthermore, Dad should praise John for his helpfulness.

As well as being constantly encouraged, the child should be included in a weekly family meeting. During this meeting, such family issues as bedtimes, the appropriateness of between-meal snacks, assignment of chores, and family fun are discussed. The meeting is democratic; each family member has a vote. Participation in family decision making is designed to enhance the self-concept of each child. By allowing each child to vote on family decisions, parents respect the child as a person as well as the child's needs and feelings.

Resolutions to conflicts with the child might also be framed in terms of choices the child can make. "You can go outside and play only in the backyard, or you can play in the house" gives the child a choice. If the child strays from the backyard, he or she can be brought in and told, "You can try again later." Such a framework teaches responsibility for and consequences of one's choices.

Finally, Driekurs suggested that the parents let natural consequences occur for their child's behavior. If a daughter misses the school bus, she walks or is charged taxi fare out of her allowance. If she won't wear a coat and boots in bad weather, she gets cold and wet. Of course, parents are to arrange logical consequences when natural consequences will not occur or would be dangerous if they did. For example, if a child leaves the television on overnight, access might be taken away the next day or so.

Criticisms of the Socioteleological Approach The socioteleological approach is sometimes regarded as impractical, since it teaches the importance of letting children take the natural consequences for their actions. Such a principle may be interpreted to let the child develop a sore throat if he or she wishes to go out in the rain without a raincoat. In reality, advocates of the method would not let the child make a dangerous decision. Rather, they would give the child a choice with a logical consequence such as "You can go outside wearing a raincoat, or you can stay inside—it is your choice."

Attachment Parenting

Dr. William Sears along with his wife, Martha Sears, developed an approach to parenting called attachment parenting (Sears & Sears, 1993). This "commonsense parenting" approach focuses on parents connecting with their baby.

Basic Perspective Sears identified three parenting goals: to know your child, to help your child feel right, and to enjoy parenting. He also suggested five concepts or tools (identified in the following section) that comprise attachment parenting that will help parents to achieve these goals. Overall, the ultimate goal is for parents to get connected with their baby. Once parents are connected, it is easy for parents to figure out what works for them and to develop a parenting style that fits them and their baby. Meeting a child's needs early in life will help him or her form a secure attachment with parents. This secure attachment will help the child to gain confidence and independence as he or she grows up.

Considerations for Childrearing The first attachment tool is for parents to connect with their baby early. The initial months of parenthood are a sensitive time for bonding with your baby and starting the process of attachment.

The second tool is to read and respond to the baby's cues. Parents should spend time getting to know their baby and learn to recognize his or her unique cues. Once a parent gets in tune with the baby's cues, it is easy to respond to the child's needs. Sears encourages parents to be open and responsive and to pick their baby up if he or she cries. Responding to a baby's cries helps the baby to develop trust and encourages good communication between child and parent. Eventually, babies who are responded to will internalize their security and will not be as demanding.

The third attachment tool is for mothers to breast-feed their babies and to do this on demand rather than trying to follow a schedule. He emphasizes the important role that fathers also play in successful breast-feeding by helping to create a supportive environment.

The fourth concept of attachment parenting is for parents to wear their baby by using a baby sling or carrier. The closeness is good for the baby and it makes life easier for the parent. Wearing your baby in a sling or carrier allows parents to engage in regular day-to-day activities and makes it easier to leave the house.

Finally, Sears advocates that parents let the child sleep in their bed with them since it allows parents to stay connected with their child throughout the night. However, some parents and babies often sleep better if the baby is in a separate crib, and Sears recognizes that wherever parents and their baby sleep best is best. Wherever you choose to have your baby sleep, Sears is clear on one thing—it is never acceptable to let your baby cry when he or she is going to sleep! Parents need to parent their child to sleep rather than leaving him or her to cry.

Criticisms of Attachment Parenting Some parents feel that responding to their baby's cries, carrying or wearing their baby, and sharing sleep with their baby will lead to a spoiled baby who is overly dependent. Some parents may feel more tied down using this parenting approach and may find it difficult to get their child on a schedule. Many women return to work after the baby is born and find some of the concepts difficult to follow. Some women choose not to breast-feed their children for a variety of reasons. Finally, the idea of sharing sleep has resulted in a lot of criticism. Some parents might be nervous that they might roll over on the child, that the child might disturb their sleep or intimacy with their partner, or that sharing a bed will mean that they will never get their child to sleep on his or her own. However, children who grow up in an emotionally secure environment and have strong attachment report less behavioral and substance abuse problems as adolescents (Elgar et al., 2003).

SUMMARY

What are the basic roles of parents?

Parenting includes providing physical care for children, loving them, being an economic resource, providing guidance as a teacher/model, and protecting them from harm.

What is a choices perspective of parenting?

Although both genetic and environmental factors are at work, it is the choices parents make that have a dramatic impact on their children. Parents who don't make a choice about parenting have already made one. The five basic choices parents make include deciding whether to have a child, deciding the number of children, deciding the interval between children, deciding one's method of discipline and guidance, and deciding the degree to which one will be invested in the role of parent.

What is the transition to parenthood like for women, men, and couples?

Transition to parenthood refers to that period of time from the beginning of pregnancy through the first few months after the birth of a baby. The mother, father, and couple all undergo changes and adaptations during this period. Most mothers relish their new role; some may experience the transitory feelings of baby blues; a few report postpartum depression.

The father's involvement with his children is sometimes predicted by the quality of the parents' romantic relationship. If the father is emotionally and physically involved with the mother, he is more likely to take an active role in the child's life. In recent years there has been a renewed cultural awareness of fatherhood.

Researchers disagree over whether children increase or decrease marital happiness. Though they may decrease happiness, they increase the likelihood that the couple will stay together.

What are several facts about parenthood?

Parenthood will involve about 40 percent of the time a couple live together, parents are only one influence on their children, each child is unique, and parenting styles differ. Research suggests that support, monitoring, and avoidance of harsh punishment are the crucial elements of a parenting style that has positive outcomes for children.

What are some of the principles of effective parenting?

Giving time, love, praise, and encouragement; monitoring the activities of one's child; setting limits; encouraging responsibility; and providing sexuality education are aspects of effective parenting.

What are five theoretical approaches to childrearing?

There are several approaches to childrearing, including the developmental-maturational approach (children are influenced by their genetic inheritance), behavioral approach (consequences influence the behaviors children learn), socioteleological approach (children seek to gain attention from parents to overcome their feelings of powerlessness), and attachment parenting (the most important goal of parents is to become emotionally attached to their children).

KEY TERMS

baby blues
behavioral approach
demandingness
developmental-maturational training approach
oxytocin
parent effectiveness training
parenting
postpartum depression
responsiveness
socioteleological approach
time out
transition to parenthood

RESEARCHING MARRIAGE AND THE FAMILY WITH INFOTRAC COLLEGE EDITION

InfoTrac College Edition, an online library, allows you to perform research online anywhere, anytime. Following are two suggested search terms and related questions to help you extend your understanding of the topics covered in this chapter. Go to www.infotrac-college.com to begin your search.

Keyword: **Discipline.** Locate articles that discuss corporal punishment as a method of child discipline. What are arguments for and against the use of corporal punishment?

Keyword: **Sex education.** Locate articles that discuss sex education in the public school system. What is the content of such education and what should the schools be providing?

The Companion Web Site for Choices in Relationships: An Introduction to Marriage and the Family, Eighth Edition

http://sociology.wadsworth.com/knox_schacht/choices8e

Supplement your review of this chapter by going to the companion Web site to take one of the Tutorial Quizzes, use the flash cards to master key terms, and check out the many other study aids you'll find there. You'll also find special features such as the Marriage and Family Resource Center, Census 2000 information, and other data and resources at your fingertips to help you with that special project or to do some research on your own.

WEBLINKS

Children, Youth and Family Consortium
http://www.cyfc.umn.edu/

National Center for Missing & Exploited Children
http://www.missingkids.org

The Children's Partnership Online
http://www.childrenspartnership.org

Mayberry USA
http://www.mbusa.net/

Parenthood.com
http://www.parenthoodweb.com/

Authors

CHAPTER 12

I learned that happiness on earth ain't just for high achievers.

Brooks and Dunn, Red Dirt Road

Balancing Work and Family Life

Contents

True or False?

1. A cultural reversal is underway whereby employed mothers are experiencing more satisfaction with employment than with parenting; they would rather be at work than be with family.
2. Employment of the wife in an unhappy marriage increases her risk of ending the marriage.
3. Men report a greater financial motivation for attending college than women.
4. The workweek in the United States is shorter than those in European countries such as France.
5. Day care for infants in terms of hygiene and safety is significantly better than day care for children at later ages.

Answers: **1.** F **2.** T **3.** T **4.** F **5.** F

Jane Swift became governor of Massachusetts in mid-2001. At the time she was pregnant with twins. Her husband and father of their existing child chose to stay in western Massachusetts, which resulted in his wife's commuting three hours (one way) back and forth to Boston. Their situation gave high visibility to the issue of balancing work and family life and the respective gender expectations. Indeed, the governor was criticized for putting her children before the needs of the state and criticized for not being a good mother when she did put the state's affairs first.

This chapter is based on the premise that families are organized around work—where the couple lives is determined by where the spouses/parents can get jobs. What time they eat, which family members eat with whom, when they go to bed, and when, where, and for how long they vacation are all influenced by the job. We also emphasize that involvement in the workforce impacts the power and roles of family members. We begin this chapter by examining the meanings of money.

MEANINGS OF MONEY

Money, it turned out, was exactly like sex—you thought of nothing else if you didn't have it and thought of other things if you did.

James Baldwin, *Novelist*

Symbolic interactionists emphasize the social meanings associated with money, including security and avoiding poverty, self-esteem, power in relationships, love, and conflict.

Security—Avoiding Poverty

College students do not want to live in poverty. While a primary goal of college students is "being happy" and "being in love" (Abowitz & Knox, 2003), a national survey of first-year students at two- and four-year universities in the United States revealed that almost three-fourths (73%) of the men and two-thirds (66%) of the women agreed that "to make more money" was very important in their deciding to go to college (American Council on Education and University of California, 2004). Money represents security most people want. Oscar Wilde once said, "When I was young, I used to think that money was the most important thing in life; now that I am older, I know it is."

Poverty has traditionally been defined as the lack of resources necessary for material well-being. The lack of resources that leads to hunger and physical deprivation is known as **absolute poverty.** Poverty is devastating to families and particularly to children. Those living in poverty have poorer physical and mental health, experience more punitive discipline styles, and are more likely to be exposed to substance abuse, domestic abuse, child abuse and neglect, and divorce (Seccombe, 2001).

NATIONAL DATA

Seven percent, 19%, and 21% of white, Hispanic, and Black families, respectively, are classified as living in poverty (*Statistical Abstract of the United States: 2003,* Tables 39, 46).

The realities of poverty were examined by Barbara Ehrenreich (2001), a Ph.D. journalist who experienced minimum-wage life as a waitress cleaning tables, a maid scrubbing showers in motel rooms, and an employee restacking clothes in the women's wear department at Wal-Mart. She worked for seven dollars an hour at the latter and discovered that it takes two jobs to afford a decent place to live "and still have time to shower between them" (p. 39). These roles are held by the working poor, who "neglect their own children so the children of others will be cared for; they live in substandard housing so that other homes will be shiny and perfect. . . ." (p. 221).

Table 12.1 2003 HHS Poverty Guidelines

Size of Family Unit	48 Contiguous States and D.C.	Alaska	Hawaii
1	$ 9,310	$11,630	$10,700
2	12,490	15,610	14,360
3	15,670	19,590	18,020
4	18,850	23,570	21,680
5	22,030	27,550	25,340
6	25,210	31,530	29,000
7	28,390	35,510	32,660
8	31,570	39,490	36,320
For each additional person, add	3,180	3,980	3,660

Source: *Federal Register,* vol. 69, no. 30, February 13, 2004, pp. 7336–7338. http://aspe.hhs.gov/poverty/04poverty.shtml

What is the definition of poverty in terms of actual dollars? Table 12.1 reflects the Department of Health and Human Services' various poverty level guidelines by size of family and where the family lives. A significant proportion of families in the United States continue to be characterized by unemployment and low wages.

Self-Esteem

Money affects self-esteem because in our society human worth, particularly for men, is often equated with financial achievement. Persons who make $10,000 a year may think of themselves as worth less than those who make $100,000 a year. Persons inadvertently compare themselves with those who make more and less money.

Power in Relationships

Money is a central issue in relationships because of its association with power, control, and dominance. Generally, the more money a partner makes, the more power that person has in the relationship. Males make considerably more money than females and generally have more power in relationships.

National Data

The average annual income of a male with some college who is working full- time is $44,911, compared with $29,273 for a female with the same education, also working full-time, year-round (*Statistical Abstract of the United States 2002,* Table 666).

When the wife earns an income, her power increases in the relationship. The authors know of a married couple in which the wife has recently begun to earn an income. Before she did this, her husband's fishing boat was in the protected carport. With her new job, and increased power in the relationship, her car is parked in the carport and her husband's fishing boat is sitting underneath the trees to the side of the house. Stier and Lewin-Epstein (2000) note that a woman who works full-time (in contrast to part-time) has increased power not only in terms of how money is spent but also in terms of getting the husband to become more involved in domestic labor.

Money also provides the employed woman the power to be independent and to leave an unhappy marriage. Indeed, the higher a wife's income, the more likely she is to leave an unhappy relationship (Schoen et al., 2002). Similarly, since adults are generally the only source of money in a family, they have considerable power over children, who have no money.

Love

To some individuals, money also means love. While admiring the engagement ring of her friend, a woman said, "What a big diamond! He must really love you." The assumption is that big diamond equals high price equals deep love.

Similar assumptions are often made when gifts are given or received. People tend to spend more money on presents for the people they love, believing that the value of the gift symbolizes the depth of their love. People receiving gifts may make the same assumption. "She must love me more than I thought," mused one man. "I gave her a CD for Christmas, but she gave me a CD player. I felt embarrassed."

Conflict

Money can also be a source of conflict in relationships. Couples argue about what to spend money on (new car? vacation? pay off credit card?) and how much money to spend. One couple in marriage therapy reported that they argued over whether to buy a new air conditioner for their car. The husband thought it necessary; the wife thought they could roll down the windows. As noted earlier, conflicts over money in a relationship often signify conflict over power in the relationship.

Marital conflicts over money may arise when there are religious disparities over what to spend money on (Curtis & Ellison, 2002). One employed wife in the authors' classes said her husband, an evangelical Christian, told her she could not buy beer. They fought vigorously over this issue with the result that she ended up buying beer—"but it's a problem in our marriage" she said.

DUAL-EARNER MARRIAGES

Two-earner marriages have become more common. Sixty-two percent of all U.S. wives are in the labor force. Most have children. Mothers least likely to be in the labor force have children under three years of age—58%. Wives most likely to be in the labor force have children between the ages of 14 and 17 (81% of wives), when their children are relatively independent and the food and clothing expenses of teenagers are highest (*Statistical Abstract of the United States: 2003,* Table 598). The stereotypical family consisting of the husband who earns the income and the wife who stays at home with two or more children is no longer the norm.

Although many low wage earners need two incomes to afford basic housing and a minimal standard of living, others have two incomes to afford expensive homes, cars, vacations, and educational opportunities for their children. As previously noted, the cultural emphasis on equality for women and their own desire to have equal power in the relationship with their husbands are important factors influencing the choice of an increasing percentage of wives to be employed. More married women seek careers and economic independence, goals traditionally reserved for men.

Because women still bear most of the child-care and other household responsibilities, women in dual-earner marriages are more likely than men to want to be employed part-time rather than full-time. If this is not possible, many women prefer to work only a portion of the year (the teaching profession allows employees to work about ten months and to have two months in the summer free). We have noted that working part time (rather than full time) decreases the power of the wife in the marital relationship.

A dual-career marriage is defined as one in which the spouses both pursue careers and maintain a life together that may or may not include dependents. A career is different from a job in that the former usually involves advanced education or training, full-time commitment, working nights and weekends "off the clock," and a willingness to relocate. Dual-career couples operate without a "wife"—a person who stays home to manage the home and care for dependents.

Nevertheless, three types of dual-career marriages are those in which the husband's career takes precedence, the wife's career takes precedence, or both careers are regarded equally. These career types are sometimes symbolized as HIS/her, HER/his, HIS/HER, and THEIR career. When couples hold traditional gender role attitudes, the husband's career is likely to take precedence (**HIS/her career**). This situation translates into the wife's being willing to relocate and to disrupt her career for the advancement of her husband's career. For couples who do not hold traditional gender role attitudes, the wife's career may take precedence (**HER/his career**). In such marriages, the husband is willing to relocate and to disrupt his career for his wife's. Such a pattern is also likely to occur when the wife earns considerably more money than her husband. When the careers of both the wife and husband are given equal status in the relationship (**HIS/HER career**), they may have a commuter marriage in which they follow their respective careers wherever they lead. Alternatively, they hire domestic/child-care help so that neither spouse functions in the role of the wife.

Authors

This father is dedicated to his career but also communicates to his son great pride for the son's accomplishments.

Finally, some couples have the same career (**THEIR career**) and may literally travel and work together. Some news organizations hire both spouses to travel abroad to cover the same story. Indeed, 35 percent of foreign correspondents are married couples who serve together. Both the news organization and the couple may benefit (Blumenthal, 1998). In the following sections we look at the effects on women, men, their marriage, and their children when both spouses are employed outside the home.

Effects of the Wife's Employment on the Wife

A woman's work satisfaction is related to whether she has children, the age of her children, the nature of her work, and the support of her husband. In general, women without children or older children, working in jobs they enjoy and married to egalitarian husbands who support their employment are much happier than their counterpart. Employed women in dual earner households who perceive inequality in the division of labor are less likely to feel happy in their marriage and are more likely to divorce.

While women have sacrificed their careers for their families, men have sacrificed their families for their careers.

Marieke Van Willigen, *Sociologist*

Employment for wives is typically associated with enhanced psychological well-being (Hochschild, 1997). Employed wives report higher self-esteem and greater feelings of independence, as well as increased social interaction with a wider network of individuals. Some wives and mothers view work as an alternative sphere that they prefer. However, Kiecolt (2003) noted that most women with young children much prefer to be at home and view home, not work, as their primary haven of satisfaction.

Indeed, family and work become spheres to manage and sometimes there is **role overload**—not having the time or energy to meet the demands of one's role (wife/parent/worker) responsibilities. Because women have traditionally been responsible for most of the housework and child care, employed women come home from work to what Hochschild (1989) calls the **second shift:** the housework and child care that employed women do when they return home from their jobs.

> *As a result, women tend to talk more intently about being overtired, sick, and "emotionally drained." Many women I could not tear away from the topic of sleep. They talked about how much they could "get by on" . . . six and a half, seven, seven and a half, less, more. . . . Some apologized for how much sleep they needed. . . . They talked about how to avoid fully waking up when a child called them at night, and how to get back to sleep. These women talked about sleep the way a hungry person talks about food.* (p. 9)

A team of researchers (Phillips et al., 2000) studied married veterinarians and found that female veterinarians reported significantly more marital/family stress in reference to their career, and they perceived less spousal support for their career, than their male counterparts. Husbands often talk of being very egalitarian and supportive of their wives' having careers, but the data backing up these assertions are sometimes lacking.

Another stressful aspect of employment for employed mothers in dual-earner marriages is **role conflict**—being confronted with incompatible role obligations. For example, the role of the employed mother is to stay late and get a report ready for tomorrow. But the role of the mother is to pick up the child from day care at five P.M. When these roles collide, there is role conflict. While most women resolve their role conflict by giving preference to the mother role, some give priority to the career role and feel guilty about it. Mary Tyler Moore wrote in her biography that she spent more time with her TV son than her real son.

Role strain, the anxiety that results from being able to fulfill only a limited number of role obligations, occurs for both women and men in dual-earner marriages. There is no one at home to take care of housework and children while they are working, and they feel strained at not being able to do everything.

Effects of the Wife's Employment on Her Husband

Husbands also report benefits from their wives' employment. These include being relieved of the sole responsibility for the financial support of the family and having more freedom to quit jobs, change jobs, or go to school. Traditionally men had no options but to work full-time. Indeed, Zuo and Tang (2000) found in a national longitudinal sample of married individuals that men were not threatened by their wives' earning an income but had positive views of their doing so in that they benefited materially from their wives' financial contributions to the family. Men also benefit by having a spouse with whom to share the daily rewards and stresses of employment. And, to the degree that women find satisfaction in their work role, men benefit by having a happier partner. Finally, men benefit from a dual-earner marriage by increasing the potential to form a closer bond with their children through active child care. Some prefer the role of househusband and stay-at-home dad. Though such husbands and dads are clearly the minority, their visibility is increasing.

However, not all husbands want their wives to be employed. Indeed, violence may erupt if the wife has a job, particularly if the husband is unemployed

Up Close 12.1 When the Wife Earns More Money

Thirty percent of working wives earn more than their husbands (Tyre & McGinn, 2003). This situation affects the couple's marital happiness in several ways. Marriages where the spouses view the provider role as the man's responsibility, where the husband cannot find employment, and where the husband is jealous of his wife's employment are vulnerable to dissatisfaction. Cultural norms typically dictate that the man is supposed to earn more money than his partner and that something is wrong with him if he doesn't. Couples who buy into this traditional norm are more likely to report marital unhappiness. Laura Doyle (2003) became the sole supporter in their marriage and reported, "Our marriage barely survived" (p. 53).

> *I was earning all the money, so I figured I was the boss. My paycheck covered the mortgage, bought the food, provided some entertainment and paid every other bill with our name on it, so I thought I was entitled to a little control over the household— and my husband. . . . I said things like, "So why aren't the dishes done? What did you do all day?"* (p. 53)

Some men express their anger at their perceived provider displacement with violence and abuse. Anderson (1997) studied the demographic characteristics of a national sample of individuals involved in domestic violence and found that men who earn less money than their partners are more likely to be violent toward them. "Disenfranchised men then must rely on other social practices to construct a masculine image. Because it is so clearly associated with masculinity in American culture, violence is a social practice that enables men to express a masculine identity" (p. 667). Fox et al. (2002) found that husbands are particularly likely to express their frustration violently if their wives' income is increasing while theirs is decreasing. Jalovaara (2003) found that the divorce rate was higher among couples where the wife's income exceeded her husband's.

On the other hand, some men want an ambitious, economically independent woman who makes a high income. Over a third of men in a national *Newsweek Poll* said that if their wife earned more money, they'd consider quitting their job or reducing their hours (Tyre & McGinn, 2003). Some want to pursue a hobby; others enjoy the role of househusband and stay-at-home parent. Indeed, lower marital satisfaction due to the wife's higher income is not inevitable (Schoen, 2002; Vannoy & Cubbins, 2001). Some husbands prefer to be "at-home-dads" and enjoy the role. Ono (2003) found that higher income on the part of the woman (in industrialized countries) actually increased her chance of getting married.

Sources

Anderson, Kristin L. 1997. Gender, status, and domestic violence: An integration of feminist and family violence approaches. *Journal of Marriage and the Family* 59:655–69.

Doyle, L. 2003. Our marriage barely survived. *Newsweek.* 12 May, 53.

Fox, G. L., M. L. Benson, A. A. DeMaris, and J. V. Wyk. 2002. Economic distress and intimate violence: Testing family stress and resources theory. *Journal of Marriage and the Family* 64:793–807.

Jalovaara, M. 2003. The joint effects of marriage partners' socioeconomic positions on the risk of divorce. *Demography* 40:67–81.

Ono, H. 2003. Women's economic standing, marriage timing, and cross-national contexts of gender. *Journal of Marriage and the Family* 65:275–86.

Schoen, R., N. M. Astone, K. Rothert, N. J. Standish, and Y. J. Kim. 2002. Women's employment, marital happiness, and divorce. *Social Forces* 81:643–62.

Tyre, P., and D. McGinn. 2003. She works, he doesn't. *Newsweek.* 12 May, 45–52.

Vannoy, D., and L. A. Cubbins. 2001. Relative socioeconomic status of spouses, gender attitudes, and attributes, and marital quality experienced by couples in metropolitan Moscow. *Journal of Comparative Family Studies* 32:195–217.

(Macmillan & Gartner, 2000). Other outcomes husbands may perceive as disadvantages from being married to a wife who has a career are feeling more pressure to engage in domestic work and less time for their own career (Rosenbluth, Steil, & Witcomb, 1998).

Effects on the Couple's Marriage of Having Two Earners

Are marriages in which both spouses earn an income more vulnerable to divorce? Not if the wife is happy; but if she is unhappy, her income will provide her a way to take care of herself when she leaves. Schoen et al. (2002) write, "Our results provide clear evidence that, at the individual level, women's employment does not destabilize happy marriages but increases the risk of disruption in unhappy marriages" (p. 643). Hence employment won't affect a happy marriage but it can do an unhappy one in.

Couples are particularly vulnerable when the wife earns more money than her husband. See Up Close 12.1.

Some dual-earner parents also wonder if the increased income is worth their sacrifices to earn it. Sefton (1998) calculated that the value of the stay-at-home mother in terms of what a dual-income family spends to pay for all services

Up Close 12.3 Day-Care Costs—It's Only the Beginning

Day-care costs vary widely—from nothing, where friends trade off taking care of their children, to very expensive institutionalized day care in large cities. The authors e-mailed a dual-earner metropolitan couple (who use day care for their two children) to inquire about the institutional costs of day care in their area (Baltimore). The response from the father is below:

There is a sliding scale of costs based on quality of provider, age of child, full-time or part-time. Full-time infant care can be hard to find in this area. Many people put their names on a waiting list as soon as they know they are pregnant. We had our first child on a waiting list for about nine months before we got him into our first choice of providers.

Today, I think infant care runs $225+ per week for high-quality day care. That works out to $11,700 per year—and you thought college was expensive.

The cost goes down when the child turns 2—to around $150–$180 per week. In Maryland, this is because the required ratio of teachers to children gets bigger at age 2. This cost break doesn't last long. Preschool programs (more academic in structure than day care) begin at age 3 or 4. Our second child turns 4 this May and begins an academic preschool in September. The school year lasts nine and a half months, and it costs about $9,000. Summer camps or summer day-care costs must be added to this amount to get the true annual cost for the child. My wife and I have budgeted about $20,000 total for our two children to attend private school and summer camps this year.

The news only gets worse when the children get older. A good private school for grades K–5 runs $10,000 per school year. Junior high (6–8) is about $11,000, and private high school here goes for $12,000 to $18,000 per nine-month school year.

Religious private schools run about half the costs above. However, they require that you be a member in good standing in their congregation and, of course, your child undergoes religious indoctrination.

There is always the argument of attending a good public school and thus not having to pay for private school. However, we have found that home prices in the "good school neighborhoods" were out of our price range. Some of the better-performing public schools also now have waiting lists—even if you move into that school's district.

In most urban and suburban areas, cost is secondary to admissions. Getting into any good private school is hard. In order to get into the good high schools, it's best to be in one of the private elementary or middle schools that serves as a "feeder" school. Of course, to get into the "right" elementary school, you must be in a good feeder kindergarten. And of course to get into the right kindergarten, you have to get into the right feeder preschool. Parents have A LOT of anxiety about getting into the right preschool—because this can put your child on the path to one of the better private high schools.

Getting into a good private preschool is not just about paying your money and filling out applications. Yes, there are entrance exams. Both of our children underwent the following process to get into preschool: First you must fill out an application. Second, you child's day-care records/transcripts are forwarded for review. If your child gets through this screening, you and your child are called in for a visit. This visit with the child lasts a few hours and your child goes through evaluation for physical, emotional, and academic development. Then you wait several agonizing months to see if your child has been accepted

Though quality day care is expensive, parents delight in the satisfaction that they are doing what they feel is best for their child.

Authors

whether a low-income mother seeks employment since the cost can absorb her paycheck (Baum, 2002). Even for dual-earner families, cost is a factor in choosing a day-care center. See Up Close 12.3 for a two-earner (high-income) couple's experience of providing quality care for their children.

State licensing regulations of day-care centers are supposed to provide a "floor of quality" for day-care programs. However, these regulations often fall short of recommended standards. In a survey of four hundred day-care centers in California, Colorado, North Carolina, and Connecticut, a team of researchers rated only 14 percent as good or excellent (Helburn et al., 1995). In these centers, the children's health and safety needs were met, the children received warmth and support from their caretakers, and learning was encouraged. Twelve percent of the centers were rated minimal—the children were ignored, their health and safety were compromised, and no learning was encouraged. The remaining (74%) received a rating just above the minimal, suggesting that the health and safety needs, nurturing needs, and educational needs were barely met.

Care for infants was particularly lacking. Of 225 infant or toddler rooms observed, 40 percent were rated less than minimal with regard to hygiene and safety. Because of the low pay and stress of the occupation, the rate of turnover for family child-care providers is very high (estimated at between 33 and 50%) (Walker, 2000). Wishard et al. (2003) emphasized the need for careful criteria in determining "quality" care: "quality child care does not involve a static or fixed way of providing care for children. Rather, individual communities vary significantly in the particular goals they hold for children" (p. 69).

Jay Belsky (1995), recognized as a leading expert on day-care centers over the past twenty-five years, makes the following observations:

1. Day care works best when children receive "care from individuals who will remain with them for a relatively long period of time (low staff turnover), who are knowledgeable about child development, and who provide care that is sensitive and responsive to their individualized needs" (p. 36).

2. Since few day-care centers are characterized by the above description, children with "early, extensive, and continuous day care experience" (that is, children whose care begins in the first year or so of life, on a full-time or near full-time basis—more than twenty to thirty hours per week—and continues through their entry into public school) are the ones who seem most susceptible to negative outcomes. Such negative outcomes include insecure attachment bonds to parents, aggressiveness toward age mates, and disobedience toward adults.

Ahnert and Lamb (2003) emphasized that any potential negative outcomes in day care can be mitigated by attentive, sensitive, loving parents. "Home remains the center of children's lives even when children spend considerable amounts of time in child care. . . . [A]lthough it might be desirable to limit the amount of time children spend in child care, it is much more important for children to spend as much time as possible with supportive parents" (pp. 1047–48).

3. Federal mandates for quality day-care standards must be established and tax provisions must be made for more economic support to allow families with children to increase their child-care options.

In the meantime, parents should be cautious in choosing day-care facilities for their children. To help them, the National Organization for the Education of Young Children has recommended criteria for accrediting day-care centers. Such recommendations include having a 1:3 to 1:4 adult-child ratio for children from 0 to 24 months, with from six to twelve children per group. For 2-year-olds, the ratio is 1:4 to 1:6, with from eight to twelve children per group (Helburn & Howes, 1996, 67). Data from the National Institute of Child Health and Human Development (NICHD) reveal that child-care quality may affect child outcomes, particularly in the cognitive domain area. "Specifically, child-care quality, defined in terms of provision of intellectual experiences and stimulation for the child, may affect cognitive development, as indexed by language ability, memory, and preacademic achievement" (NICHD, 2003, 466).

Some parents want to be aware of what is happening throughout the day while their child is in day care. Some day-care centers offer video tracking of their child. A Safer Start Child University in Cranston, Rhode Island, allows parents to log onto www.SaferStart.com and click on a room where they can see their child.

BALANCING DEMANDS OF WORK AND FAMILY

One of the major concerns of employed parents and spouses is how to juggle the demands of work and family simultaneously and achieve a sense of accomplishment and satisfaction in each area.

I have yet to hear a man ask for advice on how to combine marriage and a career.

Gloria Steinem, *Feminist writer and founder of Ms. Magazine*

Men are more likely to receive cultural support for solving the conflict between work and family by giving precedence to work. Women are more likely to resolve the conflict by giving precedence to family. Kiecolt (2003) examined national data and concluded that employed women with young children are "more likely to find home a haven, rather than finding work a haven or having high work-home satisfaction" (p. 33). As Faith Hill, the country-and-western superstar has said, "Family comes first—always—what else is there?" Hochschild (1997) had previously noted that there was a "cultural reversal" underway whereby work was becoming more satisfying than home for women. Kiecolt found "no evidence" for this reversal. Indeed, increasingly mothers are quitting their jobs and returning home to rear their children. In effect there has been "a switch in loyalties from paycheck to playpen" (Clark, 2002).

Nevertheless, the conflict between work and family is substantial and various strategies are employed to cope with the stress of role overload and role conflict, including (1) the superperson strategy, (2) cognitive restructuring, (3) delegation of responsibility, (4) planning and time management, and (5) role compartmentalization (Stanfield, 1998).

Superperson Strategy

The superperson strategy involves working as hard and as efficiently as possible to meet the demands of work and family. The person who uses the superperson strategy often skips lunch and cuts back on sleep and leisure in order to have more time available for work. Women are particularly vulnerable because they feel that if they give too much attention to child-care concerns, they will be sidelined into lower-paying jobs with no opportunities (Beck, 1998).

Hochschild (1989) noted that the term **superwoman** or **supermom** is a cultural label that allows the woman to regard herself as very efficient, bright, and confident. However, Hochschild noted that this is a "cultural cover-up" for an overworked and frustrated woman. Not only does the woman have a job in the workplace (first shift), she comes home to another set of work demands in the form of house care and child care (second shift). Finally, there is the "third shift" (Hochschild, 1997).

The **third shift** is the expense of emotional energy by a spouse or parent in dealing with various issues in family living. An example is the emotional energy needed for children who feel neglected by the absence of quality time.While young children need time and attention, responding to conflicts and problems with teenagers also involves a great deal of emotional energy—the third shift.

Cognitive Restructuring

Another strategy used by some women and men experiencing role overload and role conflict is cognitive restructuring, which involves viewing a situation in positive terms. Exhausted dual-career earners often justify their time away from their children by focusing on the benefits of their labor—their children live in a nice house in a safe neighborhood and attend the best of schools. Whether these outcomes offset the lack of "quality time" may be irrelevant—the beliefs serve simply to justify the two-earner lifestyle.

Delegation of Responsibility/Limiting Commitments

A third way couples manage the demands of work and family is to delegate responsibility to others for performing certain tasks. Because women tend to bear most of the responsibility for child care and housework, they may choose to ask their partner to contribute more or to take responsibility for these tasks. Two researchers found that spouses who help each other perceive that their relationships are more equitable and that women who receive more help from their husbands are less angry (Van Willigen & Drentea, 2001).

Parents may also involve their children in household tasks, which not only relieves the parents but also benefits children by requiring them to learn domestic skills and the value of contributing to the family. This may sound like a good idea, but when twenty married self-employed mothers (Mary Kay consultants) were interviewed, they reported that their children contributed little to the household labor, and only a few of the mothers mentioned that they would like for their children to do more (Berke, 2000).

Another form of delegating responsibility involves the decision to reduce one's current responsibilities and not take on additional ones. For example, women and men may give up or limit agreeing to volunteer responsibilities or commitments. One woman noted that her life was being consumed by the responsibilities of her church; she had to change churches since the demands were relentless. In the realm of paid work, women and men can choose not to become involved in professional activities beyond those that are required.

People are not going to work until they drop anymore.

Thomas Neff, *Executive recruiter*

Planning and Time Management

The use of planning and time management is another strategy for minimizing the conflicting demands of work and family (Grzywacz & Bass, 2003). This involves prioritizing and making lists of what needs to be done each day. Time planning also involves trying to anticipate stressful periods, planning ahead for them, and dividing responsibilities with the spouse. Such division of labor allows each spouse to focus on an activity that needs to get done (grocery shopping, picking up children at day care) and results in a smoothly functioning unit.

Having flexible jobs and/or careers is particularly beneficial for two-earner couples. Being self-employed, telecommuting, or working in academia permits flexibility of schedule so that individuals can cooperate on what needs to be done. Alternatively, some dual-earner couples attempt to solve the problem of child care by **shift work**—having one parent work during the day and the other parent work at night so that one parent can always be with the children. Shift workers often experience sleep deprivation and fatigue, which may make it difficult for them to fulfill domestic roles as a parent or spouse. Similarly, shift work may have a negative effect on a couple's relationship because of their limited time together. Presser (2000) studied the work schedules of 3,476 married couples and found that recent (married less than five years) husbands who had children and who worked at night were six times more likely to divorce than husbands/parents who worked days.

Role Compartmentalization

Some spouses use **role compartmentalization**—separating the roles of work and home so that they do not think about or dwell on the problems of one when they are at the physical place of the other. Spouses unable to compartmentalize their work and home feel role strain, role conflict, and role overload, with the result that their efficiency drops in both spheres.

While the various mechanisms identified in this section are helpful in coping with job demands, regular exercise is also effective. We discuss exercise in more detail in the chapter on stress.

Authors

This career wife attempts to balance her work and play. She walks several miles each day and spends weekends with her husband fishing.

BALANCING WORK AND LEISURE TIME WITH FAMILY

The workplace affects the family (more than the family affects the workplace) (Crawford, 2000), and one of the ways it does so is by determining the amount of leisure available to spouses and their children. Indeed, the family is a place where exhausted workers often seek leisure to recuperate from work and to enjoy life. **Leisure** refers to the use of time to engage in freely chosen activity perceived as enjoyable and satisfying, including exercise. Homemakers and those who have lower-status occupations report having very limited time for leisure and exercise, particularly aerobic exercise (Salmon et al., 2000).

Importance of Leisure

Leisure is becoming more important to people. In a nationwide poll conducted by Yankelovich Partners, almost half (48%) of the respondents reported that they would be willing to trade an increase in pay for more vacation time (Haralson & Mullins, 2000). Over one-quarter of the respondents (28%) reported that in the past five years they had voluntarily made changes in their lifestyle that resulted in making less money. Women, those with children, and those under 40 were more likely to do so. The positive value of leisure to a couple's marriage interaction and satisfaction is clear. Doumas, Margolin, and John (2003) studied forty-nine dual-earner couples who reported more positive marital interaction on days they worked less.

I wanted the gold, and I got it—
Came out with a fortune last fall,
Yet somehow life's not what I thought it,
And somehow the gold isn't all.

Robert W. Service

Functions of Leisure

Leisure fulfills important functions in our individual and interpersonal lives. Leisure activities may relieve work-related stress and pressure; facilitate social interaction and family togetherness; foster self-expression, personal growth, and skill development; and enhance overall social, physical, and emotional well-being.

Though leisure represents a means of family togetherness and enjoyment, it may also represent an area of stress and conflict. Some couples function best when they are busy with work and child care so that there is limited time to interact or for the relationship. Indeed, some prefer not to have a lot of time alone together.

BARRIERS TO LEISURE

Many individuals feel they do not have enough leisure. What factors have contributed to the "leisure shortage" that many women and men complain about? Barriers to leisure include the rising demands of the workplace, materialistic values, traditional gender roles, the "commodification of leisure," and treating leisure as work.

Demands of the Workplace

A major barrier to leisure has been the rising demands of the workplace. Technology has spawned laptops, which allow employers to get employees to expand their workdays into the night by working at home. One salesperson for a pharmaceutical company noted that "the company requires you to log in every evening after seven P.M. Their computers keep a record of how long I am on-line at night. The result is I work a full day AND evening. I put in about sixty hours a week, every week. If I don't, my colleagues will and I'll be fired if I don't put in the same unpaid hours."

Workers in the United States have the dubious distinction of spending more hours on the job than workers in any other industrial nation (Polatnick, 2000). Over a third (37%) of U.S. workers want to reduce their hours (Reynolds, 2003). Long hours and the associated limited leisure have effects on relationships. Crouter et al. (2000) studied men in the workforce and observed that long hours and high levels of role overload consistently predicted less positive relationships with both their spouses and their adolescent children.

Workers have also become distrustful of their employers. The Enron scandal, in which upper management lied to workers to inflate stock prices and then left the employees with no jobs and no retirement fund, has caused the American worker to reevaluate the wisdom of trusting any employer. Indeed, the traditional organizational career is rapidly becoming obsolete. In the past, an employee would work until retirement in one firm. Now, most individuals must change jobs, and sometimes careers, to advance. Even within the same firm, seniority no longer means security, and organizational ladders are disappearing. Even tenured university professors can be jobless when they discover their entire list of courses has been dropped from the curriculum and their department has been discontinued.

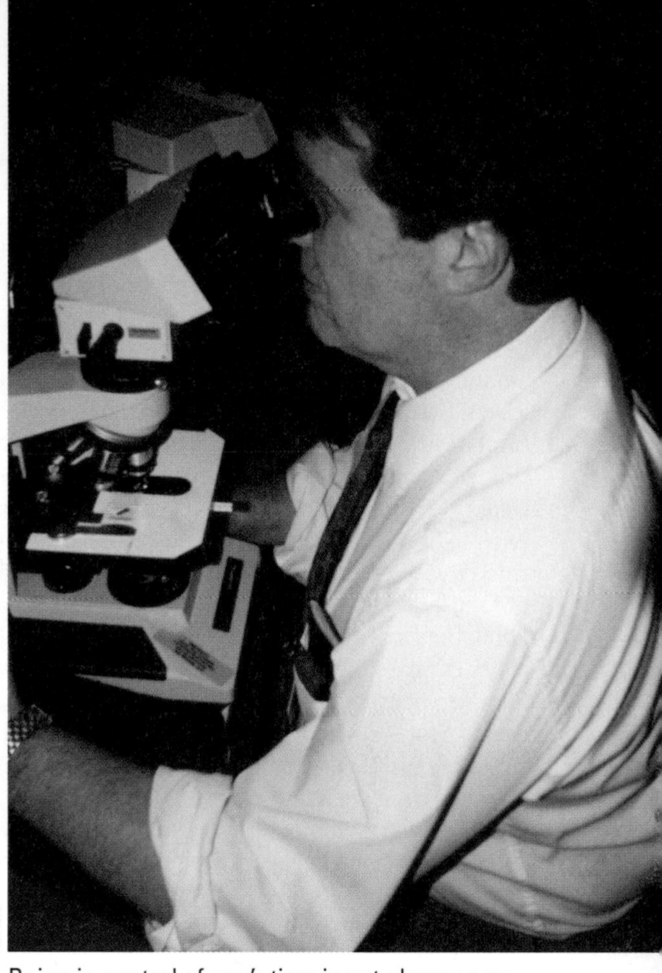
Authors

Being in control of one's time is not always possible. The fact that this pathologist must be available to determine if cells are cancerous while a patient is in surgery means that he may be called from home when he is relaxing with his family.

Some jobs demand that employees work overtime; others pay so little that an employee must take a second job or work two shifts to make ends meet. The result is exhausted workers who have very little time left over after the responsibilities of work and home have been met. High work stress is also related to problem drinking (Grzywacz & Marks, 2000). In addition, individuals may have very little energy to participate in leisure activities (including exercise) after working nine to sixteen hours a day. The Social Policy section on page 300 examines ways in which government and corporations are family-friendly.

Materialistic Values

The media are saturated with advertisements to own things, and we are encouraged to have a strong work ethic to be able to buy things. So we work long and hard hours to achieve a certain standard of living. However, once we attain the goods that we desire, we often find we want more. Thus, many couples get caught up in a vicious cycle of working long hours to achieve a certain standard of living, only to find that this standard of living no longer satisfies them. Children may be learning the materialistic values of their parents. In a *Time* study of almost one thousand children, aged 9 through 14, almost a quarter (23%) reported that they would rather be rich and unhappy than poor and happy. This percentage almost doubled from the previous year (Goldstein, 2000).

Table 12.2 shows the percentage of U.S. families at various income levels. Most families are struggling economically.

Diversity in Other Countries

Denmark has a workweek of thirty-two hours. Short work "time is seen mainly as a way to improve the well-being of workers and increase equality between women and men." France has a thirty-five-hour week. Indeed, the United States ranks near the top on most indicators of working time for couples (Jacobs & Cognick, 2002).

Government and Corporate Work-Family Policies and Programs

SOCIAL POLICY

Workers clearly value being away from work. In a study by Peter D. Hart Research Associates, almost two thirds (64%) of the adult workers reported they would prefer a workweek that consisted of four ten-hour days rather than five eight-hour days (preferred by 31%) (Carey & Mullins, 2000). The U.S. Government is an exception in regard to supporting families versus work. More than a hundred countries already provide for paid maternity leaves. Swedish law grants the parents of a newborn child a total of fifteen months of paid leave. Maternity leave in the United States is *unpaid* under federal law (Polatnick, 2000). Under the Family and Medical Leave Act all companies with twenty-five or more employees are required to provide each worker with up to twelve weeks of unpaid leave for reasons of family illness, birth, or adoption of a child. As amended in 2000, the Family Leave Act permits states to provide unemployment pay to workers who take unpaid time off to care for a newborn child or a sick relative. Nevertheless, the United States still lags behind other countries in providing paid time off for new parents. For instance, Germany provides fourteen weeks off with 100 percent salary.

Aside from government-mandated work-family policies, corporations and employers have begun to initiate policies and programs that address the family concerns of their employees. Employer-provided assistance with child care, assistance with elderly parent care, options in work schedules, and job relocation assistance are becoming more common. Examples of the rare companies providing such benefits include those provided to employees at SAS Institute, a computer software company in Cary, North Carolina, where employees may eat meals with family members and guests at the company cafeteria. The older children from the company preschool can join their parents for lunch. Another example is Pepsi-Co, Inc., in Purchase, New York, which has an on-site concierge to do personal chores for its eight hundred employees. Work-family policies benefit both employees and their families and the corporations they work for. Replacing a well-trained employee can run as much as three and a half times that person's salary (Bounds, 2000).

Not all "family-friendly" corporate policies are welcome, particularly among persons who do not have children. These workers often feel exploited in that they are transferred more often, asked more often to work weekends, and asked to work longer hours. Since they do not have families, they are assumed to be more available. "If you don't have kids, it's assumed you don't have a life outside the office" noted one employee (Grossman, 1998, p. 101).

Some corporations are responding to the concerns of childfree workers. The Quaker Oats Company gives its childfree workers an annual credit of $300 because they generate lower medical insurance costs. In addition, Spiegel Inc. of Downers Grove, Illinois, offers flexible hours to everyone, not just those who have children (Grossman, 1998).

There is disagreement over whether businesses should cater to families. Go to http://sociology.wadsworth.com/knox_schacht/choices8e, select the Opposing Viewpoints Resource Center on the left navigation bar, and enter your passcode. Conduct a search by subject, using the search term "Family," and read the articles on the Viewpoints tab that discuss family-friendly businesses. What is your evaluation of this issue?

Sources

Bounds, W. 2000. Give me a break. *Wall Street Journal.* 5 May, W1–W4.

Carey, A. R., and M. E. Mullins. 2000. Workers want fewer days at the office. *USA Today.* 11 September, B1.

Grossman, L. M. 1998. Family-friendly corporate policies can be counterproductive. In *The family: Opposing viewpoints*, edited by B. Leone. San Diego, Calif.: Greenhaven Press.

Polatnick, M. R. 2000. Working parents. *Phi Kappa Phi Journal* 80:38–41.

Traditional Gender Roles

Traditional gender roles may also create barriers to leisure in the lives of women and men. Whether employed or unemployed, women bear most of the responsibility for childrearing, housework, and general care of the family. The result is that women have little time for leisure. An exception to this is the growing number (still small) of men who not only co-parent but function in the primary role of caretaker. While their wives may be heavily involved in their careers, these fathers enjoy the day-to-day role of being with and nurturing their children.

Women also tend to spend their leisure time engaged in hobbies related to household tasks, such as cooking, preserving fruits and vegetables, and sewing. In addition, women's leisure is often combined with household chores. For example, women often watch TV while they iron clothes, talk on the phone with a friend while cooking in the kitchen, socialize with neighbors while looking after the children, listen to music while cleaning house, or clip coupons and make a grocery list while riding a stationary bicycle. In addition, women's leisure activities are often constrained by the leisure needs of the family. For example, a woman may spend an afternoon at the pool but may do so primarily to provide amusement for her children. Similarly, a woman may walk in the park primarily so that her children may enjoy the playground or feed the ducks at the pond. (It is also true that some fathers schedule their leisure activities in reference to their children.)

Table 12.2 Distribution of Income Level in U.S. Families

Income Level of Family	Percentage at This Level
Less than $15,000	9.7
$15,000–$24,999	11.3
$25,000–$34,999	11.9
$35,000–$49,999	15.7
$50,000–$79,999	20.8
$75,000–$99,999	13.1
$100,000 or more	17.6

Median family income = $51,407

Statistical Abstract of the United States: 2003. 123rd ed. Washington, D.C.: U.S. Bureau of the Census, Table 686.

Holidays provide another example of how gender roles constrain women's leisure. After the Thanksgiving holidays, many women are exhausted by the hours they spent cooking to achieve the perfect Thanksgiving meal. Christmas holidays also represent a great deal of work for women. Traditionally, women in our society who celebrate Christmas have learned it is their primary responsibility to do the Christmas shopping, mail the Christmas presents and cards, put up decorations, and bake Christmas cookies. Many women spend considerable time and energy to ensure that Christmas is a positive experience for the family. But many drive themselves to exhaustion during the process.

Women's leisure is also constrained by safety concerns. Women are less likely to use parks at certain hours or to participate in outdoor activities if they are fearful of their safety.

Men's leisure is also affected by gender role expectations that equate a man's self-worth with his income and put men in the role of primary breadwinner. Men sometimes feel that they don't have time to play when there are bills to be paid.

Leisure as a Commodity

Many leisure activities cost money that families simply do not have in their budget. One father in the authors' classes said, "I used to take the family to a movie every weekend. Not anymore. After you add the cost of popcorn and drinks, I just can't afford it. We've gone exclusively to renting videos." Schor (1991) described the **commodification of leisure** in our society:

> *Private corporations have dominated the leisure "market" encouraging us to think of free time as a consumption opportunity. Vacations, hobbies, popular entertainment, eating out, and shopping itself are all costly forms of leisure. How many of us, if asked to describe an ideal weekend, would choose activities that cost nothing? How resourceful are we about doing things without spending money?* (p. 2)

Leisure as Work

Roberts (1995) emphasized that leisure has become work because we are "using it as a means to other ends—stress reduction, therapy, fitness, and self-actualization" (p. 36). He suggests that true playfulness involves a sense of forgetting time and being completely absorbed or focused, "something that we no longer know how to do."

Some people are reluctant to take time off for a vacation. A nationwide poll of over one thousand adults revealed that 11 percent of full-time workers who were entitled to a vacation did not take one. Another 32 percent took fewer days than earned. Sixty percent of those who did not use their vacation time said that work was more important than taking a vacation; 27 percent feared something would go wrong at work if they went on vacation (Grossman, 1995).

SUMMARY

What are the various meanings of money?

Money is a means of being secure and avoiding poverty. It is also a source of self-esteem, conveys power in relationships, is viewed as an expression of love, and can be a point of conflict over how it is spent.

What are the effects of the wife's employment on her, her husband, and the marriage?

In dual-earner relationships, benefits for the woman's employment include enhanced self-esteem, more power in the relationship, greater economic independence, and a wider set of social relationships. Negatives for the woman include exhaustion as a result of role overload and frustration or guilt caused by role conflict. Women all over the world do more housework than men, even if they are employed outside the home.

Men benefit from their wives' employment by being relieved of full responsibility for the financial support of the family, having the freedom to quit or to change jobs, and having the opportunity to be a house husband and or full-time at-home parent. Men who report being dissatisfied with their wives' employment interpret such employment as a reflection of their own inadequacy to support the family. Some men are also torn between their work and family responsibilities.

Employed wives in unhappy marriages are more likely to leave the marriage. Marriages in which the wife earns more money, the husband is unemployed, and his masculinity is threatened are also more vulnerable.

Children reared in dual-earner families may have less time with the busy and frustrated parents and may experience less supervision of their behavior. Young children put in day-care centers for most of the day may be at risk.

What are the various strategies of balancing the demands of work and family?

Strategies used for balancing the demands of work and family include the superperson strategy, cognitive restructuring, delegation of responsibility, planning and time management, and role compartmentalization. Government and corporations have begun to respond to the family concerns of employees by implementing work-family policies and programs.

What is the importance of leisure and what are its functions?

Spouses and couples are increasingly beginning to value their leisure time. Leisure helps to relieve stress, facilitate social interaction and family togetherness, and foster personal growth and skill development. However, leisure time may also create conflict over how to use it.

What are the various barriers to leisure?

Barriers to having enough leisure time include incessant demands of the workplace, materialistic values, traditional gender roles, commodification of leisure, and treating leisure as work.

KEY TERMS

absolute poverty
commodification of leisure
HER/his career
HIS/her career
HIS/HER career
leisure
poverty
role compartmentalization
role conflict
role overload
role strain
second shift
shift work
supermom
superwoman
THEIR career
third shift

RESEARCHING MARRIAGE AND THE FAMILY WITH INFOTRAC COLLEGE EDITION

InfoTrac College Edition, an online library, allows you to perform research online anywhere, anytime. Following are two suggested search terms and related questions to help you extend your understanding of the topics covered in this chapter. Go to your search. www.infotrac-college.com to begin.

Keyword: **Maternal employment.** Locate articles that discuss the effect on children of the mother's employment. What factors are associated with positive and negative outcomes for children?

Keyword: **Day care.** Locate articles that discuss quality day care. What factors are associated with quality day care?

The Companion Web Site for Choices in Relationships: An Introduction to Marriage and the Family, Eighth Edition

http://sociology.wadsworth.com/knox_schacht/choices8e

Supplement your review of this chapter by going to the companion Web site to take one of the Tutorial Quizzes, use the flash cards to master key terms, and check out the many other study aids you'll find there. You'll also find special features such as the Marriage and Family Resource Center, Census 2000 information, and other data and resources at your fingertips to help you with that special project or to do some research on your own.

WEBLINKS

Career.com (career planning)
http://www.career.com/

The Down To Earth Dad
http://www.downtoearthdad.com/

Slowlane.com (stay-at-home dads)
http://slowlane.com/

© Fogel François/CORBIS SYGMA

CHAPTER 13

Stress is an ignorant state. It believes that everything is an emergency. Nothing is that important. Just lay down.

Natalie Goldberg

Stress and Crisis in Relationships

Contents

True or False?

1. Not one of fifty couples defined as "successful marriage couples" said they would automatically end their marriage if their partner was having an extramarital affair.
2. Most of the couples in one study of grief reported "chronic sorrow"; the sadness surrounding the death of their beloved was an issue that continued to surface over time.
3. The greater the number of roles a person has (e.g., employee, spouse, parent, etc.), the lower the depression and the higher the psychological well-being of the person.
4. Over 60 percent of adults going through the middle years report having a "midlife crisis."
5. Changing basic values and one's point of view is one of the most helpful strategies reported by families experiencing a crisis.

Answers: **1.** T **2.** T **3.** T **4.** F **5.** T

The slamming of two American Airlines planes into the World Trade Towers and another to follow on the Pentagon on September 11, 2001, was a crisis event for American society and set in motion a series of aftershocks in the social, economic, and political spheres worldwide. The tragedy was also a crisis for individuals (increasing workplace anxiety, Bosco & Harvey, 2003), marriages, and families and humbled all of us by reinforcing the idea that none of us is immune to sudden, dramatic, shocking change. In this chapter we review some crisis events that are not uncommon to marriages and families. These include physical illness and disability, unemployment, drug abuse, extramarital affairs, and death.

Ann Landers was once asked what she would consider the single most useful bit of advice all people could profit from. She replied, "Expect trouble as an inevitable part of life, and when it comes, hold your head high, look it squarely in the eye and say, 'I will be bigger than you.'" Life indeed brings both triumphs and tragedies. Ozer et al. (2003) noted that nearly half of all adults report experiencing at least one traumatic event at some point in their lives. This chapter is about experiencing and coping with these events and the day-to-day stress we feel during the in-between time.

NATIONAL DATA

When mental health is defined as a syndrome of symptoms of positive feelings and positive functioning in life and viewed on a continuum from flourishing to languishing, of 3,032 adults ages 25 to 74, 17 percent fit the criteria for flourishing, 57 percent for being moderately mentally healthy, 12 percent for languishing, and 14 percent for suffering major depressive symptoms (Keyes, 2002).

PERSONAL STRESS AND CRISIS EVENTS

In this section we review the definitions of crisis and stressful events, the characteristics of resilient families, and a framework for viewing a family's reaction to a crisis event.

Definitions of Stress and Crisis Events

Stress is a reaction of the body to substantial or unusual demands (physical, environmental, or interpersonal) made on it. Stress often involves irritability, high blood pressure, and depression and may result from a wide range of situations. Regardless of its source, stress is a frequent precursor to and concomitant of depression and marital discord (Prather & Darling, 1998).

Stress is a process rather than a state. For example, a person will experience different levels of stress throughout a divorce—the stress involved in acknowledging that one's marriage is over, telling the children, leaving the home, getting the final decree, and becoming remarried will vary across time.

A **crisis** is a crucial situation that requires changes in normal patterns of behavior. A family crisis is a situation that upsets the normal functioning of the family and requires a new set of responses to the stressor. Sources of stress and crises can be external (e.g., hurricane, tornado, downsizing, military separation) or internal (e.g., alcoholism, extramarital affair, Alzheimer's disease). Stressors or crises may also be categorized as expected or unexpected. Examples of expected family stressors include the need to care for aging parents and the death of one's parents. Unexpected stressors include contracting HIV, a miscarriage, or the suicide of one's teenager.

SELF-ASSESSMENT

Internality, Powerful Others, and Chance Scales

People have different feelings about their vulnerability to crisis events. The following scale addresses the degree to which you feel you have control, or feel others have control, or feel that chance has control of what happens to you.

Internality, Powerful Others, and Chance Scales

To assess the degree to which you believe that you have control over your own life (I = Internality), the degree to which you believe that other people control events in your life (P = Powerful Others), and the degree to which you believe that chance affects your experiences or outcomes (C = Chance), read each of the following statements and select a number from minus 3 to plus 3.

−3	−2	−1	+1	+2	+3
Strongly Disagree	Disagree	Slightly Disagree	Slightly Agree	Agree	Strongly Agree

Subscale

I **1.** Whether or not I get to be a leader depends mostly on my ability.

C **2.** To a great extent my life is controlled by accidental happenings.

P **3.** I feel like what happens in my life is mostly determined by powerful people.

I **4.** Whether or not I get into a car accident depends mostly on how good a driver I am.

I **5.** When I make plans, I am almost certain to make them work.

C **6.** Often there is no chance of protecting my personal interests from bad luck happenings.

C **7.** When I get what I want, it's usually because I'm lucky.

P **8.** Although I might have good ability, I will not be given leadership responsibility without appealing to those in positions of power.

I **9.** How many friends I have depends on how nice a person I am.

C **10.** I have often found that what is going to happen will happen.

P **11.** My life is chiefly controlled by powerful others.

C **12.** Whether or not I get into a car accident is mostly a matter of luck.

P **13.** People like myself have very little chance of protecting our personal interests when they conflict with those of strong pressure groups.

C **14.** It's not always wise for me to plan too far ahead because many things turn out to be a matter of good or bad fortune.

P **15.** Getting what I want requires pleasing those people above me.

C **16.** Whether or not I get to be a leader depends on whether I'm lucky enough to be in the right place at the right time.

P **17.** If important people were to decide they didn't like me, I probably wouldn't make many friends.

I **18.** I can pretty much determine what will happen in my life.

I **19.** I am usually able to protect my personal interests.

P **20.** Whether or not I get into a car accident depends mostly on the other driver.

I **21.** When I get what I want, it's usually because I worked hard for it.

P **22.** In order to have my plans work, I make sure that they fit in with the desires of other people who have power over me.

I **23.** My life is determined by my own actions.

C **24.** It's chiefly a matter of fate whether or not I have a few friends or many friends.

Scoring

Each of the subscales of Internality, Powerful Others, and Chance is scored on a six-point Likert format from minus 3 to plus 3. For example, the eight Internality items are 1, 4, 5, 9, 18, 19, 21, 23. A person who has strong agreement with all eight items would score a plus 24; strong disagreement, a minus 24. After adding and subtracting the item scores, add 24 to the total score to eliminate negative scores. Scores for Powerful Others and Chance are similarly derived.

Norms

For the Internality subscale, means range from the low 30s to the low 40s, with 35 being the modal mean (SD values approximating 7). The Powerful Others subscale has produced means ranging from 18 through 26, with 20 being characteristic of normal college student subjects (SD = 8.5). The Chance subscale produces means between 17 and 25, with 18 being a common mean among undergraduates (SD = 8).

Source

From "Differentiating Among Internality, Powerful Others and Chance" by H. Levenson, 1981. In H. M. Lefcourt (ed.), *Research with the Locus of Control.* Vol. 1, pp. 57–59. Copyright © by Academic Press, Inc. Used by permission.

Both stress and crises are a normal part of family life and sometimes reflect a developmental sequence. Pregnancy, childbirth, job changes or loss, children leaving home, retirement, and widowhood are all stressful and predictable for most couples and families. Crisis events also have a cumulative effect— the greater their number, the greater the need for coping and adjustment skills (Forehand, Giggar, & Kotchick, 1998).

Resilient Families

Just as the types of stress and crisis events vary, individuals and families who are particularly resilient tend to have certain characteristics. **Resiliency** refers to a family's strengths and ability to respond to a crisis in a positive way. Several characteristics associated with resilient families include having a joint cause or purpose, emotional support for each other, good problem-solving skills, the ability to delay gratification, flexibility, accessing residual resources, communication, and commitment (Walsh, 2003; DeHaan, 2002; Patterson, 2002). A family's ability to bounce back from a crisis (from loss of one's job to the death of a family member) reflects its level of resiliency. Resiliency may also be related to individuals' perception of the degree to which they are in control of their destiny. The Self-Assessment on page 306 measures this perception.

When it is dark enough, you can see the stars.

Charles Beard

A Family Stress Model

Various theorists have explained how individuals and families experience and respond to stressors.

Burr and Klein (1994) reviewed the ABC-X model of family stress, developed by Reuben Hill in the 1950s. The model can be explained as follows:

A = stressor event

B = family's management strategies, coping skills

C = family's perception, definition of the situation

X = family's adaptation to the event

A is the stressor event, which interacts with B, which is the family's coping ability, or crisis-meeting resources. Both A and B interact with C, which is the family's appraisal or perception of the stressor event. X is the family's adaptation to the crisis. Thus, a family that experiences a major stressor (e.g., spinal cord-injured spouse) but has great coping skills (e.g., high spirituality, love, communication, commitment) and perceives the event to be manageable will experience a moderate crisis. But a family that experiences a less critical stressor event (e.g., child makes Cs and Ds in school) but has minimal coping skills (e.g., everyone blames everyone else) and perceives the event to be catastrophic will experience an extreme crisis. Hence, how a family experiences and responds to stress depends not only on the event but also on the family's coping resources and perceptions of the event.

POSITIVE STRESS MANAGEMENT STRATEGIES

Researchers Burr and Klein (1994) administered an eighty-item questionnaire to seventy-eight adults to assess how families experiencing various stressors such as bankruptcy, infertility, disabled child, and troubled teen used various coping strategies and how useful they evaluated these strategies. Below we detail some helpful stress-management strategies.

Authors

Exercise is one of the most effective ways of remaining healthy and reducing stress.

Changing Basic Values and Perspective

The strategy that the highest percentage of respondents reported as being helpful was changing basic values as a result of the crisis situation. After a fall from a horse paralyzed him, Christopher Reeve reported a complete change in his basic values. Previously, as an avid scuba diver, equestrian, and athlete, he had placed great value on his athleticism. Now his limbs were dead weight. "You begin to see," he said, "that your body is not you and that your mind and spirit must take over."

In responding to the crisis of bankruptcy, people may reevaluate the importance of money and conclude that relationships are more important. In coping with unemployment, people may decide that the amount of time they spend with family members is more valuable than the amount of time they spent making money.

Baldwin, Chambliss, and Towler (2003) found that, among a sample of African-American college students, those with an optimistic attitude toward the future tended to report less perceived and global stress than their pessimistic counterparts. Hence, one's basic view of life has its benefits independent of any particular event.

Exercise

Exercise has also been associated with successful crisis coping and better health (Finucane et al., 1997). Exercise is also an effective stress reducer. Williams and Lord (1997) studied 187 older women who became involved in a twelve-month program of group exercise and found that anxiety reduction was associated with their exercise. Physiological and cognitive benefits were also realized. The Centers for Disease Control and Prevention (CDC) and the American College of Sports Medicine (ACSM) recommend that people aged 6 and older engage regularly, preferably daily, in light to moderate physical activity for at least thirty minutes at a time.

Biofeedback

Biofeedback is a process in which information that is relayed back to the brain enables a person to change his or her biological functioning. Biofeedback treatment teaches a person to influence biological responses such as heart rate, nervous system arousal, muscle contractions, and even brain wave functioning. Biofeedback is used at about 1,500 clinics and treatment centers worldwide. A typical session lasts about an hour and costs $60 to $150.

There are several types of biofeedback:

1. Electromyographic (EMG) biofeedback, which measures electrical activity created by muscle contractions, is often used for relaxation training, stress, and pain management.

2. Thermal or temperature biofeedback uses a temperature sensor to detect changes in temperature of the fingertips or toes. Stress causes blood vessels in the fingers to constrict, reducing blood flow, leading to cooling. Thermal biofeedback, which trains people to quiet the nervous system arousal mechanisms that produce hand and or foot cooling, is often used for stress, anxiety, and pain management.

3. Galvanic skin response (GSR) biofeedback utilizes a finger electrode to measure sweat gland activity. This measure is very useful for relaxation and stress management training, and is also used in the treatment of attention deficit/hyperactivity disorder.

4. ***Neurofeedback,*** also called *neurobiofeedback* or *EEG (electroencephalogram) biofeedback,* may be particularly helpful for individuals coping with a crisis. It trains people to enhance their brain wave functioning and has been found to be effective in treating a wide range of conditions, including anxiety, stress, depression, tension and migraine headaches, addictions, and high blood pressure.

As neurofeedback is the fastest-growing field in biofeedback, we take a closer look at this treatment modality. Neurofeedback involves a series of sessions in which a client sits in a comfortable chair facing a specialized game computer. Small sensors are placed on the scalp to detect brain-wave activity and transmit this information to the computer. The neurofeedback therapist (in the same room) also sits in front of a computer that displays the client's brain-wave patterns in the form of an electroencephalogram (EEG). After a clinical assessment of the client's functioning, the therapist determines what kinds of brain-wave patterns are optimal for the client. During neurofeedback sessions, clients learn to produce desirable brain waves by controlling a computerized game or task, similar to playing a videogame, but instead of a joystick, it is the client's brain waves that control the game.

Neurofeedback is like an exercise of the brain, helping it to become more flexible and effective. But unlike body exercise, which when training is stopped will lose its benefits over time, brain-wave training generally does not. Once the brain is trained to function in its optimal state (which may take an average of twenty to twenty-five sessions), it generally remains in this more healthy state. Neurofeedback therapists liken the process to that of learning to ride a bicycle—once you learn to ride a bike, you can do so even if it has been years since you have ridden one. One neurofeedback therapist explains that "clients speak often of their disorders—panic attacks, chronic pain, etc.—as if they were stuck in a certain pattern of response. Consistently in clinical practice, EEG biofeedback helps 'unstick' people from these unhealthy response patterns" (Carlson-Catalano, 2003).

Sleep

Getting an adequate amount of sleep is also associated with low stress levels. Fourteen percent of a sample of adults who slept seven to eight hours each night reported feeling stress, in contrast to 43 percent who slept fewer than six hours each night (Kate, 1994). In spite of the need for sleep, most Americans feel sleep-deprived.

Love

A love relationship also helps an individual cope with stress. Forty-four percent of 351 undergraduates responded "a lot" or "some" when asked how important a love relationship was in helping them cope with life. Five percent said "a great deal" (Langehough et al., 1997). Similarly, intimacy in one's marriage (Harper, Schaalje, and Sandberg, 2000), as well as being able to talk with family members (Walker, 2000), is associated with stress reduction.

Religion and Spirituality

One hundred African-American males and 103 African-American females identified spiritual support as their primary method of coping with life's difficulties (Blake & Darling, 1998). Spirituality, defined as having purpose and meaning in life, having inner resources, feeling connected to others, and being able to transcend one's physical or

Neurofeedback can be helpful in reducing anxiety, stress, and tension related to crisis. It can be effective with children as well as adults.

Authors

Authors

This couple has been married over fifty years and use humor to cope with various health issues. They point out that the jackass is the one in the middle.

psychological condition (McGee, Nagel, & Moore, 2003) may also be positively related to reducing stress and coping with crisis events.

Friends and Relatives

Reaching out to friends and relatives was a successful strategy employed by a sample of adults who reported their reaction to 9/11. The authors of the study noted, "by doing so, people do not feel so alone and vulnerable in the world and perhaps this enables them to prepare themselves for the worst (another effective strategy) (Alexandrian et al., 2002).

Multiple Roles

Another factor that helps individuals cope with stress is to be involved in multiple roles. Two researchers studied the psychological consequences of multiple roles of 461 mothers over the age of 55 who occupied an average of five roles—parent of a nonhandicapped child, spouse, relative, friend, and parent of an adult child with mental retardation (Hong & Seltzer, 1995). They found that the greater the number of roles, the lower the depression and the higher the psychological well-being. Starting a new job (adding even another role) also had positive psychological effects.

He who laughs, lasts.
Mary Poolegive

If your needs are not being met, drop some of your needs.
George Carlin, *Comedian*

Humor

A sense of humor may be defined as the ability to find something amusing in a situation and laugh. It is associated with a number of positive outcomes, including stress reduction, physical health, mental well-being, and life satisfaction. In general, the greater the humor, the more satisfied people are with their lives

Other Helpful Strategies

Wallerstein and Blakeslee (1995) identified other management strategies that spouses in fifty successful marriages found helpful in responding to a crisis.

1. Intervened early in a crisis. When the spouses saw a crisis coming, they intervened early. Whether the crisis is depression or alcoholism or violence, early intervention helps to prevent the problem from mushrooming into a much larger issue. Cano et al. (2002) noted that stressor events may create marital dissatisfaction that if not dealt with could lead to continued dissatisfaction. They encouraged couples to seek therapy if early resolution is not forthcoming.

2. Avoided blame. The spouses did not blame each other for the crisis. Phrases such as "If you had done what I told you, you wouldn't have lost your job" or "If you had gotten x-rayed earlier, you would have had time to treat the cancer" were not spoken. Rather, the spouses were careful not only to avoid blaming each other but also to protect each other from self-blame and self-reproach.

3. Kept destructive impulses in check. Recognizing that fear can lead to anxiety and helpless rage, the spouses sought to avoid "driving away the people they needed and loved the most" (p. 123). Rather than exacerbate the crisis by responding to it with anger, drug/alcohol abuse, or emotional withdrawal, spouses made a great effort to "keep destructive tendencies from getting out of control and harming the marriage."

4. Sought opportunities for fun. In spite of the tragedy the couples were coping with, they sought to enjoy life as best they could. They took the view of seizing opportunities for fun rather than spiraling downward into a deep depression. Whether taking trips, going out to eat, or renting a movie, they still sought to enjoy life. "They tried their best not to let the tragedy totally dominate their world" (p. 123).

Christakis and Iwashyna (2003) studied the health impact on spouses who had very sick partners and found that those who used hospice during the final months of their partner's life were less likely to die themselves eighteen months after the death of their partners. Their finding emphasizes the stress toll of caregiving and the benefit of seeking help.

HARMFUL STRATEGIES

Some coping strategies not only are ineffective for resolving family problems but also add to the family's stress by making the problem worse. Respondents in the Burr and Klein (1994) research identified several strategies they regarded as harmful to overall family functioning. These included keeping feelings inside, taking out frustrations on others, and denying or avoiding the problem.

Burr and Klein's research also suggests that there are differences between women and men in their perceptions of the usefulness of various coping strategies. Women were more likely than men to view such strategies as sharing concerns with relatives and friends, becoming more involved in religion, and expressing emotions as helpful. Men were more likely than women to view potentially harmful strategies such as using alcohol, keeping feelings inside, and keeping others from knowing how bad the situation was as helpful coping strategies. Bouchard et al. (1998) also found that men were more likely than women to use denial as a coping mechanism.

FIVE FAMILY CRISES

Some of the more common crisis events faced by spouses and families include physical illness, extramarital affair, unemployment, alcohol/drug abuse, and death. Before we examine each of these, we take a look at the concept of "midlife crisis" in Up Close 13.1 on page 312.

Physical Illness and Disability

"Major illness/injury to self" was ranked the number 3 most stressful life event (from a list of 51) by over 3,000 adult respondents (death of spouse and death of close family member were numbers 1 and 2). Most who promise "in sickness and in health" on their wedding day never anticipate that such illness and disability will happen to them, yet such an event in the marriage of a couple is not unusual. Almost a quarter (23 percent) of 1,393 adults aged 18 to 55 reported that they had had a major illness or accident that required them to spend a week or more in the hospital (Turner & Lloyd, 1995).

Although short-term illness and disability often produce stress in the family, long-term illness and disability have profound and enduring effects on family members and family life. Smith and Soliday (2001) studied the effects of chronic kidney disease on the family. Among the concerns of their 123 parents who were either on dialysis or had had a kidney transplant were financial problems ("My job interfered with my dialysis, so I had to quit"), gender role loss ("I can't cook for my family"/"I am no longer a provider"), and perceptions by their children ("I feel they have seen me sick for so long they are tired of trying to deal with it and me").

Up Close 13.1 Midlife Crisis?

The stereotypical explanation for the 45-year-old person who buys a convertible sports car, has an affair, marries a 20-year-old or adopts a baby is "they are having a midlife crisis." The label conveys that the person feels old, thinks that life is passing him or her by, and seizes one last great chance to do what he or she has always wanted. However, a ten-year study of close to eight thousand U.S. adults aged 25 to 74 by the MacArthur Foundation Research Network on Successful Midlife Development revealed that, for most respondents, the middle years were no crisis at all but a time of good health, productive activity, and community involvement. Fewer than a quarter (23%) reported a "crisis" in their lives. Those who did experience a crisis were going through a divorce. Two-thirds were accepting of their getting older; one-third did feel some personal turmoil related to the fact that they were aging (Goode, 1999).

Reference

Goode, E. 1999. New study finds middle age is prime of life. *New York Times*. 17 July, D6.

Lou Gehrig's disease is an example of an illness that has severe long-term effects on the family. One wife described her family's experience:

> *After years of a marriage in which we were both healthy and had two terrific children, my husband was diagnosed with what is commonly known as Lou Gehrig's disease. It is a disease of the muscles which is progressive, degenerative, and fatal. There is no cure. My husband went from an independent, strong, head baseball coach at East Carolina University to a frail man who cannot feed himself and needs twenty-four-hour care. He can only "speak" with the use of a computer, which he activates, with the movement of his eyes.*
>
> *The effect on our lives, marriage, and family has been enormous. But our religious faith has helped us to pull together to rise above this crisis. We feel blessed. Life is hard but we were never promised a rose garden. (Authors' files)*

This couple reflect the true meaning of "in sickness and in health." Several years after this photo was taken, the husband developed Lou Gehrig's disease and was confined to a wheelchair.

Authors

Booth and Johnson (1994) studied a national sample of 1,298 spouses with regard to how the change in the health of one partner affected the quality of the marital relationship over a three-year period. They found that spouses of sick partners experienced a greater decline in marital quality than did the spouses who themselves became ill. "Changes in the financial circumstances, shifts in the division of labor, declines in marital interaction, and problematic behavior by the afflicted individual account for much of the health-marital quality relationship" (p. 222).

In their study of fifty couples with successful marriages, Wallerstein and Blakeslee (1995) noted that a major developmental task is to confront the inevitable and unpredictable adversities of life in ways that enhance the relationship despite suffering. Rolland (1994) identified several issues that spouses might consider discussing when one of them becomes seriously ill.

Intimacy and Threatened Loss Although the definitions of intimacy differ by social class, intimacy often includes emotional, intellectual, sexual, spiritual, and recreational aspects. The diagnosis of a serious condition can threaten the continuation of such intimacy, which may result in partners' either "pulling away from one another or clinging to each other in a fused way" (Rolland, 1994, 329).

Hence, rather than retreat into their respective thoughts of doom and gloom, spouses might benefit from talking about the disorder in a way that frames it in a positive light. Researchers have consistently shown that the choice to view something positively, as a challenge, is a major coping mechanism. For example, rather than viewing cancer as debilitating one partner, the partners can choose to regard it as a challenge to face adversity together.

While some spouses cope with the disability of each other, others cope with the disability of their children. A profound source of stress for parents is caring for a disabled child and the concern associated with who will care for the child upon the parents' death. Life insurance on both parents is one way of coping with such stress (King, 1998).

Establishing Healthy Boundaries Couples coping with physical illness may need to express emotions and discuss concerns related to the illness. However, if these emotions and concerns permeate every discussion and activity in the marriage, then the illness becomes the focus of the relationship. Thus, couples are challenged to establish healthy boundaries between illness-related emotions and concerns and other aspects of the relationship. For example, couples might agree to limit talks related to illness to "before noon" and to the "living room."

Togetherness and Separateness Couples must also balance the need to be together with the need to spend time away from each other. "Couples adapt best to chronic disorders when they can transform their understanding of 'we-ness' to include a new version of separateness that acknowledges different needs and realities" (Rolland, 1994, 336). Hence, while the partner with the malady may have fears of disability and death that may translate into the desire for more closeness, the other partner may "pull away in detachment that represents preparation for the final separation of death" (p. 336). The respective needs to become closer and to separate are normative, and an openness between the partners about their feelings is preferable to silence. Cathy Crimmins (2000) noted the importance of maintaining relationships with others where the caretaker role is absent. Her husband suffered a traumatic brain injury and she became his constant caretaker. Though she loved and was committed to her husband and his rehabilitation, she noted the necessity of keeping a part of her life separate from the caretaker role. Indeed, all couples, even those not experiencing a crisis, need to seek a balance between togetherness and separateness. A crisis event such as illness or disability may bring this need into focus.

In those cases where the illness is fatal, **palliative care** is helpful. This term describes the health care (the focus is on relief of pain and suffering) of the individual who has a life-threatening illness and support for them and their loved ones. Such care may involve the person's physician or a palliative care specialist who works with the physician, nurse, social worker, and chaplain. Pharmacists or rehabilitation specialists may also be involved. The effects of such care are to approach the end of life with planning (how long should life be sustained on machines?) and forethought, to relieve pain, and to provide closure.

We have been discussing the management of physical illness. But not all illnesses are physical. Some are mental. Everyone experiences problems in living from time to time, including relationship problems, criminal or domestic victimization, work-related problems, and low self-esteem. However, some conditions or mental disorders are more intense, persistent, and debilitating. Many are manageable. Hyman (2000) studied thirty-two physicians who had been diagnosed with bipolar disorder (4 million individuals in the United States live with this disorder) and observed their information strategies (they were selective in whom they told of their disorder), time management strategies (some carefully structured every minute of their days), stress management (exercise, yoga, swimming), and symptom management. To manage symptoms they stayed alert to the warning signs predictive of a relapse (decrease in sleep, forgetfulness, hearing voices) and sought help immediately.

National Data

In a study of more than eight thousand respondents between the ages of 15 and 54, almost half (48%) reported having experienced at least one psychiatric disorder at some time in their lives, and nearly 30 percent reported experiencing at least one disorder in the previous twelve months. The most common disorders were depression, alcohol dependence, and social phobias (Kessler et al., 1994).

The toll of mental illness on a relationship can be immense. A major initial attraction of partners to each other includes intellectual and emotional qualities. When these are affected, the partner may feel that the mate has already died psychologically, since he or she is, literally, "not the same person I married." Chambless et al. (2002) studied couples in which the wife was agoraphobic (fear of being in open public places with the result that the person prefers to stay at home) reported higher levels of marital distress, higher rates of negative nonverbal behavior, and longer sequences of negative exchanges.

While we have focused this section on the physical and mental disability of spouses, the physical and mental disability of a child represents a similar crisis event for a couple and a family. Physical disabilities can begin at conception through the prenatal period and include cerebral palsy (10,000 babies per year), fetal alcohol syndrome (5,000 babies per year), and spina bifida (2 of every 10,000 births). Mental disabilities may involve autism (three to four times more common in boys), attention-deficit/hyperactivity disorder (4% of schoolchildren), and antisocial behavior. These difficulties can stress a couple and family to the limit of their coping capacity not only at onset but also across time (Burke, Eakes, & Hainsworth, 1999). Initially, parents may blame themselves or each other for the disorder and this may destroy the marriage (Patterson, 1988).

In the rest of this chapter, we examine how spouses cope with the crisis events of an extramarital affair, unemployment, drug abuse, and death. Each of these events can be viewed either as devastating and the end of meaning in one's life or as an opportunity and challenge to rise above.

Extramarital Affair

Extramarital affair refers to the emotional and sexual involvement of a spouse with someone other than the mate.

National Data

Twenty-three percent of U.S. adult husbands and 12 percent of U.S. adult wives report ever having had intercourse with someone to whom they were not married (Wiederman, 1997).

International Data

The percentages of male respondents from the United States and Britain who reported having had extramarital sex in the previous twelve months were 4.7 and 4.8, respectively. The percentage of female respondents from the United States and Britain who reported having had extramarital sex in the past twelve months were 1.9 and 1.6, respectively (Michael et al., 2001).

Extradyadic involvement refers to all pair-bonded individuals who are emotionally and sexually involved with someone other than the partner. College students disapprove of extradyadic involvements. In a sample of 620 undergraduates, 69 percent said that they would "end a relationship with someone who cheated on me." Almost half (45%) reported that they had ended such a relationship. Though there were no significant differences between women and men, those who were in love, and those who had been emotionally and physically abused were more likely to have ended such a relationship (Knox et al., 2000).

An extramarital affair is ranked by marriage and family therapists as the second most stressful crisis event for a couple (physical abuse is number 1) (Olson et al. 2002). Characteristics associated with spouses who are more likely to have extramarital sex include being male, having a strong interest in sex, permissive sexual values, low subjective satisfaction in their existing relationship, employment outside the home, low church attendance, and greater sexual opportunities (Olson et al., 2002; Tres & Giesen, 2000).

Extramarital affairs range from brief sexual encounters to full-blown romantic affairs. Becoming more common today is the computer affair. Although legally an extramarital affair does not exist unless two persons (one being married) have intercourse, an on-line computer affair can be just as disruptive to a marriage or a couple's relationship. Computer friendships may move to feelings of intimacy, involve secrecy (one's partner does not know the level of involvement), include sexual tension (even though there is no overt sex), and take time, attention, energy, and affection away from one's partner. Schneider (2000) studied ninety-one women who had experienced serious adverse consequences from their partner's cybersex involvement, including loss of interest in relational sex. These women noted that the cyber affair was as emotionally painful as an off-line affair.

Gender Differences in Views of an Extramarital Affair Symbolic interactionists note that men and women have different views of an extramarital affair. In a *Psychology Today* survey, 26 percent of the men reported having extramarital sex without being emotionally involved, versus 3 percent of the women (Marano, 2003).

Reasons for Extramarital Involvements Spouses report a number of reasons why they become involved in an emotional or sexual encounter outside their marriage. Some of these reasons are discussed in the following subsections.

I was married, and so I was one of those cheating husbands. I was full of pot. I was full of cocaine. I'm not blaming the drugs but they certainly facilitate things.
George Carlin, *Comedian*

1. Variety, novelty, and excitement. Extradyadic sexual involvement may be motivated by the desire for variety, novelty, and excitement. One of the characteristics of sex in long-term committed relationships is the tendency for it to become routine. Early in a relationship, the partners cannot seem to have sex often enough. But with constant availability, the partners may achieve a level of satiation, and the attractiveness and excitement of sex with the primary partner seem to wane.

The **Coolidge Effect** is a term used to describe this waning of sexual excitement and the effect of novelty and variety on sexual arousal:

> *One day President and Mrs. Coolidge were visiting a government farm. Soon after their arrival, they were taken off on separate tours. When Mrs. Coolidge passed the chicken pens, she paused to ask the man in charge if the rooster copulated more than once each day. "Dozens of times," was the reply. "Please tell that to the President," Mrs. Coolidge requested. When the President passed the pens and was told about the rooster, he asked, "Same hen every time?" "Oh no, Mr. President, a different one each time." The President nodded slowly and then said, "Tell that to Mrs. Coolidge."*
> (Bermant, 1976, 76–77)

Whether or not individuals are biologically wired for monogamy continues to be debated. Monogamy among mammals is rare (from 3 to 10 percent) and monogamy tends to be the exception more often than the rule (Morell, 1998). Equally debated is whether, even if such biological wiring for plurality of partners does exist, such wiring justifies nonmonogamous behavior—that individuals are responsible for their decisions.

2. Workplace friendships. A common place for extramarital involvements to develop is the workplace (Treas & Giesen, 2000). Coworkers share the same world eight to ten hours a day and over a period of time may develop good feelings for each other that eventually lead to a sexual relationship. It is no coincidence that the Lewinsky/Clinton scandal developed in the workplace.

3. Relationship dissatisfaction. It is commonly believed that people who have affairs are not happy in their marriage. Spouses who feel misunderstood, unloved, and ignored sometimes turn to another who offers understanding, love, and attention. Treas and Giesen (2000) found in a nationally representative survey that lower subjective relationship satisfaction, among both cohabitants and marrieds, was associated with infidelity.

One source of relationship dissatisfaction is an unfulfilling sexual relationship. Some spouses engage in extramarital sex because their partner is not interested in sex. Others may go outside the relationship because their partners will not engage in the sexual behaviors they want and enjoy. The unwillingness of the spouse to engage in oral sex, anal intercourse, or a variety of sexual positions sometimes results in the other spouse's looking elsewhere for a more cooperative and willing sexual partner.

4. Revenge. Some extramarital sexual involvements are acts of revenge against one's spouse for having an affair. When partners find out that their mate has had or is having an affair, they are often hurt and angry. One response to this hurt and anger is to have an affair to get even with the unfaithful partner.

5. Homosexual relationship. Some individuals marry as a front for their homosexuality. Cole Porter, known for "I've Got You Under My Skin," "Night and Day," and "Easy to Love," was a homosexual who feared no one would buy or publish his music if his sexual orientation were known (*De-lovely* is a film biography released in 2004). He married Linda Lee Porter (alleged to be a lesbian), and they had an enduring marriage for thirty years.

Other gay individuals marry as a way of denying their homosexuality. These individuals are likely to feel unfulfilled in their marriage and may seek involvement in an extramarital homosexual relationship. Other individuals may marry and then discover later in life that they desire a homosexual relationship. Such individuals may feel that (1) they have been homosexual or bisexual all along, (2) their sexual orientation has changed from heterosexual to homosexual or bisexual, (3) they are unsure of their sexual orientation and want to explore a homosexual relationship, or (4) they are predominately heterosexual but wish to experience a homosexual relationship for variety.

6. Aging. A frequent motive for intercourse outside marriage is the desire to return to the feeling of youth. Ageism, discrimination against the elderly, promotes the idea that it is good to be young and bad to be old. Sexual attractiveness is equated with youth, and having an affair may confirm to an older partner that he or she is still sexually desirable. Also, people may try to recapture the love, excitement, adventure, and romance associated with youth by having an affair. The Lewinsky/Clinton affair makes the point—she was thirty years younger than he.

7. Absence from partner. One factor that may predispose a spouse to an affair is prolonged separation from the partner. Some wives whose husbands are away for military service report that the loneliness can become unbearable. Some husbands who are away say it is difficult to be faithful. Partners in commuter relationships may also be vulnerable to extradyadic sexual relationships.

Reactions to the knowledge that one's spouse has been unfaithful vary. While unusual, some spouses become enraged to the point of murder. Clara Harris of Houston, Texas, ran over her husband with her Mercedes-Benz after discovering him and his former receptionist in a hotel (she was sentenced to twenty years in prison) (Associated Press, 2003).

More often spouses suffer in private.

Forget sleeping for a while. Forget friendly conversations. Forget peace of mind. You will literally begin the process of grieving as if a death had occurred. And in reality a death has occurred. The total happiness with which you have embraced your spouse and your marriage is now dead. The innocence with which you have always pictured your spouse is dead. Trust is long gone and you now doubt every word said. You will consider yourself violated and your spouse has now become somewhat "unclean." You will have wide, sweeping mood changes within a five-minute period. (Author'files)

PERSONAL CHOICES

Should You Seek a Divorce If Your Partner Has an Affair?

In light of her husband's infidelities, a nation wondered why Hillary Clinton stayed with him. About 20 percent of spouses face the decision of whether to stay with a mate who has had an extramarital affair. One alternative is to end the relationship immediately on the premise that trust has been broken and can never be mended. Persons who take this position regard fidelity as a core element of the marriage that if violated requires a divorce. Not only has trust been abused, but the fear that the partner will repeat the behavior dictates that the marriage must end. Charney and Parnass (1995) analyzed sixty-two cases in which one spouse had had an affair. Of these, 34 percent of the marriages ended in divorce attributable to the affair.

Other couples respond to a partner's emotional and sexual involvement with acceptance. "It was a sin of weakness, not of meanness" noted Hillary Clinton (2003). Other couples have open relationships with others. In Chapter 1 we discussed polyamory, in which partners are open and encouraging of multiple relationships at the same time. The term *infidelity* does not exist for polyamorous couples.

Even for traditional couples, infidelity need not be the end of a couple's marriage but the beginning of a new, enhanced, and more understanding relationship. Healing takes time, commitment on the part of the straying partner not to repeat the behavior, forgiveness by the partner (and not bringing it up again), and a new focus on improving the relationship. Couples who have been through the crisis of infidelity admonish, "Don't make any quick decisions" (Olson et al., 2002, 433). In spite of the difficulty of adjusting to an affair, most spouses are reluctant to end a marriage. Not one of fifty successful couples said that they would automatically end their marriage over adultery (Wallerstein & Blakeslee, 1995).

The spouse who chooses to have an affair is often judged as being unfaithful to the vows of the marriage, as being deceitful to the partner, and as inflicting enormous pain on the partner (and children). What is often not considered is that when an affair is defined in terms of giving emotional energy, time, and economic resources to something or someone outside the primary relationship, other types of "affairs" may be equally as devastating to a relationship. Spouses who choose to devote their lives to their children, careers, parents, friends, or recreational interests may deprive the partner of significant amounts of emotional energy, time, and money and create a context in which the partner may choose to become involved with a person who provides more attention and interest.

Diversity in Other Countries

A man of the Masai tribe (200,000 of whom live in Arusha, Tanzania, East Africa), coming back from a long journey in the forest herding animals, can put a spear outside the house of another man's wife to alert the husband that he is inside having sex with his wife. Seeing the spear, the husband does not bother the visitor, since to do so would be impolite (Knox, 2001).

Societies also differ as to how serious a violation of morality they regard adultery. In the United States, the penalty ranges from nothing to alimony. In countries that practice the Islamic religion (in Northern Africa and parts of the Middle East), adultery is considered a major sin, and execution by stoning is considered appropriate. Confirmed adulterers (four eyewitnesses are needed to confirm their shameless behavior) are buried from the waist down and stoned to death (Dar, 2003).

The United States and France also view extramarital involvements differently. In the Unites States, the Clinton/Lewinsky scandal was big news and threatened the presidency. Knowledge of the adulterous relationship of France's president François Mitterand in which he fathered a child "was not seen as a matter worthy of much publicity or warranting critical comment" (Allan & Harrison, 2002, 46).

When spouses do stay together after an affair, the price is high. In the Charney and Parnass (1995) study, 43 percent of those who stayed together (66%) were judged to have difficulty coping: "[T]he majority of the betrayed husbands and wives suffered significant damage to their self-image, personal confidence, or sexual confidence, feelings of abandonment, attacks on their sense of belonging, betrayals of trust, enraged feelings, and/or a surge of justification to leave their spouses" (p. 100).

Among the respondents of researchers Charney and Parnass (1995) only 14.5 percent of those who remained together after an affair described their relationship as characterized by improvement and growth. Of these, almost half were relationships in which the affair was a one-night stand, not a long-term romantic and sexual involvement. (Interestingly, 89 percent of the betrayed spouses "really knew" that their partner had had an affair before being told.)

Another issue in deciding whether to take the spouse back following extramarital sex is the concern over HIV. One spouse noted that though he was willing to forgive and try to forget his partner's indiscretion, he required that she be tested for HIV and that they use a condom for six months. She tested negative, but their use of a condom was a reminder, he said, that sex outside one's bonded relationship in today's world has a life-or-death meaning. Related to this issue is that wives are more likely to develop cervical cancer if their husbands have other sexual partners.

There is no one way to respond to a partner who has an extramarital relationship. Most are hurt and think of ways to work through the crisis. Some succeed.

Sources

Charney, I. W., and S. Parnass. 1995. The impact of extramarital relationships on the continuation of marriages. *Journal of Sex and Marital Therapy* 21:100–15.

Clinton, H. 2003. *Living history*. New York: Simon and Schuster.

Olson, M. M., C. S. Russell, M. Higgins-Kessler, and R. B. Miller. 2002. Emotional processes following disclosure of an extramarital affair. *Journal of Marital and Family Therapy* 28:423–34.

Wallerstein, J. S., and S. Blakeslee. 1995. *The good marriage*. Boston: Houghton Mifflin.

Unemployment

Corporate America continues to downsize. The result is massive layoffs and insecurity in the lives of American workers. Unemployment and/or the threatened loss of one's job are major stressors for individuals, couples, and families. Also, when spouses or parents lose their jobs as a result of physical illness or disability, the family experiences a double blow—loss of income combines with high medical bills. Unless the unemployed spouse is covered by the partner's medical insurance, unemployment can also result in loss of health insurance for the family. Insurance for both health care and disability is very important to help protect the family from an economic disaster. Long-term health care coverage may also be important for the elderly couple who is no longer employed.

The effects of unemployment may be more severe for men than for women. Our society expects men to be the primary breadwinners in their families and equates masculine self-worth and identity with job and income. Stress, depression, suicide, alcohol abuse, and lowered self-esteem are all associated with unemployment. Macmillan and Gartner (2000) also observed that men who are unemployed are more likely to be violent toward their working wives.

Women tend to adjust more easily to unemployment than men. Women are not burdened with the cultural expectation of the provider role and their identity is less tied to their work role. Hence, women may view unemployment as an opportunity to spend more time with their families; many enjoy doing so.

Regardless of which partner is unemployed, the fact that one spouse loses a job will impact the balance of power in the relationship and the expected division of labor—e.g., the unemployed husband is expected to do more housework. However, some couples report that the loss of a job by one spouse is associated with an improved marital relationship (Burr & Klein, 1994). Sharing the event, talking about it, and feeling greater commitment to each other helped to produce a positive outcome for a third of their respondents adjusting to unemployment.

While unemployment can be stressful, increasing numbers of workers are experiencing job stress. With layoffs and downsizing, workers are given the work of two and told "we'll hire someone soon." Meanwhile, the employer learns it can pay once and get the work of two. The result is lowered morale, exhaustion, and stress that can escalate into violence. Workers note that taking frequent breaks or a day off is their best way of coping with job stress (Armour, 2003).

Drug Abuse

Spouses, parents, and children who abuse drugs contribute to the stress and conflict experienced in their respective marriages and families (Keller et al., 2002). Although some individuals abuse drugs to escape from physically or sexually abusive relationships (Bartholomew & Klein, 2000) or the stress of family problems, drug abuse inevitably adds to the family's problems when it results in health and medical problems, legal problems, loss of employment, financial ruin, school failure, relationship conflict, and even death. Country-and-western singer Lorrie Morgan said of her husband, Keith Whitley, an alcoholic, who died of alcohol poisoning—with twenty shots of 100-proof whiskey in his system. She said of his alcoholism, "You can't love someone into sobriety. It just doesn't work." Collins, Grella, and Hser (2003) found that involvement in the role of parent, particularly for fathers, was associated with lower levels of addiction. The researchers suggested that relationship enhancement with children was one avenue to explore in the treatment of drug addiction.

Family crises involving alcohol and/or drugs are not unusual. Alcohol is also a major problem on campus (over half of 772 college students who reported drinking alcohol reported having blacked out—White, Jamieson-Drake, & Swartzwelder, 2002). Univerisity attempts (mostly unsuccessful) have been made to curb excessive drinking (see the Social Policy on page 320).

As indicated in Table 13.1, drug use is most prevalent among 18- to 25-year-olds. Drug use among teenagers under age 18 is also high. Because teenage drug use is more common than drug use among older adults, we focus on the problems parents confront when their teenagers abuse drugs.

Teenage Drug Abuse Parents are usually unaware of drug use on the part of their teenagers. Brogenschneider et al. (1998) studied 199 white mother-adolescent dyads and 144 white father-adolescent dyads in regard to what parents knew

Table 13.1 Drug Use (Ever), by Type of Drug and Age Group

Type of Drug	Age 12 to 17	Age 18 to 25	Age 26 to 34
Marijuana and hashish	20%	50%	48%
Cocaine	2	13	16
Alcohol	43	85	no data
Cigarettes	34	69	no data

Source: Adapted from *Statistical Abstract of the United States: 2003.* 123rd ed. Washington D.C.: U.S. Bureau of the Census, Table 201.

Alcohol Abuse on Campus

SOCIAL POLICY

Alcohol is the drug most frequently used by college students. University towns promote drinking by offering "All you can drink" on Sunday nights or "happy hours." About half of college students report having drunk wine or liquor in the previous year (American Council on Education and the University of California, 2002). And in a study of 72,000 18- to 25-year-olds (some of whom were college students), almost 40 percent (38.7) reported "binge" drinking in the last year—defined as drinking five or more drinks on the same occasion (*Statistical Abstract of the United States: 2003*). Between a half and 60 percent report having engaged in a "drinking game" (e.g., taking a drink whenever a certain word is mentioned in a song or television) (Borsari & Bergen-Cico, 2003). College students who drink alcohol heavily are more likely to miss class, make lower grades, have auto accidents, have low self-esteem, and be depressed (Williams et al., 2002). Nevertheless, unless a college student drinks every day, he or she is unlikely to define himself or herself as having a drinking problem (Lederman et al., 2003).

Campus policies include alcohol bans, enforcement and sanctions, and peer support and education. Banning alcohol is infrequent since administrators fear that students will attend other colleges where they are allowed to drink. And some alumni want to drink at football games and view such university banning as intrusive. Some attorneys think colleges and universities can be held liable for not stopping dangerous drinking patterns, but other people argue that college is a place for students to learn how to behave responsibly. However, universities are considering ways to limit alcohol use on campus (McKinnon, O'Rourke, & Byrd, 2003).

The use of enforcement and sanctions is also debated. Officials suggest that enforcement must be balanced. If police are too restrictive, drinking will go underground, where it is more difficult to detect.

Peer support and education may be the most helpful. Students at the University of Buffalo in New York created a video illustrating responsible drinking and discussions by students of their own alcohol poisoning. Such educational interventions are associated with more accurate knowledge of alcohol and its effects. However, whether there is subsequent reduced drinking is unknown (Williams et al., 2002). Rutgers regularly conducts campus focus groups and recommends replacement of the term "binge drinking" with "dangerous drinking" since the latter emphasizes outcomes (Lederman et al., 2003).

Some universities have hired full-time alcohol-education coordinators. Providing nonalcoholic ways to meet others and to socialize has also been suggested (Borsari & Bergen-Cico, 2003). Policies will continue to shift in reference to parental pressure to address the issue. The latest approach has been to inform parents of their son's or daughter's alcohol or drug abuse on campus. An alcohol or drug infraction is reported to the parents, who may intervene early in curbing abuse. Kuo et al. (2002) emphasized that parents may play a potentially important role in prevention efforts.

Sources

American Council on Education and University of California. 2002. *The American freshman: National norms for fall, 2002.* Los Angeles: Higher Education Research Institute. U.C.L.A. Graduate School of Education and Information Studies.

Borsari, B., and D. Bergen-Cico. 2003. Self-reported drinking-game participation of incoming college students. *Journal of American College Health* 51:149–54.

Kuo, M., E. M. Adlaf, H. Lee, L. Gliksman, A. Demers, and H. Wechsler. 2002. More Canadian students drink but Amerian students drink more: Comparing college alcohol use in two countries. *Addiction* 97:1583–92.

Lederman, L. C., L. P. Stewart, F. W. Goodhart, and L. Laitman. 2003. A case against "binge" as a term of choice: Convincing college students to personalize messages about dangerous drinking. *Journal of Health Communication* 8:79–91.

McKinnon, S., K. O'Rourke, and T. Byrd. 2003. Increased risk of alcohol abuse among college students living on the US-Mexico border: Implications for prevention. *Journal of American College Health* 51:163–67.

Statistical Abstract of the United States: 2003. 123rd ed. Washington D.C.: U.S. Bureau of the Census, Table 201.

Williams, D. J., A. Thomas, W. C. Buboltz, Jr., and M. McKinney. 2002. Changing the attitudes that predict underage drinking in college students: A program evaluation. *Journal of College Counseling* 5:39–49.

about their adolescent's alcohol behavior. Whereas *all* adolescents in the study reported regular alcohol use, less than a third of their parents were aware of such use. The title of the researchers' article was "Other Teens Drink, but Not My Kid."

Various factors associated with teenage alcohol and drug abuse include peers who use drugs, low self-esteem, alcoholic parents, low grades, low IQ, being male, being from a single-parent home, lacking support from parents, and being from a higher social class (Keller et al., 2002; Kylin, Meschke, & Borden, 2000). For the drug abuser, the use of alcohol or other drugs may cause a problem in many areas of life—health, school, work, and social relationships. Parents can best prevent their children from using drugs by creating and maintaining close, healthy, supportive relationships with them (Tuttle, 1995). Parker and Benson (2000) emphasized the importance of parents' being "watchful" of their adolescents and being supportive of their interests and activities. Other preventive measures on the part of parents include not abusing alcohol or other drugs themselves and keeping communication channels open with their children.

Drug Abuse Support Groups If the substance-abuse problem is alcohol, Alcoholics Anonymous (AA) is an appropriate support group (national headquarters mailing address: AA General Service Office, P.O. Box 459, Grand Central Station, New York, NY 10017). There are over 15,000 AA chapters nationwide; the one in your community can be found through the Yellow Pages. The only requirement for membership is the desire to stop drinking.

Former drug (other than alcohol) abusers meet regularly in local chapters of Narcotics Anonymous (NA), patterned after Alcoholics Anonymous, to help each other continue to be drug-free. As with AA, the premise of NA is that the best person to help someone stop abusing drugs is someone who once abused drugs. NA members of all ages, social classes, and educational levels provide a sense of support for each other to remain drug-free.

Al-Anon is an organization that provides support for family members and friends of alcohol abusers. Such support is also often helpful for parents coping with a teenage drug abuser (O'Farrell & Fals-Stewart, 2003).

In an attempt to improve parent-child communication in families where parents use drugs, the Strengthening Families Program provides specific social skills training for both parents and children (Wyman, 1997). After families attend a five-hour retreat, parents and children are involved in face-to-face skills training over a four-month period. A twelve-month follow-up revealed that parenting skills remained improved and reported heroin and cocaine use had declined.

Death

Even more devastating than drug abuse are family crises involving death—of one's child, parent, or loved one (we discuss the death of one's spouse in Chapter 17, Aging in Marriage and Family Relationships). The crisis is particularly acute when the death is a suicide.

Grief fills up the room with my absent child.
William Shakespeare

Death of One's Child A parent's worst fear is the death of a child. Most people expect the death of their parents but not the death of their children. Brotherson (2000) reports that the death of an infant has a more severe impact than any other type of loss, "leaving one life ended and another indelibly changed." Jiong Li et al. (2003) found that the loss is particularly devastating to mothers, who experience a higher mortality risk after their child's death.

Many parents experience the loss of their child even before it is born. About 15 percent of pregnancies end in a miscarriage (Nielsen & Hahlin, 1995). Although some parents may be relieved by a miscarriage if the pregnancy was unwanted, many parents feel sadness, frustration, disappointment, and anger. Some women blame themselves for the miscarriage and believe that they are being punished for something they have done in the past.

Among infants under 1 year of age, the leading causes of death in the United States are congenital anomalies, sudden infant death syndrome (SIDS), disorders relating to short gestation and low birth weight, and respiratory distress syndrome. Between the ages of 1 and 24, accidents are the primary cause of death. Types of accidental deaths include drowning and poisoning from ingesting household products or medication. But the most common cause of accidental death among youth is motor vehicle accidents (Anderson & Smith, 2003).

Brotherson (2000) interviewed nineteen bereaved parents (thirteen mothers and six fathers) and observed that both parents reacted to the loss of their child as the loss of a role that was focused on protecting and taking care of their child. Fathers were angry and mothers were vulnerable and "out of control."

Mothers and fathers sometimes respond to the death of their child in different ways. When they do, the respective partners may interpret these differences in negative ways, leading to relationship conflict and unhappiness. For example, after the death of their 17-year-old son, one wife accused her husband of not sharing in her grief. The husband explained that while he was deeply grieved he poured his grief into working more as a way of distraction. To deal with these differences, spouses might need to be patient and practice tolerance in allowing each other to grieve in his or her own way.

Sobieski (1994) suggests that spouses who experience the loss of a child also experience changes in their marital relationship. Their marriage may change in emphasis, direction, or quality. It may grow stronger or it may deteriorate. Previous differences in values and beliefs of spouses may become more apparent following the death of their child. For example, one partner may begin to attend church and seek solace through religion. The other partner may question religion and lose faith in God for allowing such a tragedy to happen. Sobieski (1994, 16) advises couples who have experienced the death of a child to discuss the following issues:

- *Intimacy and sexual needs*
- *Views and feelings about having other children*
- *Methods of childrearing to be used for the surviving child or children*

Death of One's Parent Terminally ill parents may be taken care of by their children. Such care over a period of years can be emotionally stressful, financially draining, and exhausting. Hence, by the time the parent dies, a crisis has already occurred.

Reactions to the death of one's parents include depression, loss of concentration, and anger (Ellis & Granger, 2002). Umberson (1995) studied the marital effects of the death of a parent on forty-two individuals and compared them with a group of spouses who had not lost a parent. In addition to individual grief, she observed that marital quality declined with the death of one of the spouse's parents:

> *Results suggest that this decline may occur because the partner fails to provide desired emotional support, the partner cannot comprehend the significance and meaning of the loss, the partner is disappointed in the bereaved individual's inability to recover quickly, because individuals have new support needs following the death of a parent and these needs may be difficult to meet, and because some partners feel imposed upon—particularly by continuing distress and depression in the bereaved person.* (P. 721)

Whether the death is that of a child or a parent, Burke, Eakes, and Hainsworth (1999) noted that grief is not a one-time experience that people adjust to and move on. Rather, for some, there is "chronic sorrow" where grief-related feelings occur periodically throughout the lives of those left behind. Paul Newman, the aging Hollywood celebrity, was asked how he got over the death of his son who overdosed. He replied, "You never get over it." Grief feelings may be particularly acute on the anniversary of the death or when the bereaved individual thinks of what might have been had the person lived. Burke, Eakes, and Hainsworth (1999) noted that 97 percent of the individuals in one study who had experienced the death of a loved one two to twenty years earlier met the criteria for chronic sorrow. Field, Gal-Oz, and Bananno (2003) also observed bereavement-related distress five years after the death of a spouse.

Surviving the Suicide of a Loved One Worldwide, about 800,000 individuals elect to commit suicide annually (Mercy et al., 2003). Research suggests that those who experience the death of a friend or loved one by suicide (i.e., suicide survivors) tend to experience different grief reactions than those whose loved ones die from accidents or natural causes (Stillion, 1996). Many family members of suicide victims experience higher levels of guilt, shame, rejection, and anger. Families of suicide victims are often left with questions about why their loved ones killed themselves and what could have been done to prevent the suicide. Although such questions are generally unanswerable, they often linger for years, prolonging the grieving process. In addition, survivors of suicide victims struggle with how others will view them and their family.

Because some religions consider suicide a mortal sin, surviving family members may also struggle with what the fate of their deceased loved one might be in the hereafter. Suicide among young people is particularly devastating not only because a young life is ended but because of the impact on the victims' family members and friends. A team of researchers identified 15,555 suicides among 15–24-year-olds in thirty-four countries in a one-year period and found an association between divorce rates and suicide rates (Johnson, Krug, & Potter, 2000).

Family members who are left behind after the suicide of a loved one often feel that there is something they could have done to prevent the death. Singer Judy Collins lost her son to suicide and began to attend a support group for persons who had a loved one who had committed suicide. One of the members in the group answered the question of whether there was something she could have done with a resounding no.

> *I was sitting on his bed saying "I love you, Jim. Don't do this. How can you do this?" I had my hand on his hand, my cheek on his cheek. He said excuse me, reached his other hand around, took the gun from under the pillow, and blew his head off. My face was inches from his. If somebody wants to kill themselves, there is nothing you can do to stop them.* (Collins, 1998, 210)

MARRIAGE AND FAMILY THERAPY

Couples might consider marriage and family therapy rather than continue to drag through dissatisfaction related to a crisis event. Signs to look for in your own relationship that suggest you might consider seeing a therapist include feeling distant and not wanting or being unable to communicate with your partner, avoiding each other, drinking heavily or taking other drugs, privately contemplating separation, being involved in an affair, and feeling depressed.

If you are experiencing one or more of these problems, it may be wise not to wait until the relationship reaches a stage beyond which repair is impossible. Relationships are like boats. A small leak will not sink them. But if left unattended, the small leak may grow larger or new leaks may break through. Marriage therapy sometimes serves to mend relationship problems early by helping the partners to sort out values, make decisions, and begin new behaviors so that they can start feeling better about each other.

> *The average total cost of marital therapy is about a thousand dollars for eleven sessions. Even twice this amount seems certainly less than the cost of a divorce and pales in comparison to the cost of many medical procedures.*
>
> James Bray, Ernest Bray, *Psychologists*

Availability of Marriage and Family Therapists

There are around 50,000 marriage and family therapists in the United States. About 40 percent are clinical members of the American Association for Marriage and Family Therapy (AAMFT). Currently there are fifty-seven masters, nineteen doctoral, and sixteen postgraduate programs accredited by the AAMFT. All but six states (Delaware, Montana, New York, North Dakota, Ohio, and West Virginia) and the District of Columbia regulate (e.g., require a license) for a person to practice marriage and family therapy (Northey, 2002).

Therapists holding membership in AAMFT have had graduate training in marriage and family therapy and a thousand hours of direct client contact (two hundred hours of supervision). The phone number of AAMFT is 1-202-452-0109; address is 1133 15th Street, N.W., Suite 300, Washington, DC 20005-2710. Clients are customers and should feel comfortable with their therapists and the progress they are making. If they don't, they should switch therapists.

Managed care has resulted in private therapists' lowering their fees so as to compete with what insurance companies will pay (about $80 per session). Many mental health centers offer marital and family therapy on a sliding-fee basis so that spouses, parents, and their children can be seen for as little as $5. The average number of sessions per couple varies. Four was the median number of sessions (in a range of one to eighteen) for 152 couples who were seen at a university marriage and family therapy clinic (Poche, White, & Smith, 1997). Marriage therapists generally see both the husband and the wife together (called conjoint therapy); about 60 percent will involve the children as necessary but often not as active participants (Lund, Zimmerman, & Haddock, 2002).

To what degree do spouses, parents, and children benefit from marriage and family therapy? Shadish and Baldwin (2003) synthesized twenty intervention studies on marriage and family therapy and concluded that such interventions were "clearly efficacious compared to no treatment" (p. 566).

Whether a couple in therapy remain together will depend on their motivation to do so, how long they have been in conflict, the severity of the problem, and whether one or both partners are involved in an extramarital affair. Two moderately motivated partners with numerous conflicts over several years are less likely to work out their problems than a highly motivated couple with minor conflicts of short duration.

Some couples come to therapy with the goal of separating amicably. The therapist then discusses the couple's feelings about the impending separation, the definition of the separation (temporary or permanent), the "rules" for their interaction during the period of separation (e.g., see each other, date others), and whether to begin discussions with a divorce mediator or attorneys.

One alternative for enhancing one's relationship is the Association for Couples in Marriage Enrichment (ACME). This organization provides conferences for couples to enrich their marriage. The address of ACME is P.O. Box 10596, Winston-Salem, NC 27108.

Brief Solution-Based Therapy

There are over twenty different treatment approaches used by members of AAMFT (Northey, 2002). The largest percentage (27%) report that they use a "cognitive-behavioral" approach, which means they focus on the cognitions or assumptions that underlie a marriage or family with the goal of ensuring that these are accurate and functional. The therapist also examines what behaviors family members want increased or decreased, initiated or terminated, and try to negotiate ways to accomplish these behavioral goals (hence the cognitive-behavioral label).

Other therapists (Imago relationship therapists) may conceptualize relationship and communication difficulties as rooted in family dynamics in which one was reared, specify how these patterns are operative in one's present relationship, and look for ways of understanding and improvement (Gelbin, 2003).

In the past, marriage and family therapy models were not constrained by the economics of managed care. As the bottom line has become the driving force de-

Up Close 13.2 Some Caveats about Marriage and Family Therapy

In spite of the potential benefits of marriage therapy, some valid reasons exist for not becoming involved in such therapy. Not all spouses who become involved in marriage therapy regard the experience positively. Some feel that their marriage is worse as a result. Saying things the spouse can't forget, feeling hopeless at not being able to resolve a problem "even with a counselor," and feeling resentment over new demands made by the spouse in therapy are reasons some spouses cite for negative outcomes.

Therapists also may give clients an unrealistic picture of loving, cooperative, and growing relationships in which partners always treat each other with respect and understanding, share intimacy, and help each other become whoever each wants to be. In creating this idealistic image of the perfect relationship, therapists may inadvertently encourage clients to focus on the shortcomings in their relationship and to expect more of their marriage than is realistic. Dr. Robert Sammons (2004) calls this his first law of therapy, "That spouses always focus on what is missing rather than what they have . . . indeed the only thing that is important to couples in therapy is that which is missing."

Couples and families in therapy must also guard against assuming that therapy will be a quick and easy fix. Changing one's way of viewing a situation (cognitions) and behavior requires a deliberate, consistent, relentless commitment to make things better. Without it, couples are wasting their time and money.

Couples who become involved in marriage counseling may also miss work, have to pay for child care, and be "exposed" at work if they use their employer's insurance policy to cover the cost of the therapy. Though these are not reasons to decide against seeing a counselor, they are issues that concern some couples.

Reference

Sammons, R. A., Jr. 2004. First law of therapy. Personal communication. Grand Junction, Colorado (Dr. Sammons is a psychiatrist in private practice).

termining what services insurance companies will pay for, increasingly marriage and family therapy has become more time-bound, brief, and solution-focused. However, marriage therapy is usually ineffective where one or both spouses are depressed (Marchand & Hock, 2000), and the treatment of such depression must precede effective marital intervention. A global concern among therapists is that managed-care economics is dictating therapy models and treatment alternatives.

On-Line Marriage Counseling

The Internet now features over 200 "on-line therapy" sites promising access to over 350 on-line counselors (*not* marriage counselors). Though these may be helpful for getting e-mail answers to e-mail questions, ongoing on-line marital therapy is virtually unknown. Since effective marriage counseling requires the participation and involvement of both spouses, on-line marital therapy is made difficult since both partners would need to be on-line at the same time. In addition, nonverbal interaction behaviors cannot be observed by the therapist. If these concerns are not enough to dissuade spouses from trying therapy on-line, they should be aware that e-mail communications are not always safe and the therapist might not be available in times of a crisis. One client noted that his counselor simply disappeared for three weeks only to reemerge with the explanation that "my e-mail was jammed and messages were not getting through" (Segall, 2000, 43).

SUMMARY

What is stress and what is a crisis event?

Stress is a reaction of the body to substantial or unusual demands (physical, environmental, or interpersonal) made on it. Stress is a process rather than a state. A crisis is a situation that requires changes in normal patterns of behavior. A family crisis is a situation that upsets the normal functioning of the family and requires a new set of responses to the stressor. Sources of stress and crises can be external (e.g., hurricane, tornado, downsizing, military separation) or internal (e.g., alcoholism, extramarital affair, Alzheimer's disease).

Resiliency refers to a family's strengths and ability to respond to a crisis in a positive way. Characteristics associated with resilient families include having a joint cause or purpose, emotional support for each other, good problem-solving skills, the ability to delay gratification, flexibility, accessing residual resources, communication, and commitment.

What are positive stress management strategies?

Changing one's basic values and perspective is the most helpful strategy in reacting to a crisis. Viewing ill health as a challenge, bankruptcy as an opportunity to spend time with one's family, and infidelity as an opportunity to improve communication are examples. Other positive coping strategies are exercise, adequate sleep, love, religion, friends/relatives, multiple roles, and humor. Still other strategies include intervening early in a crisis, not blaming each other, keeping destructive impulses in check, and seeking opportunities for fun.

What are harmful strategies for reacting to a crisis?

Some harmful strategies include keeping feelings inside, taking out frustrations on others, and denying or avoiding the problem.

What are five of the major family crisis events?

Some of the more common crisis events faced by spouses and families include physical illness, extramarital affair, unemployment, alcohol/drug abuse, and the death of one's spouse and children. The existence of a "midlife crisis" is reported by less than a quarter of adults in the middle years. Those who did experience a crisis were going through a divorce.

What help is available from marriage and family therapists?

There are around 50,000 marriage and family therapists in the United States. About 40 percent are clinical members of the American Association for Marriage and Family Therapy (AAMFT). Whether a couple in therapy remain together will depend on their motivation to do so, how long they have been in conflict, the severity of the problem, and whether one or both partners are involved in an extramarital affair. Two moderately motivated partners with numerous conflicts over several years are less likely to work out their problems than a highly motivated couple with minor conflicts of short duration.

Brief solution-based therapies are provided that focus on behavioral and/or cognitive change. On-line marriage therapy is virtually nonexistent.

KEY TERMS

Al-Anon
biofeedback
Coolidge Effect
crisis
extradyadic involvement
extramarital affair
palliative care
resiliency
stress

RESEARCHING MARRIAGE AND THE FAMILY WITH INFOTRAC COLLEGE EDITION

InfoTrac College Edition, an online library, allows you to perform research online anywhere, anytime. Following are two suggested search terms and related questions to help you extend your understanding of the topics covered in this chapter. Go to www.infotrac-college.com to begin your search.

Keyword: **Alcohol on campus.** Locate articles that discuss what college and university campus policies are being tried to eliminate deaths on campus due to alcohol consumption. Which of these do you feel will and will not be effective?

Keyword: **Extramarital affair.** Locate articles that discuss ending or continuing a marriage in the wake of a spouse's affair. What would you do if your spouse had an affair and you found out about it?

The Companion Web Site for Choices in Relationships: An Introduction to Marriage and the Family, Eighth Edition

http://sociology.wadsworth.com/knox_schacht/choices8e

Supplement your review of this chapter by going to the companion Web site to take one of the Tutorial Quizzes, use the flash cards to master key terms, and check out the many other study aids you'll find there. You'll also find special features such as the Marriage and Family Resource Center, Census 2000 information, and other data and resources at your fingertips to help you with that special project or to do some research on your own.

WEBLINKS

Dear Peggy.com (coping with an affair)
http://www.vaughan-vaughan.com/

Association for Applied and Therapeutic Humor
http://www.aath.org

Association for Couples in Marriage Enrichment
http://www.bettermarriages.org/

All State Investigations, Inc. (help investigating infidelity)
http://www.infidelity.com/

American Association for Marriage and Family Therapy
http://www.aamft.org/

Association for Applied Psychophysiology and Biofeedback
www.aapb.org

© Royalty-Free/CORBIS

CHAPTER 14

Even if I did do this [kill my wife], it would be because I loved her very much, right?

O. J. Simpson of his murdered wife, Nicole Brown Simpson

Violence and Abuse in Relationships

Contents

True or False?

1. The longer individuals have been involved in a relationship, the more likely the relationship is to include violence and abuse.
2. Fathers are more likely than mothers to use spanking as a method of discipline.
3. Men are most likely to express violence by pushing a woman while women are most likely to slap a man.
4. Younger children are more likely than older children to disclose being sexually abused.
5. Physical abuse by perpetrators who use drugs only is more severe than that by perpetrators who use alcohol only.

Answers: **1.** T **2.** F **3.** T **4.** T **5.** T

There is considerable cultural awareness that violence may occur in relationships. In early 2004, there was the newspaper story of 70-year-old soul singer James Brown being jailed on a charge of domestic violence. He was alleged to have pushed his 33-year-old wife to the floor during an argument (Associated Press, 2004). Television programs like *American Justice, City Confidential,* and *Cold Case Files,* not to speak of the nightly news, are replete with stories of domestic violence, abuse, and murder. What these newspaper articles and television stories have in common is that they reveal a frightening reality—the person most likely to be violent and abusive toward you (even kill you) is the person you are married to, living with, or hanging out with. We expect our intimate relationships to be havens where we will be loved, protected, and cared for. The reality is sometimes different. In this chapter we look at intimate partners who hurt each other. We begin by defining terms.

The one thing that is unforgivable is deliberate cruelty.

Blanche, in *A Streetcar Named Desire* (Tennessee Williams)

DEFINTIONS OF VIOLENCE/PHYSICAL ABUSE AND EMOTIONAL ABUSE

Violence (also referred to as **physical abuse**) may be defined as the intentional infliction of physical harm by either partner on the other. Examples of physical violence include pushing, throwing something at the partner, slapping, hitting, and forcing sex on the partner. The physically abusive behavior men are most likely to engage in is to push, grab, or shove the woman; women are most likely to slap the man (Lloyd & Emery, 2000).

National Data

The actual number of physically abusive acts is unknown. Dugan, Nagin, & Rosenfeld (2003) reported results from a national sample of women who had been physically assaulted. Almost three-fourths (73%) did not report the incident to the police—the leading reason was that they believed the police could not help.

International Data

At least one in three women worldwide has been beaten, coerced into sex, or abused in some way—most often by someone she knows (United Nations Population Fund, 2000).

Intimate-partner violence is an all-inclusive term that refers to crimes committed against current or former spouses, boyfriends, or girlfriends. **Battered-woman syndrome** refers to the general pattern of battering that a woman is subjected to and is defined in terms of the frequency, severity, and injury she experiences. Battering is severe if the person's injuries require medical treatment or the perpetrator could be prosecuted. The syndrome is sometimes used (with limited success) as a defense by women who kill their husbands (e.g., "He was an abusive husband so I killed him") (Terrance & Matheson, 2003).

Being slapped, pushed and shoved, or grabbed, though startling, does not qualify as being battered. Emotional abuse also does not qualify as battering. The fact that these less dramatic forms of abuse do not qualify as battering is a limitation of the term. Battering in any form is the denigration of a person, disrespect for their integrity and is abuse.

In addition to being physically violent or abusive, partners may also engage in **emotional abuse** (also known as **psychological abuse, verbal abuse,** or **symbolic aggression**), which may be even more damaging to the individual. Though such abuse does not involve physical harm, it is designed to denigrate the partner, reduce the partner's status, and make the partner vulnerable, thereby giving the abuser more control. Though there is debate about what constitutes psychological abuse, examples include

calling the partner obese, stupid, crazy, ugly, and repulsive

telling the partner she or he is pitiful/pathetic and lucky to have anyone

threatening to leave the partner to enjoy being with someone else

controlling the partner's time with friends, siblings, and parents

controlling the money of the partner to ensure dependence

refusing to talk to or to touch the partner

accusing the partner of infidelity

threatening to harm oneself if the partner leaves the relationship

threatening to harm one's children or take them away

demeaning or insulting the partner in front of others

restricting the partner's mobility—e.g., use of the car

criticizing the partner's child care, food preparation, or job performance

threatening to harm the partner, the partner's relatives, or the partner's pets

telling the partner that he or she is a terrible lover

threatening to have the partner committed to a mental institution

demanding the partner does as he or she is told

EXPLANATIONS FOR VIOLENCE/ ABUSE IN RELATIONSHIPS

Research suggests that numerous factors contribute to violence and abuse in intimate relationships. These factors include those that occur at the cultural, community, and individual and family levels.

Cultural Factors

In many ways, American culture tolerates and even promotes violence. Violence in the family stems from the acceptance of violence in our society as a legitimate means of enforcing compliance and solving conflicts at interpersonal, familial, national, and international levels. Violence and abuse in the family may be linked to cultural factors, such as violence in the media, acceptance of corporal punishment, gender inequality, and the view of women and children as property. The context of stress is also conducive to violence.

Violence in the Media A report from the Federal Trade Commission accused the movie, music, and video game industries of "peddling" sex and violent content to kids (Seiler, 2000). One need only watch the evening news to see accounts of school, domestic, and global violence. In addition, feature films and TV movies regularly reflect themes of violence. As a society, we are inundated with violence in the media.

Corporal Punishment **Corporal punishment** is defined as the use of physical force with the intention of causing a child to experience pain, but not injury, for the purpose of correction or control of the child's behavior (Straus, 2000). Violence has become a part of our cultural heritage through the corporal punishment of children. Straus (2000) reviewed the literature on the use of corporal punishment and found that 94 percent of parents of toddlers reported using corporal punishment. Spankers are more likely to be young parents and mothers and to have been hit when they were a child (Walsh, 2002). Undergraduates testify to the fact that their parents used corporal punishment. Eighty-three percent of over 11,000 undergraduate students at the University of Iowa reported that they had experienced some form of physical punishment during their childhood (Knutson & Selner, 1994).

In his book *Beating the Devil Out of Them,* Murray Straus (1994) noted that one of the effects of corporal punishment on children is that the child has an increased chance of being violent in his or her own life (p. 151). Children who are victims of corporal punishment display more antisocial behavior, are more violent, and have an increased incidence of depression as adults. Straus (2000) recommended an end to corporal punishment to reduce the risk of physical abuse and other harm to children. Sweden passed a law in 1979 that effectively abolished corporal punishment as a legitimate childrearing practice.

Gender Inequality Domestic violence and abuse may also stem from traditional gender roles. Nayak et al. (2003) found that individuals in countries espousing very restrictive roles for women (e.g., Kuwait) tend to be more accepting of violence against women. Traditional male gender roles have taught men to be aggressive. Traditionally, men have also been taught that they are superior to women and that they may use their aggression toward women, believing that women need to be "put in their place." Traditional female gender roles have also taught women to be submissive to their male partners' control. Demaris et al. (2003) found that when a nontraditional woman is paired with a traditional male, violence in their relationship is more likely.

Some occupations lend themselves to contexts of gender inequality. Men in military and police roles are less likely to treat women as their equals and may not separate their work roles from their domestic roles. The effect is spillover from the work role into the domestic role. Melzer (2002) found more domestic violence among men in military and police occupations.

It is not surprising that marital violence is found to be higher among the less educated (Verma, 2003), which is another context for inequality. Similarly, women who have higher incomes than their husbands report a higher frequency of wife beatings (Verma, 2003).

View of Women and Children as Property Prior to the late nineteenth century, a married woman was considered the property of her husband. A husband had a legal right and marital obligation to discipline and control his wife through the use of physical force.

Stress Our culture is also a context of stress. The stress associated with getting and holding a job, rearing children, staying out of debt, and paying bills may predispose one to lash out at others. Haskett et al. (2003) found that parenting stress was particularly predictive of child abuse. Persons who have learned to handle stress by being abusive toward others are vulnerable to becoming abusive partners.

Community Factors

Community factors that contribute to violence and abuse in the family include social isolation, poverty, and inaccessible or unaffordable health-care, day-care, elder-care, and respite-care services and facilities.

Diversity in Other Countries

In their study on marital power, conflict, and violence in South Korea (where the patriarchal authority power structure is in place), Kim and Emery (2003) found that violence against women who were not subservient was considered justified. They recommended that the only way to mute such abusiveness was for roles between women and men to become more egalitarian.

Diversity in Other Countries

In cultures where a man's "honor" is threatened if his wife is unfaithful, there is tolerance toward the husband's violence toward her. In some cases there are formal, legal traditions defending a man's right to beat or even to kill his wife in response to her infidelity (Vandello & Cohen, 2003).

Authors

Crime is higher in poverty contexts, where families struggle for the basic necessities of housing.

Social Isolation Living in social isolation from extended family and community members increases the risk of being abused. Isolated spouses are removed from the emotional support of extended family and community members. Spouses whose parents live nearby are least vulnerable.

Poverty Abuse in adult relationships occurs among all socioeconomic groups. However, poverty and low socioeconomic development are associated with crime and higher incidences of violence. This violence may spill over into interpersonal relationships as well as the frustration of living in poverty (Tolan, Golan-Smith, & Henry, 2003).

Inaccessible or Unaffordable Community Services Failure to provide medical care to children and elderly family members sometimes results from the lack of accessible or affordable health-care services in the community. Failure to provide supervision for children and adults may result from inaccessible day-care and elder-care services. Without elder-care and respite-care facilities, families living in social isolation may not have any help with the stresses of caring for elderly family members and children.

Lack of Violence Prevention Programs Foshee et al. (1998) collected baseline and follow-up data on 1,700 adolescents who participated in a study to assess the impact of exposure to a "safe date program." Less psychological abuse, less physical violence, and less sexual violence were reported among those who were involved in the program than among similarly aged adolescents who were not involved. Because such programs are not widely available, the opportunity for early intervention is limited. Browning (2002) found that communities that gave high visibility to positive domestic relations sent a signal for low tolerance of domestic abuse. Lower violence was found in such communities.

Individual Factors

Individual factors associated with domestic violence and abuse include psychopathology, personality characteristics, and alcohol or substance abuse.

Psychopathology Some abusing spouses and parents have psychiatric conditions that predispose them to abusive behavior. Symptoms of psychiatric conditions that are related to violence and abuse include low frustration tolerance, emotional distress, inappropriate expression of anger, and antisocial personalities (McBurnett et al., 2001). Rosenbaum and Leisring (2003) found that men who abuse women report having had limited love from their mothers, more punishment from their mothers, and less attention from their fathers than men who do not abuse their partners.

Personality Factors A number of personality characteristics have been associated with persons who are abusive in their intimate relationships. Some of these characteristics follow:

1. Dependency. Therapists who work with batterers have observed that they are extremely dependent on their partners. Because the thought of being left by their partners induces panic and abandonment anxiety, batterers use physical aggression and threats of suicide to keep their partners with them.

2. Jealousy. Along with dependence, batterers exhibit jealousy, possessiveness, and suspicion. An abusive husband may express his possessiveness by isolating his wife from others; he may insist she stay at home, not work, and not socialize with others. His extreme, irrational jealousy may lead him to accuse his wife of infidelity and beat her for her presumed affair. Indeed, jealousy is sometimes viewed as a sign of love (Puente & Cohen, 2003) (take another look at the opening quote of this chapter!).

*3. **Need to control.*** Batterers are often described as individuals who have an excessive need to exercise power and control over their partners (Dutton & Starzomski, 1997). They do not let their partners make independent decisions, and they want to know everything their partners do. They like to be in charge of all aspects of family life, including finances and recreation. Men who are dissatisfied with the power they have in their relationships are particularly vulnerable to being abusive (Ronfeldt, Kimmerling, & Arias, 1998).

*4. **Unhappiness and dissatisfaction.*** Abusive partners often report being unhappy and dissatisfied with their lives, both at home and at work. Many abusers have low self-esteem and high levels of anxiety, depression, and hostility. They may expect their partner to make them happy.

*5. **Anger and aggressiveness.*** Abusers tend to have a history of interpersonal aggressive behavior. They have poor impulse control and can become instantly enraged and lash out at the partner. Battered women report that episodes of violence are often triggered by minor events, such as a late meal or an unironed shirt.

*6. **Quick involvement.*** Because of feelings of insecurity, the potential batterer will move his partner quickly into a committed relationship. If the woman tries to break off the relationship, the man will often try to make her feel guilty for not giving him and the relationship a chance.

*7. **Blaming others for problems.*** The abuser takes little responsibility for his problems and blames everyone else. For example, when he makes a mistake, he will blame the woman for upsetting him and keeping him from concentrating on his work. He becomes upset because of what she said, hits her because she smirked at him, and kicks her in the stomach because she poured him too much alcohol.

*8. **Jekyll-and-Hyde personality.*** The abuser has sudden mood changes so that his partner is continually confused. One minute he is nice, and the next minute angry and accusatory. Explosiveness and moodiness are typical.

*9. **Isolation.*** The abusive person will try to cut off the person from all family, friends, and activities. Ties with anyone are prohibited. Isolation may reach the point at which the abuser tries to stop the victim from going to school, church, or work.

*10. **Other factors.*** Alcohol or drug abuse (see section below), having grown up in a violent family, having traditional role relationship ideology and not being able to express their feelings easily (Umberson et al., 2003) are other factors associated with the potential to be abusive.

While alcohol may provide a context for relaxation and enjoyment, it is also associated with the escalation of violence in a relationship.

Authors

Alcohol and Other Drug Use McBurnett et al. (2001) found that alcohol and drug use was associated with men's violence toward women. Whether alcohol acts to inhibit one's use of violence, acts to allow one to avoid responsibility for being violent, or increases one's aggression is still being debated. But the relationship between using alcohol and violence was clear in this sample of over 1,000 high school students in eight schools in the Los Angeles area. Shook et al. (2000) also studied 395 female and 177 male college students and confirmed the association between alcohol and violence. Finally, Demaris et al. (2003) studied over 4,000 couples and found an association between violence and substance abuse.

Family Factors

Family factors associated with domestic violence and abuse include being abused as a child, having parents who abused each other, and not having a father in the home.

Child Abuse in Family of Origin Individuals who were abused as children are more likely to report being abused in an adult domestic relationship (Heyman & Slep, 2002: Salter et al., 2003).

Parents Who Abused Each Other Men who witnessed their fathers abusing their mothers are more likely to become abusive partners themselves (Heyman & Slep, 2002). Similarly, women who witnessed their mothers being violent toward their fathers are more likely to be abusive toward their intimates and others (Babcock, Miller, & Siard, 2003). However, a majority of those who witnessed abuse do not continue the pattern. A family history of violence is only one factor out of many that may be associated with a greater probability of adult violence. Three researchers (Simmons, Kuei-Hsiu, & Gordon, 1998) provided data to support the hypothesis that it is not the imitation of abusive behavior by parents toward each other but the lack of support and emotional involvement with one's parents that predicts antisocial behavior and drug use in offspring. These factors, in turn, predict involvement in interpersonal violence in one's own relationships.

ABUSE IN DATING RELATIONSHIPS

Violence in dating relationships begins in grade school, is mutual, and escalates with emotional involvement. Ten percent of the 7,824 twelfth-grade females in Howard and Wang's (2003) study reported at least one aspect of dating violence. Gray and Foshee (1997) studied violence in relationships among 185 adolescents from the sixth through twelfth grades and found that 60 percent of the students reporting violence said that it was mutual. Moreover, 63 percent were still dating the partner even though mutual violence was a part of their relationship (p. 136).

The woman in this photo was previously involved in an emotionally and physically abusive relationship, withdrew from the relationship, and is now involved with a gentle and loving partner (pictured).

Authors

Estimates of violence in dating relationships among college students range from 20 percent to 82 percent (Riggs & Caulfield, 1997; Shook et al., 2000). Up Close 14.1 reveals the experiences of one university student who was involved in a relationship that had become abusive.

The Self-Assessment, Abusive Behavior Inventory, allows you to assess the degree to which abuse is occurring in your current or most recent relationship.

Hanley and O'Neill (1997) found that violence occurred more often among couples in which the partners were likely to disagree about each other's level of emotional commitment. Similarly, Lloyd and Emery (2000) found that physical abuse often occurs when the relationship is being threatened and when the perpetrator has been drinking alcohol or using drugs. In this regard, a team of researchers (Willson et al., 2000) found that physical abuse by perpetrators who had been using drugs only was more severe than abuse by perpetrators who had been using alcohol only.

Most research on abuse in relationships has been conducted on heterosexuals. An exception is Freedner et al. (2002), who found similar prevalence rates of dating violence among gay, lesbian, and bisexual dating couples.

Up Close 14.1 Broken-Heart Land

A broken-heart land where no one smiles,
is where I lived for a while.

He came one day and made me glad,
but it all changed and got really bad.

He started criticizing me in every way,
taunting and hurting me more each day.

I would cry and cry but he didn't care,
he'd just sit and laugh at the clothes I'd wear.

I was so torn apart by the things he'd say;
like how bad I looked and how much I weighed.

I will never forget the night so clear—
the night he hit me and I went into total fear.

I played it off the next day
I was afraid that if I told, somehow he'd make me pay.

He'd changed so much into this monster I didn't know,
I thought he might hurt me again if I let anything show.

So I cried and I thought about him and me,
and decided in the end that it shouldn't be.

So I told him how I felt and he called me a whore,
and made me feel worse than I had before.

I stayed with him, I guess because he made me feel that I should,
Even though he said he was too good for me and would dump me when he could.

And so in this world I lived for a while,
in this land of broken hearts, where no one smiles.

By Erin Brantley

Individuals in emotionally and physically abusive relationships may be more vulnerable to being stalked when the relationship ends. **Stalking** is defined as the willful, repeated, and malicious following or harassment of another person. Coleman (1997) studied data from 144 undergraduate college women and found that those who had been in romantic but abusive relationships were more likely to have been stalked than those who had not been in such abusive relationships. Such stalking is usually designed either to seek revenge or to win the partner back.

Acquaintance and Date Rape

The word *rape* often evokes images of a stranger jumping out of the bushes or a dark alley to attack an unsuspecting victim. However, most rapes are perpetrated not by strangers but by persons who have a relationship with the victim. About 85 percent of rapes are perpetrated by someone the woman knows. This type of rape is known as **acquaintance rape,** which is defined as nonconsensual sex between adults who know each other. One type of acquaintance rape is **date rape,** which refers to nonconsensual sex between people who are dating or on a date. Women who were date raped identified the primary reason for the rape as sexual satisfaction: "He just wanted to get laid" (Lloyd & Emery, 2000, 96). In a university sample of 275 men and 381 women, 70 percent of the sample reported that they had been subjected to at least one "tactic" of pressure to get them to have sex after they had already specified that they were not interested. One-third of the sample said that they had used a tactic to get a partner to have sex. The most frequent tactics were sexual arousal, emotional manipulation, and lies. An example of the use of sexual arousal by a woman forcing sex on a man follows:

> *I locked the room door that we were in. I kissed and touched him. I removed his shirt and unzipped his pants. He asked me to stop. I didn't. Then I sat on top of him. He had had two beers but wasn't drunk.* (Struckman-Johnson, Struckman-Johnson, & Anderson, 2003, 84)

Though date rapes do occur, it is important to keep in mind that friends, coworkers, and neighbors also rape and that these are trusted acquaintances who

SELF-ASSESSMENT

Abusive Behavior Inventory

Circle the number that best represents your closest estimate of how often each of the behaviors happened in your relationship with your partner or former partner during the previous six months.

1 Never
2 Rarely
3 Occasionally
4 Frequently
5 Very frequently

1. Called you a name and/or criticized you. **1 2 3 4 5**
2. Tried to keep you from doing something you wanted to do (e.g., going out with friends, going to meetings). **1 2 3 4 5**
3. Gave you angry stares or looks. **1 2 3 4 5**
4. Prevented you from having money for your own use. **1 2 3 4 5**
5. Ended a discussion with you and made the decision himself/herself. **1 2 3 4 5**
6. Threatened to hit or throw something at you. **1 2 3 4 5**
7. Pushed, grabbed, or shoved you. **1 2 3 4 5**
8. Put down your family and friends. **1 2 3 4 5**
9. Accused you of paying too much attention to someone or something else. **1 2 3 4 5**
10. Put you on an allowance. **1 2 3 4 5**
11. Used your children to threaten you (e.g., told you that you would lose custody, said he/she would leave town with the children). **1 2 3 4 5**
12. Became very upset with you because dinner, housework, or laundry was not done when he/she wanted it or done the way he/she thought it should be. **1 2 3 4 5**
13. Said things to scare you (e.g., told you something "bad" would happen, threatened to commit suicide). **1 2 3 4 5**
14. Slapped, hit, or punched you. **1 2 3 4 5**
15. Made you do something humiliating or degrading (e.g., begging for forgiveness, having to ask his/her permission to use the car or to do something). **1 2 3 4 5**
16. Checked up on you (e.g., listened to your phone calls, checked the mileage on your car, called you repeatedly at work). **1 2 3 4 5**
17. Drove recklessly when you were in the car. **1 2 3 4 5**
18. Pressured you to have sex in a way you didn't like or want. **1 2 3 4 5**
19. Refused to do housework or child care. **1 2 3 4 5**
20. Threatened you with a knife, gun, or other weapon. **1 2 3 4 5**
21. Spanked you. **1 2 3 4 5**
22. Told you that you were a bad parent. **1 2 3 4 5**
23. Stopped you or tried to stop you from going to work or school. **1 2 3 4 5**
24. Threw, hit, kicked, or smashed something. **1 2 3 4 5**
25. Kicked you. **1 2 3 4 5**
26. Physically forced you to have sex. **1 2 3 4 5**
27. Threw you around. **1 2 3 4 5**
28. Physically attacked the sexual parts of your body. **1 2 3 4 5**
29. Choked or strangled you. **1 2 3 4 5**
30. Used a knife, gun, or other weapon against you. **1 2 3 4 5**

Scoring

Add the numbers you circled and divide the total by 30 to find your score. The higher your score, the more abusive your relationship.

The inventory was given to 100 men and 78 women equally divided into groups of abusers/abused and nonabusers/nonabused. The men were members of a chemical dependency treatment program in a veterans' hospital and the women were partners of these men. Abusing or abused men earned an average score of 1.8; abusing or abused women earned an average score of 2.3. Nonabusing/abused men and women earned scores of 1.3 and 1.6, respectively.

Source

Melanie F. Shepard and James A. Campbell. The abusive behavior inventory: A measure of psychological and physical abuse. *Journal of Interpersonal Violence,* September 1992, 7, no. 3, 291–305. Inventory is on pages 303–304. Used by permission of Sage Publications, 2455 Teller Road, Newbury Park, CA 91320.

do not fit into the category of dates. Homosexual acquaintance rape also occurs. Hodge and Canter (1998) studied these rapes and noted that they were not likely to be reported to the police. If they were reported, the likelihood of the perpetrator's being convicted was remote.

A more inclusive concept than rape is feeling pressure to have sex. Ten percent of 248 women reported that they had used aggressive strategies (physical force, exploitation of a man's incapacitated state, and verbal pressure) to make a man engage in physical touch, sexual intercourse, or oral sex against his will (Krahe, Waizenhofer, & Moller, 2003).

Shapiro and Schwarz (1997) studied the effects of date rape on forty-one college women. They reported higher levels of anxiety, depression, anger and irritability, and sexual dysfunctions than for college women who had not experienced date rape. They also had lower levels of sexual self-esteem.

Rophypnol—The Date Rape Drug

Rophypnol, known as "the date rape drug," causes profound, prolonged sedation and short-term memory loss. Individuals who ingest this drug are particularly vulnerable. Alcohol may also cause a person to lose control. In California, consistent with the Alcohol Law of 1995, if the victim is intoxicated and this fact is reasonably known to the perpetrator of the sex act, the perpetrator can be charged with rape.

ABUSE IN MARRIAGE RELATIONSHIPS

The chance of abuse in a relationship increases with marriage.

O.J. is going to kill me and get away with it.

Nicole Brown Simpson, *murdered wife of O.J. Simpson*

General Abuse in Marriage

The ways in which spouses are abusive toward each other resemble the abusive behavior of unmarried couples. The difference is in the context where the husband may feel "ownership" of the wife and feels he must control her by any means necessary. This attitude is known as patriarchal terrorism (Johnson, 1995).

Rape in Marriage

Marital rape, now recognized in all states, is forcible rape by one's spouse. Abuse in marital relationships may take the form of rape and sexual assault. Ten percent of married women in a Boston survey reported that they had been raped by their husbands (Finkelhor & Yllo, 1988). Such rapes may have included not only intercourse but also other types of sexual activities in which the wife did not want to engage, most often fellatio and anal intercourse.

Sexual violence against women in an intimate relationship is often repeated. Rand (2003) analyzed data from the National Violence Against Women survey and noted that about half of the women raped by an intimate partner and two-thirds of the women physically assaulted by an intimate partner had been victimized multiple times.

EFFECTS OF ABUSE

Abuse affects the physical and psychological well-being of the victim. Abuse between parents also affects the children.

Effects of Partner Abuse on Victims

The most obvious effect of physical abuse by an intimate partner is physical injury. Indeed, former surgeon general Antonia Novella noted that battering is the single major cause of injury to women in the United States. As many as 35 percent of women who seek hospital emergency room services are suffering from injuries incurred by battering (Novello, 1992). Other less obvious effects of abuse by one's intimate partner include fear, feelings of helplessness, confusion, isolation, humiliation, anxiety, stress-induced illness, symptoms of post-traumatic stress disorder (PTSD), and suicide attempts (Lloyd & Emery, 2000). Depression and substance abuse have also been associated with partner violence (Anderson, 2002).

When the abuse by a partner is sexual, it may be more devastating than sexual abuse by a stranger. The primary effect on a woman is to destroy her ability to trust men in intimate interpersonal relationships. In addition, the woman raped by her husband lives with her rapist and may be subjected to repeated assaults.

Effects of Partner Abuse on Children

Abuse between adult partners affects children. Some women are abused during their pregnancy, resulting in a high rate of miscarriage and birth defects. The March of Dimes has concluded that the physical abuse of pregnant women causes more birth defects than all the diseases put together for which children are usually immunized (Gibbs, 1993).

Negative effects may also accrue to children who just witness domestic abuse. Kitzmann et al. (2003) analyzed 118 studies to identify outcomes on children who were and were not exposed to interparental violence. The researchers found more negative outcomes (arguing, withdrawing, avoidance, overt hostility) among children who had witnessed such behavior than those who had not. Howard et al. (2002) found that adolescents who had witnessed violence reported "intrusive thoughts, distraction, and feeling a lack of belonging" (p. 455). Children who witness high levels of parental violence are also more likely to blame themselves for the violence (Grych, Harold, & Miles, 2003), to be violent toward their parents (particularly their mother) (Ulman & Straus, 2003), and to engage in aggressive delinquent behavior (Kernic et al., 2003).

It is not unusual for children to observe and to become involved in adult domestic violence. One-fourth of 114 battered mothers noted that their children yelled, called for help, or intervened when they (the mothers) were being physically abused by their adult partners (Edleson, 2003).

PERSONAL CHOICES

Should You End an Abusive Marital Relationship?

Students in the authors' classes were asked whether they would end a marriage if the spouse hit or kicked them.

Some said, "Seek a divorce."

Those opting for divorce felt they couldn't live with someone who had abused or would abuse them. "I abhor violence of any kind, and since a marriage should be based on love, kicking is certainly unacceptable. I would lose all respect for my mate and I could never trust him again. It would be over."

Most said, "Don't overreact and try to work it out."

Most felt that marriage was too strong a commitment to end if the abuse could be stopped.

I would not divorce my spouse if she hit or kicked me. I'm sure that there's always room for improvement in my behavior, although I don't think it's necessary to assault me. I recognize that under certain circumstances, it's the quickest way to draw my attention to the problems at hand. I would try to work through our difficulties with my spouse.

The physical contact would lead to a separation. During that time, I would expect him to feel sorry for what he had done and to seek psychiatric help. My anger would be so great, it's quite hard to know exactly what I would do.

I wouldn't leave him right off. I would try to get him to counseling. If we could not work through the problem, I would leave him. If there was no way we could live together, I guess divorce would be the answer.

I would not divorce my husband, because I don't believe in breaking the sacred vows of marriage. But I would separate from him and let him suffer!

I would tell her I was leaving but that she could keep me if she would agree for us to see a counselor to ensure that the abuse never happened again.

Some therapists emphasize that a pattern of abuse develops and continues if such behavior is not addressed immediately. The first time abuse occurs, the couple should seek therapy. The second time it occurs, they should separate. The third time, they might consider a divorce. Earlier in the chapter we discussed disengaging from an abusive relationship.

THE CYCLE OF ABUSE

Up Close 14.2 reflects a stage in the cycle of abuse—flowers.

The cycle of abuse begins when a person is abused and the perpetrator feels regret, asks for forgiveness, and starts acting nice (e.g., gives flowers; see Up Close 14.2). The victim, who perceives few options and feels guilty terminating the relationship with the partner who asks for forgiveness, feels hope for the relationship at the contriteness of the abuser and does not call the police or file charges. There is usually a makeup or honeymoon period, during which the person feels good again about his or her partner. But tensions mount in the relationship, again resulting in stress, anger, and tension release in the form of violence. Such violence is followed by the familiar sense of regret and pleadings for forgiveness accompanied by being nice (a new bouquet of flowers, etc.).

Up Close 14.2 I Got Flowers Today

I got flowers today. It wasn't my birthday or any other special day. We had our first argument last night, and he said a lot of cruel things that really hurt me. I know he is sorry and didn't mean the things he said, because he sent me flowers today.

I got flowers today. It wasn't our anniversary or any other special day. Last night, he threw me into a wall and started to choke me. It seemed like a nightmare. I couldn't believe it was real. I woke up this morning sore and bruised all over. I know he must be sorry, because he sent me flowers today.

Last night, he beat me up again. And it was much worse than all the other times. If I leave him, what will I do? How will I take care of my kids? What about money? I'm afraid of him and scared to leave. But I know he must be sorry, because he sent me flowers today.

I got flowers today. Today was a very special day. It was the day of my funeral. Last night, he finally killed me. He beat me to death.

If only I had gathered enough courage and strength to leave him, I would not have gotten flowers today.

Author unknown

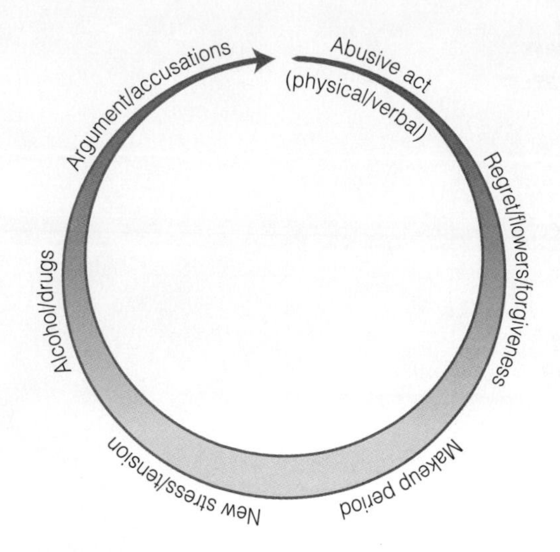

Figure 14.1
The Cycle of Abuse

Figure 14.1 illustrates this cycle, which occurs in clockwise fashion. In the rest of this section we discuss reasons why people stay in an abusive relationship and how to get out of such a relationship.

As the cycle of abuse reveals, some wives do not prosecute their partners who abuse them. To deal with this problem, Los Angeles has adopted a "zero tolerance" policy toward domestic violence. Under the law, an arrested person is required to stand trial and his victim required to testify against the perpetrator. The fine in Los Angeles County for partner abuse is up to six months in jail and a fine of $1,000.

Why People Stay in Abusive Relationships

One of the most frequently asked questions of people who are abused by their partners is, "Why don't you get out of the relationship?" The following are various reasons why abused partners stay in the abusive relationships; they include love, emotional dependency, commitment to the relationship, hope, view of violence as legitimate, guilt, fear, economic dependence, and isolation. Hendy et al. (2003) found that "fear of loneliness" was the most important reason why women stayed.

It is easier to stay out than to get out.
Mark Twain, *Humorist*

1. Love. Despite the physical and emotional pain of the abuse, abused partners often feel love for the abuser. Love feelings may be maintained by the fact that abusive partners do not always behave in abusive ways; they may also act in positive and loving ways. In a study of abused women in dating relationships, those who were more likely to stay in the relationship had partners who engaged in a high frequency of positive behaviors (Kasian & Painter, 1992). The researchers hypothesized that "the presence of positive behaviors maintains a relationship regardless of the level of negative experiences" (p. 361).

2. Emotional dependency. Abused partners may feel they cannot live without the abuser. Such codependency was expressed by one woman who said: "I know my boyfriend treats me badly, but I wouldn't know what to do without him. I need him. I would rather put up with the abuse than be alone without him" (authors' files). Some battered women also have very low self-esteem and feel that they would be incapable of attracting a new partner, so they feel that staying in an abusive relationship is not such a terrible alternative.

3. Commitment to the relationship. Abused partners, especially those in marital relationships, may feel committed to the relationship. Some abused spouses stay in the relationship because they don't believe in divorce; they believe that marriage, no matter what its quality, is a permanent relationship.

4. Hope. Abused partners may stay in the relationship because they hope it will improve. Abused women who stay in an abusive relationship often feel that things will get better, that one more chance is what their partner and the relationship need. Many feel forgiving toward their husbands and take them back usually seven times before they finally leave.

5. View of violence as legitimate. Some abused partners stay in the relationship because they accept violence as a legitimate part of intimate relationships. This may be the result of growing up in a home in which the parents abused each other, which may convey the message that marital violence is natural, inevitable, and to be expected. Some abused partners may feel that the violence directed toward them is legitimate because it is their fault; they either caused the abuse or deserved it. Some abused partners feel that if only they were a better partner or a better person, they would not be abused.

6. Guilt. Other abused partners stay in the abusive relationship because leaving would produce guilt. They may feel guilty about breaking up a family, especially if children are involved. Some abusers threaten suicide if their partner leaves or use other verbal pleas to induce their partner into staying because of guilt.

7. Fear. Abused partners may stay in a relationship because they fear that the abuser will become even more violent if they leave. Often, such fear is the result of threats made by the abuser. One woman in the authors' classes reported that she stayed with an abusive husband for eleven years because he threatened to "slit my throat if I tried to leave." When he became abusive toward her children, she took them and left the state.

8. Economic dependence. Some spouses stay in an abusive relationship because they are economically dependent on the partner. This is most often true of a wife who has limited economic resources. Leaving the husband would mean giving up not only his monthly income but retirement benefits and health insurance as well. The woman referred to in the preceding paragraph also noted that her husband kept the checkbook and gave her money only as he wished. When she finally left, she went to a women's shelter, which provided food for her and her children.

9. Isolation. Battered women may be physically and socially isolated from family, friends, and community resources. Isolation may be even greater in the case of victims who live in rural areas or for foreign-born wives. A team of researchers studied battered women in women's shelters and identified the primary reasons they stayed in abusive relationships.

Battered women stay because they rarely have escape routes related to educational or employment opportunities, relatives were critical of plans to leave the relationship, parenting responsibilities impeded escape, and abusive situations contributed to low self-esteem and negative emotions—especially anxiety and depression. Such troubles and the dangerousness of the abuser probably undermined rebellious and self-enhancing actions (Forte et al., 1996, 69).

Disengaging from an Abusive Relationship

The abused woman may have difficulty breaking free because of limited resources—poor health, limited material resources, and inadequate social supports (Carlson, 1997). Rosen and Stith (1997) also noted the difficulty of breaking through the cultural mandate to "stand by your man." Rather, women in abusive relationships must manage to come to their senses, "break free and reconnect with themselves and healthy support systems" (p. 181). The catalyst for breaking free combines the sustained aversiveness of staying, the perception that she and her children will be harmed by doing so, and the awareness of an alternative path.

Once a decision is made to withdraw from an abusive relationship, often the woman not only has to call the police and have the man arrested but also has to take out a restraining order restricting the man from contacting her or coming within a certain distance of her. Women who fear that the abuser will harm them (or their children) sometimes leave town or leave the state. Other women hide in the home of a friend or seek refuge in a women's shelter. In either case, disengagement from the abusive relationship takes a great deal of courage. Calling 800-799-7233, the national domestic hotline, is a point of beginning. She must also find work or a way to take care of her children since those who do not often return to their spouses (Anderson, 2003).

Strategies to Prevent Domestic Abuse

Family violence and abuse prevention strategies are focused at three levels: the general population, specific groups thought to be at high risk for abuse, and families who have already experienced abuse. Public education and media campaigns aimed at the general population convey the criminal nature of domestic assault, suggest ways to prevent abuse (seek therapy for anger, jealousy, or dependency), and identify where abuse victims and perpetrators can get help.

Preventing or reducing family violence through education necessarily involves altering aspects of American culture that contribute to such violence. For example, violence in the media must be curbed or eliminated, and traditional gender roles and views of women and children as property must be replaced with egalitarian gender roles and respect for women and children.

Another important cultural change is to reduce violence-provoking stress by reducing poverty and unemployment and by providing adequate housing, nutrition, medical care, and educational opportunities for everyone. Integrating families into networks of community and kin would also enhance family well-being and provide support for families under stress.

Treatment of Partner Abusers

Treatment for men who abuse their partners involves teaching them to take responsibility for their own abusive behavior, developing empathy for their partner's victimization, reducing their dependency on their partners, and improving their communication skills (Scott & Wolfe, 2000). These new behaviors can be learned in individual or group therapy. Since alcohol or drug abuse and violence toward a partner are often related, addressing one's alcohol or substance abuse problem is often a prerequisite for treating partner abuse. O'Farrell et al. (2003) found that partner abuse decreased after abusers received treatment for alcoholism.

In addition, some men stop abusing their partners only when their partners no longer put up with it. One abusive male said that his wife had to leave him before he learned not to be abusive toward women. "I've never touched my second wife," he said.

This concludes our discussion of abuse in romantic relationships. In the pages to follow we discuss other forms of abuse, including child abuse and parent, sibling, and elder abuse.

GENERAL CHILD ABUSE

NATIONAL DATA

The number of children investigated by protective services who are alleged to have been abused is around 3 million (*Statistical Abstract of the United States: 2003*, Table 345).

The most prevalent form of child abuse in America today is neglect. People are working 18 hours a day and our children are being neglected.

Laura Schlessinger, *Radio personality*

Child abuse may take many forms—physical abuse, neglect, and sexual abuse.

Physical Abuse and Neglect

A 25-pound 8-year-old daughter was found emaciated and near death in a locked closet in January of 2002. Her mother, 30-year-old Barbara Atkins of Dallas, Texas, pleaded guilty to the offense and was sentenced to prison for life; she will be eligible for parole in thirty years. She said that her daughter was being punished and that she did not love her daughter as much as her other five children. The stepfather was also charged with sexual assault and serious bodily injury to a child (Mother gets life for abuse of girl, 2002).

Child abuse can be defined as any interaction or lack of interaction between a child and his or her parents or caregiver that results in nonaccidental harm to the child's physical or psychological well-being. Child abuse includes physical

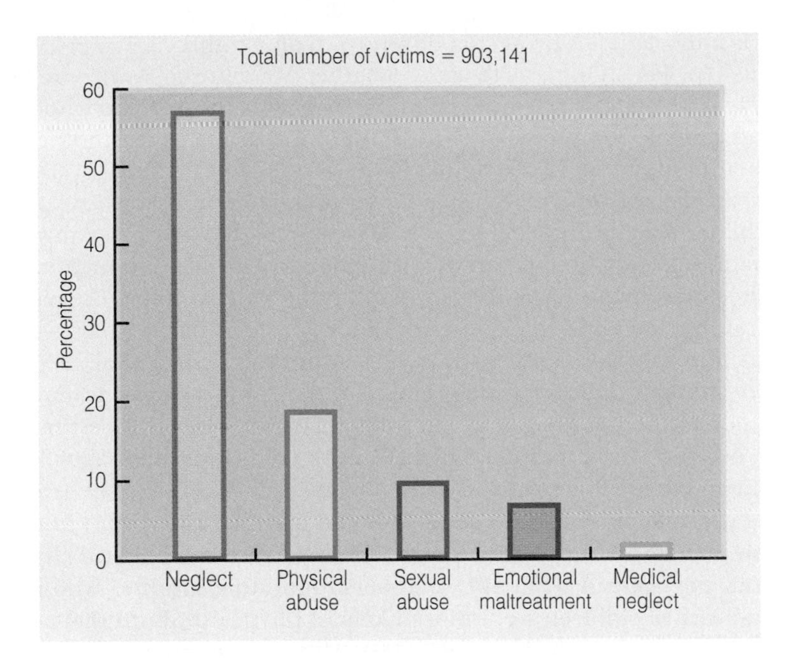

Figure 14.2
Child Abuse and Neglect Cases: 2001
Source: *Statistical Abstract of the United States 2003.* 123rd ed. Washington, D.C.: U.S. Bureau of the Census, Table 344.

abuse, such as beating and burning; verbal abuse, such as insulting or demeaning the child; and neglect, such as failing to provide adequate food, hygiene, medical care, or adult supervision for the child. A child can also experience emotional neglect by his or her parents. Infants and children more likely to be abused are the result of unintended pregnancies and have poor health and developmental problems (Sidebotham et al., 2003).

The percentages of various types of child abuse in substantiated victim cases are illustrated in Figure 14.2. Notice that "neglect" is the largest category of abuse. The children most likely to be neglected are young—between infancy and age 5 (Connell-Carrick, 2003). De Paul and Arruabarrena (2003) observed that these families are the most difficult to treat and have the highest dropout rates.

Although our discussion will focus on physical abuse, it is important to keep in mind that children often are not fed, are not given medical treatment, and are left to fend for themselves.

Factors Contributing to General Child Abuse

A variety of factors contribute to child abuse.

1. Parental psychopathology. Andrea Yates, who drowned all five of her children in Houston, Texas, in 2001, used mental illness as a defense for her behavior. Symptoms of parental psychopathology that may predispose a parent to abuse or neglect children include low frustration tolerance, inappropriate expression of anger, and alcohol or substance abuse.

Munchausen syndrome by proxy (MSP) is a rare form of child abuse whereby a parent (usually the mother) takes on the sick role indirectly (hence, by proxy) by inducing illness or sickness in her child. The parent may suffocate the child to the point of unconsciousness or scrub the child's skin with sandpaper to induce a rash. The goal of the caretaker (often the mother) is to find a fulfilling role, to get attention from friends and family for being heroic, or to get money from insurance companies. Sheridan (2003) analyzed 451 cases from 154 journals and found that the victims may be either males or females and are usually under four years of age. One-fourth of the victims' known siblings are *dead* and 61 percent of the siblings had similar illnesses. Not much is understood about MSP, and accuracy is important "to protect children from abuse and caretakers from false allegations" (p. 444). One way to assess whether Manchusen syndrome by proxy is operative is to separate the child from the mother. The children who return to health and show no symptoms of ill health in the absence of the mother are victims of their mother's abuse (Wright, 2004).

2. Unrealistic expectations. Abusing parents often have unrealistic expectations of their children's behavior. For example, a parent might view the crying of a baby as a deliberate attempt on the part of the child to irritate the parent. **Shaken baby syndrome**—whereby the caretaker, most often the father, shakes the baby to the point of causing the child to experience brain or retinal hemorrhage—most often occurs in response to a baby who won't stop crying (Ricci et al., 2003). Most victims are younger than 6 months (Smith, 2003). **Abusive head trauma** (AHT) refers to nonaccidental head inury in infants and toddlers. It is estimated that there are over one thousand cases of AHT each year (Ricci et al., 2003, p. 279). Head injury is the leading cause of death in abused children (Rubin et al., 2003). Managing parents' unrealistic expectations and teaching the parent to cope with frustration is an important part of an overall plan to reduce physical child abuse.

3. History of abuse. A team of researchers found that mothers who had been sexually abused as children were more likely to physically abuse their own children (Di Lillo, Tremblay, & Peterson, 2000). Although the majority do not, some parents who were themselves physically or verbally abused or neglected may duplicate these patterns in their own families. Although parents who were abused as children are somewhat more likely to repeat that behavior than parents who were not abused, we would like to emphasize that the majority of parents who were abused do *not* abuse their own children. Indeed, many parents who were abused as children are dedicated to ensure nonviolent parenting of their own children precisely because of their own experience of abuse.

4. Displacement of aggression. One cartoon shows several panels consisting of a boss yelling at his employee, the employee yelling at his wife, the wife yelling at their child, and the child kicking the dog, who chases the cat up a tree. Indeed, frustration may spill over from the adults to the children, the latter being less able to defend themselves.

5. Social isolation. The saying "it takes a village to raise a child" is relevant to child abuse. Unlike inhabitants of most societies of the world, many Americans rear their children in closed and isolated nuclear units. In extended kinship societies, other relatives are always present to help with the task of childrearing. Isolation means no relief from the parenting role as well as no supervision by others who might interrupt observed child abuse.

6. Fatherless homes. Living in a home where the father is absent increases a child's risk for being abused. Numerous studies show that children are more likely to be sexually abused by a stepfather or a mother's boyfriend than by their biological father (Blankenhorn, 1995). This is largely because stepfathers and mothers' boyfriends are not constrained by the cultural incest taboo that prohibits fathers from having sex with their children.

7. Other factors. In addition to the factors just mentioned, the following factors are associated with child abuse and neglect:

- **a.** *The pregnancy is premarital or unplanned, and the father or mother does not want the child.*
- **b.** *Mother-infant attachment is lacking.*
- **c.** *The child suffers from developmental disabilities or mental retardation.*
- **d.** *Childrearing techniques are harsh, with little positive reinforcement.*
- **e.** *The parents are unemployed.*
- **f.** *Abuse between the husband and wife is present.*
- **g.** *The children are adopted or are foster children.*

Effects of General Child Abuse

How does being abused in general affect the victim as a child and later as an adult? In general, the effects are negative and vary according to the intensity and frequency of the abuse. Reviews of research have found that children who have been abused are more likely to display the following (Muller & Lemieux, 2000; Windom, 1999):

1. *Few close social relationships and an inability to love or trust*
2. *Communication problems and learning disabilities*
3. *Aggression, low self-esteem, depression, and low academic achievement*
4. *Physical injuries that may result in disfigurement, physical disability, or death*
5. *Increased risk of alcohol or substance abuse and suicidal tendencies as adults*
6. *Post-traumatic stress disorder (PTSD)*

What is the relative impact of parental emotional abuse, child sexual abuse, and parental substance abuse on a child's development as an adult? Melchert (2000) studied the psychological distress of 255 college students and discovered that the largest amount of distress was explained by previous emotional abuse and neglect.

CHILD SEXUAL ABUSE

Often found in combination with other forms of child abuse is child sexual abuse (Dong et al., 2003). Two types are extrafamilial and intrafamilial.

Extrafamilial Child Sexual Abuse—The Catholic Example

In extrafamilial child sexual abuse the perpetrator is someone outside the family who is not related to the child. Extrafamilial child sexual abuse received national attention with the 2002 accusation that an estimated two thousand priests (McGeary et al., 2002) in the Catholic Church had had sex with young children; more than three thousand children have claimed abuse (Hewitt et al., 2002).

Defrocked priest John Geoghan is thought to have molested 130 children and was sentenced to nine to ten years in prison for abusing a 10-year-old boy (Geoghan was subsequently murdered by a fellow inmate). More than $30 million has already been paid to settle more than eighty lawsuits. It is reported that priests told their victims that they had a "special relationship" with them—an effort to "convince them that molestation is some kind of holy secret that must never be disclosed" (Hewitt et al., 2002). In the wake of this horrific set of abuses, "lives have been hurt, trust damaged, and the credibility of the Church to speak on social issues tainted" (McGeary et al., 2002, 52).

I was only 9 years old when I was raped by my 19-year-old cousin. He was the first of three family members to sexually molest me.
Oprah Winfrey, *TV talk show host*

Intrafamilial Child Sexual Abuse

A more frequent type of child sexual abuse is **intrafamilial child sexual abuse** (formerly referred to in professional literature as incest). This refers to exploitive sexual contact or attempted sexual contact between relatives before the victim is 18. Sexual contact or attempted sexual contact includes intercourse, fondling of the breasts and genitals, and oral sex. Relatives include biologically related individuals but may also include stepparents and stepsiblings.

Female children are more likely than male children to be sexually abused. Prevalence rates suggest that one out of four girls and one out of ten boys experience sex abuse (Fieldman & Crespi, 2002). However, Goodman et al. (2003)

suggested that nearly 40 percent of abused adults fail to report their own documented child sexual abuse. About 60 percent of sexually abused children disclose their abuse to someone within two weeks after it happens—40 percent within forty-eight hours. Children who are younger, who did not feel responsible for the abuse, and who were abused by someone external to the family are more likely to disclose the abuse sooner than older children, those who felt partly responsible, and those who were abused by someone inside the family (Goodman-Brown et al., 2003).

Intrafamilial child sexual abuse, particularly when the perpetrator is a parent, involves an abuse of power and authority. The following describes the experience of one woman who was forced to have sexual relations with her father:

> *I was around 6 years old when I was sexually abused by my father. He was not drinking at that time; therefore, he had a clear mind as to what he was doing. On looking back, it seemed so well planned. For some reason, my father wanted me to go with him to the woods behind our house to help him saw wood. Once we got there, he looked around for a place to sit and wanted me to sit down with him. He said, "Susan, I want you to do something for Daddy. I want you to lie down, and we are going to play Mama and Daddy." Being a child, I said, "Okay," thinking it was going to be fun. I don't know what happened next and I can't remember if there was pain or whatever. I was threatened not to tell, and remembering how he beat my mother, I didn't want the same treatment. It happened a few more times. I remember not liking this at all.*
>
> *But what could I do? Until age 18, I was constantly on the run, hiding from him when I had to stay home alone with him, staying out of his way so he wouldn't touch me by hiding in the cornfields all day long, under the house, in the barns, and so on until my mother got back home, then getting punished by her for not doing the chores she had assigned to me that day. It was a miserable life, growing up in that environment.* (Authors' files)

When an ex-spouse accuses her husband of child sex abuse, more often than not, the sexual abuse did occur. In a study of nine thousand families embroiled in contested divorce proceedings, 169 cases involved an allegation of child sexual abuse. Of these, only 14 percent were deliberately false allegations. Most were proven to be legitimate (Goldstein & Tyler, 1998).

Conviction of a child sex abuser is rare. In one study of 323 court cases, only 15 went to trial and in only 6 cases was the offender convicted (Faller & Henry, 2000).

Effects of Child Sexual Abuse

Child sexual abuse may have serious negative long-term consequences. The most devastating effects of being sexually abused occur when the sexual abuse is forceful, is prolonged, and involves intercourse, and when it is perpetrated by a father or stepfather (Beitchman et al., 1992). Not only has the child been violated physically, but he or she has lost an important social support. Adult-child sexual contact is, in most cases, a child's first introduction to adult sexuality. The sexual script acquired during such relationships forms the basis on which other sexual experiences are assimilated. Even if an adult-child sexual experience is recognized to be "wrong," the process of acquiring new and more appropriate sexual scripts may be difficult for a maturing adolescent (Browning & Laumann, 1998, 557). Researchers (Koss et al. 2003; Walrath et al. 2003) on the effects of being sexually abused have found the following:

1. Early forced sex is associated with being withdrawn, anxious, and depressed and with delinquency and substance abuse (Walrath et al. 2003). Anderson et al. (2003) noted that women who experienced childhood sexual abuse were 7.75 times more likely to attempt suicide. PTSD is also frequently associated with childhood sexual abuse (Koss et al., 2003).

2. Daughters of mothers who have been sexually abused are 3.6 times more likely to be sexually victimized than are daughters whose mothers have not been abused (McCloskey & Bailey, 2000). It is possible that women who have been sexually abused develop an "internal working model" of sexual relationships that encompasses exploitative, coercive, and domineering behavior among men. If such a relationship "template" results from early exposure to sexual abuse, then these women might be more tolerant of men either in their households or in their social spheres who are potential abusers of their daughters (p. 1032).

3. Spouses who were physically and sexually abused as children report lower marital satisfaction, higher individual stress, and lower family cohesion than do couples with no abuse history (Nelson & Wampler, 2000).

4. Adult males who were sexually abused as children tend to develop negative self-perceptions, anxiety disorders, sleep and eating disturbances, and sexual dysfunctions such as decreased sexual desire, rapid ejaculation, and difficulty with ejaculation (Elliott & Briere, 1992).

Strategies to Reduce Child Sexual Abuse

Strategies to reduce child sexual abuse include regendering cultural roles, providing specific information to children on sex abuse, improving the safety of neighborhoods, providing healthy sexuality information for both teachers and children in the public schools at regular intervals, and promoting public awareness campaigns.

From a larger societal preventive perspective, Bolen (2001) recommended that child sexual abuse should be viewed as a gendered problem. Indeed, "child sexual abuse is endemic within society and may be a result of the unequal power of males over females" (p. 249). Evidence for this claim includes the facts that females are at greater risk of abuse than males, males are more likely to offend, and child sexual abuse is most frequently heterosexual. The implication here is that one way to discourage child sexual abuse is to change traditional notions of gender role relationships, masculinity, and male sexuality so that men respect the sexuality of women and children and take responsibility for their sexual behavior.

The public schools are also helping children acquire specific knowledge and skills to protect themselves from sexual abuse. A survey of four hundred school districts in the United States revealed that 85 percent had offered a prevention program in the past year (Davis & Gidyez, 2000). Through various presentations in the elementary schools, children are taught how to differentiate between appropriate and inappropriate touching by adults or siblings, to understand that it is okay to feel uncomfortable if they do not like the way someone else is touching them, to say no in potentially exploitative situations, and to tell other adults if the offending behavior occurs. Although children who participate in these prevention programs have greater knowledge of sexual abuse and learn strategies to circumvent it, "it cannot be assumed that children participating in abuse prevention programs are at lower risk for sexual abuse" (Davis & Gidyez, 2000, 263).

Helping to ensure that children live in safe neighborhoods where it is known if neighbors are former convicted child molesters is the basis of Megan's Law (see Social Policy section).

Megan's Law and Beyond

SOCIAL POLICY

In 1994, Jesse Timmendequas lured 7-year-old Megan Kanka into his Hamilton Township house in New Jersey to see a puppy. He then raped and strangled her and left her body in a nearby park. Prior to his rape of Megan, Timmendequas had two prior convictions for sexually assaulting girls. Megan's mother, Maureen Kanka, argued that she would have kept her daughter away from her neighbor if she had known about his past sex offenses. She campaigned for a law, known as **Megan's Law,** requiring that communities be notified of a neighbor's previous sex convictions. New Jersey and forty-five other states have enacted similar laws. President Clinton signed a federal version in 1996.

The law requires that convicted sexual offenders register with local police in the communities in which they live. It also requires the police to go out and notify residents and certain institutions (such as schools) that a dangerous sex offender has moved into the area. It is this provision of the law that has been challenged on the belief that individuals should not be punished forever for past deeds. Critics of the law argue that convicted child molesters who have been in prison have paid for their crime. To stigmatize them in communities as sex offenders may further alienate them from mainstream society and increase their vulnerability for repeat offenses.

In many states, Megan's Law is not operative because it is on appeal. Although parents ask, "Would you want a convicted sex offender, even one who has completed his prison sentence, living next door to your 8-year-old daughter?" the reality is that little notification is afforded parents in most states. Rather, the issue is tied up in court and will likely remain so until the Supreme Court decides it. A group of concerned parents (Parents for Megan's Law) are trying to implement Megan's Law nationwide. See Weblinks at the end of this chapter for their Web site address.

In July 2002, Parents for Megan's Law sought to enact legislation for the "civil commitment for a specific group of the highest risk sexual predators who freely roam our streets, unwilling or unable to obtain proper treatment. This kind of predator commitment follows a criminal sentence and generally targets repeat sex offenders who then remain in a sexual predator treatment facility until it is safe to release them to a less restrictive environment or into the community." Indeed, this group not only wants parents to be notified of criminal sex offenders but also wants the offenders to be housed in treatment centers on release from prison.

PARENT, SIBLING, AND ELDER ABUSE

As we have seen, intimate partners and children may be victims of relationship violence and abuse. Parents, siblings, and the elderly may also be abused by family members.

Parent Abuse

Some people assume that because parents are typically physically and socially more powerful than their children, they are immune from being abused by their children. But parents are often targets of their children's anger, hostility, and frustration. It is not uncommon for teenage and even younger children to physically and verbally lash out at their parents. In a national survey of family violence, almost 10 percent of the parents reported that they had been hit, bitten, or kicked at least once by their children (Gelles & Straus, 1988). The same researchers found that 3 percent of parents reported that they had been victimized at least once by a severe form of violence inflicted by a child aged 11 or older. Ulman (2003) noted that more children were violent toward mothers than fathers and that boys were more violent toward parents than girls.

The older the child, the less frequent the violence (Ulman & Straus, 2003). Background characteristics of children who abuse their parents include observing parents' violence toward each other and having parents who used corporal punishment on their children as a method of discipline (Ulman & Straus, 2003).

Children have been known to push parents down stairs, set the house on fire while their parents are in it, and use weapons such as guns or knives to inflict serious injuries or even to kill a parent. Heide (1992) reported that about three hundred parents in the United States are killed each year by their children. However, according to one attorney who specializes in defending adolescents who have killed a parent, over 90 percent of youths who kill their parents have been abused by them. "In-depth portraits of such youths have frequently shown that they killed because they could no longer tolerate conditions at home" (p. 6). This was the defense unsuccessfully used by the Menendez brothers in California, who were convicted of murdering their parents. Nevertheless, Ulman (2003) found that in most cases where children were violent toward their parents, parents had been violent toward the children.

Authors

While sibling relationships are often kind and loving in childhood, some turn into very abusive relationships in adolescence and young adulthood.

Sibling Abuse

Observe a family with two or more children and you will likely observe some amount of sibling abuse. Even in "well-adjusted" families, some degree of fighting among the children is expected. Most incidents of sibling violence consist of slaps, pushes, kicks, bites, and punches. What passes for "normal," "acceptable," or "typical" behavior between siblings would often be regarded as violent and abusive behavior outside the family context.

Sibling abuse is said to be the most prevalent form of abuse. Ninety-eight percent of the females and 89 percent of the males in one study reported having received at least one type of emotionally aggressive behavior from a sibling; 88 percent of the females and 71 percent of the males reported having received at least one type of physically aggressive behavior from a sibling (Simonelli et al., 2002). Sibling abuse may include sexual exploitation, whereby an older brother will coerce younger female siblings into nudity or sex. Though some sex between siblings is consensual, often it is not. "He did me *and* all my sisters before he was done," reported one woman. Another woman reported that (as a child and young adolescent) she performed oral sex on her brother three times a week for years since he told her that the man's semen was the only way a woman would be able to have babies as an adult. Even milder forms of sibling abuse seem to feed the cycle of abuse, as persons who were abused by their siblings report abusing others in their adulthood (Simonelli et al., 2002).

Elder Abuse

As increasing numbers of the elderly end up in the care of their children, abuse, though infrequent, is likely to increase. Currently it is estimated that about half a million elderly are abused (Bergeron & Gray, 2003). The forms of such abuse are numerous.

1. *Neglect—failing to buy or give the elderly needed medicine, failing to take them to receive necessary medical care, or failing to provide adequate food, clean clothes, and a clean bed. Neglect is the most frequent type of domestic elder abuse.*
2. *Physical abuse—inflicting injury or physical pain or sexual assault.*
3. *Psychological abuse—verbal abuse, deprivation of mental health services, harassment, and deception.*
4. *Social abuse—unreasonable confinement and isolation, lack of supervision, abandonment.*
5. *Legal abuse—improper or illegal use of the elder's resources.*

Another type of elder abuse that has received recent media attention is **granny dumping.** Adult children or grandchildren who feel burdened with the care of their elderly parent or grandparent drive the elder to the entrance of a hospital and leave him or her there with no identification. If the hospital cannot identify responsible relatives, it is required by state law to take care of the abandoned elder or transfer the person to a nursing-home facility, which is paid for by state funds. Relatives of the dumped granny, hiding from financial responsibility, never visit or see granny again.

Adult children who are most likely to dump or abuse their parents tend to be under a great deal of stress and to use alcohol or other drugs. In some cases, parent abusers are getting back at their parents for mistreating them as children. In other cases, the children are frustrated with the burden of having to care for their elderly parents. Such frustration is likely to increase. As baby boomers age, they will drain already limited resources for the elderly, and their children will be forced to care for them with little governmental support. Prevention of elder abuse involves reducing the stress for caregivers by linking caregivers to community services (Bergeron & Gray, 2003).

SUMMARY

How are violence and abuse defined?

Violence/physial abuse may be defined as the intentional infliction of physical harm by either partner on the other. *Intimate partner violence* is an all-inclusive term that refers to crimes committed against current or former spouses, boyfriends, or girlfriends. Battered-woman syndrome refers to the general pattern of battering that a woman is subjected to and is defined in terms of the frequency, severity, and injury she experiences.

Emotional abuse (also known as psychological abuse, verbal abuse, or symbolic aggression) is designed to denigrate the partner, reduce the partner's status, and make the partner vulnerable, thereby giving the abuser more control.

What are explanations for violence in relationships?

Cultural explanations for violence include violence in the media, corporal punishment in childhood, gender inequality, and stress. Community explanations involve social isolation of individuals and spouses from extended family, poverty, inaccessible community services, and lack of violence prevention programs. Individual factors include psychopathology of the person (antisocial), personality (dependent/jealous), and alcohol abuse. Family factors include child abuse by one's parents and observing parents who abuse each other.

How does abuse in dating relationships manifest itself?

Violence in dating relationships begins in grade school, is mutual, and escalates with emotional involvement. Violence occurs more often among couples in which the partners disagree about each other's level of emotional commitment and when the perpetrator has been drinking alcohol or using drugs. Acquaintance rape is defined as nonconsensual sex between adults who know each other. One type of acquaintance rape is date rape, which refers to nonconsensual sex between people who are dating or on a date.

How does abuse in marriage relationships manifest itself?

Abuse in marriage is born out of the need to control the partner and may include repeated rape. About half of the women raped by an intimate partner and two-thirds of the women physically assaulted by an intimate partner have been victimized multiple times.

What are the effects of abuse?

The most obvious effect of physical abuse by an intimate partner is physical injury. As many as 35 percent of women who seek hospital emergency room services are suffering from injuries incurred by battering. Other less obvious effects of abuse by one's intimate partner include fear; feelings of helplessness, confusion, isolation, and humiliation; anxiety; stress-induced illness; symptoms of post-traumatic stress disorder; and suicide attempts.

When the abuse by a partner is sexual, it may be more devastating than sexual abuse by a stranger. The primary effect on a woman is to destroy her ability to trust men in intimate interpersonal relationships. In addition, the woman raped by her husband lives with her rapist and may be subjected to repeated assaults.

Abuse between adult partners affects children. Some women are abused during their pregnancy, resulting in a high rate of miscarriage and birth defects. The March of Dimes has concluded that the physical abuse of pregnant women causes more birth defects than all the diseases put together for which children are usually immunized.

What is the cycle of abuse and why do people stay in an abusive relationship?

The cycle of abuse begins when a person is abused and the perpetrator feels regret, asks for forgiveness, and starts acting nice (e.g., gives flowers). The victim, who perceives few options and feels guilty terminating the relationship with the partner who asks for forgiveness, feels hope for the relationship at the contriteness of the abuser and does not call the police or file charges. There is usually a makeup or honeymoon period, during which the person feels good again about his or her partner. But tensions mount again and are released in the form of violence. Such violence is followed by the familiar sense of regret and pleadings for forgiveness accompanied by being nice (a new bouquet of flowers, etc.).

The reasons people stay in abusive relationships include love, emotional dependency, commitment to the relationship, hope, view of violence as legitimate, guilt, fear, economic dependency, and isolation. The catalyst for breaking free combines the sustained aversiveness of staying, the perception that they and their children will be harmed by doing so, and the awareness of an alternative path.

What is child abuse and what factors contribute to it?

Child abuse can be defined as any interaction or lack of interaction between a child and his or her parents or caregiver that results in nonaccidental harm to the child's physical or psychological well-being. Child abuse includes physical abuse, such as beating and burning; verbal abuse, such as insulting or demeaning the child; and neglect, such as failing to provide adequate food, hygiene, medical care, or adult supervision for the child. A child can also experience emotional neglect by his or her parents.

Some of the factors that contribute to child abuse include parental psychopathology, a history of abuse, displacement of aggression, and social isolation.

What are the effects of general child abuse?

The negative effects of child abuse include impaired social relationships, difficulty in trusting others, aggression, low self-esteem, depression, low academic achievement, and post-traumatic stress disorder (PTSD). Physical injuries may result in disfigurement, physical disability, and even death.

One of the most devastating types of child abuse is child sexual abuse, which has serious negative long-term consequences. The effects include being withdrawn, anxious, and depressed; delinquency; suicide attempts; and substance abuse. Post-traumatic stress disorder, which means recurrent experiencing of the event, is common, as is heavy alcohol or substance abuse. Strategies to reduce child sexual abuse include regendering cultural roles, providing specific information to children on sex abuse, improving the safety of neighborhoods, providing healthy sexuality information for both teachers and children in the public schools at regular intervals, and promoting public awareness campaigns.

What is the nature of parent, sibling, and elder abuse?

Parent abuse is the deliberate harm (physical or verbal) of parents by their children. Ten percent of parents report that they have been hit, bitten, or kicked at least once by their children. About three hundred parents are killed by their children annually. The Menendez brothers of California brutally murdered both parents.

Sibling abuse is the most severe prevalent form of abuse.What passes for "normal," "acceptable," or "typical" behavior between siblings would often be regarded as violent and abusive behavior outside the family context.

Elder abuse is another form of abuse in relationships. Granny dumping is a new form of abuse in which children or grandchildren who feel burdened with the care of their elderly parents or grandparents drive them to the emergency entrance of a hospital and dump them. If the relatives of the elderly patient cannot be identified, the hospital will put the patient in a nursing home at state expense.

KEY TERMS

abusive head trauma
acquaintance rape
battered-woman syndrome
battering rape
child sexual abuse
corporal punishment
date rape
emotional abuse
granny dumping
intimate partner violence
intrafamilial child sexual abuse
marital rape
Megan's Law
Munchausen syndrome by proxy
physical abuse
psychological abuse
Rophypnol
shaken baby syndrome
stalking
symbolic aggression
verbal abuse
violence

RESEARCHING MARRIAGE AND THE FAMILY WITH INFOTRAC COLLEGE EDITION

InfoTrac College Edition, an online library, allows you to perform research online anywhere, anytime. Following are two suggested search terms and related questions to help you extend your understanding of the topics covered in this chapter. Go to www.infotrac-college.com to begin your search.

Keyword: **Abuse in marriage or marital abuse.** Locate articles that provide a discussion of marital abuse and take a position on how much abuse warrants terminating a marriage. A harsh word, a slap, a hit, a push—where is the line?

Keyword: **Megan's Law.** Locate articles that discuss Megan's Law. What are the arguments for and against informing individuals in a neighborhood that a convicted sex offender now lives in the neighborhood?

The Companion Web Site for Choices in Relationships: An Introduction to Marriage and the Family, Eighth Edition

http://sociology.wadsworth.com/knox_schacht/choices8e

Supplement your review of this chapter by going to the companion Web site to take one of the Tutorial Quizzes, use the flash cards to master key terms, and check out the many other study aids you'll find there. You'll also find special features such as the Marriage and Family Resource Center, Census 2000 information, and other data and resources at your fingertips to help you with that special project or to do some research on your own.

WEBLINKS

Childabuse.com
    http://www.childabuse.com/

Male Survivor
    http://www.malesurvivor.org/

Minnesota Center Against Violence and Abuse
    http://www.mincava.umn.edu

Parents for Megan's Law
    http://www.parentsformeganslaw.com

Rape, Abuse & Incest National Network (RAINN)
    http://www.rainn.org/

Stop It Now! The Campaign to Prevent Child Sexual Abuse
    http://www.stopitnow.com/

V-Day
    http://www.vday.org/

Authors

CHAPTER 15

And make no mistake—divorce is a death. It kills the dreams of your youth, those innocent beliefs that your marriage can weather sickness as it can weather health, that life will be kind and fair, that the joys will be shared and the vicissitudes bring you closer.

Wendy Swallow, Breaking Apart

Divorce

Contents

True or False?

1. The divorce rate continues to increase, and there is no sign that this trend will be reversed.
2. Because of a strong extended family, Blacks have a lower divorce rate than whites.
3. The large majority of children of divorce drop out of school, get arrested, abuse drugs, and suffer long-term emotional distress.
4. Parental alienation syndrome refers to the fact that when spouses become parents they are more likely to become alienated from each other.
5. A study of seventy-two family law attorneys revealed that most felt that the divorce and custody system is biased against fathers.

Answers: **1.** F **2.** F **3.** F **4.** F **5.** T

Couples in love rarely acknowledge on their wedding day the possibility that their marriage will end in divorce. Endings such as this (we think) occur only to "other" couples ("who lack strong love and commitment values"). Just as the denial that divorce won't happen to us is intense and entrenched, the reality is shocking—it can and often does happen to us. In this chapter we review the stark reality of divorce (about 40 percent of today's marriages will end in divorce), the societal and individual reasons for divorce, and the impact on the spouses and children.

DIVORCE: LEGAL DECREE AND RATES IN THE UNITED STATES

College students are no strangers to divorce. About a quarter (24%) have parents who are divorced (American Council on Education and the University of California, 2004.). **Divorce** is the legal ending of a valid marriage contract.

Legal Decree

The legal document filed with the court by an attorney is similar to the one below.

IN THE CIRCUIT COURT OF YOUR COUNTY, YOUR STATE

YOUR NAME,

Plaintiff,

vs.

YOUR SPOUSE,

Defendant.

COMPLAINT

1. Plaintiff and Defendant are both over the age of nineteen (19) years and are both bona fide resident citizens of YOUR COUNTY AND STATE, and have been for more than six months next preceding the filing of this complaint.

2. Plaintiff and Defendant were married to each other in YOUR COUNTY/ STATE, on or about the 29th day of MONTH, YEAR, and lived together as husband and wife until on or about MONTH, YEAR at which time they separated and have not lived together as husband and wife since that date.

3. There were no children born to the parties as a result of this marriage. The Defendant is not pregnant at this time.

4. The Parties have accumulated personal property during their marriage which should be equitably divided.

5. Plaintiff avers that since their marriage, the temperaments, habits, personalities and personal preferences of the Parties, along with their overall life styles, have been in constant conflict; that the Parties have made repeated efforts to reconcile their differences but those differences have become more distinct; that constant conflict and discord have become the lot of the Parties hereto in their domestic affairs and what love they held for each other has ceased to exist; that the conflict of temperaments between the Parties is irreconcilable and irremediable and that the Parties are incompatible to such an extent that it is impossible for them to live together as husband and wife and that their incompatibility is so great that there is no possibility of reconciliation between them and Plaintiff avers that he/she is entitled to a divorce on the grounds of incompatibility;

WHEREFORE, Plaintiff demands judgment against the Defendant as follows:

1. An absolute divorce.

2. An equitable division of personal property.

3. General relief.

Kathy Moore
Attorney for Plaintiff

****
****

Divorce Rates

NATIONAL DATA

There are about 1 million divorces each year (*Statistical Abstract of the United States: 2003*, Table 83).

There are different ways of expressing divorce rates, ratios, and probabilities. The **crude divorce rate** is a statement of how many divorces have occurred for every 1,000 people in the population. In 2002, there were 3.9 divorces per 1,000 population. This rate reflects a stabilization (and slight drop) of the divorce rate (as it was 4.0 the two previous years) (Sutton, 2003).

More meaningful is the **refined divorce rate,** which is an expression of the number of divorces and annulments in a given year divided by the number of married women in the population times 1,000. For example, there are ap-

Diversity in Other Countries

Compared with other countries, the United States has one of the highest divorce rates in the world. Only Belarus of Eastern Europe comes close when crude divorce rates are compared. However, one of the reasons for a high U.S. divorce rate is a very high marriage rate (Blossfeld & Muller, 2002).

proximately 1.1 divorces/annulments per year and 58 million married women, which translates into .0189 times 1,000, or 18.9 = the refined divorce rate.

A final way of estimating divorce is to identify the percentage of people who are married who eventually get divorced. The problem with this statistic is the period of time considered in identifying those who divorced. The percentage of those who divorced within five years would be lower than the percentage of those who divorced within a ten-year period. Current estimates suggest that about 40 percent of those who married in the past couple of decades will divorce (Hawkins et al., 2002). Goodwin (2003) noted that 20 percent of first marriages will end in divorce within five years, 33 percent within 10 years, and 43 percent within fifteen years of marriage.

Diversity in the United States

A higher percentage of Blacks are divorced than Whites (11.2 versus 10) (*Statistical Abstract of the United States: 2003*, Table 61). Contributing to this higher percentage of divorce among Blacks are the economic independence of the Black female and the difficulty for the Black male of finding and keeping a stable job.

Asian-Americans and Mexican-Americans have lower divorce rates than whites or African-Americans because they consider the family unit to be of greater value than their individual interests. Personal unhappiness is less likely to result in movement toward divorce for these spouses (Mindel et al. 1998).

Regardless of how one measures the rate of divorce in the United States, "divorce rates have been stable or dropping for two decades" (Coltrane & Adams, 2003, 363). The principal factor for such decline is that persons are delaying marriage so that they are older at the time of marriage. According to Heaton (2002), "age at marriage plays the greatest role in accounting for trends in marital dissolution. . . . [W]omen who marry at older ages have more stable marriages."

Divorce, with its consequences, has been one of the most researched issues in marriage and the family. We now review its causes, both macro and micro.

MACRO FACTORS CONTRIBUTING TO DIVORCE

Sociologists emphasize that social context creates outcomes. This is nowhere better illustrated than in the statistic that there was an average of only one divorce per year in Massachusetts among the Puritans from 1639 to 1760 (Morgan, 1944). The social context, reflected in strict divorce laws and strong social pressure to stay married, kept couples married. In contrast, divorce today is a frequent phenomenon. Various structural and cultural factors, also known as macro factors, help to account for our relatively high divorce rate.

Increased Economic Independence of Women

In the past, the unemployed wife was dependent on her husband for food and shelter. No matter how unhappy her marriage was, she stayed married because she was economically dependent on her husband. Her husband literally represented her lifeline. Finding gainful employment outside the home made it possible for the wife to afford to leave her husband if she wanted to. Now that almost 70 percent of all wives are employed (and this number is increasing), fewer and fewer wives are economically trapped in an unhappy marriage relationship. As we noted earlier, the wife's employment does not increase the risk of divorce in a happy marriage. But it does provide an avenue of escape for women in an unhappy or abusive marriage.

Employed wives are also more likely to require an egalitarian relationship; while some husbands prefer this role relationship, others are unsettled by it. Another effect of the wife's employment is that it may result in meeting someone new so that the wife becomes aware of alternatives to her current partner. Finally, unhappy husbands may be more likely to divorce if their wives are employed and able to be financially independent.

Diversity in Other Countries

Two researchers revealed the conditions under which a national survey of Canadians reported that divorce was justified. Ninety-five percent reported that "abusive behavior from the partner" justified divorce. Eighty-eight percent reported that "infidelity" and "lack of love and respect from the partner" justified divorce. Only 43 percent said that they would stay for the children (Frederick & Hamel, 1998).

Changing Family Functions and Structure

Many of the protective, religious, educational, and recreational functions of the family have largely been taken over by outside agencies. Family members may now look to the police, the church or synagogue, the school, and commercial recreational facilities rather than to each other for fulfilling these needs. The result is that although meeting emotional needs remains a primary function of the family, fewer reasons exist to keep the family together.

In addition to the changing functions of the family brought on by the Industrial Revolution, the family structure has changed from that of larger extended families in rural communities to smaller nuclear families in urban communities. In the former, individuals could turn to a lot of people in times of stress; in the latter, more stress necessarily fell on fewer shoulders. Cohen and Savaya (2003) documented an increased divorce rate among Muslim Palestinian citizens of Israel due to an increased acceptance of the modern views of marriage— increased emphasis on happiness and compatibility.

Liberalized Divorce Laws

No-fault divorce was first made available in California in 1970. All states now recognize some form of no-fault divorce. Although the legal grounds for it are irreconcilable differences and incompatibility, the reality is that spouses can get a divorce if they want to without having to prove that one of the partners is at fault (e.g., through adultery). The effect of **no-fault divorce** laws (neither party is assigned blame for the divorce) has been "to make divorce less acrimonious and restrictive, rendering the legal environment neutral and noncoercive" (Vlosky &Monroe,2002,317). However, a backlash has occurred in response to the fact that divorce has become too easy to obtain, and a movement is afoot to make divorce harder to get. One divorced spouse said, "I should have stayed married when the hard times hit—it was just too easy to walk out" (author's files).

Social structure will ensure that these young Masai wives of Arusha, Tanzania (East Africa) remain married since divorce would result in their return to their respective homes, where they would bring shame on their parents. In effect, they have no other role than that of wife and mother—the role of divorcee is not an option. Hence, though these women may be model wives in terms of taking care of their husbands and children, they will be perceived by their family and by the Masai group as worthy of shame if they divorce.

Authors

Fewer Moral and Religious Sanctions

Many priests and clergy recognize that divorce may be the best alternative in a particular marital relationship and attempt to minimize the guilt that members of their congregation may feel at the failure of their marriage. Increasingly, marriage is viewed in secular rather than religious terms. Hence, divorce has become more acceptable.

Starter Marriages

In her book *The Starter Marriage and the Future of Matrimony,* Paul (2002) noted that marriage is sometimes perceived as a growth experience whereby people marry, mature, and divorce before children arrive. Evidence for the phenomenon is that the highest chance of divorce is in the early years, when couples bail out if their expectations are not fulfilled. Such a view of marriage reflects our sometimes cultural disregard for marriage as a permanent life-sustaining relationship. Indeed, the label of "starter" suggests something temporary, like a starter house from which the owners will eventually move.

More Divorce Models

As the number of divorced individuals in our society increases, the probability increases that a person's friends, parents, siblings, or children will be divorced. The more divorced people a person knows, the more normal divorce will seem to that person. The less deviant the person perceives divorce to be, the greater the probability the person will divorce if that person's own marriage becomes strained. Divorce has become so common that there is now a magazine, *Divorce Magazine*, especially for divorced people.

Mobility and Anonymity

That individuals are highly mobile results in fewer roots in a community and greater anonymity. Spouses who move away from their respective family and friends are surrounded by strangers who don't care if they stay married or not. Divorce thrives when pro-marriage social expectations are not operative. In addition, the factors of mobility and anonymity also result in the removal of a consistent support system to help spouses deal with the difficulties they may encounter in marriage.

Individualistic Cultural Goal of Happiness

Unlike familistic values in Asian cultures, individualistic values in American culture emphasize the goal of personal happiness in marriage. When spouses stop having fun (when individualistic goals are no longer met), they sometimes feel that there is no reason to stay married. Only 8 percent of 620 never-married undergraduate students agreed with the statement "There are no conditions under which I would get a divorce." (Knox & Zusman, 1998). Reflecting an individualistic philosophy, Geraldo Rivera asked in 2003, "Who cares if I've been married five times?"

MICRO FACTORS CONTRIBUTING TO DIVORCE

Although macro factors may make divorce a viable cultural alternative to marital unhappiness, they are not sufficient to "cause" a divorce. One spouse must make a choice to divorce and initiate proceedings. Such a view is micro in that it focuses on the individual decisions and interactions within specific family units. The following subsections discuss some of the micro factors that may be operative in influencing a couple toward divorce.

Loss of Love

Deciding to divorce is related to both the absence of positives and the presence of negatives. Previti and Amato (2003) noted that couples who no longer viewed themselves as being in love were much more likely to divorce than those who had other issues. Indeed, love, respect, friendship, and good communication were mentioned by 60 percent of the respondents as reasons for staying married. Huston et al. (2001) also found that couples who report "weaker love feelings" and "being ambivalent" about the relationship are more likely to divorce.

Negative Behavior

People marry because they anticipate greater rewards from being married than from being single. During courtship, each partner engages in a high frequency of positive verbal (compliments) and nonverbal (eye contact, physical affection) behavior toward the other. The good feelings the partners experience as a result of these positive behaviors encourage them to marry to "lock in" these feelings across time.

Just as love feelings are based on positive behavior from the partner, negative feelings are created when the partner engages in a high frequency of negative behavior. Amato and Rogers (1997) interviewed a national sample of spouses (1,748) and identified some of these behaviors. Negative behaviors that wives reported their husbands engaging in included getting angry easily, being moody, not talking, displaying irritating habits, not being home enough, and spending money foolishly.

Negative behaviors that husbands reported their wives engaged in included getting their feelings hurt easily, being moody, not talking, being critical, and getting angry easily. Specific behaviors associated with a couple's subsequent divorce included sexual infidelity, jealousy, drinking, spending money, moodiness, not communicating, and anger (p. 622). Shumway and Wampler (2002) emphasized that the absence of positive behaviors such as "small talk," "reminiscing about shared times together," and "encouragement" increases a couple's marital dissatisfaction.

When a spouse's negative behavior continues to the point of creating more costs than rewards in the relationship, either partner may begin to seek a more reinforcing situation. Divorce (being single again) or remarriage may appear to be a more attractive alternative to being married to the present spouse. This is certainly true of a marriage in which one's spouse is an alcoholic or substance abuser or physically or emotionally abuses the partner.

Lack of Conflict Resolution Skills

Though every relationship experiences conflict, not every couple has the skills to effectively resolve conflict. Some partners respond to conflict by withdrawing emotionally from their relationship; others respond by attacking, blaming, and failing to listen to their partner's point of view. Without skills to resolve conflict in their relationships, partners drift into patterns of communication that may escalate rather than resolve conflict. When partners don't reduce conflict, they become vulnerable to divorce (Gottman et al., 1998). Ways to negotiate differences and reduce conflict were discussed in Chapter 9, Communication in Relationships.

Value Changes

Both spouses change throughout the marriage. "He's not the same person I married" is a frequent observation of persons contemplating divorce. People may undergo radical value changes after marriage. One minister married and decided seven years later that he did not like the confines of the marriage role. He left the ministry, earned a Ph.D., and began to drink and have affairs. His wife, who had married him when he was a minister, now found herself married to a clinical psychologist who spent his evenings at bars with other women. The couple divorced.

Because people change throughout their lives, the person that one selects at one point in life may not be the same partner one would select at another. Margaret Mead, the famous anthropologist, noted that her first marriage was a student marriage; her second, a professional partnership; and her third, an intellectual marriage to her soul mate, with whom she had her only child. At each of several stages in her life, she experienced a different set of needs and selected a mate who fulfilled those needs.

Satiation

Satiation, also referred to as habituation, refers to the state in which a stimulus loses its value with repeated exposure. Spouses may tire of each other. Their stories are no longer new, their sex is repetitive, and their presence no longer stimulates excitement as it did in courtship. Some persons, feeling trapped by the boredom of constancy, divorce and seek what they believe to be more excitement by a return to singlehood and, potentially, new partners. A developmental task of marriage is for couples to be creative to maintain excitement in their marriage. Going new places, doing new things, and making time for intimacy in the face of rearing children and sustaining careers become increasingly important across time. Alternatively, spouses need to be realistic and not expect every evening to be a New Year's Eve. Kurdek (2002) found that initial high levels of marital satisfaction were more resistant to negative outcomes (divorce).

Extramarital Relationship

Spouses who feel mistreated by their partners or bored and trapped sometimes consider the alternative of a relationship with someone who is good to them, who is exciting and new, and who offers an escape from the role of spouse to the role of lover. Extramarital involvements sometimes hurry a decaying marriage toward divorce because the partner begins to contrast the new lover with the spouse. The spouse is often associated with negatives (bills, screaming children, nagging); the lover, almost exclusively with positives (clandestine candlelight dinners, new sex, emotional closeness). The choice is stacked in favor of the lover. Although most spouses do not leave their mates for a lover, the existence of an extramarital relationship may weaken the emotional tie between the spouses so that they are less inclined to stay married.

I was dumbfounded, heartbroken, and outraged that I'd believed him at all.

Hillary Clinton of her husband's confession of his affair with Monica Lewinsky

Perception That Being Divorced Is Better Than Being Married

Brinig and Allen (2000) noted that two-thirds of the applications for divorce are filed by women. Their behavior is rational and based on the perception that they will achieve greater power over their own life, money (in the form of child support and/or alimony) without having the liability of dealing with an unsupportive husband on a daily basis, and greater control over their children, since in 80 percent of the cases custody is awarded to the woman. The researchers argue that "who gets the children" is by far the greatest predictor of who files for divorce, and they contend that if the law presumes that joint custody will follow divorce, there will be a reduction in the number of petitions for divorce, since there will be less to gain.

Top Twenty Factors Associated with Divorce

Researchers have identified the characteristics of those most likely to divorce (Goodwin, 2003; Orbuch et al., 2002; Wills, 2002; Teachman, 2002; Huston et al., 2001; Amato, 2001). Some of the more significant associations include the following:

1. *Courtship of less than two years (partners know less about each other)*
2. *Having little in common (similar interests serve as a bond between people)*
3. *Marrying in teens (associated with low education and income and lack of maturity)*
4. *Not being religiously devout (less bound by traditional values)*
5. *Differences in race, education, age, religion, social class, values (widens gap between spouses)*

6. *A cohabitation history (established history of breaking social norms)*
7. *Previous marriage (less fearful of divorce)*
8. *No children (less reason to stay married)*
9. *Spending little leisure time together (interests may be elsewhere)*
10. *Urban residence (more anonymity, less social control in urban environment)*
11. *Infidelity (trust broken, emotional reason to leave relationship) and ambivalence about marriage (look for alternative)*
12. *Divorced parents (models for ending rather than repairing relationship; may have inherited traits such as alcoholism that are detrimental to staying married) or parents who never married and never lived together (being unmarried is normative)*
13. *Poor communication skills (issues go unresolved and accumulate)*
14. *Unemployment of husband (his self-esteem lost, loss of respect by wife who expects breadwinner)*
15. *Employment of wife (wife more independent and can afford to leave an unhappy marriage; husband feels threatened)*
16. *Depression, alcoholism, or physical illness of spouse (partners undergo grave change from earlier in relationship)*
17. *Having seriously ill child (impacts stress, finances, couple time)*
18. *Low self-esteem of spouses (associated with higher jealousy, less ability to love and accept love)*
19. *Race (African-Americans live under more oppressive conditions, which increases stress in the marital relationship; 32 percent of European-American marriages, in contrast to 47 percent of African-American marriages, will end within ten years (Goodwin, 2003)).*
20. *Limited education (associated with lower income, more stress, less happiness)*

The more of these factors that exist in a marriage, the more vulnerable a couple is to divorce. Regardless of the various factors associated with divorce, there is debate about the character of people who divorce. Are they selfish, amoral people who are incapable of making good on a commitment to each other and who wreck the lives of their children? Or are they individuals who care a great deal about relationships and won't settle for a bad marriage? Indeed, they may divorce because they want to rescue their children from being reared in an unhappy, dysfunctional family.

CONSEQUENCES FOR SPOUSES WHO DIVORCE

Divorce is definitely not a single event but a long-lasting process of radically changing family relationships that begins in the failing marriage, continues through the often chaotic period of the marital rupture and its immediate aftermath, and extends even further, often over many years of disequilibrium.

Judith S. Wallerstein

Divorce is often an emotional and financial disaster (see Table 15.1, which identifies stages and issues of the divorce process) for both women and men (Amato, 2001). Aside from the death of a spouse, separation and divorce are the most difficult of life crisis events.

Physical, Emotional, Psychological, and Sexual Consequences

In spite of the prevalence of divorce and the suggestion that it may be the path to greater self-actualization or fulfillment, a comparison of married and unmarrieds (including the divorced) reveals that the divorced report being the least healthy and happy and the most depressed and suicidal (Waite & Gallagher, 2000). These characteristics are particularly present during the early stage of divorce when divorcing spouses are sad and angry, blame each other, and fight over children and property.

Besides feeling angry and bitter, the divorced may experience feelings of despair in reference to three basic changes in their life: termination of a major

Table 15.1 Stages and Issues of Divorce Adjustment

Stage 1: Pre-separation

Issues

Personal. Consider seeing a marriage counselor with your spouse to improve your marriage. If you must seek a divorce, get a formal/legal separation agreement drawn up and signed before you and your spouse begin to live in separate residences.

Spouse: If divorce is inevitable, adopt the perspective that you will nurture as positive a relationship as possible with your soon-to-be-former spouse. You, your former spouse, and your children benefit from such a relationship.

Children: Tell your children nothing until you make a definite decision to get divorced and have begun to develop a separation agreement.

Relatives: Same as for your children.

Finances: Anticipate a drop in income. Look for alternative housing. If unemployed, get a job.

Legal: Contact a divorce mediator if your relationship with your spouse is civil to develop the terms of your separation agreement. If mediation is not an option, each spouse needs to hire the best attorney he/she can afford.

Stage 2: Separation

Issues

Personal: Consider seeing a therapist alone to help you through the emotional devastation of divorce. Don't separate (move out) until you have a formal/legal signed agreement.

Spouse: Endeavor to cooperate/be civil with your former partner.

Children: Tell your children of your decision to divorce. Make clear to them that they are not to blame and that the divorce will not change your love and care for them.

Relatives: Tell your parents/friends of your decision to divorce and tell them you will need their support during this period of transition. Reach out to the parents of your soon-to-be-former partner and tell them that in spite of the divorce you would like to maintain a positive relationship with them—which your children will benefit from.

Finances: Your divorce mediator or attorney will instruct you to develop an inventory of what you own. Open up a separate savings and checking account.

Legal: Complete divorce mediation with a mediator. If this is not possible, ask your attorney to develop a legal separation agreement you can live with. Be reasonable. The more you and your spouse can agree on, the more time and money you save in legal fees. Mediated divorces cost about $1,500 and take about a month. A litigated divorce costs $15,000 plus and takes about three years.

Stage 3: Divorce

Issues

Personal: Nurture your relationships with friends and family who will provide support during the process of your divorce.

Spouse: Continue to nurture as civil a relationship as possible with your soon-to-be-former partner.

Children: Ensure that they have frequent and regular contact with each parent. Nurture their relationship with the other parent and their grandparents on both sides.

Relatives: Continue to have regular contact with friends/family who provide emotional support.

Finances: Be frugal.

Legal: Do what your attorney tells you.

Stage 4: Postdivorce

Issues

Personal: Seek other relationships but go slow in terms of commitment to a new partner. Give yourself at least 18 months before making a new commitment to a new partner.

Former Spouse: Continue as civil a relationship as possible. Encourage his/her involvement in another relationship. Be positive about a new partner in his/her life. You will want the same from your former partner some day.

Children: Spend individual time with each child. Do not require your children to like/enjoy your new partner.

Relatives: Provide both sets of grandparents access to your children.

Finances: Continue to be frugal. Move toward getting out of debt.

Legal: Make a payment plan with your attorney—begin to reduce this debt.

Adapted from The Divorce Room at Heartchoice.com. Used by permission.

source of intimacy, disruption of their daily routine, and awareness of a new status—divorced person. Going through a divorce can also lead to feelings of failure or defeat, especially if the other partner initiated the divorce. But the partner who initiates the divorce also undergoes considerable stress and guilt in making the decision to separate.

Divorce also shatters one's daily routine and emphasizes one's aloneness, particularly for men (Pinquart, 2003). Eating alone, watching TV alone, and driving alone to a friend's house for companionship are role adaptations made necessary by the end of one's marital role. In addition, the support system of the ex-spouse's family, with whom the individual may have been close, may no longer be available.

Although we tend to think of divorce as an intrinsically stressful event, one researcher suggested that the amount of stress involved in a divorce is determined by how stressful the marriage was (Wheaton, 1990). In marriages that were very stressful, going through a divorce may actually reduce more stress than it creates.

How do women and men differ in their emotional and psychological adjustment to divorce? Researchers (Siegler & Costa, 2000; Arendell 1995) noted that women fare better emotionally after separation and divorce than men do. Women are more likely than men not only to have a stronger network of supportive relationships but also to profit from divorce by developing a new sense of self-esteem and confidence, since they are thrust into a more independent role. On the other hand, men are more likely to have been dependent on their wives for domestic and emotional support and to have a weaker external emotional support system. As a result, divorced men are more likely than divorced women to date more partners sooner and to remarry more quickly. Hence, gender differences in long-term divorce adjustment are minimal. In the meantime, resources are available to assist individuals going through divorce.

We have been discussing the personal emotional consequences of divorce for the respective spouses, but the extended family and friends also feel the impact of a couple's divorce. Whereas some parents are happy and relieved that their offspring are divorcing, others grieve. Either way, the relationship with their grandchildren may be jeopardized. Friends often feel torn and divided in their loyalties. The courts divide the property and identify who gets the kids. But who gets the friends?

Aside from the emotional and psychological consequences for the spouses, their children, extended family, and friends, divorce means the loss of a regular sex partner. While some divorcing spouses stopped having sex months before the separation, others continue to enjoy sex up to the day of separation (and some continue even when separated). We noted in Chapter 5 that spouses report high levels of physical and emotional satisfaction with their sexual relationship. The divorced are on the other end of the continuum.

Financial Consequences

Both women and men experience a drop in income following divorce, but women may suffer more (Amato, 2001). African-American divorced women are especially vulnerable economically. Since men usually have greater financial resources, they may take all they can with them when they leave. The only money they may continue to give to the ex-wife is court-ordered in the form of child support or spousal support. The latter is rare. Some states, such as Texas, do not provide alimony at all. However, most states do provide for an equitable distribution of property, whereby property is divided according to what seems fairest to each party, on the basis of a number of factors (like ability to earn a living, fault in breaking up the marriage, etc.).

Although 56 percent of custodial mothers are awarded child support, the amount is usually inadequate, infrequent, and not dependable, and the woman is forced to work (sometimes at more than one job) to take financial care of her children.

Fathers' Separation from Children

About 5 million divorced dads wake up every morning in an apartment or home while their children are waking up in another place. These are noncustodial fathers who may be allowed to see their children only at specified times (e.g., two weekends a month). Baskerville (2003) emphasized that the state benefits economically from removing the father and is intent on doing so:

> *Divorce courts and their huge entourage of personnel depend for their existence on broken, single-parent homes. The first principle of family court is therefore: remove the father. So long as fathers remain with their families, the divorce practitioners earn nothing. This is why the first thing a family court does when it summons a father on a divorce petition—even if he has done nothing wrong and not agreed to the divorce— is to strip him of custody of his children.*
>
> *Once the father is eliminated, the state functionally replaces him as protector and provider. By removing the father, the state also creates a host of problems for itself to solve: child poverty, child abuse, juvenile crime, and other problems associated with single-parent homes. In this way the divorce machinery is self-perpetuating and self-expanding.*

Most divorced fathers separated from their children feel a sense of loss and believe that the courts have little interest in protecting their relationship with their children. A study of 72 family law attorneys revealed that most felt that the divorce and custody system is biased against fathers (Braver et al., 2002).

Dudley (1991) identified eighty-four divorced fathers (divorced for an average of six years) who reported infrequent contact with their children. Forty percent of the fathers who saw their children infrequently reported the primary reason was interference on the part of the former spouse, who refused to provide access to the children or who talked negatively to the children about their father. Two-thirds of these fathers reported that they had to return to court over visitation issues. This conflict had a negative impact on the father's relationship with his children, resulting in less involvement (Leite & McKenry, 2002). Fathers who want to stay connected with their children must make this a conscious goal (making it a priority in their lives), try to get along with the child's mother, and hire an attorney as necessary to ensure regular and frequent contact or custody. *The Divorced Dad's Survival Book* (Knox, 2000) is a resource for noncustodial dads.

Some divorced fathers seek joint custody, which ensures regular contact with their children. Increasingly, the courts will become more gender-neutral in making custody decisions so that the mother will no longer be presumed to be the primary parent. Coleman et al. (1998) noted, "Part of the myth of motherhood is that mothers are naturally better parents than are fathers and that the mother is the best care provider for the child" (p. 19).

Although we have been discussing the situation in which fathers are separated from their children, it sometimes happens that the father is given full custody. Coles (2003) studied Black single custodial fathers and found that a major motivating factor for their seeking custody was "the desire to embody the kind of father they themselves did not have" (p. 247).

Shared Parenting Dysfunction

Shared parenting dysfunction refers to the set of behaviors on the part of each parent (embroiled in a divorce) that are focused on hurting the other parent and are counterproductive for the child's well-being. Examples identified by Turkat (2002) include

> *a parent that forced the children to sleep in a car to prove the other parent had bankrupted them; a non-custodial parent who burned down the house of the primary residential parent after losing a court battle over custody of the children; and a parent who bought a cat for the children because the other parent was highly allergic to cats. Finally, some of the most destructive displays of Shared Parenting Dysfunction may include kidnapping, physical abuse, and murder* (p. 390).

Children who are deprived of regular and consistent access to the other parent may eventually resent the custodial parent for such deprivation. In addition, they may feel deceived if they are told negative things about the other parent and later learn that these were designed to foster a negative relationship with that parent. The result is often a strained and distanced relationship with the custodial parent when the child grows up—an unintended consequence (Walz, 1998).

Parental Alienation Syndrome

Parental alienation syndrome is a disturbance in which children are obsessively preoccupied with deprecation and/or criticism of a parent, denigration that is unjustified and/or exaggerated (Gardner, 1998).

Although the "alienators" are fairly evenly balanced between fathers and mothers (Gardner, 1998), the custodial parent (more often the mother) has more opportunity and control to alienate the child from the other parent. Schacht (2000) identified several types of behavior that either parent may engage in to alienate the child from the other parent:

1. *Minimizing the importance of contact and the relationship with the other parent*
2. *Excessively rigid boundaries: rudeness, refusal to speak to or inability to tolerate the presence of the other parent, even at events important to the child; refusal to allow the other parent near the home for drop-off or pick-up visitations*
3. *No concern about missed visits with the other parent*
4. *No positive interest in the child's activities or experiences during visits with the other parent*
5. *Granting autonomy to the point of apparent indifference ("It's up to you, I don't care.")*
6. *Overt expressions of dislike of visitation ("OK, visit, but you know how I feel about it . . .")*
7. *Refusal to discuss anything about the other parent ("I don't want to hear about . . . ") or selective willingness to discuss only negative matters*
8. *Innuendo and accusations against the other parent, including statements that are false*
9. *Portraying the child as an actual or potential victim of the other parent's behavior*
10. *Demanding that the child keep secrets from the other parent*
11. *Destruction of gifts or memorabilia of the other parent*
12. *Promoting loyalty conflicts (such as by offering an opportunity for a desired activity that conflicts with scheduled visitation)*

The most telling sign of a child who has been alienated from a parent is the irrational behavior of the child, who for no properly explained reason says that he or she wants nothing further to do with one of the parents.

Children who are alienated from the other parent are sometimes unable to see through the alienation process and regard their negative feelings as natural. Such children are similar to those who have been brainwashed by cult leaders to view outsiders negatively. Consequences are devastating and include long-term depression, uncontrollable guilt, hostility, and alcoholism. Successful reversal of parental alienation is difficult and can be achieved only by complete separation from the alienating parent for a substantial period of time (minimum of six months to as much as two years) (Gardner, 1998).

PERSONAL CHOICES

Choosing to Maintain a Civil Relationship with Your Former Spouse

Spouses who divorce must choose the kind of relationship they will have with each other. This choice is crucial in that it affects not only their own but also their children's lives. Constance Ahrons identified four types of ex-spouse relationships, including "fiery foes," which 25 percent of her divorcing spouses exemplified. Another 25 percent were categorized as "angry associates." Hence, half of her respondents had adversarial relationships with their former partners. Other patterns included "perfect pals" (12%) and cooperative colleagues (38%). (Stark, 1986).

Everyone loses when the "fiery foes" and "angry associates" pattern develops and continues. The parents continue to harbor negative feelings for each other, and the children are caught in the crossfire. They aren't free to develop or express love for either parent out of fear of disapproval from the other. Ex-spouses might consider the costs to their children of continuing their hostility and do whatever is necessary to maintain a civil relationship with their former partner. We emphasize the benefits of coparenting after divorce in Chapter 16.

Source

Stark, E. 1986. Friends through it all. *Psychology Today.* May, 54–60.

EFFECTS OF DIVORCE ON CHILDREN

One million children annually experience the divorce of their parents (Wallace & Koerner, 2003). The Children's Beliefs about Parental Divorce Scale provides a way to measure the perceived effects of divorce on children.

Children of divorce report feeling less close to their biological mother and father than children whose parents are still married.

Angela Decuzzi

Wallerstein (2000) conducted a longitudinal study of ninety-three children from divorced homes and found negative effects that surfaced in the child's 20s and 30s in the form of difficulty in relationship formation and fears of loss. Not only were the children of divorce more likely to have difficulty deciding whom they wanted to marry, but they also expressed concerns over whether to have children. Wallerstein's study has been criticized since she did not compare her subjects with those of similar age from intact homes. Hence, one cannot assume that children from intact homes do not also have concerns about whom to marry and whether to have children. In addition, some research contradicts Wallerstein's predictions. King (2002) found that parental divorce was unrelated to offspring trust in romantic relationships.

However, a review of the literature (Knox, 2000) on the effects of divorce on children revealed that children whose parents divorce (and particularly where the father drops out of the child's life) are more vulnerable to loss of self-esteem

SELF-ASSESSMENT

Children's Beliefs About Parental Divorce Scale

The following are some statements about children and their separated parents. Some of the statements are true about how you think and feel, so you will want to check YES. Some are NOT TRUE about how you think or feel, so you will want to check NO. There are no right or wrong answers. Your answers will just tell us some of the things you are thinking now about your parents' separation.

1. It would upset me if other kids asked a lot of questions about my parents. ____Yes ____No
2. It was usually my father's fault when my parents had a fight. ____Yes ____No
3. I sometimes worry that both my parents will want to live without me. ____Yes ____No
4. When my family was unhappy it was usually because of my mother. ____Yes ____No
5. My parents will always live apart. ____Yes ____No
6. My parents often argue with each other after I misbehave. ____Yes ____No
7. I like talking to my friends as much now as I used to. ____Yes ____No
8. My father is usually a nice person. ____Yes ____No
9. It's possible that both my parents will never want to see me again. ____Yes ____No
10. My mother is usually a nice person. ____Yes ____No
11. If I behave better I might be able to bring my family back together. ____Yes ____No
12. My parents would probably be happier if I were never born. ____Yes ____No
13. I like playing with my friends as much now as I used to. ____Yes ____No
14. When my family was unhappy it was usually because of something my father said or did. ____Yes ____No
15. I sometimes worry that I'll be left all alone. ____Yes ____No
16. Often I have a bad time when I'm with my mother. ____Yes ____No
17. My family will probably do things together just like before. ____Yes ____No
18. My parents probably argue more when I'm with them than when I'm gone. ____Yes ____No
19. I'd rather be alone than play with other kids. ____Yes ____No
20. My father caused most of the trouble in my family. ____Yes ____No

("Don't my parents love me enough to stay together?"), drop in school grades (due to the stress associated with the divorce and distracted parents), and increase in drug use (supervision of children drops with divorce). Increased delinquency was also thought to be related to divorce, but new research suggests that association with deviant peers (lack of adult supervision?) may be the culprit (Rebellon, 2002).

Though research on gender of the child and adjustment to divorce is inconsistent (Amato 2001), sons may experience more negative consequences than daughters (Nielsen, 1999). If the mother talks negatively about the father, the son is likely to have negative feelings about himself since he identifies with the father. Also, the son who does not maintain a close relationship with his father after the divorce is more likely to develop "serious social, sexual, emotional, or psychological problems" (p. 545). Hence, it is not the divorce but the low father involvement that has a negative impact on the father-child relationship (Ahrons & Tanner, 2003).

In this regard, the age of the child becomes relevant to the timing of divorce. Woodward, Ferguson, & Belsky (2000) observed that the younger the age of the

21. I feel that my parents still love me. ___Yes ___No

22. My mother caused most of the trouble in my family. ___Yes ___No

23. My parents will probably see that they have made a mistake and get back together again. ___Yes ___No

24. My parents are happier when I'm with them than when I'm not. ___Yes ___No

25. My friends and I do many things together. ___Yes ___No

26. There are a lot of things about my father I like. ___Yes ___No

27. I sometimes think that one day I may have to go live with a friend or relative. ___Yes ___No

28. My mother is more good than bad. ___Yes ___No

29. I sometimes think that my parents will one day live together again. ___Yes ___No

30. I can make my parents unhappy with each other by what I say or do. ___Yes ___No

31. My friends understand how I feel about my parents. ___Yes ___No

32. My father is more good than bad. ___Yes ___No

33. I feel my parents still like me. ___Yes ___No

34. There are a lot of things about my mother I like. ___Yes ___No

35. I sometimes think that once my parents realize how much I want them to they'll live together again. ___Yes ___No

36. My parents would probably still be living together if it weren't for me. ___Yes ___No

Scoring

The CBAPS identifies problematic responding. A "yes" response on items 1, 2, 3, 4, 6, 9, 11, 12, 14–20, 22, 23, 27, 29, 30, 35, 36 and a "no" response on items 5, 7, 8, 10, 13, 21, 24–26, 28, 31–34 indicate a problematic reaction to one's parents divorcing. A total score is derived by summing the number of problematic beliefs across all items, with a total score of 36. The higher the score, the more problematic the beliefs about parental divorce.

Norms: A total of 170 schoolchildren, 84 boys and 86 girls, with a mean age of 11 whose parents were divorced, completed the scale. The mean for the total score was 8.20, with a standard deviation of 4.98.

Source

Kurdek, L. A., and B. Berg. 1987. Children's beliefs about parental divorce scale: Psychometric characteristics and concurrent validity. *Journal of Consulting and Clinical Psychology* 55: 712–18. Copyright ©, Professor Larry Kurdek, Department of Psychology, State University, Dayton, OH 45435-0001. Used by permission of Dr. Kurdek.

child when separated from a parent, the lower the subsequent parental attachment and the more likely the child to perceive the parent as less caring. Hence, if parents divorce when the child is young, and the father becomes the noncustodial parent, the risk to the father-child relationship is greater.

The daughter's relationship with her father is also damaged by the mother who does not actively support that relationship. Indeed, "fathers and children usually remain close only if the mother actively encourages and facilitates their relationship" (Nielsen, 1999, p. 553).

Jekielek (1998) found that children from marriages characterized by high conflict that ended in divorce actually experience an improvement in their emotional well-being. The fact that they are no longer exposed to consistent conflict is experienced as a sense of relief. Sun and Li (2002) studied children three years before their parents' divorce and three years after and found that they fared less well both before and after on most well-being measures. Hence, children are impacted by the whole process—both before and after.

In discussing the effects of divorce on children, it is important to emphasize that while some children may be more vulnerable to negative outcomes from the divorce of their parents, the large majority of children of divorce do not experience severe or long-term problems: Most do not drop out of school, get arrested, abuse drugs, or suffer long-term emotional distress. Kelly and Emery (2003) noted that 75 percent to 85 percent of children/young adults whose parents divorce," do not suffer from major psychological problems, including depression; have achieved their education and career goals; and retain close ties to their families (pp. 357–58). Indeed, Coltrane and Adams (2003) emphasized that claiming that divorce seriously damages children is a "symbolic tool used to defend a specific moral vision for families and gender roles within them" (p. 369). They go on to state that "Understanding this allows us to see divorce not as the universal moral evil depicted by divorce reformers, but as a highly individualized process that engenders different experiences and reactions among various family members . . ." (p. 370).

Minimizing Negative Effects of Divorce on Children

Researchers have identified the conditions under which a divorce has the fewest negative consequences for children (Wallerstein, 2003; Stewart et al., 1997; Ahrons, 1995). Some of these follow.

1. Healthy parental psychological functioning. To the degree that parents remain psychologically fit and positive, and socialize their children to view the divorce as "something that happens" and a "challenge to learn from," children benefit. Parents who nurture self-pity, abuse alcohol or drugs, and socialize their children to view the divorce as a tragedy from which they will never recover create negative outcomes for their children. Some parents can benefit from therapy during the divorce period as a method of stabilizing their feelings, keeping a clear head, and making choices in the best interest of their children.

Some parents also enroll their children in the "new beginnings program"—"an empirically driven prevention program designed to promote child resilience during the postdivorce period" (Hipke et al., 2002, 121). The program focuses on improving the quality of the primary residential mother-child relationship, ensuring continued discipline, reduced exposure to parental conflict, and access to the nonresidential father. Outcome data reveal that not all children benefit, particularly those with "poor regulatory skills" and "demoralized" mothers (p. 127).

2. A cooperative relationship between the parents. The most important variable in a child's positive adjustment to divorce is that the child's parents continue to maintain a cooperative relationship throughout the separation, divorce, and post-divorce period. In contrast, bitter parental conflict places the children in the middle. One daughter of divorced parents said: "My father told me, 'If you love me, you would come visit me,' but my mom told me, 'If you love me, you won't visit him.'" Baum (2003) confirmed that the longer and more conflictual the legal proceedings, the worse the coparental relationship. Bream and Buchanan (2003) noted that children of conflictual divorcing parents are "children in need."

Mandatory parenting classes for divorcing parents now exist in forty-one states. Over 1,600 programs are in operation (Griffin, 1998) covering over fifty topic areas (Geasler & Blaisure, 1998). Data comparing divorcing parents in divorce education programs with divorcing parents not exposed to such programs show that these programs reduce child exposure to parental conflict.

3. Parents' attention to the children and allowing them to grieve. Both the custodial and the noncustodial parent continue to spend time with the children and to communicate to them that they love them and are interested in them. Parents also need to be aware that their children do not want the divorce and to allow them to grieve over the loss of their family as they knew it. Indeed, children do not want their parents to separate. Holroyd and Sheppard (1997) studied twenty-eight children whose parents had separated. Not one of them wanted their parents to separate and all wanted their parents to get back together. Some children are devastated to the point of suicide. A team of researchers identified 15,555 suicides among 15–24-year-olds in thirty-four countries in a one-year period and found an association between divorce rates and suicide rates (Johnson, Krug, & Potter, 2000).

*4. **Encouragement to see noncustodial parent.*** Children who usually live with custodial mothers following divorce are encouraged by the mother to maintain a regular and stable visitation schedule with their father.

*5. **Attention from the noncustodial parent.*** Noncustodial parents, usually the fathers, establish frequent and consistent times to be with the children. Noncustodial parents who do not show up at regular intervals exacerbate their children's emotional insecurity by teaching them, once again, that parents cannot be depended on. Parents who show up often and consistently teach their children to feel loved and secure. Sometimes joint custody solves the problem of children's access to their parents. Hsu et al. (2002) noted the devastating effect on children who grow up without a father.

PERSONAL CHOICES

Is Joint Custody a Good Idea?

Traditionally, sole custody to the mother was the only option considered by the courts for divorcing parents. The presumption was made that the "best interests of the child" were served if they were with their mother (the presumed more involved, more caring parent). As men have become more involved in the nurturing of their children, the courts no longer assume that "parent" means "mother." Indeed, a new **family relations doctrine** is emerging that suggests that even nonbiological parents may be awarded custody or visitation rights if they have been economically and emotionally involved in the life of the child (Holtzman 2002). A stepparent is an example.

Given that fathers are no longer routinely excluded from custody considerations, over half of the states have enacted legislation authorizing joint custody. About 16 percent of separated and divorced couples actually have a joint custody arrangement. In a typical joint physical custody arrangement, the parents continue to live in close proximity to each other. The children may spend part of each week with each parent or may spend alternating weeks with each parent.

New terminology is being introduced in the lives of divorcing spouses and in the courts. The term *joint custody,* which implies ownership, is being replaced with *shared parenting,* which implies cooperation in taking care of children.

There are several advantages of joint custody-shared parenting. Ex-spouses may fight less if they have joint custody because there is no inequity in terms of their involvement in their children's lives. Children will benefit from the resultant decrease in hostility between parents who have both "won" them. Unlike sole-parent custody, in which one parent (usually the mother) wins and the other parent loses, joint custody allows children to continue to benefit from the love and attention of both parents. Children in homes where joint custody has been awarded might also have greater financial resources available to them than children in sole-custody homes—fathers awarded joint custody are more likely to pay child support (Arditti, 1991).

Joint physical custody may also be advantageous in that the stress of parenting does not fall on one parent but rather is shared. One mother who has a joint custody arrangement with her ex-husband said, "When my kids are with their Dad, I get a break from the parenting role, and I have a chance to do things for myself. I love my kids, but I also love having time for myself." Another joint-parenting father said, "When you live with your kids every day, you're just not always happy to be with them. But after you haven't seen them for three days, it feels good to see them again."

A disadvantage of joint custody is that it tends to put hostile ex-spouses in more frequent contact with each other, and the marital war continues. Children do not profit from being subjected to bickering, yet relationships between children and bickering parents are not significantly different from those between children and parents who do not have joint custody. In a national study of children whose parents had a joint custody arrangement, the researchers found no evidence of less conflict or better relationships with their parents than if the children lived with one parent and saw the other on a visitation basis (Donnelly & Finkelhor, 1992). California, confronted with evidence that joint custody is not always in the best interest of children, has rescinded its 1979 law of "presumptive joint custody."

Depending on the level of hostility between the ex-partners, their motivations for seeking sole or joint custody, and their relationship with their children, any arrangement could have positive or negative consequences for the ex-spouses as well as for the children. In those cases in which the spouses exhibit minimal hostility toward each other, have strong emotional attachments to their children, and want to remain an active influence in their children's lives, joint custody may be the best of all possible choices.

Sources

Arditti, J. A. 1991. Child support noncompliance and divorced fathers: Rethinking the role of paternal involvement. *Journal of Divorce and Remarriage* 14:107–20.

Donnelly, D., and D. Finkelhor. 1992. Does equality in custody arrangement improve the parent-child relationship? *Journal of Marriage and the Family* 54:837–45.

Holtzman, M. 2002. The "family relations" doctrine: Extending Supreme Court precedent to custody disputes between biological and nonbiological parents. *Family Relations* 51:335–43.

6. Assertion of parental authority. Both parents continue to assert their parental authority with their children and continue to support the discipline practices of each other to their children.

7. A temperament on the part of the child that allows the child to adjust to change easily. Kalter (1989) found that the reaction of a child to divorce is influenced by the child's temperament. Some children are not easily frustrated and readily adapt to change; others have difficulty with even minor changes.

8. Regular and consistent child support payments. Support payments (usually from the father to the mother) are associated with enhanced well-being of the children of divorced parents (King, 1994).

9. Stability. The parents don't move the children to a new location. Moving them causes them to be cut off from their friends, neighbors, and teachers. It is important to keep their life as stable as possible during a divorce.

10. No new children in a new marriage. Manning and Smock (2000) found that divorced noncustodial fathers who remarried and who had children in the new marriages were more likely to shift their emotional and economic resources to the new family unit than were fathers who did not have new biological children. Fathers might be alert to this potential and consider each child, regardless of when he or she was born and with which wife in which marriage, as worthy of a father's continued love and support.

Because the greatest damage to children from a divorce is a continuing hostile and bitter relationship between their parents, some states require **divorce mediation** as a mechanism to encourage civility in working out differences and to clear the court calendar from protracted court battles. The Social Policy of this chapter focuses on divorce mediation.

CONDITIONS OF A "SUCCESSFUL" DIVORCE

Sometimes you have to take the leap and build your wings on the way down.

Kobi Yamada

While acknowledging that divorce is usually an emotional and economic disaster, it is possible to have a successful divorce. Indeed, most people are resilient and "are able to adapt constructively to their new life situation within 2–3 years following divorce, a minority being defeated by the marital breakup, and a substantial group of women being enhanced" (Hetherington, 2003, p. 318). The following are some of the behaviors spouses can engage in to achieve this.

1. Mediate rather than litigate the divorce. Divorce mediators encourage a civil, cooperative, compromising relationship while moving the couple toward an agreement on the division of property, custody, and child support. By contrast, attorneys make their money by encouraging hostility so that spouses will prolong the conflict, thus running up higher legal bills. In addition, money spent on attorneys (average is more than $15,000) cannot be divided by the couple. Ricci (2000) noted that "the best advice is still to do as much as you can to settle the issues about children outside the court where parents can maintain control of the process and keep their discussions private, and where they, not a judge, decide what is best for their child" (p. F12).

2. Coparent with your ex-spouse. Setting negative feelings about your ex-spouse aside so as to cooperatively coparent not only can reduce your personal stress but also can facilitate a successful adjustment on the part of the children going through the divorce (Blau, 1995). Although it may be unsettling for children to go through the transition from a nuclear to a binuclear family, Ahrons (1995) observed that "it doesn't follow that you have damaged your children for life." You can provide your children with love and nurturance no matter what form your family takes. Not coparenting with your ex-spouse works against such a possibility.

3. Take some responsibility for the divorce. Since marriage is an interaction between the spouses, one person is seldom totally to blame for a divorce. Rather, both spouses share in the demise of the relationship. Take some responsibility for what went wrong.

4. Learn from the divorce. View the divorce as an opportunity to improve yourself for future relationships. What did you do that you might consider doing differently in the next relationship?

5. Create positive thoughts. Divorced people are susceptible to feeling as though they are failures. They see themselves as Divorced persons with a capital D, a situation sometimes referred to as "hardening of the categories" disease. Improving their self-esteem is important for divorced persons. They can do this by systematically thinking positive thoughts about themselves. One technique is to write down twenty-one positive statements about yourself ("I am honest," "I am a good cook," "I am a good parent," etc.) and transfer them to seven three-by-five cards, each containing three statements. Take one of the cards with you each day and read the thoughts at three regularly spaced intervals (e.g., seven in the morning, one in the afternoon, seven at night). This ensures that you are thinking good things about yourself and are not allowing yourself to drift into a negative set of thoughts ("I am a failure"; no one will love me").

Should Divorce Mediation Be Required before Litigation?

SOCIAL POLICY

Divorce mediation is a process in which spouses who have decided to separate or divorce meet with a neutral third party (mediator) to negotiate the issues of (1) child custody and visitation, (2) child support, (3) property settlement, and (4) spousal support. Mediation is not for everyone. It does not work where there is a history of spouse abuse, where the parties do not disclose their financial information, where one party is controlled by someone else (e.g., a parent), where there is the desire for revenge, or where the mediator is biased. The latter situation can be mitigated by selecting a professional who has specific training and experience in divorce mediation.

Benefits of Mediation

1. Better relationship. Spouses who choose to mediate their divorce have a better chance for a more civil relationship because they cooperate in specifying the conditions of their separation or divorce. Mediation emphasizes negotiation and cooperation between the divorcing partners. Such cooperation is particularly important if the couple has children in that it provides a positive basis for discussing issues in reference to the children across time.

2. Economic benefits. Mediation is less expensive than litigation. The cost of hiring an attorney and going to court over issues of custody and division of property is around $15,000. A mediated divorce costs about $1,500. What the couple spend in legal fees they cannot keep as assets to later divide.

3. Less time-consuming process. A mediated divorce takes two to three months versus two to three years if the case is litigated.

4. Avoidance of public exposure. Some spouses do not want to discuss their private lives and finances in open court. Mediation occurs in a private and confidential setting.

5. Greater overall satisfaction. Mediation results in an agreement developed by the spouses, not one imposed by a judge or the court system. A comparison of couples who chose mediation with couples who chose litigation found that those who mediated their own settlement were much more satisfied with the conditions of their agreement. In addition, children of mediated divorces are exposed to less marital conflict, which may facilitate their adjustment to divorce.

Basic Mediation Guidelines

1. Children. What is best for a couple's children is a primary concern of the mediator. Children of divorced parents adjust best under three conditions: (1) the noncustodial parent is allowed regular and frequent access, (2) the children see the parents relating in a polite and positive way, and (3) each parent talks positively about the other parent and neither parent talks negatively about the other to the children.

2. Fairness. It is important that the agreement be fair, with neither party being exploited or punished. It is fair for both parents to contribute financially to the children. It is fair for the noncustodial parent to have regular access to his or her children.

3. Open disclosure. The spouses will be asked to disclose all facts, records, and documents to ensure an informed and fair agreement regarding property, assets, and debts.

4. Other professionals. During mediation the spouses may be asked to consult an accountant regarding tax laws. In addition, spouses are encouraged to consult an attorney throughout the mediation and to have the attorney review the written agreements that result from the mediation. However, during the mediation sessions, all forms of legal action by the spouses against each other should be stopped.

5. Confidentiality. The mediator will not divulge anything the spouses say during the mediation sessions without their permission. The spouses are asked to sign a document stating that should they not complete mediation, they agree not to empower any attorney to subpoena the mediator or any records resulting from the mediation for use in any legal action. Such an agreement is necessary for the spouses to feel free to talk about all aspects of their relationship without fear of legal action against them for such disclosures.

Divorce mediation is not for every couple getting divorced. Divorcing couples in which at least one of the spouses is abusive, is hiding assets, or wants to punish the partner will not benefit from mediation. Mediation requires that the partners not feel afraid of each other, be open and honest with each other about finances, and be conciliatory in demeanor toward each other.

6. Avoid alcohol and other drugs. The stress and despair that some people feel following a divorce make them particularly vulnerable to the use of alcohol or other drugs. These should be avoided because they produce an endless negative cycle. For example, stress is relieved by alcohol; alcohol produces a hangover and negative feelings; the negative feelings are relieved by more alcohol, producing more negative feelings, etc.

Up Close 15.1 Divorce—Uniquely Yours: You *Can* Do It Your Way

After exhausting all efforts to save our twenty-two-year marriage, my husband and I made a decision to divorce. Our two main objectives were the well-being of our children and the preservation of our family financial resources.

We accomplished both goals because we worked together instead of against each other. We saved thousands of dollars in legal fees, made important family decisions instead of letting strangers (lawyers) do so, and no one suffered emotionally or financially as a result.

You, too, can have a nondestructive, inexpensive divorce. Here's our advice:

1. Your children must be your top priority. Children love both parents. Adults must accept this. Never play one against the other.

2. Minimize the involvement of professionals, extended family members, and friends. The only people who have to like the choices are the spouses and the children. Only you know what works best for you. No one else has to like it; and remember everything is negotiable. Stand firm on this.

3. Joint custody and house sharing—it's legal to be different. We agreed on joint custody. We modified the standard joint custody arrangement by house sharing. We had our house, and we leased an apartment, and every Sunday at 6 P.M. instead of the children switching residences, the parents switched off between the house and the apartment. This way the children stayed in their home, which provided much more stability. This gave each parent equal opportunity to be fully present in the children's lives, plus during the one parent's week off, there was an opportunity for that parent to begin to get to know themselves again. All responsibility for raising the children was shared equally. We divided the costs fairly, based on our individual incomes.

4. Have family meetings to hear everyone's concerns and desires. Incorporate the input of each family member. Family members must feel safe to express themselves, without fear of criticism. Each family member must be supportive and respectful of the others at all times. Listening is essential. Always be available for your children—no excuses.

We agreed on childrearing, joint custody, and division of property without the advice of lawyers, child support officials, or well-meaning family and friends. Together we designed our own new lives. We negotiated our own terms of divorce, along with how our children would be raised and who would pay for what, and our divorce cost us less than $500. Our family nest egg was kept in our family. We did this without written agreements. We were very determined to keep our money in our family. This required total trust. Fortunately, we had that, and you can, too. If you feel a need for written agreements, you can write your own agreements, have them notarized and include them as part of your divorce decree that is filed with the courts. Keep it simple. Law firms charge hundreds of dollars to do what you can do yourself. Every time you call an attorney, it will cost you at least $100 to $200 an hour, and your nest egg disappears quickly.

The home and the family provide children with stability and foundation. Keeping them stable and secure is significant in how children learn to approach difficult issues in their own lives. If the adults are unstable, it is more difficult for the children to stabilize. Our children's lives changed very little through this adjustment in our family, because we worked together.

It takes more energy to conjure up negative and revengeful approaches than it does to simply handle business in a positive, mature manner. Your children are observing and learning from you. Setting a healthy example is your gift to them.

You may discover in the end that even though you will no longer be married, by working together, your friendship with your ex-spouse will be strengthened, providing a continued stability for your children similar to the one they had within the family unit, and everyone still feels human.

Written for this text by Kathy Moyer.

Kathy Moyer and Matt, her former husband, with whom she parted on the best of terms.

You can't get ahead while getting even.

Unknown

7. Relax without drugs. Deep muscle relaxation can be achieved by systematically tensing and relaxing each of the major muscle groups in the body. Alternatively, yoga, transcendental meditation, and getting a massage can induce a state of relaxation in some people. Whatever the form, it is important to schedule a time each day to get relaxed.

8. Engage in aerobic exercise. Exercise helps one not only to counteract stress but also to avoid it. Jogging, swimming, riding an exercise bike, or other similar exercise for thirty minutes every day increases the oxygen to the brain and helps facilitate clear thinking. In addition, aerobic exercise produces endorphins in the brain, which create a sense of euphoria ("runner's high").

9. Engage in fun activities. Some divorced people sit at home and brood over their "failed" relationship. This only compounds their depression. Doing what they have previously found enjoyable—swimming, horseback riding, skiing, sporting events with friends—provides an alternative to sitting on the couch alone.

10. Continue interpersonal connections. Adjustment to divorce is easier when intimate interaction with friends and family is continued (Bursik, 1991). This is particularly true for individuals who divorce past the age of 45 (Goodman, 1992) and for men. When divorced women and men are compared regarding life satisfaction, men report more dissatisfaction.

11. Let go of anger for your ex-partner. Former spouses who stay negatively attached to their ex by harboring resentment and trying to get back at the ex and who do not make themselves available to new relationships limit their ability to adjust to a divorce (Tschann et al., 1989). The old adage that venom does more damage to the person who holds it than the person for whom it is intended is true.

12. Allow time to heal. Since self-esteem usually drops after divorce, a person is often vulnerable to making commitments before working through feelings about the divorce. The time period most people need to adjust to divorce is between twelve and eighteen months. Although being available to others may help to repair one's self-esteem, getting remarried during this time should be considered cautiously. Two years between marriages is recommended.

Up Close 15.1 on page 375 emphasizes the importance of taking charge of your own divorce and not letting it unravel into a disaster. It was written by the divorced parent of a student who benefitted from her parents' "friendly" divorce.

ALTERNATIVES TO DIVORCE

Divorce is not the only means of terminating a marriage. Others include annulment, separation (legal or informal), and desertion.

Annulment

An **annulment** returns the spouses to their premarital status. In effect, an annulment means that the marriage never existed in the first place. The concept of annulment is both religious and civil. As a religious matter, annulment is a technical mechanism that allows Catholics to remarry. The Roman Catholic Church views marriage as a sacrament that is indissoluble except by death. Hence, while the Church does not recognize divorce, it does recognize annulment.

The basis used by the Catholic Church for granting an annulment is usually "lack of due discretion," which means that one or both parties lacked the ability needed to consent to the "essential obligations of matrimony." Personality or psychiatric disorders, premarital pregnancy, and problems in one's family of origin that one is trying to escape are examples of factors that are said by the Catholic Church to interfere with a person's ability to fulfill the essential matrimonial obligations, such as a permanent partnership, faithfulness, and sharing.

Since, when the Church annuls a marriage, it says that the marriage between the parties never existed, the individuals are allowed to marry someone else as though they had never been married. Although 90 percent of Catholics who want to end their marriage ignore the annulment process and get divorced and remarried, over 50,000 seek formal annulments from the Catholic Church annually (Rauch-Kennedy, 1997). Celebrities who have had their marriages annulled by the Church include Joseph P. Kennedy II, the eldest son of Robert Kennedy. His former wife, Sheila Rauch-Kennedy (1997), bitterly opposed the annulment of her 1979 marriage to him—they had been married twelve years and had two children.

Over 95 percent of petitions for annulment are granted. The average cost of the process is around $1,000 (Wallace, 1994). However, no one is denied having his or her petition processed because of lack of funds, and many annulments cost considerably less or cost just enough to cover court costs. Catholic couples who get divorced (rather than have their marriage annulled) and who remarry are not allowed full participation in the Catholic Church because divorce is not recognized.

Even though a Catholic couple's marriage may be annulled by the Church, the couple must still get a civil annulment (or a divorce) for the action to be legal and for the couple to legally remarry. Annulment as conceptualized by the Catholic Church is different from a civil annulment, which can be granted only by a civil court in the state in which the couple resides. An annulment granted by a civil court specifies that no valid marriage ever existed and returns both parties to their premarital status. Any property the couple exchanged as part of the marriage arrangement is returned to the original owner. Neither party is obligated to support the other economically.

Common reasons for granting civil annulments are fraud, bigamy, being under legal age, erectile failure, and insanity. As an example of fraud, a university professor became involved in a relationship with one of his colleagues. During courtship, he promised her that they would rear a "houseful of babies." But after the marriage, the woman discovered that the man had had a vasectomy several years earlier and had no intention of having more children. The marriage was annulled on the basis of fraud—the man's misrepresentation of himself to the woman. Most annulments are for fraud.

Bigamy is another basis for annulment. In our society, a person is allowed to be married to only one spouse at a time. If another marriage is contracted at the time a person is already married, the new spouse can have the marriage annulled. All 104 wives of confessed and convicted bigamist Giovanni Vigliotto were entitled to have their marriage to him annulled.

Most states have age requirements for marriage. When individuals younger than the minimum age marry without parental consent, the marriage may be annulled if either set of parents does not approve of the union. However, if neither set of parents or guardians disapproves of the marriage, the marriage may be regarded as legal; it is not automatically annulled.

Intercourse is a legal right of marriage. In some states, if a spouse is impotent, refuses to have intercourse, or is unable to do so for physical or psychological reasons, the other spouse can seek and may be granted an annulment.

Insanity and a lack of understanding of the marriage agreement are also reasons for annulment. Someone who is mentally deficient and incapable of understanding the meaning of a marriage ceremony can have a marriage annulled. Britney Spears (pop singer) and Jason Allen Alexander married in early 2004 and sought an annulment (on grounds of that she "lacked understanding of her actions") less than twelve hours after their marriage (Chen, 2004).

Separation

There are two types of separation—formal and informal. Typical items in a **formal separation** agreement include the following: (1) the husband and wife live separately; (2) their right to sexual intercourse with each other is ended; (3) the economic responsibilities of the spouses to each other are limited to those in the separation agreement; and (4) custody of the children is specified in the agreement, with visitation privileges granted to the noncustodial parent. The spouses may have emotional and sexual relationships with others, but neither party has the right to remarry. Although some couples live under this agreement until the death of one spouse, others draw up a separation agreement as a prelude to divorce. In some states, being legally separated for one year is a ground for divorce.

An **informal separation** (which is much more common) is similar to a legal separation except that no lawyer is involved in the agreement. The husband and wife settle the issues of custody, visitation, alimony, and child support between themselves. Because no legal papers are drawn up, the couple is still married from the state's point of view.

Attorneys advise against an informal separation (unless it is temporary) to avoid subsequent legal problems. For example, after three years of an informal separation, a mother decided that she wanted custody of her son. Although the father would have been willing earlier to sign a separation agreement that would have given her legal custody of her son, he was now unwilling to do so. Each spouse hired a lawyer, and a bitter and expensive court fight ensued.

Desertion

Desertion differs from informal separation in that the deserter walks out and breaks off all contact. Although either spouse may desert, it is usually the husband who does so. A major reason for deserting is to escape the increasing financial demands of a family. Desertion usually results in nonsupport, which is a crime.

The sudden desertion by a husband sometimes has more severe negative consequences for the wife than divorce. Unlike the divorced woman, the deserted woman is usually not free to remarry for several years. In addition, she receives no child support or alimony payments, and the children are deprived of a father.

Desertion is not unique to husbands. Wives and mothers also leave their husbands and children, although this happens infrequently. Their primary reason for doing so is to escape an intolerable marriage and the sense of being trapped in the role of mother. "I'm tired of having to think about my children and my husband all the time—I want a life for myself," said one woman who deserted her family. "I want to live, too." But such desertion is not without its consequences. Most mothers who desert their children feel extremely guilty.

DIVORCE PREVENTION

We have to find ways to make marriage work, if for no other reason than marriage is extraordinarily important to children.

David Popenoe, *Sociologist*

Divorce remains stigmatized in our society, as evidenced by the term **divorcism**—the belief that divorce is a disaster. In view of this cultural attitude, a number of attempts have been made to reduce it. Schacht (2000) identified a number of divorce prevention strategies, including a divorce tax. Under this proposal, all property settlements in divorces would be subject to an additional tax.

Another attempt at divorce prevention is covenant marriage now available in Louisiana, Arizona, and Arkansas. In these states a couple agrees to the following when they marry: (1) marriage preparation (meeting with a counselor who discusses marriage and their relationship), (2) full disclosure of all information that could reasonably affect the partner's decision to marry (e.g., one's homosexuality), (3) an oath that their marriage is a lifelong commitment, (4) an agreement to consider divorce only for "serious" reasons such as abuse, adultery, and imprisonment on a felony or separation of more than two years, and (5) an agreement to see a marriage counselor if problems threaten the marriage (Hawkins et al., 2002).

While most of a sample of 1,324 adults in a telephone survey in Louisiana, Arizona, and Minnesota were positive about covenant marriage (Hawkins et al., 2002), fewer than 3 percent of marrying couples elected covenant marriages when given the opportunity to do so (Licata, 2002). And, though already married couples can convert their standard marriages to covenant marriages, there are no data on how many have done so (Hawkins et al., 2002).

SUMMARY

What is the nature of divorce in the United States?

Divorce is the legal ending of a valid marriage contract. About 40 percent of couples who have married in the last two decades will eventually divorce. A quarter of college students experience the divorce of their parents.

What are macro factors contributing to divorce?

Macro factors contributing to divorce include increased economic independence of women, liberal divorce laws, fewer religious sanctions, more divorce models, and the individualistic cultural goal of happiness.

What are micro factors contributing to divorce?

Micro factors include negative behavior, lack of conflict resolution skills, satiation, and extramarital relationships. Having a courtship of less than two years, having little in common, and having divorced parents are all associated with subsequent divorce.

What are the consequences of divorce for spouses?

When compared with never-marrieds and marrieds, the divorced report being the least healthy and happy and the most depressed and suicidal. These characteristics are particularly present during the early stage of divorce when divorcing spouses are sad and angry, blame each other, and fight over children and property.

The amount of stress involved in a divorce is determined by how stressful the marriage was. In marriages that were very stressful, going through a divorce may actually reduce more stress than it creates. Women tend to fare better emotionally after separation and divorce than do men. Women are more likely than men not only to have a stronger network of supportive relationships but also to profit from divorce by developing a new sense of self-esteem and confidence, since they are thrust into a more independent role.

Factors associated with a quicker adjustment on the part of both spouses include mediating rather than litigating the divorce, coparenting their children, avoiding alcohol or other drugs, reducing stress through exercise, engaging in enjoyable activities with friends, and delaying a new marriage for two years.

What are the effects of divorce on children?

Although researchers agree that a civil, cooperative, coparenting relationship between ex-spouses is the greatest predictor of a positive outcome for children, researchers disagree on the long-term negative effects of divorce on children. Divorce mediation encourages civility between divorcing spouses who negotiate the issues of division of property, custody, visitation, child support, and spousal support.

What are alternatives to divorce?

Alternatives to divorce include annulment, separation, and desertion. Annulment returns the parties to their premarital state and is both a religious and civil concept. The Catholic Church does not recognize divorce but does recognize annulment, which in effect says that the parties were never married. However, the parties are still legally married and must have their marriage legally annulled, or they must get a divorce. Reasons for legal annulment include bigamy and being underage.

What are strategies to prevent divorce?

In response to the alarming incidence of divorce and the devastation to spouses and children that follows, divorce prevention strategies include a divorce tax whereby all property settlements in divorces would be subject to an additional tax. In addition, three states (Louisiana, Arizona, and Arkansas) offer covenant marriages, in which spouses agree to divorce only for serious reasons such as imprisonment on a felony or separation of more than two years. They also agree to see a marriage counselor if problems threaten the marriage. When given the option to choose a covenant marriage, few couples do so.

KEY TERMS

annulment
bigamy
crude divorce rate
desertion
divorce
divorce mediation
divorcism
family relations doctrine
formal separation
informal separation
no-fault divorce
parental alienation syndrome
refined divorce rate
satiation
shared parenting dysfunction

RESEARCHING MARRIAGE AND THE FAMILY WITH INFOTRAC COLLEGE EDITION

InfoTrac College Edition, an online library, allows you to perform research online anywhere, anytime. Following are two suggested search terms and related questions to help you extend your understanding of the topics covered in this chapter. Go to www.infotrac-college.com to begin your search.

Keyword: **Divorce laws.** Locate articles that discuss divorce laws in the United States. Do you think divorce laws make it too difficult or too easy to divorce? What should be the function of divorce laws?

Keyword: **Divorce mediation.** Locate articles that discuss divorce mediation. In what cases is divorce mediation *not* a good idea?

The Companion Web Site for Choices in Relationships: An Introduction to Marriage and the Family, Eighth Edition

http://sociology.wadsworth.com/knox_schacht/choices8e

Supplement your review of this chapter by going to the companion Web site to take one of the Tutorial Quizzes, use the flash cards to master key terms, and check out the many other study aids you'll find there. You'll also find special features such as the Marriage and Family Resource Center, Census 2000 information, and other data and resources at your fingertips to help you with that special project or to do some research on your own.

WEBLINKS

Divorce at Heartchoice.com
http://heartchoice.com/divorce/index.php

Divorcesource.com
http://www.divorcesource.com/

Divorcesupport.com
http://www.divorcesupport.com/index.html

Divorceinfo.com (divorce Weblinks)
http://www.divorceinfo.com/

Parental Alienation Syndrome
http://www.coeffic.demon.co.uk/pas.htm

National Center for Health Statistics (marriage and divorce data)
http://www.cdc.gov/nchs/

Authors

CHAPTER 16

Stepfamilies are not blended! Healthy ones recognize that children from prior relationships have two families and do not blend solely into one family.

Stepfamily Association of America

Remarriage and Stepfamilies

Contents

TRUE OR FALSE?

1. For a man who is divorced, his support obligations to the children in his first marriage are reduced if he has children in a second marriage.
2. Persons who have remarried are more likely to divorce in the early years than persons who have married for the first time.
3. There is a movement away from the use of the term *blended* since stepfamilies really do not blend.
4. At least two years is the recommended interval between marriages.
5. African-Americans and Hispanics are more likely to remarry than whites.

Answers: **1.** F **2.** T **3.** T **4.** T **5.** F

September 16 of each year is Stepfamily Day. Most governors issue Stepfamily Day proclamations commemorating the event as a time to recognize the pervasiveness of stepfamilies. About 40 percent of marriages today are remarriages, and most of these involve children or the creation of stepfamilies. We begin with a discussion of remarriage.

REMARRIAGE

Persons who left their previous marriage rather than persons who were left by their spouse are more likely to remarry first (Sweeney, 2002). Men who were domestically dependent on their partners are also more likely to repartner (DeGraaf and Kalmijn, 2003). When the divorced who have remarried and the divorced who have not remarried are compared, the remarried report greater personal and relationship happiness. The positive effect of remarriage on divorce adjustment was also observed by Wang and Amato (1998), who analyzed longitudinal data collected between 1980 and 1997. Like persons in first marriages, the newly remarried are in love and looking forward to a fulfilling life with their new spouse.

The biggest mistake of remarried spouses with children is that they try to use the nuclear family as their map . . . and that's like trying to put a square block in a round hole and it doesn't work . . . even if you shave off the edges.

James Bray, *Stepfamilies*

Remarriage for the Divorced

Ninety percent of remarriages consist of persons who are divorced rather than widowed. The principle of homogamy is illustrated in the selection of remarriage partners. Sixty percent of divorced people marry other divorced people (Ganong & Coleman, 1994). The majority of the divorced get remarried for many of the same reasons as those getting married for the first time—love, companionship, emotional security, and a regular sex partner. Other reasons are unique to remarriage and include financial security (particularly for the wife), help in rearing one's children, the desire to provide a "social" mother or father for one's children, avoidance of the stigma associated with the label "divorced person," and legal threats regarding the custody of children. With regard to the latter, a parent seeking custody of a child is viewed more favorably by the courts if he or she is married.

Diversity in Other Countries

Although getting remarried is regarded as an option by both women and men in the United States, such is not the case in all societies. Divorced women living in urban centers in central India are not likely to want to remarry since they are only allowed to remarry widowers. Divorced men are allowed to remarry any single woman of their choice.

Diversity in the United States

Remarriage rates are lower among Hispanics and African-Americans. Only about a third of Hispanics and 20 percent of African-Americans will eventually remarry. Remarriage rates among Hispanics may be depressed because Hispanics are predominantly Catholic (the Catholic Church opposes remarriage). Lower remarriage rates among African-Americans reflect "the lesser place of marriage in the African-American family" (greater focus on mother-child relationship) (Cherlin, 1996, 384).

Regardless of the reason for remarriage, it is best to proceed slowly into a remarriage. Grieving over the loss of a first spouse and developing a relationship with a new partner takes time. At least two years is the recommended interval between marriages (Marano, 2000). Older divorced women (over 40) are less likely than younger divorced women to remarry. Not only are there fewer available men, but the mating gradient whereby men tend to marry women younger than themselves is operative. In addition, older women are more likely to be economically independent, enjoy the freedom of singlehood, and want to avoid the restrictions of marriage. If they remarry, they are more likely to do so after their children have left home (Sweeney, 1998).

Divorced persons getting remarried are usually about ten years older than those marrying for the first time. Persons in their mid-30s who are considering remarriage have usually finished school and are established in a job or career. Courtship is usually short and takes into account the individuals' respective work schedules and career commitments. Because each partner may have children, much of the couple's time together includes their children. The practice of going out on an expensive dinner date during courtship before the first marriage is replaced with eating pizza at home and renting a DVD to watch with the kids.

Preparation for Remarriage

It is not uncommon for persons who are divorced to live together with a new partner before remarriage (Ganong & Coleman, 1994). Like other cohabitants, they drift into living together by gradually spending more time together. Aside from living together, they (like most couples in courtship, whether first or second marriage) do little else to prepare for their new marriage. Ceglian and Gardner (1998) found that women in second marriages were more likely to deny problems and to avoid discussing relationship issues than women in first marriages. Timmer and Veroff (2000) emphasized the importance of getting to know the partner's family (particularly wives getting to know their husbands' families) and nurturing such relationships; they pointed out that doing so is related to increases in reported marital happiness of remarried partners.

Lamaro (1997) emphasized the importance of pre-second marriage education to help prepare couples for both the new marriage and the stepfamily they are about to enter. Persons getting remarried are also more likely to sign a prenuptial agreement. Those who have considerable assets from their first marriages and who want to ensure that those assets go to their children rather than to their spouses may insist on the signing of a prenuptial agreement.

PERSONAL CHOICES

Should a Woman Marry a Divorced Man with Children?

Women considering marriage to a divorced man with children should be cautious. Knox and Zusman (2001) analyzed 274 questionnaires of second wives—women who married men who had been married before. Almost four in ten (39%) reported having thought about divorcing their husbands, and one in four reported that, with what she now knew, she would not have married him. Implications of the study included the following:

1. Acknowledge that a second marriage is vulnerable. Be realistic about the degree to which his children will accept you and/or your children. And how do you really feel about his children?

2. Question whether living together is beneficial to future marital success. Eighty-three percent of second wives reported that they had lived with their husbands before they married. Abundant evidence suggests that living together is *not* associated with positive relationship outcomes.

3. Delay marriage to a person who has been married before. Over a quarter (27.4%) of the wives reported marrying their current husband less than a year after his divorce became final. We have previously noted the importance of knowing a person at least two years before getting married.

4. Consider a fresh start in a new home. Eighty-two percent of the second wives reported that their new husbands had moved into their homes—homes the former husband of the new wife had vacated in the prior divorce. Not one of the second wives reported that she currently lived in a newly bought or rented place with her new husband. Only 16 percent reported that they (the second wives) had moved into their husband's home or condo.

Source

Knox, D., and M E. Zusman. 2001. Marrying a man with "baggage": Implications for second wives. *Journal of Divorce and Remarriage* 35:67–80.

Stages of Remarriage

People who remarry go through several stages (Berger, 1998; Goetting, 1982; Ganong & Coleman, 1994).

Boundary Maintenance Movement from divorce to remarriage is not a static event that happens in a brief ceremony and is over. Rather, ghosts of the first marriage in terms of the ex-spouse and, possibly, the children must be dealt with. A parent must decide how to relate to his or her ex-spouse in order to maintain a good parenting relationship for the biological children while keeping an emotional distance to prevent problems from developing with the new partner. Some spouses continue to be emotionally attached to the ex-spouse and have difficulty breaking away. These former spouses have what Masheter (1999) terms a "negative commitment" in that they

> *have decided to remain [emotionally] in this relationship and to invest considerable amounts of time, money, and effort in it . . . [T]hese individuals do not take responsibility for their own feelings and actions, and often remain "stuck," unable to move forward in their lives* (p. 297).

Emotional Remarriage Remarriage involves beginning to trust and love another person in a new relationship. Such feelings may come slowly as a result of negative experiences in the first marriage.

Psychic Remarriage Divorced individuals considering remarriage may find it difficult to give up the freedom and autonomy of being single again and to develop a mental set conducive to pairing. This transition may be particularly difficult for people who sought a divorce as a means to personal growth and autonomy. These individuals may fear that getting remarried will put unwanted constraints on them.

Remarriage for the formerly married is like marriage the first time—it is a time of hope and promise.

Authors

Community Remarriage This stage involves a change in focus from single friends to a new mate and other couples with whom the new pair will interact. The bonds of friendship established during the divorce period may be particularly valuable because they have lent support at a time of personal crisis. Care should be taken not to drop these friendships.

The community also involves religion. The Catholic Church requires that a first marriage be annulled before the Church will acknowledge a remarriage. A divorced person is restricted from participating in the sacraments of the Church, including communion. An annulment (discussed in the previous chapter) permits the person to take the sacraments again.

Parental Remarriage Because most remarriages involve children, people must work out the nuances of living with someone else's children. Since mothers are usually awarded primary physical custody, this translates into the new stepfather's adjusting to the mother's children and vice versa. If a person has children from a previous marriage who do not live primarily with him or her, the new spouse must adjust to these children on weekends, holidays, vacations, or other visitation times.

Economic and Legal Remarriage The second marriage may begin with economic responsibilities to the first marriage. Alimony and child support often threaten the harmony and sometimes even the economic survival of second marriages. Although legally the income of a new wife is not used to decide the amount her new husband is required to pay in child support for his children of a former marriage, his ex-wife may petition the court for more child support. The premise of her doing so is that his living expenses are reduced with the new wife and therefore he can afford to pay more child support. While the ex-wife is not likely to win, she can force the new wife to court and a disclosure of her income (all with considerable investment of time and legal fees for the newly remarried couple).

Economic issues in a remarriage may become evident in another way. A remarried woman who was receiving inadequate child support from her ex-spouse needed money for her child's braces and wrestled with how much money to ask her new husband for.

Remarriage for the Widowed

Only 10 percent of remarriages consist of widows or widowers. Nevertheless, remarriage for the widowed is usually very different from remarriage for the divorced. Unlike the divorced, they are usually much older and their children are grown. A widow or widower may marry someone of similar age or someone who is considerably older or younger. Marriages in which one spouse is considerably older than the other are referred to as May-December marriages (discussed in Chapter 8, Marriage Relationships). Here we will discuss only **December marriages,** in which both spouses are elderly.

A study of twenty-four elderly couples found that the need to escape loneliness or the need for companionship was the primary motivation for remarriage (Vinick, 1978). The men reported a greater need to remarry than the women. Most of the spouses (75 percent) met through a mutual friend or relative and married less than a year after their partner's death (63 percent).

The children of the couples had mixed reactions to their parent's remarriage. Most of the children were happy that their parent was happy and felt relieved that the companionship needs of their elderly parent would now be met by someone on a more regular basis. But some children

Diversity in Other Countries

Though stepparents in the United States have limited or no legal decision-making authority (e.g., signing medical releases for children in their family) (Mason et al., 2002), England has created a guardianship role that allows stepparents to assume some legal responsibility for stepchildren (Pasley, 2000).

also disapproved of the marriage out of concern for their inheritance rights. "If that woman marries Dad," said a woman with two children, "she'll get everything when he dies. I love him and hope he lives forever, but when he's gone, I want the farm." Though children may be less than approving of the remarriage of their widowed parent, adult friends of the couple, including the kin of the deceased spouses, are usually very approving (Ganong & Coleman, 1994).

Stability of Remarriages

National data reflect that remarriages are more likely than first marriages to end in divorce in the early years of remarriage (Clarke & Wilson, 1994). Having been married before does not make one immune to the delusionary effect of love. Indeed, among those about to remarry, love continues to delude them into thinking that "this time it will be different." The divorced also tend to blame their ex-spouse for the divorce (and thus fail to learn from the experience) and tend to feel that they "just married the wrong person" (and that the new person is right for them) (Marano, 2000). Individuals who have been divorced once are also less fearful of divorce if something goes wrong in the second marriage.

Though remarried persons are more vulnerable to divorce in the early years of their subsequent marriage, after fifteen years of staying in the second marriage, they are less likely to divorce than those in first marriages (Clarke & Wilson, 1994). Hence, these spouses are likely to remain married because they want to, not because they fear divorce.

STEPFAMILIES

Stepfamilies, also known as blended, binuclear, step, remarried, or reconstituted families represent the fastest-growing type of family in the United States. A **blended family** is one in which the spouses in a new marriage relationship are blended with the children of at least one of the spouses from a previous marriage. (Another popular term is **binuclear,** which refers to a family that spans two households—when a married couple with children divorce, their family unit spreads into two households.)

There is a movement away from the use of the term *blended* since stepfamilies really do not blend. The term **stepfamily** is the term currently in vogue. This section examines how stepfamilies differ from nuclear families; how they are experienced from the viewpoints of women, men, and children; and the developmental tasks that must be accomplished to make a successful stepfamily.

Definition and Types of Stepfamilies

Although a stepfamily can be created when a never-married or a widowed parent with children marries a person with or without children, most stepfamilies today are composed of spouses who were once divorced. This is different from stepfamilies characteristic of the early twentieth century, which more often were composed of spouses who had been widowed.

There are several types of stepfamilies (Berger, 1998):

1. *Stepfamilies in which a child lives with his or her married parent and new spouse*
2. *Stepfamilies in which the children from a previous marriage visit with their remarried parent and new spouse*
3. *An unmarried couple living together in which at least one of the partners has children from a previous relationship who live with or visit them*

4. *A remarried couple in which each of the spouses brings children into the new marriage from the previous marriage*
5. *A couple who not only bring children from a previous marriage but also have a child or children of their own*
6. *A remarried couple, both of whom have children from a previous marriage. The children may live in another state and have very little contact with the remarried couple.*

Not all stepfamilies are heterosexual. Wright (1998) observed that lesbian stepfamilies model gender flexibility and that mothers and stepmothers share parenting—both traditional mothering and fathering—tasks. This allows the biological mother some freedom from motherhood as well as support in it.

Just as there are different types of stepfamilies, theorists view the stepfamily constellation from various perspectives. These are reviewed in Up Close 16.1.

Unique Aspects of Stepfamilies

Stepfamilies differ from nuclear families in a number of ways. To begin with, the children in a nuclear family are biologically related to both parents, whereas the children in a stepfamily are biologically related to only one parent. Also, in a nuclear family, both biological parents live with their children, whereas only one biological parent in a stepfamily lives with the children. In some cases, the children alternate living with each parent.

Though nuclear families are not immune to loss, everyone in a stepfamily has experienced the loss of a love partner, which results in grief. About 70 percent of children are living without their biological father (who some children desperately hope will reappear and reunite the family). The respective spouses may also have experienced emotional disengagement and physical separation from a once-loved partner. Stepfamily members may also experience losses because of having moved away from the house in which they lived, their familiar neighborhood, and their circle of friends.

Stepfamily members also are connected psychologically to others outside their unit. Bray and Kelly (1998) referred to these relationships as the "Ghosts at the table":

> *Children are bound to absent parents, adults to past lives and past marriages. These invisible psychological bonds are the Ghosts at the table and because they play on the most elemental emotions—emotions like love and loyalty and guilt and fear—they have the power to tear a marriage and stepfamily apart.* (p. 4)

Children in nuclear families have also been exposed to a relatively consistent set of beliefs, values, and behavior patterns. When children enter a stepfamily, they "inherit" a new parent, who may bring a new set of values and beliefs and a new way of living into the family unit. Likewise, the new parent now lives with children who may have been reared differently from the way in which the stepparent would have reared them if he or she had been their parent all along. One stepfather explained:

> *It's been a difficult adjustment for me living with Molly's kids. I was reared to say "Yes sir" and "Yes ma'am" to adults and taught my own kids to do that. But Molly's kids just say "yes" or "no." It rankles me to hear them say that, but I know they mean no wrong with "yes" and "no" as long as it is said politely, and that it is just something that I am going to have to live with.*

Another unique aspect of stepfamilies is that the relationship between the biological parent and the children has existed longer than the relationship between the adults in the remarriage. Jane and her twin children have a nine-year relationship and are emotionally bonded to each other. But Jane has known her

Up Close 16.1 Stepfamilies in Theoretical Perspective

Structural functionalists, conflict theorists, and symbolic interactionists view stepfamilies from different points of view.

Structural-Functional Perspective
To the structural functionalist, integration or stability of the system is highly valued. The very structure of the stepfamily system can be a threat to the integration and stability of the family system. The social structure of stepfamilies consists of a stepparent, a biological parent, biological children, and stepchildren. Functionalists view the stepfamily system as vulnerable to an alliance between the biological parent and the biological children who have a history together.

In 75 percent of the cases, the mother and child create the alliance. The stepfather, as an outsider, may view this alliance between the mother and her children as the mother's giving the children too much status or power in the family. While the mother relates to her children as equals, the stepfather relates to the children as unequals whom he attempts to discipline. The result is a fragmented parental subsystem whereby he accuses her of being too soft and she accuses him of being too harsh and impartial. Structural family therapists suggest that parents should have more power than children and that they should align themselves with each other. Not to do so is to give children family power, which they may use to splinter the parents off from each other and create another divorce.

Conflict Perspective
Conflict theorists view conflict as normal, natural, and inevitable as well as functional in that it leads to change. Conflict in the stepfamily system is seen as desirable in that it leads to equality and individual autonomy.

Conflict is a normal part of stepfamily living. The spouses, parents, children, and stepchildren are constantly in conflict for the limited resources of space, time, and money. Space refers to territory (rooms) or property (television, CD player) in the house that the stepchildren may fight over. Time refers to the amount of time that the parents will spend with each other, with their biological children, and with their stepchildren. Money must be allocated in a reasonably equitable way so that each member of the family has a sense of being treated fairly.

Problems arise when space, time, and money are limited. Two new spouses who each bring a child from a former marriage into the house have a situation fraught with potential conflict. Who sleeps in which room? Who gets to watch which channel on television? To further complicate the situation, suppose the couple have a baby. Where does the baby sleep? And since both parents may have full-time jobs, the time they have for the three children is scarce, not to speak of the fact that the baby will require a major portion of their available time. As for money, the cost of the baby's needs, such as formula and disposable diapers, will compete with the economic needs of the older children. Meanwhile, the spouses may need to spend time alone as well as together and may want to spend money as they wish. All these conflicts are functional, since they increase the chance that a greater range of needs will be met within the stepfamily.

Interactionist Perspective
Symbolic interactionists emphasize the meanings and interpretations that members of the stepfamily develop for events and interactions in the family. Children may blame themselves for their parents' divorce and feel that they and their stepfamily are stigmatized; parents may view stepchildren as spoiled.

Stepfamily members also nurture certain myths. Stepchildren sometimes hope that their parents will reconcile and that their nightmare of divorce and stepfamily living will end. This is the myth of reconciliation. Another is the myth of instant love, usually held by each biological parent, who hopes that his or her children will instantly love the new partner. Although this does happen, particularly if the child is young and has no negative influences from the other parent, it is unlikely.

new partner only a year, and although her children like their new stepfather, they hardly know him.

In addition, the relationship between the biological parent and his or her children is of longer duration than that of the stepparent and stepchildren. The short history of the relationship between the child and the stepparent is one factor that may contribute to increased conflict between these two during the child's adolescence. Children may also become confused and wonder whether they are disloyal to their biological parent if they become friends with the stepparent.

Another unique feature of stepfamilies is that unlike children in the nuclear family, who have one home they regard as theirs, children in stepfamilies have two homes they regard as theirs. In some cases of joint custody, children spend part of each week with one parent and part with the other; they live with two sets of adult parents in two separate homes.

Money, or lack of it, from the ex-spouse (usually the husband) may be a source of conflict. In some stepfamilies, the ex-spouse (usually the father) is expected to send child support payments to the parent who has custody of the children. Fewer than one-half of these fathers send any money; those who do may be irregular in their payments. Fathers who pay regular child support tend to have higher incomes, to have remarried, to live close to their children, and to visit them regularly. They are also more likely to have legal shared or joint custody, which helps to ensure that they will have access to their children.

Fathers who do not voluntarily pay child support and are delinquent by more than one month may have their wages garnished by the state. Some fathers change jobs frequently and move around to make it difficult for the government to keep up with them. Such dodging of the law is frustrating to custodial mothers who need the child support money. Added to the frustration is the fact that fathers are legally entitled to see their children even though they do not pay court-ordered child support. This angers the mother who must give up her child on weekends and holidays to a man who is not supporting his child financially. Such distress on the part of the mother is probably conveyed to the child.

New relationships in stepfamilies experience almost constant flux. Each member of a new stepfamily has many adjustments to make. Issues that must be dealt with include how the mate feels about the partner's children from a former marriage, how the children feel about the new stepparent, and how the newly married spouse feels about the spouse's sending alimony and child support payments to an ex-spouse. In general, families in the Bray and Kelly (1998) study did not begin to think and act like a family until the end of the second or third year. These early years are the most vulnerable, with a quarter of the stepfamilies ending during that period.

Stepfamilies are also stigmatized. **Stepism** is the assumption that stepfamilies are inferior to biological families. Stepism, like racism, heterosexism, sexism, and ageism, involves prejudice and discrimination. Social changes need to be made to give support to stepfamilies. For example, "If there's a banquet, is there an opportunity for a child to invite all four of his or her parents?"(Everett, 2000). More often, children have to make a choice, which usually results in selecting the biological parent and ignoring the stepparents. The more parents children have who love and support them, the better, and our society should support this.

We are all familiar with the wicked stepmother in "Cinderella." The fairy tale certainly gives us the impression that to be in a stepfamily with stepparents is a disaster. Even textbooks in marriage and the family emphasize a deficit model when discussing stepfamilies. After reviewing twenty-six such introductory texts, researchers concluded that the books "contain more sources of stress than potential strengths/benefits, and positive outcome variables are ignored in favor of a focus on problems" (Coleman, Ganong, & Goodwin, 1994, 289).

Stepparents also have no childfree period. Unlike the newly married couple in the nuclear family, who typically have their first child about two and one-half years after their wedding, the remarried couple begins their marriage with children in the house.

Profound legal differences exist between nuclear and blended families. Whereas biological parents in nuclear families are required in all states to support their children, only five states require stepparents to provide financial support for their stepchildren. However, when there is a divorce, this and other discretionary types of economic support usually stop (Ganong, Coleman, & Mistina, 1995). Other legal matters with regard to nuclear families versus stepfamilies involve inheritance rights and child custody. Stepchildren do not automatically inherit from their stepparents, and courts have been reluctant to give stepparents legal access to stepchildren in the event of a divorce. Without legal support to ensure such access, these relationships tend to dwindle and become nonfunctional.

Finally, extended family networks in nuclear families are smooth and comfortable, whereas those in stepfamilies often become complex and strained. Table 16.1 summarizes the differences between nuclear families and stepfamilies.

Table 16.1 Differences Between Nuclear Families and Stepfamilies

Nuclear Families	Stepfamilies
1. Children are (usually) biologically related to both parents.	1. Children are biologically related to only one parent.
2. Both biological parents live together with children.	2. As a result of divorce or death, one biological parent does not live with children. In the case of joint physical custody, the children may live with both parents, alternating between them.
3. Beliefs and values of members tend to be similar.	3. Beliefs and values of members are more likely to be different because of different backgrounds.
4. Relationship between adults has existed longer than relationship between children and parents.	4. Relationship between children and parents has existed longer than relationship between adults.
5. Children have one home they regard as theirs.	5. Children may have two homes they regard as theirs.
6. The family's economic resources come from within the family unit.	6. Some economic resources may come from ex-spouse.
7. All money generated stays in the family.	7. Some money generated may leave the family in the form of alimony or child support.
8. Relationships are relatively stable.	8. Relationships are in flux: new adults adjusting to each other; children adjusting to stepparent; stepparent adjusting to stepchildren; stepchildren adjusting to each other.
9. No stigma is attached to nuclear family.	9. Stepfamilies are stigmatized.
10. Spouses had childfree period.	10. Spouses had no childfree period.
11. Inheritance rights are automatic.	11. Stepchildren do not automatically inherit from stepparents.
12. Rights to custody of children are assumed if divorce occurs.	12. Rights to custody of stepchildren are usually not considered.
13. Extended family networks are smooth and comfortable.	13. Extended family networks become complex and strained.
14. Nuclear family may not have experienced loss.	14. Stepfamily has experienced loss.

Stages in Becoming a Stepfamily

Just as a person must pass through various developmental stages in becoming an adult, a stepfamily goes through a number of stages as it overcomes various obstacles. Researchers such as Bray and Kelly (1998) and Papernow (1988) have identified various stages of development in stepfamilies. These stages include the following.

Stage 1: Fantasy Both spouses and children bring rich fantasies into the new marriage. Spouses fantasize that their new marriage will be better than the previous one. If the person they are marrying has adult children, they assume that these children will not be part of the new marriage. Young children have their own fantasy—they hope that their biological parents will somehow get back together, that the stepfamily is temporary.

You can't depend on your judgment when your imagination is out of focus.
Mark Twain

Stage 2: Reality Instead of realizing their fantasies, new spouses find that stepchildren and ex-spouses interfere with their new life together. Stepparents feel that they are outsiders in an already functioning unit (the biological parent and child).

Stage 3: Being Assertive Initially the stepparent assumes a passive role and accepts the frustrations and tensions of stepfamily life. Eventually, however, the resentment reaches a level where the stepparent is driven to make changes. The stepparent makes the partner aware of the frustrations and suggests that the marital relationship should have priority some of the time. The stepparent may also make specific requests, such as reducing the number of conversations the partner has with the ex-spouse, not allowing the dog on the furniture, or requiring the stepchildren to use better table manners. This stage is successful to the degree that the partner supports the recommendations for change.

Stage 4: Strengthening Pair Ties During this stage the remarried couple solidify their relationship by making it a priority. At the same time, the biological parent must back away somewhat from the parent-child relationship so that the new partner can have the opportunity to establish a relationship with the stepchildren. This relationship is the product of small units of interaction and develops slowly across time. Many day-to-day activities, such as watching television together, eating meals together, and riding in the car together, provide opportunities for the stepparent-stepchild relationship to develop. It is important that the stepparent not attempt to replace the relationship that the stepchildren have with their biological parents.

Stage 5: Recurring Change A hallmark of all families is change, but this is even more true of stepfamilies. Bray and Kelly (1998) note that even though a stepfamily may be functioning well when the children are preadolescent, when they become teenagers, a new era has begun as the children may begin to question how the family is organized and run. But such questioning by adolescents is not unique to stepchildren.

Michaels (2000) noted that spouses who become aware of the stages through which stepfamilies pass report that they feel less isolated and unique. Involvement in stepfamily discussion groups such as the Stepfamily Enrichment Program provides enormous benefits.

STRENGTHS OF STEPFAMILIES

Stepfamilies have both strengths and weaknesses. Strengths include children's exposure to a variety of behavior patterns, their observation of a happy remarriage, adaptation to stepsibling relationships inside the family unit, and greater objectivity on the part of the stepparent.

Exposure to a Variety of Behavior Patterns

Children in stepfamilies experience a variety of behaviors, values, and lifestyles. They have had the advantage of living on the inside of two families. One 12-year-old said: "My real mom didn't like sports and rarely took me anywhere. My stepmother is different. She likes to take me thrift stores and to rock concerts. She recently bought me a tent and is going to take me camping this summer."

Happier Parents

Single parenting can be a demanding and exhausting experience. Remarriage can ease the stress of solo parenting and provide a happier context for the parent. Research by Kurdek and Fine (1991) suggests that wives are happier in stepfamilies than husbands and are more optimistic about stepfamily living. One daughter said:

> *Looking back on my parents' divorce, I wish they had done it long ago. While I miss my dad and am sorry that I don't see him more often, I was always upset listening to my parents argue. They would yell and scream, and it would end with my mom cry-*

ing. It was a lot more peaceful (and I know my mom was a lot more happy) after they got divorced. Besides, I like my stepdaddy. Although he isn't my real dad, I know he cares about me.

Opportunity for New Relationship with Stepsiblings

Though some children reject their new stepsiblings, others are enriched by the opportunity to live with a new person to whom they are now "related." One 14-year-old remarked, "I have never had an older brother to do things with. We both like to do the same things and I couldn't be happier about the new situation." Some stepsibling relationships are maintained throughout adulthood. Two researchers (White & Riedmann, 1992) analyzed national data and assessed the degree to which full and step/half-siblings keep in touch as adults. They found that though step/half-siblings see each other less often, "adults report substantial contact with their step/half-siblings and only 0.5% of step-siblings were so estranged that they did not at least know where their step/half-siblings lived" (p. 206). Characteristics of those more likely to maintain contact included being female, African-American, younger, and geographically closer.

More Objective Stepparents

Because of the emotional tie between a parent and a child, some parents have difficulty discussing certain issues or topics. A stepparent often has the advantage of being less emotionally involved and can relate to the child at a different level. One 13-year-old said of the relationship with her father's new wife: "She went through her own parents' divorce and knows what it's like for me to be going through my dad's divorce. She is the only one I can really talk to about this issue. Both my dad and mom are too emotional about the subject to be able to talk about it."

The foregoing represent only a few of the many strengths of stepfamilies that are often given limited visibility. Other stepfamily strengths include the fact that the adults are often more mature (and therefore better role models) than they were in their first marriage, the children are more adaptable (good for their self-esteem), and the spouses adhere less rigidly to stereotyped gender roles (also good modeling) (Coleman, Ganong, & Goodwin, 1994). Still other strengths are that the circle of resource adults for children widens and that children observe a variety of parenting styles.

WOMEN IN STEPFAMILIES

The French have stopped using the pejorative word for stepmother (*marâtre*) and replaced it with a new word—*belle-mère,* which literally means "beautiful mother." In a study of eighty stepmothers, Church (1994) observed that most regarded the "good mother" and the "good stepmother" as the same. Furthermore, the stepmothers' ideal of being the good stepmother was often at odds with the reality—most of the women were frequently frustrated in their attempt to be the good mother/stepmother (Church, 1994).

That's why we are called STEPMOTHERS. I have footprints all over my body.
A stepmother

Specific sources of frustration for the woman in the role of stepmother include accepting and being accepted by her new partner's children, adjusting to alimony and child support payments made by her husband to an ex-wife, having the new partner accept her children and having her children accept him, and having another child in the new marriage.

Accepting Partner's Children

"She'd better think a long time before she marries a guy with kids," said one 29-year-old woman who had done so. This stepmother went on to explain:

> *It's really difficult to love someone else's children. Particularly if the kid isn't very likable. A year after we were married, my husband's 9-year-old daughter visited us for a summer. It was a nightmare. She didn't like anything I cooked, was always dragging around making us late when we had to go somewhere, kept her room a mess, and acted like a gum-chewing smart aleck. I hated her, but felt guilty because I wanted to have feelings of love and tenderness. Instead, I was jealous of the relationship she had with her father, and I wanted to get rid of her. I began counting how many days until she would be gone.*
>
> *You can hide your dislike for a while, but eventually you must tell your partner how you feel. I was lucky. My husband also thought his daughter was horrible to live with and wasn't turned off by my feelings. He told her if she couldn't act more civil, she couldn't come back. The message to every woman about to marry a guy with kids is to be aware that your man is a package deal and that the kid is in the package.*

Partner's Children Accepting Stepmother

The negative view of stepmothers is reflected in the prefix *step,* which in Old English referred to a family relationship caused by death. A stepchild, in essence, was an orphan child, and a stepmother was one who took care of an orphan. This view was enhanced by the structure of remarriages in early European households. A father whose wife had died had to remarry quickly so as to have a companion to share the work of farming, household tasks, and caring for children. A new wife found that one of the ways to establish her value with the new husband was to have a biological child. And since her own child had to compete with "his" children for resources, she tended to favor her own.

> *One could imagine her desire to advance her biological children's interests within the household and her jealousy over the advancement of her stepchildren. One could also imagine that her stepchildren would be angry at her and resent her biological children. And one could imagine the appeal of a story in which the stepmother wants her biological daughters, not her stepdaughter, to meet the prince at the ball.* (Cherlin, 1996, 395)

Such a legacy of stepmother folklore may have contributed to the fact that children have more difficulty accepting a new stepmother than a new stepfather. (It may also be that she is the more active parent, which provides greater opportunity for conflict.) In addition, a stepchild may feel the need to keep an emotional distance in the relationship with the stepmother so as not to incur the anger of the biological mother.

Lou Everett (2000), a specialist who works with stepfamilies, recommends that children remember their stepmother on Mother's Day. "The card doesn't have to say 'Mom' on it. Just something to express your appreciation . . . how great the cookies were that she baked or how happy she makes your dad. Those things mean so much to the stepmother."

Resenting Alimony and Child Support

In addition to the potential problems of not liking the partner's children, there may be problems of alimony and child support. As noted earlier, it is not unusual for a wife to become upset when her husband mails one-quarter or one-third of his income to a woman with whom he used to live. This amount of money is often equal to the current wife's earnings. Some wives in this position see themselves as working for their husband's ex-wife—a perception that is very likely to create negative feelings.

Her New Partner and How Her Children Accept Him

For some remarried wives, two main concerns are how her new husband accepts her children and how her children accept him. The ages of the children are important in these adjustments. If the children are young (age 3 or younger), they will usually accept any new adult in the natural parent's life. On the other hand, if the children are in adolescence, they are most likely struggling for independence from their natural parents and may not want any new authority figures in their lives.

In regard to whether the new spouse will accept and invest in her children, Hofferth and Anderson (2003) found that the new husband was more likely to do so if he did not already have and was not already supporting nonresidential children from a prior relationship. The authors hypothesized that, from a biological perspective, such a father did not invest as heavily in the current family to facilitate future reproduction.

Having Another Child

A national study of women in second marriages revealed that about two-thirds have a baby within six years after remarriage (Wineberg, 1992). Those women most likely to have a baby in second marriage do not have a biological child of their own (Stewart, 2002).

When this baby is a first biological baby, stepmothers report greater dissatisfaction with their stepchildren. This is also true for first-time fathers and may be a function of having to choose between one's biological children and stepchildren in making certain decisions. "Evolutionary views of parental psychology suggest that stepparents have a genetic propensity to express greater solicitude toward their biological children than toward stepchildren" (MacDonald & DeMaris, 1996, 23).

In spite of the decreased satisfaction with one's stepchildren, having a child in a second marriage is associated with increasing the stability of the relationship and reducing the probability of divorce (Wineberg, 1992). The researcher suggested that "couples with a mutual birth may tolerate more marital stress before considering divorce than couples with no mutual children" (p. 885). Remarried spouses who have another child soon discover how biased our society is against remarriages and the children born into these marriages. Our Social Policy of this chapter discusses this issue.

Although we have been discussing the stepmother and her adjustment to her stepchild, the biological mother has her own unique feelings and adjustments. One such mother lamented that she now had to share her daughter with the woman her husband left her for and that this was a very difficult adjustment. "I had a particularly hard time when she took my daughter shopping with her on their vacation during Christmas. . . . I wanted to share that with her. And his new wife's name was 'Star' and almost my daughter's age, to top it off" (author's files).

Child Support and Second Families

SOCIAL POLICY

The court system is biased against men who have children in second marriages. A man who has been married before and who has children from that marriage is required to pay child support for those children at the expense of children he has in a second marriage. For example, a divorced father of two children earning $40,000 per year would normally be required to pay 25 percent of his gross income, or $10,000, annually in child support (and his ex-wife may still get the tax deduction of the children as dependents). Should he remarry and have two additional children with his new wife, the needs of those children in the second marriage are irrelevant—the man must still pay $10,000 in child support for his first two children.

Moreover, if the man's income increases or the needs of the children in his first marriage increase (excessive medical bills), the man's ex-wife can petition the court and have his child support payments increased. Again, the fact that the man has other children in his second marriage who have clothing, food, and medical needs is not recognized by the courts. Indeed, "second family children are considered hardships and their needs are given consideration only after the needs of the children in the first family are satisfied," notes Dianna Thompson. She is the founder of Second Wives Crusade, an organization for women married to men who have children from previous marriages. She discovered, to her dismay, that her husband's child support obligations to the children in his first marriage were calculated without consideration for their own children. This is the law in all fifty states. Ms. Thompson is lobbying legislators to recognize the bias against children in second marriages and encouraging them to adopt new laws to protect all children—not just those who happen to be born of first marriages.

Source

Dianna Thompson, founder of Second Wives Crusade.

MEN IN STEPFAMILIES

Men in stepfamilies may or may not have children from a previous relationship. Three possible stepfamily combinations include a man with children married to a woman without children, a man with children married to a woman with children, and a man without children married to a woman with children. Data from national studies have suggested that the experiences for African-American stepfathers and white stepfathers are similar (Fine et al., 1992). Men in stepfamilies are, in some ways, like men in biological families. They tend to be less involved in child care and spend less time with children than the stepmother or the biological mother (Kyungok, 1994). Men in stepfamilies are also different from those in nuclear families in that stepchildren often want not a new dad but a stepdad who will respect their relationship with their biological dad (Bray & Kelly, 1998).

Man with Children Married to a Woman without Children

Men with biological children enter stepfamilies with an appreciation for the role of parent, some skills (one would hope) in reference to the role, and a bond with a child or children who usually live in another house. The latter may mean a grieving father who not only misses his children but also experiences a number of push-pull factors that influence how involved he is in his children's lives.

Men with children typically want their new wife to bond with their children and actively participate in their care. Such an expectation places the new wife in the role of the "instant mother," which may backfire as a result of the children's rejecting her. Since the new wife in the role of the new mother has typically spent limited time with her partner's children, it is important that he allow sufficient time (two to seven years) for a relationship to develop.

Another concern is whether the new wife will want children of her own. Often she does, but the man who already has children may be less interested. Nevertheless, it is not unusual for men to have additional children with a new wife.

Man with Children Married to a Woman with Children

As more men are awarded custody of their children, an increasing number of stepfamilies will include two sets of children. The number of relationships to manage increases with each new person added to the family, and the role of the stepfather appears to be enhanced when the father has his biological children living in the household. This was the conclusion of a team of researchers who studied the family context of sixty stepfathers:

> *[S]tepfathers may be drawn closer to their stepchildren, and they may have fewer negative attitudes toward them because of the strategies they adopt in striving to treat both sets of children in an equitable manner. The presence of their biological children in the household may, in effect, force them to parent to a greater extent than if they had merely been absorbed into a pre-existing family. It becomes incumbent upon them to constitute a viable living pattern and take a more active role with regard to all children in the household. This may lead them to minimize negative thoughts and feelings about their stepchildren and also to exaggerate positive attitudes about stepchildren in the interests of fairness.* (Palisi et al., 1991, 102)

MacDonald and DeMaris (2002) also found that the less involved the biological father, the more involved the stepfather. Moreover, while stepfathers tend to expect their stepchildren to be "obedient" (Bernhardt et al., 2002), stepfathers also report increased satisfaction in their role as stepfathers when they frequently engage in authoritative parenting behaviors. Hence, the more engaged the stepfather in disciplining his stepchildren, the greater the level of reported satisfaction (Fine, Ganong, & Coleman, 1997).

Men who become committed to women with children soon begin to spend time in the company of the children. Over half (56%) of divorced mothers who were dating again reported that they included their children in the activities with their future spouse at least once a week (Montgomery et al., 1992).

Man without Children Married to a Woman with Children

The adjustments of the never-married, divorced, or widowed man without children who marries a woman with children are primarily related to her children, their acceptance of him, and his awareness that his wife is emotionally bonded to her children. Unlike childfree marital partners, who are bonded only to each other, the husband entering a relationship with a woman who has children must accept her attachment with her children from the outset. New partners may view such bonding differently. One man said that such concern for one's own children was a sign of a caring and nurturing person. "I wouldn't want to live with anyone who didn't care about her kids." But another said, "I feel left out and that she cares more about her kids than me. I don't like the feeling of being an outsider."

Lou Everett (1998) interviewed six stepfathers and identified two primary factors that contribute to a positive stepfather-stepchild relationship:

1. Active involvement in teaching the stepchild something mutually valued. Just spending time with the stepchild (eating meals, watching television) had no positive effect on the relationship. Teaching the stepchild (how to skate, fish, fly a kite) made the stepfather feel as though he was contributing to the development of the stepchild and endeared the stepchild to the stepfather.

2. An intense love relationship between the stepfather and the biological mother of the stepchild. "Men who love their wives are more tolerant of their stepchildren," observed Dr. Everett. Marsiglio (1992) analyzed data from 195 men in stepfamilies and observed that the men, like the stepfathers in the Everett study, reported more positive relationships with stepchildren to the degree that they were happy with their partner. In addition, 55 percent reported that it was "somewhat true" or "definitely true" that "having stepchildren is just as satisfying as having your own children" (p. 204). Men who were living with their partners had perceptions of their stepfathering role similar to those of married men.

Here are some questions a man without children might ask a woman who has children:

1. How do you expect me to relate to your children? Am I supposed to be their friend, daddy, or something else? Men who develop a relationship first with their stepchild before they start disciplining the child report more positive outcomes (Bray & Kelly, 1998).

2. How do you feel about having another child? How many additional children are you interested in having?

3. How much money do you get in alimony and child support from your ex-husband? How long will you be receiving the money?

4. What expenses of the children do you expect me to pay for? Who is going to pay for their college expenses? In essence, the mother receiving child support from the biological father creates the illusion for the potential stepfather that the ex-husband will take care of the children's expenses. In reality, child-support payments cover only a fraction of what is actually spent on a child; consequently, the new stepfather may feel burdened with more financial responsibility for his stepchildren than he bargained for. This may engender negative feelings toward the wife in the new marriage relationship.

CHILDREN IN STEPFAMILIES

NATIONAL DATA

About 10 percent of all children live in stepfamilies (Furukawa, 1994). Of these children in stepfamilies, 86 percent live with their biological mother and stepfather; 14 percent live with their biological father and stepmother (Rutter, 1994).

Research confirms that of the children growing up in stepfamilies (see National Data) "80 percent . . . are doing well. Children in stable stepfamilies look very much like those raised in stable first families" (Pasley, 2000, 6).

Stepchildren have viewpoints and must make adjustments of their own when their parents remarry. They have experienced the transition from living in a family with both biological parents to either living alone with one parent (usually the mother) or alternating between the homes of both parents. When their parents remarry, they must adjust again to living with two new sets of stepparents/ stepsiblings.

"The biggest source of problems for kids in stepfamilies is parental conflict left over from the first marriage" (Rutter, 1994, 32). "Study after study shows that divorce and remarriage do not harm children—parental conflict does" (p. 33). Children caught in the conflict between their parents find it impossible to be loyal to both parents. The child temporarily resolves the conflict by siding with one parent at the expense of the relationship with the other parent. No one wins—the child feels bad for abandoning a loving parent, the parent who has been tossed aside feels deprived of the opportunity for a close parent-child relationship, and the custodial parent runs the risk that the children, as adults, may resent being prevented from developing or continuing a relationship with the other parent.

Divided Loyalties, Discipline, Stepsiblings

Other problems experienced by children in stepfamilies often revolve around feeling abandoned, having divided loyalties, discipline, and stepsiblings. Some stepchildren feel that they have been abandoned twice—once when their parents got divorced and again when the parents turned their attention to their new marital partners. One adolescent explained:

> *It hurt me when my parents got divorced and my dad moved out. I really missed him and felt he really didn't care about me. But we adjusted with just my mom, and when everything was going right again, she got involved with this new guy and we were left with baby-sitters all the time. I feel like I've lost both parents in two years.*

Coping with feelings of abandonment is not easy. It is best if the parents assure the children that the divorce was not their fault and that they are loved a great deal by both parents. In addition, the parents should be careful to find a balance between spending time with their new partner and spending time with their children. This translates into spending some alone time with their children.

Some children experience abandonment yet again if their parents' second marriage ends in divorce. They may have established a close relationship with the new stepparent only to find the relationship disrupted. Relationships with the stepgrandparents may also become strained. Stepgrandparents may be an enormous source of emotional support for the child, but a divorce can make it stillborn.

Divided loyalties represent another issue children must deal with in stepfamilies. Sometimes children develop an attachment to a stepparent that is more positive than the relationship with the natural parent of the same sex. When these feelings develop, the children may feel they are in a bind. One adolescent boy explained:

> *My real dad left my mother when I was 6, and my mom remarried. My stepdad has always been good to me, and I really prefer to be with him. When my dad comes to pick me up on weekends, I have to avoid talking about my stepdad because my dad doesn't like him. I guess I love my dad, but I have a better relationship with my stepdad.*

For some adolescents, the more they care for the stepparent, the more guilty they feel, so they may try to hide their attachment. The stepparent may be aware of both positive and negative feelings coming from the child. Ideally, both the biological parent and the stepparent should encourage the child to have a close relationship with the other parent.

Discipline is another issue for stepchildren. "Adjusting to living with a new set of rules from your stepparent," "accepting discipline from a stepparent," and "dealing with the expectations of your stepparent" are situations that 80 percent of more than a hundred adolescents in stepfamilies said they had experienced (Lutz, 1983). At least two studies have identified the stepmother as the more difficult stepparent to adjust to from the child's point of view (Fine & Kurdek, 1992). This is probably because the woman is more often in the role of the active parent, which increases the potential for conflict with the child. There are also data that stepfather families monitor children less than stepmother families, so the opportunity for conflict with stepfathers may be less (Fisher et al., 2003).

Adolescents respond most favorably to a supportive stepparenting style that includes warmth, acceptance, and nurturance along with supervision and discipline (Crosbie-Burnett & Giles-Sims, 1994). Pauk (1996) found that lower conflict was associated with stepparents who spent time alone with their stepchildren.

Siblings can also be a problem for stepchildren. Children in stepfamilies experience higher levels of stress if they have stepsiblings than if they do not (Lutz, 1983). The stress seems to be a result of more arguments among the adults when both sets of children are present and of the perception that parents are more fair with their own children: "I could bounce the ball in the den and my stepdad would jump all over me. But let my stepsister bounce it and he wouldn't say a word. All I want is to be treated fairly, and that's not what's happening in this family."

Stepsiblings also compete for space and TV access, which leads to bitter territorial squabbles. The children who are already in the house may feel imposed upon and threatened. The entering children may feel out of place and that they do not belong there.

Moving Between Households

Other issues unique to stepchildren including moving back and forth between households. Though sometimes this is a structured transition, it may also be a mechanism used by the child to manipulate the parents. In effect they communicate to each parent that if they don't get their way, they will move in with the other parent. Parents in conflict with each other are easy prey for this destructive ploy.

Ambiguity of the Extended Family

A final issue for children in stepfamilies is their ambiguous place in the extended family system of the new stepparent. While some parents and siblings of the new stepparent welcome the new stepchild into the extended family system, others may ignore the child since they are busy enough with their own grandchildren and children, respectively.

Some evidence supports the belief that children tend to have greater well-being in nuclear families than in stepfamilies. Data from a nationally representative sample of 17,110 children under age 18 revealed that children living with mothers and stepfathers were more likely than those living with both biological parents to have repeated a grade of school, to have been expelled, to have been treated for emotional or behavioral problems, and to report more health problems (Dawson, 1991). In regard to dropping a grade, when parents are going through a divorce, they may be so preoccupied with their own concerns and trying to handle work demands, too, that they spend less time helping children with homework.

Stepfamily living is often difficult for everyone involved: remarried spouses, children, even ex-spouses, grandparents, and in-laws. Though some of the problems begin to level out after a few years, many continue for ten or more. Many couples become impatient with the unanticipated problems that are slow to abate, and they divorce.

DEVELOPMENTAL TASKS FOR STEPFAMILIES

A **developmental task** is a skill that, if mastered, allows the family to grow as a cohesive unit. Developmental tasks that are not mastered will edge the family closer to the point of disintegration. Some of the more important developmental tasks for stepfamilies are discussed in this section.

Acknowledge Losses and Changes

As noted earlier, each stepfamily member has experienced the loss of a spouse or a biological parent in the home. These are losses of an attachment figure and are significant (Marano, 2000). These losses are sometimes compounded by home, school, neighborhood, and job changes. Feelings about these losses and changes should be acknowledged as important and consequential. In addition, children should not be required to love their new stepparent or stepsiblings (and vice versa). Such feelings will develop only as a consequence of positive interaction over an extended period of time.

Preserving original relationships is helpful in reducing the child's grief over loss. It is sometimes helpful for the biological parent and child to take time to nurture their relationship apart from stepfamily activities. This will reduce the child's sense of loss and any feelings of jealousy toward new stepsiblings.

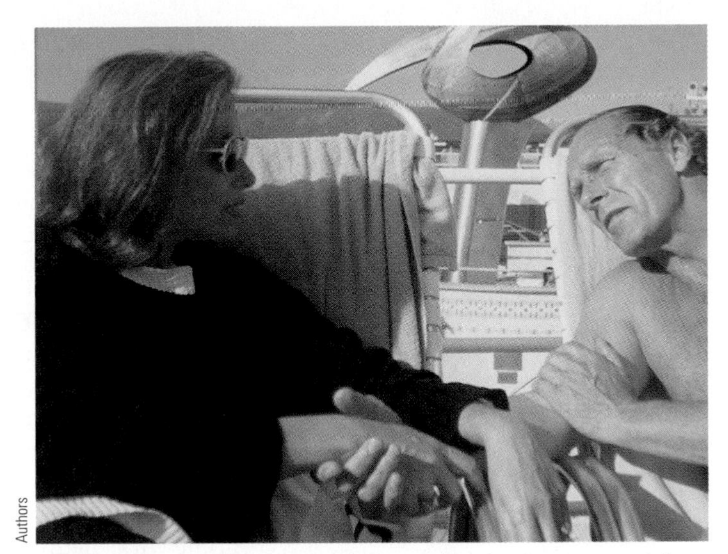

Authors

It is important that the remarried couple take time to nourish their relationship by spending some time alone with each other.

Nurture the New Marriage Relationship

It is critical to the healthy functioning of a new stepfamily that the new spouses nurture each other and form a strong unit. From this base, the couple can communicate, cooperate, and compromise with regard to the various issues in their new blended family. Too often spouses become child-focused and neglect the relationship on which the rest of the family depends. Such nurturing translates into spending time alone with each other, sharing each other's lives, and having fun with each other. One remarried couple goes out to dinner at least once a week without the children. "If you don't spend time alone with your spouse, you won't have one," says one stepparent. Two researchers studied 115 stepfamily couples and found that the husbands tended to rank "spouse" as their number one role (over parent or employee) while wives tended to rank "parent" as their number one role (Degarmo & Forgatch, 2002).

The principal challenge of stepfamily life is building an emotionally satisfying marriage.

James Bray and John Kelly

Integrate the Stepfather into the Child's Life

Stepfathers who become interested in what their stepchild is interested in and who spend time alone with the stepchild report greater integration into the life of the stepchild and the stepfamily. The benefits to both the father and the stepchild in terms of emotional bonding are enormous. In addition, the mother of the child feels closer to her new husband if he has bonded with her offspring.

Allow Time for Relationship between Partner and Children to Develop

In an effort to escape single parenthood and to live with one's beloved, some individuals rush into remarriage without getting to know each other. Not only do they have limited information about each other, but their respective children may have spent little or no time with their future stepparent. One stepdaughter remarked, "I came home one afternoon to find a bunch of plastic bags in the living room with my soon-to-be-stepdad's clothes in them. I had no idea he was moving in. It hasn't been easy." Both adults and children should have had meals together and spent some time in the same house before becoming bonded by marriage as a family.

They myth of instant love and its antithesis of never-ending, unsolvable problems need to be dispelled.

Stepfamily Association of America

Have Realistic Expectations

Because of the complexity of meshing the numerous relationships involved in a stepfamily, it is important to be realistic. Dreams of one big happy family often set up stepparents for disappointment, bitterness, jealousy, and guilt. As noted earlier, stepfamily members do not begin to feel comfortable with each other until the third year (Bray & Kelly, 1998). Just as nuclear and single-parent families do not always run smoothly, neither do stepfamilies.

Accept Your Stepchildren

Rather than wishing your stepchildren were different, it is more productive to accept them. All children have positive qualities; find them and make them the focus of your thinking. Stepparents may communicate acceptance of their stepchildren through verbal praise and positive or affectionate statements and gestures. In addition, stepparents may communicate acceptance by engaging in pleasurable activities with their stepchildren and participating in daily activities such as homework, bedtime preparation, and transportation to after-school activities. Funder (1991) studied 313 parents who had been separated five to eight years and who had become involved with new partners. In general, the new partners were very willing to be involved in the parenting of their new spouses' children. Such involvement was highest when the children lived in the household.

Establish Your Own Family Rituals

One of the bonding elements of nuclear families is its rituals. Stepfamilies may integrate the various family members by establishing common rituals, such as summer vacations, visits to and from extended kin, and religious celebrations. These rituals are most effective if they are new and unique, not mirrors of rituals in the previous marriages and families.

Decide about Money

Money is an issue of potential conflict in stepfamilies because it is a scarce resource and several people want to use it for their respective needs. The father wants a new computer; the mother wants a new car; the mother's children want bunk beds, dance lessons, and a satellite dish; the father's children want a larger room, clothes, and a phone. How do the newly married couple and their children decide how money should be spent?

Some stepfamilies put all their resources into one bank and draw out money as necessary without regard for whose money it is or for whose child the money is being spent. Others keep their money separate; the parents have separate incomes and spend them on their respective biological children. Although no one pattern is superior to the other, it is important for remarried spouses to agree on whatever financial arrangements they live by.

In addition to deciding how to allocate resources fairly in a stepfamily, remarried couples may face decisions regarding sending the children and stepchildren to college. Remarried couples may also be concerned about making a will that is fair to all family members.

Give Parental Authority to Your Spouse

How much authority the stepparent will exercise over the children should be discussed by the adults before they get married. Some couples divide the authority—each spouse disciplining his or her own children. But children may test the stepparent in such an arrangement when the biological parent is not around. One stepmother said, "Joe's kids are wild when he isn't here because I'm not supposed to discipline them."

Support Child's Relationship with Absent Parent

A continued relationship with both biological parents is critical to the emotional well-being of the child. Ex-spouses and stepparents should encourage children to have a positive relationship with both biological parents. Respect should also be shown for the biological parent's values. Bray and Kelly (1998) note that this is "particularly difficult" to exercise. "But asking a child about an absent parent's policy on movies or curfews shows the child that, despite their differences, his/her mother and father still respect each other" (p. 92).

Cooperate with the Child's Biological Parents and Coparent

It is well established that a cooperative, supportive, and amicable coparenting relationship following divorce is beneficial to children, but Richmond and Christensen (1998), in a study of 144 divorced individuals (52 men, 92 women), found that developing such a relationship with one's ex also has positive physical and psychological benefits for the adults. Hence, both children and parents experience enormous benefits from a coparenting relationship.

Support Child's Relationship with Grandparents

It is important to support children's continued relationships with their natural grandparents on both sides of the family. This is one of the more stable relationships in the child's changing world of adult relationships. Regardless of how ex-spouses feel about their former in-laws, they should encourage their children to have positive feelings for their grandparents. One mother said, "Although I am uncomfortable around my ex-in-laws, I know my children enjoy visiting them, so I encourage their relationship."

Anticipate Great Diversity

Stepfamilies are as diverse as nuclear families. It is important to let each family develop its own uniqueness. One remarried spouse said, "The kids were grown when the divorces and remarriages occurred, and none of the kids seem particularly interested in getting involved with the others" (Rutter, 1994, 68). Ganong et al. (1998) also emphasized the importance of resisting the notion that stepfamilies are not "real" families and that adoption makes them a real family. Indeed, social policies need to be developed that allow for "the establishment of some legal ties between the stepparent and stepchild without relinquishing the biological parent's legal ties" (p. 69).

We end this chapter with the Stepfamily Success Scale, which allows individuals contemplating a stepfamily to assess the degree to which they might have a successful stepfamily.

SELF-ASSESSMENT

Stepfamily Success Scale

This scale is designed to measure the degree to which you and your partner might expect to have a successful stepfamily. There are no right or wrong answers. After carefully reading each sentence, circle the number that best represents your feelings.

1 Strongly disagree
2 Mildly disagree
3 Undecided
4 Mildly agree
5 Strongly agree

	SD	MD	U	MA	SA
1. I am a flexible person.	1	2	3	4	5
2. I am not a jealous person.	1	2	3	4	5
3. I am a patient person.	1	2	3	4	5
4. My partner is a flexible person.	1	2	3	4	5
5. My partner is not a jealous person.	1	2	3	4	5
6. My partner is a patient person.	1	2	3	4	5
7. My partner values our relationship as much as the relationship with his or her children.	1	2	3	4	5
8. My partner understands that it is not easy for me to love someone else's children.	1	2	3	4	5
9. I value the relationship with my partner as much as the relationship with my children.	1	2	3	4	5
10. I understand that it will be difficult for my partner to love my children as much as I do.	1	2	3	4	5
11. My partner and I have amicable relationships with our ex-spouses.	1	2	3	4	5
12. My children and my partner's children live (except on alternate weekends) with the respective ex-spouse.	1	2	3	4	5
13. We will have plenty of money in our stepfamily.	1	2	3	4	5
14. I feel positive about my partner's children.	1	2	3	4	5
15. My partner feels positive about my children.	1	2	3	4	5
16. My children and those of my partner feel positive about each other.	1	2	3	4	5
17. My partner and I will begin our stepfamily in a place neither of us has lived before.	1	2	3	4	5
18. My partner and I agree on how to discipline our children.	1	2	3	4	5
19. My children feel positive about my new partner.	1	2	3	4	5
20. My partner's children feel positive about me.	1	2	3	4	5

Scoring

Add the numbers you circled. 1 (strongly disagree) is the most negative response you can make, and 5 (strongly agree) is the most positive response you can make. The lower your total score (20 is the lowest possible score), the greater the number of potential problems and the lower the chance of success in a stepfamily with this partner. The higher your total score (100 is the highest possible score), the greater the chance of success in a stepfamily with this partner. A score of 60 places you at the midpoint between the extremes of having a difficult or an easy stepfamily experience.

Note: This self-assessment is intended to be thought-provoking and suggestive; it is not a clinical diagnostic instrument.

SUMMARY

What is the nature of remarriage in the United States?

About 40 percent of marriages today are remarriages, and most of these involve children or the creation of stepfamilies. When the divorced who have remarried and the divorced who have not remarried are compared, the remarried report greater personal and relationship happiness. Ninety percent of remarriages consist of persons who are divorced rather than widowed.

It is not uncommon for persons who are divorced to live together with a new partner before remarriage. Aside from living together, they (like most couples in courtship, whether first or second marriage) do little else to prepare for their new marriage. National data reflect that remarriages are more likely than first marriages to end in divorce in the early years of remarriage.

What is the nature of stepfamilies in the United States?

Stepfamilies represent the fastest-growing type of family in the United States. A blended family is one in which the spouses in a new marriage relationship are blended with the children of at least one of the spouses from a previous marriage. There is a movement away from the use of the term *blended,* since stepfamilies really do not blend.

Although a stepfamily can be created when a never-married or a widowed parent with children marries a person with or without children, most stepfamilies today are composed of spouses who were once divorced.

Stepfamilies differ from nuclear families: the children in a nuclear family are biologically related to both parents, whereas the children in a stepfamily are biologically related to only one parent. Also, in a nuclear family, both biological parents live with their children, whereas only one biological parent in a stepfamily lives with the children. In some cases, the children alternate living with each parent.

Stepism is the assumption that stepfamilies are inferior to biological families. Stepism, like racism, heterosexism, sexism, and ageism, involves prejudice and discrimination.

What are the strengths of stepfamilies?

The strengths of stepfamilies include exposure to a variety of behavior patterns, a happier parent, and greater objectivity on the part of the stepparent.

What can women do to ease their transition into stepfamily living?

Learning to get along with the husband's children, not being resentful of his relationship with his children, and adapting to the fact that one-third to one-half of his net income may be sent to his ex-wife as alimony or child support are skills the new wife must develop. The new wife may also want children with her new partner or may bring her own children into the marriage. In the latter case, she is anxious that her new husband will accept her children.

What can men do to ease their transition into stepfamily living?

Getting along with their wife's children, paying for many of the expenses of their stepchildren, having their new partner accept their own children, and dealing with the issue of having more children are among the issues to be confronted by the new stepfather.

What are the challenges for children in stepfamilies?

Children must cope with feeling abandoned and with problems of divided loyalties, discipline, and stepsiblings. They may also move between households and their role in the extended stepfamily may be ambiguous.

What are the developmental tasks of stepfamilies?

Developmental tasks for stepfamilies include nurturing the new marriage relationship, allowing time for partners and children to get to know each other, deciding whose money will be spent on whose children, deciding who will discipline the children and how, and supporting the child's relationship with both parents and natural grandparents. Both sets of parents and stepparents should form a parenting coalition in which they cooperate and actively participate in childrearing.

KEY TERMS

binuclear
blended family
December marriage
developmental task
stepfamily
stepism

RESEARCHING MARRIAGE AND THE FAMILY WITH INFOTRAC COLLEGE EDITION

InfoTrac College Edition, an online library, allows you to perform research online anywhere, anytime. Following are two suggested search terms and related questions to help you extend your understanding of the topics covered in this chapter. Go to www.infotrac-college.com to begin your search.

***Keyword:* Remarriage stability.** Locate articles that compare first and second marriages on marital happiness and stability. What factors would influence more and less happiness and stability in first and second marriages?

***Keyword:* Stepchildren.** Locate articles that discuss the relationships parents have with stepchildren. Why are these more difficult relationships than those with biological children?

The Companion Web Site for Choices in Relationships: An Introduction to Marriage and the Family, Eighth Edition

http://sociology.wadsworth.com/knox_schacht/choices8e

Supplement your review of this chapter by going to the companion Web site to take one of the Tutorial Quizzes, use the flash cards to master key terms, and check out the many other study aids you'll find there. You'll also find special features such as the Marriage and Family Resource Center, Census 2000 information, and other data and resources at your fingertips to help you with that special project or to do some research on your own.

WEBLINKS

Second Wives Club
http://www.secondwivesclub.com/

Stepfamily Association of America
http://www.stepfam.org

Stepmothers.org/Stepmothers International
http://www.saafamilies.org/smi/old/resources.htm

Authors

CHAPTER 17

It doesn't interest me how old you are. I want to know if you will risk looking like a fool for love, for your dream, for the adventure of being alive.

Oriah, Indian elder

Aging in Marriage and Family Relationships

Contents

True or False?

1. Having a strong religious faith is the single most important variable affecting one's successful aging.
2. The process of death, rather than death itself, is the primary fear of the elderly.
3. Failing health is the criterion used by the elderly to define themselves as old.
4. The Supreme Court has ruled that grandparents have the right to see their grandchildren even though their biological children do not want them to.
5. Sexual orientation is an irrelevant variable when deciding to put a parent in a nursing home.

Answers: **1.** F **2.** T **3.** T **4.** F **5.** F

Tuck Everlasting is a 2002 film whose theme dealt with never aging. The Tuck family had drunk an anti-aging water they found in the woods and never aged. Chronologically they were over 100 years old but the parents were still in their 40s and the children were in their teens. The film asks the question, "If you could remain the age you are now forever, would you?"

Of course, the question is pure fantasy. Contemporary concerns such as the cost of even a brief stay in the hospital, the cost of prescription medication, and care for the elderly in a retirement or long-term care facility are a major focus in U.S. society. Although most couples getting married today rarely give a thought to their own aging, the care of their aging parents is an issue that looms ahead in their future. In this chapter, we focus on both the factors that confront couples as they age and the dilemma of how to care for their aging parents. We begin by looking at the concept of age.

AGE AND AGEISM

All societies have a way to categorize their members by age. And all societies provide social definitions for particular ages.

The Concept of Age

A person's **age** may be defined chronologically, physiologically, psychologically, sociologically, and culturally. Chronologically, an "old" person is defined as one who has lived a certain number of years. How many years it takes to be regarded as old varies with one's own age. A child of 12 may regard a sibling of 18 as old—and his or her parents as "ancient." But the teenagers and the parents may regard themselves as "young" and reserve the label "old" for their grandparents' generation.

Chronological age has obvious practical significance in everyday life. Bureaucratic organizations and social programs identify chronological age as a criterion of certain social rights and responsibilities. One's age determines the right to drive, vote, buy alcohol or cigarettes, and receive Social Security and Medicare. Age also determines retirement. Federal law requires airline pilots to retire at age 65, the bureaucratic chronological definition of "old."

In ancient Greece or Rome, where the average life expectancy was 20 years, one was old at 18; similarly, one was old at 30 in medieval Europe and at age 40 in the United States in 1850. In the United States today, however, people are usually not considered old until they reach age 65. Current life expectancy is shown in Table 17.1.

Our society is moving toward new chronological definitions of "old." Three groups of the elderly are the "young-old," the "middle-old" and the "old-old." The young-old are typically between the ages of 65 and 74; the middle-old, 75–84, and the old-old, 85 and beyond. Individuals aged 18–35 identify 50 as the age when the average man or woman becomes "old." However, those between the ages of 65 and 74 define "old" as 80 (Cutler, 2002).

Old age is always twenty years away.
Jack LaLanne, *Fitness expert, at age 83*

Ballard and Morris (2003) found that interest in family life education topics varies by age, with those 85 and beyond least likely to be interested in family and relationship issues. Where an interest does exist (widowhood was of particular interest to this age group), print material in the form of newsletters, brochures and self-help books were most preferred (Morris & Ballard, 2003).

Some individuals seem to have been successful in delaying the aging process. Frank Lloyd Wright, the famous architect, enjoyed the most productive period

Table 17.1 Life Expectancy

Year	Caucasoid Males	African-American Males	Caucasoid Females	African-American Females
2000	74.9	68.3	80.1	75.2
2005	75.4	69.9	81.1	76.8
2010	76.1	70.9	81.8	77.8

Source: *Statistical Abstract of the United States: 2003*. 123rd ed. Washington D.C.: U.S. Bureau of the Census, 2003, Table 105.

of his life between 80 and 92. George Burns and Bob Hope were active into their 90s.

Physiologically, a person is old when his or her auditory, visual, respiratory, and cognitive capabilities decline significantly. Persons who need full-time nursing care for eating, bathing, and taking medication properly and who are placed in nursing homes are thought of as being old. Failing health is the criterion used by the elderly to define themselves as old (O'Reilly, 1997) and successful aging is typically defined as maintaining one's health, independence, and cognitive ability. Jorm et al. (1998) observed that the prevalence of successful aging declines steeply from age 70–74 to age 80+.

Persons who have certain diseases are also regarded as old. Although younger individuals may suffer from Alzheimer's, arthritis, and heart problems, these ailments are more often associated with aging. As medical science conquers more diseases, the physiological definition of aging changes so that it takes longer for people to be defined as "old."

Psychologically, a person's self-concept is important in defining how old that person is. Two researchers compared young (mean age 19 years) and older adults (mean age 74 years) and found that the latter expressed less positive attitudes about their facial attractiveness. "These differences are in line with the structural changes that occur in the face as people age, moving them further from cultural beauty standards" (Franzoi & Koehler, 1998, 1).

Sociologically, people are viewed as old when they occupy social roles that have traditionally been associated with the elderly. Grandparent, widow, Medicare patient, and retiree are roles typically associated with someone who is old.

Culturally, the society in which an individual lives defines when and if a person becomes old and what being old means. In U.S. society, the period from age 18 through 64 is generally subdivided into young adulthood, adulthood, and middle age. Cultures also differ in terms of how they view and take care of their elderly. Spain is particularly noteworthy in terms of care for the elderly, with eight of ten elderly persons receiving care from family members and other relatives. The elderly in Spain report very high levels of satisfaction in the relationships with their children, grandchildren, and friends (Fernandez-Ballesteros, 2003).

Though chronologically this man is viewed by U.S. society as "old old," he is psychologically young since he views himself as such. He is 94.

Authors

Ageism

Every society has some form of **ageism**—the systematic persecution and degradation of people because they are old. Ageism is similar to sexism, racism, and heterosexism. The elderly are shunned,

Table 17.2 Theories of Aging

Name of Theory	Level of Theory	Theorists	Basic Assumptions	Criticisms
Disengagement	Macro	Elaine Cumming William Henry	The gradual and mutual withdrawal of the elderly and society from each other is a natural process. It is also necessary and functional for society that the elderly disengage so that new people can be phased in to replace them in an orderly transition.	Not all people want to disengage; some want to stay active and involved. Disengagement does not specify what happens when the elderly stay involved.
Activity	Macro	Robert Havighurst	People continue the level of activity they had in middle age into their later years. Though high levels of activity are unrelated to living longer, they are related to reporting high levels of life satisfaction.	Ill health may force people to curtail their level of activity. The older a person, the more likely the person is to curtail activity.
Conflict	Macro	Karl Marx Max Weber	The elderly compete with youth for jobs and social resources such as government programs (Medicare).	The elderly are presented as disadvantaged. Their power to organize and mobilize political resources such as the American Association of Retired Persons is underestimated.
Age stratification	Macro	M. W. Riley	The elderly represent a powerful cohort of individuals passing through the social system that both affect and are affected by social change.	Too much emphasis is put on age, and little recognition is given to other variables within a cohort such as gender, race, and socioeconomic differences.
Modernization	Macro	Donald Cowgill	The status of the elderly is in reference to the evolution of the society toward modernization. The elderly in premodern societies have more status because what they have to offer in the form of cultural wisdom is more valued. The elderly in modern technologically advanced societies have low status since they have little to offer.	Cultural values for the elderly, not level of modernization, dictate the status of the elderly. Japan has high respect for the elderly and yet is highly technological and modernized.
Symbolic interaction	Micro	Arlie Hochschild	The elderly socially construct meaning in their interactions with others and society. Developing social bonds with other elderly can ward off being isolated and abandoned. Meaning is in the interpretation, not in the event.	The power of the larger social system and larger social structures to affect the lives of the elderly is minimized.
Continuity	Micro	Bernice Neugarten	The earlier habit patterns, values, attitudes of the individual are carried forward as the person ages. The only personality change that occurs with aging is the tendency to turn one's attention and interest on the self.	Other factors than one's personality affect aging outcomes. The social structure influences the life of the elderly rather than vice versa.

discriminated against in employment, and sometimes victims of abuse. Media portrayals contribute to the negative image of the elderly. They are portrayed as difficult, complaining, and burdensome and are often underrepresented in commercials and comic strips.

Negative stereotypes and media images of the elderly engender **gerontophobia**—a shared fear or dread of the elderly, which may create a self-fulfilling prophecy. For example, an elderly person forgets something and attributes his or her behavior to age. A younger person, however, engaging in the same behavior, is unlikely to attribute forgetfulness to age, given cultural definitions surrounding the age of the onset of senility.

The negative meanings associated with aging underlie the obsession of many Americans to conceal their age by altering their appearance. With the hope of holding on to youth a little bit longer, aging Americans spend millions of dollars each year on exercise equipment, hair products, facial creams, and plastic surgery. The latest attempt to reset the aging clock is to have regular injections of HGH, human growth hormone, which promises to lower blood pressure, build muscles without extra exercise, increase the skin's elasticity, thicken hair, heighten sexual potency, etc. It is part of the regimen of such clinics as Lifespan (Beverly Hills) which costs $1,000 a month after an initial workup of $5,000. Many physicians are skeptical and repeat their contention that there is no fountain of youth (Oldenburg, 2000).

Theories of Aging

Gerontology is the study of aging. Table 17.2 identifies several theories, the level (macro or micro) of the theory, the theorists typically associated with the theory, assumptions, and criticisms. As will be noted, there are diverse ways of conceptualizing the elderly. Currently in vogue are the age stratification and life course perspectives (Mitchell, 2003).

CAREGIVING FOR THE ELDERLY—THE "SANDWICH GENERATION"

There are many married adults who have both children still living at home and elderly parents whom they are responsible for taking care of. These adults are known as members of the "sandwich generation" since they are in the middle of taking care of the needs of both children and parents.

Family caregiving of the elderly may include daily bathing and feeding, handling of finances (balancing checkbook), and transporting the elderly to the grocery store or physician. Caregiving can be informal (nonpaid and provided by family and friends) or formal (paid and provided by professionals) (Coleman & Pandya, 2002). Most (80%) of the caregiving of the elderly is informal care provided by family members.

The typical caregiver is a middle-aged married woman who works outside the home. Brewer (2000) found that high levels of stress, fatigue, and exhaustion, as well as guilt, anger, and grief were associated with taking care of one's elders. Martire and Stephens (2003) noted even higher levels of fatigue and competing demands among women who were both employed and caring for an aging parent.

Fifty-five percent of a sample of women at midlife (most of whom had children) reported that they were providing care to their mothers; 34 percent were caring for their fathers (Peterson, 2002). The number of individuals in the sandwich generation will increase for the following reasons:

*1. **Longevity.*** The over-85 age group, the segment of the population most in need of care, is the fastest-growing segment of our population.

*2. **Chronic disease.*** In the past, diseases took the elderly quickly. Today, diseases such as arthritis and Alzheimer's offer not an immediate death sentence but a lifetime of managing the illness and being cared for by others. Nancy Reagan has publicly acknowledged the challenge of taking care of her husband, former President Reagan.

*3. **Fewer siblings to help.*** The current generation of elderly had fewer children than the elderly in previous generations. Hence, the number of adult siblings to help look after parents is more limited. Only children are more likely to feel the weight of caring for elderly parents alone. Elderly spouses who regret having no children are most likely to feel distress (Koropeckyj-Cox, 2002) and to end up in formal residential care (Wenger, Scott, & Patterson, 2000).

*4. **Commitment to parental care.*** Contrary to the myth that adult children in the United States abrogate responsibility for taking care of their elderly parents, most children institutionalize their parents only as a last resort. Furthermore, Wells (2000) identified some benefits of caregiving, including a closer relationship to the dependent person and a feeling of enhanced self-esteem. Most of Peterson's (2002) sample of women caring for their parents did not view doing so as a burden. Asian children, specifically Chinese children, are socialized to expect to take care of their elderly. Zhan et al. (2003) observed, "Children were raised for the security of old age" (p. 209).

*5. **Lack of support for the caregiver.*** Caring for a dependent, aging parent requires a great deal of effort, sacrifice, and decision making on the part of over 14 million adults in the United States who are challenged with this situation. The emotional toll on the caregiver may be heavy. Guilt (over not doing enough), resentment (over feeling burdened), and exhaustion (over the relentless care demands) are common feelings that are sometimes mixed. One caregiver adult child said, "I must be an awful person to begrudge taking my mother supper, but I feel that my life is consumed by the demands she makes on me, and I have no time for myself, my children or my husband." Marks, Lambert, & Choi (2002) noted an increase in symptoms of depression among a national sample of caregivers (of a child, parent, or spouse). Older caregivers also risk their own health in caring for their aging parents (Wallsten, 2000). Caregiving can also be expensive and cut into the family budget.

Some reduce the strain of caring for an elderly parent by arranging for home health care. This involves having a nurse go to the home of the parent and provide such services as bathing the parent and giving medication. Other services may include taking meals to the elderly (e.g., through Meals on Wheels). The National Family Caregiver Support Program, enacted in 2000, provides support services for individuals (including grandparents) who provide family caregiving services. Such services might include elder-care resource and referral services, caregiver support groups, and classes on how to care for an aging parent. In addition, increasingly, states are providing family caregivers a tax credit or deduction.

Offspring who have no help may become overwhelmed and frustrated. Elder abuse, an expression of such frustration, is not unheard of (we discussed this in the chapter on abuse). Many wrestle with the decision to put their parents in a nursing home or other long-term care facility. We discuss this issue in the Personal Choices section.

Diversity in Other Countries

Whereas female children in the United States have the most frequent contact and are more involved in the caregiving of their elderly parents than male children, it is the daughter-in-law in Japan who offers the most help to elderly individuals. The female child who is married gives her attention to the parents of her husband (Ikegami, 1998). Eastern cultures emphasize filial piety, which is love and respect toward their parents. **Filial piety** involves respecting parents, bringing no dishonor to parents, and taking good care of parents (Jang & Detzner, 1998). Western cultures are characterized by **filial responsibility** emphasizing duty, protection, care, and financial support.

PERSONAL CHOICES

Should I Put My Parents in a Long-Term-Care Facility?

Over 1.1 million individuals (almost three times as many women as men) over the age of 65 are in a nursing home (*Statistical Abstract of the United States: 2003*, Table 77). Factors relevant in deciding whether to care for an elderly parent at home, arrange for nursing home care, or provide another form of long-term care include the following.

1. Level of care needed. As parents age, the level of care that they need increases. An elderly parent who cannot bathe, dress, prepare meals, or be depended on to take medication responsibly needs either full-time in-home care or a skilled nursing facility that provides twenty-four-hour nursing supervision by registered or licensed vocational nurses. Commonly referred to as "nursing homes" or "convalescent hospitals," these facilities provide medical, nursing, dietary, pharmacy, and activity services.

An intermediate-care facility provides eight hours of nursing supervision per day. Intermediate care is less extensive and expensive and generally serves patients who are ambulatory and who do not need care throughout the night.

A skilled nursing facility for special disabilities provides a "protective" or "security" environment to persons with mental disabilities. Many of these facilities have "locked" areas where patients reside for their own protection.

An assisted living facility is for individuals who are no longer able to live independently but who do not need the level of care that a nursing home provides. Although nurses and other health care providers are available, assistance is more typically in the form of meals and housekeeping.

Retirement communities involve a range of options from apartments where residents live independently to skilled nursing care. These communities allow older adults to remain in one place and still receive the care they need as they age.

2. Temperament of parent. Some elderly parents have become paranoid, accusatory, and angry at their caregivers. Family members no longer capable of coping with the abuse may arrange for their parents to be taken care of in a nursing home or other facility.

3. Philosophy of adult child. Most children feel a sense of filial responsibility—a sense of personal obligation for the well-being of aging parents. Theoretical explanations for such responsibility include the norm of reciprocity (adult children reciprocate the care they received from their parents), attachment theory (caring results from positive emotions for one's parents), and a moral imperative (caring for one's elderly parents is the right thing to do).

One only child promised his dying father that he would take care of the father's spouse (the child's mother) and not put her in a nursing home. When the mother became 82 and unable to care for herself, the son bought a bed and made the living room of his home the place for his mother to spend the last days of her life. "No nursing home for my mother," he said. This man also had the same philosophy for his mother-in-law and moved her into his home when she was 94.

Whereas some adult children have the philosophy that they must care for their elderly parents themselves, others prefer to hire the help needed. Their philosophy in combination with the amount of care needed, the amount of help from other family members, and the commitment to other responsibilities (e.g., work) influences their decision.

A crisis in care for the elderly may be looming. In the past, women have taken care of their elderly parents. But these were women who lived in traditional families where one paycheck took care of a family's economic needs. Women today work out of economic necessity, and quitting work to take care of an elderly parent is becoming less of an option. As more women enter the labor force, less free labor is available to take care of the elderly. Government programs are not in place to take care of the legions of elderly Americans. Who will care for them when both spouses are working full-time?

4. Length of time for providing care. Offspring must also consider how long they will be in the role of caring for an aging parent. Eighteen years is the average number of years an elderly parent will need care (Family Caregiver Alliance Clearinghouse, 1998).

5. Privacy needs of caregivers. Some spouses take care of their elderly at home but note the effect on their own marital privacy. One of the spouses interviewed by Agee and Blanton (2000) described the care of their Alzheimer parent:

> *The biggest inconvenience was just the lack of privacy and the fact that you can't get away from—whatever. If we're on the telephone, if she needs help with her checkbook or something, then she'll stand there while we're trying to have a conversation, and try to talk to us about whatever the problem is. And in the middle of the night, she'll come in and turn on a light and not shut a door. Right now she's right by our bedroom, so coming in and going out, we have to pass her, and she can catch us. (p. 5)*

Chadiha, Rafferty, & Pickard (2003) found that lower levels of caretaking burden were associated with higher levels of marital functioning. The impact on one's marriage needs to be considered.

6. Cost. For private full-time nursing home care, including room, board, medical care, etc., count on spending $1,000 to $1,500 a week. **Medicare,** a federal health insurance program for persons 65 and older, was developed for short-term acute hospital care. Medicare generally does not pay for long-term nursing care. In practice, adult children who arrange for their aging parent to be cared for in a nursing home end up paying for it out of the elder's own funds. After all of these economic resources are depleted, **Medicaid,** a state welfare program for low-income individuals, will pay for the cost of care. A federal law prohibits offspring from shifting the assets of an elderly parent so as to become eligible for Medicaid.

7. Sexual orientation. Homosexual elders may be resistant to going to a nursing home because they fear prejudice and discrimination from workers and patients at the facility. Some feel they have to "go back in the closet" (Gallanis, 2002).

The elderly should be included in the decision to be cared for in a long-term-care facility. Wielink, Huijsman, & McDonnell (1997) found that the more frail the elderly person, the more willing the person was to go to a nursing home. Once a decision is made for nursing home care, it is important to assess several facilities. Taking a tour of the facility, eating a meal at the facility, and meeting staff are also helpful in making a decision and a smooth transition. Since there is an acute nursing shortage, it is important to find out the ratio of registered nurses per patient. Some facilities hide their number of registered nurses by lumping them with those without training in a category called "nursing staff" (Jacoby, 2003).

Some elderly adapt well to living in a residential facility, such as a nursing home. They enjoy the community of others of similar age, enjoy visiting others in the nursing home, and reach out to others to make new friends. In essence, they find positive meaning in the nursing home experience (Reid, 2000). Ritblatt and Drager (2000) found that the elderly who were in age-segregated living arrangements reported higher levels of satisfaction and social support.

8. Other issues. Whether or not the decision is made to put one's parent (or spouse) in a nursing home, a document detailing the conditions under which life support measures should be used (do you want your parent to be sustained on a respirator?), called **advance directive** (also known as a **living will**) should be completed by the elderly person or those with power of attorney. These decisions, made ahead of time, spare the adult children the responsibility of making them in crisis contexts and give clear directives to the medical staff in charge of the elderly person. A **durable power of attorney,** which gives the adult children complete authority to act on behalf of the elderly, is also advised. These documents also help to save countless legal hours, time, and money for those responsible for the elderly. Copies of the living will and durable power of attorney are in Appendixes D and E.

Finally, adult children may consider buying long-term-care insurance (LTCI) to cover what Medicare and many private health care plans do not—"nonmedical" day-to-day care such as bathing or eating for an Alzheimer's parent and nursing home costs. Cost of LTCI is about $1,600 a year for a good policy from a good company (Cloud, 2000). However, the prices of these policies vary a great deal depending on the age and health of the insured.

Sources

Agee, A., and P. Blanton. 2000. Listening to the stories of Alzheimer's families: A collective case study approach. Poster session at the 62nd Annual Conference of the National Council on Family Relations, Minneapolis, November 12.

Cloud, J. 2000. A kinder, gentler death. *Time*. 18 September, 59 passim.

Gallanis, T. P. 2002. Aging and the nontraditional family. *The University of Memphis Law Center* 32:607–42.

Jacoby, S. 2003. The nursing squeeze. *AARP Bulletin*. May, 6 passim.

Reid, J. Living creatively in a nursing home. *Journal of the American Society on Aging* 23:51–55.

Ritblatt, S. N., and L. M. Drager. 2000. The relationship between living arrangement and perception of social support, depression, and life satisfaction among the elderly. Poster session at the 62nd Annual Conference of the National Council on Family Relations, Minneapolis, November 12.

Statistical Abstract of the United States: 2003. 123rd ed. Washington, D.C.: U.S. Bureau of the Census.

Wielink, G., R. Huijsman, and J. McDonnell. 1997. A study of the elders living independently in the Netherlands. *Research on Aging* 19:174–98.

Some elderly become so physically debilitated that questions about the quality of life sometimes lead to consideration of physician-assisted suicide. Some debilitated elderly ask their physicians to end their lives. Spouses or adult children may also be asked by the debilitated elderly for help in ending their life. Physician-assisted suicide is the topic of this chapter's Social Policy.

ISSUES CONFRONTING THE ELDERLY

Numerous issues become concerns as a person ages. In middle age, the issues are early retirement (sometimes forced), job layoffs (recession-related cutbacks), **age discrimination** (older persons are not hired and younger workers are hired to take their place), separation or divorce from spouse, and adjustment to children's leaving home. For some in middle age, grandparenting is an issue if they become the primary caregiver for their grandchildren.

Though these may be difficult issues during middle age, there are positive counterbalances such as relief from worrying about pregnancy, voluntary grandparenting, opportunities to try new things, and freedom now that the children have left home (Glazer et al., 2002). As the couple moves from the middle to the later years, the issues become more focused on income, housing, health, retirement, and sexuality.

This woman is in a nursing home and enjoys a visit from family.

Authors

Physician-Assisted Suicide for Terminally Ill Family Members

SOCIAL POLICY

One's parents may experience a significant drop in quality of life and a total loss of independence due to illness, accident, or, more typically, aging. Adult children or spouses are often asked their recommendations about withdrawing life support (food, water, or mechanical ventilation), starting medications to end life (intravenous vasopressors), or withholding certain procedures to prolong life (cardiopulmonary resuscitation). Sometimes the patient asks for death. The top reasons cited by patients for wanting to end their lives are losing autonomy (84%), decreasing ability to participate in activities they enjoyed (84%) and losing control of bodily functions (47%) (Chan, 2003). One elderly individual said of the end of life:

> *I came into this world as a human being and I wish to leave in the same manner. Being able to walk, to communicate, to take care of my own needs, to think, to feel. . . . There is no need for me to re-experience my first few months of life though my last months in this world. . . . I do not wish to be once again in a diaper. Just the thought about me losing control over my body frightens me.*
> (Leichtenritt & Rettig, 2000, 3)

Fourteen percent of a sample of 1,255 physicians in New Zealand (where euthanasia is illegal) reported that they had taken actions to hasten the death of a patient. Those doing so were typically older and less likely to report a religious affiliation (Mitchell & Owens, 2003). In a study of 1,117 physicians in Tennessee, almost half (47%) did not favor euthanasia or physician-assisted suicide (PAS) and would oppose the legalization of such procedures (Essinger, 2003). Where the practice did exist, the physicians were in favor of restrictions and safeguards to protect vulnerable patients.

Though all fifty states now have living-will statutes permitting individuals who are not terminally ill (or family members) to refuse artificial nutrition (food) and hydration (water), there is no *constitutional* right of a family member or physician to end the life of another. **Euthanasia** is from Greek words meaning "good death" or dying without suffering. Euthanasia may be passive, where medical treatment is withdrawn and nothing is done to prolong the life of the patient, or active, which involves deliberate actions to end a person's life. Active euthanasia received nationwide attention through the actions of Dr. Jack Kevorkian (convicted and now in prison), who was involved in over 130 physician-assisted suicides at the patient's request.

The Supreme Court has ruled that state law will apply in regard to physician-assisted suicide. Only Oregon recognizes the right of PAS with its Death with Dignity Act. Two physicians must agree that the patient is terminally ill and is expected to die within six months, the patient must ask three times for death both orally and in writing, and the patients must swallow the barbiturates themselves rather than be injected with a drug by the physician. The number of physician-assisted suicides increased from twenty-one in 2001 to thirty-eight in 2002 (an 81% increase). Most patients had cancer and were more likely to be white, male, and well educated (Chan, 2003).

However, the U.S. attorney general is seeking to declare the prescribing of federally controlled drugs to assist suicide an illegitimate medical practice. The courts continue to debate the issue.

The official position of the American Medical Association is that physicians must respect the patient's decision to forgo life-sustaining treatment but that they should not participate in patient-assisted suicide: "PAS is fundamentally incompatible with the physician's role as healer." The AMA disagrees with Dr. Kevorkian. Arguments against PAS emphasize that the practice can be abused by people who want to end the life of those they feel burdened by (or worse, to get the person's money) and that since physicians make mistakes, what is diagnosed as "terminal" may not in fact be terminal.

Euthanasia remains controversial. Kevorkian forced the issue by videotaping his administration of drugs to end the life of a terminally ill man (who gave his written consent). Showing of the videotape occurred nationwide on *60 Minutes.* A jury found him guilty of murder.

Other examples of euthanasia have already occurred. Between 1935 and 1994, 144 people have been charged with mercy killing. One of these was Roswell Gilbert, who shot his wife of fifty-one years twice in the head, allegedly to end her suffering. He was convicted of first-degree murder but was paroled after five years in prison.

Physician-assisted suicide has been legal in Holland for fifteen years. A concern has been the potential to misuse the law. But a study of twenty-five years of euthanasia and physician-assisted suicide in Dutch general practice in which there were about five thousand requests per year concluded, "Some people feared that the lives of increasing numbers of patients would end through medical intervention, without their consent and before all palliative options were exhausted. Our results, albeit based on requests only, suggest that this fear is not justified" (Marquet et al., 2003, 202).

Sources

Chan, S. 2003. Rates of assisted suicides rise sharply in Oregon. *Student BMJ* 11:137–38.

Essinger, D. 2003. Attitudes of Tennessee physicians toward euthanasia and assisted death. *Southern Medical Journal* 96:427–35.

Leichtentritt, R. D., and K. D. Rettig. 2000. Conflicting value considerations for end-of-life decisions. Poster session at the 62nd Annual Conference of the National Council on Family Relations, Minneapolis, November 12.

Marquet, R., A. Bartelds, G. J. Visser, P. Spreeuwenberg, and I. Peters. 2003. Twenty-five years of requests for euthanasia and physician assisted suicide in Dutch practice: Trend analysis. *British Medical Journal.* 327:201–2.

Mitchell, K. and R. G. Owens. 2003. National survey of medical decisions at end of life made by New Zealand practitioners. *British Medical Journal* 327:202–3

Income

For most individuals, the end of life is characterized by reduced income. Social Security and pension benefits, when they exist, are rarely equal to the income a retired person formerly earned.

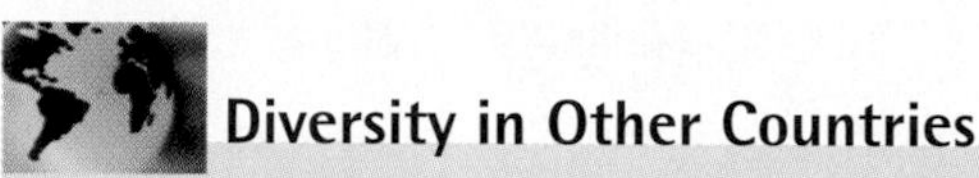

Diversity in Other Countries

Loss of income is minimized among the elderly Chinese who receive money from their children. Logan and Bian (2003) noted that money from "children accounts for nearly a third of parents' incomes" (p. 85).

NATIONAL DATA

The median income of men aged 65 and older is $19,688; women, $11,313 (*Statistical Abstract of the United States: 2003,* Table 693)

Those at greatest risk for poverty are those that live the longest, are widowed, and live alone (Golden & Saltz, 1997). Women are particularly disadvantaged since their work history has often been discontinuous, part-time, and low-paying. Social Security and private pension plans favor those with continuous full-time work histories.

Housing

The idea that most elderly individuals live in a nursing home is a myth. Most elderly Americans, about 94 percent, live in a noninstitutionalized setting (Cockerham, 1997). Almost 80 percent (78.1) over the age of 75 own their own home (*Statistical Abstract of the United States: 2002,* Table 939). These homes are typically those the individual has lived in for years in the same neighborhood. Even those who do not live in their own home are likely to live in a family setting.

For the most part, the physical housing of the elderly is adequate. Indeed, only 6 percent of the housing units inhabited by the elderly are inadequate. Where deficiencies exist, the most common are inadequate plumbing and heating. However, as individuals age, they find themselves living in homes that are not "elder-friendly"—bathroom doors wide enough for a wheelchair, grab bars in the bathrooms, and the absence of stairs (Nicholson, 2003).

Though most elderly own and live in their own homes, as they age, their need for care increases. The most recent trend in housing for the elderly is home health care (mentioned earlier) as an alternative to nursing home care. In this situation the elderly person lives in a single-family dwelling, and other people are hired to come in regularly to help with various needs. Other elderly live in group living or shared housing arrangements.

Physical Health

Good physical health is the single most important determinant of an elderly person's reported happiness (Smith et. al., 2002). Most elderly individuals, even those of advanced years, continue to define themselves as being in good health.

Even those who have a chronic, debilitating illness maintain a perception of good health as long as they are able to function relatively well. However, when the elders' vision, hearing, physical mobility, and strength are markedly diminished, there is a significant impact on their sense of well-being (Smith et al., 2002). For many, the ability to experience the positive side of life seems to become compromised after age 80 (Smith et al., 2002). Driving accidents also increase with aging. However, unlike teenagers, who also have a high percentage of automobile accidents, the elderly are less likely to die in an accident since they are usually driving at a much slower speed (Pope, 2003).

When health care is obtained for the elderly, those with spouses are advantaged. They are more likely to be hospitalized in teaching hospitals and those ranked as being among the "best" hospitals (Iwashyna & Christakis, 2003).

I'm 64 now and trying to stay calm while senility makes its incursions. The loss of sex drive is nothing. It's the loss of memory and general slowing-down of mental coordination that's demoralizing. Apparently each of us has to find all this out ourselves. I can only say there's a whole new world waiting out there for the young. Alas!

Name withheld by request

SELF-ASSESSMENT

Alzheimer's Quiz

	True	False	Don't Know
1. Alzheimer's disease can be contagious.	____	____	____
2. People will almost certainly get Alzheimer's if they just live long enough.	____	____	____
3. Alzheimer's disease is a form of insanity.	____	____	____
4. Alzheimer's disease is a normal part of getting older, like gray hair or wrinkles.	____	____	____
5. There is no cure for Alzheimer's disease at present.	____	____	____
6. A person who has Alzheimer's disease will experience both mental and physical decline.	____	____	____
7. The primary symptom of Alzheimer's disease is memory loss.	____	____	____
8. Among persons older than age 75, forgetfulness most likely indicates the beginning of Alzheimer's disease.	____	____	____
9. When the husband or wife of an older person dies, the surviving spouse may suffer from a kind of depression that looks like Alzheimer's disease.	____	____	____
10. Stuttering is an inevitable part of Alzheimer's disease.	____	____	____
11. An older man is more likely to develop Alzheimer's disease than an older woman.	____	____	____
12. Alzheimer's disease is usually fatal.	____	____	____
13. The vast majority of persons suffering from Alzheimer's disease live in nursing homes.	____	____	____
14. Aluminum has been identified as a significant cause of Alzheimer's disease.	____	____	____
15. Alzheimer's disease can be diagnosed by a blood test.	____	____	____
16. Nursing-home expenses for Alzheimer's disease patients are covered by Medicare.	____	____	____
17. Medicine taken for high blood pressure can cause symptoms that look like Alzheimer's disease.	____	____	____

Answers: 1–4, 8, 10, 11, 13–16 False; remaining items True.

Source

Copyright by Neal E. Cutler, Boettner/Gregg Professor of Financial Gerontology, Widener University. Originally published in *Psychology Today,* 20th Anniversary Issue, "Life Flow: A Special Report—The Alzheimer's Quiz" 1987, Vol. 21, No. 5, pp. 89, 93. Used by permission of Neal E. Cutler. The scale was completed by sixty-nine undergraduates at East Carolina University in 1998. Forty percent of the respondents identified less than 50 percent of the items correctly.

Mental Health

Mental processes are also affected by aging. Elderly persons (particularly those 85 and older) more often have a reduced capacity for processing information quickly, for cognitive attention to a specific task, for retention, and for motivation to focus on a task. However, judgment may not be affected, and experience and perspective are benefits to decision making.

Mental health may worsen for some elderly. Mood disorders, with depression being the most frequent, are more common among the elderly. Such depression is usually related to the fact that the older a person gets, the more likely it is that he or she will experience chronic health problems and related disabilities. It is these events and not age itself that result in a higher incidence of depression (Miller, 1998). Elderly women who have not had children and who have not accepted their childlessness also report more mental distress than those who had children or those who have accepted their childfree status (Wu & Hart, 2002). The mental health of elderly men seems unaffected by their parental status.

Regardless of the source of depression, it has a negative impact on the elderly couple's relationship. Sandberg, Miller, & Harper (2002) studied depression in twenty-six elderly couples and concluded, "The most striking finding was the frequent mention of marital conflict and confrontation among the depressed couples and the almost complete absence of it among the nondepressed couples" (p. 261).

NATIONAL DATA

Depression among the elderly may be linked to suicide. Suicide rates are among the highest for white males aged 85 or older. Eighteen percent of all suicides in 2000 were among individuals aged 65 and over (National Institute of Mental Health, 2003).

Dementia, which includes Alzheimer's disease, is the mental disorder most associated with aging. In spite of the association, only 3 percent of the aged population experience severe cognitive impairment—the most common symptom is loss of memory. It can be devastating to the individual and the partner. An 87-year-old woman, who was caring for her 97-year-old demented husband, said, "After 56 years of marriage, I am waiting for him to die, so I can follow him. At this point, I feel like he'd be better off dead. I can't go before him and abandon him" (Johnson & Barer, 1997, 47).

The self-assessment on page 418 reflects some of the misconceptions about Alzheimer's disease.

Retirement

Retirement represents a rite of passage through which most elderly pass. In 1983 Congress raised the retirement age at which individuals can receive full Social Security benefits from 65 (for those born before 1938) to 67 (for those born after 1960). It is possible to take early retirement at age 62 with reduced benefits. Retirement affects the individual's status, income, privileges, power, and prestige. Persons least likely to retire are not married, widowed, single-parent women who need to continue working because they have no pension or even Social Security benefits—if they don't work, they have no income.

Boris Karloff, actor of the forties, was 81 years old, minus one lung and inhaling oxygen as he sat in his wheelchair with a steel brace, ready to do four more films.

Patricia Fox-Sheinwold

Some workers experience what is called **blurred retirement** rather than a clear-cut one. A blurred retirement means the individual works part-time before completely retiring or takes a bridge job that provides a transition between a lifelong career and full retirement.

Individuals who have a positive attitude toward retirement are those who have a pension waiting for them, are married (and thus have social support for the transition), have planned for retirement, are in good health, and have high self-esteem. Those who regard retirement negatively have no pension waiting for them, have no spouse, gave no thought to retirement, have bad health, and have negative self-esteem (Mutran, Reitzes, & Fernandez 1997). In general, most individuals have thought about retirement and enjoy it. Few regard continuing to work as a privilege. Although they still prefer to be active, they want to pick their own activity and pace, as they could not do when they were working. Summarizing the literature, Cockerham (1997) noted: "Apparently, only a minority view retirement negatively, and this seems especially so if it results in a significant loss of income. Otherwise, it appears that most people reach a point in their lives when retirement is desired" (p. 163).

Retirement may have positive consequences for the marriage. Szinovacz and Schaffer (2000) analyzed national data and concluded that husbands viewed the retirement of their wives as associated with a reduction in heated arguments. And if both spouses are retired, there is a perceived reduction of disagreements. Webber, Scott, & Wampler (2000) found that spouses had an easier time adjusting to retirement if they had similar retirement goals or expectations in the areas of leisure, family relationships, friendships, and finances. Kupperbusch, Levenson, and Ebling (2003) found that husbands who were physiologically relaxed and who had a good relationship with their wives were happier in retirement.

Some individuals experience disenchantment during retirement. It is not enjoyable and does not live up to their expectations. Lee Iacocca, former president of Chrysler Corporation, reported that he was bored with retirement, missed "the action," and warned others "never to retire" but to stay busy and involved. Some heed his advice and either "die in harness" or go back to work.

Sexuality

Levitra, Cialis, and Viagra (prescription drugs that help a man obtain and maintain an erection) are advertised regularly on television and have given cultural visibility to the issue of sexuality among the elderly. Though there are physiological changes for the elderly (both men and women) that impact sexuality, older adults often continue their interest in sexual activity. Table 17.3 describes the physiological changes that elderly men experience during the sexual response cycle

Table 17.4 describes the physical changes elderly women experience during the sexual response cycle. Gelfand (2000) summarized these changes in women as the result of decreasing levels of estrogen and testosterone, which are associated with decreased sexual libido, sensitivity, and response.

There may be ethnic differences in sexual interest in midlife. A third of 3,262 women 42 to 52 reported that sex was "very important" (Cain et al., 2003). African-American women were most likely to report that sex was "very important"; Japanese and Chinese women were least likely to report this. Caucasian women were between the extremes.

Overall, though there is a decline in sexual activity with age, many elderly continue sexual activity, with women being less sexually active. The less frequent activity for women is due to fewer available men. Indeed, the biggest sexual complaint of elderly women over age 80 is the absence of men (Michael et al., 1994).

Although sex may occur less often, it remains physically satisfying for most elderly. Over 60 percent of both men and women reported that sex was as physically satisfying now as in their 40s. Health issues such as disability and side effects of medication were the primary causes of reduced satisfaction. A quarter of the elderly men complained of erectile dysfunction (AARP, 1999). Mackey and O'Brien (2000) observed that there are important balances that contribute to marital happiness. "These balances include the deepening of psychological intimacy, increases in mutual decision making and improved skills in managing conflict and adapting to differences" (p. 593). Winterich (2003) found that in spite of vaginal, libido, and orgasm changes past menopause, women of both sexual orientations reported that they continued to enjoy active, enjoyable sex lives due to open communication with their partners.

The debate continues about whether menopausal women should become involved in estrogen and estrogen-progestin replacement therapy (ERT). Schairer et al. (2000) studied 2,082 cases of breast cancer and concluded that the estrogen-progestin regimen increased the risk of breast cancer beyond that associated with estrogen alone. As of 2003, women were no longer routinely encouraged to take ERT, and the effects of their not doing so on their physical and emotional health were minimal. A major study (Hays et al., 2003) of over 16,608 postmenopausal women 50 to 79 found no significant benefits from ERT in terms of

Table 17.3 Physiological Sexual Changes in Elderly Men

Phases of Sexual Response	Changes in Men
Excitement phase	As men age, it takes them longer to get an erection. While the young man may get an erection within 10 seconds, elderly men may take several minutes (10 to 30). During this time, they usually need intense stimulation (manual or oral). Unaware that the greater delay in getting erect is a normal consequence of aging, men who experience this for the first time may panic and have erectile dysfunction.
Plateau phase	The erection may be less rigid than when the man was younger, and there is usually a longer delay before ejaculation. This latter change is usually regarded as an advantage by both the man and his partner.
Orgasm phase	Orgasm in the elderly male is usually less intense, with fewer contractions and less fluid. However, orgasm remains an enjoyable experience, as over 70 percent of older men in one study reported that having a climax was very important when having a sexual experience.
Resolution phase	The elderly man loses his erection rather quickly after ejaculation. In some cases, the erection will be lost while the penis is still in the woman's vagina and she is thrusting to cause her own orgasm. The refractory period is also increased. Whereas the young male needs only a short time after ejaculation to get an erection, the elderly man may need considerably longer.

Source: Adapted from W. Boskin, G. Graf, and V. Kreisworth. *Health dynamics: Attitudes and behaviors.* St. Paul: West, 1990, p. 209. Used by permission.

Table 17.4 Physiological Sexual Changes in Elderly Women

Phases of Sexual Response	Changes in Women
Excitement phase	Vaginal lubrication takes several minutes or longer, as opposed to 10 to 30 seconds. Both the length and the width of the vagina decrease. Considerable decreased lubrication and vaginal size are associated with pain during intercourse. Some women report decreased sexual desire and unusual sensitivity of the clitoris.
Plateau phase	Little change occurs as the woman ages. During this phase, the vaginal orgasmic platform is formed and the uterus elevates.
Orgasm phase	Elderly women continue to experience and enjoy orgasm. Of women aged 60 to 91, almost 70 percent reported that having an orgasm made for a good sexual experience. With regard to their frequency of orgasm now as opposed to when they were younger, 65 percent said "unchanged," 20 percent "increased," and 14 percent "decreased."
Resolution phase	Defined as a return to the preexcitement state, the resolution phase of the sexual response phase happens more quickly in elderly than in younger women. Clitoral retraction and orgasmic platform disappear quickly after orgasm. This is most likely a result of less pelvic vasocongestion to begin with during the arousal phase.

Source: Adapted from W. Boskin, G. Graf, and V. Kreisworth. *Health dynamics: Attitudes and behaviors.* St. Paul: West, 1990, p. 210. Used by permission.

quality-of-life outcomes. Those with severe symptoms (e.g., hot flashes, sleep disturbances) do seem to benefit without negative outcomes.

SUCCESSFUL AGING

Researchers who worked on the Landmark Harvard Study of Adult Development (Valliant, 2002) followed 824 men and women from their teens into their 80s and identified those factors associated with successful aging. These include not smoking or quitting early, developing a positive view of life and life's crises, avoiding alcohol and substance abuse, maintaining healthy weight, exercising daily, continuing to educate oneself, and having a happy marriage. Indeed those who were identified as "happy and well" were six times more likely to be in a good marriage than those who were identified as "sad and sick."

Crosnoe and Elder (2002b) identified success in one's career as a factor associated with successful aging. Indeed, one of the regrets a person is most likely to have is not pursuing a specific career. Almost two-thirds of a sample of individuals reported that they regretted not pursuing a particular degree or profession (Wrosch & Heckhausen, 2002).

Not smoking is "probably the single most significant factor in terms of health" according to Valliant (2002) of the Landmark Harvard Study on Adult Development. Smokers who quit before age 50 were as healthy at 70 as those who had never smoked.

RELATIONSHIPS AT AGE 85 AND BEYOND

Relationships continue into old age. Here we examine relationships with one's spouse, siblings, and children.

Relationship with Spouse at Age 85 and Beyond

Marriages that survive into late life are characterized by little conflict, considerable companionship, and mutual supportiveness. All but one of the thirty-one spouses in the Johnson and Barer (1997) study reported "high expressive rewards" from their mate.

Field and Weishaus (1992) reported interview data on seventeen couples who had been married an average of fifty-nine years and found that the husbands and wives viewed their marriages very differently. Men tended to report more marital satisfaction, more pleasure in the way their relationships had been across time, more pleasure in shared activities, and closer affectional ties. Whereas club activities, including church attendance, were related to marital satisfaction, financial stability, amount of education, health, and intelligence were not. Sex was also more important to the husbands. Every man in the study reported that sex was always an important part of the relationship with his wife, but only four of the seventeen wives had the same report.

The wives, according to the researchers, presented a much more realistic view of their marriage. For this generation of women, "it was never as important for them to put the best face on things" (Field & Weishaus, 1992, 273). Hence, these wives were not unhappy; they were just more willing to report disagreements and changes in their marriages across time.

Only a small percentage (8%) of individuals over 100 are married. Most married centenarians are men in their second or third marriage. Many have outlived some of their children. Marital satisfaction in these elderly marriages is related to a high frequency of expressing love feelings to one's partner. Though it is assumed that spouses who have been married for a long time should know how their partners feel, this is often not the case. Telling each other "I love you" is very important to these elderly spouses.

Relationship with Siblings at Age 85 and Beyond

Relationships with siblings are primarily emotional (enjoying time together) rather than functional (the sibling provides money or services). Only 7 percent of siblings are in the role of caregiver (Bedford, 1996). Almost 70 percent of the elderly report emotional benefits of having a sibling, but only 9 percent report monthly or more contact with a sibling. Other relatives such as nieces, nephews, cousins, and in-laws are also characterized by affection but infrequent contact.

Relationship with One's Own Children at Age 85 and Beyond

In regard to relationships of the elderly with their children, emotional and expressive rewards are high. Actual caregiving is rare. Only 12 percent of the Johnson and Barer sample of older adults over 85 lived with their children. Most preferred to be independent and to live in their own residence. "This independent stance is carried over to social supports; many prefer to hire help rather than bother their children. When hired help is used, children function more as mediators than regular helpers, but most are very attentive in filling the gaps in the service network" (Johnson & Barer, p. 86).

Relationships among multiple generations will increase. Whereas three-generation families have been the norm, increasingly four- and five-generation families will become the norm. These changes have already become visible.

GRANDPARENTHOOD

Another significant set of relationships for individuals in later life is with grandchildren. Among adults aged 40 and older who had children, close to 95 percent are grandparents and most have on average five or six grandchildren. The average age of becoming a grandparent is 48; the age at which grandparents become caregivers of their grandchildren is from 53 to 59 (Landry-Meyer, 2000). Fifteen percent of grandparents provide child-care services for their grandchildren while the parents work (Davies, 2002).

Most grandparents see their grandchildren two to four times a month (Davies, 2002). However, an increasing number of grandparents are becoming primary caregivers of their grandchildren. About 5 percent of all grandchildren are being reared in grandparent households full-time (Blackburn, 2000).

Perceptions of Grandparenting

Neugarten and Weinstein (1964) assessed the significance of the grandparent role for a group of seventy grandparent couples. Examples of grandmother and grandfather perceptions included the following:

1. *Biological renewal—feeling young again through one's grandchildren. Grandmothers were almost twice (42% versus 23%) as likely as grandfathers to report this effect.*
2. *Biological continuity—seeing one's life continue into the future through one's grandchildren.*
3. *Emotional self-fulfillment—becoming involved in their grandchildren's lives, as they had not been in their children's lives because they were too busy with their work and careers. These grandparents felt as though they had a second chance at parenthood. However, some grandparents still had to work, which limited their time with their grandchildren.*
4. *Resource person—offering life experience and financial aid to their grandchildren.*

One of the great joys of this woman's life is her grandbaby.

Authors

Styles of Grandparenting

Grandparents also have different styles of relating to grandchildren. Whereas some grandparents are formal and rigid, others are informal and playful, and authority lines are irrelevant. Still others are surrogate parents providing considerable care for working mothers and/or single parents. Some grandparents have regular contact with their grandchildren; others are distant and show up only for special events like birthdays. E-mail is helping grandparents to stay connected to their grandchildren. Davies (2002) reported that 35 percent of grandparents use e-mail to communicate with their grandchildren.

Age seems to be a factor in determining how grandparents relate to their grandchildren. Grandparents over the age of 65 are less likely to be playful and fun-seeking than those under 65. This may be because the older grandparents are less physically able to engage in playful activities with their grandchildren. Indeed, according to Johnson & Barer (1997),

> *members of the oldest generation in our study place more emphasis on their relationship with their own children over their grandchildren. This situation could stem from the fact that as grandchildren reach adulthood, they become more independent from their own parent. That parent then is freed up to strengthen the relationship with their oldest old parent, at the same time they, as the middle generation, maintain a lineage bridge linking their parent to their child.* (p. 89)

Finally, the quality of the grandparent-grandchild relationship is affected by the parents' relationship to their own parents. If a child's parents are estranged from their parents, it is unlikely that the child will have an opportunity to develop a relationship with the grandparents.

Effect of Divorce on Grandparent-Child Relationship

The degree of involvement of grandparents in the life of a grandchild is sometimes related to whether the grandparents are divorced (King, 2003). "Divorced grandparents have less contact with grandchildren and participate in fewer shared activities with them" (p. 180).

In addition, grandparental involvement with grandchildren is also related to whether their own children are divorced and whether the grandchild is on the mother's or father's side. Since mothers end up with custody in about 85 percent of the cases, the maternal grandparents may escalate their involvement with their grandchildren. However, since divorcing fathers may have less access to their children, the paternal grandparents may find that their time with their grandchildren is radically reduced. Sometimes a custodial parent will purposefully try to sever the relationship in an attempt to seek revenge or vent hostility against his or her former spouse (Knox, 2000).

Indeed, when their children divorce, some grandparents are not allowed to see their grandchildren. Although all fifty states have laws granting grandmothers and grandfathers the right to petition for visitation with their grandchildren over the protests of the parent or parents (Hill, 2000), the role of the grandparent has limited legal and political support. By a vote of six to three the Supreme Court in 2000 *(Troxel v. Granville)* sided with the parents and virtually denied 60 million grandparents the right to see their grandchildren. The court viewed parents as having a fundamental right to make decisions about their children (Willing & McMahon, 2000). Stepgrandparents have no legal rights to their stepgrandchildren.

Benefits to Grandchildren

Grandchildren report enormous benefits from having a close relationship with grandparents, including development of a sense of family ideals, moral beliefs, and a work ethic. Kennedy (1997) focused on the memories grandchildren had of their grandparents and found that "love and companionship" was identified most frequently by the grandchildren. In addition, a theme running through a fourth of the memories was that the grandchild felt that he or she was regarded as "special" by the grandparent, either because of being the first or last grandchild, having personality characteristics similar to the grandparent's, or being the child of a favorite son or daughter of the grandparent. Two researchers found that the relationship between the grandparents and the offspring is enhanced when the latter enters college (Crosnoe & Elder, 2002a).

THE END OF ONE'S LIFE

The end of one's life sometimes involves the death of one's spouse.

Death of One's Spouse

The death of one's spouse is the most stressful life event individuals experience (Hobson et al., 1998). Because women tend to live longer than men, and because women are often younger than their husbands, women are more likely than men to experience the death of their marital partner. Antonucci et al. (2002) noted that women in several countries—e.g., Germany, Japan, France—as well as in the United States, are all more likely than men to experience widowhood, illness, and financial strain.

Widows are divided into two classes—the relieved and the bereaved.

Anonymous

Although individual reactions and coping mechanisms for dealing with the death of a loved one vary, several reactions to death are common. These include shock, disbelief and denial, confusion and disorientation, grief and sadness, anger, numbness, physiological symptoms such as insomnia or lack of appetite, withdrawal from activities, immersion in activities, depression, and guilt. Eventually, surviving the death of a loved one involves the recognition that life must go on, the need to make sense out of the loss, and the establishment of a new identity. Grief and sadness often return on the anniversary of the death, on the deceased's birthday, and on other special occasions.

Women and men tend to have different ways of reacting to and coping with the death of a loved one. Women are more likely than men to express and share feelings with family and friends and are also more likely to seek and accept help, such as attending support groups of other grievers. Initial responses of men are often cognitive rather than emotional. From early childhood, males are taught to be in control, to be strong and courageous under adversity, and to be able to take charge and fix things. Showing emotions is labeled weak.

Men sometimes respond to the death of their spouse in behavioral rather than emotional ways. Sometimes they immerse themselves in work or become involved in physical action in response to the loss. For example, a widower immersed himself in repairing a beach cottage he and his wife had recently bought. Later, he described this activity as crucial to getting him through those first two months. Another coping mechanism is the increased use of alcohol and other drugs, particularly among men.

Women's response to the death of their husbands may necessarily involve practical considerations. Johnson and Barer (1997) identified two major problems of widows—the economic effects of losing a spouse and the practical problems of maintaining a home alone. The latter involves such practical issues as cleaning the gutters, painting the outside, and changing the filters in the furnace.

Whether a spouse dies suddenly or after a prolonged illness has an impact on the reaction of the remaining spouse. The sudden rather than prolonged death of one's spouse is associated with being less at peace with death, being more angry, and not having discussed impending death with the spouse. Widows of spouses who die a "painful" death report higher anxiety and intrusive thoughts (Carr, 2003).

The age at which one experiences the death of a spouse is also a factor in adjustment to it. Persons in their 80s may be so consumed with their own health and disability concerns that they spend little emotional energy on the death of their partner. On the other hand, a team of researchers (Hansson, Berry, & Berry, 1999) noted that death does not end the relationship with the deceased. Some widows and widowers report a year after the death of their beloved a feeling that their spouses are with them at times and are watching out for them. They may also dream of them, talk to their photographs, and remain interested in carrying out their wishes. Such continuation of the relationship may be adaptive by providing meaning and purpose for the living or maladaptive in that it may prevent one from establishing new relationships.

Involvement with New Partners at Age 80 and Beyond

Most women who live to age 80 have lost their husbands. At age 80 there are only fifty-three men for every one hundred women. Patterns women use to adjust to this lopsided man-woman ratio include dating younger men, romance without marriage, and "share-a-man" relationships. Most individuals in their 80s who have lost a partner do not remarry (Stevens, 2002). Those who do report as the primary reasons a need for companionship and to provide meaning in life. Stevens (2002) noted "evidence of continuing loyalty to the deceased" in all the late-life partnerships she studied, suggesting that new partners do not simply "replace" former partners.

These individuals are both widowed from previous relationships. They have no interest in marrying each other and do not feel it proper to live together. But they travel the world together. In this photo they are on a cruise up the Ohio River.

Authors

Over 3 million women are married to men who are at least ten years younger than they. Although traditionally women were socialized to seek older, financially established men, the sheer shortage of men has encouraged many women to seek younger partners.

Faced with a shortage of men but reluctant to marry those who are available, some elderly women are willing to share a man. In elderly retirement communities such as Palm Beach, Florida, women count themselves lucky to have a man who will come for lunch, take them to a movie, or be an escort to a dance. They accept the fact that the man may also have lunch, go to a movie, or go dancing with other women.

But finding a new spouse is usually not a goal. Women in their later years have also moved away from the idea that they must remarry and have become more accepting of the idea that they can enjoy the romance of a relationship without the obligations of a marriage. Ken Dychtwald (1990) in his book *Age Wave* observed that many elderly women were interested in romance with a man but were not interested in giving up their independence. "Many say they do not have the same family-building reasons for marriage that young people do. For women especially, divorce or widowhood may have marked the first time in their lives that they have been on their own, and many now enjoy their independence" (p. 222). In addition, many elders are reluctant to marry because of a new mate's deteriorating health. "They would not want to become the caretaker of an ill spouse, especially if they had been through an emotionally draining ordeal before" (p. 222).

To avoid marriage, some elderly couples live together. Parenthood is no longer a goal and many do not want to entangle their assets. For some, marriage would mean the end of their Social Security benefits or other pension moneys.

Preparing for One's Own Death

What is it like for those near the end of life to think about death? To what degree do they go about actually "preparing" for death? Johnson and Barer (1997) interviewed forty-eight individuals with an average age of 93 to find out their perspective on death. Most were women (77%) who lived alone (56%), but most had some sort of support in terms of children or one or more social support services (73%). The findings to follow are specific to those who died within a year after the interview.

It is nothing to die, but it is frightful not to live.

Jean Valjean
Les Miserables

Thoughts the Last Year of Life Most had thought about death and saw their life as one that would soon end. Most did so without remorse or anxiety. With their spouses and friends dead and their health failing, they accepted death as the next stage in life. Some comments follow (Johnson & Barer, 1997, 205).

- *If I die tomorrow, it would be all right. I've had a beautiful life, but I'm ready to go. My husband is gone, my children are gone, and my friends are gone.*
- *That's what is so wonderful about living to be so old. You know death is near and you don't even care.*
- *I've just been diagnosed with cancer, but it's no big deal. At my age, I have to die of something.*

The major fear expressed by these respondents was not the fear of death but the dying process. Dying in a nursing home after a long illness is a dreaded fear. Sadly, almost 60 percent of the respondents died after a long progressive illness. They had become frail, fatigued, and burdened by living. They identified dying in their sleep as the ideal way to die. Some hasten their death by no longer taking their medications or wish they could terminate their own life. "I'm feeling kind of useless. I don't enjoy anything anymore. . . . What the heck am I living for? I'm ready to go anytime—straight to hell. I'd take lots of sleeping pills if I could get them" (Johnson & Barer, 1998, 204).

Behaviors the Last Year of Life Aware that they are going to die, most simplify their life, disengage from social relationships, and leave final instructions. In simplifying their life, they sell their home and belongings and move to smaller quarters. One 81-year-old woman sold her home, gave her car away to a friend, and moved into a nursing home. The extent of her belongings became a chair, lamp, and TV.

Just think! some night the stars will gleam
Upon a cold, grey stone,
And trace a name with silver beam,
And lo! 'twill be your own.

Robert W. Service

Disengaging from social relationships is selective. Some maintain close relationships with children and friends but others "let go." Christmas cards are no longer sent out, letters stop, and phone calls become the source of social connections.

Some leave final instructions in the form of a will or handwritten note expressing wishes of where to be buried, handling costs associated with disposal of the body, and what to do about pets. One of Johnson's and Barer's (1997) respondents left $30,000 to specific caregivers to take care of each of several pets (p. 204).

SUMMARY

***What is meant by the terms* age *and* ageism?**

Age is defined chronologically (time), physiologically (capacity to see, hear etc.) psychologically (self-concept), sociologically (social roles), and culturally (value placed on elderly). Ageism is the denigration of the elderly, and gerontophobia is the dreaded fear of being elderly. Theories of aging range from disengagement (individuals and societies mutually disengage from each other) to continuity (the habit patterns of youth are continued in old age). Age stratification and life course theoretical perspectives are currently in vogue.

What is the "sandwich generation"?

Elder care combined with child care is becoming common among the **sandwich generation,** adult children responsible for the needs of both their parents and their children. Guilt over not doing enough, resentment over feeling burdened, and exhaustion over the relentless demands are among the feelings reported by members of the sandwich generation.

Deciding whether to arrange for an elderly parent's care by a nursing home requires attention to a number of factors, including the level of care needed by the parent, the philosophy and time availability of the adult child, and the resources of the adult children and other siblings. Full-time nursing care is expensive. Elderly parents who are dying from terminal illnesses incur enormous medical bills. Some want to die and ask for help. Our society continues to wrestle with physician-assisted suicide and euthanasia. Dr. Jack Kevorkian brought these issues to a legal confrontation by his televised lethal injection of a terminally ill patient.

What issues confront the elderly?

The elderly have various concerns, including housing, health, retirement, and sexuality. Most elderly live in their own homes, which they have paid for. Most housing of the elderly is adequate, although repair becomes a problem with the age of the person.

Health concerns are paramount for the elderly. Good health is the single most important factor associated with an elderly person's perceived life satisfaction. Hearing and visual impairments, arthritis, heart conditions, and high blood pressure are all common to the elderly. Mental problems may also occur with mood disorders; depression is the most common.

Though the elderly are thought to be wealthy and living in luxury, most are not. The median household income of persons over the age of 65 is less than half of what the couple earned in the prime of their lives. The most impoverished elderly are those who have lived the longest, who are widowed, and who live alone. Women are also particularly disadvantaged since their work history has often been discontinuous, part-time, and low-paying.

Sexuality among the elderly involves, for most women and men, lower reported interest, activity, and capacity. Fear of the inability to have an erection is the most frequently reported sexual problem by elderly men. The absence of a sexual partner is the most frequently reported sexual problem among elderly women.

What factors are associated with successful aging?

Factors associated with successful aging include not smoking or quitting early, developing a positive view of life and life's crises, avoiding alcohol and substance abuse, maintaining healthy weight, exercising daily, continuing to educate oneself, and having a happy marriage. Indeed, those who were identified as "happy and well" were six times more likely to be in a good marriage than those who were identified as "sad and sick." Success in one's career is also associated with successful aging.

What are the relationships of those 85 and older like?

Marriages that survive into old age (beyond age 85) tend to have limited conflict, considerable companionship, and mutual supportiveness. Relationships with siblings are primarily emotional rather than functional. In regard to relationships of the elderly with their children, emotional and expressive rewards are high. Caregiving help is available but rare. Only 12 percent of one sample of older adults over 85 lived with their children.

What is grandparenthood like?

Among adults aged 40 and older who had children, close to 95 percent are grandparents. There is considerable variation in role definition and involvement. Whereas some delight in seeing their lineage carried forward in their grandchildren and provide emotional and economic support, others are focused on their own lives or on their own children and relate formally and at a distance to their grandchildren. When grandparents are involved in their lives, grandchildren benefit in terms of positive psychological and economic benefits.

How do the elderly face the end of life?

The end of life involves adjusting to the death of one's spouse and to the gradual decline of one's health. Most elderly are satisfied with their life, relationships, and health. Declines begin when persons are their 80s.

KEY TERMS

advance directive
age
age discrimination
ageism
blurred retirement
Cialis
dementia
durable power of attorney
euthanasia
family caregiving
filial piety
filial responsibility
gerontology
gerontophobia
Levitra
living will
Medicaid
Medicare
sandwich generation
Viagra

RESEARCHING MARRIAGE AND THE FAMILY WITH INFOTRAC COLLEGE EDITION

InfoTrac College Edition, an online library, allows you to perform research online anywhere, anytime. Following are two suggested search terms and related questions to help you extend your understanding of the topics covered in this chapter. Go to www.infotrac-college.com to begin your search.

Keyword: **Grandparents.** Locate articles that discuss the role of grandparents. To what degree do grandparents enjoy their role instead of feeling burdened by it?

Keyword: **Physician-assisted suicide.** Locate articles that discuss this controversial issue. What is your opinion?

The Companion Web Site for Choices in Relationships: An Introduction to Marriage and the Family, Eighth Edition

http://sociology.wadsworth.com/knox_schacht/choices8e

Supplement your review of this chapter by going to the companion Web site to take one of the Tutorial Quizzes, use the flash cards to master key terms, and check out the many other study aids you'll find there. You'll also find special features such as the Marriage and Family Resource Center, Census 2000 information, and other data and resources at your fingertips to help you with that special project or to do some research on your own.

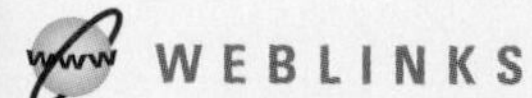

WEBLINKS

AARP (widowed persons services)
http://www.aarp.org

ElderWeb
http://www.elderweb.com

Foundation for Grandparenting
http://www.grandparenting.org/

Generations United
http://www.gu.org

GROWW (Grief Recovery Online)
http://www.groww.com

Nolo: Law for All (wills and legal issues)
http://www.nolo.com

The Future of Marriage and the Family

What can we predict about marriage and the family as we pass midway into the first decade of the new century? We have no crystal ball but think it reasonable to suggest the following:

> ***Marriage*** *will continue to be the lifestyle of choice for most (over 95%) of U.S. adults. Though there is evidence that individuals are putting off getting married until their mid-20s, there is no evidence that they intend to avoid marriage completely* (Statistical Abstract of the United States: *2003). For the almost 5 million annually who marry (Sutton, 2003), those who elect this lifestyle will be among the happiest, healthiest, and most sexually fulfilled in our society.*

Children will be a focus of marriage less often than the adult relationship. Whitehead and Popenoe (2003) emphasized that

> *though Americans aspire to marriage, they are ever more inclined to see it as an intimate relationship between adults rather than as a necessary social arrangement for rearing children. . . . Indeed, if there is a story to be told about marriage over recent decades, it is not that it is withering away for adults but that it is withering away for children.*

According to Whitehead and Popenoe (2003), almost 80 percent of young adults disagree with the statement, "The main purpose of marriage is children."

> ***Singlehood*** *will (in the cultural spirit of diversity) lose its stigma, slightly more will choose this option, and most of those who do will find satisfaction in it.*

> ***Gay marriage*** *in the United States will continue to be a controversial issue with equally powerful opposing forces. The Supreme Court striking down an anti-sodomy law in Texas, Massachusetts giving gay and lesbian couples the same legal status as traditional heterosexual married couples, and the legalization of gay marriage in two Canadian provinces are changes reflecting a momentum toward increased acceptance of same-sex marriage. Among all university students, 67 percent of women and 50 percent of men agree that "same sex couples should have the right to legal marital status" (American Council on Education and the University of California, 2004).*
>
> *However, national data on U.S. adults (not just college students) reflect that only a minority (24%) favor a law allowing homosexual couples to get legally married (34% favor civil unions of gay couples) (Grossman, 2004). In addition, President Bush (reflecting the mood of conservative right-wing America) labeled homosexuality a sin (McQuillan, 2003) and emphasized in his 2004 State of the Union message that "Our nation must defend the sanctity of marriage." The federal government and thirty-eight states have enacted laws barring the recognition of gay marriages.*

Living together *will become an accepted and predictable stage of courtship. The link between cohabitation and subsequent divorce will dissolve as more individuals elect to cohabit before marriage. Previously, only risk takers and persons willing to abandon traditional norms lived together before marriage. In the future, mainstream individuals will increasingly cohabit.*

Children *will continue to be desired by both individuals and couples. While about 65 percent will continue to be born into married, two-parent homes, there will be increased acceptability for conceiving and rearing children outside legal marriage in single-parent families (both heterosexual and homosexual). Children born to married couples will not be immune to being reared in single-parent families since around 40–45 percent of their parents will divorce. Those whose parents remain together will enjoy the stability of and benefits of living in a two-parent family (lower risk for poverty, economic insecurity, emotional and school problems, and unwed teen pregnancy).*

Dual-earner relationships *will increase. There will be no return to the one-income family. The prices of goods and services are such that it takes two incomes to pay for housing, food, cars, etc. Allowing time with the family will continue to take precedence over work time, and job flexibility will therefore remain an important factor in selecting a career and a job. Women will continue to give greater priority to family life than men and pay a greater cost in terms of decreased wages and career advancement (the motherhood penalty) (Avellar & Smock, 2003).*

Day-care *concerns will continue. As more children will spend more time in day care, increasing pressure will be put on the industry to provide "quality" care. But the sheer demand will not be followed by the income to pay for trained staff in smaller classes, with the result that more children will spend time in substandard day-care facilities. Concerned parents will put pressure on Congress to address this national crisis.*

Violence and abuse *will continue to occur behind closed doors. But cultural visibility of abusive relationships, support for leaving them, and the availability of hotlines and shelters will cause increasing numbers of individuals to bravely leave these relationships.*

Divorce *will continue to claim more marriages than the death of a spouse. Between 40 and 45 percent of persons beginning their lives together as spouses will end up as ex-spouses haggling over custody, child support, and visitation. Divorce mediation will become a fortunate alternative to some litigated endings.*

The stigma of divorce and attempts to reduce it will continue. "Marriage education" courses focusing on improving communication skills, managing conflict, and developing empathy and showing love for the partner will continue. As part of the government's welfare reform proposal, the Senate continues to consider a $1.5 billion plan over five years (including matching funds from states) for pro-marriage programs (Peterson, 2004).

Elderly *individuals, particularly those in their 80s, will continue to find the end of life difficult. Health problems, lower incomes, and lack of health care alternatives with few solutions are the reality for the elderly in the United States today. As our population continues to age, attention to problems of the elderly will continue and give hope to improvement of the final days.*

"Nothing endures but change" is the summary statement for the future of marriage and the family. We embrace the future.

Authors

SPECIAL TOPIC

1

Marriage is like twirling a baton, turning handsprings, or eating with chopsticks. It looks easy till you try it.

E. C. McKenzie

Careers in Marriage and the Family

Contents

Students who take courses in marriage and the family sometimes express an interest in working with people and ask what careers are available if they major in marriage and family studies. In this Special Topics section we review some of these career alternatives, including family life education, marriage and family therapy, child and family services, and family mediation. These careers often overlap so that you may engage in more than one of these at the same time. For example, you may work in family services but do family life education as part of your job responsibilities.

For all the careers discussed in this section, it is helpful to have a bachelor's degree in a family-related field such as family science, sociology, or social work. Family science programs are the only academic programs that have a focus specifically on families and approach working with people from a family systems perspective. These programs have many different names, including child and family studies, human development and family studies, child development and family relations, and family and consumer sciences. Marriage and family programs are offered through sociology departments; family service programs are typically offered through departments of social work as well as through family science departments. While some jobs are available at the bachelor's level, others require a master's or Ph.D. degree. More details on the various careers available to you in marriage and the family follow.

Appreciation is expressed to Sharon Ballard, Ph.D., CFLE, for the development of Special Topic 1. Dr. Ballard is an assistant professor of Child Development and Family Relations at East Carolina University. She is also a certified family life educator through the National Council on Family Relations.

FAMILY LIFE EDUCATION

Family life education (FLE) is an educational process that focuses on prevention and on strengthening and enriching individuals and families. The family life educator empowers family members by providing them with information that will help prevent problems and enrich their family well-being. There are different ways that this education may be offered to families: a newsletter, one on one, or through a class or workshop. Examples of family life education programs include parent education for parents of toddlers through a child care center, a brown-bag lunch series on balancing work and family in a local business, a premarital or marriage enrichment program at your local church, a class on sexuality education in a high school classroom, and a workshop on family finance and budgeting at a local community center. Your role as a family life educator would involve your making presentations in a variety of settings, including schools, churches, and even prisons. As a family life educator, you may also work with military families on military bases, within the business world with human resources or employee assistance programs, and within social service agencies or cooperative extension programs. Some family life educators develop their own business providing family life education workshops and presentations.

To become a family life educator, you need a minimum of a bachelor's degree in a family-related field such as family science, sociology, or social work. You can become a **certified family life educator** (CFLE) through the National Council on Family Relations (NCFR). The CFLE credential offers you credibility in the field and shows that you have competence in conducting programs in all areas of family life education. These areas are Families in Society, Internal Dynamics of the Family, Human Growth and Development, Interpersonal Relationships, Human Sexuality, Parent Education and Guidance, Family Resource Management, Family Law and Public Policy, and Ethics. In addition, you must show competence in planning, developing, and implementing family life education programs.

Your academic program at your college or university may be approved for provisional certification. In other words, if you follow a specified program of study at your school, you may be eligible for a provisional CFLE certification. Once you gain work experience, you can then apply for full certification.

MARRIAGE AND FAMILY THERAPY

While family life educators help prevent the development of problems, marriage and family therapists help spouses, parents, and family members resolve existing interpersonal conflicts and problems. The range of problems they treat include communication, emotional and physical abuse, substance abuse, sexual dysfunctions, and parent-child relationships. They work in a variety of contexts, including mental health clinics, social service agencies, schools, and private practice.

Currently, forty-two states license or certify marriage and family therapists. While an undergraduate degree in sociology, family studies, or social work is a good basis for becoming a marriage and family therapist, a master's degree is required in one of these areas. Some universities offer accredited master's degree programs specific to marriage and family therapy; these involve courses in marriage and family relationships, family systems, and human sexuality as well as numerous hours of clinical contact with couples and families under supervision. Full certification involves clinical experience of one thousand hours of direct client/couple/family contact; two hundred of these hours must be under the direction of a supervisor approved by the American Association of Marriage and Family Therapists (AAMFT). In addition, most states require a licensure examination. The AAMFT is the organization that certifies marriage and family therapists.

CHILD AND FAMILY SERVICES

In addition to work as a family life educator or marriage and family therapist, careers are available in agencies and organizations that work with families, often referred to as social service agencies. The job titles within these agencies include family interventionist, family specialist, and family services coordinator. Your job responsibilities in these roles might involve your helping clients over the telephone, coordinating services for families, conducting intake evaluations, performing home visits, facilitating a support group, or participating in grant-writing activities. In addition, family life education is often a large component of child and family services. You may develop a monthly newsletter, conduct workshops or seminars on particular topics, or facilitate regular educational groups.

Some agencies or organizations focus on helping a particular group of people. If you are interested in working with children, youth, or adolescents, you might find a position with Head Start, youth development programs such as the Boys and Girls Club, after-school programs (e.g., pregnant or parenting teens), child-care resource or referral agencies or early intervention services. Child-care resource/ referral agencies assist parents in finding child care, provide training for child-care workers, and serve as a general resource for parents and for child-care providers. Early intervention services focus on children with special needs. If you worked in this area, you might work directly with the children or you might work with the families and help to coordinate services for them.

Other agencies focus more on specific issues that confront adults or families as a whole. Domestic violence shelters, family crisis centers, and employee assistance programs are examples of employment opportunities. In many of these positions, you will function in multiple roles. For example, at a family crisis center, you might take calls on a crisis hotline, work one on one with clients to help them find resources and services, and offer classes on sexual assault or dating violence to high school students.

Another focus area in which jobs are available is aging. There are opportunities within residential facilities such as assisted living facilities or nursing homes, senior centers, organizations such as the Alzheimer's Association, or agencies such as Area Agencies on Aging. There is also a need for elder-care resource and referral as more and more families are finding that they have caregiving responsibilities for an aging family member. These families have a need for resources, support, and assistance in finding residential facilities and/or other services for their aging family member. Many of the available positions with these types of agencies are open to individuals with bachelor's degrees. However, if you get your master's degree in a program emphasizing the elderly, you might have increased opportunity and will be in a position to compete for various administrative positions.

FAMILY MEDIATION

In the chapter on divorce, we emphasized the value of divorce mediation. It is also known as family mediation and involves a neutral third party negotiating with divorcing spouses the issues of child custody, child support, spousal support, and division of property. The purpose of mediation is not to reconcile the partners but to help the couple to make decisions about children, money, and property as amicably as possible. A mediator does not make decisions for the couple, but supervises communication between the partners offering possible solutions.

While some family and divorce mediators are attorneys, family life professionals are becoming more common. Specific training is required that may include numerous workshops or a master's degree offered at some universities (e.g., University of Maryland). Most practitioners conduct mediation in conjunction with their role as a family life educator, marriage and family therapist, or other professional. In effect, you would be in business for yourself as a family or divorce mediator.

Students interested in any of the above career paths can profit from getting initial experience in working with people through volunteer or internship agencies. Most communities have crisis centers, mediation centers, and domestic abuse centers that permit students to work for them and gain experience. Not only can you provide a service, but you can assess your suitability for the "helping professions" as well as discover new interests. Talking with persons already in the profession you want to enter is also a good idea for new insights. Your teacher may already be in the marriage and family profession you would like to pursue or can refer you to someone who is.

Authors

SPECIAL TOPIC 2

My latest survey shows that people don't believe in surveys.

Laurence J. Peter

Evaluating Research in Marriage and the Family

Contents

"New Research Study" is a frequent headline in popular magazines (e.g., *Cosmopolitan*) promising accurate information about "hooking up," "what women want," "what men want," or other relationship/marriage/family issues. As you read such articles, as well as the research in such texts as this, be alert to their potential flaws. Following are specific issues to keep in mind when evaluating research in marriage and the family.

SAMPLE

Some of the research on marriage and the family is based on random samples. In a **random sample,** each individual in the population has an equal chance of being included in the sample. Random sampling involves selecting individuals at random from an identified population. Studies that use random samples are based on the assumption that the individuals studied are similar to and therefore representative of the population that the researcher is interested in. For example, suppose you want to know the percentage of unmarried seniors (US) on your campus who are living together. Although the most accurate way to get this information is to secure an anonymous yes or no response from every US, doing so is not practical. To save yourself time, you could ask a few USs to complete your questionnaire and assume that the rest of them would say yes or no in the same proportion as those who answered the questions. To decide who those few USs would be, you could put the name of every US on campus on a separate note card, stir these cards in your empty bathtub, put on a blindfold, and draw one hundred cards. Because each US would have an equal chance of having his or her card drawn from the tub, you would obtain a random sample. After administering the questionnaire to this sample and adding the yes and no answers, you would have a fairly accurate idea of the percentage of USs on your campus who are living together.

The term *random sample,* however, may not always mean "random." For example, in the preceding study of unmarried seniors, not all the names you put in the bathtub to select from would have addresses and phone numbers. Hence, even if you drew the person's name, finding her or him to complete a questionnaire could be difficult. In addition, some people refuse to complete a questionnaire.

Because of the trouble and expense of obtaining random samples, most researchers study subjects to whom they have convenient access. This often means students in the researchers' classes. The result is an overabundance of research on "convenience" samples consisting of white, Protestant, middle-class college students. Because college students cannot be assumed to be similar to their noncollege peers or older adults in their attitudes, feelings, and behaviors, research based on college students cannot be generalized beyond the base population. To provide a balance, this text included data that reflected people of different ages, marital statuses, racial backgrounds, lifestyles, religions, and social classes. When only data on college samples are presented, it is important not to generalize the findings too broadly.

In addition to having a random sample, it is important to have a large sample. The random study of first-semester undergraduates at over four hundred colleges and universities referred to in Chapter 1 represented a large national sample, which provides credible data on first-year college students. If only fifty college students had been in the sample, the results would have been very unreliable in terms of generalizing beyond that sample. Be alert to the sample size of the research you read. Most studies are based on small samples. Other researchers have emphasized the limitations of norms based on populations of college students only (Meyers & Shurts, 2002).

CONTROL GROUPS

Any study that concludes that divorce (or any independent variable) is associated with lower grades (or any dependent variable) for children of divorced parents must necessarily include two groups: (1) children whose parents are divorced and (2) children whose parents are still married. The latter would serve as a **control group**—the group not exposed to the independent variable you are studying. Hence, if you find that children in both groups make low grades, you know that divorce cannot be the potential cause. Be alert to the existence of a control group, which is usually *not* included in research studies. An exception is Wallerstein (2000), who included a control group of 44 children (parents divorced) and compared them with children in the experimental group of 131 (parents still together) in her study of divorce. However, although twenty-five years later she reinterviewed the children whose parents had divorced, she did not reinterview children in the control group, thus weakening her conclusions. Yet the Wallerstein study has achieved massive public visibility for suggesting that some children are permanently damaged by divorce. It is also accurate that some children of divorce are flourishing, and there are some children in homes with intact alcoholic or abusive marriages who are floundering.

AGE AND COHORT EFFECTS

In some research designs, different cohorts or age groups are observed and/or tested at one point in time. One problem that plagues such research is the difficulty—even impossibility—of discerning whether observed differences between the subjects studied are due to the research variable of interest, cohort differences, or some variable associated with the passage of time (e.g., biological aging). A good illustration of this problem is found in research on changes in marital satisfaction over the course of the family life cycle. In such studies, researchers may compare the level of marital happiness reported by couples who have been married for different lengths of time. For example, a researcher may compare the marital happiness of two groups of people—those who have been married for fifty years and those who have been married for five years. But differences between these two groups may be due to (1) differences in age (age effect), (2) the different historical time period that the two groups have lived through (cohort effect), or (3) being married different lengths of time (research variable). It is helpful to keep these issues in mind when you read studies on marital satisfaction over time.

TERMINOLOGY

In addition to being alert to potential shortcomings in sampling and control groups, you should consider how the phenomenon being researched is defined. For example, in a preceding illustration of unmarried seniors living together, how would you define *living together?* How many people, of what sex, spending what amount of time, in what place, engaging in what behaviors will constitute your definition? Indeed, researchers have used more than twenty definitions of what constitutes living together.

What about other terms? What is meant by *marital satisfaction, commitment, interpersonal violence,* and *sexual fulfillment?* Even the term *romantic behavior* has a number of referents. Quiles (2003) specified fifteen such behaviors, including kissing, making love, flowers, saying "I love you," hugging, candlelight dinner, slow dancing, cuddling, love cards or letters, holding hands, etc. Before reading too far in a research study, be alert to the definitions of the terms being used. Exactly what is the researcher trying to measure?

RESEARCHER BIAS

Although one of the goals of scientific studies is to gather data objectively, it may be impossible for researchers to be totally objective. McGraw et al. (2000) emphasized that marriage and family research is inherently political in content and method. Researchers are human and have values, attitudes, and beliefs that may influence their research methods and findings. It may be important to know what the researcher's bias is in order to evaluate that researcher's findings. For example, a researcher who does not support abortion rights may conduct research that focuses only on the negative effects of abortion. Occasionally, researchers are not just biased but outright deceptive (Pound, 2000).

In addition, some researchers present an interpretation of what other researchers have done. Two layers of bias may be operative here: (1) when the original data were collected and interpreted and (2) when the second researcher read the study of the original researcher and made his or her own interpretation. Much of this text was based on interpretations of other researchers' studies. As a consumer you should be alert to the potential bias in reading such secondary sources. To help control for this bias, we have provided references to the original sources for your own reading.

Even the particular topics selected by researchers can reflect a gender bias. For example, as a result of male bias in the scientific community, research on women's issues has not been a major focus. Thus research on male contraception is almost nonexistent, since male researchers focused on the "female pill."

With regard to sexual dysfunctions, women are often presented as dysfunctional because of their difficulty in achieving an orgasm, whereas in fact the lack of adequate stimulation provided by the male might be a more accurate emphasis. A more gendered approach to research in marriage and the family would include more qualitative studies based on women's experiences, recognition that gender is a socially created category, and commitment to design research with the aim of eliminating bias and improving the lives of women.

TIME LAG

Typically, a two-year lag exists between the time a study is completed and the study's appearance in a professional journal. Because textbooks take even longer to develop than getting an article printed in a professional journal, they do not always present the most cutting-edge research, especially on topics in flux. In addition, even though a study may have been published recently, the data on which the study was based may be old. For example, Pluhar and her colleagues (2003) published a study in 2003 on college student behavior regarding contraception and prevention of sexually transmitted diseases. But the data were collected in 1996, making them almost ten years old in the text (published in 2005) you are reading. Be aware that the research you read in this or any other text may not reflect the most current, cutting-edge research.

DISTORTION AND DECEPTION

Our society is no stranger to distortion and deception—corporate CEOs (e.g., Enron), Wall Street analysts, and celebrities (e.g., Martha Stewart) have been indicted for giving fake information. Similarly, writers at the prestigious *New York Times* have been fired when they were discovered fabricating articles. Distortion and deception, deliberate or not, also exist in marriage and family research. Marriage is a very private relationship that happens behind closed doors; individual respondents to questionnaires/interviews have been socialized not to reveal to strangers the intimate details of their lives. Hence, they are prone to distort, omit, or exaggerate information, perhaps unconsciously, to cover up what they may feel is no one else's business. Thus, the researcher sometimes obtains inaccurate information. Marriage and family researchers know more about what people say they do than about what they actually do.

An unintentional and probably more frequent form of distortion is inaccurate recall. Sometimes researchers ask respondents to recall details of their relationships that occurred years ago. Time tends to blur some memories, and respondents may not relate what actually happened but will relate only what they remember to have happened, or, worse, what they wish had happened.

OTHER RESEARCH PROBLEMS

Nonresponse on surveys and the discrepancy between attitudes and behaviors are other research problems. With regard to nonresponse, not all individuals who complete questionnaires or agree to participate in an interview are willing to provide information about such personal issues as date rape and partner abuse. Such individuals leave the questionnaire blank or tell the interviewer they would rather not respond. Others respond but give only socially desirable answers. The implications for research are that data gatherers do not know the nature or extent to which something may be a problem because people are reluctant to provide accurate information.

The discrepancy between the attitudes people have and their behavior is another cause for concern about the validity of research data. It is sometimes assumed that if a person has a certain attitude (for example, extramarital sex is wrong), then his or her behavior will be consistent with that attitude (avoid extramarital sex). However, this assumption is not always accurate. People do indeed say one thing and do another. This potential discrepancy should be kept in mind when reading research on various attitudes.

Finally, most research reflects information provided by volunteers. But volunteers may not represent nonvolunteers when they are completing surveys.

In view of the research cautions identified here, you might ask, "Why bother to report the findings?" The quality of some family science research is excellent. For example, articles published in *Journal of Marriage and the Family* (among other journals) reflect the high level of methodologically sound articles that are being published. Even less sophisticated journals provide useful information on marital, family, and other relationship data. Particularly when there are multiple replications of a study to assess consistency of results, there is increasing confidence in the research process (Riniolo & Schmidt, 2000). Table ST.2 summarizes potential inadequacies of any research study.

Table ST2.1 Potential Inadequacies of Research Studies

Weakness	Consequences	Example
Sample not random	Cannot generalize findings	Opinions of college students do not reflect opinions of other adults.
No control group	Inaccurate conclusions	Study on the effect of divorce on children needs control group of children whose parents are still together.
Age differences between groups of respondents	Inaccurate conclusions	Effect may be due to passage of time or to cohort differences.
Unclear terminology	Inability to measure what is not clearly defined	What is living together, marital happiness, sexual fulfillment, good communication, quality time?
Researcher bias	Slanted conclusions	Male researcher may assume that since men usually ejaculate each time they have intercourse, women should have an orgasm each time they have intercourse.
Time lag	Outdated conclusions	Often-quoted Kinsey sex research is over fifty years old.
Distortion	Invalid conclusions	Research subjects exaggerate, omit information, and/or recall facts or events inaccurately. Respondents may remember what they wish had happened.

Unmarried, childfree college students probably do not need life insurance. No one is dependent on them for economic support. However, the argument used by some insurance agents who sell campus policies is that college students should buy life insurance while they are young when the premiums are low and when insurability is guaranteed. Still, consumer advocates typically suggest that life insurance for unmarried, childfree college students is not necessary.

When considering income protection for dependents, there are two basic types of life insurance policies: (1) term insurance and (2) insurance plus investment. As the name implies, term insurance offers protection for a specific time period (usually one, five, ten, or twenty years). At the end of the time period, the protection stops. Although a term insurance policy offers the greatest amount of protection for the least cost, it does not build up cash value (money the insured would get upon surrendering the policy for cash).

Insurance-plus-investment policies are sold under various names. The first is straight life, ordinary life, or whole life, in which the individual pays a stated premium (based on age and health) as long as the individual lives. When the insured dies, the beneficiary is paid the face value of the policy (the amount of insurance originally purchased). During the life of the insured, the policy also builds up a cash value (which is tax free), which permits the insured to borrow money from the insurance company at a low rate of interest. A second type of life insurance is a limited payment policy, in which the premiums are paid up after a certain number of years (usually twenty) or when the insured reaches a certain age (usually 60 or 65). As with straight life, ordinary life, or whole life policies, limited payment policies build up a cash value, and the face value of the policy is not paid until the insured dies. The third type of life insurance is endowment insurance, in which the premiums are paid up after a stated number of years and can be cashed in at a stated age.

Regardless of how they are sold, insurance-plus-investment policies divide the premium paid by the insured. Part pays for the actual life insurance, and part is invested for the insured, giving the policy a cash value. Unlike term insurance, insurance-plus-investment policies are not canceled at age 65. Which type of policy, term or insurance-plus-investment, should you buy? An insurance agent is likely to suggest the latter and point out the advantages of cash value, continued protection beyond age 65, and level premiums. But the agent has a personal incentive for your buying an insurance-plus-investment policy: the commission on this type of policy is much higher than it is on a term insurance policy.

A strong argument can be made for buying term insurance and investing the additional money that would be needed to pay for the more expensive insurance-plus-investment policy.

The annual premium for $50,000 worth of renewable term insurance at age 25 is about $175. The same coverage offered in an ordinary life policy—the most common insurance-plus-investment policy—costs $668 annually, so the difference is $493 per year. If you invested this money and were able to increase its value by 5 percent, at the end of five years you would have $2,860.32. In contrast, the cash value of an ordinary life policy after five years would be $2,350. But to get this money, you have to pay the insurance company interest to borrow it. If you don't want to pay the interest, the company will give you this amount but cancel your policy. In effect, you lose your insurance protection if you receive the cash value of your policy. With term insurance, you have the $2,860.32 for your trouble and you can use the money whenever you want. And you can do so without affecting your insurance program.

It should be clear that for term insurance to be cheaper, you must invest the money you would otherwise be paying for an ordinary life insurance policy. If you can't discipline yourself to save (and if your investments turn out to be unlucky ones), you might wish you had bought an insurance-plus-investment policy to ensure savings.

Finally, what about the fact that term insurance stops when you are 65, just as you are moving closer to death and needing the protection more? Again, by investing the money that you would otherwise have spent on an insurance-plus-investment policy, you will have as much money for your beneficiary as your insurance-plus-investment policy would earn, or even more.

Whether you buy a term policy, an insurance-plus-investment policy, or both, there are three options to consider: guaranteed insurability, waiver of premiums, and double or triple indemnity. All are inexpensive and generally should be included in a life insurance policy.

Guaranteed insurability means that the company will sell you more insurance in the future regardless of your medical condition. For example, suppose you develop cancer after you buy a policy for $10,000. If the guaranteed insurability provision is in your contract, you can buy additional insurance. If not, the company can refuse you more insurance.

Waiver of premiums provides that your premiums will be paid by the company if you become disabled for six months or longer and are unable to earn an income. Such an option ensures that your policy will stay in force because the premiums will be paid. Otherwise, the company will cancel your policy.

Double or triple indemnity (rent the video of the classic movie *Double Indemnity* to see how this works) means that if you die as the result of an accident, the company will pay your beneficiary twice or three times the face value of your policy.

An additional item you might consider adding to your life insurance policy is a disability income rider. If the wage earner becomes disabled and cannot work, the financial consequences for the family are the same as though the wage earner were dead. With disability insurance, the wage earner can continue to provide for the family up to a maximum of $3,500 per month or two-thirds of the individual's salary, whichever is smaller. If the wage earner is disabled by accidental injury, payments are made for life. If illness is the cause, payments may be made only to age 65. A 27-year-old spouse and parent who was paralyzed in an automobile accident said, "It was the biggest mistake of my life to think I needed only life insurance to protect my family. Disability insurance turned out to be more important."

In deciding to buy life insurance, it might be helpful to consider the following:

1. Decide whether you really need insurance. People with no dependents rarely need it.

2. Decide how much coverage you need. The cost of rearing a middle-class child from conception through college (public) is approximately $300,000.

3. Compare prices. Not all policies and prices are the same. In some cases, the higher premiums are for lower coverage.

4. Select your agent carefully. Only one in ten life insurance agents stays in the business. The person you buy life insurance from today may be in the real estate business tomorrow. Choose an agent who has been selling life insurance for at least ten years.

5. Select your company carefully. The big names in insurance such as John Hancock, Prudential, and State Farm may not always offer you the best protection at the most affordable prices.

6. Seek group rates. Group life insurance is the least expensive coverage. See whether your employer offers a group plan.

7. Proceed slowly. Don't rush into buying an insurance policy. Consult several agents, read *Consumer Reports,* and talk with friends to find out what they are doing about their insurance needs.

8. Select features carefully. Don't buy features you don't need. A waiver of premium, for which you are assessed a percentage of the yearly cost of the policy as a surcharge, continues to pay your premiums if you become permanently disabled. But you may be better off purchasing more disability insurance.

Although this section emphasizes life insurance, health coverage is equally important. With hospital costs of over $1,000 a day, life savings can be eliminated within a short time if you suffer a serious injury or illness. The least expensive health policies are group policies available through some employers. If such policies are not available, major medical policies with high deductibles are relatively inexpensive and will protect the insured in case of major illness.

CREDIT

College students might be mindful that their economic future can be wrecked by using credit unwisely. The "free" credit cards they receive in the mail are a Trojan horse and can plunge them into massive debt from which it will take years to recover.

Types of Credit Accounts

You use credit when you take an item (CD, beer, clothes, etc.) home today and pay for it later. The amount you pay later will depend on the arrangement you make with the seller. Suppose you want to buy a big-screen color television set that costs $600. Unless you pay cash, the seller will set up one of three types of credit accounts with you: installment, revolving charge, or open charge.

Under the installment plan, you make a down payment and sign a contract to pay the rest of the money in monthly installments. You and the seller negotiate the period of time over which the payments will be spread and the amount you will pay each month. The seller adds a finance charge to the cash price of the television set and remains the legal owner of the set until you have made your last payment. Most department stores, appliance and furniture stores, and automobile dealers offer installment credit. The cost of buying the $600 big-screen color TV is calculated in Table ST3.2.

Table ST3.2 Calculating the Cost of Installment Credit

Amount to be financed	
Cash price	$600.00
− down payment (if any)	−50.00
Amount to be financed	$550.00
Amount to be paid	
Monthly payments	$ 35.00
× number of payments	×18
Total amount repaid	$630.00
Cost of credit	
Total amount repaid	$630.00
− amount financed	−550.00
Cost of credit	$ 80.00
Total cost of TV	
Total amount repaid	$630.00
+ down payment (if any)	+50.00
Total cost of TV	$680.00

Instead of buying your $600 television set on the installment plan, you might want to buy it on the revolving charge plan. Most credit cards, such as Visa and MasterCard, represent revolving charge accounts that permit you to buy on credit up to a stated amount during each month. At the end of the month, you may pay the total amount you owe, any amount over the stated minimum payment due, or the minimum payment. If you choose to pay less than the full amount, the cost of the credit on the unpaid amount is approximately 1.5 percent per month, or 18 percent per year. For instance, if you pay $100 per month for your television for six months, you will still owe $31.62 to be paid the next month, for a total cost (television plus finance charges) of $631.62. It is estimated that 70 percent of college students in four-year universities have at least one credit card (Joo, Grable, & Blackwell, 2003). In a sample of 242 undergraduates, about 10 percent paid only the minimum monthly balance each month. Almost half paid the balance on their cards in full at the end of each month. The average balance for those who held a credit card was $890.42; of those who paid only the monthly minimum, the average balance was $1,769.85 (Joo, Grable, & Blackwell, 2003).

It is important to protect your credit card numbers. ID or identity theft is the number one concern of consumers today (10 million annually report having been victimized): someone gains access to your Social Security number and credit card numbers and presents themselves to others as though they are you. Once they convince someone else that they are you, they can buy things on credit in your name and never pay. They get the goods and you get stuck with the bad credit. It can take months and years to clear your name and credit rating.

You can also purchase items on an open charge (thirty-day) account. Under this system you agree to pay in full within thirty days. Since there is no direct service charge or interest for this type of account, the television set would cost only the purchase price. For example, Sears and J.C. Penney offer open charge (thirty-day) accounts. If you do not pay the full amount in thirty days, a finance charge is placed on the remaining balance. The use of both revolving charge and open charge accounts is wise if you pay off the bill before finance charges begin. In deciding which type of credit account to use, remember that credit usually costs money; the longer you take to pay for an item, the more the item will cost you.

Three Cs of Credit

Whether you can get credit will depend on the rating you receive on the three Cs: character, capacity, and capital. Character refers to your honesty, sense of responsibility, soundness of judgment, and trustworthiness. Capacity refers to your ability to pay the bill when it is due. Such issues as the amount of money you earn and the length of time you have held a job will be considered in evaluating your capacity to pay. Capital refers to such assets as bank accounts, stocks, bonds, money market funds, and real estate.

It is particularly important that married individuals establish credit ratings in their own name in case they become widowed or divorced. Otherwise, their credit will depend on their spouse; if one spouse dies, the other spouse will have no credit of his or her own. Similarly, if a couple divorce, each ex-spouse will want to have established his or her own credit during the marriage.

It is also important that individuals not depend on credit to pay for necessities such as food, rent, and utilities. Continually spending more than one's income and taking all credit cards to the limit can lead to financial trouble and eventual bankruptcy.

IDENTITY THEFT

When someone poses as you and uses your credit history to buy goods and services, they have stolen your identity. Over ten million Americans are victims of **identity theft,** which can destroy your credit, plunge you into debt, and keep you awake at night with lawsuits from creditors. It happens when someone gets access to personal information such as your social security number, credit card number, or bank account number and goes online to pose as you to buy items or services.

Safeguards include never giving the above information over the phone or online unless you initiate the contact and shredding (a shredder can cost as little as $20) bank /credit card statements and preapproved credit card offers. Also, don't pay your bills by putting an envelope in your mailbox with the flag up—use a locked box or the post office. Finally, check your credit reports, scrutinize your bank statements, and guard your PIN number at ATMs. If you use the Internet, protect your safety by installing firewall software.

BUYING A HOME

This text has been about choices. In reference to housing, the number of choices are seemingly endless. Such choices begin with what you can afford.

Figuring Out How Much You Can Afford

A prerequisite for owning a house or condominium is being able to afford one. The median price of a single-family home in 2003 was $169,900. In San Francisco it was $558,100. A financial officer in a bank or a real estate agent can determine the value of the dwelling you can afford. Such a determination is based on your net worth—what you own minus what you owe, how much you can afford as a down payment, and how much you can afford in monthly payments. In general, the more you pay down, the lower your monthly payments. Some loans involve paying only 5 percent of the value of the dwelling down, and others may require 10 percent or 20 percent.

Deciding What Type of Housing You Want

If you are buying a house, you must decide whether to buy a house that has already been built, buy one that is to be built identical to a model already available, or build your own custom-made dream house.

Many would-be homeowners believe that custom houses cost more than those already built or those based on a model. This is not necessarily so since custom houses avoid real estate agent fees. But be careful. The investment in a custom-built home is in terms of your own time (and frustration).

An alternative to buying a single-family house (old, new, custom-built) is to buy a condominium, a cooperative apartment, or a mobile home. A condominium involves your owning the space in which you live and sharing the ownership of common areas such as sidewalks, grounds, parking lot, and elevators. A cooperative apartment means that you own a share of the corporation that owns the apartments and your privilege to live there is based on your ownership. Mobile homes, some of which are referred to as double-wides, are rarely mobile, since 90 percent stay on their original site. Some individuals buy land on which they put their mobile home. Others lease the land on which they place their home. Such an arrangement makes the homeowner vulnerable, since the owner of the land can force you to move your home.

The type of housing that individuals select is based on what they can afford and what they need. The following are some of the factors to be considered:

Location—neighborhood and proximity to work, school, grocery stores, etc.

Space—number of bedrooms, baths, basement, workshop areas, garage, etc.

Floor plan—one or two levels, separation of activities, kitchen placement, etc.

Other—need for fenced-in backyard, land for garden, etc.

The lot on which the house sits is particularly important for persons who enjoy gardening, who have small children, or who have a pet. The best house on an undesirable lot is not a house to buy. Other considerations are how much privacy you want and whether you will maintain the grounds yourself. Finally, identify your personal objections, such as small bathrooms, fake brick siding, heavy traffic, no trees, or too little sunlight.

Deciding on a House

Having identified what you can afford and what you want, look at prospective houses advertised in the newspaper or recommended by a real estate agent or a buyer's broker. The latter is a person you hire at a fixed price to find the house you want at the least cost. The typical real estate agent gets a 6 percent commission on the house you buy. Hence, the more expensive the house you buy, the more money the agent makes. You can find the name of a buyer's broker from your local Board of Realtors. About 10 percent of home buyers hire a buyer's broker.

As you begin to look at houses, look at a lot of them so that you become aware of the range of possibilities. Once you have identified the house that you want, consider having it inspected by a person listed in the yellow pages under "Building Inspection Services" or "Real Estate Services." Such a person is knowledgeable about houses and will make a written report to you on issues such as water drainage, exterior walls, roofing, basement, electrical system, plumbing system, heating/cooling systems, and kitchen and bathrooms in terms of ventilation and flooring.

After the house passes inspection, have the house appraised. This involves hiring a professional residential appraiser who is not invested in selling you the house to tell you what the house is actually worth. This may prevent you from paying too much or make you aware that you have found a good deal. Once you have decided what you are willing to pay, make an offer in writing to the seller and identify all conditions, such as gutters to be fixed by seller within a certain period of time, a positive termite inspection report, etc. Many asking prices have a good bit of padding built into them, to see if anyone will take the bait. "Don't feel you have to offer the full asking price, or even something close to it, just because that's what the owner is seeking," noted Knight Kiplinger, financial adviser.

Finding the Right Mortgage

Once you have signed a contract for the house, find the best deal on a mortgage—the amount of money you will pay a lender for the house. Shop around—you are looking for a loan, not a lender. A difference of .5 percent on a thirty-year mortgage can mean thousands you save or pay unnecessarily. Paying 5.5 instead of 6.0 percent on a thirty-year, fixed-rate $100,000 loan will save you thousands of dollars. When comparing rates, ask for the true annual percentage rate (APR).

Also decide whether you want a fixed rate or an adjustable rate. The latter means that your monthly payment will go up or down depending on how the economy is doing. Conventional wisdom would have buyers lock in low interest rates with a fixed-rate loan when it appears that interest rates will be heading up over the next few years. On the other hand, if it looks as if rates will go down or stay about the same, an adjustable-rate mortgage might be better.

Regardless of the rate you decide on, be aware of the enormous price you are paying for the loan. The first payment on a $100,000 loan at 7 percent (in 2005 you should be able to get a loan for much less than 7 percent) for thirty years is $665.25. Only $50.00 of this amount is applied to the principal (the amount that actually goes toward your owning the house). The remainder ($615.25) is interest. Over the thirty-year period, you will pay $239,490 for the $100,000 loan. To shorten the number of payments, and therefore the total amount you must pay, consider making extra payments each month directed specifically toward the principal. The sooner the principal is paid off, the sooner all payments will stop.

WEBLINKS

Ms.Money.com (budgeting documents)
http://www.msmoney.com

Countrywide Financial
http://www.countrywide.com

FinanCenter
http://www.financenter.com

HomeGain
http://www.homegain.com

MBA Online (Mortage Bankers Association)
http://www.mbaa.org

Quotesmith (Insure.com)
http://www.quotesmith.com

© Tim Thompson/CORBIS

SPECIAL TOPIC 4

HIV and Sexually Transmitted Diseases

Contents

HIV refers to the human immunodeficiency virus that attacks the immune system and can lead to **AIDS**—acquired immunodeficiency syndrome. **STD** (sometimes known as STI, for sexually transmitted infection) refers to a more general category of sexually transmitted diseases. The focus of this Special Topic section on STDs emphasizes that sexual encounters necessitate making choices about the level of risk one is willing to take in reference to becoming infected. The wrong choice might lead to an early death.

Media information about the value of abstinence, safer sex, and the regular use of latex and polyurethane condoms may not translate into consistent condom use. Denial expressed as "It won't happen to me" is the primary reason for failure to use condoms.

HUMAN IMMUNODEFICIENCY VIRUS (HIV) INFECTION

HIV attacks the white blood cells (T-lymphocytes) in human blood, impairing the immune system and a person's ability to fight other diseases. Of all the diseases that can be transmitted sexually, HIV infection is the most life-threatening. Symptoms, if they occur at all, surface between

Appreciation is expressed to Beth C. Burt, M.A.Ed., CHES, a health education specialist, for updating Special Topic 4. She is Program Consultant for the HIV/STD Branch of the North Carolina Department of Health and Human Services.

two and six weeks after infection and are often dismissed because they are similar to symptoms of influenza. Antibodies may appear in the blood in two months but more often take three to six months before they reach reliable detectable levels.

International Data

An estimated 42 million are now infected worldwide with HIV. About half of the people living with HIV are women (Stephenson, 2003).

Before the HIV virus or its antibodies are detectable, infected individuals will test negative for HIV, making them silent carriers of the virus (this period is called the "window period"). Although not all persons who have HIV get AIDS (half of those infected with HIV will develop AIDS within ten years after the infection), all are infectious and are able to transmit the virus to others. Hence, even though your partner tested negative for HIV, she or he could still transmit the virus to you. If HIV progresses to AIDS, the person's body is vulnerable to opportunistic diseases that would be resisted if the immune system were not damaged. The two most common diseases associated with AIDS are a form of cancer called Kaposi's sarcoma (KS) and Pneumocystis carinii pneumonia (PCP), a rare form of pneumonia. Seventy percent of all HIV deaths result from PCP. HIV can also invade the brain and nervous system, producing symptoms of neurological impairment and psychiatric illness.

Transmission of HIV and High-Risk Behaviors

The human immunodeficiency virus can be transmitted in several ways:

1. Sexual contact. HIV is found in several body fluids of infected individuals, including blood, semen, and vaginal secretions. During sexual contact with an infected individual, the virus enters a person's bloodstream through the rectum, vagina, penis (an uncircumcised penis is at greater risk because of the greater retention of the partner's fluids), and possibly the mouth during oral sex. Saliva, sweat, and tears are not body fluids through which HIV is transmitted.

2. Intravenous drug use. Drug users who are infected with HIV can transmit the virus to other drug users with whom they share needles, syringes, and other drug-related implements.

3. Blood transfusions. HIV can be transmitted through receiving HIV-infected blood or blood products. Currently, all blood donors are screened, and blood is not accepted from high-risk individuals. Blood that is accepted from donors is tested for the presence of HIV. However, prior to 1985, donor blood was not tested for HIV. Individuals who received blood or blood products prior to 1985 may have been infected with HIV.

4. Mother-child transmission of HIV. A pregnant woman infected with HIV has a 40 percent chance of transmitting the virus through the placenta to her unborn child. These babies will initially test positive for HIV as a consequence of having the antibodies from their mother's bloodstream. However, AZT taken by the mother twelve weeks before birth seems to reduce by two-thirds the transmission of HIV by the mother. Although rare, HIV may also be transmitted from mother to infant through breast-feeding.

5. Organ or tissue transplants and donor semen. Receiving transplant organs and tissues, as well as receiving semen for artificial insemination, could involve risk of contracting HIV if the donors have not been HIV-tested. Such testing is essential, and recipients should insist on knowing the HIV status of the organ, tissue, or semen donor.

6. Other methods of transmission. For health care professionals, HIV can also be transmitted through contact with amniotic fluid surrounding a fetus, synovial fluid surrounding bone joints, and cerebrospinal fluid surrounding the brain and spinal cord.

Sexual Orientation and HIV Infection

In the United States, HIV infection was first seen among homosexual and bisexual men having multiple sex partners. Homosexual transmission is the predominant mode of HIV infection among U.S. males. The predominant mode of transmission of HIV for women is heterosexual contact (Stephenson, 2003).

Prevalence of HIV/AIDS

There are an estimated 886,575 people diagnosed with AIDS in the United States. About 80 percent of these are male, 20 percent female. These figures reflect data collected through 2002 (HIV/AIDS Surveillance Report, 2003). *Risk group* is a term that implies that a certain demographic trait determines who has a higher chance of becoming infected with HIV. However, anyone who is exposed can become infected.

Tests for HIV Infection

Early medical detection of HIV has decided benefits, including taking medications to reduce the growth of HIV and preventing the development of some life-threatening conditions. An example of the latter is pneumonia, which is more likely to develop when one's immune system has weakened.

HIV counselors recommend that individuals who answer yes to any of the following questions should definitely seek testing:

If you are a man, have you had sex with other men?

Have you had sex with someone you know or suspect was infected with HIV?

Have you had an STD?

Have you shared needles or syringes to inject drugs or steroids?

Did you receive a blood transfusion or blood products between 1978 and 1985?

Have you had sex with someone who would answer yes to any of these questions?

Additionally, counselors suggest that if you have had sex with someone whose sexual history you do not know or if you have had numerous sexual partners, your risk of HIV infection is increased, and you should seriously consider being tested. Finally, if you plan to become pregnant or are pregnant, the American Medical Association recommends being tested.

The newest method of HIV testing provides the fastest accurate results. In November 2002, Orasure Technologies, Inc. received approval from the U. S. Food and Drug Adminstration for its OraQuick® Rapid HIV-1 Antibody Test. OraQuick is the first rapid, point-of-care test designed to detect antibodies to HIV-1 within approximately twenty minutes. Other forms of testing may take up to two weeks for results. This test requires taking a blood sample through a finger stick.

Another collection method that was approved in 1994 by the Food and Drug Administration that is achieving growing popularity is the OraSure® Oral Specimen Collection Device. This device allows for HIV testing with twenty-minute results, using a simple, two-minute collection procedure that can be performed by trained health care providers. Both methods are 99 percent effective for detecting HIV antibodies when present.

Home HIV testing is available by calling 1-800-HIV-TEST (800-448-8378). The cost is $49. You mail in a sample of blood and call seven days later to get anonymous test results.

Treatment for HIV and Opportunistic Diseases

While research to find a vaccine for HIV continues (Sternberg, 2003), several drugs (AZT, 3TC, Indinavir, Ritonavir, and Saquinavir) used in various combinations have demonstrated the most efficacy in the treatment of HIV. We emphasize "the most efficacy" since some of the drugs have proven to be less potent and more toxic than previously thought. Patients are not cured, but the progress of the disease is slowed and the survival rate increased. About twenty-five other drugs are used to treat AIDS-related illnesses. Drug therapy for AIDS and associated illnesses is expensive. The cost for just the drugs used by one AIDS patient in one month is over $1,000.

Not all persons who become HIV-infected progress to AIDS. Indeed, about 3 percent are referred to as long-term nonprogressors—people who have not suffered any apparent damage to their immune system in twenty years.

Treatment for HIV and AIDS is not just medical. Human interaction networks, including spouses, parents, and siblings, are affected by the person who is HIV-positive. Getting the secret out and establishing a supportive network are essential in managing the psychological trauma of being diagnosed with HIV. The HBO miniseries *Angels in America* dealt with various HIV issues.

The following sections consider other sexually transmissible diseases. The relationship of HIV infection to other STDs is being given increased attention by researchers. HIV infection leads to altered manifestations of other STDs and thereby probably promotes their spread. Genital and some herpes ulcers normally heal within one to three weeks, but they can persist for months as highly infectious ulcers in persons with HIV infection.

OTHER SEXUALLY TRANSMISSIBLE DISEASES

There are numerous other sexually transmitted diseases. Young women, formerly married women, persons with little education who are smokers, persons with a high number of sexual partners, and persons who do not use condoms regularly are more likely to be among those who contract HIV and other STDs (Lane & Althaus, 2002; Capaldi et al., 2003). Some of the more common ones include HPV, chlamydia, genital herpes, gonorrhea, and syphilis.

Human Papilloma Virus (HPV)

Human papilloma virus (HPV) is the most common STD. The more sexual partners an individual has, the more likely the person is to contract HPV (Sellors et al., 2003).

There are more than seventy types of HPV. More than a dozen of these types can cause warts (called **genital warts,** or *condyloma*) or more subtle signs of infection in the genital tract. The virus infects the skin's top layers and can remain inactive for months or years before any obvious signs of infection appear. Often warts appear within three to six months after infection. However, some types of HPV produce no visible warts. Fewer than 1–2 percent of people who are infected with HPV develop symptoms. Any sexual partners of an infected individual should have a prompt medical examination.

HPV can be transmitted through vaginal or rectal intercourse, through fellatio and cunnilingus, and through other skin-to-skin contact. Genital warts are small bumps that are usually symptomfree, but they may itch. In women, genital warts most commonly develop on the vulva, in the vagina, or on the cervix. They can also appear on or near the anus. In men, the warts appear most often on the penis but can appear on the scrotum or anus or within the rectum. Incidence of infection radically increases as the number of sexual partners increases.

Health care providers disagree regarding the efficacy of treating HPV when there are no detectable warts. However, when the warts can be seen, either by visual inspection or by colposcope, providers do typically advise treatment. A number of treatment options are available. Choosing among them depends upon the number of warts and their location, availability of equipment, training of health care providers, and the preferences of the patient. Most of the treatments are at least moderately effective, but many are quite expensive. Treatments range from topical application of chemicals to laser surgery. Treatment of warts destroys infected cells, but not all of them, as HPV is present in a wider area of skin than just the precise wart location. To date, no therapy has proved effective in eradicating HPV, and relapse is common.

A danger for women exposed to certain strains of HPV is a higher risk for cervical cancer. The majority of cervical cancers (80%) are caused by just four types of HPV. Women who are diagnosed with HPV should carefully follow recommendations for cervical cancer screening and have Pap smears as directed by their health care providers.

Chlamydia

Chlamydia trachomatis (CT) is a bacterium that can infect the genitals, eyes, and lungs. **Chlamydia** (clah-MID-ee-uh) is the most frequently occurring bacterial STD on college campuses. Indeed, it is estimated that 3 million new cases occur annually, and chlamydia is the most frequently reported infectious disease in the United States. Worldwide, chlamydial infections are even more extensive. Trachoma inclusion conjunctivitis, a chlamydial infection that occurs rarely in the United States, is the leading cause of blindness in Third World countries. Untreated, chlamydia can lead in rare cases to pelvic inflammatory disease, sterility, spontaneous abortion, and premature birth.

CT is easily transmitted directly from person to person via sexual contact or by sharing sex toys. The microorganisms are most often found in the urethra of the man, the cervix, uterus, and fallopian tubes of the woman, and in the rectum of either men or women.

Genital-to-eye transmission of the bacteria can also occur. If a person with a genital CT infection rubs his or her eye or the eye of a partner after touching infected genitals, the bacteria can be transferred to the eye, and vice versa. Finally, infants can get CT as they pass through the cervix of their infected mother during delivery. CT rarely shows obvious symptoms, which accounts for its being known as "the silent disease." Both women and men who are infected with CT usually do not know that they have the disease. The result is that they infect new partners unknowingly, who affect others unknowingly—unendingly.

Genital Herpes

Herpes refers to more than fifty viruses related by size, shape, internal composition, and structure. One such herpes is **genital herpes.** Whereas the disease has been known for at least two thousand years, media attention to genital herpes is relatively new. Also known as **herpes simplex virus type 2** (HSV-2), genital herpes is a viral infection that is almost always transmitted through sexual contact. Symptoms occur in the form of a cluster of small, painful blisters or sores at the point of infection, most often on the penis or around the anus in men. In women, the blisters usually appear around the vagina but can also develop inside the vagina, on the cervix, and sometimes on the anus. Pregnant women can transmit the herpes virus to their newborn infants, causing brain damage or death.

Another type of herpes originates in the mouth. **Herpes simplex virus type 1** (HSV-1) is a biologically different virus with which people are more familiar as cold sores on the lips. These sores can be transferred to the genitals by the fingers or by oral-genital contact. In the past, genital and oral herpes had site specificity: HSV-1 was always found on the lips or in the mouth, and HSV-2 was always found on the genitals. But because of the increase in fellatio and cunnilingus, HSV-1 herpes can be found in the genitals and HSV-2 can be found on the lips.

Herpes symptoms range from no symptoms at all to painful ulcers or blisters. Other symptoms that may occur include discharge, itching or burning sensation during urination, back pain, leg pain, stiff neck, sore throat, headache, fever, aches, swollen glands, fatigue, and heightened sensitivity of the eyes to light. The outbreaks with herpes last about ten to fourteen days on average, although they can last for as long as six weeks if not treated. "I've got herpes," said one sufferer, "and it's a very uneven discomfort. Some days I'm okay, but other days I'm miserable."

As with syphilis, the sores associated with genital herpes subside (the sores dry up, scab over, and disappear), and the person feels good again. But the virus settles in the nerve cells in the spinal column and may cause repeated outbreaks of the symptoms in about one-third of those infected.

Stress, menstruation, sunburn, fatigue, and the presence of other infections seem to be related to the reappearance of herpes symptoms. Although such recurrences are usually milder and of shorter duration than the initial outbreak, the resurfacing of the symptoms can occur throughout the person's life. "It's not knowing when the thing is going to come back that's the bad part about herpes," said one woman.

The herpes virus is usually contagious during the time that a person has visible sores but not when the skin is healed. However, infected people may have a mild recurrence yet be unaware that they are contagious. Aside from visible sores, itching, burning, or tingling sensations at the sore site also suggest that the person is contagious. Using a latex or polyurethane condom reduces the risk of transmitting or acquiring herpes, since the virus doesn't permeate the condom. However, if the condom doesn't cover the site of the sore, the virus may be spread through skin-to-skin contact.

At the time of this writing, there is no cure for herpes. Because it is a virus, herpes does not respond to antibiotics as do syphilis and gonorrhea. A few procedures that help to relieve the symptoms and promote healing of the sores include seeing a physician to look for and treat any other genital infections near the herpes sores, keeping the sores clean and dry, taking hot sitz baths three times a day, and wearing loose-fitting cotton underwear to enhance air circulation. Proper nutrition, adequate sleep and exercise, and avoiding physical or mental stress help people to cope better with recurrences.

Acyclovir, marketed as Zovirax, is an ointment that can be applied directly on the sores that helps to relieve pain, speed healing, and reduce the amount of time that live viruses are present in the sores. A more effective tablet form of acyclovir, which significantly reduces the rate of recurring episodes of genital herpes, is also available. Once acyclovir is stopped, the herpetic recurrences resume. Acyclovir seems to make the symptoms of first-episode genital herpes more manageable, but it is less effective during subsequent outbreaks. ImmuVir—an alternative to acyclovir—is primarily for use by persons who have frequent outbreaks of genital herpes (once a month or more). This ointment is designed to reduce pain, healing time, and number of outbreaks. The drug has no known side effects.

Coping with the psychological and emotional aspects of having genital herpes is often more difficult than coping with the physical aspects of the disease.

Gonorrhea

Also known as the clap, the whites, morning drop, and the drip, **gonorrhea** is a bacterial infection that is sexually transmissible. Individuals contract gonorrhea through having sexual contact with someone who is carrying Neisseria gonorrhea bacteria. The gonorrhea bacteria, and most other STDs, cannot live long outside the human body—outside mucous membranes—so they could not survive on a toilet seat unless fluid were present, and even then they would not survive long. These bacteria thrive in warm, moist cavities, including the urinary tract, cervix, rectum, mouth, and throat. A pregnant woman can transmit gonorrhea to her infant at birth, causing eye infection. Many medical experts recommend gonorrhea testing for all pregnant women and antibiotic eyedrops for all newborns.

Although some infected men show no signs, 80 percent exhibit symptoms between three and eight days after exposure. They may begin to discharge a thick, yellowish pus from the penis and to feel pain or discomfort during urination. They may also have swollen lymph glands in the groin. Women are more likely to show no signs (70 percent to 80 percent have no symptoms) of the infection, but when they do, the symptoms are sometimes a yellowish discharge from the vagina along with a burning sensation or spotting between periods or after sexual intercourse. More often, a woman becomes aware of gonorrhea only after she feels extreme discomfort, which results when the untreated infection travels up into her uterus and fallopian tubes, causing pelvic inflammatory disease (PID). Salpingitis (inflammation of the fallopian tube) occurs in 10 percent to 20 percent of infected women and can cause infertility or ectopic pregnancy.

Undetected and untreated, gonorrhea can do permanent damage. Not only can the infected person pass the disease on to the next partner, but other undesirable consequences can result. Untreated gonorrhea commonly causes long-term reproductive system complications, such as blocking and/or scarring of the urethra and possible infertility in men and pelvic inflammatory disease (infection of the fallopian tubes), infertility, and damage to the uterus, fallopian tubes, and ovaries in women. In rare cases, the untreated bacteria can affect the brain, heart valves, and joints. Both men and women could develop endocarditis (an infection of the heart valves) or meningitis (inflammation of the tissues surrounding the brain and spinal column), arthritis, and sterility. Infected pregnant women could have a spontaneous abortion or a premature or stillborn infant.

A physician can detect gonorrhea by analyzing penile or cervical discharge under a microscope. A major problem with new cases of gonorrhea is the emergence of new strains of the bacteria that are resistant to penicillin. Because of high rates of resistance to penicillin and tetracycline, the current recommended treatment for gonorrhea is a single shot of ceftriaxone or a single dose of such oral medications as oflotaxin, cefixime, and ciprofloxacin.

Syphilis

Syphilis is caused by bacteria that can be transmitted through sexual contact with an infected individual. Syphilis can also be transmitted by an infected pregnant woman to her unborn baby. Although syphilis is less prevalent than gonorrhea, its effects are more devastating and include mental illness, blindness, heart disease—even death. The spirochete bacteria enter the body through mucous membranes that line various body openings. With your tongue, feel the inside of your cheek. This is a layer of mucous membrane—the substance in which spirochetes thrive. Similar membranes are in the vagina and urethra of the penis. If you kiss or have genital contact with someone harboring these bacteria, the bacteria can be absorbed into your mucous membranes and cause syphilitic infection. Syphilis progresses through at least three stages, plus a latency stage before the final stage.

In stage one (primary-stage syphilis), a small sore, or chancre, will appear at the site of the infection between ten and ninety days after exposure. The chancre, which can show up anywhere on the man's penis, in the labia, vaginal membranes, or cervix of the woman, or in either partner's mouth or rectum, neither hurts nor itches and, if left untreated, will disappear in three to five weeks. The disappearance leads infected people to believe that they are cured—one of the tricky aspects of syphilis. In reality, the disease is still present and doing great harm, even though there are no visible signs.

During the second stage (secondary-stage syphilis), beginning from two to twelve weeks after the chancre has disappeared, other signs of syphilis appear in the form of a rash all over the body or just on the hands or feet. Welts and sores can also occur, as well as fever, headaches, sore throat, and hair loss. Syphilis has been called the great imitator because it mimics so many other diseases (for example, infectious mononucleosis, cancer, and psoriasis). Whatever the symptoms, they, too, will disappear without treatment. The person may again be tricked into believing that nothing is wrong.

Following the secondary stage is the latency stage, during which there are no symptoms and the person is not infectious. However, the spirochetes are still in the body and can attack any organ at any time.

Tertiary syphilis—the third stage—can cause serious disability or even death. Heart disease, blindness, brain damage, loss of bowel and bladder control, difficulty in walking, and erectile dysfunction can result.

Early detection and treatment are essential. Blood tests and examination of material from the infected site can help to verify the existence of syphilis. But such tests are not always accurate. Blood tests reveal the presence of antibodies, not spirochetes, and it sometimes takes three months before the body produces detectable antibodies. Sometimes there is no chancre anywhere on the person's body.

Treatment for syphilis is similar to that for gonorrhea. Penicillin or other antibiotics (for those allergic to penicillin) are effective. Infected persons treated in the early stages can be completely cured with no ill effects. If the syphilis has progressed into the later stages, any damage that has been done cannot be repaired.

GETTING HELP FOR HIV AND OTHER STDS

If you are engaging in unprotected sex or are having symptoms, you should get tested for HIV and other STDs. There are different tests for different STDs. Women must specifically ask for such tests from their gynecologist, since only a few doctors routinely perform them. If you do not know whom to call to get tested, call your local health department or the national STD hotline at 1-800-227-8922. You will not be asked to identify yourself but will be given the name and number of local STD clinics that offer confidential, free treatment. In addition, Duke University in Durham, North Carolina, has opened an AIDS clinic to treat

AIDS patients (919-684-2660). For the Centers for Disease Control National AIDS hotline, call 1-800-342-2437. Students can also obtain information and assistance about HIV and STDs from their student health care facility on campus.

PREVENTION OF HIV AND STD TRANSMISSION

There is no assurance that education can prevent infection. Pedlow and Carey (2003) reviewed twenty-three intervention programs and found that education alone is not sufficient.

The best way to avoid getting a sexually transmitted disease is to avoid sexual contact or to have contact only with partners who are not infected. This means restricting your sexual contacts to those who limit their relationships to one person. The person most likely to get a sexually transmitted disease has sexual relations with a number of partners or with a partner who has a variety of partners.

Even if you are in a mutually monogamous relationship, you may be at risk for acquiring or transmitting an STD. This is because health officials suggest that when you have sex with someone, you are having sex (in a sense) with everyone that person has had sexual contact with in the past ten years.

Partners may believe that they are in a mutually monogamous relationship when they are not. It is not uncommon for partners in "monogamous" relationships to have extradyadic sexual encounters that are not revealed to the primary partner. Partners may also lie about how many sexual partners they have had and whether or not they have been tested for STDs.

In addition to restricting sexual contacts, putting on a latex or polyurethane condom (natural membrane condoms do not block the transmission of STDs) before the penis touches the partner's body will make it difficult for sexually transmitted diseases to pass from one person to another. It is important to withdraw the penis while it is erect to prevent fluid from leaking from the base of the condom into the partner's genital area. If a woman is receiving oral sex, she should wear a dental dam, a flat latex device that is held over the vaginal area, preventing direct contact between the woman's genital area and her partner's mouth. Condoms should be used for vaginal, anal, and oral sex and should never be reused.

WEBLINKS

American Social Health Association
http://www.ashastd.org

Centers for Disease Control and Prevention
http://www.cdc.gov/
http://www.cdc.gov/hiv/rapid_testing (HIV testing)

HIV Testing
Home Access
www.homeaccess.com

OraSure Technologies, Inc.
http://www.orasure.com/products/default.asp?cid=16&subx=2&sec=3

Food and Drug Administration
http:www.fda.gov/bbs/topics/NEWS/NEW00503.html

gure ST5.1
kternal Female Genitalia

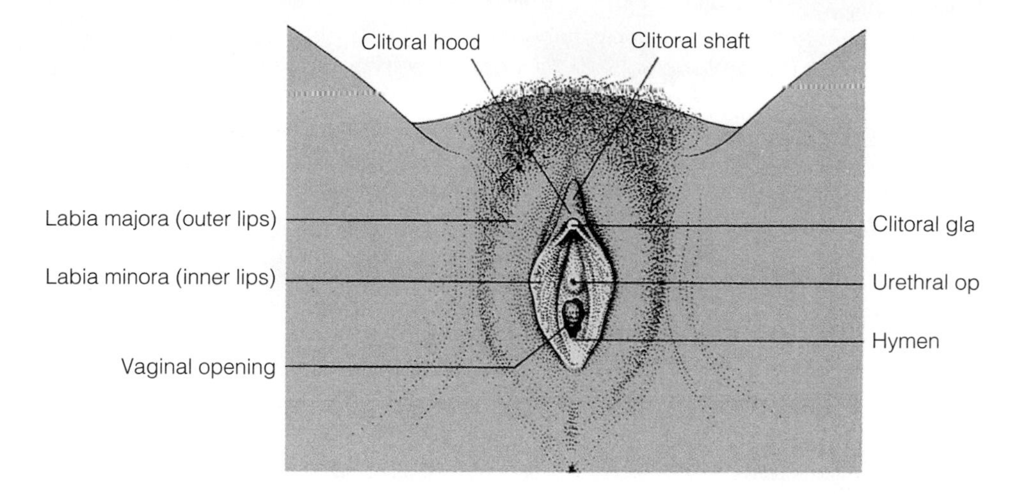

Mons Veneris

The soft cushion of fatty tissue overlying the pubic bone is called the *mons veneris* (mahns vuh-NAIR-ihs), also known as the *mons pubis.* This area becomes covered with hair at puberty and has numerous nerve endings. The purpose of the mons is to protect the pubic region during sexual intercourse.

Labia

In the sexually unstimulated state, the urethral and vaginal openings are protected by the *labia majora* (LAY-bee-uh muh-JOR-uh), or outer lips—two elongated folds of fatty tissue that extend from the mons to the *perineum,* the area of skin between the opening of the vagina and the anus. Located between the labia majora are two additional hairless folds of skin, called the *labia minora* (muh-NOR-uh), or inner lips, that cover the urethral and vaginal openings and join at the top to form the hood of the clitoris. Some contend that the clitoral hood provides clitoral stimulation during intercourse. Both sets of labia—particularly the inner labia minora—have a rich supply of nerve endings that are sensitive to sexual stimulation.

Clitoris

At the top of the labia minora is the *clitoris* (KLIHT-uh-ruhs), which also has a rich supply of nerve endings. The clitoris is a very important site of sexual excitement and, like the penis, becomes erect during sexual excitation.

Vaginal Opening

The area between the labia minora is called the *vestibule.* This includes the urethral opening and the vaginal opening, or *introitus* (ihn-TROH-ih-tuhs), neither of which is visible unless the labia minora are parted. Like the anus, the vaginal opening is surrounded by a ring of sphincter muscles. Although the vaginal opening can expand to accommodate the passage of a baby at childbirth, under conditions of tension these muscles can involuntarily contract, making it difficult to insert an object, including a tampon, into the vagina. The vaginal opening is sometimes covered by a *hymen,* a thin membrane.

Probably no other body part has caused as much grief to so many women as the hymen, which has been regarded throughout history as proof of virginity. A newlywed woman who was thought to be without a hymen was often returned to

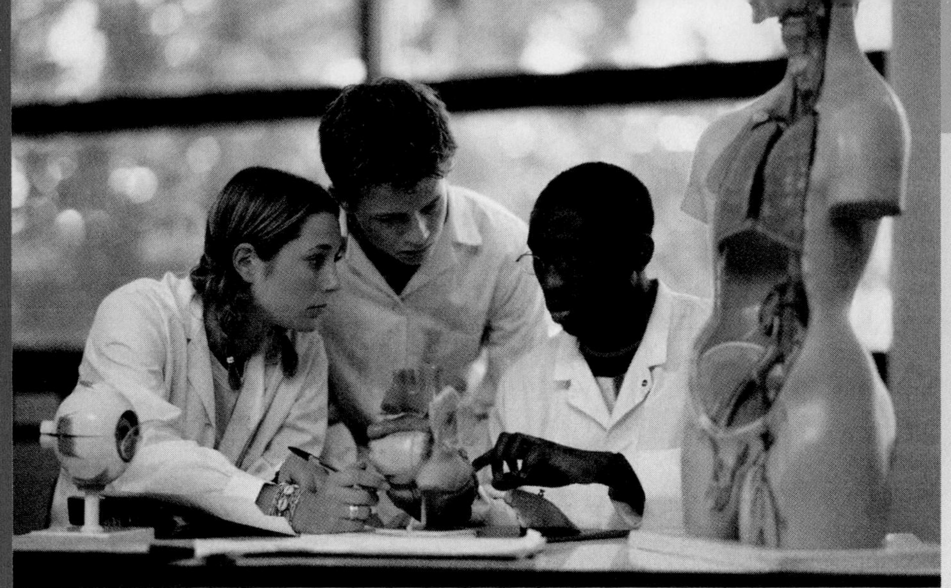

© John Davis/ImageState-Pictor/PictureQuest

SPECIAL TOPIC

In the culture in wh[...] live, it is the custom [...] informed upon that [...] concerning which e[...] individual should k[...] most— namely, the [...] and function of his [...] body.

Ashley Montague

Sexual Anatomy and Physiology

If we think of the human body as a special type of machine, *anatomy* refers to that machine's part and *physiology* refers to how the parts work. This Special Topic reviews the sexual anatomy and physiology of women and men and the reproductive process.

CONTENTS

FEMALE EXTERNAL ANATOMY AND PHYSIOLOGY

The external female genitalia are collectively known as the *vulva* (VUHL-vuh), a Latin term meaning "covering." The vulva consists of the mons veneris, the labia, the clitoris, and the vaginal and urethral openings (see Figure ST5.1). The female genitalia differ in size, shape, and color, resulting in considerable variability in appearance.

Terms for male genitalia (e.g., *penis, testicles*) are more commonly known than are terms for female genitalia. Some women do not even accurately name their genitals. At best, little girls are taught that they have a vagina, which becomes the word for everything "down there"; they rarely learn they also have a vulva, a clitoris, and labia.

her parents, disgraced by exile, or even tortured and killed. It has been a common practice in many societies to parade a bloody bedsheet after the wedding night as proof of the bride's virginity. The anxieties caused by the absence of a hymen persist even today; in Japan and other countries, sexually experienced women may have a plastic surgeon reconstruct a hymen before marriage. Yet the hymen is really a poor indicator of virtue. Some women are born without a hymen or with incomplete hymens. In others, the hymen is accidentally ruptured by vigorous physical activity or insertion of a tampon. In some women, the hymen may not tear but only stretch during sexual intercourse. Even most doctors cannot easily determine whether a woman is a virgin.

Urethral Opening

Just above the vaginal opening is the urethral opening, where urine passes from the body. A short tube, the *urethra,* connects the bladder (where urine collects) with the urethral opening. Because of the shorter length of the female urethra and its close proximity to the anus, women are more susceptible than men to cystitis, a bladder inflammation.

FEMALE INTERNAL ANATOMY AND PHYSIOLOGY

The internal sex organs of the female include the vagina, pubococcygeus muscle, uterus, and paired fallopian tubes and ovaries (see Figure ST5.2).

Vagina

Some people erroneously believe that the *vagina* is a dirty part of the body. In fact, the vagina is a self-cleansing organ. The bacteria that are found naturally in the vagina help to destroy other potentially harmful bacteria. In addition, secretions from the vaginal walls help to maintain the vagina's normally acidic environment. The use of feminine hygiene sprays, as well as excessive douching, can cause irritation, allergic reactions, and, in some cases, vaginal infection by altering the normal chemical balance of the vagina.

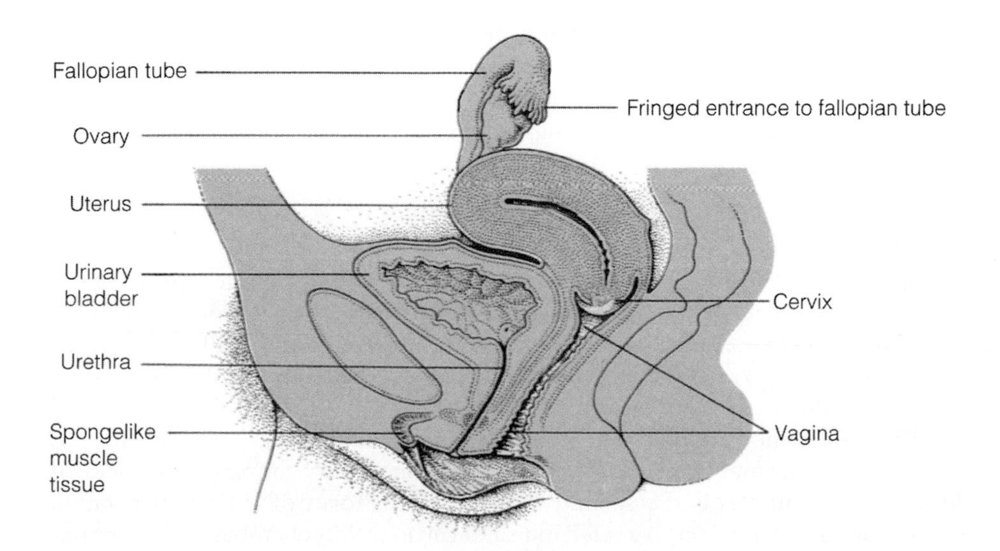

Figure ST5.2
Internal Female Sexual and Reproductive Organs

Figure ST5.3
Alleged Grafenberg Spot

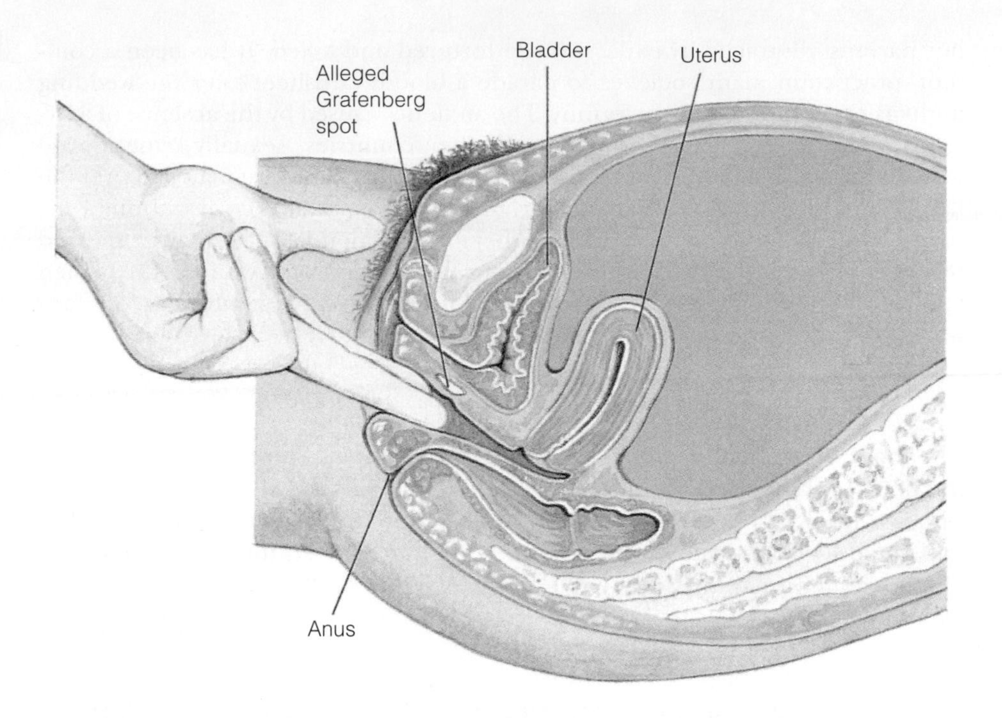

Some researchers have reported that some women experience extreme sensitivity in an area in the front wall of the vagina one to two inches into the vaginal opening. The spot (or area) may swell during stimulation, and although a woman's initial response may be a need to urinate, continued stimulation generally leads to orgasm. The area was named the *Grafenberg spot,* or *G spot,* for gynecologist Ernest Grafenberg, who first noticed the erotic sensitivity of this area over forty years ago. Not all women report a G spot.

Pubococcygeus Muscle

Also called the *PC muscle,* the *pubococcygeus muscle* is one of the pelvic floor muscles that surround the vagina, the urethra, and the anus. To find her PC muscle, a woman is instructed to voluntarily stop the flow of urine after she has begun to urinate. The muscle that stops the flow is the PC muscle.

A woman can strengthen her PC muscle by performing the Kegel exercises, named after the physician who devised them. The Kegel exercises involve contracting the PC muscle several times for several sessions per day. Kegel exercises are often recommended after childbirth to restore muscle tone to the PC muscle, which is stretched during the childbirth process, and help prevent involuntary loss of urine.

Uterus

The *uterus* (YOOT-uh-ruhs), or *womb,* resembles a small, inverted pear, which measures about three inches long and three inches wide at the top in women who have not given birth. A fertilized egg becomes implanted in the wall of the uterus and continues to grow and develop until delivery. At the lower end of the uterus is the *cervix,* an opening that leads into the vagina.

All adult women should have a pelvic exam, including a Pap test, each year. A Pap test is extremely important in the detection of cervical cancer. Cancer of the cervix and uterus is the second most common form of cancer in women. Some women may neglect to get a Pap test because they feel embarrassed or anxious about it or because they think they are too young to worry about getting cancer. For all women over age 20, however, having annual Pap tests may mean the difference between life and death.

Fallopian Tubes

The *fallopian* (ful-LOH-pee-uhn) *tubes* extend about four inches laterally from either side of the uterus to the ovaries. Fertilization normally occurs in the fallopian tubes. The tubes transport the ovum, or egg, by means of *cilia* (hairlike structures) down the tube and into the uterus.

Ovaries

The *ovaries* (OH-vuhr-eez) are two almond-shaped structures, one on either side of the uterus. The ovaries produce eggs (ova) and the female hormones estrogen and progesterone. At birth, a woman has a total of about 400,000 immature ova in her ovaries. Each ovum is enclosed in a thin capsule forming a follicle. Some of the follicles begin to mature at puberty; only about 400 mature ova will be released in a woman's lifetime.

MALE EXTERNAL ANATOMY AND PHYSIOLOGY

Although they differ in appearance, many structures of the male (see Figure ST5.4) and female genitals develop from the same embryonic tissue (the penis and the clitoris, for example).

Penis

The *penis* (PEE-nihs) is the primary male sexual organ. In the unaroused state, the penis is soft and hangs between the legs. When sexually stimulated, the penis enlarges and becomes erect, enabling penetration of the vagina. The penis functions not only to deposit sperm in the female's vagina but also as a passageway from the male's bladder to eliminate urine. In cross-section, the penis can be seen to consist of three parallel cylinders of tissue containing many cavities, two *corpora cavernosa* (cavernous bodies), and a *corpus spongiosum* (spongy body) through which the urethra passes. The penis has numerous blood vessels; when stimulated, the arteries dilate and blood enters faster than it can leave. The cavities of the cavernous and spongy bodies fill with blood, and pressure against the fibrous membranes causes the penis to become erect. The head of the penis is called the glans. At birth, the glans is covered by *foreskin.*

Circumcision, the surgical procedure in which the foreskin of the male is pulled forward and cut off, has been practiced for at least six thousand years. About 80 percent of men in the United States have been circumcised.

Circumcision is a religious rite for members of the Jewish and Muslim faiths. To Jewish people, circumcision symbolizes the covenant between God and Abraham. In the United States, the procedure is generally done within the first few days after birth. Among non-Jewish people, circumcision first became popular in the United States during the nineteenth century as a means of preventing masturbation. Such effectiveness has not been demonstrated.

Today, the primary reason for performing circumcision is to ensure proper hygiene and to maintain tradition. The smegma that can build up under the foreskin is a potential breeding ground for infection. But circumcision may be a

Figure ST5.4
External Male Sexual Organs

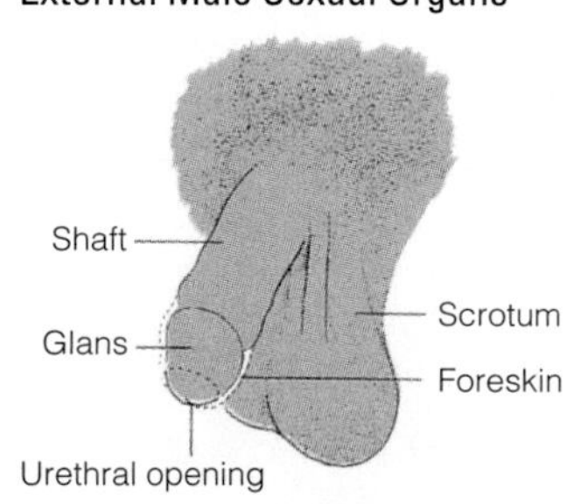

rather drastic procedure merely to ensure proper hygiene, which can just as easily be accomplished by pulling back the foreskin and cleaning the glans during normal bathing. However, being circumcised is associated with having a lower risk of contracting penile HPV; similarly, female partners of men who engage in risky sexual behavior have a reduced likelihood of having cervical cancer if the man is circumcised (Lane, 2002). Circumcision is a relatively low-risk surgical procedure. And although the male does feel pain, it can be minimized by administering local anesthesia.

Scrotum

The *scrotum* (SCROH-tuhm) is the sac located below the penis that contains the *testes.* Beneath the skin covering the scrotum is a thin layer of muscle fibers that contract when it is cold, helping to draw the testes (testicles) closer to the body to keep the temperature of the sperm constant. Sperm can be produced only at a temperature several degrees lower than normal body temperature; any prolonged variation can result in sterility.

MALE INTERNAL ANATOMY AND PHYSIOLOGY

The male internal organs, often referred to as the reproductive organs, include the testes, where the sperm are produced; a duct system to transport and propel the sperm out of the body; and some additional structures that produce the seminal fluid in which the sperm are mixed before ejaculation (see Figure ST5.5)

Testes

The male gonads—the paired testes, or testicles—develop from the same embryonic tissue as the female gonads (the ovaries). The two oval-shaped testicles are suspended in the scrotum by the *spermatic cord* and enclosed within a fibrous sheath. The function of the testes is to produce spermatozoa and male hormones, primarily testosterone.

Figure ST5.5
Internal and External Male Sexual Organs

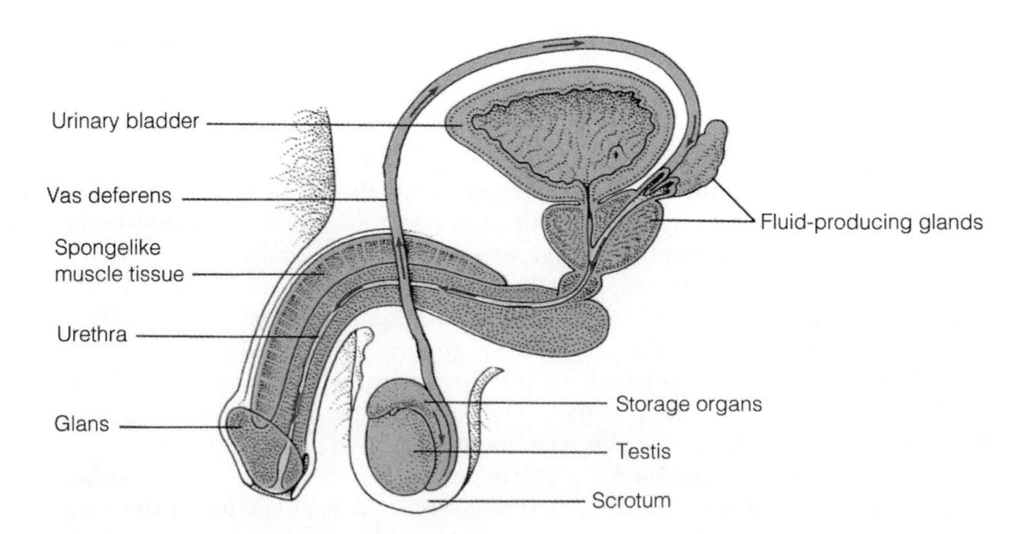

Duct System

Several hundred *seminiferous tubules* come together to form a tube in each testicle called the *epididymis* (ehp-uh-DIHD-uh-muhs), the first part of the duct system that transports sperm. If uncoiled, each tube would be twenty feet long. Sperm spend from two to six weeks traveling through the epididymis as they mature and are reabsorbed by the body if ejaculation does not occur. One ejaculation contains an average of 360 million sperm cells.

The sperm leave the scrotum through the second part of the duct system, the *vas deferens* (vas-DEF-uh-renz). These fourteen- to sixteen-inch paired ducts transport the sperm from the epididymis up and over the bladder to the prostate gland. Rhythmic contractions during ejaculation force the sperm into the paired ejaculatory ducts that run through the prostate gland. The entire length of this portion of the duct system is less than one inch. It is here that the sperm mix with seminal fluid to form semen before being propelled to the outside through the urethra.

Seminal Vesicles and Prostate Gland

The *seminal vesicles* resemble two small sacs, each about two inches in length, located behind the bladder. These vesicles secrete their own fluids, which empty into the ejaculatory duct to mix with sperm and fluids from the prostate gland.

Most of the seminal fluid comes from the *prostate gland,* a chestnut-sized structure located below the bladder and in front of the rectum. The fluid is alkaline and serves to protect the sperm in the more acidic environments of the male urethra and female vagina. Males over 45 should have a rectal exam annually to detect the presence of prostate cancer.

A small amount of clear, sticky fluid is also secreted into the urethra before ejaculation by two pea-sized *Cowper's,* or *bulbourethral,* glands located below the prostate gland. This protein-rich fluid alkalizes the urethral passage, which prolongs sperm life.

The fluid secreted by the Cowper's glands can often be noticed on the tip of the penis during sexual arousal. Since it may contain sperm, withdrawal of the penis from the vagina before ejaculation is a risky method of birth control.

© Rachel Epstein/PhotoEdit

SPECIAL TOPIC 6

Resources and Organizations

ABORTION—Pro Choice
Religious Coalition for Reproductive Choice
1025 Vermont Avenue NW, Suite 1130
Washington, DC 20005
Phone: 202-628-7700
www.rcrc.org

ABORTION—Pro Life
National Right to Life Committee
419 Seventh Street NW
Washington, DC 20004
Phone: 202-626-8800
www.nrlc.org/Unborn_Victims/

ADOPTION
Dave Thomas Foundation for Adoption
1-800-275-3832
Email: adoption@wendys.com

Evan B. Donaldson Adoption Institute
120 Wall Street, 20th Floor
New York, New York 10005
212-269-5080
www.adoptioninstitute.org/

National Council for Adoption
1930 Seventeenth Street NW
Washington, DC 20009
202-328-1200

AL-ANON FAMILY GROUPS
1600 Corporate Landing Parkway
Virginia Beach, VA 23454
Phone: 757-563-1655
www.al-anon.org

CHILD ABUSE
National Committee for Prevention of Child Abuse
332 S. Michigan Avenue
Chicago, Illinois 60604-4357
Phone: 312-663-3520

CHILDREN
National Association for the Education of Young Children
1509 Sixteenth Street NW
Washington, DC 20036-1426
Phone: 800-424-2460
www.naeyc.org

COMMUNES/INTENTIONAL COMMUNITIES
Intentional Communities
www.ic.org

Twin Oaks
138 Twin Oaks Road
Louisa, VA 23093
www.twinoaks.org/

Sandhill
Route 1, Box 155
Rutledge, Mo 63563
Phone: 660-883-5543

DIVORCED FATHERS
American Coalition for Fathers and Children
22994 El Toro Road, Suite 114
Lake Forest, CA 92630
Phone: 1 800 978 3237
www.acfc.org

Fathers are Capable Too (FACT)
Telephone (416) 410-FACT
www.fact.on.ca

DIVORCE MEDIATION
Academy of Family Mediators
5 Militia Drive
Lexington, MA 02421
Phone: 781-674-2663

DIVORCE RECOVERY
Heartchoice.com
The divorce room
http://heartchoice.com/divorce/index.php
Information for divorced fathers
http://heartchoice.com/contrib/men_and_divorce.html
Information for divorced mothers
http://heartchoice.com/contrib/womenanddiv.html

DOMESTIC VIOLENCE
National Coalition against Domestic Violence
P.O. Box 18749
1201 East Colfax 385
Denver, CO 80218
Phone: 303-839-1852
www.ncadv.org

National Toll Free Number for Domestic Violence
Phone: 1-800-799-7233

FAMILY PLANNING
Planned Parenthood Federation of America
810 Seventh Avenue
New York, New York 10019
www.plannedparenthood.org

GRANDPARENTING
The Foundation for Grandparenting
108 Farnham Road
Ojai, CA 93023
www.grandparenting.org

AARP (grandparent information center)
601 E St. NW
Washington, DC 20049
www.aarp.org/grandparents/

HEALTHY BABY
National Healthy Mothers, Healthy Babies Coalition
121 N. Washington Street
Alexandria, VA 22314
Phone: 703-836-6110
www.hmhb.org

HOMOSEXUALITY
National Gay and Lesbian Task Force
2320 Seventeenth Street NW
Washington, DC 20009
Phone: 202-332-6483
www.ngltf.org

Parents, Families and Friends of Lesbians and Gays (PFLAG)
P.O. Box 96519
Washington, DC 20009-6519
www.pflag.org

INFERTILITY
Infertility and Reproductive Technology
American Fertility Society
1209 Montgomery Highway
Birmingham, Alabama 35216-2809

MARRIAGE
National Marriage Project
Rutgers, The State University of New Jersey
54 Joyce Kilmer Avenue, Lucy Stone Hall A347
Piscataway, NJ 08854-8045
732-445-7922
http://marriage.rutgers.edu

Groves Conference on Marriage and the Family
3005 Hidden Lane
Knoxville, TN 37920-7300
Phone: 865-579-3549

Coalition for Marriage, Family and Couples Education
5310 Belt Rd NW
Washington, DC 20015-1961
202-362-3332
www.smartmarriages.com/

The Couples Place: Successful Marriages
www.couples-place.com/marriage.asp

MARRIAGE AND FAMILY THERAPY
American Association for Marriage and Family Therapy
1133 Fifteenth Street NW, Suite 300
Washington, DC 20005
Phone: 202-452-0109
http://www.aamft.org/index_nm.asp

MARRIAGE ENRICHMENT
ACME (Association for Couples in Marriage Enrichment)
56 Windsor Court
New Brighton, MN 55112
Phone: 651-604-9345
E-mail: hamb1001@tc.umn.edu
www.bettermarriages.org

MATE SELECTION
Right Mate
http://heartchoice.com/rightmate

MEN'S AWARENESS
American Men's Studies Association
22 East Street
Northampton, MA 01060

MOTHERHOOD
Centre for Research on Mothering
726 Atkinson
York University
4700 Keele Street
Toronto, Ontario Canada
www.yorku.ca/crm

PARENTAL ALIENATION SYNDROME
The National Parental Alienation Foundation
816 Connecticut Ave. NW, Suite 900
Washington, DC 20006
Phone: 202-466-7778
Fax: 202-466-7779
info@pasresearch.org

PARENTING EDUCATION
Parenting Education
P.O. Box 390744
Edina, MN 55439-0744
Phone: 612-713-0898
E-mail: parentips@aol.com

REPRODUCTIVE HEALTH
Association of Reproductive Health Professionals
2401 Pennsylvania Avenue, NW
Suite 350
Washington, DC 20037-1718
202-466 3825
www.arhp.org

SEX ABUSE
VOICES in Action, Inc. (Victims of Incest Can Emerge Survivors)
P.O. Box 148309
Chicago, Illinois 60614
Phone: 1-800-7-VOICE-8
www.voices-action.org

National Clearinghouse on Marital and Date Rape
2325 Oak Street
Berkeley, California 94708-1697
Phone: 510-524-1582
www.ncmdr.org

SEX EDUCATION
Sexuality Information and Education Council of the United States
New York University
32 Washington Plaza
New York, New York 10003
www.siecus.org

SEX THERAPY
Masters and Johnson Institute
24 S. Kings Highway
St. Louis, MO 63108
www.mastersandjohnson.com

SEXUAL INTIMACY
http://www.heartchoice.com/sex_intimacy/

SEXUALLY TRANSMISSIBLE DISEASES
American Social Health Association
(Herpes Resource Center and HPV Support Program)
P.O. Box 13827
Research Triangle Park, NC 27709
Phone: 919-361-8400
www.ashastd.org

National AIDS Hotline
800-342-AIDS

National Herpes Hotline
919-361-8488
800-230-6039

National STD Hotline
800-227-8922

STD/AIDS Information
919-361-8400

SINGLEHOOD
Alternatives to Marriage Project
P.O. Box 991010
Boston, MA 02199
Phone: 781-793-0296
Fax: 781-394-6625
www.unmarried.org

SINGLE PARENTHOOD
Parents without Partners
8807 Colesville Road
Silver Spring, Maryland 20910
Phone: 301-588-9354
www.parentswithoutpartners.org

Single Mothers by Choice
1642 Gracie Square Station
New York, New York 10028
Phone: 212-988-0993
www.singlemothersbychoice.com

STEPFAMILIES
Stepfamily Association of America
650 J Street, Suite 205
Lincoln, Nebraska 68508
Phone: 800-735-0329
www.saafamilies.org

Stepfamily Associates
1368 Beacon Street
Suite 108
Brookline, MA 02146
Phone: 617-734-8831
www.stepfamilyboston.com

Stepfamily Foundation
333 West End Ave
New York, New York 10023
Phone: 800-759-7837
www.stepfamily.org

TRANSGENDER
Tri-Ess: The Society for the Second Self, Inc.
8880 Bellaire Blvd
B2, PMB 104
Houston, TX 77036
Phone: 903-813-3398
www.tri-ess.org

WIDOWHOOD
Widowed Person's Service
American Association of Retired Persons
609 E St. NW
Washington, DC 20049

WOMEN'S AWARENESS
National Organization for Women
1000 Sixteenth Street NW
Suite 700
Washington, DC 20036
Phone: 202-331-0066
www.now.org

SPECIAL TOPICS KEY TERMS

AIDS
certified family life educator
chlamydia
control group
genital herpes
genital warts
gonorrhea
herpes simplex virus type 1
herpes simplex virus type 2
HIV
identity theft
random sample
STD

APPENDIX A

Individual Autobiography Outline

Your instructor may ask you to write a paper that reflects the individual choices you have made that have contributed to your becoming who you are. Check with your instructor to determine what credit (if any) is assigned to your completing this autobiography. Use the following outline to develop your paper. Some topics may be too personal, and you may choose to avoid writing about them. Your emotional comfort is important, so skip any questions you want and answer only those questions you feel comfortable responding to.

I. Choices: Free Will Versus Determinism

Specify the degree to which you feel that you are free to make your own interpersonal choices versus the degree to which your choices are determined by social constraints and influences. Give examples.

II. Relationship Beginnings

A. *Interpersonal context into which you were born.* How long had your parents been married before you were born? How many other children had been born into your family? How many followed your birth? Describe how these and other parental choices affected you before you were born and the choices you will make in regard to family planning that will affect the lives of your children.

B. *Early relationships.* What was your relationship with your mother, father, and siblings when you were growing up? What is your relationship with each of them today? Who took care of you as a baby? If this person was other than your parents or siblings, who was the person (e.g., grandmother) and what is your relationship with that person today? How have your experiences in the family in which you were reared impacted who you are today? In regard to decision making, how easy or difficult is it for you to make decisions and how have your parents influenced this capacity?

C. *Early self-concept.* How did you feel about yourself as a child, an adolescent, and a young adult? What significant experiences helped to shape your self-concept? How do you feel about yourself today? What choices have you made that have resulted in your feeling good about yourself? What choices have you made that have resulted in your feeling negatively about yourself?

III. Subsequent Relationships

A. *First love.* When was your first love relationship with someone outside your family? What kind of love was it? Who initiated the relationship? How long did it last? How did it end? How did it affect you and your subsequent relationships? What choices did you make in this first love relationship you are glad you made? What choices did you make that you now feel were a mistake?

B. *Subsequent love relationships.* What other significant love relationships (if any) have you had? How long did they last and how did they end? What choices did you make in these relationships you are glad you made? What choices did you make that you now feel were a mistake?

C. *Lifestyle preferences.* What are your preferences for remaining single, being married, or living with someone? How would you feel about living in a commune? What do you believe is the ideal lifestyle? Why?

IV. Communication Issues

A. *Parental models.* Describe your parents' relationship and their manner of communicating with each other. How are your interpersonal communication patterns similar to and different from theirs?

B. *Relationship communication.* How comfortable do

you feel talking about relationship issues with your partner? How comfortable do you feel telling your partner what you like and don't like about his or her behavior? To what degree have you told your partner your feelings for him or her? Your desires for the future?

C. Sexual communication. How comfortable do you feel giving your partner feedback about how to please you sexually? How comfortable are you discussing the need to use a condom with a potential sex partner? How would you approach this topic?

D. Sexual past. How much have you disclosed to a partner about your previous relationships? How honest were you? Do you think you made the right decision to disclose or withhold? Why?

V. Sexual Choices

A. Sex education. What did you learn about sex from your parents, peers, and teachers (both academic and religious)? What choices did your parents, peers, and teachers make about your sex education that had positive consequences? What decisions did they make that had negative consequences for you?

B. Sexual experiences. What choices have you made about your sexual experiences that had positive outcomes? What choices have you made that had negative outcomes?

C. Sexual values. To what degree are your sexual values absolutist, legalist, or relativistic? How have your sexual values changed since you were an adolescent?

D. Safer sex. What is the riskiest choice you have made with regard to your sexual behavior? What is the safest choice you have made with regard to your sexual behavior? What is your policy about asking your partner about his or her previous sex history and requiring that both of you be tested for STDs and HIV before having sex? How comfortable are you buying and using condoms?

VI. Violence and Abuse Issues

A. Violent/abusive relationship. Have you been involved in a relationship where your partner was violent or abusive toward you? Give examples of the violence or abuse (verbal and nonverbal) if you have been involved in such a relationship. How many times did you leave and return to the relationship before you left permanently? What was the event that triggered your decision to leave the first and last time? Describe the context of your actually leaving (e.g., left when partner was at work). To what degree have you been violent or abusive toward a partner in a romantic relationship?

B. Family, sibling abuse. Have you been involved in a relationship where your parents or siblings were violent or abusive toward you? Give examples of any violence or abuse (verbal or nonverbal) if such experiences were part of your growing up. How have these experiences affected the relationship you have with the abuser today?

C. Forced sex. Have you been pressured or forced to participate in sexual activity against your will by a parent, sibling, partner, or stranger? How did you react at the time and how do you feel today? Have you pressured or forced others to participate in sexual experiences against their will?

VII. Reproductive Choices

A. Contraception. What is your choice for type of contraception? How comfortable do you feel discussing the need for contraception with a potential partner? In what percentage of your first-time intercourse experiences with a new partner did you use a condom? (If you have not had intercourse, this question will not apply).

B. Children. How many children (if any) do you want and at what intervals? How important is it to you that your partner wants the same number of children as do you? How do you feel about artificial insemination, sterilization, abortion, and adoption? How important is it to you that your partner feel the same way?

VIII. Childrearing Choices

A. Discipline. What are your preferences for your use of "time out" or "spanking" as a way of disciplining your children? How important is it to you that your partner feel as you do on this issue?

B. Day care. What are your preferences for whether your children grow up in day care or whether one parent will stay home with and rear the children? How important is it to you that your partner feel as do you on this issue?

C. Education. What are your preferences for whether your children attend public or private school or are "home schooled"? What are your preferences for whether your child attends a religious school? To what degree do you feel it is your responsibility as parents to pay for the college education of your children? How important is it to you that your partner feel as do you on these issues?

IX. Education/Career Choices

A. Own educational/career choices. What is your major in school? How important is it to you that you finish undergraduate school? Earn a master's degree? Earn a Ph.D., M.D. ,or law degree? To what degree do you want to be a stay-at-home mom or stay-at-home dad? How important is it to you that your partner be completely supportive of your educational, career, and family aspirations?

B. Expectations of partner. How important is it to you that your partner have the same level of education that you have? To what degree are you willing to be supportive of your partner's educational/career aspirations?

APPENDIX B

Family Autobiography Outline

Your instructor may want you to write a paper that reflects the influence of your family on your development. Check with your instructor to determine what credit (if any) is assigned to your completing this family autobiography. Use the following outline to develop your paper. Some topics may be too personal, and you may choose to avoid writing about them. Your emotional comfort is important, so skip any questions you like and answer only those you feel comfortable responding to.

I. Family Background

A. Describe yourself, including age, gender, place of birth, and additional information that helps to identify you. On a scale of 0 to 10 (10 = highest), how happy are you? Explain this number in reference to the satisfaction you experience in the various roles (offspring, sibling, partner in a relationship, employee, student, parent, roommate, friend, etc.) you currently occupy.

B. Identify your birth position; give the names and ages of children younger and older than you. How did you feel about your "place" in the family? How do you feel now?

C. What was your relationship with and how did you feel about each parent and sibling when you were growing up?

D. What is your relationship and how do you feel today about each of these family members?

E. Which parental figure or sibling are you most like? How? Why?

F. Who else lived in your family (e.g., grandparent, spouse of sibling?) and how did they impact family living?

G. Discuss the choice you made before attending college that you regard as the wisest choice you made during this time period. Discuss the choice you made prior to college that you regret.

H. Discuss the one choice you made since you began college that you regard as the wisest choice you made during this time period. Discuss the one choice you made since you began college that you regret.

II. Religion and Values

A. In what religion were you socialized as a child? Discuss the impact of religion on you as a child and as an adult. To what degree will you choose to teach your own children similar religious values? Why?

B. Explain what you were taught in your family in regard to each of the following values: intercourse outside of marriage, need for economic independence, manners, honesty, importance of being married, qualities of a desirable spouse, children, alcohol and drugs, safety, elderly family members, persons of other races and religions, persons with disabilities, homosexuals, persons with less or more education, persons with and without "wealth," importance of having children, occupational role (in regard to the latter, what occupational role were you encouraged to pursue?). To what degree will you choose to teach your own children similar values? Why?

C. What was the role relationship between your parents in terms of dominance, division of labor, communication, affection, etc? How has your observation of the parent of the same sex and opposite sex influenced the role you display in your current relationships with intimate partners? To what degree will you choose to have a similar relationship with your own partner that your parents had with each other?

D. How close were your parents emotionally? How emotionally close were/are you with your parents? To what degree did you and your parents discuss feelings? On a scale of 0 to 10, how well do your parents "know" how you think and feel? To what degree will you choose to have a similar level of closeness or distance with you own partner and children?

E. Did your parents have a pet name for you? How did you feel about this?

F. How did your parents resolve conflict between themselves? How do you resolve conflict with partners in your own relationship?

III. Economics and Social Class

A. Identify your "social class" (lower, middle, upper), the education/jobs/careers of your respective parents, and the economic resources of your family. How did your social class and economic well-being affect you as a child? How has the economic situation in which you were reared influenced your own choices of what you want for yourself? To what degree are you economically self-sufficient?

B. How have the career choices of your parents influenced your own?

IV. Parental Plus and Minuses

A. What is the single most important thing your mother and father, respectively, said or did that has affected your life in a positive way?

B. What is the single biggest mistake your mother and father, respectively, made in rearing you? Discuss how this impacted you negatively. To what degree are you still affected by this choice?

V. Personal Crisis Events

A. Everyone experiences one or more crisis events that have a dramatic impact on his or her life. Identify and discuss each event or events you have experienced and your reaction and adjustment to them.

B. How did your parents react to this crisis event you were experiencing? To what degree did their reaction help or hinder your adjustment? What different choice or choices (if any) could they have made to assist you in ways you would have regarded as more beneficial?

VI. Family Crisis Events

Identify and discuss each crisis event your family has experienced. How did each member of your family react and adjust to each event? An example of a family crisis event would be unemployment of the primary breadwinner, prolonged illness of a family member, aging parent coming to live with the family, death of a sibling, alcoholism, etc.

VII. Future

Describe yourself two, five, and ten years from now. What are your educational, occupational, marital, and family goals? How has the family in which you were reared influenced each of these goals? What choices might your parents make to assist you in achieving your goals?

APPENDIX C

Prenuptial Agreement of a Remarried Couple

Pam and Mark are of sound mind and body, have a clear understanding of the terms of this contract and of the binding nature of the agreements contained herein; they freely and in good faith choose to enter into the PRENUPTIAL AGREEMENT and MARRIAGE CONTRACT and fully intend it to be binding upon themselves.

Now, therefore, in consideration of their love and esteem for each other and in consideration of the mutual promises herein expressed, the sufficiency of which is hereby acknowledged, Pam and Mark agree as follows:

Names

Pam and Mark affirm their individuality and equality in this relationship. The parties believe in and accept the convention of the wife's accepting the husband's name, while rejecting any implied ownership.

Therefore, the parties agree that they will be known as husband and wife and will henceforth employ the titles of address: Mr. and Mrs. Mark Stafford, and will use the full names of Pam Hayes Stafford and Mark Robert Stafford.

Relationships with Others

Pam and Mark believe that their commitment to each other is strong enough that no restrictions are necessary with regard to relationships with others.

Therefore, the parties agree to allow each other freedom to choose and define their relationships outside this contract, and the parties further agree to maintain sexual fidelity each to the other.

Religion

Pam and Mark reaffirm their belief in God and recognize He is the source of their love. Each of the parties has his/her own religious beliefs.

Therefore, the parties agree to respect their individual preferences with respect to religion and to make no demands on each other to change such preferences.

Children

Pam and Mark both have children. Although no minor children will be involved, there are two (2) children still at home and in school and in need of financial and emotional support.

Therefore, the parties agree that they will maintain a home for and support these children as long as is needed and reasonable. They further agree that all children of both parties will be treated as one family unit, and each will be given emotional and financial support to the extent feasible and necessary as determined mutually by both parties.

Careers and Domicile

Pam and Mark value the importance and integrity of their respective careers and acknowledge the demands that their jobs place on them as individuals and on their partnership. Both parties are well established in their respective careers and do not foresee any change or move in the future.

The parties agree, however, that if the need or desire for a move should arise, the decision to move shall be mutual and based on the following factors:

1. The overall advantage gained by one of the parties in pursuing a new opportunity shall be weighed against the disadvantages, economic and otherwise, incurred by the other.
2. The amount of income or other incentive derived from the move shall not be controlling.

3. Short-term separations as a result of such moves may be necessary.

Mark hereby waives whatever right he might have to solely determine the legal domicile of the parties.

Care and Use of Living Spaces

Pam and Mark recognize the need for autonomy and equality within the home in terms of the use of available space and allocation of household tasks. The parties reject the concept that the responsibility for housework rests with the woman in a marriage relationship while the duties of home maintenance and repair rest with the man.

Therefore, the parties agree to share equally in the performance of all household tasks, taking into consideration individual schedules, preferences, and abilities.

The parties agree that decisions about the use of living space in the home shall be mutually made, regardless of the parties' relative financial interests in the ownership or rental of the home, and the parties further agree to honor all requests for privacy from the other party.

Property, Debts, Living Expenses

Pam and Mark intend that the individual autonomy sought in the partnership shall be reflected in the ownership of existing and future-acquired property, in the characterization and control of income, and in the responsibility for living expenses. Pam and Mark also recognize the right of patrimony of children of their previous marriages.

Therefore, the parties agree that all things of value now held singly and/or acquired singly in the future shall be the property of the party making such acquisition. In the event that one party to this agreement shall predecease the other, property and/or other valuables shall be disposed of in accordance with an existing will or other instrument of disposal that reflects the intent of the deceased party.

Property or valuables acquired jointly shall be the property of the partnership and shall be divided, if necessary, according to the contribution of each party. If one party shall predecease the other, jointly owned property or valuables shall become the property of the surviving spouse.

Pam and Mark feel that each of the parties to this agreement should have access to monies that are not accountable to the partnership.

Therefore, the parties agree that each shall retain a mutually agreeable portion of their total income and the remainder shall be deposited in a mutually agreeable banking institution and shall be used to satisfy all jointly acquired expenses and debts.

The parties agree that beneficiaries of life insurance policies they now own shall remain as named on each policy. Future changes in beneficiaries shall be mutually agreed on after the dependency of the children of each party has been terminated. Any other benefits of any retirement plan or insurance benefits that accrue to a spouse only shall not be affected by the foregoing.

The parties recognize that in the absence of income by one of the parties, resulting from any reason, living expenses may become the sole responsibility of the employed party, and in such a situation, the employed party shall assume responsibility for the personal expenses of the other.

Both Pam and Mark intend their marriage to last as long as both shall live.

Therefore, the parties agree that should it become necessary, due to the death of either party, the surviving spouse shall assume any last expenses in the event that no insurance exists for that purpose.

Pam hereby waives whatever right she might have to rely on Mark to provide the sole economic support for the family unit.

Evaluation of the Partnership

Pam and Mark recognize the importance of change in their relationship and intend that this CONTRACT shall be a living document and a focus for periodic evaluations of the partnership.

The parties agree that either party can initiate a review of any article of the CONTRACT at any time for amendment to reflect changes in the relationship. The parties agree to honor such requests for review with negotiations and discussions at a mutually convenient time.

The parties agree that in any event, there shall be an annual reaffirmation of the CONTRACT on or about the anniversary date of the CONTRACT.

The parties agree that in the case of unresolved conflicts between them over any provisions of the CONTRACT, they will seek mediation, professional or otherwise, by a third party.

Termination of the Contract

Pam and Mark believe in the sanctity of marriage; however, in the unlikely event of a decision to terminate this CONTRACT, the parties agree that neither shall contest the application for a divorce decree or the entry of such decree in the county in which the parties are both residing at the time of such application.

In the event of termination of the CONTRACT and divorce of the parties, the provisions of this and the section on "Property, Debts, Living Expenses" of the CONTRACT as amended shall serve as the final property settlement agreement between the parties. In such event, this CONTRACT is intended to effect a complete settlement of any and all claims that either party may have against the other, and a complete settlement of their respective rights as to property rights, homestead rights, inheritance rights, and all other rights of property otherwise arising out of their partnership. The parties further agree that in the event of termination of this CONTRACT and divorce of the parties, neither party shall require the other to pay maintenance costs or alimony.

Decision Making

Pam and Mark share a commitment to a process of negotiations and compromise that will strengthen their equality in the partnership. Decisions will be made with respect for individual needs. The parties hope to maintain such mutual decision making so that the daily decisions affecting their lives will not become a struggle between the parties for power, authority, and dominance. The parties agree that such a process, while sometimes time-consuming and fatiguing, is a good investment in the future of their relationship and their continued esteem for each other.

Now, therefore, Pam and Mark make the following declarations:

1. They are responsible adults.
2. They freely adopt the spirit and the material terms of this prenuptial and marriage contract.
3. The marriage contract, entered into in conjunction with a marriage license of the State of Illinois, County of Wayne, on this 12th day of June 2004, hereby manifests their intent to define the rights and obligations of their marriage relationship as distinct from those rights and obligations defined by the laws of the State of Illinois, and affirms their right to do so.
4. They intend to be bound by this prenuptial and marriage contract and to uphold its provisions before any Court of Law in the Land.

Therefore, comes now, Pam Hayes Stafford, who applauds her development that allows her to enter into this partnership of trust, and she agrees to go forward with this marriage in the spirit of the foregoing PRENUPTIAL and MARRIAGE CONTRACT.

Therefore, comes now, Mark Robert Stafford, who celebrates his growth and independence with the signing of this contract, and he agrees to accept the responsibilities of this marriage as set forth in the foregoing PRENUPTIAL and MARRIAGE CONTRACT.

This CONTRACT AND COVENANT has been received and reviewed by the Reverend Ray Brannon, officiating.

Finally, come Vicki Oliver and Rodney Oliver, who certify that Pam and Mark did freely read and sign this MARRIAGE CONTRACT in their presence, on the occasion of their entry into a marriage relationship by the signing of a marriage license in the State of Illinois, County of Wayne, at which they acted as official witnesses. Further, they declare that the marriage license of the parties bears the date of the signing of this PRENUPTIAL and MARRIAGE CONTRACT.

Living Will

I, [Declarant], ("Declarant" herein), being of sound mind, and after careful consideration and thought, freely and intentionally make this revocable declaration to state that if I should become unable to make and communicate my own decisions on life-sustaining or life-support procedures, then my dying shall not be delayed, prolonged, or extended artificially by medical science or life-sustaining medical procedures, all according to the choices and decisions I have made and which are stated here in my Living Will.

It is my intent, hope, and request that my instructions be honored and carried out by my physicians, family, and friends, as my legal right.

If I am unable to make and communicate my own decisions regarding the use of medical life-sustaining or life-support systems and/or procedures, and if I have a sickness, illness, disease, injury, or condition which has been diagnosed by two (2) licensed medical doctors or physicians who have personally examined me (or more than two (2) if required by applicable law), one of whom shall be my attending physician, as being either (1) terminal or incurable certified to be terminal, or (2) a condition from which there is no reasonable hope of my recovery to a meaningful quality of life, which may reasonably be referred to as hopeless, although not necessarily "terminal" in the medical sense, or (3) has rendered me in a persistent vegetative state, or (4) a condition of extreme mental deterioration, or (5) permanently unconscious, then in the absence of my revoking this Living Will, all medical life-sustaining or life-support systems and procedures shall be withdrawn, unless I state otherwise in the following provisions.

Unless otherwise provided in this Living Will, nothing herein shall prohibit the administering of pain-relieving drugs to me, or any other types of care purely for my comfort, even though such drugs or treatment may shorten my life, or be habit forming, or have other adverse side effects.

I am also stating the following additional instructions so that my Living Will is as clear as possible:

[Resuscitation (CPR)]

[Intravenous and Tube Feeding]

[Life-Sustaining Surgery]

[New Medical Developments]

[Home or Hospital]

In the event that any terms or provisions of my Living Will are not enforceable or are not valid under the laws of the state of my residence, or the laws of the state where I may be located at the time, then all other provisions which are enforceable or valid shall remain in full force and effect, and all terms and provisions herein are severable.

IN WITNESS WHEREOF, I have read and understand this Living Will, and I am freely and voluntarily signing it on this the ____________ day of ____________ (month), ____________ (year) in the presence of witnesses.

Signed:

[Declarant]

Street Address:

County:

City and State:

Witness:

We, the undersigned witnesses, certify by our signatures below, that we are adult (at least 18 years old), mentally competent persons; that we are not related to the Declarant by blood, marriage, or adoption; that we do not stand to inherit anything from the Declarant by any means, including will, trust, operation of law or the laws of intestate succession, or by beneficiary designation, nor do we stand to benefit in any way from the death of the Declarant; that we are not directly responsible for the health or medical care, or general welfare of the Declarant; that neither of us signed the Declarant's signature on this document; and that the Declarant is known to us.

We hereby further certify that the Declarant is over the age of 18; that the Declarant signed this document freely and voluntarily, not under any duress or coercion; and that we were both present together, and in the presence of the Declarant to witness the signing of this Living Will on this the ____________ day of ____________ (month), ____________ (year).

Witness signature:

Residing at:

Witness signature:

Residing at:

Notary Acknowledgment

This instrument was acknowledged before me on this the ____________ day of ____________ (month), ____________ (year) by

[Declarant], the Declarant herein, on oath stating that the Declarant is over the age of 18, has fully read and understands the above and foregoing Living Will, and that the Declarant's signing and execution of same is voluntary, without coercion, and is intentional.

Notary Public

My commission or appointment expires:

Durable Power of Attorney

KNOW ALL MEN BY THESE PRESENTS

That I, ____________________, as principal ("Principal"), a resident of the State and County aforesaid, have made, constituted, appointed and by these presents do make, constitute, and appoint ____________________ and ____________________, either or both of them, as my true and lawful agent or attorney-in-fact ("Agent") to do and perform each and every act, deed, matter, and thing whatsoever in and about my estate, property, and affairs as fully and effectually to all intents and purposes as I might or could do in my own proper person, if personally present, including, without limiting the generality of the foregoing, the following specifically enumerated powers which are granted in aid and exemplification of the full, complete and general power herein granted and not in limitation or definition thereof:

1. To forgive, request, demand, sue for, recover, elect, receive, hold all sums of money, debts due, commercial paper, checks, drafts, accounts, deposits, legacies, bequests, devises, notes, interest, stock of deposit, annuities, pension, profit sharing, retirement, Social Security, insurance, and all other contractual benefits and proceeds, all documents of title, all property and all property rights, and demands whatsoever, liquidated or unliquidated, now or hereafter owned by me, or due, owing, payable, or belonging to me or in which I have or may hereafter acquire an interest, to have, use and take all lawful means and equitable and legal remedies and proceedings in my name for collection and recovery thereof, and to adjust, sell, compromise, and agree for the same, and to execute and deliver for me, on my behalf, and in my name all endorsements, releases, receipts or other sufficient discharges for the same.

2. To buy, receive, lease as lessor, accept, or otherwise acquire; to sell, convey, mortgage, grant options upon, hypothecate, pledge, transfer, exchange, quitclaim, or otherwise encumber or dispose of; or to contract or agree for the acquisition, disposal, or encumbrance of any property whatsoever or any custody, possession, interest, or right therein for cash or credit and upon such terms, considerations, and conditions as Agent shall think proper, and no person dealing with Agent shall be bound to see to the application of any monies paid.

3. To take, hold, possess, invest, or otherwise manage any or all of the property or any interest therein; to eject, remove, or relieve tenants or other persons from, and recover possession of, such property by all lawful means; and to maintain, protect, preserve, insure, remove, store, transport, repair, build on, raze, rebuild, alter, modify, or improve the same or any part thereof, and/or to lease any property for me or my benefit, as lessee without option to renew, to collect and receive any receipt for rents, issues, and profits of my property.

4. To invest and reinvest all or any part of my property in any property and undivided interest in property, wherever located, including bonds, debentures, notes secured or unsecured, stock of corporations regardless of class, interests in limited partnerships, real estate or any interest in real estate whether or not productive at the time of the investment, interest in trusts, investment trusts, whether of the open and/or closed funds types, and participation in

common, collective, or pooled trust funds or annuity contracts without being limited by any statute or rule of law concerning investment by fiduciaries.

5. To make, receive, and endorse checks and drafts; deposit and withdraw funds; acquire and redeem certificates of deposit in banks, savings and loan associations, or other institutions; and execute or release such deeds of trust or other security agreements as may be necessary or proper in the exercise of the rights and powers herein granted.

6. To pay any and all indebtedness of mine in such manner and at such times as Agent may deem appropriate.

7. To borrow money for any purpose, with or without security or on mortgage or pledge of any property.

8. To conduct or participate in any lawful business of whatsoever nature for me and in my name; execute partnership agreements and amendments thereto; incorporate, reorganize, merge, consolidate, recapitalize, sell, liquidate, or dissolve any business; elect or employ officers, directors, and agents; carry out the provisions of any agreement for the sale of any business interest or stock therein; and exercise voting rights with respect to stock either in person or by proxy, and to exercise stock options.

9. To prepare, sign, and file joint or separate income tax returns or declarations of estimated tax for any year or years; to prepare, sign, and file gift tax returns with respect to gifts made by me for any year or years; to consent to any gift and to utilize any gift-splitting provision or other tax election; and to prepare, sign, and file any claims for refund of tax.

10. To have access at any time or times to any safe deposit box rented by me, wheresoever located, and to remove all or any part of the contents thereof, and to surrender or relinquish said safety deposit box in any institution in which such safety deposit box may be located shall not incur any liability to me or my estate as a result of permitting Agent to exercise this power.

11. To execute any and all contracts of every kind or nature.

As used herein, the term "property" includes any property, real or personal, tangible or intangible, wheresoever situated.

The execution and delivery by Agent of any conveyance paper instrument or document in my name and behalf shall be conclusive evidence of Agent's approval of the consideration therefore, and of the form and contents thereof, and that Agent deems the execution thereof in my behalf necessary or desirable.

Any person, firm, or corporation dealing with Agent under the authority of this instrument is authorized to deliver to Agent all considerations of every kind or character with respect to any transactions so entered into by Agent and shall be under no duty or obligation to see to or examine into the disposition thereof.

Third parties may rely upon the representation of Agent as to all matters relating to any power granted to Agent, and no person who may act in reliance upon the representation of Agent or the authority granted to Agent shall incur liability to me or my estate as a result of permitting Agent to exercise any power. Agent shall be entitled to reimbursement for all reasonable costs and expenses incurred and paid by Agent on my behalf pursuant to any provisions of this durable power of attorney, but Agent shall not be entitled to compensation for services rendered hereunder.

Notwithstanding any provision herein to the contrary, Agent shall not satisfy any legal obligation of Agent out of any property subject to this power of attorney, nor may Agent exercise this power in favor of Agent, Agent's estate, Agent's creditors, or the creditors of Agent's estate.

Notwithstanding any provision hereto to the contrary, Agent shall have no power or authority whatever with respect to (a) any policy of insurance owned by me on the life of Agent, and (b) any trust created by Agent as to which I am Trustee.

When used herein, the singular shall include the plural and the masculine shall include the feminine.

This power of attorney shall become effective immediately upon the execution hereof.

This is a durable power of attorney made in accordance with and pursuant to Section (this section to be completed in reference to laws of the state in which the document is executed).

This power of attorney shall not be affected by disability, incompetency, or incapacity of the principal.

Principal may revoke this durable power of attorney at any time by written instrument delivered to Agent. The guardian of Principal may revoke this instrument by written instrument delivered to Agent.

IN WITNESS WHEREOF, I have executed this durable power of attorney in three (3) counterparts, and I have directed that photostatic copies of this power be made, which shall have the same force and effect as an original.

DATED THIS THE _____ day of ____________ (month), ________ (year).

WITNESS:

____

Name of Principal here ____________________________

THE STATE OF___________________

and ______________________ (County)

I, a Notary Public in and for said County, in said State, hereby certify that _________________________, whose name is signed to the foregoing durable power of attorney, and who is known to me, acknowledged before me on this day that being informed of the contents of the durable power, she executed the same voluntarily on the day the same bears date.

GIVEN under my hand and seal this _______ day of ____________ (month), ____________ (year).

____

NOTARY PUBLIC

My Commission Expires:

Glossary

A

Abortion rate the number of abortions per 1,000 women aged 15–44.

Abortion ratio the number of abortions per 1,000 live births.

Absolute poverty the lack of resources that leads to hunger and physical deprivation.

Absolutism sexual value system based on unconditional allegiance to the authority of science, law, tradition, or religion (e.g., sexual intercourse before marriage is wrong).

Abusive head trauma nonaccidental head inury in infants and toddlers.

Accommodating style of conflict conflict style in which the respective partners are not assertive in their positions but are cooperative. Each attempts to soothe the other and to seek a harmonious solution.

Acquaintance rape nonconsensual sex between adults who know each other.

Advance directive a legal document detailing the conditions under which a person wants life-support measures to be used for him or her in the event of a medical crisis.

Agape love style love style characterized by a focus on the well-being of the love object with little regard for reciprocation. The love of parents for their children is agape love.

Age term defined chronologically, physiologically, sociologically, and culturally.

Age discrimination discriminating against a person because of age (e.g., not hiring a person because he or she is old or hiring a person because he or she is young).

Ageism systematic persecution and degradation of people because they are old.

AIDS acquired immunodeficiency syndrome, the last stage of HIV infection, in which the immune system of a person's body is so weakened that it becomes vulnerable to disease and infection.

Al-Anon an organization that provides support for family members and friends of alcohol abusers.

Androgyny a blend of traits that are stereotypically associated with masculinity and femininity.

Annulment a mechanism that returns the parties to their premarital status. An annulment states that no valid marriage contract ever existed. Annulments are both religious and civil.

Antinatalism Opposition to having children.

Asceticism the belief that giving in to carnal lusts is wrong and that one must rise above the pursuit of sensual pleasure to a life of self-discipline and self-denial.

Avoiding style of conflict conflict style in which the partners are neither assertive nor cooperative.

B

Baby blues transitory symptoms of depression twenty-four to forty-eight hours after a baby is born.

Battered-woman syndrome general pattern of battering that a woman is subjected to, defined in terms of the frequency, severity, and injury she experiences.

Battering rape a marital rape that occurs in the context of a regular pattern of verbal and physical abuse.

Behavioral approach an approach to childrearing based on the principle that behavior is learned through classical and operant conditioning.

Beliefs definitions and explanations about what is thought to be true.

Bigamy marriage to more than one person at the same time.

Binuclear family family in which the members live in two households. Most often one parent and offspring live in one household, and the other parent lives in a separate household. They are still a family even though the members live in separate households.

Biofeedback in reference to stress management, a process in which information that is fed back to the brain helps a person to experience the desired brain wave state.

Biphobia negative beliefs about and stigmatization of bisexuality and those identified as bisexuals.

Bisexuality a sexual orientation that involves cognitive, emotional, and sexual attraction to members of both sexes.

Blended family a family created when two individuals marry and at least one of them brings with him or her a child or children from a previous relationship or marriage. Also referred to as a stepfamily.

Blind marriage practiced in traditional China, a marriage in which the bride and groom were prevented from seeing each other until their wedding day.

Blurred retirement the process of retiring gradually so that the individual works part-time before completely retiring or takes a bridge job that provides a transition between a lifelong career and full retirement.

Brainstorming suggesting as many alternatives as possible without evaluating them.

Branching in communication, going out on different limbs of an issue rather than staying focused on the issue.

Bundling a courtship custom practiced by the Puritans whereby the would-be groom slept in the future bride's bed. Both were fully clothed and had a board between them.

C

Certified family life educator (CFLE) a credential offered through the National Council on Family Relations that provides credibility for one's competence in conducting programs in all areas of family life education.

Cervical cap a contraceptive device that fits over the cervix.

Child sexual abuse exploitative sexual contact or attempted sexual contact by an adult with a child before the victim is 18. Sexual contact or attempted sexual contact includes intercourse, fondling of the breasts and genitals, and oral sex.

Chlamydia known as the silent disease, a sexually transmitted disease caused by bacteria that can be successfully treated with antibiotics.

Cialis known as the "weekend Viagra" (effective for 36 hours), a medication that allows the aging male (when stimulated) to get and keep an erection as desired.

Civil union a pair-bonded relationship given legal significance in terms of rights and privileges (more than a domestic relationship and less than a marriage). Vermont recognizes civil unions of same-sex individuals.

Cohabitation two unrelated adults (by blood or by law) involved in an emotional and sexual relationship who sleep in the same residence at least four nights a week.

Cohabitation effect the research finding that couples who cohabit before marriage have greater marital instability than couples who do not cohabit.

Coitus interruptus the ineffective contraceptive practice whereby the man withdraws his penis from the vagina before he ejaculates.

Collaborating style of conflict conflict style in which the partners are both assertive and cooperative. Each expresses his or her view and cooperates to find a solution.

Commensality eating with others. Most spouses eat together and negotiate who joins them.

Commitment an intent to maintain a relationship.

Commodification of leisure the perception of free time as a consumption opportunity whereby one expects to spend money (e.g., on vacations) to enjoy leisure.

Common-law marriage a marriage by mutual agreement between a cohabiting man and woman without a marriage license or ceremony (recognized in some, but not all, states). A common-law marriage may require a legal divorce if the couple breaks up.

Communication the process of exchanging information and feelings between two people.

Compersion the "opposite of jealousy," feeling good about and being supportive of a partner's emotional and physical involvement with and enjoyment of another person.

Competing style of conflict conflict style in which the partners are both assertive and uncooperative. Each tries to force his or her way on the other so that there is a winner and a loser.

Complementary-needs theory mate selection theory that persons with complementary needs are attracted to each other. Also known as the "opposites attract" theory.

Compromising style of conflict conflict style in which there is an intermediate solution so that both partners find a middle ground they can live with.

Conception see Fertilization.

Conflict the interaction that occurs when the behavior or desires of one person interfere with the behavior or desires of another.

Congruent message one in which the verbal and nonverbal behaviors match: the message sent by what a person says is the same as that conveyed by what he or she does.

Conjugal love the love between married people characterized by companionship, calmness, comfort, and security. This is in contrast to romantic love, which is characterized by excitement and passion.

Control group group used to compare with the experimental group that is not exposed to the independent variable being studied.

Coolidge Effect term used to describe the waning of sexual excitement and the effect of novelty and variety on sexual arousal.

Corporal punishment the use of physical force with the intention of causing a child to experience pain, but not injury, for the purpose of correction or control of the child's behavior.

Covenant marriage type of marriage that permits divorce only under conditions of fault (such as abuse, adultery, or imprisonment on a felony).

Crisis a sharp change for which typical patterns of coping are not adequate and new patterns must be developed.

Cross-dresser a generic term for individuals who may dress or present themselves in the gender of the other sex (e.g., a heterosexual male will dress as a woman).

Crude divorce rate the number of divorces that have occurred in a given year for every thousand people in the population.

Cunnilingus the oral stimulation of a woman's genitals by her partner.

D

Date rape nonconsensual sex (sexual intercourse, anal sex, oral sex) between two people who are dating or on a date.

Dating a mechanism whereby some men and women pair off for recreational purposes, which may lead to exclusive, committed relationships for the reproduction, nurturing, and socialization of children.

December marriage a marriage in which both spouses are elderly.

Defense mechanisms unconscious techniques that function to protect individuals from anxiety and minimize emotional hurt.

Defense of Marriage Act legislation passed by Congress denying federal recognition of homosexual marriage and allowing states to ignore same-sex marriages licensed elsewhere.

Demandingness the degree to which parents place expectations on children and use discipline to enforce the demands.

Dementia the mental disorder most associated with aging whereby the normal cognitive functions are slowly lost.

Depo-Provera® also referred to as "dep," a contraceptive shot taken by the woman every six months.

Desertion a separation in which one spouse leaves the other and breaks off all contact with him or her.

Developmental-maturational training approach an approach to childrearing that views what children do, think, and feel as being influenced by their genetic inheritance.

Developmental task skills developed at one stage of life that are helpful at later stages.

Diaphragm a barrier method of contraception that involves a rubber dome over the uterus to prevent sperm from moving into the uterus.

Disenchantment the change in a relationship from a state of newness and high expectation to a state of mundaneness and boredom in the face of reality.

Displacement shifting one's feelings, thoughts, or behaviors from the person who evokes them onto someone else who is a safer target.

Divorce legal ending of a valid marriage contract.

Divorce mediation process in which divorcing parties make agreements with a third party about custody, visitation, child support, property settlement, and spousal support. Divorce mediation is quicker and less expensive than litigation.

Divorcism the belief that divorce is a disaster.

Domestic partnership a relationship in which individuals who live together are emotionally and financially interdependent and are given some kind of official recognition by a city or corporation so as to receive partner benefits (e.g., health insurance).

Double standard the idea that there is one standard for women and another for men (e.g., having numerous sexual partners suggests promiscuity in a woman and manliness in a man).

Durable power of attorney a legal document that gives one person the power to act on behalf of another.

E

Ecosystem the interaction of families with their environment.

Emergency contraception also referred to as postcoital contraception, the various types of morning-after pills that are used in three circumstances: when a woman has unprotected intercourse, when a contraceptive method fails (such as condom breakage), and when a woman is raped.

Emotional abuse the denigration of an individual with the purpose of reducing the victim's status and increasing the victim's vulnerability so that he or she can be more easily controlled by the abuser. Also known as verbal abuse or symbolic aggression.

Endogamy in mate selection, the cultural expectation to marry within one's own social group in terms of race, religion, and social class.

Erectile dysfunction a man's inability to get and maintain an erection.

Eros love style love style characterized by passion and romance.

Escapism the simultaneous denial and withdrawal from a problem.

Euthanasia derived from Greek words meaning "good death," or dying without suffering/pain. Euthanasia may be passive, where medical treatment is withdrawn and nothing is done to prolong the life of the patient, or active, which involves deliberate actions to end a person's life.

Exchange theory theory that emphasizes that relations are formed and maintained between individuals offering the greatest rewards and least costs to each other.

Exogamy in mate selection, the social expectation that individuals marry outside their family group (e.g., avoid sex and marriage with a sibling or other close relative).

Extended family the nuclear family or parts of it plus other relatives such as parents, grandparents, aunts, uncles, cousins, and siblings.

Extradyadic relationship (involvement) emotional/sexual involvement between a member of a pair and someone other than the partner. The term *extradyadic,* external to the couple, is broader than *extramarital,* which refers specifically to spouses.

Extramarital affair emotional/sexual involvement of a spouse with someone other than the mate.

F

Familism philosophy in which decisions are made in reference to what is best for the family as a collective unit.

Family as defined by the U.S. Census Bureau, a group of two or more persons related by blood, marriage, or adoption. Broader definitions include individuals who live together who are emotionally and economically interdependent.

Family career the various stages and events that occur within the family.

Family caregiving activities to help the elderly, which may include daily bathing and feeding, balancing checkbooks, and transportation to the grocery store or physician.

Family life course development the process through which families change over time.

Family of orientation the family of origin into which a person is born.

Family of origin the family into which an individual is born or reared, usually including a mother, father, and children.

Family of procreation the family a person begins by getting married and having children.

Family relations doctrine an emerging doctrine holding that even nonbiological parents may be awarded custody or visitation rights if they have been economically and emotionally involved in the life of the child.

Fellatio the oral stimulation of a man's genitals by his partner.

Female circumcision sometimes referred to as genital cutting or mutilation, the cultural practice (e.g., in parts of Africa and some Middle Eastern countries) of cutting off the clitoris to reduce the woman's libido and make her marriageable.

Female condom a condom that fits inside the woman's vagina to protect her from pregnancy, HIV infection, and other STDs.

Female genital mutilation see Female circumcision.

Female genital operations see Female circumcision.

Feminization of poverty the idea that poverty is experienced disproportionately by women.

Fertilization also known as conception, the fusion of the egg and sperm.

Filial piety love and respect toward one's parents.

Filial responsibility feeling a sense of duty to take care of one's elderly parents.

Formal separation a separation between spouses based on a legal agreement drawn up by an attorney and specifying the rights and responsibilities of the parties, including custody issues.

Foster parent, also known as a family caregiver, a person who at home either alone or with a spouse takes care of and fosters a child taken into custody.

Friends with benefits a relationship consisting of nonromantic friends who also have a sexual relationship.

Functionalists structural functionalist theorists who view the family as an institution with values, norms, and activities meant to provide stability for the larger society.

G

Gender the social and psychological behaviors that women and men are expected to display in society.

Gender dysphoria the condition in which one's gender identity does not match one's biological sex.

Gender identity the psychological state of viewing oneself as a girl or a boy, and later as a woman or a man.

Gender role ideology the proper role relationships between women and men in a society.

Gender roles behaviors assigned to women and men in a society.

Gender role transcendence abandoning gender frameworks and looking at phenomena independent of traditional gender categories.

Genital herpes also known as herpes simplex virus type 2, a sexually transmissible viral infection. Can also be transmitted to a newborn during birth.

Genital warts sexually transmitted lesions that commonly appear on the cervix, vulva, or penis, or in the vagina or rectum.

Gerontology the study of aging.

Gerontophobia fear or dread of the elderly.

Gonorrhea also known as the clap, the whites, and morning drop, a bacterial infection that is sexually transmitted.

Granny dumping a situation in which adult children or grandchildren who feel burdened with the care of their elderly parent or grandparent drive the elder to the entrance of a hospital and leave him or her there with no identification.

H

Hedonism sexual value system emphasizing the pursuit of pleasure and the avoidance of pain.

HER/his career wife's career given precedence over husband's career.

Hermaphrodites (also **intersexed individuals**) persons with mixed or ambiguous genitals.

Herpes simplex virus type 1 a viral infection that can cause blistering, typically of the lips and mouth; it can also infect the genitals.

Herpes simplex virus type 2 see Genital herpes.

Heterosexism the denigration and stigmatization of any behavior, person or relationship that is not heterosexual.

Heterosexuality the predominance of cognitive, emotional, and sexual attraction to those of the opposite sex.

HIS/her career husband's career given precedence over wife's career.

HIS/HER career both husband's and wife's career given equal precedence.

HIV human immunodeficiency virus that attacks the immune system and can lead to AIDS.

Homogamy in mate selection, selecting someone with similar characteristics, such as interests, values, age, race, religion, and education.

Homonegativity a construct that refers to antigay responses such as negative feelings (fear, disgust, anger), thoughts (homosexuals are HIV carriers), and behavior (homosexuals deserve a beating).

Homophobia used to refer to negative attitudes toward homosexuality.

Homosexuality the predominance of cognitive, emotional, and sexual attraction to those of a person's own sex.

Hooking up meeting someone and becoming sexually involved at that first meeting with no commitment or expectation beyond the encounter.

Human ecology the study of ecosystems or the interaction of families with their environment.

Hysterectomy removal of a woman's uterus.

I

Individualism philosophy in which decisions are made on the basis of what is best for the individual as opposed to the family (familism).

Induced abortion the deliberate termination of a pregnancy through chemical or surgical means.

Infatuation a state of passion or attraction that is not based on reality.

Infertility the inability to achieve a pregnancy after at least one year of regular sexual relations without birth control, or the inability to carry a pregnancy to a live birth.

Informal separation similar to a formal separation except that no lawyer is involved in the separation agreement; the husband and wife settle issues of custody, visitation, alimony, and child support between themselves.

Inhibited female orgasm inability to achieve an orgasm after a period of continuous stimulation.

Intentional community group of people living together on the basis of shared values and worldview (e.g., Twin Oaks in Louisa, Virginia).

Intersexed individuals see Hermaphrodite.

Intimate partner violence an all-inclusive term that refers to crimes committed against current or former spouses, boyfriends, or girlfriends.

Intrafamilial child sexual abuse sex by an adult or older family member with a child in the family.

Intrauterine device (IUD) a small object that is inserted by a physician into a woman's uterus through the vagina and cervix for the purpose of preventing implantation of a fertilized egg in the uterine wall.

"I" statements statements that focus on the feelings and thoughts of the communicator without making a judgment on others.

J

Jadelle® a contraceptive consisting of rod-shaped silicone implants that are inserted under the skin in the upper inner arm and provide time-release progestin into a woman's system for contraception.

Jealousy an emotional response to a perceived or real threat to an important or valued relationship.

L

Laparoscopy a form of salpingectomy (tubal ligation) that involves a small incision through the woman's abdominal wall just below the navel.

Leisure the use of time to engage in freely chosen activity perceived as enjoyable and satisfying.

LesBiGays collective term referring to lesbians, gays, and bisexuals.

Levitra a medication taken by aging men to help them get and maintain an erection (an alternative to Viagra and Cialis).

Living together see Cohabitation.

Living will a legal document that identifies the wishes of an individual with regard to end-of-life care.

Lose-lose solution a solution to a conflict in which neither partner benefits.

Ludic love style love style in which love is viewed as a game whereby the love interest is one of several partners, is never seen too often, and is kept at an emotional distance.

M

Mania love style an out-of-control love whereby the person "must have" the love object. Obsessive jealousy and controlling behavior are symptoms of manic love.

Marital rape forcible rape (sexual intercourse, anal intercourse, oral sex) by one's spouse.

Marital success relationship in which the partners have spent many years together and define themselves as happy and in love (hence the factors of time and emotionality).

Marriage a legal contract signed by a heterosexual couple with the state in which they reside that regulates their economic and sexual relationship. (The highest court in Massachusetts ruled in 2004 that homosexuals may marry.)

Mating gradient the tendency for husbands to marry wives who are younger and have less education and less occupational success.

May-December marriage an age-discrepant marriage in which the younger woman is in the spring of her life (May) and he is in his later years (December).

Medicaid state welfare program for low-income individuals.

Medicare federal insurance program for short-term acute hospital care and other medical benefits for persons over 65.

Megan's Law Federal law requiring that convicted sex offenders register with local police when they move into a community.

Mexican-American referring to people of Mexican origin or descent who are American citizens.

Mifepristone also known as RU-486, a synthetic steroid that effectively inhibits implantation of a fertilized egg.

Miscarriage see Spontaneous abortion.

Munchausen syndrome by proxy rare form of child abuse whereby a parent (usually the mother) takes on the sick role indirectly (hence, by proxy) by inducing illness or sickness in her child so that she can gain attention and status as a caring parent.

N

Natural family planning a method of contraception that involves refraining from sexual intercourse when the woman is thought to be fertile.

No-fault divorce a divorce that assumes that neither party is to blame.

Nuclear family family consisting of an individual, his or her spouse, and his or her children or of an individual and his or her parents and siblings.

NuvaRing® a soft, flexible, and transparent ring approximately two inches in diameter that is worn inside the vagina and provides month-long pregnancy protection.

O

Occupational sex segregation the concentration of women in certain occupations and men in other occupations.

Oophorectomy removal of a woman's ovaries.

Open relationship a stable relationship in which the partners regard their own relationship as primary but agree that each may have emotional and physical relationships with others.

Ortho Evra® is a contraceptive transdermal patch that delivers hormones to a woman's body through skin absorption.

Oxytocin a hormone from the pituitary gland during the expulsive stage of labor that has been associated with the onset of maternal behavior in lower animals.

P

Palimony a take-off on the word *alimony,* referring to the amount of money one "pal" who lives with another "pal" may have to pay if the partners terminate their relationship.

Palliative care health care focused on the relief of pain and suffering of the individual who has a life-threatening illness and support for them and their loved ones.

Pantogamy group marriage where each member of the group is married to each other.

Parallel style of conflict style of conflict whereby both partners deny, ignore, and retreat from addressing a problem issue (e.g., "Don't talk about it, and it will go away").

Parent effectiveness training a model of childrearing that focuses on trying to understand what a child is feeling and experiencing in the here and now.

Parental alienation syndrome a disturbance in which children are obsessively preoccupied with deprecation and/or criticism of a parent (usually noncustodial parent); the denigration is unjustified and/or exaggerated. Found in situations where one divorced parent has "brainwashed" the child to perceive the other parent in very negative ways.

Parental investment any investment by a parent that increases the chance that the offspring will survive and thrive.

Parenting the provision by an adult or adults of physical care, emotional support, instruction, and protection from harm in an ongoing structural (home) and emotional context to one or more dependent children.

Patriarchy "rule by the father," a norm designed to ensure that women are faithful to their husbands and remain in the home as economic assets, childbearers, and child-care workers.

Periodic abstinence refraining from sexual intercourse during the one to two weeks each month when the woman is thought to be fertile.

Physical abuse intentional infliction of physical harm in the form of socking, slapping, pushing, kicking, or burning (e.g., with water or cigarettes) by one individual on another.

Polyamory three or more men and women in a committed emotional and sexual relationship. The partners may rear children in this context.

Polyandry a form of polygamy in which one wife has two or more husbands.

Polygamy a generic term referring to a marriage in which there are more than two spouses.

Polygyny a form of polygamy in which one husband has two or more wives.

Positive androgyny a view of androgyny that is devoid of the negative traits associated with masculinity (aggression, being hard-hearted, indifferent, selfish, showing off, and vindictive) and femininity (being passive, submissive, temperamental, and fragile).

POSSLQ an acronym used by the U.S. Census Bureau that stands for People of the Opposite Sex Sharing Living Quarters.

Postpartum depression a reaction more severe than the "baby blues" to the birth of one's baby, characterized by crying, irritability, loss of appetite, and difficulty in sleeping.

Poverty the lack of resources necessary for material well-being—most importantly, food and water but also housing and health care.

Power the ability to impose one's will on another and to avoid being influenced by the partner.

Pragma love style love style that is logical and rational. The love partner is evaluated in terms of pluses and minuses and regarded as a good or bad "deal."

Pregnancy a condition that begins five to seven days after conception when the fertilized egg is implanted (typically in the uterine wall).

Prenuptial agreement a contract between intended spouses specifying what assets will belong to whom and who will be responsible for paying what in the event of a divorce.

Primary group small, intimate, informal group. A family is a primary group.

Primary sexual dysfunction a dysfunction that the person has always had.

Principle of least interest principle stating that the person who has the least interest in a relationship controls the relationship.

Projection attributing one's own thoughts, feelings, and desires to someone else while avoiding recognition that these are one's own thoughts, feelings, and desires.

Pronatalism view that encourages having children.

Propinquity the tendency to date and marry someone who lives, goes to school, or works in the same area. The term *propinquity* means "nearness."

Psychological abuse see Emotional abuse.

R

Race group of individuals who have physical characteristrics that are identified and labeled as being socially significant.

Racism the attitude that one group of people is inferior to another group on the basis of physical characteristics.

Random sample sample in which each person in the population being studied has an equal chance of being included in the sample.

Rapid ejaculation persistent or recurrent ejaculation with minimal sexual stimulation before, upon, or shortly after penetration and before the partner wishes it.

Rationalization the cognitive justification for one's own behavior that unconsciously conceals one's true motives.

Refined divorce rate the number of divorces/annulments in a given year divided by the number of married women in the population times 1,000.

Reflective listening paraphrasing or restating what a person has said to indicate that the listener understands.

Relativism sexual value system whereby decisions are made in the context of the situation and the relationship (e.g., sexual intercourse is justified in a stable, caring, monogamous context).

Residential propinquity the situation in which individuals who live close to each other tend to date and marry each other.

Resiliency the ability of a family to respond to a crisis in a positive way.

Responsiveness the extent to which parents respond to and meet the needs of their children. Refers to such qualities as warmth, reciprocity, person-centered communication, and attachment.

Rite of passage event that marks the transition from one status to another (e.g., a wedding marks the rite of passage of individuals from lovers to spouses).

Role the behavior individuals in certain status positions are expected to engage in (e.g., spouses are expected to be faithful).

Role compartmentalization separating the roles of work and home so that an individual does not dwell on the problems of one role while physically being at the place of the other role.

Role conflict being confronted with incompatible role obligations (e.g., the wife is expected to work full-time and also to be the primary caretaker of children).

Role overload the convergence of several aspects of one's role resulting in having neither time nor energy to meet the demands of that role. For example, the role of wife may involve coping with the demands of employee, parent, and spouse.

Role strain the anxiety that results from not being able to take care of several role needs at once.

Romantic love an intense love whereby the lover believes in love at first sight, only one true love, and that love conquers all.

Rophypnol date rape drug that renders persons unconscious so that they have no memory of what happens when they are under the influence.

RU-486 see Mifepristone.

S

Salpingectomy tubal ligation, or tying a woman's fallopian tubes, to prevent pregnancy.

Sandwich generation individuals who attempt to meet the needs of their children and elderly parents at the same time.

Satiation the state in which a stimulus loses its value with repeated exposure. Partners sometimes get tired of each other because they are around each other all the time.

Second shift the cooking, housework, and child care that employed women do when they return home from their jobs.

Secondary group large or small group characterized by impersonal and formal interaction. A civic club is an example.

Secondary sexual dysfunction a dysfunction that the person is currently experiencing following a period of satisfactory sexual functioning.

Secondary virginity the conscious decision of a sexually active person to refrain from intimate encounters for a specified period of time.

Sensate focus an exercise whereby the partners focus on pleasuring each other in nongenital ways.

Sex the biological distinction between being female and being male (for example, having XX or XY chromosomes).

Sexism an attitude, action, or institutional structure that subordinates or discriminates against an individual or group because of their biological sex (e.g., women are discriminated against as national television news anchors).

Sex roles behaviors defined by biological constraints. Examples include wet nurse, sperm donor, and childbearer.

Sexual double standard see Double standard.

Sexual orientation the aim and object of one's sexual interests—toward members of the same sex, the opposite sex, or both sexes.

Sexual script shared interpretations and expected behaviors in sexual situations.

Sexual values moral guidelines for sexual behavior.

Shaken baby syndrome behavior in which the caretaker, most often the father, shakes the baby to the point of causing the child to experience brain or retinal hemorrhage.

Shared parenting dysfunction the set of behaviors by both parents that are focused on hurting the other parent and are counterproductive for the child's well-being.

Shift work having one parent work during the day and the other parent work at night so that one parent can always be with the children.

Single-parent family family in which there is only one parent—the other parent is completely out of the child's life through death, sperm donation, or complete abandonment, and no contact is ever made with the other parent.

Single-parent household a household in which one parent typically has primary custody of the child or children but the parent living out of the house is still a part of the child's family.

Socialization the process through which we learn attitudes, values, beliefs, and behaviors appropriate to the social positions we occupy.

Sociobiology a theory that emphasizes that there are biological explanations for social behavior.

Sociological imagination the perspective of how powerful social structure and culture are in influencing personal decision making.

Socioteleological approach an approach to childrearing that explains children's behavior as resulting from the attempt to compensate for feelings of inferiority.

Spectatoring mentally observing one's own and one's partner's sexual performance. Often interferes with performance because of the associated anxiety.

Spermicide a chemical that kills sperm.

Spontaneous abortion an unintended termination of a pregnancy.

Stalking the willful, repeated, and malicious following or harassment of another person.

Status a position a person occupies within a social group such as parent, spouse, child.

STD sexually transmitted disease that can be transmitted through sociosexual contact.

Stepfamily a family in which at least one spouse brings at least one child into a remarriage. Also referred to as a blended family.

Stepism the assumption that stepfamilies are inferior to biological families. Stepfamilies are stigmatized.

Sterilization a permanent surgical procedure that prevents reproduction.

Storge love style a love consisting of friendship that is calm and nonsexual.

Stratification the ranking of people according to socioeconomic status, usually indexed according to income, occupation, and educational attainment.

Stress a nonspecific response of the body to demands made on it.

Supermom a cultural label that allows the mother who is experiencing role overload to regard herself as particularly efficient, energetic, and confident.

Superwoman see Supermom.

Symbolic aggression see Emotional abuse.

Syphilis a sexually transmissible disease caused by spirochete entering the mucous membranes that line various body openings. Can also be transmitted by a pregnant woman to her unborn child.

T

THEIR career career shared by couple. They travel and work together (e.g., journalists).

Theoretical framework a set of interrelated principles designed to explain a particular phenomenon and to provide a point of view.

Therapeutic abortion an abortion performed to protect the life or health of the woman.

Third shift the emotional energy expended by a spouse/parent in dealing with various family issues. The job and housework are the first and second shifts, respectively.

Time out a discipline procedure whereby the child is removed from an enjoyable context to a nonreinforcing one and left there (alone) for a minute for every year of the child's age.

Transgendered a generic term that refers to a broad spectrum of individuals who express characteristics other than those of their assigned gender.

Transgenderism a political movement seeking to challenge the belief that every person can be categorized as simply a woman or a man.

Transgenderist an individual who lives in a gender role that does not match his or her biological sex, but has no desire to surgically alter his or her genitalia.

Transition to parenthood the period of time from the beginning of pregnancy through the first few months after the birth of a baby.

Transracial adoption the practice of parents adopting children of another race—for example, a white couple adopting a Korean or African-American child.

Transsexual an individual who has the anatomical and genetic characteristics of one sex but the self-concept of the other.

Transvestite a person who enjoys dressing in the clothes of the other sex.

Two-stage marriage term used by Margaret Mead whereby a couple would first live together without having children; if the relationship was stable and durable, they would marry and have children.

U

Utilitarianism the doctrine holding that individuals rationally weigh the rewards and costs associated with behavioral choices

V

Values standards regarding what is good and bad, right and wrong, desirable and undesirable.

Vasectomy male sterilization involving cutting out small portions of the vas deferens.

Verbal abuse see Emotional abuse.

Viagra a medication taken by aging men to help them get and maintain an erection (the first medication of its kind—before Levitra and Cialis).

Violence the intentional infliction of physical harm by one individual toward another.

W

Win-lose solution a solution to a conflict in which one partner benefits at the expense of the other.

Win-win relationship a relationship in which conflict is resolved so that each partner derives benefits from the resolution.

Withdrawal see Coitus interruptus.

Y

"You statements" Statements that tend to assign blame (e.g., "You made me angry") rather than "I statements" ("I got angry and blew up").

References

Chapter 1

Aborampah, O. M. 1999. Systems of kinship and marriage in Africa: Continuities and change. In *Till death do us part: A multicultural anthology on marriage,* edited by Sandra Lee Browning and R. Robin Miller. Stamford, Conn: JAI Press, 123–38.

Abowitz, D. A., and D. Knox. 2003. College student life: Gender, gender ideology, and the effects of Greek status. Paper presented at the 73rdAnnual Meeting of the Eastern Sociological Society, Philadelphia, February 28.

Al-Krenawi, A., J. R. Graham, and V. Slonim-Nevo. 2002. Mental health aspects of Arab-Israeli adolescents from polygamous versus monogamous families. *Journal of Social Psychology* 142:446–60.

American Council on Education and University of California. 2002. *The American freshman: National norms for fall, 2002.* Los Angeles: Higher Education Research Institute. U.C.L.A. Graduate School of Education and Information Studies.

Arrindell, W. A., and F. Luteijn. 2000. Similarity between intimate partners for personality traits as related to individual levels of satisfaction with life. *Personality and Individual Differences* 28:629–37.

Barrera, M., H. M. Prelow, L. E. Dumka, N. A. Gonzales, G. P. Knight, M. L. Michaels, M. W. Roosa, and J. Tien. 2002. Pathways from family economic conditions to adolescents' distress: Supportive parenting, stressors outside the family, and deviant peers. *Journal of Community Psychology* 30:135–52.

Biskupic, J. 2003. Court's opinion on gay rights reflects trends. *USA Today.* 18 July, A-1.

Blumer, H. G. 1969. The methodological position of symbolic interaction. In *Symbolic interactionism: Perspective and method.* Englewood Cliffs, N.J.: Prentice-Hall.

Bronfenbrenner, U. 1979. *The ecology of human development: Experiments by nature and design.* Cambridge, Mass.: Harvard University Press.

Bubolz, M. M., and M. S. Sontag. 1993. Human ecology theory. In *Sourcebook of family theory and methods: A contextual approach,* edited by P.G. Boss, W. J. Doherty, R. LaRossa, W. R. Schumm, and S. K. Steinmetz. New York: Plenum.

Clapp, J. D., and A. L. McDonnell. 2000. The relationship of perceptions of alcohol promotion and beer drinking norms to alcohol problems reported by college students. *Journal of College Student Development* 41:19–26.

Cooksey, E. C., F. L. Mott, and S. A. Neubauer. 2002. Friendships and early relationships: Links to sexual initiation among American adolescents born to young mothers. *Perspectives on Sexual and Reproductive Health* 34:118–126.

Cooley, C. H. 1964. *Human nature and the social order.* New York: Schocken.

Crook, J. H., and S. J. Crook. 1988. Tibetan polyandry: Problems of adaptation and fitness. In *Human reproductive behavior: A Darwinian perspective,* edited by Laura Betzig, Monique B. Bulder, and Paul Turke. Cambridge, Eng.: Cambridge University Press, 97–114.

Crosnoe, R., and G. H. Elder, Jr. 2002. Life course transitions, the generational stake, and grandparent-grandchild relationships. *Journal of Marriage and the Family* 64:1089–96.

Crowell, J. A., D. Treboux, Y. Gao, C. Fyffe, H. Pan, and E. Waters. 2002. Assessing secure base behavior in adulthood: Development of a measure, links to adult attachment representations, and relations to couples' communication and reports of relationships. *Developmental Psychology* 38:679–93.

DeGraaf, P. M., and M. Kalmijn. 2003. Alternative routes in the remarriage market: Competing-risk analysis of union formation after divorce. *Social Forces* 81:1459–98.

Frosch, C. A., S. C. Mangelsdorf, and J. L. McHale. 2000. Marital behavior and the security of preschooler-parent attachment relationships. *Journal of Family Psychology* 14:144–61.

Gillmore, M. R., M. E. Archibald, D. M. Morrison, A. Wilsdon, E. A. Wells, M. J. Hoppe, D. Nahom, and E. Murowchick. 2002. Teen sexual behavior: Applicability of the theory of reasoned action. *Journal of Marriage and the Family* 64:885–97.

Henley-Walters, L., W. Warzywoda-Kruszynska, and T. Gurko. 2002. Cross-cultural studies of families: Hidden differences. *Journal of Comparative Family Studies* 33:433–50.

Howard, Margo. 2003. *A life in letters: Ann Landers's letters to her only child.* New York: Warner.

Hunt, M. O. 2003. Race, ethnicity, and beliefs about the black/white gap in socioeconomic status, 1977–2000. Paper presented at the 73rd Annual Meeting of the Eastern Sociological Society, Philadelphia, February 28.

Jenkins, J. M., J. Rasbash, and T. G. O'Connor. 2003. The role of the shared family context in differential parenting. *Developmental Psychology* 39:99–113.

Jones, D. J., R. Forehand, G. H. Brody, and L. Armistead. 2002. Positive parenting and child psychosocial adjustment in inner-city single parent African American families. *Behavior Modification* 26:464–81.

Kanfer, S. 2000. *Groucho: The Life and Times of Julius Henry Marx.* New York: Knopf.

Kim, H. K., and P. C. McKenry. 2002. The relationship between marriage and psychological well-being. *Journal of Family Issues* 23:885–911.

Kitzmann, K. M. 2000. Effects of marital conflict on subsequent triadic family interactions and parenting. *Developmental Psychology* 36:3–13.

Knox, D., M. E. Zusman, M. Kaluzny, and L. Sturdivant. 2000. Attitudes of college students toward infidelity. *College Student Journal* 34:162–64.

LaBrie, J. W., J. Schiffman, and M. Earleywine. 2002. Expectancies specific to condom use mediate the alcohol and sexual risk relationship. *Journal of Sex Research* 39:145—52.

LaSala, M. 2002. Walls and bridges: How coupled gay men and lesbians manage their intergenerational relationships. *Journal of Marital and Family Therapy* 28:327–39.

Lorber, J. 1998. *Gender inequality: Feminist theories and politics.* Los Angeles: Roxbury.

McAdoo, H. P. 2000. Transference of values in African American families and children. *National Council on Family Relations Report* 45:5 et passim.

McCullough, D., and D. S. Hall. 2003. Polyamory—What it is and what it isn't. *Electronic Journal of Human Sexuality* 6: http://www.ejhs.org/volume6/polyamory.htm

McNeely, C., M. L. Shew, T. Beuhrign, R. Sieving, B. C. Miller, and R. W. Blum. 2002. Mothers' influence on the timing of first sex among 14- and 15-year-olds. *Journal of Adolescent Health* 31:256–65.

Murdock, G. P. 1949. *Social structure.* New York: Free Press.

Nelson, S. 2002. Sense of humor and women's psychological health. *Progress: Family Systems Research and Therapy* 11:117–24.

Neyer, F. J., and F. R. Lang. 2003. Blood is thicker than water: Kinship orientation across adulthood. *Journal of Personality and Social Psychology* 84:310–21.

Peterson, K. S. 2000. Unhappily ever after: Children of divorce grow into bleak legacy. *USA Today.* 5 September, D1.

Pimentel, E. E. 2000. Just how do I love thee? Marital relations in urban China. *Journal of Marriage and the Family* 62:32–47.

Pinsof, W. 2002. Introduction to the special issue on marriage in the 20th century in Western civilization: Trends, research, therapy, and perspectives. *Family Process* 41:133–34.

Previti, D., and P. R. Amato. 2003. Why stay married? Rewards, barriers, and marital stability. *Journal of Marriage and Family* 65:561–73.

Ramos, K., and M. Fine. 2000. Parent-child cosleeping in the context of parental belief systems. Poster session at the 62nd Annual Meeting of the National Council on Family Relations, Minneapolis, November 12.

Salmon, J., N. Owen, A. Bauman, M. K. H. Schmitz, and M. Booth. 2000. Leisure-time, occupational, and household physical activity among professional, skilled, and less-skilled workers and homemakers. *Prevention Medicine* 30:191–99.

Sasaki, K. 2002. Attachment style in childhood: Effects on our romantic relationships and parenting of children. *Progress: Family Systems Research and Therapy* 11:169–80.

Shaw, D. S., M. Gilliom, E. M. Ingoldsby, and D. S. Nagin. 2003. Trajectories leading to school-age conduct problems. *Developmental Psychology* 39:189–200.

Skolnick, A. 2000. A time of transition: Sustainable families in the new global economy. *National Council on Family Relations Report* 45:11 passim.

Statistical Abstract of the United States: 2003. 123rd ed. Washington, D.C.: U.S. Bureau of the Census.

Stuart, T. D., and M. E. B. Garrison. 2002. The influence of daily hassles and role balance on health status: A study of mothers of grade school children. *Women and Health* 36:1–11.

Sutton, P. D. 2003. Births, marriages, divorces, and deaths: Provisional data for January–March 2002. *National Vital Statistics Report* 51, no 6. Hyattsville, Md.: National Center for Health Statistics.

Tamura, T., and A. Lau. 1992. Connectedness versus separateness: Applicability of family therapy to Japanese families. *Family Process* 31:319–40.

Teachman, J. D. 2002. Childhood living arrangements and the intergenerational transmission of divorce. *Journal of Marriage and Family* 64:717–29.

Waite, L., and M. Gallagher. 2000. *The case for marriage: Why married people are happier, healthier and better off financially.* New York: Doubleday.

Walen, H. R., and M. E. Lachman. 2000. Social support and strain from partner, family, and friends: Costs and benefits for men and women in adulthood. *Journal of Social and Personal Relationships* 17:5–30.

White, J. M., and D. M. Klein. 2002. *Family theories,* 2d ed. Thousand Oaks, Calif.: Sage.

Wilmoth, J., and G. Koso. 2002. Does marital history matter? Marital status and wealth outcomes among preretirement adults. *Journal of Marriage and the Family* 64:254–68.

Yee, B. W. K. 1992. Gender and family issues in minority groups. In *Cultural diversity and families,* edited by K. G. Arms, J. K. Davidson, Sr., and N. B. Moore. Dubuque, Iowa: Bench Press, 5–10.

Young, J. R. 2003. Prozac campus. *Chronicle of Higher Education* 69:A37–38.

Chapter 2

Abowitz, D. A., and D. Knox. 2003. College student life goals: Gender, gender ideology, and the effects of Greek status. Paper presented at the 73rd Annual Meeting of the Eastern Sociological Society, Philadelphia, February 28.

Alcock, J. 2001. *The triumph of sociobiology.* New York: Oxford University Press.

Aldous, J., and R. Woodberry. 1994. Gender, marital status, and the pursuit of happiness. Paper presented at the 56th Annual Meeting of the National Council on Family Relations, Minneapolis.

American Council on Education and University of California. 2002(4). The American freshman: National norms for fall, 2001(3). Los Angeles: Higher Education Research Institute.

Baca Zinn, M., and A. Y. H. Pok. 2002. Tradition and transition in Mexican-origin families. In *Minority families in the United States: A Multicultural perspective,* edited by Ronald L. Taylor. Upper Saddle River, N. J.: Prentice –Hall, 79–100.

Basow, S. A. 1992. *Gender: Stereotypes and roles.* 3d ed. Pacific Grove, Calif.: Brooks/Cole.

Bem, S. L. 1983. Gender schema theory and its implications for child development: Raising gender-aschematic children in a gender-schematic society. *Signs* 8:596–616.

Berglund. D. M., and T. Inman. 2000. Gender role stereotypes and family roles in comic strips. Poster session presented at the 62nd Annual Meeting of the National Council on Family Relations, Minneapolis, November 12.

Bird, C. E., and A. M. Fremont. 1991. Gender, time, use, and health. *Journal of Health and Social Behavior* 32:114–29.

Blair, K. D. 2002. School social work, the transmission of culture, and gender roles in schools. *School Social Work* 24:21–33.

Brewer, N., P. Mitchell, and N. Weber. 2002. Gender role organization, status, and conflict management styles. *International Journal of Conflict Management* 13:78–94.

Buss, D. M. 1989. Sex differences in human mate preferences: Evolutionary hypotheses tested in 37 cultures. *Behavioral and Brain Sciences* 12:1–13.

Chodorow, N. 1978. *The reproduction of mothering.* Berkeley: University of California Press.

Colapinto, J. 2000. *As nature made him: The boy who was raised as a girl.* New York: Harper Collins.

Consolatore, D. 2002. What next for the women of Afghanistan? *The Humanist* 62:10–15.

Croyle, K. L., and J. Waltz. 2002. Emotional awareness and couples' relationship satisfaction. *Journal of Marital and Family Therapy* 28:435–44.

Cullen, L. T. 2003. I want your job, lady! *Time.* 12 May, 52 et passim.

Diamond, D. A. 2003. Breaking down the barricades: The admission of women at Virginia Military Institute and the United States Military Academy at West Point. Paper presented at the Annual Meeting of the Eastern Sociological Society, Philadelphia, February.

Donaghue, N., and B. J. Fallon. 2003. Gender-role self-stereotyping and the relationship between equity and satisfaction in close relationships. *Sex Roles: A Journal of Research* 48:217–31.

Elson, J. 2003. Hormonal hierarchy: Hysterectomy and stratified stigma. *Gender and Society* 17:750–70.

Fox, G. L., and V. M. Murry. 2001. Gender and families: Feminist perspectives and family research. In *Understanding families into the new millennium: A decade in review,* edited by Robert M. Milardo. Minneapolis: The National Council on Family Relations, 379–91.

Furham, A., and L. Gasson. 1998. Sex differences in parental estimates of their children's intelligence. *Sex Roles* 38:151–62.

Glenn, E. N., and G. H. Yap. 2002. Chinese American families In *Minority families in the United States: A multicultural perspective,* edited by Ronald L. Taylor. Upper Saddle River, N.J.: Prentice Hall, 134–63.

Heckert, T. M., H. E. Dorste, G. W. Farmer, P. J. Adams, J. C. Bradley, and B. M. Bonness. 2002. Effect of gender and work experience on importance of job characteristics when considering job offers. *College Student Journal* 36:344–51.

Hout, M., and C. S. Fischer. 2002. Why more Americans have no religious preference: Politics and generations. *American Sociological Review* 67:165–90.

Hynie, M., R. A. Schuller, and L. Couperthwaite. 2003. Perceptions of sexual intent: The impact of condom possession. *Psychology of Women Quarterly* 27:75–79.

Jones, D. 2003. Few women hold top executive jobs, even when CEO's are female. *USA Today.* 27 January, B1.

Joseph, R. 2000. The evolution of sex differences in language, sexuality, and visual-spatial skills. *Archives of Sexual Behavior* 29:35–66.

Kayamy, J. M., and P. Yelsma. 2000. Displacement effects of online media in the socio-technical contexts of households. *Journal of Broadcasting and Electronic Media* 44:215–29.

Kennedy, M. 2000. Gender role observations of East Africa. Written exclusively for this text.

Kimmel, M. S. 2001. Masculinity as homophobia: Fear, shame, and silence in the construction of gender identity. In *Men and masculinity: A text reader,* edited by T. F. Cohen. Belmont, Calif.: Wadsworth, 29–41.

Kissee, J. E., S. D. Murphy, G. L. Bonner, and L. C. Murley. 2000. Effects of family origin dynamics on college freshmen. *College Student Journal* 34:172–81.

Knox, D., M. E. Zusman, and H. R. Thompson. In press. Emotional perceptions of self and others: Stereotypes and data. *College Student Journal.*

Kohlberg, L. 1966. A cognitive-developmental analysis of children's sex-role concepts and attitudes. In *The development of sex differences,* edited by E. E. Macoby. Stanford, Calif.: Stanford University Press.

———. 1969. State and sequence: The cognitive-developmental approach to socialization. In *Handbook of socialization theory and research,* edited by D. A. Goslin. Chicago: Rand McNally, 347–480.

Laner, M., and N. A. Ventrone. 2000. Dating scripts revisited. *Journal of Family Issues* 21:488–500.

Lindsey, E. W., and J. Mize. 2001. Contextual differences in parent-child play: Implications for children's gender role development. *Sex Roles: A Journal of Research* 44:155–76.

Lippa, R. A. 2002. Gender-related traits of heterosexual and homosexual men and women. *Archives of Sexual Behavior* 31:83–98.

Lorber, J. 2001. "Night to his day": The social construction of gender. In *Men and masculinity: A text reader,* edited by T. F. Cohen. Belmont, Calif.: Wadsworth, 19–28.

Martin, K. A. 2003. Giving birth like a girl. *Gender and Society.* 17:54–72.

McNeely, A., D. Knox, and M. E. Zusman. 2004. Beliefs about men: Gender differences among college students. Poster presented at the Annual Meeting of the Southern Sociological Society, Atlanta, April 16–17.

Mead, Margaret. 1935. *Sex and temperament in three primitive societies.* New York: William Morrow.

Miller, A. S., and R. Stark 2002. Gender and religousness: Can socialization explanations be

saved? *American Journal of Sociology* 107:1399–1423.
Miller, E. M., and C. Costello. 2001. The limits of biological determinism. *American Sociological Review* 66:592–98.
Miller, S. 2000. When sexual development goes awry. *The World & I* 15:148–55.
Moghadam, V. M. 2002. Patriarchy, the Taliban, and the politics of public space in Afghanistan. *Women's Studies International Forum* 25:19–31.
Monro, S. 2000. Theorizing transgender diversity: Towards a social model of health. *Sexual and Relationship Therapy* 15:33–42.
Murphy, E. M. 2003. Being born female is dangerous to your health. *American Psychologist* 58:205–10.
National Opinion Research Center at the University of Chicago. 2002. Summary Report 2001: Doctorate recipients from United States Universities, Chicago, Illinois.
Peoples, J. G. 2001.The cultural construction of gender and manhood. In *Men and masculinity: A text reader,* edited by T. F. Cohen. Belmont, Calif.: Wadsworth, 9–18.
Pimentel, E. E. 2000. Just how do I love thee? Marital relations in urban China. *Journal of Marriage and the Family* 62:32–47.
Pollack, W. S. (with Shuster, T.). 2001. *Real boys' voices.* New York: Penguin Books.
Riseman, B. J., and D. Johnson-Sumerford. 1998. Doing it fairly: A study of postgender marriages. *Journal of Marriage and the Family* 60:23–40.
Roen, K. 2001. "Either/or" and "both/neither": Discursive tensions in transgender politics. *Signs: Journal of Women in Culture and Society* 27:501–22.
Ross, C. E., and M. Van Willigen. 1997. Education and the subjective quality of life. *Journal of Health and Social Behavior* 38:275–97.
Salgado de Snyder, V. N., A. Acevedo, M. D. J. Diaz-Perez, A. Saldivar-Garduno. 2000. Understanding the sexuality of Mexican-born women and their risk for HIV/AIDS. *Psychology of Women Quarterly* 24:100–110.
Skaine, R. 2002. *The women of Afghanistan under the Taliban.* Jefferson, N.C.: McFarland.
Skoe, E. E. A., A. Cumberland, N. Eisenberg, K. Hansen, and J. Perry. 2002. The influences of sex and gender-role identity on moral cognition and prosocial personality traits. *Sex Roles* 46:295–309.
Stake, J. E. 2000. When situations call for instrumentality and expressiveness: Resource appraisal, coping strategy choice, and adjustment. *Sex Roles* 42:865–85.
Statistical Abstract of the United States: 2003. 123rd ed. Washington, D.C.: U.S. Bureau of the Census.
Taylor, R. L. 2002 Black American families. In *Minority families in the United States: A Multicultural perspective,* edited by Ronald L. Taylor. Upper Saddle River, N.J.: Prentice Hall, 19–47.
Thelen, M. H., W. Vander-Wal, J. S. A. Muir-Thomas, and R. Harmon. 2000. Fear of intimacy among dating couples. *Behavior Modification* 24:223–40.
Thornhill, R., and C. T. Palmer. 2000. *A natural history of rape: Biological bases of sexual coercion.* Cambridge, Mass.: MIT Press.
Vogel, D. L., S. R. Wester, M. Heesacker, and S. Madon. 2003. Confirming gender stereotypes: A social role perspective. *Sex Roles* 48:519–28.
Waite, L. J. 2000. Trends in men's and women's well-being in marriage. In *The ties that bind,* edited by L. J. Waite. New York: Aldine de Gruyter, 368–92.
Ward, C. A. 2001. Models and measurements of psychological androgyny: A cross-cultural extension of theory and research. *Sex Roles: A Journal of Research* 43:529–52.
Weinber, M. S., I. Lottes, and F. M. Shaver. 2000. Sociocultural correlates of permissive sexual attitudes: A test of Reiss's hypothesis about Sweden and the United States. *Journal of Sex Research* 37:44–52.
Wermuth, L., and M. Ma'At-Ka-Re Monges. 2002. Gender stratification: A structural model for examining case examples of women in less developed countries. *Frontiers* 23:1–22.
West, C., and D. H. Zimmerman. 1991. Doing gender. In *The social construction of gender,* edited by J. Lorber and S. A. Forrell, 13–37. Thousand Oaks, Calif.: Sage.
Wiederman, M. W. 2000. Women's body image self-consciousness during physical intimacy with a partner. *Journal of Sex Research* 37:60–68.
Williams, J. E., and D. L. Best. 1990. *Measuring sex stereotypes: A multination study.* London: Sage.
Winkvist, A., and H. Z. Akhtar. 2000. God should give daughters to rich families only: Attitudes toward childbearing among low-income women in Punjab, Pakistan. *Social Science and Medicine* 51:73–81.
Woodhill, B. M., and C. A. Samuels. 2003. Positive and negative androgyny and their relationship with psychological health and well-being. *Sex Roles* 48:555–65.
Zimmerman, T. S., K. E. Holm, and S. A. Haddock. 2000. A decade of advice for women and men in best-selling self-help literature. Poster session at the 62nd Annual Meeting of the National Council on Family Relations, Minneapolis, November 12.

Chapter 3

Ackerman, D. 1994. *A natural history of love.* New York: Random House.
Anderson, J. E., R. Wilson, L. Doll, T. S. Jones, and P. Barker. 2000. Condom use and HIV risk behaviors among U.S. adults: Data from a national survey. *Family Planning Perspectives* 31:24–28.
Attridge, M., E. Berscheid, and S. Sprecher. 1998. Dependence and insecurity in romantic relationships: Development and validation of two companion scales. *Personal Relationships* 5:31–58.
Bankole, A., J. E. Darroch, and S. Singh. 2000. Determinants of trends in condom use in the United States, 1988–1995. *Family Planning Perspectives* 31:264–71.
Brehm, S. S. 1992. *Intimate relationships.* 2d ed. New York: McGraw-Hill.
Bulcroft, R., K. Bulcroft, K. Bradley, and C. Simpson. 2000. The management and production of risk in romantic relationships: A postmodern paradox. *Journal of Family History* 25:63–92.
Buss, D. M. 2000. Prescription for passion. *Psychology Today.* May/June, 54–61.
Buunk, B. P., A. Angleitner, V. Oubaid, and D. Buss. 1996. Sex differences in jealousy in evolutionary and cultural perspective: Tests from the Netherlands, Germany, and the United States, *Psychological Science* 7:359–63.
Cassidy, M. L., and G. Lee. 1989. The study of polyandry: A critique and synthesis. *Journal of Comparative Family Studies* 20:1–11.
DeHart, T., S. L. Murray, B. W. Pelham, and P. Rose. 2002 The regulation of dependency in parent-child relationships. *Journal of Experimental Social Psychology* 39:59–67.
Diamond, L. M. 2003. What does sexual orientation orient? A biobehavioral model distinguishing romantic love and sexual desire. *Psychological Review* 110:173–92.
Freud, S. 1938. Three contributions to the theory of sex. In *The basic writings of Sigmund Freud,* edited by A. A. Brill. New York: Random House (article originally published in 1905).
Fromm, E. 1963. *The art of loving.* New York: Bantam
Fuller, J. A., and R. M. Warner. 2000. Family stressors as predictors of codependency. *Genetic, Social, and General Psychology Monographs* 126:5–22.
Gallmeier, C. P., M. E. Zusman, D. Knox, and L. Gibson. 1997. Can we talk? Gender differences in disclosure patterns and expectations. *Free Inquiry in Creative Sociology* 25:219–25.
Gross, N., and S. Simmons 2002. Intimacy as a double-edged phenomenon? An empirical test of Giddens. *Social Forces* 81:531–55.
Harris, C. R., and N. Christenfeld. 1996. Gender, jealousy, and reason. *Psychological Science* 7:364–66.
Hendrick, S. S., C. Hendrick, and N. L. Adler. 1988. Romantic relationships: Love, satisfaction, and staying together. *Journal of Personality and Social Psychology* 54:980–88.
Huston, T. L., Caughlin J. P., Houts, R. M., Smith, S. E., & George, L. J. (2001). The connubial crucible: Newlywed years as predictors of marital delight, distress, and divorce. *Journal of Personality and Social Psychology* 80:237–52.
Inman-Amos, J., S. S. Hendrick, and C. Hendrick. 1994. Love attitudes: Similarities between parents and between parents and children. *Family Relations* 43:456–61.
Jankowiak, W. R., and E. F. Fischer. 1992. A cross-cultural perspective on romantic love. *Ethnology* 31:149–55.
Knox, D., and M. Zusman. 1998. Unpublished data on 620 undergraduates at a large southeastern university collected for this text.
Knox, D., M. E. Zusman, L. Mabon, and L. Shivar. 1999. Jealousy in college student relationships. *College Student Journal* 33:328–29.
Knox, D., M. Zusman, and W. Nieves. 1998. What I did for love: Risky behavior of college students in love. *College Student Journal* 32:203–5.
Knox, D., M. E. Zusman, and H. R. Thompson. In press. Emotional perceptions of self and others: Stereotypes and data. *College Student Journal.*
Lee, J. A. 1973. *The colors of love: An exploration of the ways of loving.* Don Mills, Ontario: New Press.
———. 1988. Love-styles. In *The Psychology of Love,* edited by R. Sternberg and M. Barnes. New Haven, Conn.: Yale University Press, 38–67.
Lewis, D. 1985. *In and out of love: The mystery of personal attraction.* London: Methuen.
Medora, N. P., J. H. Larson, N. Hortacsu, and P. Dave. 2002. Perceived attitudes towards romanticism: A cross-cultural study of American, Asian-Indian, and Turkish young adults. *Journal of Comparative Family Studies* 33:155–78.
Meyers, J. E., and M. Shurts. 2002. Measuring positive emotionality: A review of instruments assessing love. *Measurement and evaluation in counseling and development.* 34:238–254.
Moller, N. P., C. J. McCarthy, and R. T. Fouladi. (2002). Earned attachment security: Its relationship to coping resources and stress symptoms among college students following relationship breakup. *Journal of College Student Development* 43:213–30.
Neto, F., J. C. Deschamps, J. Barros. 2000. Cross-cultural variations in attitudes toward love. *Journal of Cross-Cultural Psychology* 31:626–35.
Paul, E. L., B. McManus, and A. Hayes. 2000. "Hookups": Characteristics and correlates of college students' spontaneous and anonymous sexual experiences. *Journal of Sex Research* 37:76–88.

Petrie, J., J. A. Giordano, and C. S. Roberts. 1992. Characteristics of women who love too much. *Affilia: Journal of Women and Social Work* 7:7–20.

Pimentel, E. E. 2000. Just how do I love thee? Marital relations in urban China. *Journal of Marriage and the Family* 62:32–47.

Pines, A. M. 1992. *Romantic jealousy: Understanding and conquering the shadow of love.* New York: St. Martin's.

Pines, A. M., and A. Friedman. 1998. Gender differences in romantic jealousy. *Journal of Social Psychology* 138:54–71.

Pistole, M. C., and L. C. Vocaturo. 2000. Attachment and commitment in college students' romantic relationships. *Journal of College Student Development* 40:710–20.

Radecki Bush, C. R., J. P. Bush, and J. Jennings. 1988. Effects of jealousy threats on relationship perceptions and emotions. *Journal of Social and Personal Relationships* 5:285–303.

Reik, T. 1949. *Of love and lust.* New York: Farrar, Straus, and Cudahy.

Reiss, I. L. 1960. Toward a sociology of the heterosexual love relationship. *Journal of Marriage and Family Living* 22:139–45.

Sanderson, C. A., and K. H. Karetsky. 2002. Intimacy goals and strategies of conflict resolution in dating relationships: A mediational analysis. *Journal of Social and Personal Relationships* 19:317–37.

Sharp, E. A., and L. H. Ganong. 2000. Raising awareness about marital expectations: An unrealistic beliefs change by integrative teaching? *Family Relations* 49:71–76.

Sprecher, S., and P. C. Regan. 1998. Passionate and companionate love in courting and young married couples. *Sociological Inquiry* 68:163–85.

Statistical Abstract of the United States: 2003. 123rd ed. Washington, D.C.: U.S. Bureau of the Census.

Sternberg, R. J. 1986. A triangular theory of love. *Psychological Review* 93:119–35.

Toufexis, Anastasia. 1993. The right chemistry. *Time.* 15 February, 49–51.

Walster, E., and G. W. Walster. 1978. *A new look at love.* Reading, Mass.: Addison-Wesley.

White, G. L. 1980. Inducing jealousy: A power perspective. *Personality and Social Psychology Bulletin* 6:222–27.

Chapter 4

American Council on Education and University of California. 2002. The American freshman: National norms for fall, 2002. Los Angeles: Los Angeles Higher Education Research Institute.

Bachrach, C., M. J. Hindin, and E. Thomson. 2000. The changing shape of ties that bind: An overview and synthesis. In *The ties that bind,* edited by Linda J. Waite. New York: Aldine de Gruyter, 3–16.

Barker, O. 8 minutes to a love connection. 2002. *USA Today,* 12 December, D1.

Brooks, S. 2003. Personal communication.

Bulcroft, R., K. Bulcroft, K. Bradley, and C. Simpson. 2000. The management and production of risk in romantic relationships: A postmodern paradox. *Journal of Family History* 25:63–92.

Cohan, C. L., and S. Kleinbaum.2002. Toward a greater understanding of the cohabitation effect: Premarital cohabitation and marital communication. *Journal of Marriage and the Family* 64:180–92.

Cotton, S. R. 2003. Presentation on Internet research. Department of Sociology, East Carolina University, February 28.

Gallmeier, C., D. Knox, and M. E. Zusman. 2002. Going out or hanging out: Individual and couple dating in the new millennium. *Free Inquiry in Creative Sociology* 30:1–3.

Guldner, G. T. 2003. *Long distance relationships: The complete guide.* Corona, Calif.: JFMilne Publications.

Hatfield, E., and R. L. Rapson. 1996. *Love and sex: Cross-cultural perspectives.* Boston: Allyn and Bacon.

Heimdal, K. R., and S. K. Houseknecht. 2003. Cohabiting and married couples' income organization: Approaches in Sweden and the United States. *Journal of Marriage and the Family* 65:525–38.

Hodge, A. 2003. Video chatting and the males who do it. Paper presented at the 73rd Annual Meeting of the Eastern Sociological Association, Philadelphia, February 28.

Jamieson, L., M. Anderson, D. McCrone, F. Bechhofer, R. Stewart and Y. Li. 2002. Cohabitation and commitment: partnership plans of young men and women. *The Sociological Review* 50:356–77.

Kamp Dush, C. M. K., C. L. Cohan, and P. R. Amato. 2003. The relationship between cohabitation and marital quality and stability. Change across cohorts? *Journal of Marriage and the Family* 65:539–49.

Kiernan, K. 2000. European perspectives on union formation. In *The ties that bind,* edited by Linda J. Waite. New York: Aldine de Gruyter, 40–58.

Knox, D., V. Daniels, L. Sturdivant, and M. E. Zusman. 2001. College student use of the Internet for mate selection. *College Student Journal* 35:158–60.

Knox, D., L. Gibson, M. E. Zusman, and C. Gallmeier. 1997. Why college students end relationships. *College Student Journal* 31:449–52.

Knox, D., and Schacht, C. 2000. Interview with a Hopi elder, Hopi Indian Reservation, Arizona.

Knox, D., and M.E. Zusman. 2001. Marrying a man with baggage: Implications for second wives. *Journal of Divorce and Remarriage* 35:67–80.

Knox, D., M. Zusman, V. Daniels, and A. Brantley. 2002. Absence makes the heart grow fonder? Long- distance dating relationships among college students. *College Student Journal* 36:365–67.

Lugaila, T. A. 1998. Marital status and living arrangements: March 1997 (update). *Current Population Reports.* U.S. Census Bureau, P20–506, June.

Lydon, J., T. Pierce, and S. O'Regan. 1997. Coping with moral commitment to long-distance dating relationships. *Journal of Personality and Social Psychology* 73:104–13.

McGinnis, S. L. 2003. Cohabiting, dating, and perceived costs of marriage: A model of marriage entry. *Journal of Marriage and Family* 65:105–16.

Michael, R. T., J. H. Gagnon, E. O. Laumann, and G. Kolata. 1994. *Sex in America: A definitive survey.* Boston: Little, Brown.

Miller, R. S. 1997. Inattentive and contented: Relationship commitment and attention to alternatives. *Journal of Personality and Social Psychology* 73:758–66.

Mitchell, M. 1977. *Gone with the Wind.* New York: Macmillan. (Originally published 1936.)

Moors, G. 2000. Values and living arrangements: A recursive relationship. In *The ties that bind,* edited by Linda J. Waite. New York: Aldine de Gruyter, 212–26.

Morgan, C., and S. R. Cotton. 2003. The relationship between Internet activities and depressive symptoms in a sample of college freshmen. Forthcoming in *CyberPsychology & Behavior* 6 (2).

O'Flaherty, K. M., and L. W. Eells. 1988. Courtship behavior of the remarried. *Journal of Marriage and the Family* 50:499–506.

Oppenheimer, V. K. 2003. Cohabiting and marriage during young men's career-development process. *Demography* 40:127–49.

Peterson K. S. 2003. Dating game has changed. *USA Today.* 11 February 11, 9D.

Pimentel, E. E. 2000. Just how do I love thee? Marital relations in urban China. *Journal of Marriage and the Family* 62:32–47.

Raley, R. K. 2000. Recent trends and differentials in marriage and cohabitation: The United States. In *The ties that bind,* edited by Linda J. Waite. New York: Aldine de Gruyter, 19–38.

Rohall, D., S. R. Cotton, and C. Morgan. 2002. Internet use and the self concept: Linking specific uses to global self-esteem. *Current Research in Social Psychology* 8(1):1–18.

Sigle-Rushton, W., and S. McLanahan. 2002. The living arrangements of new unmarried mothers. *Demography* 39:415–433.

Skinner, K. B., S. J. Bahr, D. R. Crane, and V. R. A. Call. 2002. Cohabitation, marriage, and remarriage. *Journal of Family Issues* 23:74–90.

Smock, P. J. 2000. Cohabitation in the United States: An appraisal of research themes, findings, and implications. *Annual Review of Sociology* 26:1–20.

Stack, S., and J. R. Eshleman. 1998. Marital status and happiness: A 17-nation study. *Journal of Marriage and the Family* 60:527–36.

Stanley, S. M., and H. J. Markman. 1997. *Marriage in the 90s: A nationwide random phone survey.* Denver: PREP, Inc.

Statistical Abstract of the United States: 2003. 123rd ed. Washington, D.C.: U.S. Bureau of the Census.

Stets, J. E. 1993. The link between past and present intimate relationships. *Journal of Family Issues* 14:236–60.

Van Horn, K. R., A. Arone, K. Nesbitt, L. Desilets, T. Sears, M. Giffin, and R. Brundi. 1997. Physical distance and interpersonal characteristics in college students' romantic relationships. *Personal Relationships* 4:25–34.

Chapter 5

Adam, B. D., A. Sears, and E. G. Shellenberg. 2000. Accounting for unsafe sex: Interviews with men who have sex with men. *Journal of Sex Research* 37:24–36.

American Council on Education and University of California. 2002. The American Freshman: National norms for fall, 2002. Los Angeles, Calif.: Los Angeles Higher Education Research Institute.

American Foundation for Urologic Disease, Reproductive Health Council (AFUD). 2003. TRT clear and simple: Basic facts about testosterone replacement therapy for men. *Family Urology* 8:9–10.

Anderson, J. E., R. Wilson, L. Doll, T. S. Jones, and P. Barker. 2000. Condom use and HIV risk behaviors among U.S. adults: Data from a national survey. *Family Planning Perspectives* 31:24–28.

Balanko, S. L. 2000. Good sex? A critical review of school sex education. *Guidance & Counseling* 17:117–214.

Bancroft, J., E. Janissen, D. Strong, and Z. Vukadinovic. 2003. The relationship between mood and sexuality in gay men. *Archives of Sexual Behavior* 32:231–45.

Bancroft, J., J. Loftus, and J. S. Long. 2003. Distress about sex: A national survey of women in heterosexual relationships. *Archives of Sexual Behavior* 32:193–208.

Bay-Cheng, L. Y. 2001. SexEd.com: Values and norms in web-based sexuality education. *Journal of Sex Research* 38:241–51.

Bleske, A. L., and D. M. Buss. 2000. Can men and women be just friends? *Personal Relationships* 7:131–51.

Bogle, K. A. 2002. From dating to hooking up: Sexual behavior on the college campus. Paper presented at the Annual Meeting of the Society for the Study of Social Problems, Summer.

———. 2003. Sex and "Dating" in college and after: A look at perception and behavior. Paper delivered at the 73rd Annual Meeting of the Eastern Sociological Society, Philadelphia, February 28.

Byers, E. S ,and G. Grenier. 2003. Premature or rapid ejaculation: Heterosexual couples' perceptions of men's ejaculatory behavior. *Archives of Sexual Behavior* 32:261–70.

Buss, D. M., and D. P. Schmitt. 1993. Sexual strategies theory: An evolutionary perspective on human mating. *Psychological Review* 100:204–32.

Carpenter, L. M. 2003. Like a Virgin . . . Again?: Understanding Secondary Virginity in Context. Paper presented at the 73rdAnnual Meeting of the Eastern Sociological Society, Philadelphia, February 28.

Centers for Disease Control and Prevention. 2000. *HIV/AIDS Surveillance Report* 12:3–43.

Civic, D. 2000. College students' reasons for nonuse of condoms within dating relationships. *Journal of Sex and Marital Therapy* 26:95–105.

Coren, C. 2003. Timing, amount of teenage alcohol or marijuana use may make future risky sex more likely. *Perspectives on Sexual and Reproductive Health* 25:49–51.

Davis, K. R., and S. C. Weller. 2000. The effectiveness of condoms in reducing transmission of HIV. *Family Planning Perspectives* 31:272–79.

Dunn, K. M., P. R. Croft, and G. I. Hackett. 2000. Satisfaction in the sex life of a general population sample. *Journal of Sex and Marital Therapy* 26:141–51.

East, P. L. 1998. Racial and ethnic differences in girls' sexual, marital, and birth expectations. *Journal of Marriage and the Family* 60:150–62.

Eisenberg, M. E. 2002. The association of campus resources for gay, lesbian, and bisexual students with college students' condom use. *Journal of American College Life* 51:109–16.

Elia, J. P. 2000. Democratic sexuality education: A departure from sexual ideologies and traditional schooling. *Journal of Sex Education and Therapy* 25:122–29.

Ellis, B. J., and D. Symons. 1990. Sex differences in sexual fantasy: An evolutionary psychological approach. *Journal of Sex Research* 27:527–56.

Eyre, R.C., G. Zheng, and A. A. Kiessling. 2000. Multiple drug resistance mutations in human immunodeficiency virus in semen but not blood of a man on antiretroviral therapy. *Urology* (online) 55:591 passim.

Gagnon, John H. 2004. Personal communication via E-mail from Dr. Gagnon, January 16. Used by permission.

Glenn, N., and E. Marquardt. 2001 *Hooking up, hanging out, and hoping for Mr. Right: College women on dating and mating today.* New York: Institute for American Values.

Haavio-Mannila, E., and O. Kontula. 1997. Correlates of increased sexual satisfaction. *Archives of Sexual Behavior* 26:399–420.

Hyde, J. S., J. D. DeLamater, and A. M. Durik. 2001. Sexuality and the dual-earner couple, Part II: Beyond the baby years. *Journal of Sex Research* 38:10–23.

Hynie, M., J. E. Lydon, and A. Taradash. 1997. Commitment, intimacy, and women's perceptions of premarital sex and contraceptive readiness. *Psychology of Women Quarterly* 21:447–64.

Jaccard, J., P. J. Dittus, and V. V. Gordon. 2000. Parent-teen communication about premarital sex: Factors associated with the extent of communication. *Journal of Adolescent Research* 15:187–208.

Kaestle, C. E., D. E. Morisky, and D. J. Wiley. 2002. Sexual intercourse and the age difference between adolescent females and their romantic partners. *Perspectives on Sexual and Reproductive Health* 34:304–30.

Katz, B. P., D. Fortenberry, G. D. Zimet, M. J. Blythe, and D. P. Orr. 2000. Partner-specific relationship characteristics and condom use among young people with sexually transmitted diseases. *Journal of Sex Research* 37:69–75.

Kimmel, M. S. 2000. *The gendered society.* New York: Oxford University Press.

Knox, D., C. Cooper, and M. E. Zusman. 2001. Sexual values of college students. *College Student Journal* 35:24–27.

Knox, D., V. Daniels, L. Sturdivant, and M. E. Zusman. 2001. College student use of the Internet for mate selection. *College Student Journal* 35:158–60.

Knox, D., M. E. Zusman, and C. Cooper. 2001. Sexual values of college students. *College Student Journal* 35:24–27.

Kornreich, J. L., K. D. Hern, G. Rodriguez, and L. F. O'Sullivan. Sibling influence, gender roles, and the sexual socialization of urban early adolescent girls. *Journal of Sex Research* 40:101–10.

Lichter, S. R., L. S. Lichter, and D. R. Amundson, D. R. 2001. Sexual imagery in popular entertainment. Center for Media and Public Affairs. http://www.cmpa.com/archive/sexpopcult.htm

Masters, W. H., and V. E. Johnson. 1970. *Human sexual inadequacy.* Boston: Little, Brown.

McGuirl, K. E., and M. W. Wiederman. 2000. Characteristics of the ideal sex partner: Gender differences and perceptions of the preferences of the other gender. *Journal of Sex and Marital Therapy* 26:153–59.

Meier, A. M. 2003. Adolescents' transition to first intercourse, religiosity, and attitudes about sex. *Social Forces* 81:1031–52.

Metz, M. F., and J. L. Pryor. 2000. Premature ejaculation: A psychophysiological approach for assessment and management. *Journal of Sex and Marital Therapy* 26:293–320.

Michael, R. T., J. H. Gagnon, E. O. Laumann, and G. Kolata. 1994. *Sex in America.* Boston: Little, Brown.

Miller, B. C. 2002. Family influences on adolescent sexual and contraceptive behavior. *Journal of Sex Research* 39:22–27.

Motley, M. T., and H. M. Reeder. 1995. Unwanted escalation of sexual intimacy: Male and female perceptions of connotations and relational consequences of resistance messages. *Communication Monographs* 62:355–82.

Mulligan Rauch, S. A., and B. Bryant. 2000. Gender and context differences in alcohol expectancies. *Journal of Social Psychology* 140:240–53.

Naimi, T. S., L. E. Lipscomb, R. D. Brewer, and B. C. Gilbert. 2003. Binge drinking in the preconception period and the risk of unintended pregnancy: Implications for women and their children. *Pediatrics* 111:1136–41.

O'Sullivan, L. F., and H. F. L. Meyer-Bahlburg. 2003. African-American and Latina inner-city girls' reports of romantic and sexual development. *Journal of Social and Personal Relationships* 20:221–38.

Osman, S. L., and C. M. Davis. 2000. Predicting perceptions of date rape based on individual beliefs and female alcohol consumption. *Journal of College Student Development* 40:701–9.

Pepper, T., and D. L. Weiss. 1987. Proceptive and rejective strategies of U.S. and Canadian college women. *Journal of Sex Research* 23:455–80.

Peterson, K. S. 2000. "For many teens, oral sex is not sex." *USA Today.* 16 November, D1.

Raley, R. K. 2000. Recent trends and differentials in marriage and cohabitation: The United States. In *The ties that bind,* edited by Linda J. Waite. New York: Aldine de Gruyter, 19–39.

Realo, A., and R. Bodwin. 2003. Family-related allo centrism and HIV risk behavior in central and eastern Europe. *Journal of Cross-Cultural Psychology* 34:690–701.

Reiss, I. L. 1967 *The social context of premarital sexual permissiveness.* New York: Holt, Rinehart and Winston.

Reiss, I. L., and H. M. Reiss. *An end to shame: Shaping our next sexual revolution.* New York: Prometheus Books.

Rempel, J. K. and B. Baumgartner. 2003. The relationship between attitudes toward menstruation and sexual attitudes, desires, and behavior in women. *Archives of Sexual Behavior* 32:155–63.

Rostosky, S. S., D. Welsh, M. C. Kawaguchi, and R. V. Galliher. 1999. Commitment and sexual behaviors in adolescent dating relationships. In *Handbook of interpersonal commitment and relationship stability,* edited by J. M. Adams and W. H. Jones. New York: Academic/Plenum Publishers, 323–38.

Sakalh-Ugurlu, N., and P. Glick. 2003. Ambivalent sexism and attitudes toward women who engage in premarital sex in Turkey. *Journal of Sex Research* 40:296–302.

Sammons, R. Personal communication, 2003. Dr. Sammons is a psychiatrist in private practice in Grand Junction, Colorado.

Sanders, S. A., and J. M. Reinisch. 1999. Would you say you "had sex" if. . . . ? *Journal of the American Medical Association* 28:275–77.

Sather, L., and K. Zinn. 2002. Effects of abstinence-only education on adolescent attitudes and values concerning premarital sexual intercourse. *Family and Community Health* 25:15–20.

Schneider, C. S., and D. A. Kenny. 2000. Cross-sex friends who were once romantic partners: Are they platonic friends now? *Journal of Social and Personal Relationships* 17:451–66.

Serovich, J. M., and K. E. Mosack. 2003. Reasons for HIV disclosure or nondisclosure to casual sexual partners. *AIDS Education and Prevention* 15:70–81.

Simon, W., and J. Gagnon. 1998. Psychosexual development. *Society* 35:60–68.

Somers, C. L., and J. H. Gleason. 2001. Does source of sex education predict adolescents' sexual knowledge, attitudes and behaviors? *Education* 121:674–82.

Sonfield, A., and R. B. Gold. 2001. States' implementation of the section 510 abstinence education program, FY 1999. *Family Planning Perspectives* 33:166–71.

Stodgill, R. 1998. Where'd you learn that? A Time/CNN poll. *Time.* 15 June, 52 passim.

Thomsen, D., and I. J. Chang. 2000. Predictors of satisfaction with first intercourse: A new perspective for sexuality education. Poster presentation at the 62nd Annual Conference of the National Council on Family Relations, Minneapolis, November.

Townsend, J. M., and G. D. Levy. 1990. Effects of potential partners' physical attractiveness and socioeconomic status on sexuality and partner selection. *Archives of Sexual Behavior* 19:149–64.

Verma, K. K., B. K. Khaitan, and O. P. Singh. The frequency of sexual dysfunctions in patients attending a sex therapy clinic in North India. *Archives of Sexual Behavior* 27:309–15.

Waite, L. J., and K. Joyner, K. 2001. Emotional and physical satisfaction with sex in married, cohabiting, and dating sexual unions: Do men and women differ? In *Sex, love, and health in America: Private choices and public policies,* edited by E. O. Laumann and R. T. Michael. Chicago: The University of Chicago Press, 239–69.

Weinberg, M. S., Lottes, I., & Shaver, F. M. 2000. Sociocultural correlates of permissive sexual attitudes: A test of Reiss's hypothesis about Sweden and the United States. *Journal of Sex Research* 37:44–52.

Whitaker, D. J., Miller, K. S., & Clark, L. F. 2000. Reconceptualizing adolescent sexual behavior: Beyond did they or didn't they? *Family Planning Perspectives* 32:111–21.

Wiederman, M. W. 1977. Extramarital sex: Prevalence and correlates in a national survey. *Journal of Sex Research* 34:167–74.

Wiley, D. C. 2002. The ethics of abstinence-only and abstinence-plus sexuality education. *Journal of School Health* 72:164–68.

Young, M., and E. S. Goldfarb. 2000. The problematic (a)–(h) in abstinence education. *Journal of Sex Education and Therapy* 25:156–61.

Chapter 6

Allen, M., & Burrell, N. 2002. Sexual orientation of the parent: The impact on the child. In M. Allen, R. Preiss, B. Gayle, & N. Burrell (Eds.). *Interpersonal Communication Research: Advances through meta-analysis* (pp. 111–124). Mahwah, NJ: Lawrence Erlbaum.

American Council on Education and University of California. 2004. The American freshman: National norms for fall, 2003. Los Angeles: Higher Education Research Institute. U.C.L.A. Graduate School of Education and Information Studies.

Beeler, J., and V. DiProva. 1999. Family adjustment following disclosure of homosexuality by a member: Themes discerned in narrative accounts. *Journal of Marital and Family Therapy* 25:443–59.

Black, D., G. Gates, S. Sanders, and L. Taylor. 2000. Demographics of the gay and lesbian population in the United States: Evidence from available systematic data sources. *Demography* 37:139–54.

Bock, J. D. 2000. Doing the right thing? Single mothers by choice and the struggle for legitimacy. *Gender and Society* 14:62–86.

Bolte, A. 2001. Do wedding dresses come in lavender? The prospects and implications of same-sex marriage. In *The gay and lesbian marriage and family reader,* edited by J. M. Lehmann. Lincoln, Neb.: Gordian Knot Books, 25–46.

Bono, Chastity, and B. Fitzpatrick (contributor). 1998. *Family outing.* Boston: Little, Brown.

Brannock, J. C., & Chapman, B. E. 1990. Negative sexual experiences with men among heterosexual women and lesbians. *Journal of Homosexuality* 19:105–10.

Buxton, A. P. 2004. Paths and pitfalls: How heterosexual spouses cope when their husbands or wives come out. *Journal of Couple and Relationship Therapy* 3.

Cantor, J. M., R. Blanchard, A. D. Paterson, & A. F. Bogaert, A. F. 2002. How many gay men owe their sexual orientation to fraternal birth order? *Archives of Sexual Behavior* 31:63–71.

Centers for Disease Control and Prevention. 2004. http://www.cdc.gov/hiv/stats.htm#exposure

Chamberlain, R. 2003. *Shattered Love.* New York: ReganBooks.

Cochran, S. D., J. G. Sullivan, and V. M. Mays. 2003. Prevalence of mental disorders, psychological distress, and mental health services use among lesbian, gay, and bisexual adults in the United States. *Journal of Consulting and Clinical Psychology* 71:53–61.

Cohen, K. M., and R. C. Savin-Williams. 1996. Developmental perspectives on coming out to self and others. In *The lives of lesbians, gays, and bisexuals: Children to adults,* edited by R. C. Savin-Williams and K. M. Cohen. Fort Worth, Tex.: Harcourt Brace, 113–51.

Cornelius-Cozzi, T. 2002. Effects of parenthood on the relationships of lesbian couples. *PROGRESS: Family Systems Research and Therapy* 11:85–94.

Curtin, S.C., and J. A. Martin. 2000. Preliminary data for 1999. *National vital statistics reports* 48 (14). Hyattsville, Md.: National Center for Health Statistics.

Dawood, K., R. C. Pillard, C. Horvath, W. Revelle, & J. M. Bailey. 2000. Familial aspects of male homosexuality. *Archives of Sexual Behavior* 29:155–63.

Demian. 2003. Legal marriage report: Global status of legal marriage. Partners Task Force for Gay & Lesbian Couples. www.buddybuddy.com

Deenen, A. A., L. Gijiis, and A. X. Van Naerssem. 1994. Intimacy and sexuality in gay male couples. *Archives of Sexual Behavior* 23:421–31.

Diamond, L. M. 2003a. Was it a phase? Young women's relinquishment of lesbian/bisexual identities over a 5 year period. *Journal of Personality and Social Psychology* 84:352–364.

———. 2003b. What does sexual orientation orient? A biobehavioral model distinguishing romantic love and sexual desire. *Psychological Review* 110:173–92.

Dickson, N., C. Paul, and P. Herbison. 2003. Same-sex attraction in a birth order cohort: Prevalence and persistence in early adulthood. *Social Science and Medicine* 56:1607–15.

Duran-Aydintug, C., & K. A. Causey. 2001. Child custody determination: Implications for lesbian mothers. In *The gay & lesbian marriage and family reader,* edited by J. M. Lehmann. Lincoln, Neb.: Gordian Knot Books, 47–64.

Dreher, R. 2002. Beds, bathhouses, and beyond: The return of public sex. *National Review* 54:14.

Eliason, M. 2001. Bi-negativity: The stigma facing bisexual men. *Journal of Bisexuality* 1:137–54.

Erera, P. I., and K. Fredericksen. 2001. Lesbian stepfamilies: A unique family structure. In *The gay and lesbian marriage and family reader,* edited by J. M. Lehmann. Lincoln, Neb.: Gordian Knot Books, 80–94.

Evans, J., and E. M. Broido. 2000. Coming out in college residence halls: Negotiation, meaning making, challenges, supports. *Journal of College Student Development* 40:658–68.

Ferguson, S. J. 2000. Challenging traditional marriage: Never married Chinese American and Japanese American women. *Gender and Society* 14:136–59.

Glenn, N. D., and C. N. Weaver. 1988. The changing relationship of marital status to reported happiness. *Journal of Marriage and the Family* 50:317–24.

Golombok, S., B. Perry, A. Burston, C. Murray, J. Money-Somers, M. Stevens, and J. Golding. 2003. Children with lesbian parents: A community study. *Developmental Psychology* 39:20–33.

Green, R. J., J. Bettinger, and E. Sacks. 1996. Are lesbian couples fused and gay male couples disengaged? In *Lesbians and gays in couples and families,* edited by J. Laird & R. J. Green. San Francisco: Jossey-Bass, 185–230.

Hemstrom, Orian. 1996. Is marriage dissolution linked to differences in mortality risks for men and women? *Journal of Marriage and the Family* 58:366–78.

Herek, G. M. 2002. Gender gaps in public opinion about lesbians and gay men. *Public Opinion Quarterly* 66:40–66.

Horowitz, S. M.,, D. L. Weis, and M. T. Laflin. 2001. Differences between sexual orientation behavior groups and social background, quality of life, and health behaviors. *Journal of Sex Research* 38:205–18.

Hyde, J. S. 2000. Becoming a heterosexual adult: The experiences of young women. *Journal of Social Issues* 56:283–96.

Kinsey, A. C., W. B. Pomeroy, C. E. Martin, and P. H. Gebhard. 1953. *Sexual behavior in the human female.* Philadelphia: Saunders.

Knox, D. (with Kermit Leggett). 2000. *The divorced dad's survival book: How to stay connected with your kids.* New York: Perseus.

Kurdek, L. A. 1994. Conflict resolution styles in gay, lesbian, heterosexual nonparent, and heterosexual parent couples. *Journal of Marriage and the Family* 56:705–22.

———. 2003. Differences between gay and lesbian couples. *Journal of Social and Personal Relationships* 20:411–36.

LaSala, M. 2000. Lesbians, gay men, and their parents: Family therapy for the coming-out crisis. *Family Process* 39:67–81.

———. 2002. Walls and bridges: How coupled gay men and lesbians manage their intergenerational relationships. *Journal of Marital and Family Therapy* 28:327–39.

LeVay, S. 1991. News and comment. *Science* 253: 956–57.

———. 1994 *The sexual brain.* Cambridge, Mass.: MIT Press.

Lever, J. 1994. The 1994 Advocate survey of sexuality and relationships: The men. *Advocate.* 23 August, 16–24.

Lever, J., D. E. Kanouse, W. H. Rogers, S. Carson, and R. Hertz. 1992. Behavior patterns and sexual identity of bisexual males. *Journal of Sex Research* 29:141–67.

Lewis, G. B. 2003. Black-white differences in attitudes toward homosexuality and gay rights. *Public Opinion Quarterly* 67:59–78.

Lichter, D. T., D. R. Graefe, and J. B. Brown. 2003. Is marriage a panacea? Union formation among economically disadvantaged unwed mothers. *Social Problems* 50:60–86.

Lisotta, C. 2003. Poll: 6 in 10 Americans OK gay unions. Gay.com 15 May.

Louderback, L. A., and B. E. Whitley. 1997. Perceived erotic value of homosexuality and sex role attitudes as mediators of sex differences in heterosexual college students' attitudes toward lesbians and gay men. *Journal of Sex Research* 34:175–82.

Lucas, R. E., A. E. Clark, Y. Georgellis, and E. Diener. 2003. Reexamining adaptation and the set point model of happiness: Reactions to changes in marital status. *Journal of Personality and Social Psychology* 84:527–39.

Mackay, J. 2001. Global sex: Sexuality and sexual practices around the world. *Sexual and Relationship Therapy* 16:71–82.

Mather, P.C. 2000. Diversity at East Carolina University: Student Perspectives. Fall.

Mattes, Jane. 1994. *Single mothers by choice.* New York: Times Books.

McLanahan, S. S. 1991. The long term effects of family dissolution. In *When families fail: The social costs,* edited by Bryce J. Christensen. New York:

University Press of America for the Rockford Institute, 5–26.

McLanahan, S., and K. Booth. 1989. Mother-only families: Problems, prospects and politics. *Journal of Marriage and the Family* 51:557–80.

Means-Christensen, A. J., D. K. Snyder, and C. Negy. 2003. Assessing nontraditional couples: Validity of the marital satisfaction inventory—revised with gay, lesbian, and cohabiting heterosexual couples. *Journal of Marital and Family Therapy* 29:69–83.

Michael, R. T. 2001. Private sex and public policy. In *Sex, love, and health in America: Private choices and public policies*, edited by E. O. Laumann and R. T. Michael. Chicago: The University of Chicago Press, 465–91.

Michand, C. 2001. Survey: Students hold mostly pro-gay views. http://www.hrc.org/familynets

Money, John. 1987. Sin, sickness, or status? Homosexual gender identity and psychoneuroendocrinology. *American Psychologist* 42:384–99.

Moss, K. 2002. Legitimizing same-sex marriages. *Peace Review* 14:101–8.

Musick, K. 2002. Planned and unplanned childbearing among unmarried women. *Journal of Marriage and Family* 64:915–29.

National survey results of gay couples in long-lasting relationships. 1990. Partners: *Newsletter for Gay and Lesbian Couples*, May/June, 1–16. Available from Stevie Bryant, Box 9685, Seattle, WA 98109.

Nobes, G., and M. Smith. 2002. Family structure and the physical punishment of children. *Journal of Family Issues* 23:349–73.

Ochs, R. 1996. Biphobia: It goes more than two ways. In *Bisexuality: The psychology and politics of an invisible minority*, edited by B. A. Firestein. Thousand Oaks, Calif.: Sage, 217–39.

Oswald, R. F. 2000. Queer country: Rural gay, lesbian, bisexual, transgender (GLBT) people and their families. Poster session presented at the National Council on Family Relations, Minneapolis, November.

Oswald, R. F., and L. S. Culton. 2003. Under the rainbow: Rural gay life and its relevance for family providers. *Family Relations* 52:72–81.

Page, S. 2003. Gay rights tough to sharpen into political "wedge issue." *USA Today*. 28 July, 10A.

Palmer, R., & Bor, R. 2001. The challenges to intimacy and sexual relationships for gay men in HIV serodiscordant relationships: A pilot study. *Journal of Marital and Family Therapy* 27:419–31.

Patel, S. 1989. Homophobia: Personality, emotional, and behavioral correlates. Master's thesis, East Carolina University.

Patel, S., T. E. Long, S. L. McCammon, and K. L. Wuensch. 1995. Personality and emotional correlates of self reported antigay behaviors. *Journal of Interpersonal Violence* 10:354–66.

Patterson, C. J. 2001. Family relationships of lesbians and gay men. In *Understanding families into the new millennium: A decade in review*, edited by R. M. Milardo. Minneapolis: National Council on Family Relations, 271–88.

Paul, J. P. 1996. Bisexuality: Exploring/exploding the boundaries. In *The lives of lesbians, gays, and bisexuals: Children to adults*, edited by R. Savin-Williams and K. M. Cohen. Fort Worth, Tex.: Harcourt Brace, 436–61.

Peplau, L. A., R. C. Veniegas, and S. N. Campbell. 1996. Gay and lesbian relationships. In *The lives of lesbians, gays, and bisexuals: Children to adults*, edited by R. C. Savin-Williams and K. M. Cohen. Fort Worth, Tex.: Harcourt Brace, 250–73.

Peterson, Karen S. 1995. Family advocates declare war on divorce. *USA Today*. 30 March, 6d.

Pinquart, M. 2003. Loneliness in married, widowed, divorced, and never-married older adults. *Journal of Social and Personal Relationships* 20:31–53.

Plummer, D. C. 2001. The quest for modern manhood: Masculine stereotypes, peer culture and the social significance of homophobia. *Journal of Adolescence* 24:15–23.

Pong, S. L., and B. Dong. 2000. The effects of change in family structure and income on dropping out of middle and high school. *Journal of Family Issues* 21:147–69.

Richman, K. 2002. Lovers, legal strangers, and parents: Negotiating parental and sexual identity in family law. *Law and Society Review* 36:285–324.

Ritter, J. 2003. Canada gives gays hope for change. *USA Today*. 30 June, 3A.

Robinson, B. E., L. H. Walters, and P. Skeen. 1989. Response of parents to learning that their child is homosexual and concern over AIDS: A national study. *Journal of Homosexuality* 18:59–80.

Roderick, T. 1994. Homonegativity: An analysis of the SBS-R. Master's thesis, East Carolina University.

Roderick, T., S. L. McCammon, T. E. Long, and L. J. Allred. 1998. Behavioral aspects of homonegativity. *Journal of Homosexuality* 36:79–88.

Rust, P. 1996. Monogamy and polyamory: Relationship issues for bisexuals. In *Bisexuality: The psychology and politics of an invisible minority*, edited by B. A. Firestein. Thousand Oaks, Calif.: Sage, 127–48.

Savin-Williams, R. C., and E. M. Dube. 1997. Parental reactions to their child's disclosure of a gay/lesbian identity. *Family Relations* 47:7–13.

Sylivant, S. 1992. The cognitive, affective, and behavioral components of adolescent homonegativity. Master's thesis, East Carolina University.

Smith, D. M., & G. J. Gates, G. J. 2001. Gay and lesbian families in the United States: Same-sex unmarried partner households: Preliminary Analysis of 2000 U.S. Census Data, A Human Rights Campaign Report, *www.hrc.org*

Statistical Abstract of the United States: 2003, 123rd ed. Washington, D.C.: U.S. Bureau of the Census.

Sugarman, Stephen D. 2003. Single-parent families. In *All our families, 2nd ed, New Policies for a new century*, edited by M. A. Mason, A. Skolnick, and S. D. Sugarman. New York: Oxford University Press, 14–39.

Theodore, P. S., and Basow, S. A. 2000. Heterosexual masculinity and homophobia: A reaction to the self? *Journal of Homosexuality 40:* 31–48.

Toufexis, A. 1996. When the ring doesn't fit. *Psychology Today* 29 (6): 52 passim.

Tyagart, C. E. 2002. Legal rights to homosexuals in areas of domestic partnerships and marriages: Public support and genetic causation attribution. *Educational Research Quarterly* 25:20–29.

Wayment, H. A., G. E. Wyatt, M. B. Tucker, G. J. Romero, J. V. Carmona, M. Newcomb, B. M. Solis, M. Riederle, and C. Mitchell-Kernan. 2003. Predictors of risky and precautionary sexual behaviors among single and married white women. *Journal of Applied Social Psychology* 33:971–816.

Welde, K. D., and E. A. Hubbard. 2003. "I'm glad I'm not gay!" Heterosexual students' emotional experience in the college classroom with a "coming out" assignment. *Teaching Sociology* 31:73–84.

White, Edmund. 1994. Sexual culture. In *The burning library: Essays by Edmund White*, edited by D. Bergman. New York: Knopf, 157–67.

Chapter 7

American Council on Education and University of California. 2002. *The American freshman: National norms for fall, 2002*. Los Angeles: Higher Education Research Institute.

Arrindell, W. A., and F. Luteijn. 2000. Similarity between intimate partners for personality traits as related to individual levels of satisfaction with life. *Personality and Individual Differences* 28:629–37.

Blackwell, D. L., and D. T. Lichter. 2000. Mate selection among married and cohabiting couples. *Journal of Family Issues* 21:275–302.

Bossard, J. H. S. 1932. Residential propinquity as a factor in marriage selection. *American Journal of Sociology* 38:219–24.

Brodsky, A. E. 2000. The role of religion in the lives of resilient, urban, African-American, single mothers. *Journal of Community Psychology* 28:199–220.

Butler, M. H., J. A. Stout, and B. C. Gardner. 2002. Prayer as a conflict resolution ritual: Clinical implications of religious couple's report of relationship softening, healing perspective, and change responsibility. *American Journal of Family Therapy* 30:19–37.

Cawley, J. 2003. Russell Crowe in Command. *Biography Magazine*. December, 50–54.

Crowell, J. A., D. Treboux, and E. Waters. 2002. Stability of attachment representations: The transition to marriage. *Developmental Psychology* 38:467–79.

Dehle, C., and R. L. Weiss 2002 Associations between anxiety and marital adjustment. *Journal of Psychology* 136:328–38.

Edwards, T. M. 2000. Flying solo. *Time*. 28 August, 47–53.

Ganong, L. W., and M. Coleman. 1992. Gender differences in expectations of self and future partner. *Journal of Family Issues* 13:55–64.

Haring, M., P. L. Hewitt, and G. L. Flett. 2003. Perfectionism, coping, and quality of relationships. *Journal of Marriage and the Family* 65:143–59.

Houts, R. M., E. Robins, and T. L. Huston. 1996. Compatibility and the development of premarital relationships. *Journal of Marriage and the Family* 58:7–20.

Huston, T. L., J. P. Caughlin, R. M. Houts, S. E. Smith, & L. J. George.2001. The connubial crucible: Newlywed years as predictors of marital delight, distress, and divorce. *Journal of Personality and Social Psychology* 80:237–52.

Ingoldsby, B., P. Schvaneveldt, and C. Uribe.2003. Perceptions of acceptable mate attributes in Ecuador. *Journal of Comparative Family Studies* 34:171–86.

Jepsen, L. K., and C. A. Jepsen. 2002. An empirical analysis of the matching patterns of same-sex and opposite-sex couples. *Demography* 39:435–53.

Kalmijn,M., and H. Flap. 2001. Assortive meeting and mating: Unintended consequences of organized settings for partner choices. *Social Forces* 79:1289–1312.

Killian, K. D. 1997. What's the difference: Negotiating race, class and gender in interracial relationships. Paper presented at the Annual Conference of the National Council on Family Relations, Crystal City, Virginia.

Knox, D., M. E. Zusman, K. McGinty, and B. Davis. 2002. College student attitudes and behaviors toward ending an unsatisfactory relationship *College Student Journal* 36:630–34.

Knox, D, M. E. Zusman, C. Buffington, and G. Hemphill. 2000. Interracial dating attitudes among college students. *College Student Journal* 34:69–71.

Knox, D., M. E. Zusman, and Wandy Nieves. 1997. College students' homogamous preferences for a date and mate. *College Student Journal* 31: 445–48.

Lewis, S. K., and V. K. Oppenheimer. 2000. Educational assortive mating across marriage markets: Non-Hispanic whites in the United States. *Demography* 37:29–40.

Licata, N. 2002. Should premarital counseling be mandatory as a requisite to obtaining a marriage license? *Family Court Review* 40:518–32.

Marcus, D. K., and R. S. Miller. 2003. Sex differences in judgments of physical attractiveness: A social relations analysis. *Personality and Social Psychology Bulletin* 29:325–35.

Medora, N. P., J. H. Larson, N. Hortacsu, and P. Dave. 2002. Perceived attitudes towards romanticism: A cross-cultural study of American, Asian-Indian, and Turkish young adults. *Journal of Comparative Family Studies* 33:155–78.

Meehan, D., and C. Negy. 2003. Undergraduate students' adaptation to college: Does being married make a difference? *Journal of College Student Development* 44:670–90.

Meyer, J. P., and S. Pepper. 1977. Need compatibility and marital adjustment in young married couples. *Journal of Personality and Social Psychology* 35:331–42.

Michael, R. T., J. H. Gagnon, E. O. Laumann, and G. Kolata. 1994. *Sex in America: A definitive survey.* Boston: Little, Brown.

Morgan, H. J., and P. R. Shaver. 1999. Attachment processes and commitment to romantic relationships. In *Handbook of interpersonal commitment and relationship stability,* edited by J. M. Adams and W. H. Jones. New York: Academic/Plenum Publishers, 109–24.

O'Hagen, S., A. Johnson, G. Lardi, and J. P. Keenan. 2003. The effect of relationship status on perceived attractiveness. *Social Behavior and Personality* 31:291–300.

Pimentel, E. E. 2000. Just how do I love thee? Marital relations in urban China. *Journal of Marriage and the Family* 62:32–47.

Regan, P. C., and A. Joshi.2003. Ideal partner preferences among adolescents. *Social Behavior and Personality* 31:13–20.

Renick, M. J., S. L. Blumberg, and H. J. Markman. 1992. The prevention and relationship enhancement program (PREP): An empirically based preventive program for couples. *Family Relations* 41:141–47.

Rosenblatt, P. C., T. A. Karis, and R. D. Powell. 1955. *Multiracial couples.* Thousand Oaks, Calif.: Sage.

Saint, D. J. 1994. Complementarity in marital relationships. *Journal of Social Psychology* 134:701–4.

Skowron, E. A. 2000. The role of differentiation of self in marital adjustment. *Journal of Counseling Psychology* 47:229–37.

Smits, J., W. Ultee, and J. Lammers. 1998. Educational homogamy in 65 countries: An explanation of differences in openness using country-level explanatory variables. *American Sociological Review* 63:64–285.

Snyder, D. K., and J. M. Regts. 1990. Personality correlates of marital dissatisfaction: A comparison of psychiatric, maritally distressed, and nonclinic samples. *Journal of Sex and Marital Therapy* 90:34–43.

Sprecher, S., and P. C. Regan. 2002. Liking some things (in some people) more than others: Partner preferences in romantic relationships and friendships. *Journal of Social and Personal Relationships* 19:463–81.

Sprecher, S. 2002. Sexual satisfaction in premarital relationships: Associations with satisfaction, love, commitment, and stability. *Journal of Sex Research* 39:190–96.

Stanley, S. M., and H. J. Markman. 1997. Marriage in the 90s: A nationwide random phone survey. Denver: PREP, Inc.

Statistical Abstract of the United States: 2003. 123rd ed. Washington, D.C.: U.S. Bureau of the Census.

Strassberg, D. S., and S. Holty. 2003. An experimental study of women's Internet personal ads. *Archives of Sexual Behavior* 32:253–61.

Toro-Morn, M., and S. Sprecher. 2003. A cross-cultural comparison of mate preferences among university students: The United States versus the People's Republic of China (PRC). *Journal of Comparative Family Studies* 34:151–62.

Waller, W., and R. Hill. 1951. *The family: A dynamic interpretation.* New York: Holt, Rinehart and Winston.

Winch, R. F. 1955. The theory of complementary needs in mate selection. Final results on the test of the general hypothesis. *American Sociological Review* 20:552–55.

Xie, Y. , J. M. Raymo, K. Govette, and A. Thornton. 2003. Economic potential and entry into marriage and cohabitation. *Demography* 40:351–64.

Zaidi, A. U., and M. Shuraydi. 2002 Perceptions of arranged marriages by young Pakistani Muslim women living in a western society. *Journal of Comparative Family Studies* 33:495–514.

Zusman, M. E. , D. Knox, J. Gescheidler, and K. McGinty. In press. Dating manners among college students. *Journal of Indiana Academy of Social Sciences.*

Chapter 8

Abowitz, D. A. 2000. A "Wedding story," or, A modern American fairy tale: Gender and romance in the construction of white weddings. Paper presented at the Eastern Sociological Society, Baltimore, March 5.

———. 2002. On the road to "happily ever after": A survey of attitudes toward romance and marriage among college students. Paper presented at the Southern Sociological Society, Baltimore, April 6.

Adams, R. G., and J. Rosen-Grandon. 2002. Mixed marriages: Music community membership as a source of marital strain. In *Inappropriate relationships: The unconventional, the disapproved, and the forbidden,* edited by R. Goodwin and D. Cramer. Mahwah, N. J,: Lawrence Erlbaum, 79–102.

Baca Zinn, M., and A. Y. H. Pok. 2002. Tradition and transition in Mexican-Origin families. In *Minority families in the United States: A multicultural perspective.* 3d ed. Upper Saddle River, N. J.: Prentice-Hall,79–100.

Becerra, R. M. 1998. The Mexican American family. In *Ethnic families in America: Patterns and variations,* 4th ed., edited by C. H. Mindel, R. W. Habenstein, and R. Wright, Jr. Upper Saddle River, N.J.: Prentice Hall, 141–59.

Billingsley, S., M. Lim, and G. Jennings. 1995. Themes of long-term, satisfied marriages consummated between 1952–1967. *Family Perspective* 29:283–95.

Boden, S. 2001. "Superbrides": Wedding consumer culture and the construction of bridal identity. Sociological Research on Line *www.socresonline .org.u.k.* Vol. 6, 1, May.

Browning, S. L. 1999. Marriage and the black male/female relationship: From culture to structure. In *Till death do us part: A Multicultural anthology on marriage,* edited by S. L. Browning and R. R. Miller. Stamford, Conn: JAI Press, 253–77.

Coburn, J., A. B. Pfeiffer, S. M. Simon, P. A. Locke, J. B. Ridley, and H. Mann. 1995. The American Indians. In *Educating for diversity,* edited by C. A. Grant. Boston: Allyn and Bacon, 225–54.

Coontz, S. 2000. Marriage: Then and now. *Phi Kappa Phi Journal* 80:16–20.

Crawford, D. W., D. Feng, J. L. Fischer, and L. K. Diana. 2003. The influence of love, equity, and alternatives on commitment in romantic relationships. *Family and Consumer Sciences Research Journal* 31:253–71.

Darroch, J. E., D. J. Landry, and S. Oslak. 2000. Age differences between sexual partners in the United States. *Family Planning Perspectives* 31:160–67.

Edwards, T. M. 2000. Flying Solo. *Time.* 28 August, 47–53.

Frye, N. E., and B. R. Karney. 2002. Being better or getting better? Social and temporal comparisons as coping mechanisms in close relationships. *Personality and Social Psychology Bulletin* 28: 1287–99.

Gaines, S. O., Jr., and J. Leaver. 2002. Interracial relationships. In *Inappropriate relationships: The unconventional, the disapproved, and the forbidden,* edited by R. Goodwin and D. Cramer. Mahwah, N. J.: Lawrence Erlbaum, 65–78.

Gottman, J., and S. Carrere. 2000. Welcome to the love lab. *Psychology Today.* September/October, 42 passim.

Harper, J. M., B. G. Schaalje, and J. G. Sandberg. 2000. Daily hassles, intimacy, and marital quality in later marriages. *American Journal of Family Therapy* 28:1–18.

Hill, M. R., and V. Thomas. 2000. Strategies for racial identity development: Narratives of Black and White women in interracial partner relationships. *Family Relations* 49:193–200.

Huyck, M. H., and D. L. Gutmann. 1992. Thirtysomething years of marriage: Understanding experiences of women and men in enduring family relationships. *Family Perspective* 26: 249–65.

Ingraham, Chrys. 1999. White weddings: Romancing heterosexuality in popular culture. New York: Routledge.

Kennedy, R. 2003. *Interracial intimacies.* New York: Pantheon.

Kirn, W., and W. Cole. 2000. Twice as nice. *Time.* 19 June, 53.

Knox, D., T. Britton, and B. Crisp. 1997. Age discrepant relationships reported by university faculty and students. *College Student Journal* 31:290–93.

Knox, D., S. O. Langehough, C. Walters, and M. Rowley. 1998. Religiosity and spirituality among college students. *College Student Journal* 32:430–32.

Knox, D., and M. E. Zusman. 1998. Unpublished data on 620 undergraduates at large southeastern university collected for this text.

Knox, D., M. E. Zusman, and V. D. Daniels. 2002 College student attitudes toward interreligious marriage. *College Student Journal* 36:84–86.

Licata, N. 2002. Should premarital counseling be mandatory as a requisite to obtaining a marriage license? *Family Court Review* 40:518–32.

Malia, J. A., and E. M. Blackwell. 1997. A study of the nature of mother-and daughter-in- law relationships using the OSR NUD-IST Program. Paper presented at the Annual Conference of the

National Council on Family Relations, Crystal City, Va.
McLain, R. 2000. The Hopi way of life. Tour to Hopi Indian reservation, 2nd Mesa near Tuba City, Arizona.
Michael, R. T., J. H. Gagnon, E. O. Laumann, and G. Kolata. 1994. *Sex in America: A definitive survey.* Boston: Little, Brown.
Miller, R. R. 1999. The implications of marital status for socially supportive ties with both extended family and community. In *Till death do us part: A Multicultural anthology on Marriage,* edited by S. L. Browning and R. R. Miller. Stamford, Conn: JAI Press, Inc., 139–58.
Murdock, G. P. 1949. *Social structure.* New York: Free Press.
National Center for Health Statistics. 2001. Births, marriages, divorces, and deaths, July 2000. *Monthly vital statistics report* 48 (19). Hyattsville, Md.: Public Health Service.
Pimentel, E. E. 2000. Just how do I love thee? Marital relations in urban China. *Journal of Marriage and the Family* 62:32–47.
Regan, P. C. 2000. The role of sexual desire and sexual activity in dating relationships. *Social Behavior and Personality* 28:51–60.
Saad, Lydia. 1995. Children, hard work taking their toll on baby boomers. *Gallup Poll Monthly.* April, 21–24.
Scovell, J. 1998. *Living in the shadows: A biography of Oona O'Neill Chaplin.* New York: Warner Books.
Serovich, J., and S. Price. 1992. In-law relationships: A role theory perspective. Paper presented at the 54th Annual Conference of the National Council on Family Relations, Orlando, Fla.
Silverthorne, Z. A., and V. L. Quinsey. 2000. Sexual partner age preferences of homosexual and heterosexual men and women. *Archives of Sexual Behavior* 29:67–76.
Sobal, J., C. F. Bove, and B. S. Rauschenbach. 2002. Commensal careers at entry into marriage: Establishing commensal units and managing commensal circles. *Sociological Review* 50:378–97.
Sousa, Lori A. 1995. Interfaith marriage and the individual and family life cycle. *Family Therapy* 22:97–104.
Stack, S., and J. R. Eshleman. 1998. Marital happiness: A 17-nation study. *Journal of Marriage and the Family* 60:527–36.
Statistical Abstract of the United States: 2003. 123rd ed. Washington, D.C.: U.S. Bureau of the Census.
Taylor, R. L. 2002. Black American families. In *Minority families in the United States: A multicultural perspective.* 3d ed. Upper Saddle River, N. J.: Prentice Hall,19–47.
Timmer, S. G., and J. Veroff. 2000. Family ties and the discontinuity of divorce in black and white newlywed couples. *Journal of Marriage and the Family* 62:349–61.
Treas, J., and D. Giesen. 2000. Sexual infidelity among married and cohabiting Americans. *Journal of Marriage and the Family* 62:48–60.
Turner, A. J. 2004. Personal communication, Huntsville, Alabama, September. Used by permission.
Vaillant, C. O., and G. E. Vaillant. 1993. Is the U-curve of marital satisfaction an illusion? A 40-year study of marriage. *Journal of Marriage and the Family* 55:230–39.
Vanlaningham, J., D. R. Johnson, and P. Amato. 2001. Marital happiness, marital duration, and the U-shaped curve: Evidence from a five-year wave panel study. *Social Forces* 79:1313–41.
Wallerstein, J., and S. Blakeslee. 1995. *The good marriage.* Boston: Houghton-Mifflin.
Walters, L. H., P. Skeen, W. Warzywoda-Krusynska, and T. Kurko. 1997. Marital happiness in young families: Similarities and differences across countries. Paper presented at the National Council on Family Relations, Crystal City, Va.
Weigel, D. J., and D. S. Ballard-Reisch. 2002. Investigating the behavioral indicators of relational commitment. *Journal of Social and Personal Relationships* 19:403–23.
Wilmoth, J., and G. Koso. 2002. Does marital history matter? Marital status and wealth outcomes among preretirement adults. *Journal of Marriage and the Family* 64:254–68.
Yee, B. W. K. 1992. Gender and family issues in minority groups. In *Cultural diversity and families,* edited by K. G. Arms, J. K. Davidson, Sr., and N. B. Moore. Dubuque, Iowa: W. C. Brown, 5–10.
Yellowbird, M., and C. M. Snipp. 2002. American Indian families. In *Minority families in the United States: A multicultural perspective.* 3d ed. Upper Saddle River, N. J.: Prentice Hall 227–49.

Chapter 9

Bargh, J. A., K. Y. A. McKenna, and G. M. Fitzsimons. 2002. Can you see the real me? Activation and expression of the "true self" on the Internet. *Journal of Social Issues* 58:33–49.
Clinton, H. 2003. 20/20 interview with Barbara Walters. ABC Television, June 8.
Davis, M., and Scott, R. S. 1988. *Lovers, doctors and the law.* New York: Harper & Row.
Derlega, V. J., S. Metts, S. Petronio, and S. T. Margulis. 1993. *Self-disclosure.* Newbury Park, Calif.: Sage.
Finkenauer, C., and H. Hazam. 2000. Disclosure and secrecy in marriage: Do both contribute to marital satisfaction? *Journal of Social and Personal Relationships* 17:245–63.
Forgatch, M. S. 1989. Patterns and outcome in family problem solving: The disrupting effect of negative emotion. *Journal of Marriage and the Family* 51:115–24.
Gable, S. L., H. T. Reis, and G. Downey. 2003. He said, she said: A Quasi-Signal detection analysis of daily interactions between close relationship partners. *Psychological Science* 14:100–105.
Gallmeier, C. P., M. E. Zusman, D. Knox, and L. Gibson. 1997. Can we talk? Gender differences in disclosure patterns and expectations. *Free Inquiry in Creative Sociology* 25:129–225.
Goodman, N. 2003. Comments on review of *Choices in Relationships.* Department of Sociology, State University of New York, Stony Brook, New York.
Gottman, John. 1994a. *What predicts divorce? The relationship between marital processes and marital outcomes.* Hillsdale, N.J.: Lawrence Erlbaum.
———. 1994b. *Why marriages succeed or fail.* New York: Simon & Schuster.
Gottman, J. M., James Coan, S. Carrere, and C. Swanson. 1998. Predicting marital happiness and stability from newlywed interactions. *Journal of Marriage and the Family* 60:5–22.
Greeff, A. P., and T. De Bruyne. 2000. Conflict management style and marital satisfaction. *Journal of Sex and Marital Satisfaction* 26:321–34.
Guerrero, L. K. 1997. Nonverbal involvement across interactions with same-sex friends, opposite-sex friends and romantic partners: Consistency or change? *Journal of Social and Personal Relationships* 14:31–54.
Hetherington, E. M. 2003. Intimate pathways: Changing patterns in close personal relationships across time. *Family Relations* 52:318–31.
Kim, M. S., and K. S. Aune. 1998. The effects of psychological gender orientations on the perceived salience of conversational constraints. *Sex Roles* 37:935–53.
Knox, D., S. Hatfield, and M. E. Zusman. 1998. College student discussion of relationship problems. *College Student Journal* 32:19–21.
Knox, D., C. Schacht, J. Holt, and J. Turner. 1993. Sexual lies among university students. *College Student Journal* 27:269–72.
Knox, D., C. Schacht, J. Turner, and P. Norris. 1995. College students' preference for win-win relationships. *College Student Journal* 29:44–46.
Knox, D., M. E. Zusman, K. McGinty, and J. Gescheidler. 2001. Deception of parents during adolescence. *Adolescence* 36:611–13.
Knox, D., M. E. Zusman, and W. Nieves. 1997. College students' homogamous preferences for a date and a mate. *College Student Journal* 31: 445–48.
Kurdek, Lawrence A. 1994. Areas of conflict for gay, lesbian, and heterosexual couples: What couples argue about influences relationship satisfaction. *Journal of Marriage and the Family* 56:923–34.
———. 1995. Predicting change in marital satisfaction from husbands' and wives' conflict resolution styles. *Journal of Marriage and the Family* 57:153–64.
Mackey, R. A., and B. A. O'Brien. 1998. Marital conflict management: Gender and ethnic differences. *Social Work* 43:128–41.
———. 1999. Adaptation in lasting marriages. *Families in Society: Journal of Contemporary Human Services* 80:587–96.
Marano, H. E. 1992. The reinvention of marriage. *Psychology Today.* January/February, 49 passim.
Marchand, J. F., and E. Hock. 2000. Avoidance and attacking conflict-resolution strategies among married couples: Relations to depressive symptoms and marital satisfaction. *Family Relations* 49:201–6.
McGinty, K., D. Knox, and M. E. Zusman. 2003. Nonverbal and verbal communication in "involved" and "casual" relationships among college students. *College Student Journal* 37:68–71.
McKenna, K. Y. A., A. S. Green, and M. E. J. Gleason. 2002. Relationship formation on the Internet: What's the big attraction? *Journal of Social Issues* 58:9–22.
Miller, K. U., and T. Abraham. 1998. Deceptive behavior in social relationships: Consequences of violated expectations. *Journal of Psychology* 122:263–73.
Moore, M. M. 1998. Nonverbal courtship patterns in women: Rejection signaling—An empirical investigation. *Semiotica* 118:201–14.
Nakanishi, M. 1986. Perceptions of self-disclosure in initial interaction: A Japanese sample. *Human Communication Research* 13:167–90.
Norwood, Chris. 1995. Mandated life versus mandatory death: New York's disgraceful partner notification record. *Journal of Community Health* 20(2): 161–70.
Notarius, C., and H. Markman. 1994. *We can work it out: Making sense of marital conflict.* New York: Putnam.
Patford, J. L. 2000. Partners and cross-sex friends: A preliminary study of the way marital and de facto partnerships affect verbal intimacy with cross-sex friends. *Journal of Family Studies* 6: 106–19.
Payn, B., K. Tanfer, J. O. G. Billy, and W. R. Grady. 1997. Men's behavior change following infection with a sexually transmitted disease. *Family Planning Perspectives* 29:152–57.
Rowatt, W. C., M. R. Cunningham, and P. B. Druen. 1998. Deception to get a date. *Personality and Social Psychology Bulletin* 24:1228–42.
Scoresby, A. L. 1977. *The marriage dialogue.* Reading, Mass.: Addison-Wesley.

Seguin-Levesque, C., M. L. N. Laliberte, L. G. Pelletier, C. Blanchard, and R. J. Vallerand. 2003. Harmonious and obsessive passion for the Internet: Their associations with the couple's relationship. *Journal of Applied Social Psychology* 33:197–221.

Sollie, D. L. 2000. Beyond Mars and Venus: Men and women in the real world. *Phi Kappa Phi Journal* 80:42–45.

Sprecher, S., and Diane Felmlee. 1993. Conflict, love and other relationship dimensions for individuals in dissolving, stable, and growing premarital relationships. *Free Inquiry in Creative Sociology* 21:115–25.

Stanley, S. M., and H. J. Markman. 1997. *Marriage in the 90s: A nationwide random phone survey.* Denver: PREP, Inc.

Stanley, S. M., H. J. Markman, and S. W. Whitton. 2002. Communication, conflict, and commitment: Insights on the foundations of relationship success from a national survey. *Interpersonal Relations* 41:659–66.

Stanley, S. M., and D. W. Trathen. 1994. Christian PREP: An empirically based model for marital and premarital intervention. *Journal of Psychology and Christianity* 13:158–65.

Tannen, D. 1990. *You just don't understand: Women and men in conversation.* London: Virago.

Turner, A. J. 2004. Communication basics. Personal communication.

Zusman, M. E., and D. Knox. 1998. Relationship problems of casual and involved university students. *College Student Journal* 32:606–9.

Chapter 10

Adler, N. E., Em J. Ozer, and J. Tschann. 2003. Abortion among adolescents. *American Psychologist* 58:211–17.

Alan Guttmacher Institute. 2002. *In their own right: Addressing the sexual and reproductive health needs of American men.* New York: Alan Guttmacher Institute.

———. 2003. Two-child families are becoming the norm. International *Family Planning Perspectives* 28:56–68.

American Council on Education and University of California. 2002. *The American freshman: National norms for fall. 2002.* Los Angeles.: Los Angeles Higher Education Research Institute.

Artz, L., Macaluso, M., Brill, I., Kelaghan, J., Austin, H., Fleenor, M. 2000. Effectiveness of an intervention promoting the female condom to patients at sexually transmitted disease clinics. *American Journal of Public Health* 90:237–44.

Assve, A. 2003. The impact of economic resources on premarital childbearing and subsequent marriage among young American women. *Demography* 40:105–26.

Avery, R. J. 1998. Information disclosure and openness in adoption: State policy and empirical evidence. *Children and Youth Services Review* 20: 57–85.

Baird, D. T. 2000. Therapeutic abortion. In *Family planning and reproductive healthcare,* edited by A. Glasier and A. Gebbie. London: Churchill Livingston, 249–62.

Bales, D., and D. Stephens. 2000. *What do teen parents need? Assessing community policies and effective services for pregnant and parenting teens.* Poster presentation at 62nd Annual Conference of the National Council on Family Relations, Minneapolis, November.

Barnett, W., N. Freudenberg, and R. Willie. 1992. Partnership after induced abortion: A prospective controlled study. *Archives of Sexual Behavior* 21:443–55.

Begue, L. 2001. Social judgment of abortion: A black-sheep effect in a Catholic sheepfold. *Journal of Social Psychology* 141:640–50.

Beksinska, M. E., Rees, H. V., Dickson-Tetteh, K. E., Mqoqi, N., Kleinschmidt, I., and McIntyre, J. A. 2001. Structural integrity of the female condom after multiple uses, washing, drying, and relubrication. *Contraception* 63:33–36.

Brothers, Z., and J. E. Maddux. 2003. The goal of biological parenthood and emotional distress from infertility: Linking parenthood to happiness. *Journal of Applied Social Psychology* 33: 248–62.

Burkman, R. 2002. Rationale for new contraceptive methods. *The Female Patient.* August, 4–13.

Centers for Disease Control and Prevention. 2002. Sexually transmitted diseases treatment guidelines 2002. MMWR 2002; 51 (No. RR-6): [1–84].

David, H. P., Z. Dytrych, and Z. Matejcek 2003. Born unwanted: Observations from the Prague study. *American Psychologist* 58:224–29.

Darroch, J. E., Singh, S., Frost, J. J., and The Study Team. 2001. Differences in teenage pregnancy rates among five developed countries: The roles of sexual activity and contraceptive use. *Family Planning Perspectives* 33: 244–50.

DeOllos, I. Y., and C. A. Kapinus, C. A. 2002. Aging childless individuals and couples: Suggestions for new directions in research. *Sociological Inquiry* 72:72–80.

Farber, M. L. Z., E. Timberlake, H. P. Mudd, and L. Cullen. 2003. Preparing parents for adoption: An agency experience. *Child and Adolescent Social Work Journal* 20:175– 96.

Finer, L. B., and S. K. Henshaw, S. K. 2003. Abortion incidence and services in the United States in 2000. *Perspectives on Sexual and Reproductive Health* 35:6–15.

Finley, G. E. 2000. Adoptive families: Dramatic changes across generations. *National Council on Family Relations Report* 45:6–7.

Fisher, A. P. 2003. A critique of the portrayal of adoption in college textbooks and readers on families, 1998–2001. *Family Relations* 52:154–60.

Flower Kim, K. M. 2003. We are family. Paper presented at the 73rd Annual Meeting of the Eastern Sociological Society, Philadelphia, February 27.

Garrett, T. M., Baillie, H. W., & Garrett, R. M. 2001. *Health care ethics.* 4th ed. Upper Saddle River, N.J.: Prentice Hall.

Geronimus, A. T. 2003. Damned if you do: Culture, identity, privilege, and teenage childbearing in the United States. *Social Science and Medicine*57:881–93.

Gillespie, R. 2003. Childfree and feminine: Understanding the gender identity of voluntarily childless women. *Gender and Society* 17:122–36.

Godecker, A. L., E. Thomson, and L. L. Bumpass. 2001. Union status, marital history and female contraceptive sterilization in the United States. *Family Planning Perspectives* 33:35–41.

Grady, W. R., J. O. G. Billy, and D. H. Klepinger. 2002. Contraception method switching in the United States. *Perspectives on Sexual and Reproductive Health* 34:135–45.

Grady, W. R., D. H. Kepinger, and A. Nelson-Wally. 2000. Contraceptive characteristics: The perceptions and priorities of men and women. *Family Planning Perspectives* 31:168–75.

Haller, D. L., D. R. Miles, and K. S. Dawson. 2003. Factors influencing treatment enrollment by pregnant substance abusers. *American Journal of Drug and Alcohol Abuse* 29:117–31.

Hart, V. A. 2002. Infertility and the role of psychotherapy. *Issues in Mental Health Nursing* 23:31–42.

Harvey, S. M., L. J. Beckman, C. Sherman, and D. Petitti. 2000. Women's experience and satisfaction with emergency contraception. *Family Planning Perspectives* 31:237–45.

Haugaard, J. J., M. Palmer, and J. C. Wojslawowicz. 1999. Single-parent adoptions. *Adoption Quarterly* 2:65–74.

Hillis, S. D. et al. 2000. Poststerilization regret: Findings from the United States collaborative review of sterilization. *Obstretics and Gynecology* 93:889–95.

Hofferth, S. L., L. Reid, and F. L. Mott. 2001. The effects of early childbearing on schooling over time. *Family Planning Perspectives* 33:259–67.

Hogan, D. P., P. R. Sun, and G. T. Cornwell. 2000. Sexual and fertility behaviors of American females aged 15–19 years: 1985, 1990, and 1995. *American Journal of Public Health* 90:1421–25.

Hollander, D. 2001. After abortion, mixed mental health. *Family Planning Perspectives* 33:1–3.

Hollingsworth, L. D. 1997. Same race adoption among African Americans: A ten year empirical review. *African American Research Perspectives* 13:44–49.

———. 2000. Sociodemographic influences in the prediction of attitudes toward transracial adoption. *Families in Society* 81:92–100.

———. 2003. When an adoption disrupts: A study of public attitudes. *Family Relations* 52:161–66.

Hubacher, D. 2002. The checkered history and bright future of intrauterine contraception in the United States. *Perspectives on Sexual and Reproductive Health* 34:98–103.

Huh, N. S., and W. J. Reid. 2000. Intercountry, transracial adoption and ethnic identity: A Korean example. *International Social Work* 43:75–87.

Isomaki, V. 2002 The fuzzy foster parenting—a theoretical approach. *Social Science Journal.* 39: 625–38.

Jones, R. K., J. E. Darroch, and S. K. Henshaw. 2002. Patterns in the socioeconomic characteristics of women obtaining abortions in 2000–2001. *Perspectives on Sexual and Reproductive Health* 34:226–35.

Judge, S. 2003. Determinants of parental stress in families adopting children from Eastern Europe. *Family Relations* 52:241–48.

Kaufman, G. 2000. Do gender role attitudes matter? Family formation and dissolution among traditional and egalitarian men and women. *Journal of Family Studies* 21:128–44.

Kennedy, R. 2003. *Interracial intimacies.* New York: Pantheon.

Kramer, L., and D. Ramsburg. 2002. Advice given to parents on welcoming a second child: A critical review. *Family Relations* 51:2–14.

Lane, T. 2003. High proportion of college men using condoms report errors and problems. *Perspectives on Sexual and Reproductive Health* 35:50–52.

Letherby, G. 2002. Childless and bereft? Stereotypes and realities in relation to "voluntary" and "involuntary" childlessness and womanhood. *Sociological Inquiry* 72:7–20.

Lewis, R. K., A. Paine-Andrews, J. Fisher, C. Custard, M. Fleming-Randle, and S. B. Fawcett. 1999. Reducing the risk for adolescent pregnancy: Evaluation of a school/community partnership in a Midwestern military community. *Family Community Health* 22:16–30.

London, S. 2003. Method-related problems account for most failures of the female condom. *Perspectives on Sexual and Reproductive Health* 35:193–99.

Macaluso, M., M. Demand, L. Artz, M. Fleenor, L. Robey, J. Kelaghan, R. Cabral, and E. W. Hook. 2000. Female condom use among women at high risk of sexually transmitted disease. *Family Planning Perspectives* 32:138–44.

Manlove, J., C. Mariner, and A. R. Papillo. 2000. Subsequent fertility among teen mothers: Longitudinal analysis of recent national data. *Journal of Marriage and the Family* 62:430–48.

Martin, J. A., B. E. Hamilton, P. D. Sutton, S. J. Ventura, F. Menacker, and M. L. Munson. 2003. Birth: Final data for 2002. *National Vital Statistics Reports* 52 (10). Hyattsville, Md.: National Center for Health Statistics.

McDonagh, E. 1996. *Breaking the abortion deadlock: From choice to consent.* New York: Oxford University Press.

Meschke, L. L., S. Bartholamae, and S. R. Zentall. 2000. Adolescent sexuality and parent-adolescent processes: Promoting healthy teen choices. *Family Relations* 49:143–54.

Michael, R. T. 2001. Abortion decisions in the United States. In *Sex, love, and health in America: Private choices and public policies,* edited by E. O. Laumann and R. T. Michael. Chicago: The University of Chicago Press, 377–438.

Mirowsky, J., and C. E. Ross. 2003. *Social causes of psychological distress.* Hawthorne, N.Y.: Aldine De Gruyter.

Mishell, D. R. 2002.The transdermal contraceptive system. *The Female Patient.* August, 14–25.

Morrison, D. M., M. R. Gillmore, M. J. Hoppe, J. Gaylord, et al. 2003. Adolescent drinking and sex: Findings from a daily diary study. *Perspectives on Sexual and Reproductive Health* 35:162–75.

Motamed, S. 2002. 100 million women can't be wrong: What most American women don't know about the IUD. Retrieved June 25, 2002, from www.plannedparenthood.org/articles/IUD/html

Naimi, T. S., L. E. Lipscomb, R. D. Brewer and B. C. Gilbert. 2003. Binge drinking in the preconception period and the risk of unintended pregnancy: Implications for women and their children. *Pediatrics* 111:1136–41.

Paulson, J. R., R. Boostanfar, P. Saadat, E. Mor, D. Tourgeman, C. C. Stater et al. 2002. Pregnancy in the sixth decade of life: Obstretic outcomes in women in advanced reproductive age. *Journal of the American Medical Association* 288:2320–24.

Pertman, A. 2000. *Adoption nation: How the adoption revolution is transforming America.* New York: Basic Books.

Peterson, B. D., C. R. Newton, and K. H. Rosen. 2003. Examining congruence between partners' perceived infertility-related stress and its relationship to marital adjustment and depression in infertile couples. *Family Process* 42:59–70.

Peterson, K. S. 2003. Adoptions reflect diversity. *USA Today.* 25 August, 7D.

Pinon, R., Jr. 2002. *Biology of Human Reproduction.* Sausalito, Calif.: University Science Books.

Population Council. 2003a. Female contraceptive development. Retrieved February 10, 2003, from *www.popcouncil.org/biomed/femalecontras.html*

———. 2003b. Jadele® Implants. Retrieved March 13, 2002, from *www.popcouncil.org/faqs/adellefaq.html*

Population Council Annual Report. 1999. Nestorone: A synthetic progestin that expands women's choices. Retrieved March 17, 2003, from http://www.popcouncil.org/about/ar99/nestorone.html

Ranjit, N., A. Bankole, J. E. Darroch, and S. Singh 2001. Contraceptive failure in the first two years of use: Differences across socioeconomic subgroups. *Family Planning Perspectives* 33:19–27.

Reardon, D. C., & P. G. Ney. 2000. Abortion and subsequent substance abuse. *American Journal of Drug and Alcohol Abuse* 26:61–73.

Regalado, A. 2002. Age is no barrier to motherhood—Study gives latest proof older women can get pregnant with donor eggs. *Wall Street Journal.* 13 November, D3.

Roberts, L. C., and P. W. Blanton. 2000. Parenting education with one-child families. Poster presentation aat the 62nd Annual Conference of the National Council on Family Relations, Minneapolis, November.

Rosenfeld, J. 1997. Postcoital contraception and abortion. In *Women's health in primary care,* edited by J. Rosenfeld. Baltimore: Williams & Wilkins, 315–29.

Rosenthal, M. B. 1997. Infertility. In *Women's health in primary care,* edited by J. Rosenfeld. Baltimore: Williams & Wilkins, 351–62.

Ross, R., D. Knox, M. Whatley and J. N. Jahangardi. 2003. Transracial adoption: Some college student data. Paper presented at the 73rd Annual Meeting of the Eastern Sociological Society. Philadelphia.

Roumen, F., D. Apter, T. Mulders, and T. Dieben. 2001. Efficacy, tolerability and acceptability of a novel contraceptive vaginal ring releasing etonogestrel and ethinyl oestradiol. *Human Reproduction.* 16:469–75.

Schwartz, J. L., and H. L. Gabelnick 2002. Current contraceptive research. *Perspectives on Sexual and Reproductive Health* 34:310–16.

Simon, R. J., and R. M. Roorda. 2000. *In their own voices: Transracial adoptees tell their stories.* New York: Columbia University Press.

Smith, P. B., R. S. Buzi, and M. L. Weinman. 2003. Targeting males for teenage pregnancy prevention in a school setting. *School of Social Work Journal* 27:23–36.

Statistical Abstract of the United States: 2003. 123rd ed. Washington, D.C. U.S. Bureau of the Census.

Stein, Z., and M. Susser. 2000. The risks of having children in later life. *British Medical Journal* 320:1681–83.

Tanfer, K., S. Wierzbicki, and B. Payn. 2000. Why are U.S. women not using long-acting contraceptives? *Family Planning Perspectives* 32:176–83.

Torres, A., and J. D. Forrest. 1988. Why do women have abortions? *Family Planning Perspectives* 20:169–76.

Townsend, J. W. 2003. Reproductive behavior in the context of global population. *American Psychologist* 58:197–204.

Trevor, L., and F. A. Althaus. 2002. The toll of unwanted pregnancies. *International Family Planning Perspectives* 28:184–85.

Urban Institute. 2003. Who will adopt the foster care child left behind? *Caring for Children* Brief 2. June 1–2.

Van Damme, L. 2000. *Advances in topical microbicides.* Paper presented at the 13th International AIDS Conference, July 9–14, Durban, South Africa.

Van Devanter, N., Gonzales, V., Merzel, C., Parikh, N. S., Celantano, D., & Greenberg, J. 2002. *American Journal of Public Health* 92:109–15.

Ventura, S. J., B. E. Hamilton, and P.D. Sutton. 2003. Revised birth and fertility rates for the United States, 2000 and 2001. *National Vital Statistics Reports* 51(4). Hyattsville, Md.: National Center for Health Statistics.

Walsh, T. L., R. G. Frezieres, K. Peacock, A. L. Nelson et al. 2003. Evaluation of the efficacy of a nonlatex condom: Results from a randomized, controlled clinical trial. *Perspectives on Sexual and Reproductive Health* 35:79–86.

Wenger, G. C., A. Scott, and N. Patterson. 2000. How important is parenthood? Childlessness and support in old age in England. *Ageing and Society* 20:161–82.

Wu, Z., and R. Hart 2002. The mental health of the childless elderly. *Sociological Inquiry* 72:21–42.

Wu, Z., and L. Macneill. 2002. Education, work and childbearing after age 30. *Journal of Comparative Family Studies* 33:191–213.

Wyeth Pharmaceuticals, Inc. 2002. Back-up contraception no longer required for women using Norplant system News release, July 26. Madison, N.J.

Zieman, M. 2002.Managing patients using the transdermal contraceptive system. *The Female Patient.* August, 26–32.

Chapter 11

Adler, A. 1992. *Understanding human nature.* New York: HarperCollins.

Aldous, J., and G. M. Mulligan.2002. Father's child care and children's behavior problems: A longitudinal study. *Journal of Family Issues* 23:624–47.

Amato, P. R., and F. Fowler. 2002 Parenting practices, child adjustment, and family diversity. *Journal of Marriage and the Family* 64:703–16.

Anderson, L. 1989. *Dear dad: Letters from an adult child.* New York: Viking Press.

Angold, A., S. C. Messer, D. Stangl, E. M. Z. Farmer, E. J. Cosello, and B. J. Burns. 1998. Perceived parental burden and service use for child and adolescent psychiatric disorders. *American Journal of Public Health* 88:75–80.

Arbona, C., and T. G. Power. 2003. Parental attachment, self-esteem, and antisocial behaviors among African American, European American, and Mexican American adolescents. *Journal of Counseling Psychology* 50:40–51.

Ball, H. L. 2002. Reasons to bed-share: Why parents sleep with their infants. *Journal of Reproductive and Infant Psychology* 20:207–22.

Baumrind, D. 1966. Effects of authoritative parental control on child behavior. *Child Development* 37:887–907.

Bigner, J. J. In press. Working with gay and lesbian parents. *Journal of Couple and Relationship Therapy.*

Booth, C. L., K. A. Clarke-Stewart, D. L. Vandell, K. McCartney, and M. T. Owen. 2002. Child-care usage and mother-infant "quality time." *Journal of Marriage and the Family* 64:16–26.

Bost, K. K., M. J. Cox, M. R. Burchinal, and C. Payne. 2002. Structural and supporting changes in couples' family and friendships networks across the transition to parenthood. *Journal of Marriage and the Family* 64:517–31.

Bushman, B. J., and J. Cantor. 2003. Media ratings for violence and sex. *American Psychologist* 58:130–41.

Casanova, M. F., D. Solursh, L. Solursh, E. Roy, E., and L. Thigpen. 2001. The history of child pornography on the Internet. *Journal of Sex Education and Therapy* 25:245–51.

Cassidy, G. L., and L. Davies. 2003. Explaining gender differences in mastery among married parents. *Social Psychology Quarterly* 66:48–61.

Coley, R. L., and J. E. Morris. 2002. Comparing father and mother reports of father involvement among low-income minority families. *Journal of Marriage and the Family* 64:982–97.

Cornelius-Cozzi, T. 2002. Effects of parenthood on the relationships of lesbian couples. *PROGRESS: Family Systems Research and Therapy* 11:85–94.

Danso, H., B. Hunsberger, and M. Pratt. 1997. The role of parental religious fundamentalism and right-wing authoritarianism child-rearing goals and practices. *Journal for the Scientific Study of Religion* 36:496–502.

Darling, N. 1999. Parenting style and its correlates. *ERIC Digest* ED427896.

Day, R. D., G. W. Peterson, and C. McCracken. 1998. Predicting spanking of younger and older children by mothers and fathers. *Journal of Marriage and the Family* 60:79–94.

Eggebeen, D. J. 2002. The changing course of fatherhood: Men's experiences with children in demographic perspective. *Journal of Family Issues* 23:486–506.

Elgar, F. J., J. Knight, G. J. Worrall, and G. Sherman. 2003. Attachment characteristics and behavioral problems in rural and urban juvenile delinquents. *Child Psychiatry and Human Development* 34:35–48.

Flouri, E., and A. Buchanan. 2003. The role of father involvement and mother involvement in adolescents' psychological well-being. *British Journal of Social Work* 33:399–406.

Flynn, C. P. 1998. To spank or not to spank: The effect of situation and age of child on support for corporal punishment. *Journal of Family Violence* 13:21–37.

Fontes, L. Anderson. 1998. Ethics in family violence research: Cross-cultural issues. *Family Relations* 47:53–61.

Galambos, N. L., E. T. Barker, and D. M. Almeida. 2003. Parents do matter: trajectories of change in externalizing and internalizing problems in early adolescent. *Child Development* 74:578–95.

Gavin, L. E., M. M. Black, S. Minor, Y. Abel, and M. E. Bentley. 2002. Young, disadvantaged fathers' involvement with their infants: An ecological perspective. *Journal of Adolescent Health* 31: 266–76.

Gennaro, S., and W. Fehder. 2000. Health behaviors in postpartum women. *Family Community Health* 22:16–26.

Gesell, A., F. L. Ilg, and L. B. Ames. 1995. *Infant and child in the culture of today.* Northvale, N.J.: Jason Aronson.

Glenn, E. N., and S. G. H. Yap. 2002. Chinese American families. In *Minority families in the United States: A multicultural perspective,* 3d ed., edited by R. L. Taylor. Upper Saddle River, N.J.: Prentice Hall, 134–63.

Golombok, S., B. Perry, A. Burston, C. Murray, J. Money-Somers, M. Stevens, and J. Golding. 2003. Children with lesbian parents: A community study. *Developmental Psychology* 39:20–33.

Gordon, T. 2000. *Parent effectiveness training: The parents' program for raising responsible children.* New York: Random House.

Gorman, J. C. 1998. Parenting attitudes and practices of immigrant Chinese mothers of adolescents. *Family Relations* 47:73–80.

Green, S. E. 2003. "What do you mean 'what's wrong with her?'" Stigma and the lives of families of children with disabilities. *Social Science & Medicine* 57:1361–74.

Greif, G. L., F. A. Hrabowski, and K. I. Maton. 1998. African-American fathers of high achieving sons: Using outstanding members of an at-risk population to guide intervention. *Families in Society.* January, 45–52.

Gross, K. H., C. S. Wells, A. Radigan-Garcia, and P. M. Dietz. 2002. Correlates of self-reports after delivery: Results from the pregnancy risk assessment monitoring system. *Maternal and Child Health Journal* 6:247–53.

Harry, A. W., and R. C. Ainslie. 1998. Marital discord and child problem behaviors. *Journal of Family Issues* 19:140–63.

Hauck, F. R., S. M. Herman, M. Donovan, C. M. Moore, S. Iyasu, E. Donoghue, R. H. Kirschner, and M. Willinger. 2003. Sleep environment and the risk of sudden infant death syndrome in an urban population: The Chicago Infant Mortality study. *Pediatrics* 111:1207–15.

Hay, C. 2003. Family strain, gender, and delinquency. *Sociological Perspectives* 46:107–35.

Hollander, D. 2003. Teenagers with the least adult supervision engage in the most sexual activity. *Perspectives on Sexual and Reproductive Health* 35:106–23.

Jacobson, K. C., and L. J. Crockett. 2000. Parental monitoring and adolescent adjustment: An ecological perspective. *Journal of Research on Adolescence* 10:65–97.

Jenkins, J. M., and J. M. Buccioni. 2000. Children's understanding of marital conflict and the marital relationship. *Journal of Child Psychology and Psychiatry and Allied Disciplines* 41:161–68.

Karnehm, Amy L. 2000. *Adolescent sexual initiation: Are parents powerless to delay it?* Paper presented at the Annual Conference of the National Council on Family Relations, Minneapolis, November.

Knox, D. (with Kermit Leggett). 2000. *The divorced dad's survival book: How to stay connected with your kids.* Reading, Mass.: Perseus Books.

Knox, D., M. E. Zusman, K. McGinty, and J. Gescheidler. 2001. Deception of parents during adolescence. *Adolescence* 36:611–13.

Laird, R. D., G. S. Pettit, J. E. Bates, and K. A. Dodge. 2003. Parents' monitoring, relevant knowledge and adolescents' delinquent behavior: Evidence of correlated developmental changes and reciprocal influences. *Child Development* 74:752–63.

Larzelere, R. E., P. R. Sather, W. N. Schneider, D. B. Larson, and P. L. Pike. 1998. Punishment enhances reasoning's effectiveness as a disciplinary response to toddlers. *Journal of Marriage and the Family* 60:388–403.

Lengua, L. J., S. A. Wolchik, I. N. Sandler, and S. G. West. 2000. The additive and interactive effects of parenting and temperament in predicting problems of children of divorce. *Journal of Clinical Child Psychology* 29:232–44.

Longmore, M. A., W. D. Manning, and P. C. Giordano. 2001. Preadolescent parenting strategies and teens' dating and sexual initiation: A longitudinal analysis. *Journal of Marriage and the Family* 63:322–355.

Lundeen, M. G. 1999. Interpersonal experience in infancy as a foundation for the capacity in adults for stable relationships. In *Handbook of interpersonal commitment and relationship stability,* edited by J. M. Adams and W. H. Jones. New York: Academic/Plenum, 91–107.

MacDermid, S. M., T. L. Huston, and S. M. McHale. 1990. Changes in marriage associated with the transition to parenthood: Individual differences as a function of sex-role attitudes and changes in the division of household labor. *Journal of Marriage and the Family* 52:475–86.

Mauer, M. and M. Chesney-Lind. 2003. *Invisible punishment: The collateral consequences of mass imprisonment.* New York: New Press.

McBride, B. A., S. J. Schoppe, and T. R. Rane. 2002. Child characteristics, parenting stress, and parental involvement: Fathers versus mothers. *Journal of Marriage and Family* 64:998–1011.

Mulsow, M., Y. M. Caldera, M. Pursley, A. Reifman, and A. C. Huston. 2002. Multilevel factors influencing maternal stress during the first three years. *Journal of Marriage and the Family* 64:944–56.

NICHD Early Child Care Research Network. 2003. Does quality of child care affect child outcomes at age $4\frac{1}{2}$? *Developmental Psychology* 39:451–69.

Newsbytes. 2001. Germans seek to centralize Internet content control. 31 August, NWSBO1, 24300e.

Pinquart, M., and R. K. Silbereisen. 2002. Changes in adolescents' and mothers' autonomy and connectedness in conflict discussions: An observation study. *Journal of Adolescence* 25:509–22.

Remez, L. 2003. Mothers exert more influence on timing of first intercourse among daughters than among sons. *Perspectives on Sexual and Reproductive Health* 35:55–56.

Riesch, S. K., S. A. Wolchik, L. Bush, C. J. Nelson, B. J. Ohm, P. A. Portz, B. Abell, M. R. Wightman, and P. Jenkins. 2000. Topics of conflict between parents and young adolescents. *Journal of the Society of Pediatric Nurses* 5:27–40.

Sammons, L. 2003. Personal communication, Grand Junction, Colorado.

Sears, W. and M. Sears. 1993. *The Baby Book.* Boston: Little, Brown.

Skinner, B. F., P. B. Dews, C. B. Ferster, C. D. Cheney, and W. H. Morse. 1997. *Schedules of reinforcement.* New York: Paul and Co. Publications Consortium.

Soltz, V., and R. Dreikurs. 1991. *Children: The challenge.* New York: Penguin.

Sprey, J. 2001 Theorizing in family studies: Discovering process. In *Understanding families into the new millennium: A decade of review,* edited by Robert M. Milardo. Minneapolis: National Council on Family Relations, 1–14.

Stanley, S. M., and H. J. Markman. 1992. Assessing commitment in personal relationships. *Journal of Marriage and the Family* 54:595–608.

Statistical Abstract of the United States: 2002. 122nd ed. Washington, D.C.: U.S. Bureau of the Census.

Straus, M. A. 1994. *Beating the devil out of them: Corporal punishment in American families.* San Francisco: Jossey-Bass.

Strom, R. D., T. E. Beckert, P. S. Strom, S. K. Strom, and D. L. Griswold. 2002. Evaluating the success of Causasian fathers in guiding adolescents. *Adolescence* 37:131–49.

Supple, Andrew J. 2000. Comparing the influence of parental support and control on African American, Mexican American, and Euro American adolescent development. Paper presented at the Annual Conference of the National Council on Family Relations, Minneapolis, November.

Troth, A., and C. C. Peterson. 2000. Factors predicting safe-sex talk and condom use in early sexual relationships. *Health Communication* 12:195–218.

Tucker, C. J., S. M. McHale, and A. C. Crouter. 2003. Dimensions of mothers' and fathers' differential treatment of siblings: Links with adolescents' sex-typed personal qualities. *Family Relations* 52:82–89.

Tucker, M. B. 1997. *Economic contributions to marital satisfaction and commitment.* Paper presented at the Annual Convention of the American Psychological Association, Chicago.

Twenge, J. M., W. Keith Campbell, and C. A. Foster. 2003. Parenthood and marital satisfaction: A meta-analytic review. *Journal of Marriage and Family* 65:574–83.

Wark, M.J., T. Kruczek, and A. Boley. 2003. Emotional neglect and family structure: Impact on student functioning. *Child Abuse and Neglect* 27:1033–43.

Wilson, L. C., C. M. Wilson, and L. Berkeley-Caines. 2003. Age, gender and socioeconomic differences in parental socialization preferences in Guyana. *Journal of Comparative Family Studies* 34:213–311.

Chapter 12

Abowitz, D., and D. Knox. 2003. Goals of college students: Some gender differences. *College Student Journal* 37:550–56.

Ahnert, L., and M. E. Lamb. 2003. Shared care: Establishing a balance between home and child care settings. *Child Development* 74:1044–49.
American Council on Education and University of California. 2004. *The American Freshman: National Norms for Fall, 2003.* Los Angeles: Higher Education Research Institute. U.C.L.A. Graduate School of Education and Information Studies.
Anderson, Kristin L. 1997. Gender, status, and domestic violence: An integration of feminist and family violence approaches. *Journal of Marriage and the Family* 59:655–69.
Baum, C. L. 2002. A dynamic analysis of the effect of child care costs on the work decisions of low income mothers with infants. *Demography* 39:139–64.
Beck, B. 1998. Women and work: At the double. *Economist* 348:s12–s15.
Belsky, Jay. 1995. A nation still at risk. *Phi Kappa Phi Journal* 75:36–38.
Berke, D. L. 2000. He does, she does, who does? Division of "family work" in couples where the wife is self-employed. Poster presentation at the Annual Conference of the National Council on Family Relations, Minneapolis, November.
Blumenthal, R. G. 1998. Foreign affairs, family affairs. *Columbia Journalism Review* 36:62–65.
Bounds, W. 2000. Give me a break. *Wall Street Journal.* 5 May, W1–W4.
Carey, A. R., and M. E. Mullins. 2000. Workers want fewer days at the office. *USA Today.* 11 September, B1.
Clark, K. 2002. Mommy's home: More parents choose to quit work to raise their kids. *U.S. News and World Report.* 25 November 25, 32 passim.
Crawford, D. W. 2000. Occupational characteristics and marital leisure involvement. *Family and Consumer Sciences Research Journal* 28:52–70.
Crouter, A. C., M. R. Head, M. F. Bumpus, and S. M. McHale. 2000. Implications of overwork and overload for the quality of men's family relationships. Poster presentation at the Annual Conference of the National Council on Family Relations, Minneapolis, November.
Curtis, K. T., and C. G. Ellison. 2002. Religious heterogamy and marital conflict: Finding from the National Survey of families and households. *Journal of Family Issues* 23:551–76.
Doumas, D. M., G. Margolin, and R. S. John. 2003. The relationship between daily marital interaction, work, and health-promoting behaviors in dual-earner couples: An extension of the work-family spillover model. *Journal of Family Studies* 24:3–20.
Doyle, L. 2003. Our marriage barely survived. *Newsweek.* 12 May, 53.
Ehrenreich, B. 2001. *Nickel and dimed: On (not) getting by in America.* New York: Henry Holt.
Fox, G. L., M. L. Benson, A. A. DeMaris, and J. V. Wyk. 2002. Economic distress and intimate violence: Testing family stress and resources theory. *Journal of Marriage and the Family* 64:793–807.
Goldstein, A. 2000. It took three dead babies. *Time.* 10 July 10, 80–81.
Grossman, Cathy L. 1995. For many, vacation is no vacation. *USA Today.* 14 July, 1D.
Grossman, L. M. 1998. Family-friendly corporate policies can be counterproductive. In *The family: Opposing viewpoints,* edited by B. Leone. San Diego, Calif.: Greenhaven Press.
Grzywacz, J. G., and B. L. Bass. 2003. Work, family, and mental health: Testing different models of work-family fit. *Journal of Marriage and the Family* 65:248–62.
Grzywacz, J. G., and N. F. Marks. 2000. Family, work, work-family spillover, and problem drinking during midlife. *Journal of Marriage and the Family* 62:336–48.
Haralson, D., and M. E. Mullins. 2000. Less pay, more play. *USA Today.* 6 November, B1.
Helburn, S., M. L. Culkin, J. Morris, N. Moran, C. Howes, L. Phillipsen, D. Bryant, R. Clifford, D. Cryer, E. Peisner-Feinberg, M. Burchinal, S. L. Kagan, and J. Rustici. 1995. Cost, quality, and child outcomes in child care centers: Key findings and recommendations. *Young Children* 50:40–44.
Helburn, S. W., and C. Howes. 1996. Day care cost and quality. *Future of Children* 6:62–82.
Hendershott, Anne. 1995. A moving story for spouses and other wage-earners. *Psychology Today,* 28:28–30.
Hochschild, A. R. 1989. *The second shift.* New York: Viking.
———. 1997. The time bind. New York: Metropolitan Books.
Jacobs, J. A., and J. C. Cognick. 2002. Hours of paid work in dual-earner couples: The United States in cross-national perspective. *Sociological Focus* 35:169–87.
Jalovaara, M. 2003. The joint effects of marriage partners' socioeconomic positions on the risk of divorce. *Demography* 40:67–81.
Kiecolt, K. J. 2003. Satisfaction with work and family life: No evidence of a cultural reversal. *Journal of Marriage and the Family* 65:23–35.
Macmillan, R., and Gartner. 2000. When she brings home the bacon: Labor-force participation and the risk of spousal violence against women. *Journal of Marriage and the Family* 61:947–58.
NICHD Early Child Care Research Network. 2003. Does quality of child care affect child outcomes at age 4½? *Developmental Psychology* 39:451–69.
Ono, H. 2003. Women's economic standing, marriage timing, and cross-national contexts of gender. *Journal of Marriage and the Family* 65: 275–86.
Perry-Jenkins, M., R. L. Repetti, and A. C. Crouter. 2001. Work and family in the 1990s. In *Understanding families into the new millennium: A decade in review,* edited by R. M. Milardo. Minneapolis: National Council on Family Relations, 200–217.
Phillips, M., L. D., N. J. Campbell, and C. R. Morrison. 2000. Work and family: Satisfaction, stress, and spousal support. *Journal of Employment and Counseling* 37:16–30.
Polatnick, M. R. 2000. Working parents. *Phi Kappa Phi Journal* 80:38–41.
Presser, H. B. 2000. Nonstandard work schedules and marital instability. *Journal of Marriage and the Family* 62:93–110.
Reynolds, J. 2003. You can't always get the hours you want: Mismatches between actual and preferred work hours in the U.S. *Social Forces* 81:1171–99.
Roberts, Paul. 1995. Goofing off. *Psychology Today.* July/August, 34–41.
Rosenbluth, S. C., J. M. Steil, and J. H. Witcomb. 1998. Marital equality: What does it mean? *Journal of Family Issues* 19:227–44.
Salmon, J., N. Owen, A. Bauman, M. K. H. Schmitz, and M. Booth. 2000. Leisure -time, occupational, and household physical activity among professional, skilled, and less-skilled workers and homemakers. *Prevention Medicine* 30: 191–99.
Schoen, R., N. M. Astone, K. Rothert, N. J. Standish, and Y. J. Kim. 2002. Women's employment, marital happiness, and divorce. *Social Forces* 81:643–62.
Schor, Juliet B. 1991. *The overworked American: The unexpected decline of leisure.* N.Y.: Basic Books.
Seccombe, K. 2001. Families in poverty in the 1990s: Trends, causes, consequences, and lessons learned. In *Understanding families into the new millennium: A decade in review,* edited by R. M. Milardo. Minneapolis: National Council on Family Relations, 313–32.
Sefton, B. W. 1998. The market value of the stay-at-home mother. *Mothering* 86:26–29.
Stanfield, J. B. 1998. Couples coping with dual careers: A description of flexible and rigid coping styles. *Social Science Journal* 35:53–62.
Statistical Abstract of the United States: 2003 123rd ed. Washington, D.C.: U.S. Bureau of the Census.
Stier, H., and N. Lewin-Epstein. 2000. Women's part-time employment and gender inequality in the family. *Journal of Family Issues* 21:390–410.
Tyre, P., and D. McGinn. 2003. She works, he doesn't. *Newsweek.* 12 May, 45–52.
Van Willigen, M., and P. Drentea. 2001. Benefits of equitable relationships: The impact of fairness, household division of labor, and decision-making power on social support. *Sex Roles* 44: 571–97.
Vannoy, D., and L. A. Cubbins. 2001 Relative socioeconomic status of spouses, gender attitudes, and attributes, and marital quality experienced by couples in metropolitan Moscow. *Journal of Comparative Family Studies* 32:195–217.
Walker, S. K. 2000. Making home work: Family factors related to stress in family child care providers. Poster presentation at the Annual Conference of the National Council on Family Relations, Minneapolis, November.
Wishard, A. G., E. M. Shivers, C. Howes, and S. Ritchie. 2003. Child care program and teacher practices: Associations with quality and children's experiences. *Early Childhood Research Quarterly* 18:65–103.
Zuo, J., and S. Tang. 2000. Breadwinner status and gender ideologies of men and women regarding family roles. *Sociological Perspectives* 43:29–43.

Chapter 13

Alexandrian, T., D. J. Alonzo, T. Ashworth, E. Berumen, H. Gyrous, P. Fabian et al. 2002. Measuring the effects of September 11, 2001. *PROGRESS: Family Systems Research and Therapy* 11:37–64.
Anderson, R.N., and B. L. Smith 2003. Leading causes of death for 2001. *National Vital Statistics Reports* 52 (1). Hyattsville, Md.: National Center for Health Statistics.
American Council on Education and University of California. 2002. The American freshman: National norms for fall, 2002. Los Angeles: Higher Education Research Institute. U.C.L.A. Graduate School of Education and Information Studies.
Armour, S. 2003. Rising job stress could affect bottom line. *USA Today.* 29 July,1B.
Associated Press. 2003. Texas woman gets 20 years in prison. *The Daily Reflector.* 15 February 15, A4.
Baldwin, D. R., L. N. Chambliss, and K. Towler. 2003. Optimism and stress: An African-American college student perspective. *College Student Journal* 37:276–83.
Bartholomew, G. S., and S. Klein. 2000. Substance-abusing women's perceptions of family life: Implications for FLE. Poster presentation at the Annual Conference of the National Council on Family Relations, Minneapolis, November.
Bermant, G. 1976. Sexual behavior: Hard times with the Coolidge Effect. In *Psychological research: The inside story,* edited by M. H. Siegel and H. P. Zeigler. New York: Harper and Row.
Blake, W. M., and C. A. Darling. 1998. Quality of life: Perceptions of African Americans. Poster

presentation at the Annual Meeting of the National Council on Family Relations, Milwaukee.
Booth, Alan, and David R. Johnson. 1994. Declining health and marital quality. *Journal of Marriage and the Family* 56:218–23.
Bosari, B., and D. Bergen-Cico. 2003. Self-reported drinking-game participation of incoming college students. *Journal of American College Health* 51:149–54.
Bosco, S. M., and D. Harvey. 2003. Effects of terror attacks on employment plans and anxiety levels of college students. *College Student Journal* 37:438–46.
Bouchard, G., J. Wright, Y. Lussier, and C. Richer. 1998. Predictive validity of coping strategies on marital satisfaction: Cross-sectional and longitudinal evidence. *Journal of Family Psychology* 12:112–31.
Brogenschneider, K., M. Wu, M. Raffaelli, and J. C. Tsay. 1998. "Other teens drink, but not my kid": Does parental awareness of adolescent alcohol use protect adolescents from risky consequences? *Journal of Marriage and the Family* 60:356–73.
Brotherson, S. E. 2000. When a child dies: Primary parenting concerns during the loss of a child. Poster presentation at the Annual Conference of the National Council on Family Relations, Minneapolis, November.
Burke, M. L., G. G. Eakes, and M. A. Hainsworth. 1999. Milestones of chronic sorrow: Perspectives of chronically ill and bereaved persons and family caregivers. *Journal of Family Nursing*. 5:387–84.
Burr, W. R., and S. R. Klein and associates.1994. *Reexamining family stress: New theory and research.* Thousand Oaks, Calif.: Sage.
Cano, A., J.Christian-Herman, K. D. O'Leary, and S. Avery-Leaf. 2002. Antecedents and consequences of negative marital stressors. *Journal of Marital and Family Therapy* 28:145–51.
Carlson-Catalano, Judy. 2003. Director, Clinical Biofeedback Services. *Health Innovations.* Greenville, NC 27878. Personal communication, June 9.
Chambless, D. L., F. J. Floyd, K. A. Wilson, A. L. Remen, J. A. Fauerbach, and B. Renneberg. 2002. Martial interaction of agoraphobic women: A controlled, behavioral observation study. *Journal of Abnormal Psychology* 111:502–11.
Charney, I. W., and S. Parnass. 1995. The impact of extramarital relationships on the continuation of marriages. *Journal of Sex and Marital Therapy* 21:100–15.
Christakis, N. A., and T. J. Iwashyna. 2003. The health impact of health care on families: A matched cohort study of hospice use by decedents and mortality outcomes in surviving widowed spouses. *Social Science & Medicine* 57:465–75.
Clinton, H. R. 2003. *Living history.* New York: Simon and Schuster.
Collins, C. C., C. E. Grella, and Y. Hser. 2003. Effects of gender and level of parental involvement among parents in drug treatment. *American Journal of Drug and Alcohol Abuse* 29:237–61.
Collins, J. 1998. *Singing lessons: A memoir of love, loss, hope, and healing.* New York: Pocket Books, Inc.
Crimmins, C. 2000. *Where is the Mango Princess?* New York: Knopf.
Dar, S. 2003. An Islamic view of sexuality. Presentation to Sociology of Human Sexuality class, Department of Sociology, East Carolina University, November 13.
DeHaan, L. 2002. Book review of *Resilient marriages* by K. J. Shirley. *FamilyRelations* 51:185–86.
Ellis, R. T., and J. M. Granger 2002. African-American adults' perceptions of the effects of parental loss during adolescence. *Child and Adolescent Social Work Journal* 19:271–86.
Field, N. P., Gal-Oz, E., and G. A. Bananno. 2003. Continuing bonds and adjustment at 5 years after the death of a spouse. *Journal of Consulting and Clinical Psychology* 71:110–17.
Finucane, P., L. C. Giles, R. T. Withers, C. A. Silagy, A. Sedgwick, P. A. Hamdorf, J. A. Halbert, L. Cobiac, M. S. Clark, and G. R. Andrews. 1997. Exercise profile and subsequent mortality in an elderly Australian population. *Australian and New Zealand Journal of Public Health* 21:155–58.
Forehand, R., H. Giggar, and B. A. Kotchick. 1998. Cumulative risk across family stressors: Short- and long-term effects for adolescents. *Journal of Abnormal Child Psychology,* 119–29.
Gelbin, R. Personal communication. 2003. (Ms. Gelbin is a Certified Professional Counselor in Arizona who specializes in Imago Therapy; see Imagocouples.com)
Goode, E. 1999. New study finds middle age is prime of life. *New York Times.* 17 July 17, D6.
Harper, J. M., B. G. Schaalje, and J. G. Sandberg. 2000. Daily hassles, intimacy, and marital quality in later marriages. *American Journal of Family Therapy* 28:1–18.
Hong, Jinkuk, and M. M. Seltzer. 1995. The psychological consequences of multiple roles: The nonnormative case. *Journal of Health and Social Behavior* 36:386–98.
Hyman, B. 2000. Resilient physicians with bipolar disorder. Poster presentation at the Annual Conference of the National Council on Family Relations, Minneapolis, November.
Jiong Li, P., H. Dorthe, P. B. Mortensen, and J. Olsen. 2003. Mortality in parents after death of a child in Denmark: A nationwide follow-up study. *Lancet* 361:363–68.
Johnson, G. R., E. G. Krug, and L. B. Potter. 2000. Suicide among adolescents and young adults: A cross-national comparison of 34 countries. *Suicide and Life-Threatening Behavior* 30:74–82.
Kate, Nancy Ten. 1994. To reduce stress, hit the hay. *American Demographics* 16:14–16.
Keller, T. E., R. F. Catalano, K. P. Haggerty, and C. B. Fleming. 2002. Parent figure transitions and delinquency and drug use among early adolescent children of substance abusers. *American Journal of Drug Alcohol Abuse* 28:399–427.
Kessler, R., K. McGonale, S. Zhao, C. Nelson, M. Hughes, S. Eshelman, H. Wittchen, and K. Kendler. 1994. Lifetime and twelve-month prevalence of DSM-III-R psychiatric disorders in the United States. *Archives of General Psychiatry* 51:8–19.
Keyes, C. L. M. 2002. The mental health continuum: From languishing to flourishing in life. *Journal of Health and Social Research* 43:207–22.
King, C. A. 1998. Special niche: Planning for families with disabled members. *National Underwriter* 102:7 passim.
Knox, D. 2001. Interview with Shaaban Mgonja, Masai guide, Arusha, Tanzania, East Africa, January 24.
Knox, D., M. E. Zusman, M. Kaluzny, and L. Shivar. 2000. Attitudes and behavior of college students toward infidelity. *College Student Journal* 34: 162–64.
Kuo, M., E. Adlaf, H. Lee, L. Gliksman, A. Demers, and H. Wechsler. 2002. More Canadian students drink but American students drink more: Comparing college alcohol use in two countries. *Addiction* 97:1583–92.
Kylin, N., L. L. Meschke, and L. M. Borden. 2000. Family matter: Family process and adolescent substance abuse. Poster presentation at the Annual Conference of the National Council on Family Relations, Minneapolis, November.
Langehough, Steven O., C. Walters, D. Knox, and M. Rowley. 1997. Spirituality and religiosity as factors in adolescents' risk for antisocial behaviors and use of resilient behaviors. Paper presented at the 59th Annual Conference of the National Council on Family Relations, Crystal City, Va.
Lederman, L. C., L. P. Stewart, F. W. Goodhart, and L. Laitman. 2003. A case against "binge" as a term of choice: Convincing college students to personalize messages about dangerous drinking. *Journal of Health Communication* 8:79–91.
Lund, L. K., T. S. Zimmerman, and S. A. Haddock. 2002. The theory, structure, and techniques for the inclusion of children in family therapy: A literature review. *Journal of Marital and Family Therapy* 28:445–54.
Macmillan, R., and Gartner. 2000. When she brings home the bacon: Labor-force participation and the risk of spousal violence against women. *Journal of Marriage and the Family* 61:947–58.
Marano, H. S. 2003. The new sex scorecard. *Psychology Today* 36:38 passim.
Marchand, J. F., and E. Hock. 2000. Avoidance and attacking conflict-resolution strategies among married couples: Relations to depressive symptoms and marital satisfaction. *Family Relations* 49:201–6.
McGee, M., L. Nagel, and M. K. Moore. 2003. A study of university classroom strategies aimed at increasing spiritual health. *College Student Journal* 37:583–94.
McKinnon, S., K. O'Rourke, and T. Byrd. 2003. Increased risk of alcohol abuse among college students living on the US-Mexico border: Implications for prevention. *Journal of American College Health* 51:163–67.
Mercy, J. A., E. G. Krug, L. L. Dahlberg, and A. B. Zwi. 2003. Violence and health: The United States in global perspective. *American Journal of Public Health* 93:256–61.
Michael, M. T., J. Wadsworth, J. A. Feinleib, A. M. Johnson, E. O. Laumann, and K. Wellings. 2001. Private sexual behavior, public opinion, and public health policy related to sexually transmitted diseases: A U.S. British comparison. In *Sex, love, and health in America: Private choices and public policies,* edited by E. O. Laumann and R. T. Michael. New York: Harper and Row,439–53.
Morell, V. 1998. A new look at monogamy. *Science* 281:1982 passim.
Nielsen, Sven, and Mats Hahlin. 1995. Expectant management of first-trimester spontaneous abortion. *Lancet* 345:84–86.
Northey, W. F. J. 2002. Characteristics and clinical practices of marriage and family therapists: A national survey. *Journal of Marriage and Family Therapy* 28:487–94.
O'Farrell, T. J., and W. Fals-Stewart. 2003. Alcohol abuse. *Journal of Marital and Family Therapy* 29:121–46.
Olson, M. M., C. S. Russell, M. Higgins-Kessler, and R. B. Miller. 2002. Emotional processes following disclosure of an extramarital affair. *Journal of Marital and Family Therapy* 28:423–34.
Ozer, E. J., S. R. Best, T. L. Lipsey, and D. S. Weiss. 2003. Predictors of posttrauamatic stress disorder and symptoms in adults: A meta-analysis. *Psychological Bulletin* 129:52–73.
Parker, J. S., and M. J. Benson. 2000. Parenting processes and adolescent problems: Delinquency, substance abuse, and peer and self-esteem deficits. Poster presentation at the Annual Conference of the National Council on Family Relations, Minneapolis, November.

Patterson, J. M. 2002. Integrating family resilience and family stress theory. *Journal of Marriage and the Family* 64:349–60.
Patterson, J.M. 1988. Chronic illness in children and the impact upon families. In *Chronic illness and disability*, edited by C. S. Chillman, E. W. Nunnally, and F. M. Cox. Beverly Hills, Calif.: Sage, 69–107.
Poche, R. S, M. B. White, and T. A. Smith, Jr. 1997. Correlates of therapeutic dropout. Paper presented at the Annual Conference of the National Council on Family Relations, Crystal City, Va.
Prather, W., and G. A. Darling. 1998. Family stress and learned helplessness: Risk factors influencing marital satisfaction. Poster session presented at the Annual Meeting of the National Council on Family Relations, Milwaukee, Wis.
Rolland, John S. 1994. In sickness and in health: The impact of illness on couples' relationships. *Journal of Marital and Family Therapy* 20:327–47.
Sammons, R. A., Jr. 2003. First law of therapy. Personal communication. Grand Junction, Colorado (Dr. Sammons is a psychiatrist in private practice).
Shadish, W. R., and S. A. Baldwin. 2003. Meta-analysis of MTF interventions. *Journal of Marriage and the Family* 29:547–70.
Schneider, J. P. 2000. Effects of cybersex addiction on the family: Results of a survey. *Sexual Addiction and Compulsivity* 7:31–58.
Segall, R. 2000. Online shrinks: The inside story. *Psychology Today*. May/June, 38–43.
Smith, S. R., and E. Soliday. 2001. The effects of parental chronic kidney disease on the family. *Family Relations* 50:171–78.
Sobieski, R. 1994. *Men and mourning: A father's journey through grief.* Mothers Against Drunk Driving (MADD), 511 E. John Carpenter Freeway, Suite 700, Irving, TX 75062-8187.
Statistical Abstract of the United States: 2003. 123rd ed. Washington, D.C.: U.S. Bureau of the Census.
Stillion, J. M. 1996. Survivors of suicide. In *Living with grief after sudden loss,* edited by K. J. Doka. Bristol, Pa.: Taylor & Francis, 41–51.
Treas, J., and D. Giesen. 2000. Sexual infidelity among married and cohabiting Americans. *Journal of Marriage and the Family* 62:48–60.
Turner, R. J., and D. A. Lloyd. 1995. Lifetime traumas and mental health: The significance of cumulative adversity. *Journal of Health and Social Adversity* 36:360–76.
Tuttle, Jane. 1995. Family support, adolescent individuation, and drug and alcohol involvement. *Journal of Family Nursing* 1:303–26.
Umberson, Debra. 1995. Marriage as support or strain? Marital quality following the death of a parent. *Journal of Marriage and the Family* 57:709–23.
Walker, S. K. 2000. Making home work: Family factors related to stress in family child care providers. Poster presentation at the Annual Conference of the National Council on Family Relations, Minneapolis, November.
Wallerstein, J. S., and S. Blakeslee. 1995. *The good marriage.* Boston: Houghton Mifflin.
Walsh, F. 2003. Clinical views of family normality, health, and dysfunction. . In *Normal Family Processes: Growing Diversity and Complexity,* edited by F. Walsh. New York/London: Guilford Press, 27–57.
White, A. M., D. W. Jamieson-Drake, and H. S. Swartzwelder. 2002. Prevalence and correlates of alcohol-induced blackouts among college students: Results of an e-mail survey. *Journal of American College Health* 51:117–32.
Wiederman, M. W. 1997. Extramarital sex: Prevalence and correlates in a national survey. *Journal of Sex Research* 34:167–74.
Williams, D. J., A. Thomas, W. C. Buboltz, Jr., and M. McKinney. 2002. Changing the attitudes that predict underage drinking in college students: A program evaluation. *Journal of College Counseling* 5:39–49.
Williams, P., and S. R. Lord. 1997. Effects of group exercise on cognitive functioning and mood in older women. *Australian and New Zealand Journal of Public Health* 21:45–52.
Wyman, J. R. 1997. Multifaceted prevention programs reach at-risk children through their families. *National Institute on Drug Abuse* 12 (3):5–7.

Chapter 14

Anderson, D. J. 2003. The impact on subsequent violence of returning to an abusive partner. *Journal of Comparative Family Studies* 34:93–107.
Anderson, K. L. 2002. Perpetrator or victim? Relationships between intimate partner violence and well-being. *Journal of Marriage and Family* 64:851–63.
Anderson, P. L., J. A. Tiro, A. W. Price, M. A. Bender, and N. J. Kaslow. 2003. Additive impact of childhood emotional, physical, and sexual abuse on suicide attempts among low-income African American woman. *Suicide and Life-Threatening Behavior* 32:131–38.
Associated Press. 2004. James Brown accused of assaulting wife. *The News and Observer.* 29 January, 2A.
Babcock, J. C., S. A. Miller, and C. Siard. 2003. Toward a typology of abusive women: Differences between partner-only and generally violent women in the use of violence. *Psychology of Women Quarterly* 27:153–61.
Beitchman, J. H., K. J. Zuker, J. E. Hood, G. A. daCosta, D. Akman, and E. Cassavia. 1992. A review of the long-term effects of child sexual abuse. *Child Abuse and Neglect* 16:101–19.
Bergeron, L. R., and B. Gray. 2003. Ethical dilemmas of reporting suspected elder abuse. *Social Work* 48:96–106.
Blankenhorn, David. 1995. *Fatherless America: Confronting our most urgent social problem.* New York: Basic Books.
Bolen, R. M. 2001. *Child sexual abuse: Its scope and our failure.* New York: Kluwer Academic/Plenum Publishers.
Browning, C. R., and E. O. Laumann. 1998. Sexual contact between children and adults: A life course perspective. *American Sociological Review* 62:540–60.
Browning, C. R. 2002. The span of collective efficacy: Extending social disorganization theory to partner violence. *Journal of Marriage and Family* 64:833–50.
Calhoun, K. S., J. A. Bernat, G. A. Clum, and C. L. Frame. 1997. Sexual coercion and attraction to sexual aggression in a community sample of young men. *Journal of Interpersonal Violence* 392–406.
Carlson, Bonnie E. 1997. A stress and coping approach to intervention with abused women. *Family Relations* 46:291–98.
Coleman, Frances L. 1997. Stalking behavior and the cycle of domestic violence. *Journal of Interpersonal Violence* 12:420–32.
Connell-Carrick, K. 2003. A critical review of the empirical literature: Identifying correlates of child neglect *Child and Adolescent Social Work Journal* 20:389–925.
Davis, M. K., and C. A. Gidyez. 2000. Child sexual abuse prevention programs: A meta-analysis. *Journal of Clinical Child Psychology* 29:257–66.
DeMaris, A., M. L. Benson, G. L. Fox, T. Hill, and J. V. Wyk. 2003. Distal and proximal factors in domestic violence: A test of an integrated model. *Journal of Marriage and Family* 65:652–67.
De Paul, J., and I. Arruabarrena. 2003. Evaluation of a treatment program for abusive and high-risk families in Spain. *Child Welfare* 82:413–42.
Di Lillo, D., G. C. Tremblay, and L. Peterson. 2000. Linking childhood sexual abuse and abusive parenting: The mediating role of maternal anger. *Child Abuse and Neglect* 24:767–79.
Dong, M., R. F. Anda, S. R. Dube, W. H. Giles, and V. J. Felitti. 2003. The relationship of exposure to childhood sexual abuse to other forms of abuse, neglect, and household dysfunction during childhood. *Child Abuse and Neglect* 27:625–39.
Dugan, L., D. S. Nagin, and R. Rosenfeld 2003. Exposure reduction or retaliation? The effects of domestic violence resources on intimate-partner homicide. *Law & Society Review* 37:169–98.
Dutton, Donald G., and Andrew J. Starzomski. 1977. Personality predictors of the Minnesota Power and Control Wheel. *Journal of Interpersonal Violence* 12:70–82.
Edleson, J. L., L. F. Mbilinyi, S. K. Beeman, and A. K. Hagemeister. 2003. How children are involved in adult domestic violence. *Journal of Interpersonal Violence* 18:18–32.
Egley, L. C. 2000. How effective are criminal justice interventions into partner abuse? *Family Focus.* September, F10–F11.
Elliott, D. M., and J. Briere. 1992. The sexually abused boy: Problems in manhood. *Medical Aspects of Human Sexuality* 26:68–71.
Faller, K. C., and J. Henry. 2000. Child sexual abuse: A case study in community collaboration. *Child Abuse and Neglect* 24:1215–25.
Feinauer, L., H. G. Hilton, E. H. Callahan. 2003. Hardiness as a moderator of shame associated with childhood sexual abuse. *American Journal of Family Therapy* 31:65–78.
Fieldman, J. P., and T. D. Crespi. 2002. Child sexual abuse: Offenders, disclosure, and school-based initiatives. *Adolescence* 37:151–60.
Finkelhor, D., and K. Yllo. 1988. Rape in marriage. In *Abuse and victimization across the life span,* edited by M. B. Straus. Baltimore: Johns Hopkins University Press, 140–52.
Forte, J. A., D. D. Franks, J. A. Forte, and D. Rigsby. 1996. Asymmetrical role-taking: Comparing battered and nonbattered women. *Social Work* 41:59–73.
Foshee, V. A., K. E. Bauman, Z. B. Arriaga, R. W. Helms, G. G. Koch, and G. F. Linder. 1998. An evaluation of safe dates, an adolescent dating violence prevention program. *American Journal of Public Health* 88:45–50.
Freedner, N., L. H. Freed, W. Yang, and S. B. Austin. 2002 Dating violence among gay, lesbian, and bisexual adolescents: Results from a community survey. *Journal of Adolescent Health* 31:469–74.
Gelles, Richard J., and M. Straus. 1988. *Intimate violence.* New York: Simon & Schuster.
Gibbs, N. 1993. Till death do us part. *Time.* 18 January, 38–45.
Goldstein, S. L., and R. P. Tyler. 1998. Frustrations of inquiry: Child sexual abuse allegations in divorce and custody cases. *FBI Law Enforcement Bulletin* 67:1–6.
Goodman, G. S., Simona Ghetti, J. A. Quas, R. S. Edelstein, K. W. Alexander, A. D. Redlich, I. M. Cordon, and D. P. H. Jones. 2003. A prospective study of memory for child sexual abuse: New findings relevant to the repressed-memory controversy. *Psychological Science* 14:113–18.

Goodman-Brown, T. B., R. S. Edelstein, G. S. Goodman, D. P. H. Jones, and D. S. Gordon. 2003. Why children tell: A model of children's disclosure of sexual abuse. *Child Abuse and Neglect* 27:525–40.

Gray, Heather M., and Vangie Foshee. 1997. Adolescent dating violence: Differences between one-sided and mutually violent profiles. *Journal of Interpersonal Violence* 12:126–41.

Grych, J. H., G. T. Harold, and C. J. Miles. 2003. A prospective investigation of appraisals as mediators of the link between interparental conflict and child adjustment. *Child Development* 74: 1176–96.

Hanley, M. Joan, and Patrick O'Neill. 1997. Violence and commitment: A study of dating couples. *Journal of Interpersonal Violence* 12:685–703.

Haskett, M. E., S. S. Scott, R. Grant, C. S. Ward, and Canby Robinson. 2003. Child-related cognitions and affective functioning of physically abusive and comparison parents. *Child Abuse and Neglect* 27:663–86.

Heide, K. M. 1992. *Why kids kill parents: Child abuse and adolescent homicide.* Columbus, Ohio: Ohio State University Press.

Hendy, H. M., D. Eggen, C. Gustitus, K. C. McLeod, and P. Ng. 2003. Decision to leave scale: Perceived reasons to stay in or leave violent relationships. *Psychology of Women Quarterly* 27:162–73.

Hewitt, B., K. Klise, L. Comander, M Schorr, A. Hardy, T. Duffy et al. 2002. Breaking the silence. *People* 57:56 passim.

Heyman, R. E., and A. M. S. Slep. 2002 Do child abuse and interpersonal violence lead to adult family violence? *Journal of Marriage and Family* 64:864–70.

Hodge, S., and D. Canter. 1998. Victims and perpetrators of male sexual assault. *Journal of Interpersonal Violence* 13:222–39.

Howard, D. E., S. Feigelman, X. Li, S. Gross, and L. Rachuba. 2002. The relationship among violence victimization, witnessing violence, and youth distress. *Journal of Adolescent Health* 31:455–62.

Howard, D.E and M. Q. Wang. 2003. Risk profiles of adolescent girls who were victims of dating violence. *Adolescence* 38:1–19.

Johnson, Michael P. 1995. Patriarchal terrorism and common couple violence: Two forms of violence against women. *Journal of Marriage and the Family* 57:283–94.

Kahn, A. S., J. Jackson, C. Kully, K. Badger, and J. Halvorsen. 2003. Calling it rape: Differences in experiences of women who do and do not label their sexual assault as rape. *Psychology of Women Quarterly* 27:233–42.

Kasian, M., and S. L. Painter. 1992. Frequency and severity of psychological abuse in a dating population, *Journal of Interpersonal Violence* 7:350–64.

Kernic, M. A., M. E. Wolfe, V. L. Holt, B. McKnight, C. E. Huebner, and F. P. Rivara. 2003. Behavioral problems among children whose mothers are abused by an intimate partner. *Child Abuse & Neglect* 27:1231–46.

Kim, J., and C. Emery. 2003. Marital power: Conflict, norm consensus, and marital violence in a nationally representative sample of Korean couples. *Journal of Interpersonal Violence* 18:197–219.

Kitzmann, K. M., N. K. Gaylord, A. R. Holt, and E. D. Kenny. 2003. Child witnesses to domestic violence: A meta-analytic review. *Journal of Clinical and Consulting Psychology* 71:339–52.

Knutson, John F., and Mary Beth Selner. 1994. Punitive childhood experiences reported by young adults over a 10-year period. *Child Abuse and Neglect* 18:155–66.

Koss, M. P., J. A. Bailey, N. P. Yaun, V. M. Herrera, and E. L. Lichter. 2003. Depression and PTSD in survivors of male violence: Research and training initiatives to facilitate recovery. *Psychology of Women Quarterly* 27:130–42.

Krahe, B., P. Habil, R. Scheinberger-Olwig, and S. Bieneck. 2003. Men's reports of nonconsensual sexual interactions with women: Prevalence and impact. *Archives of Sexual Behavior* 32:165–75.

Krahe, B., E. Waizenhofer, and I. Moller. 2003. Women's sexual aggression against men: Prevalence and predictors. *Sex Roles: A Journal of Research* 49:219–33.

Lloyd, S. A., and B. C. Emery. 2000. *The dark side of courtship: Physical and sexual aggression.* Thousand Oaks, Calif.: Sage.

McBurnett, K., C. Kerckhoff, L. Capasso, L. J. Pfiffner, P. J. Rathouz, M. McCord, and S. M. Harris. 2001. Antisocial personality, substance abuse, and exposure to parental violence in males referred for domestic violence. *Violence and Victims* 16:491–505.

McCloskey, L. A., and J. A. Bailey. 2000. The intergenerational transmission of risk for child sexual abuse. *Journal of Interpersonal Violence* 15:1019–35.

McGeary, J., R. Winters, S. Morrissey, S. Scully, M. Sieger, S. Crittle. et al. 2002. Can the church be saved? *Time Atlantic* 159:52–62.

Melchert, T. P. 2000. Clarifying the effects of parental abuse, child sexual abuse, and parental caregiving on adult adjustment. *Professional Psychology: Research and Practice* 31:64–69.

Melzer, S. A. 2002. Gender, work, and intimate violence: Men's occupational violence spillover and compensatory violence. *Journal of Marriage and the Family* 64:820–32.

Muller, R. T., and K. E. Lemieux. 2000. Social support, attachment, and psychopathology in high risk formerly maltreated adults. *Child Abuse and Neglect* 24:883–900.

Nayak, M. B., C. A. Byrne, M. K. Martin, and A. G. Abraham. 2003. Attitudes toward violence against women: A cross-nation study. *Sex Roles: A Journal of Research* 49:333–43.

Nelson, B. S., and K. S. Wampler. 2000. Systemic effects of trauma in clinic couples: An exploratory study of secondary trauma resulting from childhood abuse. *Journal of Marriage and Family Counseling* 26:171–84.

Novello, A. 1992. The domestic violence issue: Hear our voices. *American Medical News* 35 (12): 41–42.

O'Farrell, T. J., M. Murphy, W. Fals-Stewart, and C. M. Murphy. 2003. Partner violence before and after individually based alcoholism treatment for male alcoholic patients. *Journal of Consulting and Clinical Psychology* 71:92–102.

Puente, S., and D. Cohen. 2003. Jealousy and the meaning (or Nonmeaning) of violence. *Personality and Social Psychology Bulletin* 29:449–60.

Rand, M. R. 2003. The nature and extent of recurring intimate partner violence against women in the United States. *Journal of Comparative Family Studies* 34:137–46.

Ricci, L., A. Giantris, P. Merriam, S. Hodge, and T. Doyle. 2003. Abusive head trauma in Maine infants: Medical, child protective, and law enforcement analysis. *Child Abuse and Neglect* 27: 271–83.

Riggs, David S., and M. B. Caulfield. 1997. Expected consequences of male violence against their female dating partners. *Journal of Interpersonal Violence* 12:229–40.

Ronfeldt, H. M., R. Kimmerling, and I. Arias. 1998. Satisfaction with relationship power and the perception of dating violence. *Journal of Marriage and the Family* 60:70–78.

Rosen, K. H., and S. M. Stith. 1997. Surviving abusive dating relationships. In *Out of the darkness: Contemporary perspectives on family violence,* edited by G. K. Kantor and J. L. Jasinski. Thousand Oaks, Calif.: Sage, 171–82.

Rosenbaum, A., and P. A. Leisring. 2003. Beyond power and control: Towards an understanding of partner abusive men. *Journal of Comparative Family Studies* 34:7–21.

Rubin, D. M., C. W. Christian, L. T. Bilaniuk, K. A. Zaxyczny, and D. R. Durbin. 2003. Occult head injury in high-risk abused children. *Pediatrics* 111:1382–86.

Salter, D., D. McMillan, M. Richards, T. Talbot, J. Hodges, A. Bentovim, R. Hastings, J. Stevenson, and D. Skuse. 2003. *Lancet* 361:471–502.

Scott, K. L., and D. A. Wolfe. 2000. Change among batterers: Examining men's success stories. *Journal of Interpersonal Violence* 15:827–42.

Seiler, A. 2000. Entertainment marketing to children blasted. *USA Today.* 14 September, 4A.

Senn, C. Y., S. Desmarais, N. Verberg, and E. Wood. 2000. Predicting coercive sexual behavior across the lifespan in a random sample of Canadian men. *Journal of Social and Personal Relationships* 17:95–113.

Shapiro, B. L., and J. C. Schwarz. 1997. Date rape. The relationship to trauma symptoms and sexual self-esteem. *Journal of Interpersonal Violence* 12:407–19.

Sheridan, M. S. 2003. The deceit continues: An updated literature review of Manuchausen syndrome by proxy. *Child Abuse & Neglect* 27: 431–51.

Shook, N. J., D. A. Gerrity, J. Jurich, and A. E. Segrist. 2000. Courtship violence among college students: A comparison of verbally and physically abusive couples. *Journal of Family Violence* 15:1–22.

Sidebotham, P., J. Heron, and the ALSPAC study team. 2003. Child maltreatment in the "children of the nineties:" The role of the child. *Child Abuse & Neglect* 27:337–52.

Simmons, R. L., L. Kuei-Hsiu, and L. C. Gordon. 1998. Socialization in the family of origin and male dating violence: A prospective study. *Journal of Marriage and the Family* 60:467–78.

Simonelli, C. J., T. Mullis, A. N. Elliott, and T. W. Pierce. 2002. Abuse by siblings and subsequent experiences of violence within the dating relationship. *Journal of Interpersonal Violence* 17: 103–21.

Smith, J. 2003. Shaken baby syndrome. *Orthpaedic Nursing* 22:196–205.

Statistical Abstract of the United States: 2003. 123rd ed. Washington, D.C.: U.S. Bureau of the Census.

Struckman-Johnson, C., D. Struckman-Johnson, and P. B. Anderson. 2003. Tactics of sexual coercion: When men and women won't take no for an answer. *Journal of Sex Research* 40:76–86.

Straus, Murray A. 1994. *Beating the devil out of them: Corporal punishment in American families.* New York: Lexington Books/Macmillan.

———. 2000. Corporal punishment and primary prevention of physical abuse. *Child Abuse and Neglect* 24:1109–14.

Terrance, C., and K. Matheson. 2003. Undermining reasonableness: Expert testimony in a case involving a battered woman who kills. *Psychology of Women Quarterly* 27:37–45.

Tolan, P. H., D. Gorman-Smith, and D. B. Henry. 2003. The developmental ecology of urban males' youth violence. *Developmental Psychology* 39:274–91.

Ulman, A. 2003. Violence by children against mothers in relation to violence between parents and corporal punishment by parents. *Journal of Comparative Family Studies* 34:41–56.
Ulman, A., and M. A. Straus. 2003. Violence by children against mothers in relation to violence between parents and corporal punishment by parents. *Journal of Comparative Family Studies* 34:41–56.
Umberson, D. K., L. Anderson, K. Williams, and M. D. Chen. 2003. Relationship dynamics, emotional state, and domestic violence: A stress and masculine perspective. *Journal of Marriage and the Family* 65:233–47.
United Nations Population Fund. 2000. The state of world population report 2000. *www.unfpa.org/swp/2000/english/index.html*
Vandello, J. A., and D. Cohen. 2003. Male honor and female fidelity: Implicit cultural scripts that perpetuate domestic violence. *Journal of Personality and Social Psychology* 84:997–1010.
Verma, R. K. 2003. Wife beating and the link with poor sexual health and risk behavior among men in urban slums in India. *Journal of Comparative Family Studies* 34:1–61.
Walrath, C., M. Ybarra, E. W. Holden, Q. Liao, R. Santiago, and P. Leaf. 2003. Children with reported histories of sexual abuse. *Child Abuse and Neglect* 27:509–24.
Walsh, W. 2002. Spankers and nonspankers: Where they get information on spanking. *Family Relations* 51:81–88.
Willson, P., J. McFarlane, A. Malecha, K. Watson, D. Lemmey, P. Schultz, J. Gist, and N. Fredland. 2000. Severity of violence against women by intimate partners and associated use of alcohol and/or illicit drugs by the perpetrator. *Journal of Interpersonal Violence* 15:996–1008.
Windom, C. S. 1999. Posttraumatic stress disorder in abused and neglected children grown up. *American Journal of Psychiatry* 156:1223–29.
Wright, J. W. Jr. 2004. Personal communication, Monroe, La. Dr. Wright is an attorney who has been involved in litigation of Munchausen syndrome by proxy lawsuits.
Zoellner, L. A., N. C. Feeny, J. Alvarez, C. Watlington, M. L. O'Neill, R. Zager, and E. B. Foa. 2000. Factors associated with completion of the restraining order process in female victims of partner violence. *Journal of Interpersonal Violence* 15:1081–99.

Chapter 15

Ahrons, C. R., and J. L. Tanner. 2003. Adult children and their fathers: Relationship changes 20 years after parental divorce. *Family Relations* 52:340–51.
Ahrons, Constance R. 1995. *The good divorce: Keeping your family together when your marriage comes apart.* New York: HarperCollins.
Amato, Paul R. 2001. The consequences of divorce for adults and children. In *Understanding families into the new millennium: A decade in review,* edited by R. M. Milardo. Minneapolis: National Council on Family Relations, 488–506.
Amato, Paul R., and Stacy J. Rogers. 1997. A longitudinal study of marital problems and subsequent divorce. *Journal of Marriage and the Family* 59:612–24.
American Council on Education and University of California. 2004. *The American freshman: National norms for fall, 2003.* Los Angeles: Los Angeles Higher Education Research Institute.
Arditti, J. A. 1991. Child support noncompliance and divorced fathers: Rethinking the role of paternal involvement. *Journal of Divorce and Remarriage* 14:107–20.
Arendell, Terry. 1995. *Fathers and divorce.* New York: Sage.
Baskerville, S. 2003. Divorce as revolution. *The Salisbury Review.* Summer, 30–32.
Baum, N. 2003. Divorce process variables and the co-parental relationship and parental role fulfillment of divorced parents. *Family Process* 42:117–31.
Blau, Melinda. 1995. *Families apart: Ten keys to successful co-parenting.* New York: Putnam.
Blossfeld, H.P., and R. Muller. 2002. Union disruption in comparative perspective: The role of assertive partner choice and careers of couples. *International Journal of Sociology* 32:3–35.
Braver, S. L., J. T. Cookston, and B. R. Cohen. 2002. Experiences of family law attorneys with current divorce practice. *Family Relations* 51:325–34.
Bream, V., and A. Buchanan. 2003. Distress among children whose separated or divorced parents cannot agree on arrangements for them. *British Journal of Social Work* 33:227–38.
Brinig, M. F., and D. W. Allen. 2000. "These boots are made for walking": Why most divorce filers are women. *American Law and Economics Association* 2:126–69.
Bursik, K. 1991. Correlates of women's adjustment during the separation and divorce process. *Journal of Divorce and Remarriage* 14:137–62.
Chen, J. 2004. Britney-Jason union wouldn't last. *USA Today.* 5 January, D1.
Cohen, O., and R. Savaya. 2003. Lifestyle differences in traditionalism and modernity and reasons for divorce among Muslim Palestinian citizens of Israel. *Journal of Comparative Family Studies* 34:283–94.
Coleman, M., L. H. Ganong, T. S. Killian, and A. K. McDaniel. 1998. Mom's house? Dad's house? Attitudes toward physical custody changes. *Families and Society.* March–April, 112–22.
Coles, R. L. 2003. Black single custodial fathers: Factors influencing the decision to parent. *Families in Society: The Journal of Contemporary Human Services* 84:247–58.
Coltrane, S., and M. Adams. 2003. The social construction of the divorce "problem": Morality, child victims, and the politics of gender. *Family Relations* 52:363–72.
Donnelly, D., and D. Finkelhor. 1992. Does equality in custody arrangement improve the parent-child relationship? *Journal of Marriage and the Family* 54:837–45.
Dudley, J. R. 1991. Increasing our understanding of divorced fathers who have infrequent contact with their children. *Family Relations* 40:279–85.
Frederick, J. A., and J. Hamel. 1998. Canadian attitudes toward divorce. *Canadian Social Trends.* 48:6–10.
Gardner, R. A. 1998. *The parental alienation syndrome.* 2d ed. Cresskill, N.J.: Creative Therapeutics.
Geasler, M. J., and K. R. Blaisure. 1998. A review of divorce education program materials. *Family Relations* 167–75.
Goodman, C. C. 1992. Social support networks in late life divorce. *Family Perspective* 26:61–81.
Goodwin, P. Y. 2003. African American and European American women's health marital well-being. *Journal of Marriage and Family* 65:550–60.
Gottman, J. M., J. Coan, S. Carrere, and C. Swanson. 1998. Predicting martial happiness and stability from newlywed interactions. *Journal of Marriage and the Family* 60:5–22.
Griffin, H. 1998. Courts order divorcing parents to parenting classes. *Time.* 34, 100 passim.
Hawkins, A. J., S. L. Nock, J. C. Wilson, L. Sanchez, and J. D. Wright. 2002. Attitudes about covenant marriage and divorce: Policy implications from a three state comparison. *Family Relations* 51:166–75.
Heaton, T. B. 2002. Factors contributing to increasing marital stability in the United States. *Journal of Family Issues* 23:392–409.
Hetherington, E. M. 2003. Intimate pathways: Changing patterns in close personal relationships across time. *Family Relations* 52:318–31.
Hipke, K. N., S. A. Wolchik, I. N. Sandler, and S. L. Braver. 2002. Predictors of children's intervention-induced resilience in a parenting program for divorced mothers. *Family Relations* 51:121–29.
Holroyd, R., and A. Sheppard. 1997. Parental separation: Effects on children; implications for services. *Child: Care, Health and Development* 23:369–78.
Holtzman, M. 2002. The "family relations" doctrine: Extending Supreme Court precedent to custody disputes between biological and nonbiological parents. *Family Relations* 51:335–43.
Hsu, M, D. L. Kahn, and C. Huang. 2002. No more the same: The lives of adolescents in Taiwan who have lost fathers. *Family Community Health* 25:43–56.
Huston, T. L., R. M. Houts, J. P. Caughlin, and S. E. Smith. 2001. The connubial crucible: Newlywed years as predictors of marital delights, distress, and divorce. *Journal of Personality and Social Psychology* 80:237–52.
Jekielek, S. M. 1998. Parental conflict, marital disruption and children's emotional well-being. *Social Forces* 76:905–36.
Johnson, G. R., E. G. Krug, and L. B. Potter. 2000. Suicide among adolescents and young adults. A cross-national comparison of 34 countries. *Suicide and Life-Threatening Behavior* 30:74–82.
Kalter, Neil. 1989. *Growing up with divorce.* New York: Free Press/Macmillan.
Kelly, J. B., and R. E. Emery. 2003. Children's adjustment following divorce: Risk and resilience perspectives. *Family Relations* 52:352–62.
King, V. 1994. Nonresident father involvement and child well-being. *Journal of Family Issues* 15:78–96.
———. 2002. Parental divorce and interpersonal trust in adult offspring. *Journal of Marriage* and the Family 64:642–56.
Knox, David (with Kermit Leggett). 2000. *The divorced dad's survival book: How to stay connected to your kids.* New York: Perseus.
Knox, David, and M. E. Zusman. 1998. Unpublished data on 620 students at East Carolina University.
Kurdek, L. A. 2002. Predicting the timing of separation and marital satisfaction: An eight-year prospective longitudinal study. *Journal of Marriage and the Family* 64:163–79.
Leite, R. W., and P. C. McKenry. 2002. Aspects of father status and postdivorce father involvement with children. *Journal of Family Issues* 23:601–23.
Licata, N. 2002. Should premarital counseling be mandatory as a requisite to obtaining a marriage license? *Family Court Review* 40:518–32.
Manning, W. D., and P. J. Smock. 2000. "Swapping" families: Serial parenting and economic support for children. *Journal of Marriage and the Family* 62:111–22.
Mindel, C. H., R. W. Habenstein, and R. Wright, Jr. 1998. *Ethnic families in America: Patterns and variations.* Upper Saddle River, N.J.: Prentice Hall.
Moore, Martha T. 1997. CEO's ex-wife gets $20 million. *USA Today.* 4 December, 3A.
Morgan, E. S. 1944. *The Puritan family.* Boston: Public Library.

Nielsen, L. 1999. College aged students with divorced parents: Facts and Fiction. *College Student Journal* 33:543–72.

Orbuch, T. L., J. Veroff, H. Hassan, and J. Horrocks. 2002. Who will divorce: A 14-year longitudinal study of black couples and white couples. *Journal of Social and Personal Relationships* 19:179–202.

Paul, P. 2002 *The starter marriage and the future of matrimony.* New York: Random House, 2002.

Pinquart, M. 2003. Loneliness in married, widowed, divorced, and never-married older adults. *Journal of Social and Personal Relationships* 20: 31–53.

Previti, D., and P. R. Amato. 2003. Why stay married? Rewards, barriers, and marital stability. *Journal of Marriage and Family* 65:561–73.

Rauch-Kennedy, S. R. 1997. *Shattered faith: A woman's struggle to stop the Catholic Church from annulling her marriage.* New York: Pantheon.

Rebellon, C. J. 2002. Reconsidering the broken homes/delinquency relationship and exploring its mediating mechanism(s). *Criminology* 40:103–35.

Ricci, Isolina. 2000. The courts, child custody and visitation. *Family Focus.* September, F12–F13.

Schacht, T. E. 2000. Protection strategies to protect professionals and families involved in high-conflict divorce. *UALR Law Review* 22 (3): 565–92.

Shumway, S T., and R. S. Wampler. 2002. A behaviorally focused measure for relationships: The couple behavior report (CBR). *The American Journal of Family Therapy* 30:311–21.

Siegler, I., and P. Costa. 2000. Divorce in midlife. Paper presented at the Annual Meeting of the American Psychological Association, Boston.

Stark, E. Friends through it all.1986. *Psychology Today.* May, 54–60.

Statistical Abstract of the United States: 2003. 123rd ed. Washington, D.C.: U.S. Bureau of the Census.

Stewart, A.l, A. Copeland, N. L. Chester, J. Malley, and N. Barenbaum. 1997. *Separating together: How divorce transforms families.* New York: Guilford Press.

Sun, Y., and Y. Li. 2002. Children's well-being during parents' marital disruption process: A pooled time-series analysis. *Journal of Marriage and the Family* 64:472–88.

Sutton, P. D. 2003. Births, marriages, divorces, and deaths: Provisional data for January-March 2002. *National vital statistics reports* 51 (6). Hyattsville, Md.: National Center for Health Statistics.

Teachman, J. D. 2002. Childhood living arrangements and the intergenerational transmission of divorce. *Journal of Marriage and the Family* 64:717–29.

Tschann, Jeanne M., Janet R. Johnston, Marsha Kline, and Judith S. Wallerstein. 1989. Family process and children's functioning during divorce. *Journal of Marriage and the Family* 51: 431–44.

Turkat, I. D. 2002. Shared parenting dysfunction. *The American Journal of Family Therapy* 30:385–393.

Vlosky, D. A., and P. A. Monroe. 2002. The effective dates of no-fault divorce laws in the 50 states. *Family Relations* 51:317–26.

Waite, L., and M. Gallagher. 2000. *The case for marriage: Why married people are happier, healthier and better off financially.* New York: Doubleday.

Wallace, C. 1994. It never happened. *Prime Time.* American Broadcasting Company, 6 January.

Wallace, S. R., and S. S. Koerner. 2003. Influence of child and family factors on judicial decisions in contested custody cases. *Family Relations* 52:160–88.

Wallerstein, J. S. 2000. *The unexpected legacy of divorce: A 25-year landmark study.* New York: Hyperion.

———. 2003. Children of divorce: A society in search of a policy. In *All our families: New Policies for a new century,* 2d ed., edited by M. A. Mason, A. Skolnick, and S. D. Sugarman. New York: Oxford University Press, 66–95.

Walz, Bill. 1998. Personal communication, Ashville Mediation Center.

Wheaton, Blair. 1990. Life transitions, role histories, and mental health. *American Sociological Review* 55(2):209–23.

Wills, J. 2002. Research updates: Transmitting divorce across generations. *Stepfamilies* 21:1–2.

Woodward, L., D. M. Fergusson, and J. Belsky. 2000. Timing and parental separation and attachment to parents in adolescence. *Journal of Marriage and the Family* 62:162–74.

Chapter 16

Berger, R. 1998. *Stepfamilies: A multi-dimensional perspective.* New York: Haworth Press.

Bernhardt, E. M., F. K. Goldscheider, M. L. Rogers, and H. Koball. 2002. Qualities men prefer for children in the U.S. and Sweden: Differences among biological, step and informal fathers. *Journal of Comparative Family Studies* 33:235–47.

Bray, J. H., and J. Kelly. 1998. *Stepfamilies: Love, marriage and parenting in the first decade.* New York: Broadway Books.

Ceglian, C. P., and S. Gardner. 1998. Attachment style: A risk for multiple marriages. Paper presented at 60th Annual Conference of the National Council on Family Relations, Milwaukee.

Cherlin, A. J. 1996. *Public and private families.* New York: McGraw-Hill.

Church, E. 1994. What is a good stepmother? In *Families and justice: From neighborhoods to nations. Proceedings,* Annual Conference of the National Council on Family Relations, no. 4, 60.

Clarke, S. C., and B. F. Wilson. 1994. The relative stability of remarriages: A cohort approach using vital statistics. *Family Relations* 43:305–10.

Coleman, M., L. H. Ganong, and C. Goodwin. 1994. The presentation of stepfamilies in marriage and family textbooks: A re-examination. *Family Relations* 43:289–97.

Crosbie-Burnett, M., and J. Giles-Sims. 1994. Adolescent adjustment and stepparenting style. *Family Relations* 43:394–400.

Dawson, D. A. 1991. Family structure and children's health and well-being: Data from the 1988 national health interview survey on child health. *Journal of Marriage and the Family* 53:573–84.

Degarmo, D. S., and M. S. Forgatch. 2002. Identity salience as a moderator of psychological and marital distress in stepfather families. *Social Psychology Quarterly* 65:266–84.

DeGraaf, P. M., and M. Kalmijn. 2003. Alternative routes in the remarriage market: Competing-risk analysis of union formation after divorce. *Social Forces* 81:1459–98.

Everett, Lou. 1998. Factors that contribute to satisfaction or dissatisfaction in stepfather-stepchild relationships. *Perspectives in Psychiatric Care* 34(2):25–35.

———. 2000. Personal communication, East Carolina University, October 12.

Fine, M. A., L. H. Ganong, and M. Coleman. 1997. The relation between role constructions and adjustment among stepfathers. *Journal of Family Issues* 18:503–25.

Fine, M. A., and L. A. Kurdek. 1992. The adjustment of adolescents in stepfather and stepmother families. *Journal of Marriage and the Family* 54:725–36.

Fine, M. A., P. C. McKenry, B. W. Donnelly, and P. Voydanoff. 1992. Perceived adjustment of parents and children: Variations by family structure, race, and gender. *Journal of Marriage and the Family* 54:118–27.

Fisher, P. A., L. D. Leve, C. C. O'Leary, and C. Leve. 2003. Parental monitoring of children's behavior: Variation across stepmother, stepfather, and two-parent biological families. *Family Relations* 52:45–52.

Funder, K. 1991. New partners as co-parents. *Family Matters.* April, 44–46.

Furukawa, S. 1994. The diverse living arrangements of children: Summer 1991. *Current Population Reports,* Series P70, no. 38. Washington, D.C.: U.S. Bureau of the Census.

Ganong, L. H., and M. Coleman. 1994. *Remarried family relationships.* Thousand Oaks, Calif.: Sage.

Ganong, L. H., M. Coleman, M. Fine, and A. K. McDaniel. 1998. Issues considered in contemplating stepchild adoption. *Family Relations* 47:63–71.

Ganong, L. H., M. Coleman, and D. Mistina. 1995. Normative beliefs about parents' and stepparents' financial obligations to children following divorce and remarriage. *Family Relations* 44: 306–15.

Goetting, A. 1982. The six stations of remarriage: Developmental tasks of remarriage after divorce. *Family Coordinator* 31:213–22.

Hofferth, S. L., and K. G. Anderson. 2002. All dads are equal? Biology versus marriage as a basis for paternal investment. *Journal of Marriage and the Family* 65:213–32.

Knox, D., and M. E. Zusman. 2001. Marrying a man with "baggage": Implications for second wives. *Journal of Divorce and Remarriage* 35:67–80.

Kurdek, L. A., and M. A. Fine. 1991. Cognitive correlates of satisfaction for mothers and stepfathers in stepfather families. *Journal of Marriage and the Family* 53:565–72.

Kyungok, Huh. 1994. Father's child care time across family types. In *Families and justice: From neighborhoods to nations. Proceedings,* Annual Conference of the National Council on Family Relations, no. 4, 22.

Lamaro, C. 1997. Second marriage: A relationship or a stepfamily? *SCOPE: Journal of Family Services Australia* 3:27–30.

Lutz, P. 1983. The stepfamily: An adolescent perspective. *Family Relations* 32:367–75.

MacDonald, W., and A. DeMaris. 1996. Parenting stepchildren and biological children: The effects of stepparent's gender and new biological children. *Journal of Family Issues* 17:5–25.

———. 2002. Stepfather-stepchild relationship quality. *Journal of Family Issues* 23:121–37.

Marano, H. E. 2000. Divorced? Don't even think of remarrying until you read this. *Psychology Today.* March/April, 56–64.

Marsiglio, W. 1992. Stepfathers with minor children living at home. *Journal of Family Issues* 13:195–214.

Masheter, C. 1999. Examples of commitment in postdivorce relationships between spouses. In *Handbook of interpersonal commitment and relationship stability,* edited by J. M. Adams and W. H. Jones. New York: Academic/Plenum Publishers, 293–306.

Mason, M. A., S. Harrison-Jay, G. M. Svare, and N. H. Wolfinger. 2002. Stepparents: De facto parents or legal strangers? *Journal of Family Issues* 23:507–22.

Michaels, M. L. 2000. The stepfamily enrichment program: A preliminary evaluation using focus groups. *American Journal of Family Therapy* 28: 61–73.

Montgomery, M. J., E. R. Anderson, E. M. Hetherington, and W. G. Clingempeel. 1992. Patterns of courtship for remarriage: Implications for child adjustment and parent-child relationships. *Journal of Marriage and the Family* 54:686–98.
Palisi, B. J., M. Orleans, D. Caddell, and B. Korn. 1991. Adjustment to stepfatherhood: The effects of marital history and relations with children. *Journal of Divorce and Remarriage* 14:89–106.
Papernow, P. L. 1988. Stepparent role development: From outsider to intimate. In *Relative strangers,* edited by William R. Beer. Lanham, Md.: Rowman and Littlefield, 54–82.
Pasley, K. 2000. Stepfamilies doing well despite challenges. *National Council on Family Relations* Report 45:6–7.
Pauk, N. A. 1996. An exploratory study of stepparent-stepchild relationships: The stepchild's perspective. *MU McNair Journal* 4:29–33.
Richmond, L. S., and D. H. Christensen. 1998. The impact of the postdivorce coparenting relationship on parents' physical and psychological health. Paper presented at 60th Annual Conference of the National Council on Family Relations, Milwaukee.
Rutter, Virginia. 1994. Lessons from stepfamilies. *Psychology Today.* May/June, 27, 30 passim.
Stewart, S. D. 2002. The effect of stepchildren on childbearing intentions and births. *Demography* 39:181–197.
Sweeney, M. M. 1998. Remarriage of women and men after divorce: The role of socioeconomic prospects. *Journal of Family Issues* 18:479–502.
———. 2002. Remarriage and the nature of divorce: Does it matter which spouse chose to leave? *Journal of Family Issues* 23:410–40.
Timmer, S. G., and J. Veroff. 2000. Family ties and the discontinuity of divorce in black and white newlywed couples. *Journal of Marriage and the Family* 62:349–61.
Vinick, B. 1978. Remarriage in old age. *The Family Coordinator* 27:359–63.
Wang, H., and P. R. Amato. 1998. Predictors of divorce adjustment. Paper presented at the Annual Meeting of the National Council on Family Relations, Milwaukee.
White, L. K., and A. Riedmann. 1992. When the Brady Bunch grows up: Step/half- and full sibling relationships in adulthood. *Journal of Marriage and the Family* 54:197–208.
Wineberg, H. 1992. Childbearing and dissolution of the second marriage. *Journal of Marriage and the Family* 54:879–87.
Wright, J. M. 1998. *Lesbian step families.* New York: Haworth Press.

Chapter 17

AARP. 1999. AARP/Modern Maturity Sexuality Survey. Retrieved July 25, 2003 from http://research.aarp.org/health/mmsexsurvey1.html
Agee, A., and P. Blanton. 2000. Listening to the stories of Alzheimer's families: A collective case study approach. Poster session at the 62nd Annual Conference of the National Council on Family Relations, Minneapolis, November 12.
Antonucci, T. C, J. E. Lansford, H. Akiyama, J. Smith, M. M. Baltes, K. Takahaski, R. Fuhrer, and J. Francois Dartigues. 2002. Differences between men and women in social relations, resource deficits, and depressive symptomatology during later life in four nations. *Journal of Social Issues* 58:767–84.
Ballard, S. M., and M. L. Morris. 2003. The family life education needs of midlife and older adults. *Family Relations* 52:129–36.
Blackburn, M. L. 2000. America's grandchildren living in grandparent households. *Journal of Family and Consumer Sciences* 92:30–36.
Brewer, L. 2000. Gender socialization and the cultural construction of elder care givers. Poster session at the 62nd Annual Conference of the National Council on Family Relations, Minneapolis, November 12.
Cain, V. S., C. B. Johannes, N. E. Avis, B. Mohr, M. Schocken, J. Skurnick, and M. Ory. 2003. Sexual functioning and practices in a multiethnic study of midlife women: Baseline results from SWAN. *Journal of Sex Research* 40:266–76.
Carr, D. 2003. A "good death" for whom? Quality of spouses' death and psychological distress among older widowed persons. *Journal of Health and Social Behavior* 44:215–25.
Chadiha, L., J. Rafferty, and J. Pickard. 2003. The influence of caregiving stressors, social support, and caregiving appraisal on marital function among African American wife caregivers. *Journal of Marital and Family Therapy* 29:479–90.
Chan, S. 2003. Rates of assisted suicides rise sharply in Oregon. *Student MBJ* 11:137–38.
Cloud, J. 2000. A kinder, gentler death. *Time.* 18 September, 59 passim.
Cockerham, W. C. 1997. *This aging society.* Upper Saddle River, N.J.: Prentice Hall.
Coleman, B. and S. M. Pandya 2002. *Family caregiving and long-term care.* AARP: Washington, D.C.
Crosnoe, R., and G. H. Elder, Jr. 2002a. Life course transitions, the generational stake, and grandparent-grandchild relationships. *Journal of Marriage and the Family* 64:1089–96.
———. 2002b. Successful adaptation in the later years: A life course approach to aging. *Social Psychology Quarterly* 65:309–28.
Cutler, N. E. 2002. *Advising mature clients.* New York: Wiley.
Davies, C. 2002. The grandparent study 2002 report. Washington, D.C.: AARP.
Dychtwald, K. 1990. *Age wave.* New York: Bantam Books.
Essinger, D. 2003. Attitudes of Tennessee physicians toward euthanasia and assisted death. *Southern Medical Journal* 96:427–35.
Family Caregiver Alliance Clearinghouse, www.caregiver.org/factsheets/caregiver_stats.html 1998.
Fernandez-Ballesteros, R. 2003. Social support and quality of life among older people in Spain. *Journal of Social Issues* 58:645–60.
Field, D., and S. Weishaus. 1992. Marriage over half a century: A longitudinal study. In *Changing lives,* edited by M. Bloom. Columbia, S.C.: University of South Carolina Press, 269–73.
Franzoi, S. L., and V. Koehler. 1998. Age and gender differences in body attitudes: A comparison of young and elderly adults. *International Journal of Aging and Human Development* 47:1–10.
Gallanis, T. P. 2002. Aging and the nontraditional family. *The University of Memphis Law Center* 32:607–42.
Gelfand, M. M. 2000. Sexuality among older women. *Journal of Women's Health and Gender-Based Medicine.* 9:s-15–s-20.
Glazer, G., R. Zeller, L. Delumbia, C. Kalinyak, S. Hobfoll, and P. Hartman. 2002. The Ohio midlife women's study. *Health Care for Women International* 23:612–30.
Golden, R., and C. C. Saltz. 1997. The aging family. *Journal of Gerontological Social Work* 27:55–64.
Hansson, R. O., J. O. Berry, and M. E. Berry. 1999. The bereavement experience: Continuing commitment after the loss of a loved one. In *Handbook of interpersonal commitment and relationship stability,* edited by J. M. Adams and W. H. Jones. New York: Academic/Plenum Publishers, 281–91.
Hays, J., J. K. Ockene, R. L. Brunner, J. M. Kotchen, J. E. Manson, R. E. Patterson, A. K. Aragki, M. S., S. A. Shumaker, R. G. Bryzyski, et al. 2003. Effects of estrogen plus progestin on health-related quality of life. *The New England Journal of Medicine* 348:1839–54.
Hill, Twyla J. 2000. Legally extending the family: An event history analysis of grandparent visitation rights laws. *Journal of Family Issues* 21:246–61.
Hobson, C. J., J. Kamen, J. Szostek, C. M. Nethercut, J. W. Tiedmann, and S. Wojnarowiez. 1998. Stressful life events: A revision and update of the social readjustment rating scale. *International Journal of Stress Management* 5:1–23.
Ikegami, N. 1998. Growing old in Japan. *Age and Ageing* 27:277–78.
Iwashyna, T. J., and N. A. Christakis. 2003. Marriage, widowhood, and health-care use. *Social Science and Medicine* 57:2137–47.
Jacoby, S. 2003. The nursing squeeze. *AARP Bulletin.* May, 6 passim.
Jang, S., and D. F. Detzner. 1998. Filial responsibility in cross-cultural context. Poster session presented at the Annual Conference of the National Council on Family Relations, Milwaukee, Wis.
Johnson, C. L., and B. M. Barer. 1997. *Life beyond 85 years: The aura of survivorship.* New York: Springer Publishing Co.
Jorm, A. F., H. Christensen, A. S. Henderson, P. A. Jacomb, A. E. Korten, and A. Mackinnon. 1998. Factors associated with successful ageing. *Australian Journal of Ageing* 17:33–37.
Kennedy, G. E. 1997. Grandchildren's memories: A window into relationship meaning. Paper presented at the Annual Conference of the National Council on Family Relations, Crystal City, Va.
King, V. 2003. The legacy of a grandparent's divorce: Consequences for ties between grandparents and grandchildren. *Journal of Marriage and the Family* 65:170–83.
Knox, D. (with K. Leggett). 2000. *The divorced dad's survival book: How to stay connected with your kids.* Boston: Perseus.
Koropeckyj-Cox, T. 2002. Beyond parental status: Parental well-being in middle and old age. *Journal of Marriage and Family* 64:957–71.
Kupperbusch, C., R. W. Levenson, and R. Ebling. 2003. Predicting husbands' and wives' retirement satisfaction from the emotional qualities of marital interaction. *Journal of Social and Personal Relationships* 20:335–45.
Landry-Meyer, L. 2000. Grandparents as parents: What they need to be successful. *National Council on Family Relations* Report 45, 89.
Leichtentritt, R. D., and K. D. Rettig. 2000. Conflicting value considerations for end-of-life decisions. Poster session at the 62nd Annual Conference of the National Council on Family Relations, Minneapolis, November 12.
Logan, J. R., and F. Bian. 2003. Parents' needs, family structure, and regular international financial exchange in Chinese cities. *Sociological Forum* 18:85–101.
Mackey, R. A., and B. A. O'Brien. 2000. Adaptation in lasting marriages. *Families in Society: The Journal of Contemporary Human Services* 80:587–96.
Marks, N. F., J. D. Lambert, and H. Choi. 2002. Transitions to caregiving, gender, and psychological well-being: A prospective U.S. national study. *Journal of Marriage and Family* 64:657–67.
Marquet, R. ., A. Bartelds, G. J. Visser, P. Spreeuwenberg, and I. Peters. 2003. Twenty-five

years of requests for euthanasia and physician assisted suicide in Dutch practice: Trend analysis. *British Medical Journal* 327:201–2.

Martire, L. M., and M. A. P. Stephens. 2003. Juggling parent care and employment responsibilities: The dilemmas of adult daughter caregivers in the workforce. *Sex Roles: A Journal of Research* 48:167–74.

Michael, R. T., J. H. Gagnon, E. O. Laumann, and G. Kolata. 1994. *Sex in America: A definitive survey.* Boston: Little, Brown.

Miller, K. 1998. Prevalence of depression in healthy elderly persons. *American Family Physician* 57: 2238–40.

Mitchell, J. 2003. Personal communication. Center for Aging, East Carolina University, Greenville, North Carolina.

Mitchell, K., and R. G. Owens. 2003. National Survey of medical decisions at end of life made by New Zealand practitioners. *British Medical Journal* 327:202–3.

Morris, M. L., and S. M. Ballard. 2003. Instructional techniques and environmental considerations in family life education programming for midlife and older adults. *Family Relations* 52:167–73.

Mutran, E. J., D. Reitzes, and M. E. Fernandez. 1997. Factors that influence attitudes toward retirement. *Research on Aging* 19:251–73.

National Institute of Mental Health. 2003. Older adults: Depression and suicide facts. HIH Publication No. 03–4593. Retrieved on July 23, 2003 from www.nimh.nih.gov/publicat/elderlydep suicide.cfm

Neugarten, B. L., and K. K. Weinstein. 1964. The changing American grandparent. *Journal of Marriage and the Family* 26:199–204.

Nicholson, T. 2003. Homeowners fail to prepare for aging. *AARP Bulletin* 44:7.

Oldenburg, A. 2000. Boomers believe they've found a fountain of youth in a syringe. *USA Today.* 15 November, A1.

O'Reilly, E. M. 1997. *Decoding the cultural stereotypes about aging: New perspectives on aging talk and aging issues.* New York: Garland.

Peterson, B. E. 2002. Longitudinal analysis of midlife generativity, intergenerational roles, and caregiving. *Psychology and Aging* 17:161–168.

Pope, E. 2003 MIT study: Older drivers know when to slow down. *AARP Bulletin.* 44:11–12.

Reid, J. 2000. Living creatively in a nursing home. *Journal of the American Society on Aging* 23:51–55.

Ritblatt, S. N., and L. M. Drager. 2000. The relationship between living arrangement and perception of social support, depression, and life satisfaction among the elderly. Poster session at the 62nd Annual Conference of the National Council on Family Relations, Minneapolis, November 12.

Sandberg, J. G., R. B. Miller, and J. M. Harper. 2002. A qualitative study of marital process and depression in older couples. *Family Relations* 51:256–64.

Schairer, C., J. Lubin, R. Troisi, S. Sturgeon, L. Brinton, and R. Hoover. 2000. Menopausal estrogen and estrogen-progestin replacement therapy and breast cancer risk. *Journal of the American Medical Association* 283:485–91.

Smith, J., M. Borchelt, H. Maier, and D. Jopp. 2002. Health and well-being in the young old and oldest old. *Journal of Social Issues* 58:715–33.

Statistical Abstract of the United States: 2003. 123rd ed. Washington, D.C.: U.S. Bureau of the Census.

Stevens, N. 2002. Re-engaging: New partnerships in late-life widowhood. *Ageing International* 27: 27–42.

Szinovacz, M. E., and A. M. Schaffer. 2000. Effects of retirement on marital tactics. *Journal of Family Issues* 21:367–89.

Valliant, G. E. 2002. *Aging well: Surprising guideposts to a happier life from the Landmark Harvard study on adult development.* New York: Little, Brown.

Wallsten, S. S. 2000. Effects of care giving, gender, and race on the health, mutuality, and social supports of older couples. *Journal of Aging and Health* 12:90–111.

Webber, S., J. P. Scott, R. Wampler. 2000. Perecived congruency of goals as a predictor of marital satisfaction and adjustment in retirement. Poster session at the 62nd Annual Conference of the National Council on Family Relations, Minneapolis, November 12.

Wells, Y. D. 2000. Intentions to care for spouse: Gender differences in anticipated willingness to care and expected burden. *Journal of Family Studies* 5:220–34.

Wenger, G. C., A. Scott, and N. Patterson. 2000. How important is parenthood? Childlessness and support in old age in England. *Ageing and Society.* 20:161–82.

Wielink, G., R. Huijsman, and J. McDonnell. 1997. A study of the elders living independently in the Netherlands. *Research on Aging* 19:174–98.

Willing, R., and P. McMahon. 2000. Court dumps grandparents law. *USA Today.* 6 June, 3A.

Winterich, J. A. 2003. Sex, menopause, and culture: Sexual orientation and the meaning of menopause for women's sex lives. *Gender and Society* 17:627–42.

Wrosch, C. and J. Heckhausen. 2002. Perceived control of life regrets: Good for young and bad for old adults. *Psychology and Aging* 17:340–50.

Wu, Z., and R. Hart. 2002 The mental health of the childless elderly. *Sociological Inquiry* 72:21–42.

Zhan, H. J., and R. J. V. Montgomery. 2003. Gender and elder care in China: The influence of filial piety and structural constraints. *Gender and Society* 17:209–29.

Epilogue

American Council on Education and University of California. 2004. *The American freshman: National norms for fall, 2003.* Los Angeles: Higher Education Research Institute. U.C.L.A. Graduate School of Education and Information Studies.

Avellar, S., and P. J. Smock. 2003. Has the price of motherhood declined over time? A cross-cohort comparison of the motherhood wage penalty. *Journal of Marriage and Family* 65:597–607.

Grossman, C. L. 2004. Gay "civil union" not as divisive as "marriage." *USA Today.* 14 January, 6D.

McQuillan, L. 2003. President wants to "codify" marriage. *USA Today.* July 31,1A.

Peterson, K. S. 2004. Learning the dance of marriage: Couples get help from courses. *USA Today.* 5 February 5, D1.

Statistical Abstract of the United States: 2004. 124th ed. Washington, D.C.: U.S. Bureau of the Census.

Sutton, P. D. 2003. Births, marriages, divorces, and deaths: Provisional data for January–March 2002. *National Vital Statistics Reports* 51(6). Hyattsville, Md.: National Center for Health Statistics.

Special Topics

Capaldi, D. M. M. Stoolmiller, S. Clark, and L. D. Owen. 2002. Heterosexual risk behaviors in at risk young men from early adolescence to young adulthood: Prevalence, prediction, and association with STD contraction. *Developmental Psychology* 38:394–406.

HIV/AIDS Surveillance Report. 2002. Vol. 14,Cases of HIV infection and AIDS in the United States. http://www.cdc.gov/hiv/stats.htm#cumaids

Joo, S., J. E. Grable, and D. C. Bagwell. 2003. Credit card attitudes and behaviors of college students. *College Student Journal* 37:405–19.

Lane, T. 2002. Male circumcision reduces risk of both acquiring and transmitting human papillomavirus infection. *International Family Planning Perspectives.* 28:179–81.

Lane, T., and F. Althus. 2002. Who engages in sexual risk behavior? *International Family Planning Perspectives* 28:185–86.

McGraw, A., M. Zvonkovic, and A. J. Walker. 2000. Studying postmodern families: A feminist analysis of ethical tensions in work and family research. *Journal of Marriage and the Family* 62: 68–77.

Meyers, J. F., and M. Shurts. 2002 Measuring positive emotionality: A review of instruments assessing love. *Measurement and Evaluation in Counseling and Development* 34:238–54.

Pedlow, C. T., and M. P. Carey. 2003. HIV sexual risk-reduction intervention for youth: A review and methodological critique of randomized controlled trials. *Behavior Modification* 27: 135–90.

Pluhar, E. I., E. A. Frongillo, J. M. Stycos, and D. Dempster-McClain. 2003. Changes over time in college students' family planning knowledge, preference, and behavior and implications for contraceptive education and prevention of sexually transmitted infections. *College Student Journal.* 37:420–34.

Pound, E. T. 2000. University ordered to correct problems. *USA Today.* 13 July, 3A.

Quiles, J. A. 2003. Romantic behaviors of university students: A cross-cultural and gender analysis in Puerto Rico and the United States. *College Student Journal* 37:354–66.

Riniolo, T. C., and L. A. Schmidt. 2000. Searching for reliable relationships with statistical packages: An empirical example of the potential problems. *Journal of Psychology* 13:143–51.

Sellors, J. W., T. L. Karwalajtys, J. Kaczorowski, J. B. Mathony, A. Lltwyn, S. Chong, J. Sparrow, A. Lorincz, for the Survey of HPV in Ontario Women (SHOW group). 2003. Incidence, clearance and predictors of human papillomavirus infection in women. *Canadian Medical Association Journal* 168:421–25.

Stephenson, J. 2003. Growing, evolving HIV/AIDS pandemic is producing social and economic fallout. *Journal of American Medical Association* 289:31–33.

Sternberg, S. 2003. Vaccine for AIDS appears to work. *USA Today.* 24 February, A1.

Wallerstein, J. 2000. *The unexplained legacy of divorce: A 25 year landmark study.* New York: Hyperion.

Photo Credits

Chapter 1	1, 3, 7, 18, 27: Authors
Chapter 2	31: Authors; 34: Dana Danahey; 49: Elizabeth Baker; 52: John C. Newman
Chapter 3	55: Authors; 61: Rebecca Allen; 63, 70: Authors
Chapter 4	77: Authors; 79: Rebecca Allen; 81, 83: Authors
Chapter 5	99, 109, 111, 127: Authors
Chapter 6	132, 136: Authors; 142: Kate Adamson
Chapter 7	159, 161, 162, 176: Authors
Chapter 8	182, 197 *top & bottom,* 198, 200: Authors
Chapter 9	205: CORBIS; 206, 208, 211, 215: Authors
Chapter 10	225: Authors; 227: Ivor L. Livingston; 228, 235: Authors; 244: Ben Latten
Chapter 11	159, 261, 262, 270, 278: Authors
Chapter 12	285, 289, 293, 294, 298, 299: Authors
Chapter 13	304: Fogel Francois/CORBIS Sygma; 308, 309, 310, 312: Authors
Chapter 14	328: Royalty-Free/CORBIS; 332, 334, 349: Authors
Chapter 15	354, 358, 375: Authors
Chapter 16	382, 385, 401: Authors
Chapter 17	407, 409, 415, 423, 426: Authors
Special Topic 1	433, 437: Authors
Special Topic 2	437: Authors
Special Topic 3	442: Marshall Burke; 443: Authors
Special Topic 4	452: Tim Thompson/CORBIS
Special Topic 5	461: John Davis/ImageState-Pictor/PictureQuest
Special Topic 6	468: Rachel Epstein/PhotoEdit

Name Index

E

F

G

H

M

S

X

Y

Z

Subject Index

A

D

N

O

P